Particle Acceleration and Detection

The series "Particle Acceleration and Detection" is devoted to monograph texts dealing with all aspects of particle acceleration and detection research and advanced teaching. The scope also includes topics such as beam physics and instrumentation as well as applications. Presentations should strongly emphasize the underlying physical and engineering sciences. Of particular interest are

- contributions which relate fundamental research to new applications beyond the immediate realm of the original field of research
- contributions which connect fundamental research in the aforementioned fields to fundamental research in related physical or engineering sciences
- concise accounts of newly emerging important topics that are embedded in a broader framework in order to provide quick but readable access of very new material to a larger audience

The books forming this collection will be of importance to graduate students and active researchers alike.

François Méot

Understanding the Physics of Particle Accelerators

A Guide to Beam Dynamics Simulations
Using ZGOUBI

 Springer

François Méot
Department of Collider-Accelerator
Brookhaven National Laboratory
Upton, NY, USA

ISSN 1611-1052   ISSN 2365-0877 (electronic)
Particle Acceleration and Detection
ISBN 978-3-031-59978-1   ISBN 978-3-031-59979-8 (eBook)
https://doi.org/10.1007/978-3-031-59979-8

This Springer imprint is published by the registered company Springer Nature Switzerland AG
The registered company address is: Gewerbestrasse 11, 6330 Cham, Switzerland

Paper in this product is recyclable.

To my wife Sylvaine for her patient support

Foreword

The author of this book, Dr. François Méot, is an accelerator physicist at Brookhaven National Laborartory (BNL). Currently, he is promoting the acceleration of proton beams that maintain spin polarization at BNL's large accelerator RHIC, and the FFAG accelerator project, especially the non-scaling FFAG type Energy Recovery Linac (ERL) accelerator project. He has been developing for some time a numerical simulation code ZGOUBI for analyzing the dynamical behavior of charged particles including the spin motions in various types of accelerators.

Accelerators are devices that focus and accelerate charged particles using electromagnetic force, and have been developed as experimental tools for particle and nuclear physics as well as for various applications. Various types of accelerators have been invented as described in each chapter of this book. In order to understand the behaviors of the charged particle motion in these accelerators, it is necessary to solve the equation of motion using numerical methods and many numerical simulation codes have been developed. In particular, in ring accelerators where closed orbits are unknown in advance, such as scaling FFAG, it is necessary to accurately determine particle trajectories using electromagnetic field distributions obtained through actual measurements or numerical calculations. Also, when evaluating spin motion in a ring accelerator, identification and distortion of closed orbits are important points. For these reasons, there was a need for a simulation code that could evaluate the particle orbits from beam tracking based on the external electromagnetic field distribution, regardless of the type of accelerator. The code ZGOUBI answers those needs.

In the 1980s, pioneering work on spin-polarized proton beam acceleration in proton synchrotrons began with the 3 GeV proton synchrotron SATURNE 2 in Saclay, François. Dr. Méot contributed to polarized proton beam acceleration studies in SATURNE 2 by extending the ZGOUBI code, which was originally developed to analyze the orbits within magnetic field spectrometers, to the analysis of proton spin motion in synchrotrons. In the recent years, he made a major contribution to the success of polarized proton beam acceleration in RHIC.

In the late 1990s, I developed the world's first proton FFAG accelerator (POP-FFAG) with our colleagues at KEK, Japan. I first met Dr. Méot at an international conference in Paris. We decided to start the collaboration on the FFAG experimental study in Japan. Japan's monsoon summer is extremely hot and humid, and I remember the days spent doing beam experiments day after day. For François Méot, the experience brought back memories of his childhood in Southeast Asia.

Ashiya, Japan   Yoshiharu Mori
January 2024

Preface

"The history of SATURNE 1 and SATURNE 2 accelerators includes the construction of 4 spectrometers as particle detectors. The first, SPES I designed in 1968 is of high resolution. The resolution permits the selection of one nuclear level which is necessary even for elastic scattering. A program, ZGOUBI (Jean Claude Faivre, Denis Garreta) for precise orbit computing had to be written and was the first one to be fast enough to permit new spectrometer design."[1]

In the matter of raytracing accuracy, no method competes with stepwise numerical integration of the differential equations of motion. This holds for beam dynamics in accelerators and beamlines (which abide by the Lorentz force equation), for spin precession (Thomas-BMT differential equation), computation of the radiation Poynting vector, or when adding such ingredients as scattering and other stochastic energy loss.

In the era of artificial intelligence (AI) and machine learning (ML), and leaning on high performance computing (HPC) for speed and statistics, there is no reason to deprive oneself of building informed intelligence on deep understanding of the kinematics, by pushing charged particles step by step in true models of magnetic and electric fields.

Computer codes for stepwise raytracing have long been available and used. Their long lists of achievements include

- As early as the 1950s, the design of fixed field alternating gradient (FFAG) cyclic accelerators by the MURA group, from paraxial parameters to dynamical acceptance performance, using a Runge-Kutta integrator concurrently with magnetic field models including computed field maps.[2]
- The design and parameterization of nuclear physics mass spectrometers from the early times, decades ago. This is the only way that numerical simulation can

---

[1] Jacques Thirion, Preface, in "The 20 years of the synchrotron SATURNE 2". A. Boudard, P.-A. Chamouard Eds. World Scientific, May 1998.

[2] Cole, F. T.: O Camelot, a Memoir of the MURA Years (April 11, 1994); Sects. 7.1, 10.5 and 10.6. Available in the Proceedings of the Cyclotron Conference, East Lansing, USA, May 13–17, 2001. https://accelconf.web.cern.ch/c01/cyc2001/extra/Cole.pdf.

match the level of time-of-flight and momentum resolution requested from these devices.

- The simulation of radiation and bunch-radiation interaction, as part of the design of insertion devices found in synchrotron radiation rings and free electron lasers.
- The analysis of spin diffusion and polarization lifetime in electron accelerators and storage rings.
- Computation and correction of optical aberrations. They matter in all beam optics problems: imaging and image resolution, geometrical acceptance in beam lines and dynamical acceptance in rings, non-linear resonant extraction for hadrontherapy purpose, etc.

For accuracy, the representation of electrostatic and/or magnetic fields may resort to analytical modeling or to computed field maps, whereas full benefit from these models requires stepwise integration of the equations of motion. An additional benefit of the method is its allowing high resolution Monte Carlo processes, such as synchrotron radiation, in-flight decay, particle-matter interaction, etc.

Various numerical integrators have been in use for raytracing, including Taylor series, Runge-Kutta, leap-frog, etc. They are found in decades old, popular codes as RAYTRACE, GEANT, ZGOUBI, and others. Accuracy on the modeling of fields, allied with accuracy on the integration of the equations of motion, makes the numerical quality of this partnership hard to beat—and these proven 'old' codes best modern tools!

A drawback of the latter might be considered to be the longer time it takes to push particles, compared to matrix and other kick-drift mapping techniques. Computing speed is often used as a justification to jeopardize intelligence by resorting to approximations regarding kinematics, or fields, or both. However (i) better do things slowly and right than fast and wrong; (ii) with today's HPC, only a few accelerator design and operation simulation problems, e.g., space charge, dynamical aperture computation in large rings, remain affected by somewhat "long" execution times. Where paraxial optics or quasi-linear optical systems are concerned, low order approximations might, why not, fulfill requirements and accuracy might be relaxed, for the sake of computing speed ... however, several categories of beam dynamics simulations in these paraxial machines (such as multiturn bunch tracking in small rings) may anyway be performed in a reasonably short time, the more so using HPC and/or CPU clusters, without having to surrender to mapping style of field modeling and kinematics approximations.

In the era of AI and ML, there is no reason to rely on approximate integration methods, and on approximate field models: loose modeling is in patent contradiction with the goals of intelligence and learning! It would be akin to assuming $\pi = 3.1416$, or $c = 3 \times 10^8$ m/s in designing accelerators—this does not happen. Stepwise integration in realistic field models saves on the time spent figuring out and overcoming the adverse effects of approximations: *"Is what I observe due to my approximations, or is it real?"*. This allows time to be efficiently used to focus one's energy where it is needed, and of interest, i.e., the physics of phenomena and their understanding.

Table 1 Comparative advantages (+) and disadvantages (−) of numerical simulations and machine operation

	Numerical simulation	Machine operation
Studying the physics of phenomena	No limitation +	+ that's where the truth is
Explore exotic dynamics	No limitation +	− technological and principle limitations
Details of particle dynamics	No limitation +	− requires very small beam emittance
Speed	Improves with time −	+ can't be beaten!
Cost	Essentially zero +	− a lot

The truth lies in machine operation anyway. This is an additional good reason for numerical methods to stay away from approximations, as they may offset apprehension of phenomena and hamper comparisons with operation outcomes. Learning from machine operation may have limitations, compared with raytracing techniques, learning from the latter has its limitations as well, both are complementary anyway (Table 1).

This book is an introduction to the physics of particle accelerators based on beam dynamics simulations, covering accelerator concepts developed over the past century. It is as much about learning on particle accelerators, as it is about learning beam dynamics simulation techniques and tools, real life accelerator design methods, and simulation data production and treatment.

Simulation exercises are proposed, this is the "hands-on" side of the learning method. Their material is based on real life design studies, which have often been subject to tutorials in workshops and university teachings, or used in conference publications and peer-review journal articles. A lot more than proposed in this book, covering half a century of accelerator, spectrometer, and beam line design studies, can be found in the sourceforge branch https://sourceforge.net/p/zgoubi/code/HEAD/tree/branches/exemples/.

These diverse studies have most of the time been subject to laboratory tech. notes and other publications, which for some can be found in the sourceforge branch https://sourceforge.net/p/zgoubi/code/HEAD/tree/branches/publications/.

More zgoubi simulation material, guidance regarding the use of the code, its keywords, and regarding its capabilities, can be found in a general manner in

- US DOE OSTI repository [3]https://www.osti.gov/.
- PR-AB and NIM A publications,

---

[3] At the time of these writtings, "https://www.osti.gov/search/semantic:zgoubi" produces "164 Search Results".

– JACoW accelerator conference proceedings site https://www.jacow.org/Main/Pro ceedings and its search tool.[4]

Performing numerical simulations is not quite as real as being at the command/ controls in an accelerator facility, yet it may sometimes be quite close. Computer simulations closely reproduce beam manipulations done in accelerator control rooms. In fact, machine operation has much to do with beam dynamics simulation, since both make extensive use of computer models to reproduce basic beam dynamics and accel- erator optics, beam monitoring, and control. Simulations use similar post-processing tools, and deliver similar types of data: particle coordinates, bunch parameters, phase space portraits, motion spectra, machine parameters, etc.

Computer simulations proposed here manipulate, guide, and accelerate parti- cles and bunches, in most styles of particle accelerators devised since the 1920s. In performing these simulation exercises, the reader will acquire an understanding of charged particle beam optics, extend their knowledge of accelerator physics and technology, and assess the use of one or the other of the existing beam handling technologies for beam delivery in such or such particle beam application. The exer- cises take the reader in a virtual world of accelerator and beam simulations on a computer, in the way that accelerator physicists design these machines in their labo- ratories, and play with them in control rooms. They allow the basic theoretical and practical aspects of the main technological components of particle accelerators to be discovered: guiding and focusing using **E** and/or **B** fields, accelerating in voltage gaps, shaking particles in wiggler magnets, getting rid of impurities in combined **E**, **B** devices, etc. Checking the consistency of numerical results against the elements of theory introduced in the chapters, and vice-versa, is part of these exercises.

Finally, in some exercises one may think of launching a bunch tracking simulation for minutes, and why not hours, while tuning the beam and taking data as the computer is quietly pushing these bunches, repeatedly, through a virtual beam line, or around a virtual ring. Along that line, bunch tracking animations are proposed in some exercises, based on short beam lines or small rings for conveniently fast tracking.

Most accelerator species are covered in this book, in a series of chapters ordered following the historical chronology: from electrostatic systems to today's storage rings, via the classical cyclotron, relativistic cyclotron, microtron, beta- tron, synchrotron, FFAG. Without forgetting linacs, opportunistically, if not chrono- logically, introduced in racetrack microtron and in linear FFAG recirculator exer- cises. Note in passing, quite interestingly all these accelerator species, invented and developed over a century, are still in use today, in one application or another.

All the chapters are organized in a similar manner:

– a short introduction with some historical insight,
– a brief theoretical reminder, which provides recipes resorted to in the exercises,
– a series of simulation exercises.

---

[4] At the time of these writtings, looking up 'zgoubi' in JACoW advanced search tool produces about 800 results.

The solutions of the exercises are the subject of a dedicated chapter. They include zgoubi input data files, or detailed indications to build them, as well as expected results.

The reader is assumed to have a basic knowledge of charged particle beam optics in transport lines and accelerators. So the theoretical reminders are rather concise, aimed essentially at bringing the basic concepts and formulæ used in the exercises, with minimal explanations. Thus, having text books at hand when working on the simulations is a good idea. The reader may at times feel that computer code capabilities are a little (too) lightly addressed ... well ... the 400 page companion to the present opus, Zgoubi Users' Guide, happens to be indispensable as well to work out the simulations. It is available in its most recent version at https://sourceforge.net/p/zgoubi/code/HEAD/tree/trunk/guide/Zgoubi.pdf.

All details regarding the methods to work it out may not be provided when proposing a simulation exercise. For instance which keywords may be preferred (e.g., for a bend magnet: BEND, CARTEMES, DIPOLE, MULTIPOL, TOSCA, ...), which output file(s) will deliver results in an appropriate form (e.g., zgoubi.res, zgoubi.plt, zgoubi.fai, zgoubi.TWISS.out, zgoubi.MATRIX.out, ...), post-treatment to possibly apply on these files content. Thinking about it and finding the preferred ways is part of the exercise. Choices depend on the context, e.g., learning beam optics, teaching, developing AI programs, or working in a team on a particular design. Of course, this means a learning curve in order to figure out the possibilities offered by zgoubi (and, beyond, by stepwise raytracing based beam optics techniques). This is one of the goals here, and it is also why the exercise series starts with simple simulations using a very limited number of keywords: in the first two cyclotron chapters for instance, just DIPOLE or TOSCA, to bend a trajectory into a closed orbit in a magnetic field, CAVITE for resonant acceleration, and FAISCEAU to monitor particle coordinates.

So... Yes! Grab your laptop, it will be an essential tool. You'll be able to play with charged particle beams in a world of virtual accelerator optical components, beam lines, and rings, and ... have fun!

BNL, Upton, NY, USA   François Méot

Acknowledgements

This book is a product of more than 40 years of numerical simulations within the framework of tens of projects, in high energy physics and nuclear physics laboratories, CEA Saclay, BNL, CERN, FERMILAB, TRIUMF, and others. It is also the product of years of tutorials, in workshops and university courses. Not the least, it benefits from the feedbacks these collaborations, workshops, and teachings allowed.

More specifically this book has benefitted from discussions with my colleagues and friends, among whom former PhD students, Yann Dutheil, Bhawin Dhital, Malek Haj Tahar, Kiel Hock, Xiangdong Lee, Joseph Lidestri, Steve Peggs, Guillaume Robert-Demolaize, Laurent Sérani, Victor Smirnov. They provided suggestions and advice for corrections and improvements.

The work environment at the Collider-Accelerator Department at the Brookhaven National Laboratory is for a large part responsible for this undertaking. I would like to thank here the many colleagues with whom I have been in close contact during these years of collaborations on accelerator R&D and projects at BNL C-AD, and in particular Thomas Roser, Nick Tsoupas, Haixin Huang, Wolfram Fischer, and Dejan Trbojevic.

Contents

Chapter 1
Numerical Simulations

Several of the numerical simulation exercises proposed in the following chapters require step-by-step raytracing through the optical elements concerned, for diverse reasons, depending on the problem:

– some optical elements feature complicated magnetic or electrostatic fields,
– some are represented by magnetic and/or electrostatic field maps,
– fields may be time-dependent and vary while traversed,
– some processes require accuracy on field modeling: spin transport for instance, or computation of the synchrotron radiation Poynting vector from the coordinates of a relativistic particle,
– some Monte Carlo processes may require small integration steps for accuracy, such as in-flight decay, stochastic emission of radiation.

In any case, the numerical integration of the equations of motion, which is what the exercises are concerned with, requires stepwise raytracing through magnetic and/or electrostatic fields defined in 3D space, possibly including time variation, in order to ensure a faithful reproduction of the physical processes simulated.

Zgoubi [1] is resorted to here, however other well known codes use a numerical integrator to push particles step-by-step in analytical or field map based field models, and could be used in various exercises, for instance RAYTRACE [2], GEANT [3], to mention just two. Furthermore, some of the exercises involve simple optical assemblies, for which a Runge-Kutta integration may easily be written.

Various exercises lend themselves to resolution using matrix or other mapping transport techniques, however these techniques rely on approximations regarding both field models and the solution of the equations of motion, resulting *always* on questioning regarding the validity of one or the other, or both. The option of using these techniques is left to the reader, however this is not our interest here, and no such solutions to the exercises are provided: in the exercises, transport coefficients T_{ij}, T_{ijk}, ... and transport matrices $[T_{ij}]$, $[T_{ijk}]$, ... will be addressed as a subproduct

© The Author(s) 2024   1
F. Méot, *Understanding the Physics of Particle Accelerators*, Particle Acceleration
and Detection, https://doi.org/10.1007/978-3-031-59979-8_1

of stepwise raytracing, derived from particle coordinates, and mostly for comparison with theoretical expectations regarding such quantities of paraxial beam optics as magnification, optical aberrations, betatron functions, wave numbers.

1.1 Half a Century of Charged Particle Raytracing in Zgoubi

At the time of this publication, 2024, the raytracing code zgoubi celebrates its 52st anniversary, having pushed particles in accelerator laboratories over half a century.

The initial version of zgoubi was developed by D. Garetta and J. C. Faivre at CEA-Saclay in the early 1970s, for the purpose of the design and operation of magnetic spectrometers at the 3 GeV polarized ion synchrotron SATURNE 2 [4, 5], using magnetic field maps, simulated in zgoubi proper in the design approach [1, AIMANT keyword] and measured eventually. The code was used to assess, from their measured field maps, the optical properties of the magnets of the SATURNE 2 ring which was under construction in the same period.

The author of these lines inherited Zgoubi at SATURNE in the early 1980s for spectrometer design and spin dynamics studies, from Saby Valéro who was completing the design of GANIL SPEG spectrometer [6], and André Tkatchenko [7]. The diversity of the utilization of zgoubi in the following years boosted the development of analytical models of accelerator components, with today a library of more than 60 optical elements and about 50 monitoring and command keywords [1, *cf.*, Glossary of Keywords].

In September 2007 it has been made available in sourceforge, https://sourceforge. net/projects/zgoubi/ and has undergone a substantial number of downloads since, from many countries as it appears (Fig. 1.1). The sourceforge package includes

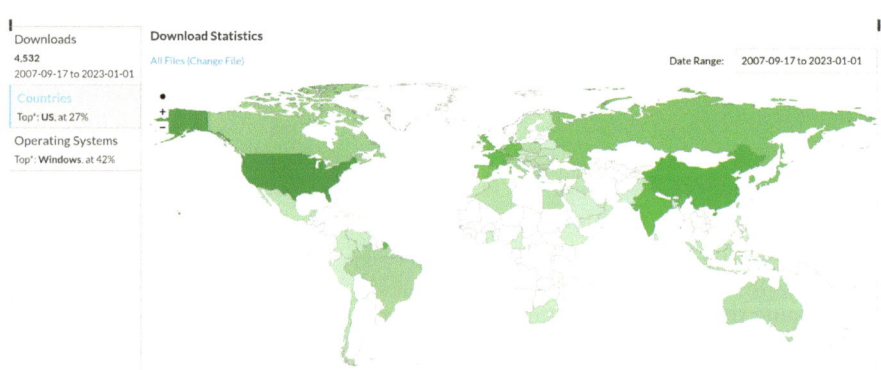

Fig. 1.1 Zgoubi downloads in sourceforge [8], 4,500+ over the period 2007–2022, from 67 different countries from USA, China, France, Germany, India, etc., to Peru, Bangladesh, Vietnam

zgoubi, as well as an ancillary data treatment/graphic companion, zpop (sometimes forsaken though, in favor of gnuplot and awk tools). The sourceforge branch https://sourceforge.net/p/zgoubi/code/HEAD/tree/branches/exemples/ offers about 300 simulations covering half a century of zgoubi usage:

- beam lines and spectrometers of all sorts,
- all possible ring accelerator methods,
- linacs, recirculating (RLA) and energy recovery linacs (ERL),
- effects of synchrotron radiation on beam and polarization,
- beam imaging using visible synchrotron radiation,
- collect and acceleration of short-lived beams,
- etc.

Most of these examples are drawn from real life R&D and teaching activities, and as such have been the subject of laboratory technical notes and other publications. It may be helpful to refer to the latter when undertaking simulations, many can be found in the sourceforge branch https://sourceforge.net/p/zgoubi/code/HEAD/tree/branches/publications/.

The first Users' Guide dates back to 1988, at CEA-Saclay [9]. Accounting for further developments, and in order to facilitate access to the program an English version of the manual was written at TRIUMF in 1990 [10]. The code, which so far had been used to push particles in beam lines and spectrometers, was introduced to the realm of cyclic accelerators in the early 1980s, for the purpose of partial Siberian snake design studies at SATURNE [11]. In the mid-1990s, the computation of synchrotron radiation electromagnetic impulse and spectra was introduced to investigate, and solved, synchrotron radiation interference issues at the LEP mini-wiggler beam profile monitor. In the mean time, several new optical elements were added, including all sorts of electro-magnetic and electrostatic bends and lenses. Zgoubi has undergone extensive developments in the recent years, for design and machine operation studies regarding high energy accelerators and storage rings, including the Neutrino Factory FFAG rings in the early 2000s, polarized beams at the Brookhaven National Laboratory 25 GeV AGS, its 3 GeV Booster, RHIC heavy ion collider, and the electron rings of the Electron-Ion Collider (EIC) complex at BNL. All this reflects in the simulation exercises proposed in this book.

1.2 Raytracing with Zgoubi—Solving the Exercises

Zgoubi is a stand alone series of Fortran files, compiling does not require any specific library. Running zgoubi requires no interface (various interfaces have been developed over the years though, and made available, see Sect. 1.3).

A beam optics problem in zgoubi consists in an ASCII input data file, its default name is zgoubi.dat. That ASCII file may actually be split, in as many ancillary files as desired, for instance according to a modular structure of an optical sequence.

Executing zgoubi.dat is as simple as this: [pathTo]/zgoubi-code/zgoubi/zgoubi *i.e.*
typing the address of the executable file. The execution produces an output ASCII listing, zgoubi.res, always. Zgoubi may produce various additional output files during
execution, according to user's requests.

One has to bear in mind that the only thing zgoubi knows to do is pushing particles: starting from an initial position and velocity, it computes particle coordinates
along an optical sequence, by stepwise integration of the Lorentz force differential
equations of motion. The input data file describes that optical sequence; it also
includes diverse commands aimed at delivering ancillary results, the latter anyway
derived from particle coordinates. As aforementioned a few things may actually happen while particles are pushed: spin motion, decay in flight, synchrotron radiation,
space charge perturbation, etc.

An optical sequence in zgoubi is a sequence of keywords, most of them followed
by one or more lines of numerical data (*e.g.*, in the case of optical elements: length,
field value, integration step size, fringe field parameters possibly, etc.), like so:

```
Title: this is my optical sequence. Particles will be
! pushed through, all the way to 'END'
'OBJET'
a few lines of data define initial particle coordinates (initial
conditions are needed to solve the differential equation of motion!)
'DIPOLE'
a few lines of parameters: field, fringe fields, etc.
                                        ! this is a comment line
'FAISCEAU'                 ! print out local particle coordinates
'QUADRUPO'
a few lines of parameters: field, fringe fields, etc.
'DIPOLE'
a second dipole
                            an empty line, not a problem
'BEND'
another type of dipole, with its own parameters and subtleties
'DRIFT'
drift length
'FAISCEAU'                 ! print out local particle coordinates
'SYSTEM'
2                                          ! 2 commands follow
echo 'this is a system call'
gnuplot < ./gnuplot_ellipses.gnu                ! some gnuplot script
'END'                                      ! execution stops here
trash                      ! whatever follows is trash, ignored
more trash
```

An optical sequence begins with a title line. And then:

OBJET: most of the time the first keyword, it defines the coordinates of particles making up the object to be transported; this is mandatory as initial conditions are needed in order to solve the Lorentz force equation.

Optical elements and commands follow, for instance

- DIPOLE: define a dipole magnet;
- EBMULT: a combined **E**, **B** multipole;
- ELCYLDEF: a cylindrical electrostatic deflector; MULTIPOL: lenses; CAVITE to accelerate; TOSCA to handle field maps; WIENFILTER; etc.

Zgoubi offers a total of about 50 magnetic and/or electrostatic optical elements [1, pp. 9, 10 and 13, 14].

Commands—which are keywords as well—are added wherever desired along the optical sequence, they include such procedures as

- FAISCEAU, FAISTORE: log local particle coordinates, respectively in zgoubi.res or in an ancillary output file;
- IMAGE[S]: compute local image density and size, etc.;
- GOTO: move the execution pointer to some arbitrary location along the sequence (useful for instance for managing beam transport amongst recirculating linacs spreader and combiner sections);
- TWISS, MATRIX: compute paraxial quantities from rays; SYSTEM: a system call;
- INCLUDE: to include ancillary input data files, a recursive command.

Keywords include switches, for instance to request

- spin tracking: SPNTRK, whose numerical data include initial spins, a necessary ingredient as initial conditions are needed in order to solve the Thomas-BMT equation;
- space charge perturbations: SPACECHARGE;
- in-flight decay: MCDESINT, synchrotron radiation: SRLOSS, etc.

Launching matching procedures resorts to FIT, FIT2 keywords, two different matching methods.

In the exercises, optical elements and procedures are most of the time referred to by their corresponding keyword, with little additional explanation: further information regarding their use and functioning is to be found in the indispensable companion to the resolution of the exercises, Zgoubi Users' Guide [1]:

- PART A of the guide describes what keywords do and how, and the physics content of the code, optical elements in particular.
- PART B details the formatting of the input data which follow most keywords (a few keywords do not require any data, for instance YMY, FAISCEAU, MARKER).
- A complete list of the available keywords can be found in the "Glossary of Keywords" sections at the beginning of both PART A and PART B.

– A quick overview of what optical elements can be simulated using `zgoubi`, and what keywords can be used for that, is given in the "Optical elements versus keywords" sections which follow the "Glossary of Keywords" sections. Note in passing, there are most of the time various ways to simulate one particular optical element, either for historical reasons, or to allow for actual and/or real life subtleties (for instance, between a gradient dipole and an offset quadrupole; between the various modes of operation of an accelerating radio-frequency system).
– The Index at the end of Zgoubi Users' Guide is a convenient tool to navigate keywords.

A concise notation KEYWORD[ARGUMENT1, OPTION, ...] is used in the exercises and solutions: it is believed that the reader will get promptly familiarized with these shortcuts, of which the main goal is to alleviate the text. The nomenclature KEYWORD[ARGUMENT1, OPTION, ...] follows the nomenclature of the Users' Guide, Part B. Three examples:

– OBJET[KOBJ = 1] stands for keyword OBJET (generating particle coordinates), and KOBJ = 1 option retained here;
– DIPOLE[IL = 2, XPAS = 2.5] stands for keyword DIPOLE (magnetic dipole); print out stepwise particle data to zgoubi.plt (this is what "IL = 2" stands for!); integration step size XPAS = 2.5 cm;
– OPTIONS[CONSTY ON, WRITE OFF] stands for keyword OPTIONS (gives access to various options), and two options retained here, (i) CONSTY (maintain constant transverse coordinates during stepwise integration through optical elements), switched ON; (ii) switch off most print outs to zgoubi.res.
– INCLUDE[NBF = N,FNAME = fileName, LBL_1A = from_A,LBL_1B = to_B] inserts locally, N times, a piece of a sequence copied from 'fileName' file, comprised between LABEL1-type MARKERS 'from_A' and 'to_B'.

Coordinate nomenclature

In the theoretical reminders, *i.e.* Sect. 1.3 in the following chapters, conventional notations are used for particle coordinates, namely,

$$\underbrace{x, \overbrace{x'}^{\text{radial}}, \overbrace{y, y'}^{\text{axial}}}_{\text{transverse coordinates}} , \underbrace{\delta s, \delta p/p}_{\text{longitudinal}}$$

δp and δs are respectively the momentum and path length offsets compared to a reference particle. These coordinates are defined in the Serret-Frénet frame, or moving frame, Fig. 1.2.

In the exercises instead, `zgoubi` coordinates are used, namely

$$\underbrace{\overbrace{Y, T}^{\text{radial}}, \overbrace{Z, P}^{\text{axial}}}_{\text{transverse coordinates}} , \underbrace{S, D}_{\text{longitudinal}}$$

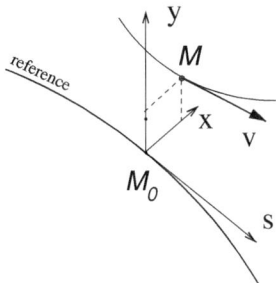

Fig. 1.2 Moving frame $(M_0; s, x, y)$ along a reference line. M_0, at path distance s from some origin, is the reference particle location

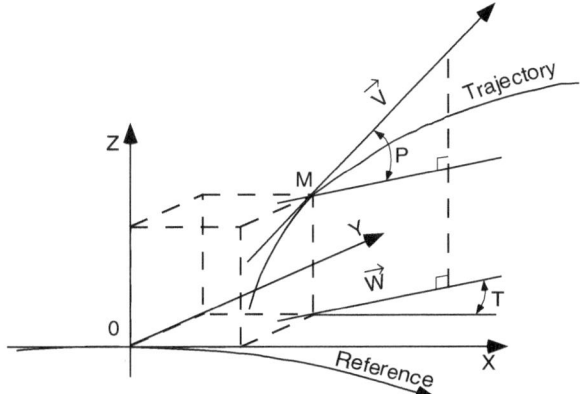

Fig. 1.3 Coordinates Y, T, Z, P in zgoubi [1, Sect. 1.1]. Reference curve: a straight axis in optical elements defined in a Cartesian frame; an arc of a circle in those defined in a cylindrical frame. OX: in the direction of motion, tangent to the reference; OY: normal to OX; OZ: orthogonal to the (X, Y) plane; $\mathbf{W}$: projection of the velocity, $\mathbf{v}$, in the (X, Y) plane; T: angle between $\mathbf{W}$ and the X-axis; P: angle between $\mathbf{W}$ and $\mathbf{v}$

The transverse coordinates are explicited in Fig. 1.3. S is the path length, D is the relative rigidity of the particle, relative to a reference rigidity specified as part of the initial object definition in zgoubi input data file. As a matter of fact, an initial object, *i.e.* the set of initial coordinates of particles to be raytraced, and possibly their spins, always has to be defined, for zgoubi to solve the differential equations of particle and spin motion.

An important additional parameter is the integration step. Figure 1.4 displays the position and velocity vectors of a particle in zgoubi frame, and a Δs push from position M_0 to position M_1. That push is performed using a Taylor expansion in Δs [1, Sect. 1.2]. The integration step size is one of the available controls on the accuracy of the integrator, when applied to the Lorentz force equation, or to the Thomas-BMT

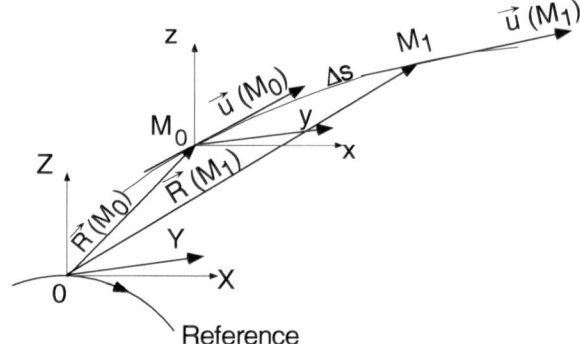

Fig. 1.4 Position vector **R** and normalized velocity vector ($\mathbf{u} = \mathbf{v}/v$) of a particle in zgoubi frame. A Δs push takes the particle from position M_0 to position M_1

spin equation. It also controls the accuracy of the simulation of events, such as photon emission, in-flight decay, etc.

Conventional and zgoubi coordinate notations may sometimes be used concurrently, for instance when equations from the main text are referred to, or resorted to, in the exercises. This is presumably in contexts exempt of ambiguity.

Reference frames of optical elements

Optical elements in zgoubi define fields in a Cartesian reference frame: this is the case for instance for MULTIPOL, BEND, EBMULT; or in a cylindrical reference frame: case of *e.g.*, DIPOLE, ELCYLDEF. And similarly for field map handling keywords: CARTEMES, TOSCA[MOD≤19], BREVOL use a Cartesian meshing, whereas POLARMES, TOSCA[MOD≥20] use polar or cylindrical meshing. Referring to Fig. 1.5: let a particle location M(X, Y, Z) project at m(X, Y) (the dashed curve figures the projected trajectory). In the case of an optical element (figured as a rectangular box) defined in Cartesian coordinates, X and Y in zgoubi.plt (columns respectively 22 and 10 [1, Sect. 8.3]) denote the coordinates taken along the fixed

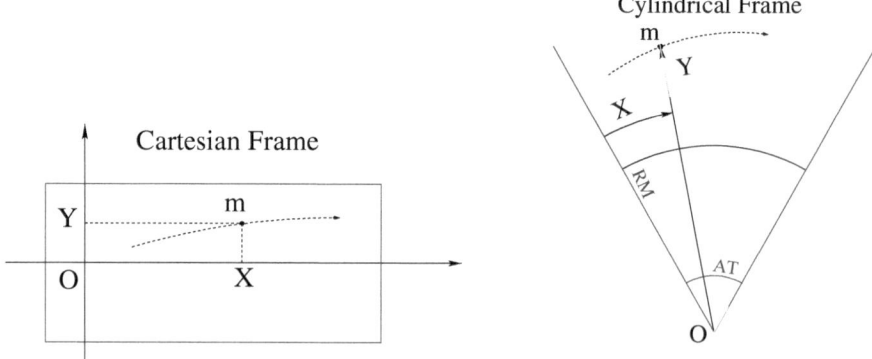

Fig. 1.5 Cartesian and cylindrical reference frames in optical elements

reference frame axes. In the case of an optical element (figured as an angular sector AT with some reference radius RM) defined in a cylindrical coordinate frame (Y, X, Z), Y is the radius, X is the polar angle, counted positive clockwise, Z is the vertical coordinate (column 12 [1, Sect. 8.3]).

1.3 Graphics, Data Treatment: zpop, gnuplot, awk, python

An execution of a beam optics problem in zgoubi produces a listing, zgoubi.res, always. However, when running a problem the user often requests logging of execution data in zgoubi.fai (produced by FAISTORE[FNAME = zgoubi.fai, or else]) and/or zgoubi.plt (produced as a result of IL = 2 flag, *e.g.* as in DIPOLE[IL = 2]).

The output file zgoubi.fai is a record of more than 50 particle data (coordinates, spin, etc.) [1, Sect. 8.2], at the location(s) where the keyword is inserted in the optical sequence.

The output file zgoubi.plt is a record of more than 50 particle data, step-by-step (coordinates, fields, step size, etc,) [1, Sect. 8.3] while integration proceeds through an optical element.

Beyond, a PRINT command available in several keywords allows specific printouts during raytracing. For instance, CAVITE[PRINT] will cause particle acceleration data to be logged in zgoubi.CAVITE.Out, which can then be accessed from gnuplot scripts, to produce graphs, data treatment, or provide debugging help. In the same line, one would get zgoubi.HISTO.out from HISTO[PRINT], zgoubi.OPTICS.out from OPTICS[PRINT], zgoubi.PICKUP.out from PICK-UPS[PRINT], zgoubi.SPNPRT.Out from SPNPRT[PRINT], etc. [1, Sect. 8].

Zpop [12], an old companion postprocessor of zgoubi's, allows handling zgoubi.fai and zgoubi.plt. It also allows brute reading of and plotting from any of the other files mentioned above. Zpop is part of the sourceforge package, portable on any linux and Mac OS. Quick to launch (in an xterm window), quick to operate. After years of development and utilization zpop allows all sorts of graphs, and various post-processing, reading particle coordinates and other data from zgoubi.plt or zgoubi.fai records.

Zpop menu 7, for instance allows plotting any variable entering the process of pushing particles step by step and element by element, against any other. There are of the order of 60 of them: particle coordinates, **E** and **B** field components, spin components, RF phase, step size, optical element number, turn number, etc., as well as derivatives or combinations [1, PART D, Sect. 1.3]. By experience, menu 7 answers most of the needs of lattice studies and beam dynamics simulations.

Zpop menu 8, allows further treatment of data read from these output files from a run, for instance drawing of synoptics with trajectories superimposed, Fourier analysis of periodic motion, matching of Enge's fringe field coefficients, etc.

Although this book is not a guide to the use of zpop, graphs found in the solutions of simulation exercises often use the latter.

When they are not produced using `zpop`, data analysis and graphic in the solutions use gnuplot, an incredibly simple yet powerful tool, even more so when added awk commands. By experience, gnuplot is quite suited as a graphic interface to `zgoubi` output data files, awk adds a powerful data analysis and treatment tool, both combined answer about any needs.

There is more, about python, following section.

1.4 Interface to Zgoubi?

Zgoubi can be run without an interfacing software, there is no need for that. Again, all that is needed is (i) an input data file, zgoubi.dat, which starts with the definition of initial coordinates, followed by a linear description of the optical sequence to be raytraced through, and with a few commands sprinkled around, and (ii) the following command:

[pathTo]/zgoubi

which is the address of the executable. Execution results are logged in output files, of which zgoubi.res *a minima*. Whatever is needed to handle the code is found in Zgoubi Users' Guide, which is part of the sourceforge package [1].

python [13]

A Zgoubi user quick startup has been written by beginners a few years ago [14]. This startup introduces to pyZgoubi, a python based interface to zgoubi developed by Sam Tygier, which has its own web site [15] and at present maintained at RAL

and BNL.  is an additional python interface, developed by a group from

Brussels university, available on internet as well [16].

Not strictly speaking python, but based on anyway, Sirepo accelerator simulation package by Radiasoft company also offers an interface to `zgoubi` [17].

References

1. F. Méot, Zgoubi Users' Guide. https://www.osti.gov/biblio/1062013-zgoubi-users-guide The latest version of the guide, on Sourceforge: https://sourceforge.net/p/zgoubi/code/HEAD/tree/trunk/guide/Zgoubi.pdf
2. S. Kowalski, H.A. Enge, *RAYTRACE. Laboratory For Nuclear Science* (MIT, Cambridge, MA, USA, 1986). http://aea.web.psi.ch/Urs_Rohrer/MyFtp/RAYTRACE/raytrace1.pdf
3. S. Agostinelli et al., GEANT4—A simulation toolkit. NIM A **506**(3), 250–303 (2003). https://geant4.web.cern.ch/
4. J. Thirion, P. Birien, *LE SPECTROMETRE II*. (Rapport Interne DPhN/ME, CEA Saclay, 1975)
5. H. Catz, *LE SPECTROMETRE SPES II* (Rapport Interne DPhN/ME, CEA Saclay, 1980)

6. P. Birien, S. Valéro, Projet de spectromìtre magnétique à haute résolution pour ions lourds. Note CEA-N-2215 (CEA-Saclay, 1981)
7. A. Tkatchenko, F. Méot, *Calculs optiques pour le spectromètre à kaons de GSI* (Rapport Interne CEA/LNS/GT/88-07, Saclay, 1988)
8. Zgoubi downloads on sourceforge: https://sourceforge.net/projects/zgoubi/files/stats/map?dates=2007-09-01 to 2023-01-01&period=monthly
9. F. Méot, S. Valéro, *Manuel d'utilisation de Zgoubi* (Rapport IRF/LNS/88-13, CEA Saclay, 1988)
10. F. Méot, S. Valéro, in collaboration with J. Doornbos, P. Stewart, Zgoubi users' guide, Int. Rep. CEA/DSM/LNS/GT/90-05, CEA Saclay (1990) & TRIUMF report TRI/CD/90-02 (1990)
11. F. Méot, A numerical method for combined spin tracking and raytracing of charged particles. NIM A313, 492, Proc. EPAC **1992**, 747 (1992)
12. Zgoubi's data treatment software zpop comes, and compiles independently, as part of the zgoubi package [8], when downloaded from sourceforge. It is available, including source files, at https://sourceforge.net/p/zgoubi/code/HEAD/tree/trunk/zpop/
13. https://www.python.org/
14. A. Pressman, K. Hock, Zgoubi. *A Startup Guide for the Complete Beginner* (2014). https://sourceforge.net/p/zgoubi/code/HEAD/tree/trunk/guide/aGuide4Beginner/arxiv.org_abs_1405.4921.pdf
15. S. Tygier, D. Kelliher, Developers: pyZgoubi. https://github.com/pyZgoubi/pyZgoubi
16. C. Hernalsteens, R. Tesse, M. Vanwelde, Zgoubidoo. https://ulb-metronu.github.io/zgoubidoo/
17. https://www.sirepo.com/en/apps/particle-accelerators/

Chapter 2
Electrostatic Systems

Abstract This chapter introduces to electrostatic systems used in beam optics, and to the theoretical material needed for the simulation exercises. It begins with a brief reminder of the historical and technological context, and continues with electrostatic optics methods which beam handling, guiding and focusing lean on. Zgoubi optical element library offers analytical modeling of several electrostatic components. For instance ELCYLDEF: an electrostatic deflector; ELMULT: a multipole, up to 20 poles; WIENFILTER: a plane condenser, possibly combining a magnetic dipole; ELMIR, ELMIRC: N-electrode mirrors and condenser lenses, with straight or circular slits. Electrostatic elements can be simulated as well using field maps, via the keywords TOSCA, MAP2D-E or ELREVOL. Running a simulation generates a variety of output files, including the execution listing zgoubi.res, always, and, on demand, such files as zgoubi.plt, zgoubi.fai, zgoubi.MATRIX.out, aimed at looking up program execution, storing data for post-treatment such as graphics, etc. Additional keywords are introduced as needed in the exercises, such as the matching procedures FIT[2]; FAISCEAU and FAISTORE to log local particle data in zgoubi.res or in a user defined ancillary file; MARKER; the 'system call' command SYSTEM; REBELOTE do-loop for multiple-pass or for parameter scans; and some more. This chapter introduces in addition to spin motion in electrostatic fields, the simulation of which is triggered by the keyword SPNTRK. SPNPRT or FAISTORE log spin vector components in respectively zgoubi.res or an ancillary file. The "IL = 2" flag logs stepwise particle data, including spin vector, in zgoubi.plt file. Simulations include deriving transport matrix, beam matrix, optical functions, from rays, using MATRIX and TWISS keywords.

Notations Used in the Text

$\mathbf{A}$	vector potential
a	electron gyromagnetic anomaly, $a = 1.15965 \times 10^{-3}$
B	magnetic field
$B\rho$	magnetic rigidity, $B\rho = p/q$

© The Author(s) 2024
F. Méot, *Understanding the Physics of Particle Accelerators*, Particle Acceleration
and Detection, https://doi.org/10.1007/978-3-031-59979-8_2

E; $m_0 c^2$; E_i	energy, $E = mc^2$; at rest; injection energy
$\mathbf{E}$; $E_{s,x,y}$; $E_\perp$	electric field vector; its components; normal component to $\mathbf{v}$
$E\rho$	electric rigidity, $E\rho/v = p/q = B\rho$
$\mathbf{F}$	Lorenz force
FOFDOD	a Focusing-drift-Focusing-Defocusing-drift-Defocusing lattice cell
G	hadron gyromagnetic anomaly. Proton: $G = 1.7928474$
m; m_0	particle mass; at rest
$(O; r, \theta, z)$	cylindrical frame
$(O; s, x, y)$	Cartesian frame
$\mathbf{p}$; $p_{s,x,y}$	momentum vector of a particle; its components
q	particle charge
R_0; r_0	condenser equipotential radii
s	path variable $[(*)' = d(*)/ds]$
T	kinetic energy
t	time variable $[(\dot{*}) = d(*)/dt]$
U	potential energy
$\mathbf{v}$; $v_{s,x,y}$	velocity vector of a particle; its components
V; V_i	voltage
α	trajectory deflection, $\alpha = \int \frac{E_\perp \, ds}{E\rho}$
β	v/c
$\delta p/p$, δ	relative momentum offset
ϕ	electrostatic scalar potential
ρ	curvature radius

2.1 Introduction

A historical electrostatic beam line is the column of electrostatic tubes which, in 1932, allowed guiding and accelerating a proton beam to a target, so producing the first artificial atom-splitting reaction, $p + {}^7Li \rightarrow 2\,{}^4He$, the Cockcroft-Walton experiment [1]. A high voltage was produced by an *ad hoc* diode and condenser cascade rectifying the AC voltage from a transformer. This high DC voltage was applied to a string of conducting cylinders (Fig. 2.1) which ensured beam guiding, (sufficient) focusing, and acceleration to $700\,\text{keV}$, a high enough energy to break the Coulomb barrier in this nuclear reaction. Which earned its authors the 1951 Nobel Prize.

Electrostatic systems allowed a landmark advancement, the first acceleration of a polarized beam, at the University of Basel in the early 1960s, a period where polarized proton and deuteron sources began operating [3, 4]. The experiment used a $200\,\text{keV}$ electrostatic accelerator. "*The Basel group [...] presented the first deuteron source in operation at the time of the first polarization conference in Basel 1960*" [5]. The convention for the sign of polarization is known as the "Basel Convention". Polarized beam acceleration at the nearby ETH Zurich $6\,\text{MV}$ Van de Graaff generator was not

Fig. 2.1 A similar tube cascade to the early 1930s Cockcroft-Walton experiment eponymous acceleration system: Fermilab's 750 keV H$^-$ injector [2]

far behind. Acceleration of polarized ions in cyclic accelerators soon followed, to way higher energy, starting with cyclotrons [4].

A landmark in physics as well: the electron column. The design of the first electron microscope and of the scanning tunneling microscope earned their authors the 1986 Nobel Prize—well, actually these designs used magnetic lenses. Nevertheless, the electron column, which combines electrostatic and magnetic components, is a widespread system since, with a number of variants such as transmission-, scanning-, or photoemission-electron microscope, and the electron-beam lithography column.

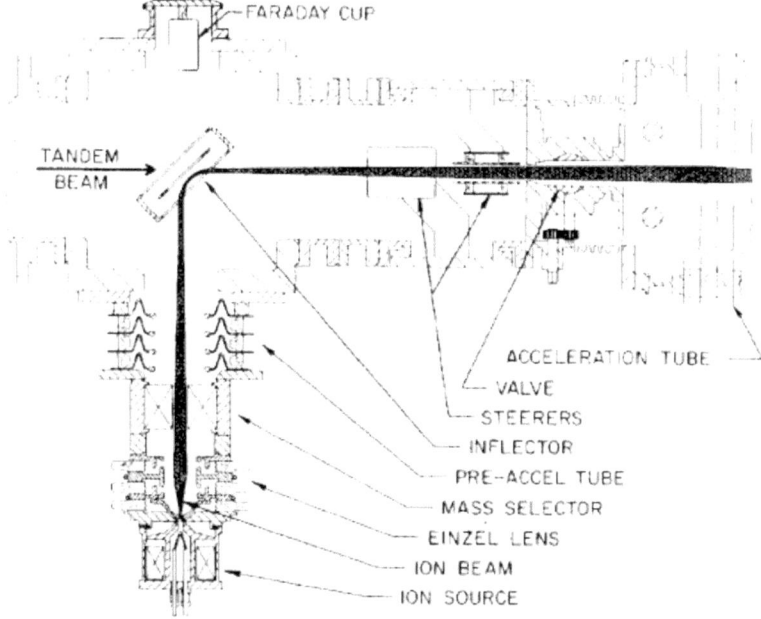

Fig. 2.2 Typical beam handling in an ion source region (BNL AGS injectors). Several electrostatic systems are at work in a short distance: a focusing Einzel lens, a Wien filter mass selector, pre-accelerating tubes, an inflector which serves as a switch with a Tandem ion line, electrostatic condensers to steer the beam, more acceleration tubes

Electron beam energies range in 0.1–1 MeV [6]. A century of design and technological refinements in electron optics have brought these systems to optical perfection.

Electrostatic optical elements are light objects. Deflectors and lenses are simple to construct, simple mechanic forms shape the required fields, electrode voltages can be up to a fraction of a MV, gradients to several MV/m, there is no remanence, power consumption is low. All reasons why electrostatic optical elements are used where energy allows, in low energy beam lines in particular (Fig. 2.2). Guiding and focusing components include prism, plane condenser, multipoles, mirrors, etc. [12], Figs. 2.3, 2.4, 2.5 and 2.6. Electrostatic components are not a specificity of low energy lines though, they span a large range of applications, with energy and size varying accordingly. On the small side are Einzel lenses used in particle source areas (Fig. 2.3). Main bends in beam lines may be of larger volume (Fig. 2.4). Even larger, in the meter range, are injection and extraction septa in GeV synchrotrons (Fig. 2.5), or pretzel orbit separators in GeV e+ e– colliders such as LEP and CESR [13, 14] (Fig. 2.6).

The electrostatic septum (Fig. 2.5) in particular is commonly used to steer beam into or out of circular accelerators. Megavolts/m gradients allow handling high beam rigidities, and achieve fraction of milliradian deflections aimed at. To give an idea of quantities at stake, the septum in Fig. 2.5 for instance is an 80 cm long device,

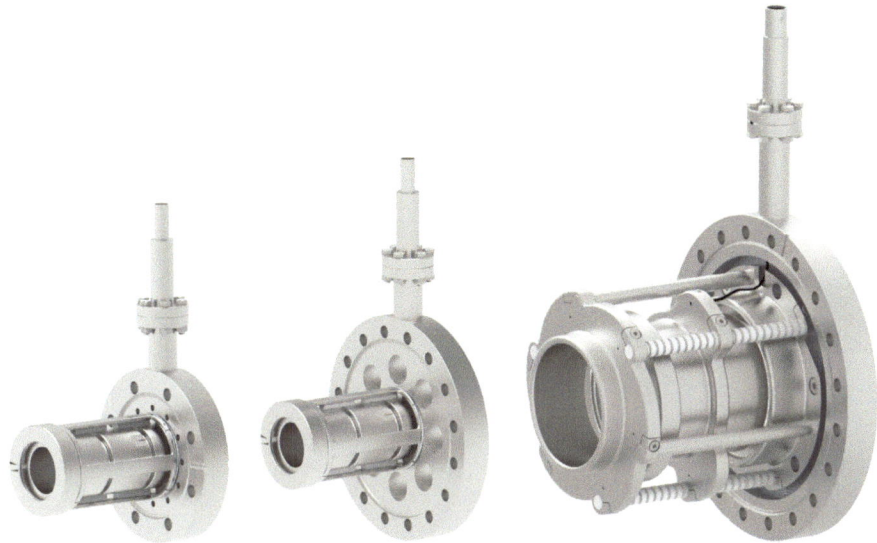

Fig. 2.3 Quite popular, the Einzel lens [7]. Three specimen here, diameters from 10 to 40 mm, operation voltage 10 to 30 kV

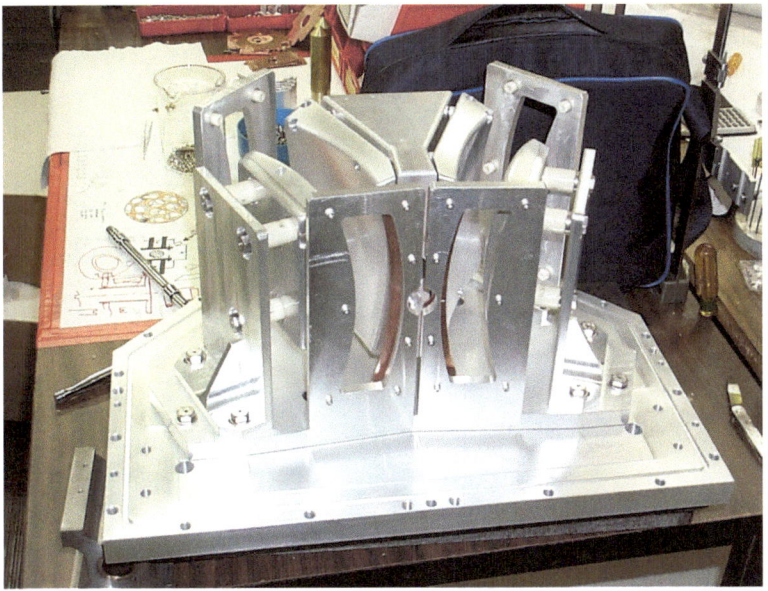

Fig. 2.4 A 3-way spherical electrostatic deflector [9]. Beam can be switched left or right, or let go straight

In-situ bake-out lamp   Deflector   HV feedthrough

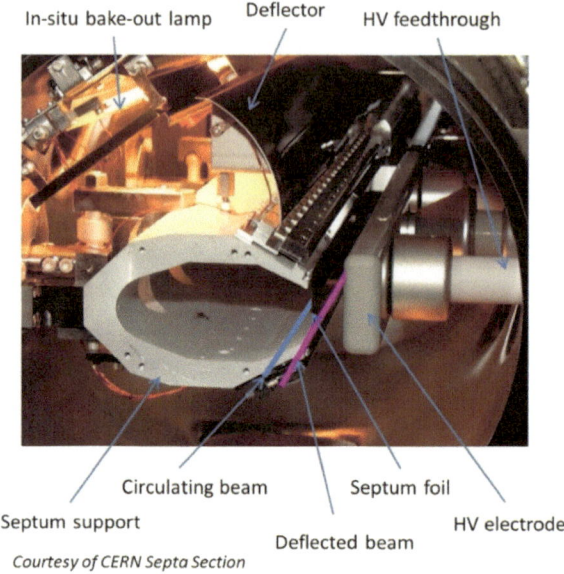

Circulating beam   Septum foil
Septum support   HV electrode
                    Deflected beam
Courtesy of CERN Septa Section

Fig. 2.5 A 250 kV septum for slow extraction at the SPS [8]. Electric field is on the extracted beam side

Fig. 2.6 Cornell ESR 3 m long horizontal pretzel separator, operating voltage ±85 kV (2 MV/m). Electrodes are split to let synchrotron radiation through

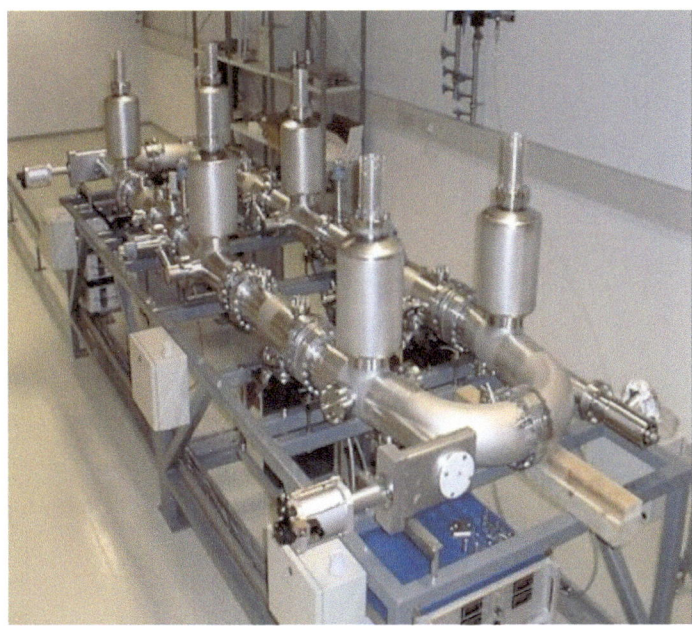

Fig. 2.7 Elisa in Aarhus, a 25 keV, 7.6 m circumference racetrack for molecular and atomic physics [10]. Its lattice combines spherical deflectors, plane deflectors and quadrupoles

Fig. 2.8 UMER ring at the University of Maryland [11]. A 10 keV, 11.5 m circumference beam optics and beam dynamics test accelerator

septum thickness is 100 μm, operating voltage 260 kV (15 MV/m over a 17 mm gap) for a deflection angle of 0.28 mrad.

Electrostatic optical elements have also invited themselves in the realm of rings. An electrostatic ring is used every once in a while for proof-of-principle purposes. The first occurrence was the "Electron Analog" (Fig. 2.9), built in 1954 to assess the

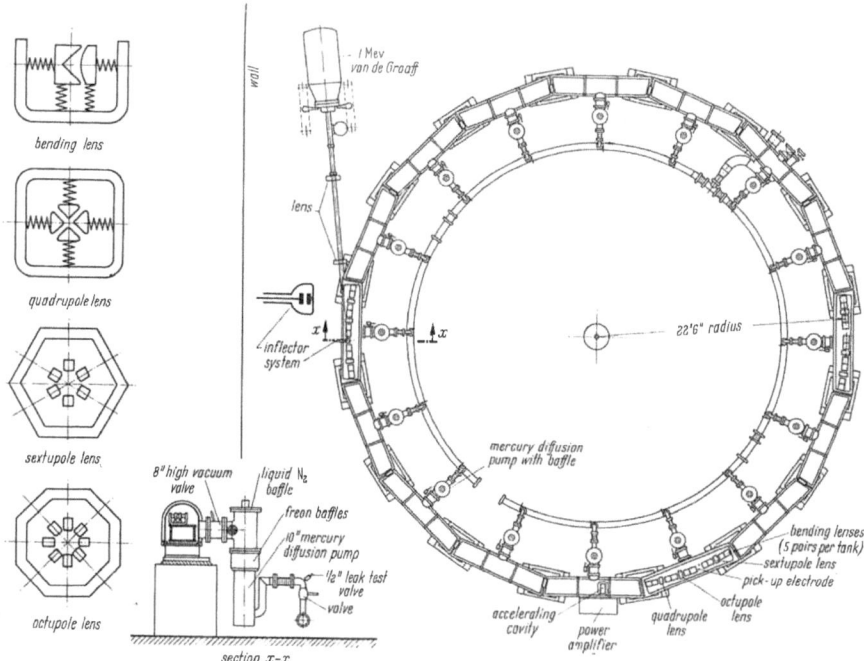

Fig. 2.9 The "Electron Analog", a prof-of-principle of BNL AGS, a strong $n = 225$ index FOF-DOD lattice, 45 ft in diameter, built in 1954 [15]. The left column shows the cross sections of the electrostatic optical elements which comprise the lattice

novel concept of strong focusing and transition-gamma crossing (cf. Chaps. 8 and 9), prior to the construction the AGS at the Brookhaven National Laboratory [15]. In the 1990s electrostatic rings were raised to the rank of tools for physics research, with energies of keVs to tens of keVs. Examples are the ion storage ring ELISA (Fig. 2.7), the beam physics ring UMER (Fig. 2.8), amongst others.

2.2 Basic Concepts and Formulæ

Mathematically speaking, electrostatic elements exploit the scalar potential component in

$$\mathbf{E} = -\mathbf{grad}\phi - \frac{\partial \mathbf{A}}{\partial t}$$

allowing local deflection and/or focusing and/or acceleration along DC voltage gaps. A fundamental aspect is that the resulting Lorentz force works. Particles exchange energy with the field, at a rate $\mathbf{F} \cdot \mathbf{v} = q\mathbf{E} \cdot \mathbf{v}$ which is in general non-zero, thus mass and velocity vary along the trajectory. This is a major difference with magnetic elements, in which $\mathbf{F} \cdot \mathbf{v} = q(\mathbf{v} \times \mathbf{B}) \cdot \mathbf{v} \equiv 0$, the magnetic force does not work, $|\mathbf{v}|$ and mass do not change.

Solving the Lorentz force differential equation $\frac{dm\mathbf{v}}{dt} = q\mathbf{E}$ requires the electric field distribution in space. The latter derives from a potential solution of the Laplace equation $\nabla^2 \phi = 0$, The necessary boundary conditions to solve it depend on the electrical properties of the device, on its shape, symmetries, and various components. For instance electrodes are equipotentials to which the electric field is normal; the electric field is along the axis in cylindrical tube; the transverse plane between two identical iso-potential tubes is a symmetry plane, etc. In simple systems, or with some *ad hoc* approximations, it is possible to find an analytical solution to the Laplace equation, from which analytical expressions of the components of the field vector $\mathbf{E} = -\mathbf{grad}\phi$ may be derived. In some complicated cases, it may still be possible to find analytical solutions for field components along a symmetry axis, or over a symmetry plane, and extrapolate from there using Taylor expansion and Maxwell's equations. With complicated geometry the easiest way may end up being the computation of a field map. Raytracing in a field map is at the expense of accuracy of the integration, though, as a result of field interpolation from a mesh. A dense mesh and an integration step size commensurate with the mesh size mitigate the issue.

2.2.1 Kinetics

Circular Motion; Rigidity

The Lorentz force on a particle of charge q and mass m in an electric field $\mathbf{E}$ is

$$\mathbf{F} = \frac{d\mathbf{p}}{dt} = \frac{d(m\mathbf{v})}{dt} = q\mathbf{E} \tag{2.1}$$

Circular motion requires velocity $\mathbf{v}$ to be normal to the electric field $\mathbf{E}$. Deflectors allow that, see below. It requires in addition, as in the cyclotron, the centripetal force to equate $\mathbf{F}$. Write it under the form $qE = -mv^2/\rho$. Ignoring the sign, this yields the electric rigidity

$$E\rho = \frac{p^2}{qm} = \frac{T}{q}\frac{1+\gamma}{\gamma} = \frac{m_0c^2}{q}\frac{\gamma^2-1}{\gamma} \tag{2.2}$$

where $T = mc^2 - m_0c^2$ is the kinetic energy of the particle. The trajectory deflection over an arc of length $\int ds$ normally to the field is

$$\alpha = \frac{\int E\,ds}{E\rho} = \frac{1}{v}\frac{\int E\,ds}{p/q} = \frac{1}{v}\frac{\int E\,ds}{(B\rho)} \tag{2.3}$$

where $(B\rho)$ denotes the particle rigidity. The velocity v appears in the expression for the deflection angle, compared to magnetic deflection $\alpha = BL/B\rho$.

Work of the Force

The work by a force $\mathbf{F}$ in the time interval t_1, t_2, over $d\mathbf{M} = \mathbf{v}dt$ is

$$\mathcal{T}_{1,2} = \int_{t_1}^{t_2} \mathbf{F}(M,t)\cdot d\mathbf{M} \tag{2.4}$$

Developing yields

$$\mathcal{T}_{1,2} = \int_{t_1}^{t_2} \frac{d}{dt}\left(\frac{m_0\mathbf{v}}{(1-v^2/c^2)^{1/2}}\right)\cdot\mathbf{v}dt = \int_{t_1}^{t_2} \frac{m_0\mathbf{v}\cdot d\mathbf{v}}{(1-v^2/c^2)^{3/2}}$$

$$= \int_{t_1}^{t_2} d\left(\frac{m_0c^2}{\sqrt{1-v^2/c^2}}\right) = \int_{t_1}^{t_2} d(mc^2) = [m_2 - m_1]c^2 \tag{2.5}$$

Thus, with kinetic energy defined as $T = mc^2 - m_0c^2 = E - m_0c^2$ the work writes

$$\mathcal{T}_{1,2} = E_2 - E_1 = T_2 - T_1 \tag{2.6}$$

If $\mathbf{F}$ derives from a time-independent potential V, namely $\mathbf{F} = -q\,\mathbf{grad}V(M, t)$, then, with $U = qV$,

$$\mathcal{T}_{1,2} = E_2 - E_1 = T_2 - T_1 = -\int_{t_1}^{t_2} \mathbf{grad}U\cdot d\mathbf{M} = U_2 - U_1 \tag{2.7}$$

thus

$$E_1 + U_1 = E_2 + U_2, \qquad T_1 + U_1 = T_2 + U_2 \tag{2.8}$$

In the non-relativistic limit $v/c \ll 1$, $\gamma \approx 1 + \beta^2/2$ so that, as expected

$$T_{1,2} = E_2 - E_1 \xrightarrow{\beta \to 0} \frac{1}{2} m_0 (v_2^2 - v_1^2) \tag{2.9}$$

Motion in a Uniform Field

Take the x axis parallel to $\mathbf{E}$, $\mathbf{E} = E_x \mathbf{x}$. The equations of motion write

$$\frac{d\mathbf{p}}{dt} = q\mathbf{E} \Rightarrow \begin{vmatrix} \frac{dp_s}{dt} = 0 \\ \frac{dp_x}{dt} = q E_x \\ \frac{dp_y}{dt} = 0 \end{vmatrix} \quad \text{thus} \quad \begin{vmatrix} p_s = p_{s0} \\ p_x = q E_x t + p_{x0} \\ p_y = p_{y0} \end{vmatrix} \tag{2.10}$$

Simplify the developments by taking the motion parallel to the s axis at time $t = 0$,

$$\mathbf{p}_0 = \begin{vmatrix} p_{s0} \\ 0 \\ 0 \end{vmatrix} \tag{2.11}$$

Integrating Eq. 2.10 is not straight forward as m is a function of v, such that

$$p_{s,x,y} = \frac{m_0 v_{s,x,y}}{\sqrt{1 - \frac{v_s^2 + v_x^2 + v_y^2}{c^2}}}$$

The difficulty can be surmounted in two steps [16]:
(i) Take $E^2 = p^2 c^2 + m_0^2 c^4$, with $p^2 = p_s^2 + p_x^2 + p_y^2 = p_{s0}^2 + (q E_x t)^2$, note $E(t = 0) = E_i$. Thus

$$E^2(t) = (m_0 c^2)^2 + p_{s0}^2 c^2 + (q E_x t)^2 c^2 = E_i^2 + (q E_x t)^2 c^2 \tag{2.12}$$

(ii) With $\mathbf{v} = \mathbf{p}/m = c^2 \mathbf{p}/E$, and $p_{s0} = \beta_i E_i/c$ as $\mathbf{p}(t = 0) = p_{s0}\mathbf{s}$, one then gets

$$\begin{vmatrix} \frac{ds}{dt} = v_s = \frac{p_{s0} c^2}{\sqrt{E_i^2 + (q E_x ct)^2}} = \frac{\beta_i E_i c}{\sqrt{E_i^2 + (q E_x ct)^2}} \\ \frac{dx}{dt} = v_x = \frac{q E_x c^2 t}{\sqrt{E_i^2 + (q E_x ct)^2}} \\ \frac{dy}{dt} = v_y = 0 \end{vmatrix} \tag{2.13}$$

An interesting result here is that the longitudinal velocity decreases with time. The transverse acceleration causes longitudinal deceleration. The radial velocity v_x increases, with c an upper limit:

$$\frac{dx}{dt} = v_x = \frac{q E_x c^2 t}{\sqrt{E_i^2 + (q E_x ct)^2}} \xrightarrow{t \to \infty} \frac{q E_x c^2 t}{\sqrt{(q E_x ct)^2}} = \pm c$$

The trajectory slope increases linearly with time,

$$\frac{dx}{ds} = \frac{dx/dt}{ds/dt} = \frac{qE_x}{p_{s0}}t = \frac{qE_xc}{\beta_i E_i}t \tag{2.14}$$

Integrate the differential Eqs. 2.13:

$$\left|\begin{array}{l} ds = \dfrac{p_{s0}c^2dt}{\sqrt{E_i^2+(qE_xct)^2}} = \dfrac{p_{s0}c}{qE_x}\dfrac{dt}{\sqrt{a^2+t^2}}, \quad \text{with } a = \dfrac{E_i}{qE_xc} \\[3mm] dx = \dfrac{qE_xc^2tdt}{\sqrt{E_i^2+(qE_xct)^2}} = \dfrac{ctdt}{\sqrt{a^2+t^2}} \\[3mm] dy = 0 \end{array}\right. \tag{2.15}$$

On the one hand $\int \frac{dt}{\sqrt{a^2+t^2}} = \mathrm{Asinh}\frac{t}{a}$; on the other hand $\int \frac{tdt}{\sqrt{a^2+t^2}} = \sqrt{a^2+t^2}$, so that

$$\left|\begin{array}{l} s = \dfrac{p_{s0}c}{qE_x}\int_0^t \dfrac{dt}{\sqrt{a^2+t^2}} = \dfrac{p_{s0}c}{qE_x}\left[\mathrm{Asinh}\dfrac{t}{a}\right]_0^t = \dfrac{p_{s0}c}{qE_x}\mathrm{Asinh}\dfrac{qE_xct}{E_i} \\[3mm] x = c\int_0^t \dfrac{tdt}{\sqrt{a^2+t^2}} = c\left[\sqrt{a^2+t^2}\right]_0^t = \dfrac{1}{qE_x}\left[\sqrt{E_i^2+(qE_xct)^2}-E_i\right] \\[3mm] y = 0 \quad \text{(motion is in (O; s, x) plane)} \end{array}\right. \tag{2.16}$$

The trajectory $x(s)$ is obtained by eliminating time between x and s using

$$qE_xct = E_i \sinh\frac{qE_xs}{p_{s0}c} \tag{2.17}$$

so that (accounting for $\cosh^2 - \sinh^2 = 1$)

$$x = \frac{E_i}{qE_x}\left(\cosh\frac{qE_xs}{p_{s0}c}-1\right) = \frac{E_i}{qE_x}\left(\cosh\frac{qE_xs}{\beta_i E_i}-1\right) \tag{2.18}$$

The motion is a catenary—the shape of a chain hanging by its two ends, under the effect of gravitation (Fig. 2.11). A paraxial approximation, valid for a small enough deflection, takes the Taylor development of cosh, yielding a parabolic trajectory

$$x_{\text{paraxial}} \approx \frac{1}{2}\frac{qE_x}{\beta_i^2 E_i}s^2 \approx \frac{s^2}{2\rho_0} \tag{2.19}$$

where $\rho_0 = \beta_i^2 E_i/qE_x$ is the radius of the tangent circle to the parabola.

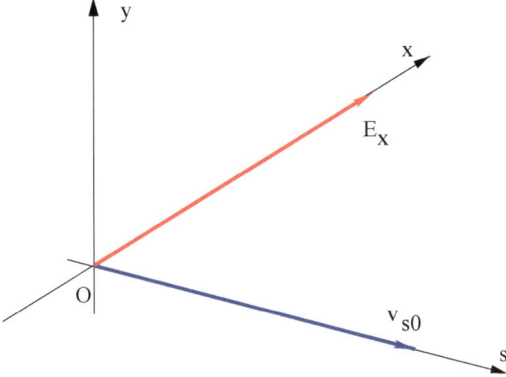

Fig. 2.10 Working frame (O; s, x, y). $\mathbf{E} \parallel \mathbf{x}$ and $\mathbf{v}(s = 0) \parallel \mathbf{s}$

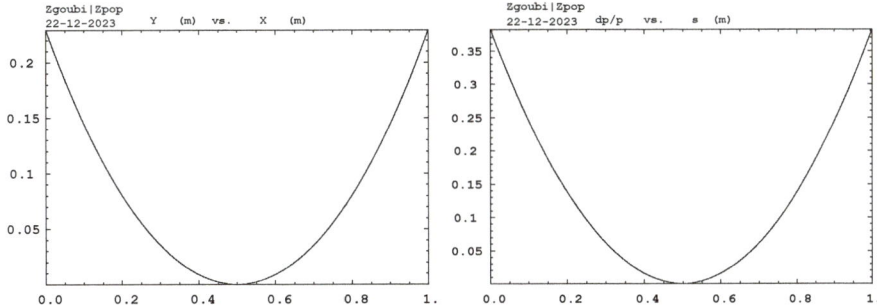

Fig. 2.11 Left: a catenary, trajectory of a 350 keV electron over 1 m in a $E_s = 980$ kV/m field. Right: the evolution of its relative momentum offset $\delta p / p_0$ from $s = 0$ to $s = 1$ m

2.2.2 Optical Components

As a particle travels in the electric field of an electrostatic element, its energy changes because the field along the path is in general not normal to the velocity, $\mathbf{F} \cdot d\mathbf{M} \neq 0$. This affects the velocity and mass (Eqs. 2.4 and 2.5).

In optical elements a reference optical axis is defined, straight or curved depending on the device. The analytical formalism in general assumes paraxial optics, i.e. trajectory angle to the optical axis remains small.

In various optical components, such as the Wien filter (see Sect. 12.2.4), quadrupoles, toroidal deflectors, the electric field is normal to the optical axis. Implications are

– the field is considered normal to trajectories as well, longitudinal velocity is assumed of constant magnitude,
– transverse excursions are small so that energy change can be ignored.

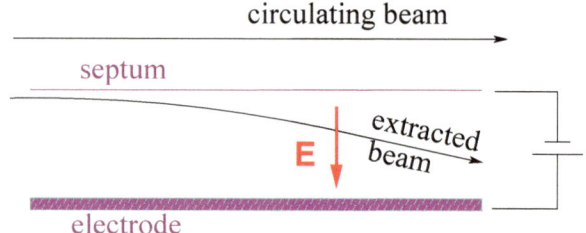

Fig. 2.12 A sketch of a plane condenser used as an electrostatic septum, a separation between a circulating beam, unaffected, and an extracted beam deflected under the effect of the field

Things are different in cylindrical lenses and mirrors, where the electric field can be near parallel, or far from normal, to trajectories.

These assumptions, aimed at allowing simplifying hypotheses for the sake of analytical modeling, are irrelevant anyway if numerical integration is used to solve the Lorentz equation.

Transverse Fields

Plane Condenser

A plane condenser is a simple concept (Fig. 2.12): a pair of parallel plates, to which a voltage is applied, allowing the deflection of a charged particle beam. The device is used in various optical systems: for beam guiding in low energy beam lines, electron columns and ion rings; for beam switching; in accelerators up to high rigidities for peeling out or switching beams; for orbit separation in high energy e+ e− colliders, to mention a few.

The paraxial approximation of the deflection $\alpha \approx \tan \alpha$ undergone over a distance L in the uniform field can be obtained from $\tan \alpha = dx/ds$ (Eq. 2.14), using Eq. 2.17 to remove time, giving

$$\alpha = \frac{qE_xL}{\beta_i p_{s0}c} = \frac{qE_xL}{\beta_i^2 E_i} \tag{2.20}$$

At this point it is interesting to compare with the equivalent effect of a force of magnetic origin, writing $qE = qc\beta B$. Thus,

$$E = c\beta B \quad \text{or} \quad E_{[\text{GV/m}]} \approx 0.3\beta B_{[\text{T}]}$$

A deflection equivalent to that due to a 1 T magnetic field could be achieved with an electric field of 9 MV/m in the case of a $\beta = 0.01$ particle, but is not doable for a $\beta \approx 1$ particle.

In the paraxial, parabolic approximation (Eq. 2.19), the radial motion writes [9]

$$x(s) = x_0 + x_0's + \left[\frac{\delta p}{p}(2 - \beta^2) - 1\right]\frac{s^2}{2\rho_0} \tag{2.21}$$

Fig. 2.13 A sketch of an electrostatic bend. A region of radial electric field is defined between concentric electrodes (equipotential surfaces) with axial and radial symmetry. The curvature radius of the reference trajectory $\rho_0 = r_0$

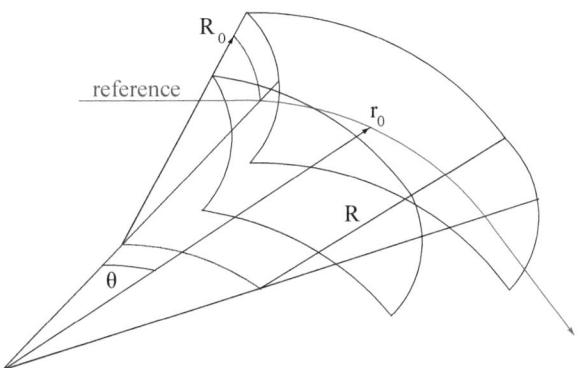

with s the longitudinal coordinate in the condenser frame (Fig. 2.10).

The length of the trajectory from the origin at $(X = 0, Y = 0)$ (assuming $\mathbf{v} \perp \mathbf{E}$ at that location), to $(X, Y(X))$ along the condenser is (Eq. 2.18)

$$l(X) = \int_0^X \left[1 + Y'^2(X)\right]^{1/2} dX \qquad (2.22)$$

$$= \int_0^X \left[1 + \left(\tfrac{1}{\beta_i} \sinh \tfrac{X}{a}\right)^2\right]^{1/2} dX = -ia\, Ei(i\tfrac{X}{a}, \beta_i^{-2})$$

$$\approx X + \tfrac{1}{6\beta^2}\tfrac{X^3}{a^2} + (\tfrac{1}{3} - \tfrac{1}{4\beta^2})\tfrac{1}{10\beta^2}\tfrac{X^5}{a^4} + \cdots$$

with $Ei(x, k)$ the elliptic integral of the second kind, i the imaginary unit and, below, a series approximation.

Section 12.2.4 further addresses the Wien filter, a combination of an electrostatic plane condenser and a magnetic dipole, featuring $\mathbf{E}$ and $\mathbf{B}$ field components normal to each other, and normal to the longitudinal axis.

Toroidal Condenser

A sketch of a toroidal condenser is given in Fig. 2.13, which also defines r_0, the radius of the reference axis and R_0, the vertical curvature radius. The reference axis is in the median plane, along an equipotential $\phi = r_0 E_0/2$ mid-way between the electrodes. This class of electrostatic bend comprises

– spherical condensers, $R_0 = r_0$, electrostatic potential $\phi = E_0 r (\tfrac{1}{2} - \ln \tfrac{r}{r_0})$, and
– cylindrical condensers, $1/R_0 = 0$, , electrostatic potential $\phi = E r (\tfrac{r_0}{r} - \tfrac{1}{2})$.

The deflection angle α along the reference axis satisfies Eq. 2.3. The energy of the ideal particle, along the optical axis, satisfies Eq. 2.2. Particle coordinates in a moving frame (see Sect. 3.2.2, Fig. 3.8) can be defined, namely, $x = (r - r_0)$ in the bend plane, y along an axis normal to the latter, and $s = r_0\theta$.

A $\delta p/p$ off-momentum particle differs from the reference one by its mass and velocity. The latter two vary as the particle travels across the bend, exchanging energy with the field. Combine these effects, appropriate approximations lead to the linear equations of motion in a cylindrical condenser $(1/R_0 = 0)$ [9]

$$\frac{d^2x}{ds^2} + \frac{2-\beta^2}{\rho_0^2}x = \frac{2-\beta^2}{\rho_0}\frac{\delta p}{p}, \quad \frac{d^2y}{ds^2} = 0 \qquad (2.23)$$

By comparison with the equations of motion in a uniform magnetic field (see Sect. 3.2.2, Eqs. 3.15 taken with a field index $k = 0$), a factor $2 - \beta^2$ appears, which tends to 1 at relativistic energy, as $\beta \to 1$.

In a toroidal condenser $(r_0/R_0 \neq 0)$, in the non-relativistic case $(\beta \approx 0)$, the equations of motion write [12]

$$\frac{d^2x}{ds^2} + \frac{2-c}{\rho_0^2}x = \frac{2}{\rho_0}\frac{\delta p}{p}, \quad \frac{d^2y}{ds^2} + \frac{c}{\rho_0^2}y = 0, \quad \text{with } c = \frac{r_0}{R_0} \qquad (2.24)$$

Quadrupole

With the force parallel to the electric field, transverse focusing requires (in an (x,y) plane transverse to the quadrupole axis)

$$E_x = -Kx = -\frac{\partial\phi}{\partial x}, \quad E_y = +Ky = -\frac{\partial\phi}{\partial y} \qquad (2.25)$$

A '$-$' sign for E_x is a convention. Thus $\mathbf{E}$ derives from the scalar potential

$$\phi = \frac{K}{2}(x^2 - y^2) \qquad (2.26)$$

In the case of a potential $\pm V/2$ applied on the electrodes (Fig. 2.14), with radius a at pole tip, then $K = -V/a^2$.

The equation of the equipotential is

$$y = \pm\sqrt{x^2 - \frac{2\phi}{K}} \qquad (2.27)$$

an hyperbola with its axes at 45 deg to the coordinate axes. As a matter of fact, pause

$$\begin{pmatrix} u \\ v \end{pmatrix} = \begin{pmatrix} \cos 45° & -\sin 45° \\ \sin 45° & \cos 45° \end{pmatrix}\begin{pmatrix} x \\ y \end{pmatrix}, \quad \text{so that} \quad \begin{pmatrix} x \\ y \end{pmatrix} = \begin{pmatrix} \cos 45° & \sin 45° \\ -\sin 45° & \cos 45° \end{pmatrix}\begin{pmatrix} u \\ v \end{pmatrix}$$

In this change of axes, ϕ changes to $\phi^* = Kuv$. Thus, an electrostatic quadrupole skewed by 45 deg achieves the same focusing as a magnetic quadrupole.

Fig. 2.14   An electrostatic quadrupole [18]. This one, a design for a 50 keV ion ring, operates in the kVolt range

The equations of motion have the same form as in a magnetic quadrupole (see Sect. 14.4.2), namely

$$
\begin{bmatrix} \frac{d^2x}{ds^2} + K_x x = 0 \\ \frac{d^2y}{ds^2} + K_y y = 0 \end{bmatrix} \quad \text{with} \quad K_x = -K_y = \frac{-q}{mv^2}\frac{V}{a^2} = \underbrace{\frac{\pm 1}{|E\rho|}}_{\text{electric rigidity (Eq. 2.2)}}\frac{V}{a^2} \quad (2.28)
$$

Relative efficiency of an electrostatic quadrupole

From $\mathbf{F} = q(\mathbf{E} + \mathbf{v} \times \mathbf{B})$ it results an equivalence between E and βcB. Technology does allow electric gradients beyond, say, 30 MV/m. For $\beta = 0.1$ for instance, the effect of such electric field is equivalent to that of $B = 30 \times 10^6/0.1c = 1$ T; for $\beta = 1$ it corresponds to $B = 30 \times 10^6/c = 0.1$ T. This relative inefficiency limits the use of electrostatic lenses to low energy beam lines.

Optical aberrations

More on the electric quadrupole can be found in [17]. Simulation outcomes reported in the latter demonstrate the accuracy which numerical integration allows on high order optical aberrations.

That publication also addresses the cancellation of second order achromatic aberrations by a lumped $(\mathbf{E}, \mathbf{B})$ quadrupole. The use of a pair of those as the final focus quadrupole doublet in a nanoprobe is the subject of Sect. 12.2.2.

Electrostatic Mirrors

Plane condensers include electrostatic mirrors [19]. These devices can be used for great trajectory deflection, including mirroring. In the latter case the longitudinal component of the velocity cancels and changes sign, a motion which stepwise ray-tracing handles efficiently.

Sketches of two such devices, available in `Zgoubi` optical element library, are given in Fig. 2.15.

The potential in the straight slit mirror can be modeled by (after [19], using the notations of Fig. 2.15)

$$V(Z, Y) = \sum_{i=2}^{N} \frac{V_i - V_{i-1}}{\pi} \arctan \frac{\sinh(\pi(Z - Z_{i-1})/D)}{\cos(\pi Y/D)} \tag{2.29}$$

where N is the number of plate pairs, D is their gap. This model assumes mid-plane symmetry, and infinite slits of negligible width. The mid-plane field component E_Z (and derivatives if needed) is obtained by differenciation, $E_Z = -dV(Z)/dZ$.

The potential of the circular slit mirror can be modeled by

$$V(r) = \sum_{i=2}^{N} \frac{V_i - V_{i-1}}{\pi} \arctan \left(\sinh \frac{\pi(r - R_{i-1})}{D} \right) \tag{2.30}$$

This model assumes mid-plane symmetry, and slits of negligible width. The mid-plane field $E_r(r)$ (and its r-derivatives if needed) are first derived by differenciation. $\mathbf{E}(r, Y)$ is then obtained by Taylor expansion in Y, using symmetries and Maxwell relations [20].

An example of a design of a time-of-flight ring for mass separation, based on these optical elements, is displayed in Fig. 2.16. More on this device can be found in [21].

Cylindrical Lenses

Cylindrical lenses are used for their focusing properties, in some cases combined with longitudinal acceleration. Focusing stems from the change of radial velocity through the gap between the tubes. It can be written

$$\Delta v_r = \int_{(gap)} \frac{q E_r(r, z)}{m v_z} dz,$$

with z the longitudinal axis, r the radial coordinate, and assuming revolution symmetry.

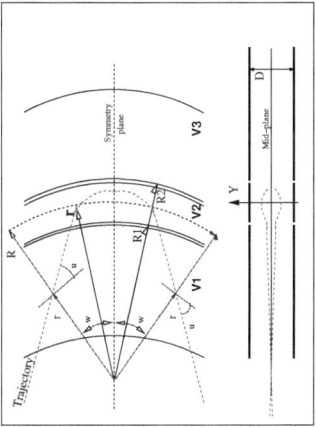

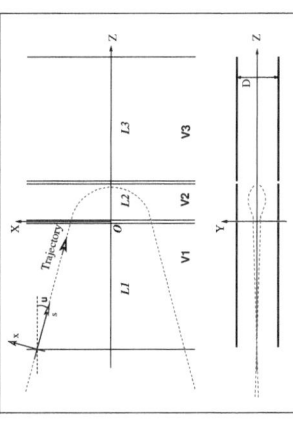

Fig. 2.15 Electrostatic N-electrode lens condensers (N = 3 electrodes, here), with straight slits on the left, circular slits on the right. They are used as deflectors in this schematic, however the device can be used as well in mirror mode, or as a focusing or defocusing lens in transmission mode. $(Z, X, Y = 0)$ is the bend plane. Parameters which define these systems include voltages V_{1-3}, plate lengths L_{1-3} and slit locations $Z_{1,2}$ or $R_{1,2}$

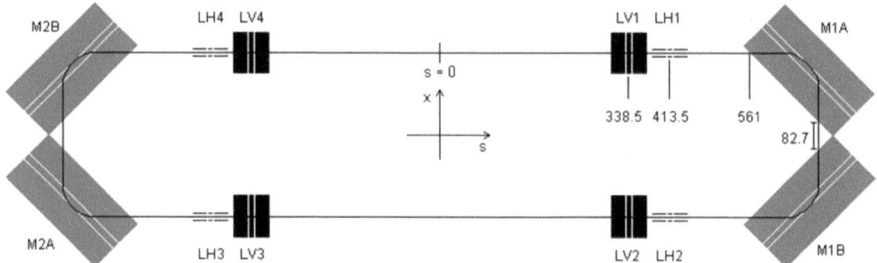

Fig. 2.16 A low energy electrostatic storage ring employed as a multiturn time-of-flight mass spectrometer. Three-plate condensers are used for both focusing (LH1-4, horizontal and LV1-4, vertical) and bending (M1A-B and M2A-B)

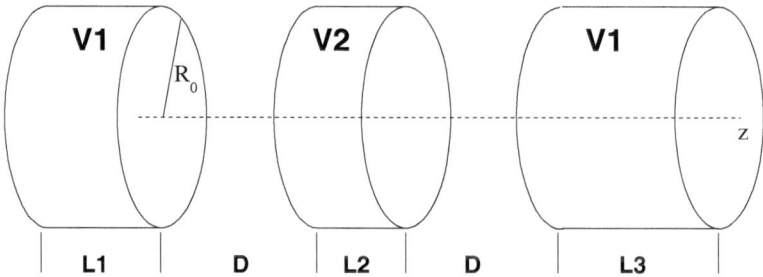

Fig. 2.17 Unipotential lens, with revolution symmetry. The tubes are distant D, they have the same diameter, $2R_0$. Their lengths and potentials are respectively L_1, V_1, L_2, V_2 and L_3, V_1

Numerical integration of the Lorentz equation along a trajectory requires knowing the potential. The electric field results, which provides the force which applies on the charged particle. Numerous publications have been dealing with the analytical modeling of cylindrical lenses, and testing of these models. Below are two examples, found in `zgoubi` optical element library.

Unipotential Lens [17]

A schematic of an unipotential lens is given in Fig. 2.17. Revolution symmetry about the z axis is assumed here. Various models for the electrostatic potential along the axis can be found in the literature, not so different anyway. A possibility is [22]

$$V(z) = \frac{V_2 - V_1}{2\omega D} \left[\ln \frac{\cosh \frac{\omega (z + 1/2L_2 + D)}{R_0}}{\cosh \frac{\omega (z + 1/2L_2)}{R_0}} + \ln \frac{\cosh \frac{\omega (z - 1/2L_2 - D)}{R_0}}{\cosh \frac{\omega (z - 1/2L_2)}{R_0}} \right]$$

(2.31)

The origin for z is taken in the middle of the central electrode, and $\omega = 1.318$.

Differenciation provides the electrostatic field component along the longitudinal axis, $E_z(z, r = 0) = -dV(z)/dz$. Radial and azimuthal field components are null

Fig. 2.18 Bipotential lens.
The tubes have the same
diameter, $2R_0$, their lengths
are respectively L_1 and L_2,
and potentials V_1 and V_2

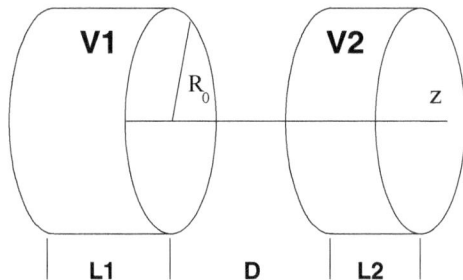

along the longitudinal axis. Their derivatives are not, off-axis Taylor expansions
provide $\mathbf{E}(r, z)$ [20, Sect 1.3.2].

More on the unipotential lens can be found in [17], including raytracing outcomes
and comparisons with numerical solution of the differential equation of radial motion
$r(z)$, namely,

$$\frac{d^2 R}{dz^2} + \frac{3}{16} \left(\frac{1}{V} \frac{dV(z)}{dz} \bigg|_{r=0} \right)^2 R = 0 \qquad (2.32)$$

with $R(z) = r(z)\, V^{1/4}(z)$ the Pitch variable.

Bipotential Lens

This is the basic optical block of a string of tubes, including multi-gap acceleration
columns. An analytical model for the potential along the axis in the geometry of
Fig. 2.18, in the case where the distance between the two tubes is negligible, is [23,
Chap. 5, Sect. 5.1.2] [20, *cf.* EL2TUB]

$$V(z) = \frac{V_2 - V_1}{2} \tanh \frac{\omega z}{R_0} + \frac{V_1 + V_2}{2} \qquad (D = 0) \qquad (2.33)$$

such that $E_z(z, r = 0) = -dV(z)/dz$. The origin for z is half-way between the
electrodes, and $\omega = 1.318$.

A second model assumes that the distance D between the two tubes is large enough
that the field fall-offs from the two lenses do not overlap. It is written

$$V(z) = \frac{V_2 - V_1}{2} \frac{1}{2\omega D / R_0} \ln \frac{\cosh \omega \dfrac{z + D}{R_0}}{\cosh \omega \dfrac{z - D}{R_0}} + \frac{V_1 + V_2}{2} \qquad (D > R_0) \quad (2.34)$$

Figure 2.19 shows examples of converging proton trajectories computed using that
model.

If a string of more than 2 tubes is modeled, an accelerating column for instance,
an upstream end lens (respectively downstream end) is modeled using $V_1 = 0$ (resp.
$V_2 = 0$).

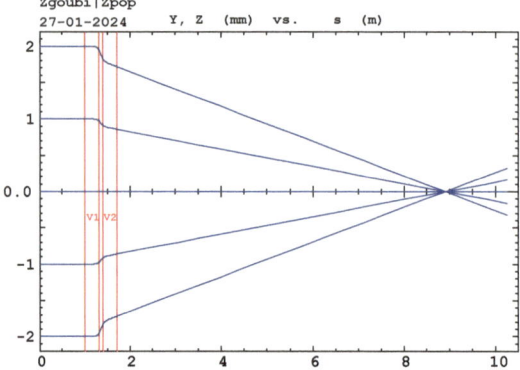

Fig. 2.19 Converging trajectories of 20 keV protons, with bipotential lens geometrical parameters $L_1 = L_2 = 36$ cm, $R_0 = 10$ cm, $D = 1.2\,R_0$, and gap voltage $V_2 - V_1 = -13$ kV

2.2.3 Periodic Structures

Periodic electrostatic structures are typically found in rings. ELISA, UMER and the Electron Analog are three examples, respectively Figs. 2.7, 2.8 and 2.9.

In the aforementioned hypotheses of paraxial optics, electric fields normal to the velocity vector, assuming negligible energy exchange between the beam and the electric field, particle motion abides by the principles of betatron motion. Basic theoretical material can be found in Chaps. 3–9.

These assumptions may however be misleading, acceleration or deceleration in the course of betatron motion may have noticeable effects. Fringe fields may also affect particle motion. This in addition translates into coupling between transverse and longitudinal motions. Stepwise raytracing is exempt of these limitations, as field models can be made as accurate as necessary, whereas numerical integration accounts for possible energy variation and coupling.

Electrostatic rings are typically synchrotron style of beam instruments. Longitudinal beam handling can use RF systems, for beam bunching, or for acceleration or deceleration. Bend and lens voltages are ramped during acceleration. Note that the latter may in principle be faster than with magnetic optics where eddy currents are a restricting factor.

Cyclic Acceleration Using an Electrostatic Field?

Is it possible to accelerate on a closed orbit using a DC voltage? The answer is 'no'.

The work of the force $\mathbf{F} = q\mathbf{E}$ over a path from A to B (top Fig. 2.20) only depends on the initial and final states, $U_A = qV_A$ and $U_B = qV_B$ (Eq. 2.7), it does not dependent on the details of the path. Thus, using an electrostatic field ($\mathbf{E} = -\mathbf{grad}\,V(\mathbf{R})$) it is not possible to accelerate a particle traveling on a closed path

Fig. 2.20 Top: the work of
the electrostatic force only
depends on voltages at A and
B, V_A and V_B, independently
of the path. Bottom: case of a
closed path, the particle loses
along (2) the energy gained
along (1)

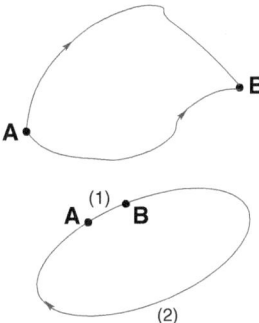

(bottom Fig. 2.20) as $\oint \mathbf{F}.d\mathbf{s} = 0$. As a consequence, a DC voltage gap in a circular
machine does not produce energy gain.

2.2.4 Spin Precession

Consider the classical model which, to the spin angular momentum $\mathbf{S}$ of a particle of
charge q and mass m, associates the magnetic moment $\mu = (1 + G)\frac{q}{2m}\mathbf{S}$ of a spinning
charge [24, Sect. 2.2]. In that model, under the effect of an ambient magnetic field
$\mathbf{B}_a$, $\mathbf{S}$ undergoes a torque

$$\frac{d\mathbf{S}}{dt} = (1 + G)\frac{q}{2m}\mathbf{S} \times \mathbf{B}_a \tag{2.35}$$

A particle traveling in the electrostatic field $\mathbf{E}$ of an optical system experiences in its
rest frame a magnetic field ($\mathbf{B}_a$) which is the Lorentz transform of the former (it also
experiences a electric field, which does not couple to μ). Expressing $\mathbf{B}_a$ in terms of
the laboratory electric field yields the differential equation of spin precession,

$$\frac{d\mathbf{S}}{ds} = \mathbf{S} \times \frac{\omega}{B\rho} \tag{2.36}$$

around a precession vector

$$\omega = \gamma \left(a + \frac{1}{1 + \gamma} \right) \frac{\mathbf{E} \times \beta}{c} \tag{2.37}$$

In these expressions, $\mathbf{S}$ is in the particle frame, it has not been Lorentz-transformed,
and all other quantities, including time, are expressed in a laboratory frame.

2.3 Exercises

2.1 Plane Condenser; Spin Motion
Solution 2.1

Electron dynamics in a parallel plate condenser is considered in this exercise. Use
WIENFILTER to simulate it, hard-edge field model.

Take condenser length 1 m, and electric field $|E| = 0.98\,\text{MV/m}$. Note: the reason
for this electric field value is to be found in Exercise 12.13 et seqs., an optimal field
setting of a Wien filter used as a spin rotator.

(a) Produce a graph of a symmetric catenary across the condenser.
    Check the transverse excursion of a particle and trajectory length, versus theory.
(b) Produce a graph of $(Y_{num} - Y_{th})/Y_{th}$ as a function of integration step size. Y_{num}:
    particle excursion at the downstream end of the condenser, from numerical inte-
    gration. Y_{th}: theoretical expectation, Eq. 2.18.
    Use REBELOTE[IOPT=1] as a do-loop, changing the integration step size,
    WIENFILTER[XPAS], at each pass.
(c) Add spin, parallel to the electric field $\mathbf{E}$ at start. In a similar manner to (b)
    produce a graph of $(\theta_{num} - \theta_{th})/\theta_{th}$ as a function of integration step size. θ_{num}:
    spin angle at the downstream end of the condenser, from numerical integration.
    θ_{th}: theoretical expectation, Eqs. 2.36 and 2.37.

For spin tracking, add SPNTRK.

2.2 Toroidal Condenser

Solution 2.2

Use ELCYLDEF to simulate a toroidal condenser.

(a) Set up a simulation showing that, in the paraxial hypothesis, in a cylindrical
    condenser, a diverging beam is re-focused after a deflection $\alpha = \pi/\sqrt{2}$.
    Test the convergence of the numerical solution versus integration step size.
(b) Produce the aberration curve $T(Y)$ at the focal plane. The moving frame can be
    shifted to the latter using AUTOREF.

2.3 A Time-of-Flight Mass Spectrometer Ring

Solution 2.3

Multiturn storage is a convenient way to achieve high resolution mass separation, in
a compact apparatus. Electrostatic mirrors are potential candidates as both deflector
and focusing optical element for a low energy storage ring. The design displayed in
Fig. 2.16 is an example. This exercise reviews various of its aspects.

(a) The parameters of the ring are given in Table 2.1. Use ELMIR with appropriate
    MOD option for both focusing lenses (LH, MOD = 22 and LV, MOD = 21) and
    bends (MA and MB, MOD = 11).

Table 2.1 Time-of-flight mass spectrometer. The parameters of a half-cell of the ring are given, as the cell is symmetric. Referring to Fig. 2.16: this parameter list starts from the center of the long drift ($s = 0$), going clockwise

Particle		N_2
Mass	uma	28
Mass	GeV	26.082
Charge	|e|	1
Kinetic energy	keV	400
Geometry		
Ring circumference[a]	cm	393.73658
Gap height in condensers	m	0.012
Number of slits in LH, LV		2
Number of slits in M		6
Length, electrode lengths		
Drift	(cm)	30.7
LV1	($3 \times$ cm)	2.525, 1.25, 2.525
Drift		1.2
LH1	($3 \times$ cm)	2.525, 1.25, 2.525
Drift	(cm)	11.6
M1	($7 \times$cm)	4.275, $5 \times$ 0.4163, 10
Drift	(cm)	6.00217933
Electrode voltages, in that order		
LV1	(V)	0, 115, 0
LH1	(V)	0, 40, 0
M1A	(V)	0, $5 \times$ 200, 400

[a] This is the length of the reference closed orbit

Build `zgoubi` input data file. Produce a synoptic of the ring in laboratory coordinates.

Produce the ring tunes, chromaticities. Produce a graph of the optical functions. TWISS can be used for that.

Produce a graph of horizontal or vertical trajectory over a few tens of turns.

(b) Produce a chromaticity scan (i.e., wave numbers as a function of momentum offset).

(c) Produce 1000-turn horizontal and vertical phase space motion, up to maximum stable amplitudes.

(d) Produce the time-of-flight histograms after 20 turns, for a bunch comprised of two masses: $M1 = 26.082$ GeV and $1.0004 \times M1$. Both bunches have a 400 keV average energy, *rms* energy spread $\delta E/E = 10^{-4}$, *rms* emittances $\epsilon_x/\pi = 0.02138 \ 10^{-6}$ m and $\epsilon_z/\pi = 0.0106 \ 10^{-6}$ m. All particles leave from s = 0 at the same time.

Use PARTICUL[M=M1,M2] to define two different masses [20, cf. PARTICUL].

Table 2.2 Parameters of the AGS electron analog [15]

Injection energy, T_i	MeV	1
Maximum energy, E_{max}	MeV	10
Physical radius, R	feet	22.5
Curvature radius, ρ	ft	15
Lattice cell		FOFDOD
Number of cells, N		40
Field index, n		225
Phase advance per cell		$\approx \pi/3$

2.4 Converging Rays in a Bipotential Cylindrical Lens

Solution 2.4

(a) Reproduce Fig. 2.19, using the analytical modeling EL2TUB.
     Check the evolution of proton rigidity across the lens.
(b) Repeat, using instead ELREVOL with a 1-D electrostatic field map.

2.5 The AGS Electron Analog

Solution 2.5

A schematic of the AGS electron analog is given in Fig. 2.9. Its parameters are given in Table 2.2. Refer to Chaps. 8 and 9 for preliminary notions regarding betatron motion.

(a) Based on these informations, build a simulation file of the electron analog. Produce a graph of its optical functions.
(b) Accelerate an electron bunch, from 1 to 10 MeV. Produce a graph of the horizontal and vertical phase spaces.

Check the betatron damping, compare with theory.

2.4 Solutions of Exercises of This Chapter: Electrostatic Systems

2.1 Plane Condenser; Spin Motion

(a) A catenary

An input data file for a catenary is given in Table 2.3. The initial coordinates under OBJET have been determined in the following way:

– perform a preliminary simulation in half the length, 50 cm, launching the electron normal to **E**. The expected transverse excursion is Y = 22.8948628 cm (Eq. 2.18), this is what raytracing yields

– that simulation also gives the electron relative rigidity, D, after 50 cm, as well as the trajectory angle, T.

The initial coordinates for the the 1 m long condenser are Y, –T and D.

The catenary and the evolution of electron momentum are displayed in Fig. 2.11. Graphs obtained using zpop: menu 7; 1/1 to open zgoubi.plt; 2/[8, 2] for Y(X), or 2/[8, 1] for D(X); 7 to plot.

The final coordinates can be found in the execution listing zgoubi.res, under DRIFT:

```
    4 Keyword, label(s) :  DRIFT
                           Drift, length =    0.00000  cm
TRAJ #1 IEX,D,Y,T,Z,P,S,time :  1  3.818208E-01  2.289511E+01  7.616725E+02 -5.839359E-15 -8.451676E-14  1.1275340E+02

Cumulative length of optical axis =    1.00  m  ;  Time  (for reference rigidity & particle) =  -4.144491E-09 s
```

It can be verified that they are quite close to the initial ones, under OBJET in the input data file, as expected.

This list includes the trajectory length, $S = 1.1275340$ m, over $X : 0 \rightarrow 1$ m, or half that value, 0.563767 m, over $X : 0 \rightarrow 0.5$ m. This is in agreement with Eq. 2.22 taken for $X = 0.5$ m, $\beta_i = 0.804837$ and $a = \beta_i E_i / E_x$ with $E_i = 860998$ eV and $E_x = 0.98$ MV.

(b, c) Sensitivity to step size. Spin precession

An input data file for a scan of the step size is given in Table 2.4. The scan is based on REBELOTE[IOPT = 1, LMNT = WIENFILT, KPRM = 8 0], which causes 1,000 times passes through the condenser, with each time the same initial conditions, and a different value of the step size XPAS (parameter number 80 under WIENFILTER). The step size value spans from 0.001 to 5 cm by $(5 - 0.001)/(1000 - 1) \approx 5 \times 10^{-3}$ cm increments

The result of the scan is displayed in Fig. 2.21. Spin outcomes are found in zgoubi.res, under "SPNTRK MATRIX", namely (an excerpt):

Table 2.3 Simulation input data file: push an electron through a condenser, along a catenary

```
Condenser. Catenary.
'OBJET'
2.3114795386518345        ! Rigidity of a 350 keV electron.
2
1  1                                     ! 1 particle.
22.89486 -7.616710E+02 0. 0. 0. 1.3818189 'o'
1                       ! Expected Y, T and D for E=980kV/m.
'PARTICUL'
ELECTRON

'WIENFILT'   S_condenser
20 ! Log particle data to zgoubi.plt, every other 10 step.
1.  -980000  0.    1
0. 0. 0.     ! 20. 5. 5.       ! Hard-edge entrance face.
0.2401  1.8639  -0.5572  0.3904 0. 0.
0.2401  1.8639  -0.5572  0.3904 0. 0.
0. 0. 0.     ! 20. 5. 5.          ! Hard-edge exit face.
0.2401  1.8639  -0.5572  0.3904 0. 0.
0.2401  1.8639  -0.5572  0.3904 0. 0.
.002
1. 0. 0. 0.
'MARKER'   E_condenser
'DRIFT'
0.    ! Gives coordinates of particle 1, with many digits.
'END'
```

A gnuplot script (*excerpt*) *to obtain Fig.* 2.21:

```
Yexp = 22.8948628 ; ttaexp = 19.5251764286/deg
stp_i = 0.001 ; stp_f = 5. ; NPASS = 1000
dStep = (stp_f-stp_i)/(NPASS-1.)
plot \
"zgoubi.fai" u ($0>2 ? stp_i + ($38-3)*dStep :1/0) \
:(abs(($10-Yexp)/Yexp) ) w lp pt 5 ps .6 lc rgb "blue" ,\
"zgoubi.fai" u ($26==1 && $0>2? stp_i+($38-3)*dStep :1/0) \
:(abs(atan($39/$38)-ttaexp)/ttaexp) axes x1y2 w lp pt 6
```

```
10  Keyword, label(s) :  SPNPRT       PRINT              MATRIX                              IPASS= 1001

                          -- 1  GROUPS  OF  MOMENTA  FOLLOW   --
                      -------------------------------------------------------------------------
                      Momentum  group  #1  (D=  1.377085E+00; particles  1  to 3 ;

                    Spin  components  of  the     3 particles,  spin  angles :
              INITIAL                    FINAL
                                                         --- angles ---
            SX SY SZ |S|      SX        SY     SZ  ||     GAMMA   |Si,Sf|  (Z,Sf_yz) (Z,Sf)
                                                            (deg.)   (deg.)    (deg.)
                                                      (Sf_yz : projection of Sf on YZ plane)
   o 1 1. 0. 0. 1.    0.942706 -0.333625  0. 1.    2.1184   19.489    90.000   90.000   1
   o 1 0. 1. 0. 1.    0.333625  0.942706  0. 1.    2.1184   19.489    90.000   90.000   2
   o 1 0. 0. 1. 1.   -0.000000  0.000000  1. 1.    2.1184    0.000     0.000    0.000   3

                 Spin transfer matrix, momentum group # 1 :
              0.942706        0.333625      -2.040746E-17
             -0.333625        0.942706       3.567997E-18
              2.042861E-17    3.571694E-18   1.00000
```

2.2 Toroidal Condenser

(a) Re-focusing at $\alpha = \pi/\sqrt{2}$ rad

The input data file to raytrace over $\alpha = \pi/\sqrt{2}$ rad in a cylindrical condenser is given in Table 2.5. Figure 2.22 shows a pair of trajectories, re-focused after a $\pi/\sqrt{2}$ deflection.

Figure 2.23 shows the sensitivity of the result to integration step size. The graph is unchanged whether $T_0 = \pm 0.1$ mrad or a hundred times less.

Table 2.4 Simulation input data file: push an electron through a condenser, along a catenary. Repeat 1,000 times with each time a different integration step size

```
E field only
'OBJET'
2.3114795386518345
2
3  1                ! 3 electrons, reason: see SPNTRK below.
0.  0. 0. 0. 0. 1. 'o'
0.  0. 0. 0. 0. 1. 'o'
0.  0. 0. 0. 0. 1. 'o'
1 1 1

'PARTICUL'
ELECTRON
'SPNTRK'   ! Allows chceking rotation of  3 spin components
4                       ! (they are computed independently).
1. 0. 0.
0. 1. 0.
0. 0. 1.

'WIENFILT'
20
0.5  -980000  0.    1
0. 0. 0.                           ! Hard-edge entrance face.
0.2401  1.8639  -0.5572  0.3904 0. 0.
0.2401  1.8639  -0.5572  0.3904 0. 0.
0. 0. 0.      ! 20. 5. 5.                    ! Hard-edge exit face.
0.2401  1.8639  -0.5572  0.3904 0. 0.
0.2401  1.8639  -0.5572  0.3904 0. 0.
.001
1. 0. 0. 0.

'SPNPRT'  MATRIX
'DRIFT'
0.
'FAISTORE'                         ! For use by gnuplot.
zgoubi.fai
1
'SPNPRT' PRINT MATRIX

'REBELOTE'
1000  1.1  0 1
1                  ! Step size, parameter #80 in WIENFILT,
WIENFILT 80 0.001:5.          ! is changed at each pass.

'SYSTEM'
1
gnuplot < ./gnuplot_scanStepSize.gnu &
'END'
```

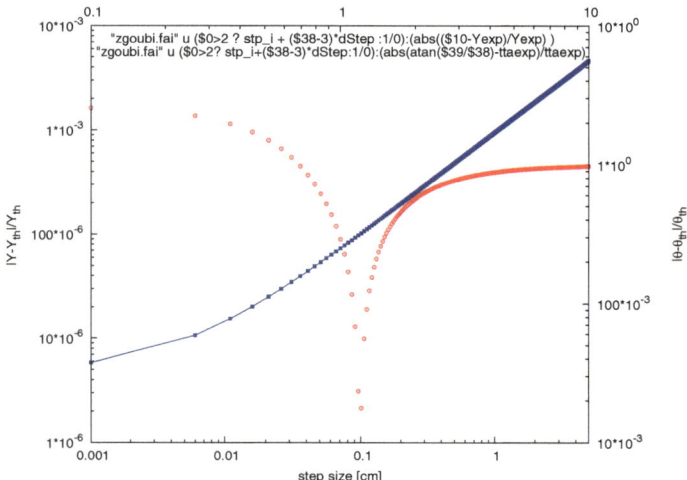

Fig. 2.21 Catenary motion. Sensitivity to step size for convergence of the numerical integration. Left vertical axis, square markers: radial excursion. Right axis, circles: spin rotation

(b) Aberration curve at the focal plane

The previous data file can be used, with the following changes:
  – use OBJET[KOBJ = 1, IMAX = 41] so to generate 41 particles launched with $T_0 \in [-20, 20]$ mrad, like so:

```
'OBJET'
2.33741826923 * 1.e3
1
1    41   1   1   1   1
.0    1. .0  0.  0.  1.
.0    0. .0  0.  0.  1.
```

  – add AUTOREF[I = 2] after ELCYLDEF, that will cause the moving frame to move to the waist formed by particles 1, 3 and 5 [20, cf. AUTOREF]
  – add FAISTORE[FNAME = zgoubi.fai] after AUTOREF, to log particle data at that location.

  The following gnuplot script produces a graph of the horizontal phase space (Fig. 2.24):

```
cm2m = 1e-2; mrd2rd = 1e-3
plot './zgoubi.fai' u ($10 *cm2m):($11 *mrd2rd) w p ps .9 pt ; pause 2
```

2.3 A Time-of-Flight Mass Spectrometer Ring
(a) TOF ring data file. Optics

  • Input data file. Synoptic.

Table 2.5 Input data file: track three positrons launched with initial horizontal angles $T_0 = 0, \pm 0.1$ mrad, through a cylindrical condenser

```
ELCYLDEF, pi/sqrt(2) bend angle.
'OBJET'
3.33564095198 * 1.e3                                    ! Case of 1GeV electron.
2
3  1
.0   0.  .0  0.  0.  1. 'o'
.0   0.1 .0  0.  0.  1. 'o'
.0  -0.1 .0  0.  0.  1. 'o'
1 1 1
'PARTICUL'
POSITRON
'ORDRE'
2                       ! In MOD=1 (default), at higher order the motion drifts dramatically, radially.

'ELCYLDEF'
200                                  ! Log stepwise data in zgoubi.plt, every 100 steps.
2.2214414691 95.492966 -10471974. 0.  3     ! Deflection/rad, r/m, E at r /V/m, index, field model.
0. 0. 5. 2.
4  .1455   2.2670  -.6395  1.1558  0. 0.  0.
0. 0. 5. 2.
4  .1455   2.2670  -.6395  1.1558  0. 0.  0.
.3 cm
2
95.492966e2 0. 95.492966e2 0.        ! No fringe field -> in & out radii normally = theortical value.

'IMAGE'                              ! Finds the location of the waist for the 3 trajectories.

'REBELOTE'
10  1.1 0 1                          ! Repeat 10 times, with step size varying from 0.31 to .4 cm.
1
ELCYLDEF  60  .31:.4
!! 5  60  .31:.4                     works as well, as ELCYLDEF is keyword #5 in the sequence.
'SYSTEM'
1
gnuplot < ./gnuplot_IMAGEvsStep.gnu
'END'
```

```
Trace =    2.8854120755;  spin precession acos((trace-1)/2) =    19.4889179609 deg
Precession axis : ( 0.,-0.,-1.) -> angle to (X,Y) plane, X axis, Z axis : -90.0000,   90.0000, 180.0000 deg
Spin precession/2pi (or Qs, fractional) :   5.4136E-02
```

The input data file for this problem is given in Table 2.6.
A synoptic of the ring can be obtained using zpop, Fig. 2.25.

● Optical parameters, optical functions.

The input data file in Table 2.6 runs a TWISS command. This produces zgoubi.
TWISS.out, which contains the optical parameters. An excerpt:

```
@ LENGTH     %le    3.311022574
@ ALFA       %le    0.7808368683
@ ORBIT5     %le             -0
@ GAMMATR    %le    1.131670108
@ Q1         %le    0.25615960E+00   2  [frac., int.]
@ Q2         %le    0.28340770E+00   3  [frac., int.]
@ DQ1        %le    2.228316236
@ DQ2        %le    36.69163200
```

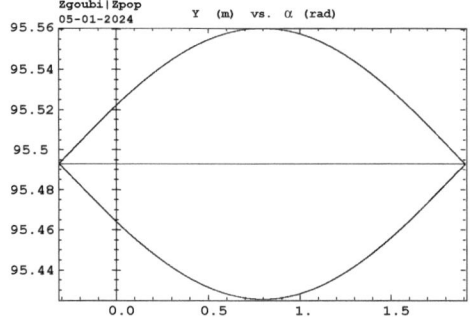

Fig. 2.22 Paraxial focusing in a cylindrical condenser. Launch angles of the three trajectories shown are 0 and ±0.1 mrad. Location of convergence is after $\alpha = \pi/\sqrt{2} = 2.22144$ rad deflection

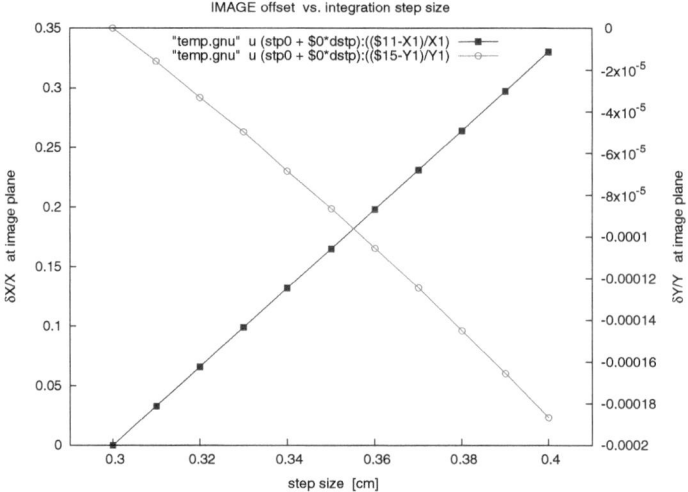

Fig. 2.23 Paraxial focusing in a cylindrical condenser. The figure shows the relative change in the distance of the waist to trajectory 1 (which leaves OBJET with all coordinates null), longitudinally (δX) and radially (δY), as a function of integration step size

@ DXMAX	%le	4.31815950E-01	-1.22847730E-08	@ DXMIN
@ DYMAX	%le	4.69095543E-25	-4.82489222E-25	@ DYMIN
@ XCOMAX	%le	1.18954794E+01	0.00000000E+00	@ XCOMIN
@ YCOMAX	%le	0.00000000E+00	0.00000000E+00	@ YCOMIN
@ BETXMAX	%le	1.02880410E+00	7.38003952E-01	@ BETXMIN
@ BETYMAX	%le	1.30020403E+00	9.01549054E-02	@ BETYMIN

That file zgoubi.TWISS.out also contains the optical functions along the structure. They are plotted using gnuplot_TWISS.gnu (Table 2.6), Fig. 2.26.

- Trajectories around the ring.

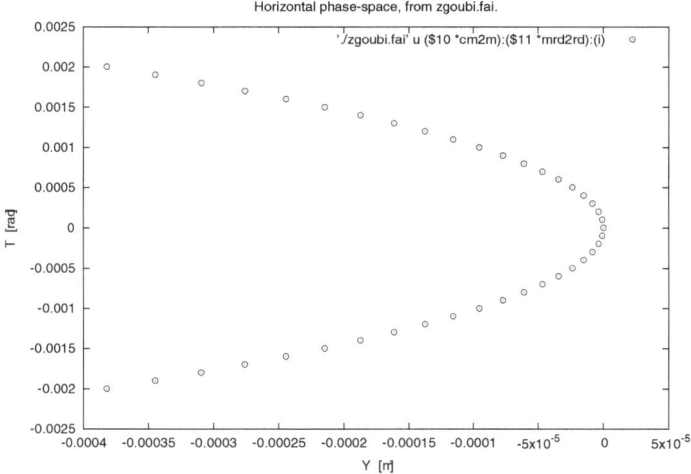

Fig. 2.24 Aberration curve at the focal point of a 127.2 deg deflection toroidal condenser: a second order (sextupole) aberration, $Y \propto T^2$, typical of geometrical non-linearities in a bending element

Fig. 2.25 A synoptic of half the TOF ring (except for the central drift, not displayed), and reference trajectories at 400 keV and $+/-20\%$. A graph obtained using zpop: menu 7, 2/[48, 42] for Y(X) coordinates, 7 to plot

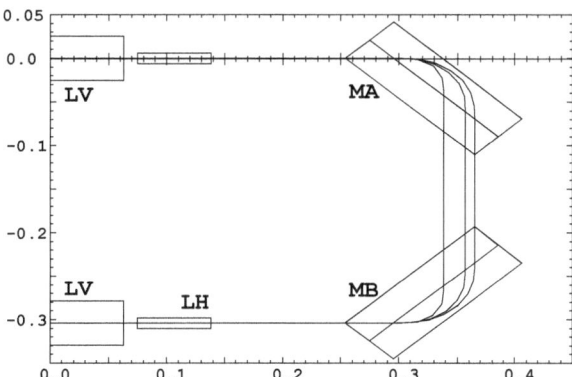

Trajectories along the structure are obtained using the previous input data file, Table 2.6, with some changes, as follows. Use the following OBJET:

Table 2.6 Simulation input data file. Left: a half of the TOF ring, from middle of long straight. This file also defines the sequence segment half-ring_S to half-ring_E, for INCLUDE in subsequent files. Right: Computation of the optical functions using a TWISS, a SYSTEM commands runs a gnuplot script taken from [pathTo]/zgoubi-code/toolbox/gnuplotFiles/gnuplot_TWISS/ folder, which plots the optical functions read from zgoubi.TWISS.out

Note: this file is available in **zgoubi** sourceforge repository at
https://sourceforge.net/p/zgoubi/code/HEAD/tree/branches/exemples/book/zgoubiMaterial/electrostaticSystems/TOF_Ring/lattice

```
'MARKER'  S_TOF-ringCell        ! Just for edition purposes.
'DRIFT'   half-ring_S
30.7
'ELMIR'          LV1                      ! Vertical lens.
0
3  0.02525  0.0125  0.02525 0.012    21
     0.       115.   0.                 ! LV.
#10|1000|10    500.010v102
1 0. 0. 0.
'DRIFT'                                  TOF-ring_TWISS.dat file. Compute & plot optical functions.
1.2                                      'OBJET'
'ELMIR'          LH1             ! Horizontal lens.   15.23683   ! Rigidity (kG.cm) ->   0.4 KeV N2+ (mass=28).
0                                        5
3  0.02525  0.0125  0.02525 0.012    22   0.01 0.001 0.01    0.001  0.  .0001
     0.       040.0  0.                 ! LH.   0. 0. 0. 0. 0. 1.
#10|1000|10    500.010v102               'PARTICUL'
1 0. 0. 0.                               26.082E3  1.60217733E-19 0. 0. 0.   ! uma=.9311->26070.8.
'DRIFT'                                  'FAISTORE'
11.6                                     zgoubi.fai #END
'ELMIR'          M1A             ! Horizontal deflector.  1
0                                        'COLLIMA'
7 .04275 .004163 .004163 .004163 .004163 .004163 .1 .012 11   1                         ! Collimator active. Elliptical,
     0.    200.    200.   200.    200.   200.   400.   2 1. 1. 0. 0.         ! H opening +\-1cm, V opening +\-1cm.
#10|3001|10    500.010v102               'INCLUDE'
3 0.   0.   .7853981634                  1
'DRIFT'                                  half-ring.inc[half-ring_S:half-ring_E]
6.00217933                               'COLLIMA'
'COLLIMA'                                1
1                         ! Collimator active. Elliptical,  2 1. 1. 0. 0.
2 1.e10 1. 0. 0.    ! H opening +\-1e10cm, V opening +\-1cm.  'INCLUDE'
'DRIFT'                                  1
6.00217933                               half-ring.inc[half-ring_S:half-ring_E]
'ELMIR'          M1B             ! Horizontal deflector.  'FAISCEAU'  #END
0
7 .04275 .004163 .004163 .004163 .004163 .004163 .1 .012 11   'TWISS'
     0.    200.    200.   200.    200.   200.   400.   2 1. 1.
#10|3001|10    500.010v102               'SYSTEM'
3 0.   0.   .7853981634                  1
'DRIFT'                                  gnuplot <./gnuplot_TWISS.gnu
11.6                                     'END'
'ELMIR'          LH1             ! Horizontal lens.
0
3  0.02525  0.0125  0.02525 0.012    22
0. 040.0  0.     ! LH
#10|1000|10    500.010v102
1 0. 0. 0.
'DRIFT'
1.2
'ELMIR'          LV1             ! Vertical lens.
0
3  0.02525  0.0125  0.02525 0.012    21
     0.     115.   0.     ! LV
#10|1000|10    500.010v102
1 0. 0. 0.
'DRIFT'
30.7
'MARKER'  half-ring_E
'MARKER'  E_TOF-ringCell         ! Just for edition purposes.
'END'
```

gnuplot_TWISS.gnu script (excerpt), from zgoubi package, to produce Fig. 2.26:

```
plot \
"zgoubi.TWISS.out" u ($13):($2) axes x1y1 w l tit "bet_x" ,\
"zgoubi.TWISS.out" u ($13):($4) axes x1y1 w l tit "bet_y" ,\
"zgoubi.TWISS.out" u ($13):($7) axes x1y2 w l tit "eta_x" ,\
"zgoubi.TWISS.out" u ($13):($9) axes x1y2 w l tit "eta_y"
```

```
'OBJET'
15.23683      Rigidity (kG.cm) ->   0.4 KeV N2+ (mass=28)
8
1  1  1
0.  0.  0.   0.   0.  1.
-1.1897385E-004  7.3901428E-001 2.5e-99
-1.1658994E-001  1.0815484E-001 120e-9
0. 1. 0.
```

Particle data are logged in zgoubi.plt, this is triggered using OPTIONS:

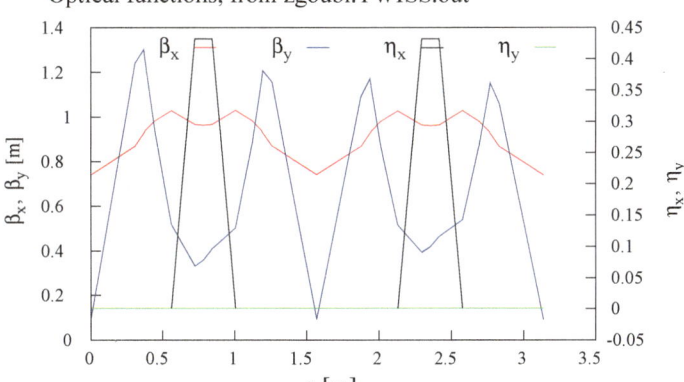

Fig. 2.26 Optical functions along the TOF ring, as read from zgoubi.TWISS.out

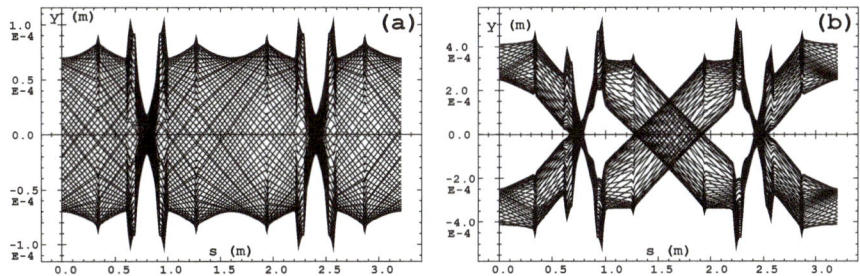

Fig. 2.27 Vertical trajectory of a particle over 80 turns around the ring. LH and LV potentials are about 38 V and 76 V, respectively. **a**: initial $z(s = 0) = 0.07$ mm. **b**: initial $z(s = 0) = 0.25$ mm, amplitude detuning brings the fractional tune near $\nu_y = 0.5$, a island configuration in phase space (see Fig. 2.29), causing the orbit closes in two turns. Graphs obtained using zpop: menu 7; 1/1 to open zgoubi.plt; 2[6, 4] to select Z(S); 7 to plot

```
'OPTIONS'
1 1
.plt 200
```

Multiturn, 80 turns here, uses REBELOTE:

```
'REBELOTE'
80 0.1 99
```

Outcomes are displayed in Fig. 2.27.

Table 2.7 Simulation input data file: a tune scan, versus momentum. This file INCLUDEs the sequence segment half-ring_S to half-ring_E defined in Table 2.6

```
scanTunes.INC.dat file. Tune scan vs. momentum.
'OBJET'
15.23683    ! Rigidity (kG.cm) ->   0.4 KeV N2+ (mass=28).
5
0.01  0.001  0.01    0.001   0.   .0001
0.  0.  0.    0.  0.  1.
'PARTICUL'
26.082E3  1.60217733E-19 0. 0. 0.     ! uma=.9311->26070.8.

'INCLUDE'
1
half-ring.inc[half-ring_S:half-ring_E]
'INCLUDE'
1
half-ring.inc[half-ring_S:half-ring_E]

'MATRIX'
1  11  PRINT      ! Tunes etc. logged in zgoubi.MATRIX.out.

'REBELOTE'  ! Change OBJET[D], and repeat 1-turn tracking,
34 1.1 0 1                         ! a total of 34 times,
1
OBJET 35 0.996:1.012   ! from D=Brho/BORO=0.996 to 1.012.

'SYSTEM'
1
gnuplot < ./gnuplot_MATRIX_Qxy.gnu
'END'
```

Fig. 2.28 Chromaticities, $dv_x/dp/p$ (red, square markers) and $dv_y/dp/p$ (blue, circles) . LH and LV voltages are 40 and 115 Volts, respectively

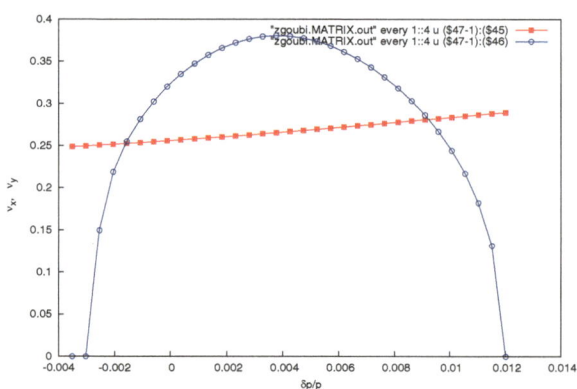

(b) Chromaticity scan.

An input data file for a scan of wave numbers as a function of momentum is given in Table 2.7. Results are displayed in Fig. 2.28.

(c) 1000-turn phase spaces.

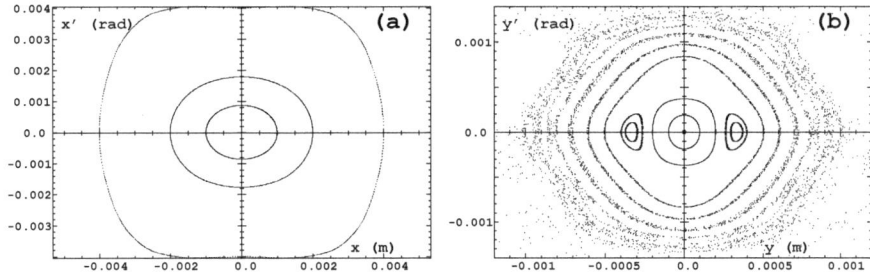

Fig. 2.29 Horizontal (**a**) and vertical (**b**) phase spaces, 1000 turns around the ring. LH and LV potentials are about 38 V and 76 V respectively. Non-linear vertical motion causes amplitude detuning towards fractional $v_y = 0.5$ and island formation

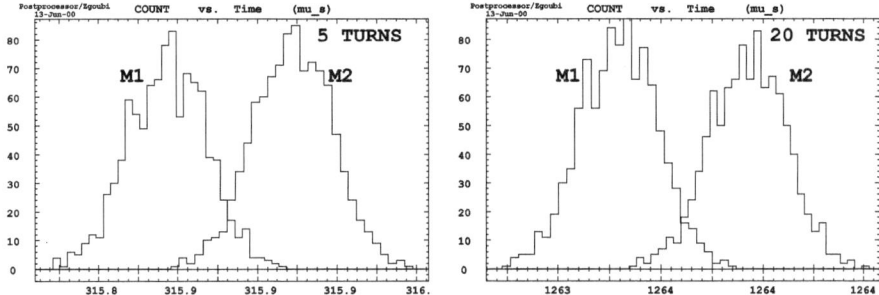

Fig. 2.30 Histograms of the time of flight of masses M1 and M2, at 5 and 20 turns, observed at s = 0

Definition of initial trajectory coordinates can use OBJET[KOBJ = 2]. Use REBE-LOTE[NPASS = 999, K = 99] for a 1000 turn loop. Outcomes are displayed in Fig. 2.29.

(d) Histograms.

Definition of initial trajectory coordinates can use MCOBJET[KOBJ = 3]. Use REBELOTE[NPASS = 19, K = 99] for a 20 turn loop. Outcomes are displayed in Fig. 2.30.

2.4 Converging Rays in a Bipotential Cylindrical Lens

(a) Using the analytical modeling EL2TUB.

The input data file used to produce Fig. 2.19 is given in Table 2.8.

The graph is obtained using `zpop`: menu 7; 1/1 to open zgoubi.plt; 2/[6, 2] for Y versus s or [6, 4] for Z versus s; 7 to plot

Proton rigidity is expected to increase from $B\rho_{in} = 0.020435\,T\,m$ (20 keV) to $B\rho_{out} = 0.026249\,T\,m$ (33 keV) across the 13 kV gap voltage, a ratio of $B\rho_{out}/B\rho_{in} = 1.2845$, confirmed in Fig. 2.31.

Table 2.8 Simulation input data file: tracking 20 keV protons in a bipotential cyclindrical lens

```
Track 20keV protons in EL2TUB
'OBJET'
20.435                                                    ! Rigidity of 20keV protons.
2                                    ! KOBJ=2: initial coordinates of the 9 particles are defined
9 1                                                           ! particle by particle.
0.2  0.  0.  0. 0. 1. 'o'                        ! Particles are launched with same Y0 and Z0 initial
0.1  0.  0.  0. 0. 1. 'o'                        ! coordinates, to show the axial symmetry: they converge
0.0  0.  0.  0. 0. 1. 'o'                        ! at the same location.
-0.1 0.  0.  0. 0. 1. 'o'
-0.2 0.  0.  0. 0. 1. 'o'
0.0  0.  0.2 0. 0. 1. 'o'
0.0  0.  0.1 0. 0. 1. 'o'
0.0  0. -0.1 0. 0. 1. 'o'
0.0  0. -0.2 0. 0. 1. 'o'
1 1 1 1 1 1 1 1 1
'PARTICUL'
PROTON            ! It is necessary to define the particle species, in order to solve \vec F=q \vec E.
'DRIFT'
100. split 5 2                                   ! IL is set to 2 for particle data storage in zgoubi.plt,
'EL2TUB'                                          ! the latter can be read to produce a graph.
2
.36 .12 .36 0.1
-20.E3 -33.E3
1.                                                ! XPAS: integration step size.
1 0 0 0
'DRIFT'
900.  split 20 2
'FAISCEAU'

'FIN'
```

Fig. 2.31 Evolution of momentum offset $(p - p_0)/p_0$ across the lens, for all the protons tracked (they all have 20 keV initial energy). A graph obtained using zpop: menu 7, 2/[6, 1] for $(p - p_0)/p_0$ versus s; 7 to plot

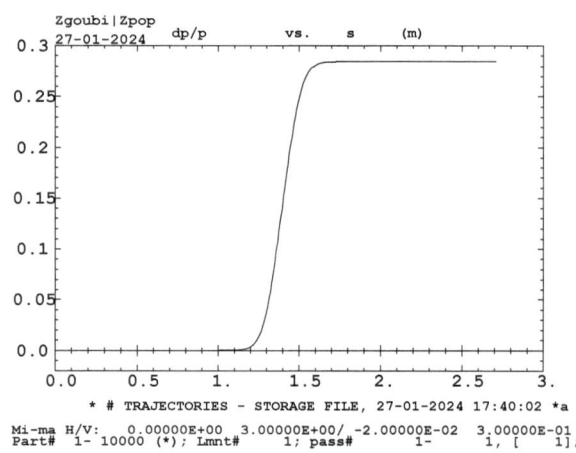

(b) Using ELREVOL with a 1-D electrostatic field map.

Proceed with the following steps:

- run the previous problem

  - with a single particle, all initial coordinates and momentum offset null
  - setting EL2TUB[IL = 2]

- read the step-by-step X coordinate of the particle and the electric component $E_X(X)$ so generated, across EL2TUB, from zgoubi.plt
- re-write that in an ascii file, say, el2tub.map, with proper format to be read by ELREVOL (two columns, col. 1 is X(step), col. 2 is E_X(step))
- change EL2TUB to ELREVOL with proper data list, with el2tub.map as field map data file (refer to [20, Lookup INDEX for EL2TUB] for ELREVOL data list and its formatting)

Run that new file. Essentially identical results are expected. Some discrepancy may arise, reasons being

- the field map mesh size (i.e., the integration step size across EL2TUB): increase mesh density as necessary for convergence of the trajectory results
- the integration step size across ELREVOL field map. It is mostly ineffective to take it smaller than the mesh size.

2.5 The AGS Electric Analog

Building the AGS Electric Analog and operating it is left to the reader. All details concerning the experiment and the ring can be found in [25, part D, Sect. 38].

ELCYLDEF can be used for the bends, WIENFILTER with null magnetic field might work since the deflection is small (thus abiding by Eq. 2.20), to be confirmed. ELMULT can be used for the two quadrupole families, focusing and defocusing.

References

1. J.D. Cockcroft, E.T.S. Walton, Experiments with High velocity positive ions. Proc. R. Soc. Lond. **A136**, 619–630 (1932)
2. Figure 2.1: Credit Reider Hahn, Fermilab
3. T. Roser, A. Zelensky, *Private Communication* (BNL, 2021)
4. A.I. Yavin, The AVF cyclotron. Phys. Today **15**(5), 19–25 (1962). https://doi.org/10.1063/1.3058175
5. G. Clausnitzer, History of polarized ion source developments, in *International Workshop on Polarized Ion Sources and Polarized Gas Jets*. ed. by Y. Mori (KEK, Tsukuba, Japan, 1990). https://inis.iaea.org/collection/NCLCollectionStore/_Public/22/051/22051667.pdf
6. W. Wan, Aberration correction in microscopes, in *TU4PBI02 Proceedings of PAC09* (Vancouver, BC, Canada)
7. Figure 2.3: © Dreebit GmbH

8. M. Paraliev, in *Proceedings of the CAS-CERN Accelerator School:Beam Injection, Extraction and Transfer*. ed. by B. Holzer (Erice, Italy, 2017); CERN Yellow Reports: School Proceedings, Vol. 5/2018, CERN-2018-008-SP (CERN, Geneva, 2018), pp. 363–394, https://doi.org/10.23730/CYRSP-2018-005 Figure 2.5: CERN, 2018. https://creativecommons.org/licenses/by/4.0; no change to the material

9. P.J. Bryant, Transverse motion and electrostatic elements, Lecture 3, in *Introduction to Particle Accelerators*. (Joint Universities Accelerator School, Archamps, 2010). Figure 2.4: copyrights under license CC-BY-3.0, https://creativecommons.org/licenses/by/3.0; no change to the material

10. S. Pape Møller, Design and first operation of the electrostatic storage ring, ELISA, in *Proceedings of EPAC'98 Accelerator Conference* (Stockholm, 1998). https://accelconf.web.cern.ch/e98/PAPERS/THZ01A.PDF Figure 2.7: copyrights under license CC-BY-3.0, https://creativecommons.org/licenses/by/3.0; no change to the material

11. R.A. Kishek et al., Benchmarking space charge codes against UMER experiments, in *WEA3MP03 Proceedings ICAP 2006* (Chamonix, France, 2006). http://accelconf.web.cern.ch/icap06/HTML/AUTHOR.HTM Figure 2.8: copyrights under license CC-BY-3.0, https://creativecommons.org/licenses/by/3.0; no change to the material

12. A. Septier (ed.), *Focusing of Charged Particles*, vol. I, II (Academic, 1967)

13. W. Kalbreier, N. Garrel, R. Guinard, R.L. Keizer, K.H. Kissler, Layout, design, and construction of the electrostatic separator system of the LEP e+ e– Collider, in *Proceedings of EPAC*, vol. 2 (1988) or CERN SPS/88-20 (ABT)

14. J.J. Welch et al., Commissioning and performance of low impedance electrostatic separators for high luminosity at CESR, in *Proceedings of the 1999 Particle Accelerator Conference* (New York, 1999). https://accelconf.web.cern.ch/p99/PAPERS/TUA156.PDF Figure 2.3: copyrights under license CC-BY-3.0, https://creativecommons.org/licenses/by/3.0; no change to the material

15. G.K. Green, E.E. Courant, The proton synchrotron. Part D, Sect. 38, The electron analog, in *Handbuch der Physik, Band XLIV* (Springer, Berlin 1959), p. 319. Figure 2.9: Springer, All rights reserved

16. G. Leleux, Accélérateurs Circulaires, in *INSTN Lectures* (Saturne, CEA Saclay, 1978). (unpublished)

17. F. Méot, Generalization of the Zgoubi method for ray-tracing to include electric fields. NIM A **340**, 594–604 (1994)

18. C.P. Welsch, Design studies of an electrostatic storage ring, in *Proceedings of the 2003 Particle Accelerator Conference*. Figure 2.14: copyrights under license CC-BY-3.0, https://creativecommons.org/licenses/by/3.0; no change to the material

19. S.P. Karetskaya et al., Mirror-bank energy analyzers, in *Advances in Electronics and Electron Physics*, vol. 89 (Academi, 1994), pp. 391–491

20. F. Méot, Zgoubi Users' Guide. https://www.osti.gov/biblio/1062013-zgoubi-users-guide. An up-to-date version of the guide can be found at: https://sourceforge.net/p/zgoubi/code/HEAD/tree/trunk/guide/Zgoubi.pdf

21. M. Baril, F. Méot, D. Michaud, Design study of a compact multiturn time of flight mass spectrometer. Internal Report CEA DSM DAPNIA/SEA-00-08 (2008)

22. A. Septier, Cours du DEA de physique des particules, optique corpusculaire, in *Electron Beams, Lenses, and Optics, Université d'Orsay 1966–1967*, vol. I, ed. by J.C. El-Kareh (Academic, New York and London, 1970), pp.38–39

23. A. Galejs, P.H. Rose, Optics of electrostatic tubes, in *Focusing of Charged Particles*, vol. 2, ed. by A. Septier (Academic, 1967)

24. F. Méot, Spin dynamics, in *Polarized Beam Dynamics and Instrumentation in Particle Accelerators, USPAS Summer 2021 Spin Class Lectures*. (Springer Nature, Open Access, 2023). https://link.springer.com/book/10.1007/978-3-031-16715-7

25. G.K. Green, E.E. Courant, The electron analog, in *Handbuch der Physik*, vol. XLIV, (Springer, Berlin, 1959), p.319

Chapter 3
Classical Cyclotron

Abstract This chapter introduces the classical cyclotron, and the theoretical material needed for the simulation exercises. It begins with a brief reminder of the historical context, and continues with beam optics and with the principles and methods which the classical cyclotron leans on, including

- ion orbit in a cyclic accelerator,
- weak focusing and periodic transverse motion,
- revolution period and isochronism,
- voltage gap and resonant acceleration,
- the cyclotron equation.

The simulation of a cyclotron dipole will either resort to an analytical model of the field: the optical element DIPOLE, or will resort to using a field map together with the keyword TOSCA to handle it and raytrace through. An additional accelerator device needed in the exercises, CAVITE, simulates a local oscillating voltage. Running a simulation generates a variety of output files, including the execution listing zgoubi.res, always, and other zgoubi.plt, zgoubi.CAVITE.out, zgoubi.MATRIX.out, etc., aimed at looking up program execution, storing data for post-treatment, producing graphs, etc. Additional keywords are introduced as needed, such as the matching procedure FIT[2]; FAISCEAU and FAISTORE which log local particle data in zgoubi.res or in a user defined ancillary file; MARKER; the "system call" command SYSTEM; REBELOTE, a 'do loop'; and some more. This chapter introduces in addition to spin motion in accelerator magnets; dedicated simulation exercises include a variety of keywords: SPNTRK, a request for spin tracking, SPNPRT or FAISTORE, to log spin vector components in respectively zgoubi.res or some ancillary file, and the "IL = 2" flag to log stepwise particle data, including spin vector, in zgoubi.plt file. Simulations include deriving transport matrices, beam matrix, optical functions and their transport, from rays, using MATRIX and TWISS keywords.

© The Author(s) 2024
F. Méot, *Understanding the Physics of Particle Accelerators*, Particle Acceleration and Detection, https://doi.org/10.1007/978-3-031-59979-8_3

Notations Used in the Text

B; B_0	magnetic field; at a reference radius R_0
$\mathbf{B}$; B_R; B_y	field vector; radial component; axial component
$BR = p/q$	magnetic rigidity
C; C_0	orbit length, $C = 2\pi R$; reference, $C_0 = 2\pi R_0$
E	ion energy, $E = \gamma m_0 c^2$
f_{rev}, f_{rf}	revolution and RF voltage frequencies
G	gyromagnetic anomaly, $G = 1.7928$ for proton, -4.184 for helion
h	harmonic number, an integer, $h = f_{\text{rf}}/f_{\text{rev}}$
$k = \frac{R}{B}\frac{dB}{dR}$	radial field index
m; m_0; M	ion mass; rest mass; in units of MeV/c^2
$\mathbf{p}$; p; p_0	ion momentum vector; its modulus; reference
q	ion charge
R; R_0; R_E	equilibrium orbit radius; reference, $R(p_0)$; at energy E
RF	Radio-Frequency
s	path variable
T_{rev}, T_{rf}	revolution and accelerating voltage periods
$\mathbf{v}$; v	ion velocity vector; its modulus
$V(t)$; $\hat{V}$	oscillating voltage; its peak value
W	kinetic energy, $W = \frac{1}{2}mv^2$
x', y'	radial and axial coordinates $\left[(*)' = \frac{d(*)}{ds}\right]$
α	trajectory deviation, or momentum compaction
$\beta = \frac{v}{c}$; β_0; β_s	normalized ion velocity; reference; synchronous
$\gamma = E/m_0 c^2$	Lorentz relativistic factor
Δp, δp	momentum offset
ε_u	Courant-Snyder invariant (u : x, r, y, l, Y, Z, s, etc.)
θ	azimuthal angle
ϕ	RF phase at ion arrival at the voltage gap

3.1 Introduction

Cyclotrons are the most widespread type of accelerator, today, used by thousands, with the production of isotopes as the dominant application. This chapter is devoted to the first cyclic accelerator: the early 1930s *classical* cyclotron which its concept limited to low energy, a few tens of MeV/nucleon. This limitation was overcome a decade later by the azimuthally varying field (AVF) technique, this is the subject of the next chapter.

The classical cyclotron is based on four main principles:

(i) the use of a cylindrical-symmetry magnetic field in the gap of an electromagnet (Fig. 3.1) to maintain ions on a circular trajectory

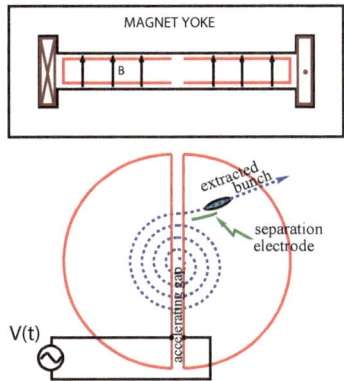

Fig. 3.1 Left: a cyclotron electromagnet, namely here that used for a model of Berkeley's 184 inch cyclotron in the early 1940s [3]. Magnetic field in the gap decreases with radius. Right: a schematic of the resonant acceleration motion; turn after turn, accelerated ions spiral out (bottom) in the quasi-uniform field (top). A double-dee (or, a variant, a single-dee facing a slotted electrode) forms an accelerating gap. The fixed-frequency oscillating voltage $V(t)$ applied is a harmonic of the revolution frequency. Ions experiencing proper voltage phase at the gap, turn by turn, are accelerated. A septum electrode allows beam extraction

(ii) transverse vertical confinement of the beam obtained by a slow radial decrease of the magnetic field. A technique known as weak focusing, applied over the years in all cyclic accelerators: microtron, betatron, synchrocyclotron, synchrotron. These weak focusing accelerator species all are still part of the landscape today

(iii) resonant acceleration by synchronization of a fixed-frequency accelerating voltage on the quasi-constant revolution time (Fig. 3.1) and

(iv) use of high voltage, to mitigate the effect of the turn-by-turn RF phase slip.

Resonant acceleration has the advantage that a small gap voltage is enough to accelerate with, in principle, no energy limitation, by contrast with the electrostatic techniques developed at the time, which required the generation of the full voltage, such as the Van de Graaff which was limited by sparking at a few tens of megavolts.

The cyclotron concept goes back to the late 1920s [1], yet it was not until the early 1930s when a cyclotron was first brought to operation [2]. The principles are summarized in Fig. 3.1: an oscillating voltage is applied on a pair of electrodes ("dees") forming an accelerating gap and placed between the two poles of an electromagnet. Ions reaching the gap during the acceleration phase of the voltage wave experience an energy boost; no field is experienced inside the dees. Under the effect of energy increase at the gap every half-revolution, they spiral out in the quasi-constant field of the dipole.

The first cyclotron achieved acceleration of H_2^+ hydrogen ions to 80 keV [2], at Berkeley in 1931. The apparatus used a dee-shaped electrode vis-à-vis a slotted electrode forming a voltage gap, the ensemble housed in a 5 in diameter vacuum chamber and placed in the 1.3 Tesla field of an electromagnet. A $\approx$ 12 MHz vacuum tube oscillator provided 1 kVolt gap voltage.

Fig. 3.2 Berkeley 27 inch cyclotron, brought to operation in 1934, accelerated deuterons up to 6 MeV. Left: a double-dee (seen in the vacuum chamber, cover off), 22 in diameter, creates an accelerating gap: 13 kV, 12 MHz radio frequency voltage is applied for deuterons for instance (through two feed lines seen at the top right corner). This apparatus was dipped in the 1.6 Tesla dipole field of a 27 in diameter, 75 ton, electromagnet. A slight decrease of the dipole field with radius, from the center of the dipole, ensures axial beam focusing. With their energy increasing, ions spiral out from the center to eventually strike a target (red arrow). Right: ionization of the air by the extracted beam (1936); the view also shows the vacuum chamber squeezed between the pole pieces of the electromagnet [3]

Fig. 3.3 Berkeley 184 in diameter, 4,000 ton cyclotron during construction [3]. The coil windings around both of the magnetic poles are clearly visible. Following the invention of longitudinal focusing it was actually operated as a synchrocyclotron, in 1946. The man on the right gives the scale

One goal foreseen in developing this technology was the acceleration of protons to MeV energy range for the study of atom nucleus. And in background, a wealth of potential applications. An 11 in cyclotron followed which delivered a 0.01 μA H_2^+ beam at 1.22 MeV [4], and a 27 in cyclotron later reached 6 MeV (Fig. 3.2) [5]. Targets were mounted at the periphery of the 11 inch cyclotron, disintegrations were observed in 1932. And, in 1933: '*The neutron had been identified by Chadwick in 1932. By 1933 we were producing and observing neutrons from every target bombarded by deuterons.*' [5, M. S. Livingston, p. 22].

A broad range of applications were foreseen: "*At this time biological experiments were started. [...] Also at about this same time the first radioactive tracer experiments on human beings were tried [...] simple beginnings of therapeutic use, coming a*

Fig. 3.4 Evolution of the number of the various cyclotron species, over the years [9]. From the 1950s on the AVF cyclotron rapidly supplanted the 1930s' classical cyclotron

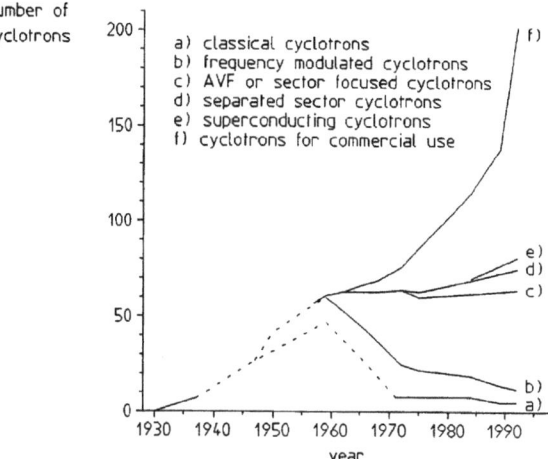

little bit later, in which neutron radiation was used, for instance, in the treatment of cancer. [...] Another highlight from 1936 was the first time that anyone tried to make artificially a naturally occurring radio-nuclide. (a bismuth isotope)" [5, McMillan, p. 26].

Berkeley's 184 in cyclotron, the largest (Fig. 3.3), commissioned in 1941, was to accelerate Deuterons to 100 MeV for meson production. Its magnet however was diverted to the production of uranium for the atomic bomb during the second world war years [1]. Re-started in 1946, as a consequence of the discovery of phase focusing the accelerator was actually operated as a synchrocyclotron (an accelerator species addressed in Chap. 7).

Limitation in energy

The understanding of the dynamics of ions in the classical cyclotron took some time, and brought two news, a bad one and a good one,

(i)   the bad one first: the energy limitation. A consequence of the loss of isochronism resulting from the relativistic increase of the ion mass so that *"[...] it seems useless to build cyclotrons of larger proportions than the existing ones [...] an accelerating chamber of 37 in radius will suffice to produce deuterons of 11 MeV energy which is the highest possible [...]"* [6], or in a different form: *"If you went to graduate school in the 1940s, this inequality ($-1 < k < 0$) was the end of the discussion of accelerator theory"* [7].

(ii) the good news now: the energy limit which results from the mass increase can be removed by shaping the magnetic pole into valley and hill field sectors. This is the azimuthally varying field (AVF) cyclotron technology, due to L.H. Thomas in 1938 [8]. It took some years to see effects of this breakthrough (Fig. 3.4). The AVF is the object of Chap. 4.

With the progress in magnet computation tools, in computer speed and in beam dynamics simulations, the AVF cyclotron ends up being essentially as simple to design and build: it has in a general manner supplanted the classical cyclotron in all energy domains (Fig. 3.4).

3.2 Basic Concepts and Formulæ

The cyclotron was conceived as a means to overcome the technological difficulty of a long series of high electrostatic voltage electrodes in a linear layout, by, instead, repeated recirculation through a single accelerating gap in synchronism with an oscillating voltage (Fig. 3.5). As the accelerated bunch spirals out in the uniform magnetic field, the velocity increase comes with an increase in orbit length; the net result is a slow increase of the revolution period T_{rev} with energy, yet, with appropriate fixed $f_{rf} \approx h/T_{rev}$ the revolution motion and the oscillating voltage can be maintained in sufficiently close synchronism, $T_{rev} \approx T_{rf}/h$, that the bunch will transit the voltage gap at an accelerating phase (Fig. 3.6) over a large enough number of turns that it acquires a significant energy boost.

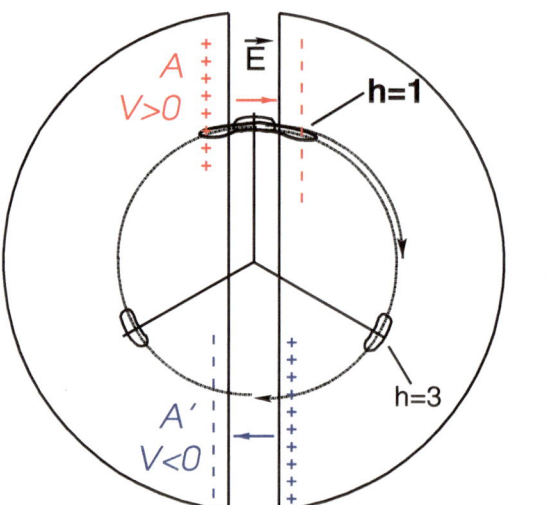

Fig. 3.5 Resonant acceleration: in an $h = 1$ configuration an ion bunch meets an oscillating field **E** across gap A, at time t, at an accelerating phase; it meets again, half a turn later, at time $t + T_{rev}/2$, the accelerating phase across gap A', and so on: the magnetic field recirculates the bunch through the gap, repeatedly. Higher harmonic allows more bunches: the next possibility in the present configuration is h = 3, and 3 bunches, 120° apart, in synchronism with **E**

Fig. 3.6 An ion which reaches the double-dee gap at the RF phase $\omega_{rf}t = \phi_A$ or $\omega_{rf}t = \phi_B$ is accelerated. If it reaches the gap at $\omega_{rf}t = \phi_C$ it is decelerated

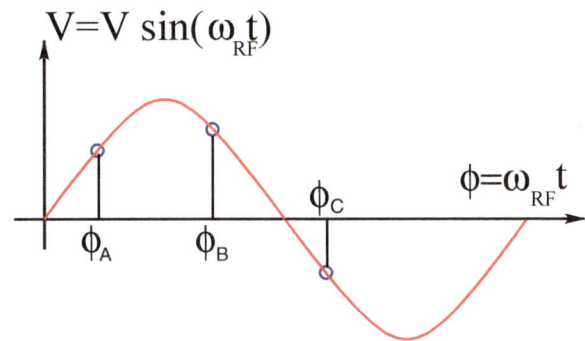

The orbital motion quantities: radius R, ion rigidity BR, revolution frequency f_{rev}, satisfy

$$BR = \frac{p}{q}, \qquad 2\pi f_{rev} = \omega_{rev} = \frac{v}{R} = \frac{qB}{m} = \frac{qB}{\gamma m_0} \qquad (3.1)$$

These relationships hold at all γ, so covering the *classical* cyclotron domain ($v \ll c$, $\gamma \approx 1$) as well as the *isochronous* cyclotron (in which the ion energy increase is commensurate with its mass). To give an idea of the revolution frequency, in the limit $\gamma = 1$, for protons, one has $f_{rev}/B = q/2\pi m = 15.25\,\text{MHz/T}$.

The cyclotron design sets the constant RF frequency $f_{rf} = \omega_{rf}/2\pi$ at an intermediate value of hf_{rev} along the acceleration cycle. The energy gain, or loss, by the ion when transiting the gap, at time t, is

$$\Delta W(t) = q\hat{V}\sin\phi(t) \quad \text{with} \quad \phi(t) = \omega_{rf}t - \omega_{rev}t + \phi_0 \qquad (3.2)$$

with ϕ its phase with respect to the RF signal at the gap (Fig. 3.6), $\phi_0 = \phi(t = 0)$, and $\omega_{rev}t$ the orbital angle. Assuming constant field B, the increase of the revolution period with ion energy satisfies

$$\frac{\Delta T_{rev}}{T_{rev}} = \gamma - 1 \qquad (3.3)$$

The mis-match so induced between the RF and cyclotron frequencies is a turn-by-turn cumulative effect and sets a limit to the tolerable isochronism defect, $\Delta T_{rev}/T_{rev} \approx$ 2–3%, or highest velocity $\beta = v/c \approx 0.22$. This results for instance in a practical limitation to ≈ 25 MeV for protons, and ≈ 50 MeV for D and α particles, a limit however dependent on energy gain per turn.

Over time multiple-gap accelerating structures where developed, whereby a "multiple-Δ" electrode pattern substitutes to a "double-D". An example is GANIL C0 injector with its 4 accelerating gaps and $h = 4$ and $h = 8$ RF harmonic operation [10].

3.2.1 Fixed-Energy Orbits, Revolution Period

In a laboratory frame (O; x, y, z), with (O; x, z) the bend plane (Fig. 3.7), assume $\mathbf{B}|_{y=0} = \mathbf{B}_y$, constant. An ion is launched from the origin with a velocity

$$\mathbf{v} = \left(\frac{dx}{dt}, \frac{dy}{dt}, \frac{dz}{dt}\right) = (v \sin \alpha, 0, v \cos \alpha)$$

at an angle α from the z-axis. Solving

$$m\dot{\mathbf{v}} = q\mathbf{v} \times \mathbf{B} \tag{3.4}$$

with $\mathbf{B} = (0, B_y, 0)$ yields the parametric equations of motion

$$\begin{cases} x(t) = \dfrac{v}{\omega_{rev}} \cos(\omega_{rev}t - \alpha) - \dfrac{v \cos \alpha}{\omega_{rev}} \\ y(t) = \text{constant} \\ z(t) = \dfrac{v}{\omega_{rev}} \sin(\omega_{rev}t - \alpha) + \dfrac{v \sin \alpha}{\omega_{rev}} \end{cases} \tag{3.5}$$

which result in

$$\left(x + \frac{v \cos \alpha}{\omega_{rev}}\right)^2 + \left(z - \frac{v \sin \alpha}{\omega_{rev}}\right)^2 = \left(\frac{v}{\omega_{rev}}\right)^2 \tag{3.6}$$

a circular trajectory of radius $R = v/\omega_{rev}$ centered at $(x_C, z_C) = (-\frac{v \cos \alpha}{\omega_{rev}}, \frac{v \sin \alpha}{\omega_{rev}})$.

Stability of the cyclic motion—The initial velocity vector defines a reference closed orbit in the median plane of the cyclotron dipole; a small perturbation in α or v results in a new orbit *in the vicinity* of the reference. An axial velocity component v_y on the other hand, causes the ion to drift away from the reference, vertically, linearly

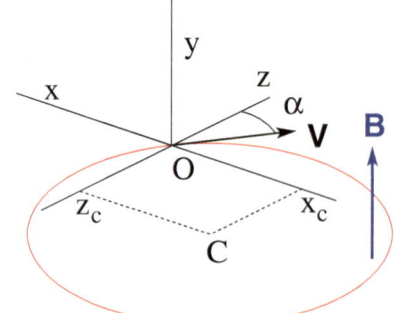

Fig. 3.7 Circular motion of an ion in the plane normal to a uniform magnetic field **B**. The orbit is centered at $x_C = -v \cos \alpha/\omega_{rev}$, $z_C = v \sin \alpha/\omega_{rev}$, its radius is v/ω_{rev}

with time, as there is no axial restoring force. The next Section will investigate the necessary field property to ensure both horizontal and vertical confinement of the cyclic motion in the vicinity of a reference orbit in the median plane.

3.2.2 Weak Focusing

In the early accelerated turns in a classical cyclotron (central region of the electromagnet, energy up to tens of keV/u), the accelerating electric field provides vertical focusing for particles with proper RF phase [11, Sect. 8], whereas a flat magnetic field with uniformity $dB/B < 10^{-4}$ is sufficient to maintain isochronism. Beyond this low energy region however, at greater radii, a magnetic field gradient must be introduced to ensure transverse stability: field must decrease with R.

Ion coordinates in the following are defined in the moving frame $(M_0; s, x, y)$ (Fig. 3.8), which moves along the reference orbit (radius R_0), with its origin M_0 the location of the reference ion on the orbit; the s axis is tangent to the latter, the x axis is normal to s, the y axis is normal to the bend plane. Median-plane symmetry of the field is assumed, thus the radial field component $B_R|_{y=0} = 0$ at all R (Fig. 3.9).

Consider small motion excursions $x(t) = r(t) - R_0 \ll R_0$; introduce Taylor expansion of the field components,

$$B_y(R_0 + x) = B_y(R_0) + x \left.\frac{\partial B_y}{\partial R}\right|_{R_0} + \frac{x^2}{2!} \left.\frac{\partial^2 B_y}{\partial R^2}\right|_{R_0} + \cdots \approx B_y(R_0) + x \left.\frac{\partial B_y}{\partial R}\right|_{R_0}$$

$$B_R(0 + y) = y \underbrace{\left.\frac{\partial B_R}{\partial y}\right|_{0}}_{= \left.\frac{\partial B_y}{\partial R}\right|_{R_0}} + \frac{y^3}{3!} \left.\frac{\partial^3 B_R}{\partial y^3}\right|_{0} + \cdots \approx y \left.\frac{\partial B_y}{\partial R}\right|_{R_0} \qquad (3.7)$$

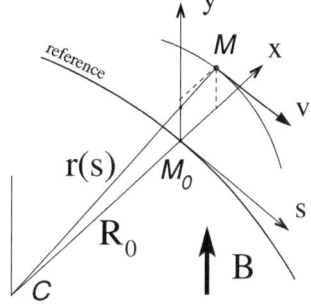

Fig. 3.8 Moving frame $(M_0; s, x, y)$ along the reference circular orbit. The curvature $1/R_0$ is constant along the orbit and $(M_0; s, x, y)$ can be considered equivalent to the cylindrical frame $(C; \theta, R_0, y)$

Fig. 3.9 Axial motion stability requires proper shaping of field lines: B_y has to decrease with radius. The Laplace force pulls a positive charge with velocity pointing out of the page, at I, toward the median plane. Increasing the field gradient (k closer to -1, gap opening up faster) increases the focusing

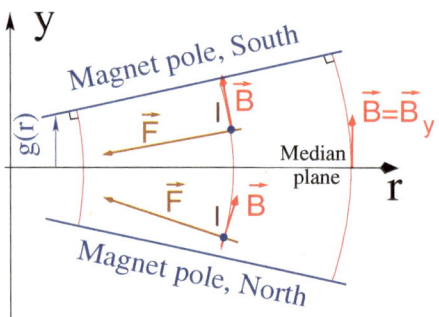

Using these, and noting $(\dot{*}) = d(*)/dt$, the linear approximation of the differential equations of motion in the moving frame writes

$$F_x = m\ddot{x} = -qvB_y(R) + \frac{mv^2}{R_0 + x} \approx -qv\left(B_y(R_0) + \frac{\partial B_y}{\partial R}\bigg|_{R_0} x\right) + \frac{mv^2}{R_0}\left(1 - \frac{x}{R_0}\right)$$

$$\rightarrow m\ddot{x} = -\frac{mv^2}{R_0^2}\left(\frac{R_0}{B_0}\frac{\partial B_y}{\partial R}\bigg|_{R_0} + 1\right)x \tag{3.8}$$

$$F_y = m\ddot{y} = qvB_R(y) = qv\frac{\partial B_R}{\partial y}\bigg|_{y=0} y + \text{higher order} \rightarrow m\ddot{y} = qv\frac{\partial B_y}{\partial R}y$$

Note $B_y(R_0) = B_0$ and introduce

$$\omega_R^2 = \omega_{\text{rev}}^2\left(1 + \frac{R_0}{B_0}\frac{\partial B_y}{\partial R}\right), \quad \omega_y^2 = -\omega_{\text{rev}}^2\frac{R_0}{B_0}\frac{\partial B_y}{\partial R} \tag{3.9}$$

substitute in Eqs. 3.8, this yields

$$\ddot{x} + \omega_R^2 x = 0 \quad \text{and} \quad \ddot{y} + \omega_y^2 y = 0 \tag{3.10}$$

A restoring force (linear terms in x and y, Eq. 3.10) arises from the radially varying field, characterized by a field index

$$k = \frac{R_0}{B_0}\frac{\partial B_y}{\partial R}\bigg|_{R=R_0, y=0} \tag{3.11}$$

Radial stability: radially this force adds to the geometrical focusing (curvature term "1" in ω_R^2, Eq. 3.9, Fig. 3.10). In the weakly decreasing field $B(R)$ an ion with momentum $p = mv$ moving in the vicinity of the R_0-radius reference orbit experiences in the moving frame a resultant force $F_t = -qvB + m\dfrac{v^2}{r}$ (Fig. 3.11) of which

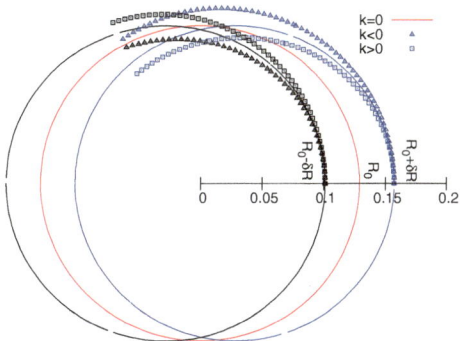

Fig. 3.10 Geometrical focusing: take k = 0; two circular trajectories which start from $r = R_0 \pm \delta R$ (solid lines, going counter-clockwise) undergo exactly one oscillation around the reference orbit $r = R_0$. A negative k (triangles), for axial focusing, decreases the radial convergence; a positive k (square markers) increases the radial convergence and increases vertical divergence

the (outward) component $f_c = m\frac{v^2}{r}$ decreases with r at a higher rate than the decrease of the Laplace (inward) component $f_B = -qvB(r)$. In other words, radial stability requires BR to increase with R, $\frac{\partial BR}{\partial R} = B + R\frac{\partial B}{\partial R} > 0$, this holds in particular at R_0, thus $1 + k > 0$.

Axial stability requires a restoring force directed toward the median plane. Referring to Fig. 3.9, this means $F_y = -a \times y$ (with a a positive quantity) and thus $B_R < 0$, at all $(r, y \neq 0)$. This is achieved by designing a guiding field which decreases with radius, $\frac{\partial B_R}{\partial y} < 0$. Referring to Eq. 3.11 this means $k < 0$.

From these radial and axial constraints the condition of "weak focusing" for transverse motion stability around the circular equilibrium orbit results, namely,

$$-1 < k < 0 \tag{3.12}$$

Note regarding the geometrical focusing: the focal distance associated with the curvature of a magnet of arc length $\mathcal{L}$ is obtained by integrating $\frac{d^2x}{ds^2} + \frac{1}{R_0^2}x = 0$ and identifying with the focusing property $\Delta x' = -x/f$, namely,

$$\Delta x' = \int \frac{d^2x}{ds^2} ds \approx \frac{-x}{R^2} \int ds = \frac{-x\mathcal{L}}{R^2}, \text{ thus } f = \frac{R^2}{\mathcal{L}} \tag{3.13}$$

Isochronism: the axial focusing constraint, B decreasing with R, contributes breaking the isochronism (in addition to the effect of the mass increase) by virtue of $\omega_{\text{rev}} \propto B$.

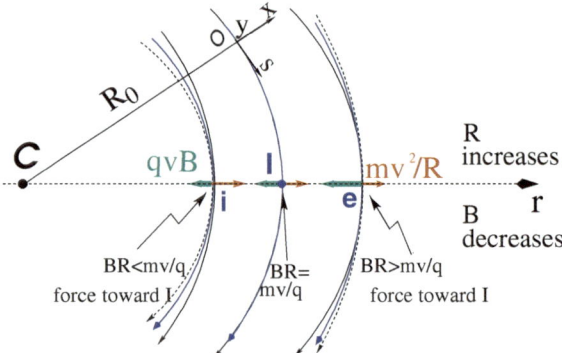

Fig. 3.11 Radial motion stability. Trajectory arcs at $p = mv$ are represented: case of $k = 0$ (thin black lines), of $-1 < k < 0$ (thick blue lines), and of $k = -1$ (dashed concentric circles). k decreasing towards -1 reduces the geometrical focusing, increases axial focusing. The resultant of the Laplace and centrifugal forces, $F_t = -qvB + mv^2/r$, is zero at I, motion is stable if F_t is toward I at i, i.e. $qvB_i < mv^2/R_i$, and toward I as well at e, i.e. $qvB_e > mv^2/R_e$

Paraxial Transverse Coordinates

Introduce the path variable s as the independent variable in Eq. 3.10 and neglect the transverse velocity components $(1 + \frac{x}{R_0} \approx 1,\ y \ll 0)$ so that

$$ds = \left[r^2(s)d\theta^2 + dr^2 + dy^2 \right]^{1/2} \approx |\mathbf{v}|dt$$

(3.14)

thus the equations of motion in the moving frame (Eq. 3.10) take the form

$$\frac{d^2x}{ds^2} + \frac{1+k}{R_0^2}x = 0 \quad \text{and} \quad \frac{d^2y}{ds^2} - \frac{k}{R_0^2}y = 0 \tag{3.15}$$

Given $-1 < k < 0$ the motion is that of a harmonic oscillator, in both planes, with respective restoring constants $(1+k)/R_0^2$ and $-k/R_0^2$, both positive quantities. The solution is a sinusoidal motion,

$$\begin{cases} r(s) - R_0 = x(s) = x_0 \cos \frac{\sqrt{1+k}}{R_0}(s - s_0) + x_0' \frac{R_0}{\sqrt{1+k}} \sin \frac{\sqrt{1+k}}{R_0}(s - s_0) \\ r'(s) = x'(s) = -x_0 \frac{\sqrt{1+k}}{R_0} \sin \frac{\sqrt{1+k}}{R_0}(s - s_0) + x_0' \cos \frac{\sqrt{1+k}}{R_0}(s - s_0) \end{cases} \tag{3.16}$$

$$\begin{cases} y(s) = y_0 \cos \frac{\sqrt{-k}}{R_0}(s - s_0) + y_0' \frac{R_0}{\sqrt{-k}} \sin \frac{\sqrt{-k}}{R_0}(s - s_0) \\ y'(s) = -y_0 \frac{\sqrt{-k}}{R_0} \sin \frac{\sqrt{-k}}{R_0}(s - s_0) + y_0' \cos \frac{\sqrt{-k}}{R_0}(s - s_0) \end{cases} \tag{3.17}$$

Radial and axial wave numbers can be introduced,

$$\nu_R = \frac{\omega_R}{\omega_{\text{rev}}} = \sqrt{1 + k} \quad \text{and} \quad \nu_y = \frac{\omega_y}{\omega_{\text{rev}}} = \sqrt{-k} \tag{3.18}$$

i.e., the number of sinusoidal oscillations of the paraxial motion about the reference circular orbit over a turn, respectively radial and axial. Both are less than 1: there is less than one sinusoidal oscillation in a revolution. In addition, as a result of the revolution symmetry of the field,

$$\nu_R^2 + \nu_y^2 = 1 \tag{3.19}$$

Off-Momentum Orbit

In a structure with revolution symmetry, the equilibrium trajectory at momentum $\begin{cases} p_0 \\ p = p_0 + \Delta p \end{cases}$ is at radius $\begin{cases} R_0 \text{ with } B_0 R_0 = \frac{p_0}{q} \\ R \text{ with } B R = \frac{p}{q} \end{cases}$, where $\begin{cases} B = B_0 + \left(\frac{\partial B}{\partial x}\right)_0 \Delta x + \cdots \\ R = R_0 + \Delta x \end{cases}$
On the other hand

$$BR = \frac{p}{q} \Rightarrow \left[B_0 + \left(\frac{\partial B}{\partial x}\right)_0 \Delta x + \cdots \right](R_0 + \Delta x) = \frac{p_0 + \Delta p}{q}$$

which, neglecting terms in $(\Delta x)^2$, and given $B_0 R_0 = \frac{p_0}{q}$, leaves $\Delta x \left[\left(\frac{\partial B}{\partial x}\right)_0 R_0 + B_0\right] = \frac{\Delta p}{q}$. With $k = \frac{R_0}{B_0}\left(\frac{\partial B}{\partial x}\right)_0$ this yields

$$\Delta x = D \frac{\Delta p}{p_0} \quad \text{with} \quad D = \frac{R_0}{1 + k} \quad \text{the dispersion function} \tag{3.20}$$

The dispersion D is an s-independent quantity as a result of the revolution symmetry of the field (k and R = p/qB are s-independent).

To the first order in the coordinates, the vertical coordinates y(s), y'(s) (Eq. 3.17) are unchanged under the effect of a momentum offset, the horizontal trajectory angle x'(s) (Eq. 3.16) is unchanged as well (the circular orbits are concentric, Fig. 3.12) whereas $x(s)$ satisfies

$$x(s, p_0 + \Delta p) = x(s, p_0) + \Delta p \left.\frac{\partial x}{\partial p}\right|_{s, p_0} = x(s, p_o) + D\frac{\Delta p}{p_0} \tag{3.21}$$

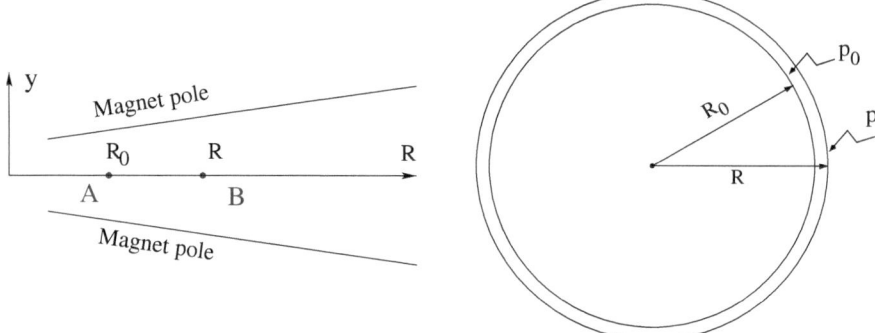

Fig. 3.12 The equilibrium radius at location A is R_0, momentum is p_0, rigidity is $B_0 R_0$. The equilibrium radius at B is R, momentum p, rigidity $B R$

Orbit and revolution period lengthening

A δp momentum offset results in (Eq. 3.20)

$$\frac{\delta C}{C} = \frac{\delta R}{R} = \frac{\delta x}{R} = \alpha \frac{\delta p}{p} \quad \text{with} \quad \alpha = \frac{1}{1+k} = \frac{1}{v_R^2} \qquad (3.22)$$

with α the momentum compaction, a positive quantity: orbit length increases with momentum. Substituting $\frac{\delta\beta}{\beta} = \frac{1}{\gamma^2}\frac{\delta p}{p}$, the change in revolution period $T_{\text{rev}} = C/\beta c$ with momentum writes

$$\frac{\delta T_{\text{rev}}}{T_{\text{rev}}} = \frac{\delta C}{C} - \frac{\delta \beta}{\beta} = \left(\alpha - \frac{1}{\gamma^2}\right)\frac{\delta p}{p} \qquad (3.23)$$

Given that $-1 < k < 0$ and $\gamma \gtrsim 1$, it results that $\alpha - 1/\gamma^2 > 0$: the revolution period increases with energy, the increase in radius is faster than the velocity increase.

3.2.3 Quasi-Isochronous Resonant Acceleration

The energy W of an accelerated ion (in the non-relativistic energy domain of the classical cyclotron) satisfies the frequency dependence

$$W = \frac{1}{2}mv^2 = \frac{1}{2}m\,(2\pi R f_{\text{rev}})^2 = \frac{1}{2}m\left(2\pi R\frac{f_{\text{rf}}}{h}\right)^2 \qquad (3.24)$$

Observe in passing: given the cyclotron size (radius R), f_{rf} and h set the limit for the acceleration range. The revolution frequency decreases with energy and the condition of synchronism with the oscillating voltage, $f_{\text{rf}} = h f_{\text{rev}}$, is only fulfilled at that

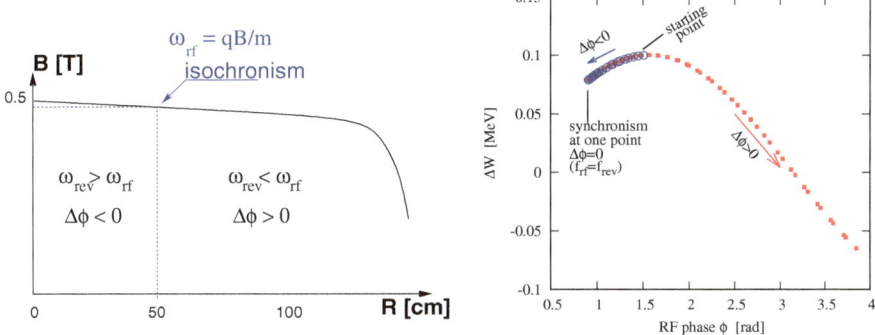

Fig. 3.13 Left: a sketch of the synchronism condition at one point (h = 1 assumed). Right: the span in phase of the energy gain $\Delta W = q\hat{V}\sin\phi$ (Eq. 3.2) over the acceleration cycle

particular radius where $\omega_{\text{rf}} = qB/m$ (Fig. 3.13-left). The out-phasing $\Delta\phi$ of the RF at ion arrival at the gap builds-up turn after turn, decreasing in a first stage (towards lower voltages in Fig. 3.13-right) and then increasing back to $\phi = \pi/2$ and beyond towards π. Beyond $\phi = \pi$ the RF voltage is decelerating.

With ω_{rev} constant between two gap passages, differentiating $\phi(t)$ (Eq. 3.2) yields $\dot\phi = \omega_{\text{rf}} - \omega_{\text{rev}}$. Between two gap passages on the other hand, $\Delta\phi = \dot\phi\Delta T = \dot\phi T_{\text{rev}}/2 = \dot\phi\frac{\pi R}{v}$, yielding a phase-shift of

$$\text{half-turn} \quad \Delta\phi = \pi\left(\frac{\omega_{\text{rf}}}{\omega_{\text{rev}}(R)} - 1\right) = \pi\left(\frac{m\omega_{\text{rf}}}{qB(R)} - 1\right) \tag{3.25}$$

The out-phasing is thus a gap-after-gap, cumulative effect. Due to this the classical cyclotron requires quick acceleration (small number of turns), which means high voltage (tens to hundreds of kVolts). As expected, with ω_{rf} and B constant, $\dot\phi$ presents a minimum ($\dot\phi = 0$) at $\omega_{\text{rf}} = \omega_{\text{rev}} = qB/m$ where exact isochronism is reached (Fig. 3.13). The upper limit to ϕ is set by the condition $\Delta W > 0$: acceleration.

The cyclotron equation determines the achievable energy range, depending on the injection energy E_i, the RF phase at injection ϕ_i, the RF frequency ω_{rf} and gap voltage $\hat{V}$. It writes [12]

$$\cos\phi = \cos\phi_i + \pi\left[1 - \frac{\omega_{\text{rf}}}{\omega_{\text{rev}}}\frac{E + E_i}{2M}\right]\frac{E - E_i}{q\hat{V}} \tag{3.26}$$

Equation 3.26 is represented in Fig. 3.14 for various values of the peak voltage and phase at injection ϕ_i. M [eV/c^2] and E [eV] are respectively the rest mass and relativistic energy, $q\hat{V}$ is expressed in electron-volts, the index i denotes injection parameters.

Fig. 3.14 A graph of the
cyclotron equation
(Eq. 3.26), for three different
accelerating voltages: 100,
200 and 400 kV/gap
(respectively square, circle
and triangle markers). The
sole settings resulting in
$-1 < \cos\phi(E) < 1, \forall E$,
allow complete acceleration
to top energy. $\phi_i = \pi/4$ at
injection for instance, does
not (upper three curves).
$\phi_i = 3\pi/4$ works (lower
three curves), with as low as
100 kV/gap

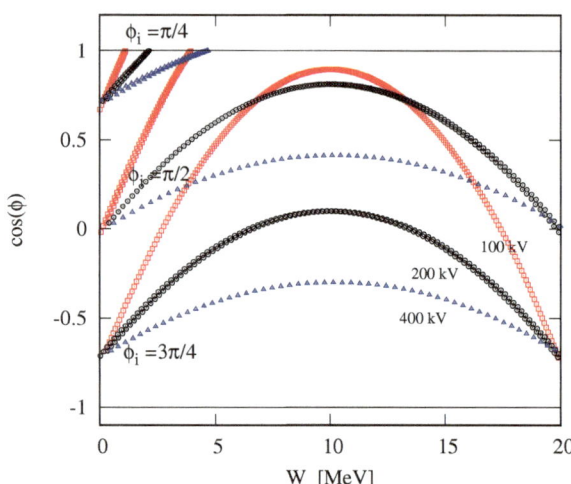

3.2.4 Beam Extraction

From $R = p/qB$ and assuming $B(R) \approx$ constant (this is legitimate as k is normally
small), in the non-relativistic approximation ($W \ll M$, $W = p^2/2M$) one gets

$$\frac{dR}{R} = \frac{1}{2}\frac{dW}{W} \tag{3.27}$$

Integrating yields

$$R^2 = R_i^2 \frac{W}{W_i} \tag{3.28}$$

with R_i, W_i initial conditions. From Eqs. 3.27 and 3.28, assuming $W_i \ll W$ and constant acceleration rate dW such that $W = n\,dW$ after n turns, one gets the scaling laws

$$R \propto \sqrt{n}, \quad dR \propto \frac{R}{W} \propto \frac{1}{R} \propto dW, \quad \frac{dR}{dn} = \frac{R}{2n} \tag{3.29}$$

The turn separation dR is proportional to the energy gain per turn and inversely
proportional to the orbit radius.

The radial distance between successive turns decreases with energy, toward zero
(Fig. 3.15), eventually resulting in insufficient spacing for insertion of an extraction
septum.

Orbit modulation

Consider an ion bunch injected in the cyclotron with some (x_0, x_0') conditions in
the vicinity of the reference orbit, and assume slow acceleration. While accelerated
the bunch undergoes an oscillatory motion around the equilibrium orbit (Eq. 3.16).

Fig. 3.15 The radial distance between successive turns decreases with energy, in inverse proportion to the orbit radius. The red and blue segments here figure the accelerating gap

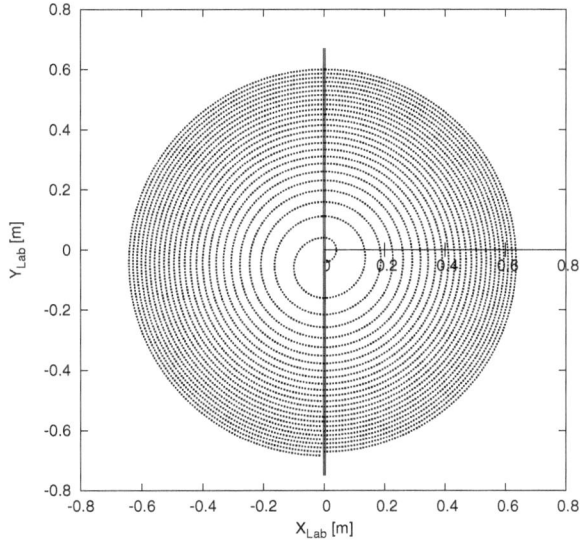

Observed at the extraction septum this oscillation modulates the distance of the bunch to the local equilibrium orbit, moving it outwards or inwards depending on the turn number, which modulates the distance between the accelerated turns. This effect can be resorted to, so to increase the separation between the final two turns and so enhance the extraction efficiency [9].

3.2.5 Spin Dance

"Much of the physics of spin motion can be illustrated using the simplest model of a storage ring consisting of uniform horizontal bending and no straight sections" [13].

By virtue of this statement, a preliminary introduction to spin motion in magnetic fields is given in the present chapter. In support to this in addition, comes the fact that cyclotrons happened to be the first circular machines to acelerate polarized beams (first acceleration of polarized beams had happened earlier in the 1960s, using electrostatic columns at voltage generators, when polarized proton and deuteron sources began operating [14]).

The magnetic field **B** of the cyclotron dipole exerts a torque on the spin angular momentum **S** of an ion, causing it to precess following the Thomas-BMT differential equation [15]

$$\frac{d\mathbf{S}}{dt} = \mathbf{S} \times \underbrace{\frac{q}{m} \left[(1 + G)\mathbf{B}_{\parallel} + (1 + G\gamma)\mathbf{B}_{\perp} \right]}_{\omega_{\mathrm{sp}}} \qquad (3.30)$$

Fig. 3.16 Spin and velocity
vector precession in a
constant field, from **S** to **S'**
and **v** to **v'** respectively. In
the moving frame the spin
precession along the arc
$\mathcal{L} = R\alpha$ is $G\gamma\alpha$, in the
laboratory frame the spin
precesses by $(1 + G\gamma)\alpha$

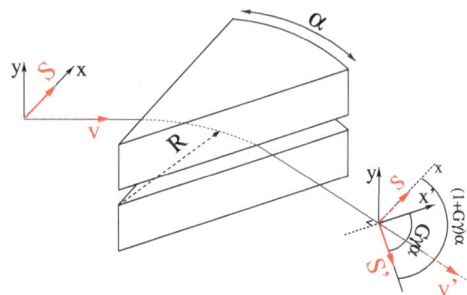

where t is the time; ω_{sp} the precession vector: a combination of $\mathbf{B}_\parallel$ and $\mathbf{B}_\perp$ components of $\mathbf{B}$ respectively parallel and orthogonal to the ion velocity vector. G is the
gyromagnetic anomaly,

   $G = 1.7928474$ (proton), -0.178 (Li), -0.143 (deuteron), -4.184 (^{3}He) ...

S in this equation is in the ion rest frame, all other quantities are in the laboratory
frame.

   In the case of an ion moving in the median plane of the dipole, $\mathbf{B}_\parallel = 0$, thus
the precession axis is parallel to the magnetic field vector, $\mathbf{B}_y$, so that $\omega_{\text{sp}} = \frac{q}{m}(1 + G\gamma)\mathbf{B}_y$. The spin precession angle over a trajectory arc $\mathcal{L}$ is

$$\theta_{\text{sp, Lab}} = \frac{1}{v}\int_{(\mathcal{L})} \omega_{\text{sp}}\, ds = (1 + G\gamma)\frac{\int_{(\mathcal{L})} B\, ds}{BR} = (1 + G\gamma)\alpha \qquad (3.31)$$

with α the velocity vector precession (Fig. 3.16). The precession angle in the moving
frame (the latter rotates by an angle α along $\mathcal{L}$) is

$$\theta_{\text{sp}} = G\gamma\alpha \qquad (3.32)$$

thus the number of 2π spin precessions per ion orbit around the cyclotron is $G\gamma$. By
analogy with the wave numbers (Eq. 3.18) this defines the "spin tune"

$$\nu_{\text{sp}} = G\gamma \qquad (3.33)$$

3.3 Exercises

Note: some of the input data files for these simulations are available in `zgoubi` sourceforge   repository   at   https://sourceforge.net/p/zgoubi/code/HEAD/tree/
branches/exemples/book/zgoubiMaterial/cyclotron_classical/

3.1 Modeling a Cyclotron Dipole: Using a Field Map
Solution 3.1

In this exercise, ion trajectories are ray-traced, various optical properties addressed in the foregoing are recovered, using a field map to simulate the cyclotron dipole. Fabricating that field map is a preliminary step of the exercise.

The interest of using a field map is that it is an easy way to account for fancy magnet geometries and fields, including field gradients and possible defects. A field map can be generated using mathematical field models, or from magnet computation codes, or from magnetic measurements. The first method is used, here. TOSCA[MOD.MOD1 = 22.1] keyword [16, cf. INDEX] is used to ray-trace through the map.

Working hypotheses: A 2-dimensional $m(R, \theta)$ polar meshing of the median plane is considered (Fig. 3.17). It is defined in a $(O; X, Y)$ frame and covers an angular sector of a few tens of degrees. The mid-plane field map is the set of values $B_Z(R, \theta)$ at the nodes of the mesh. During ray-tracing, TOSCA[MOD.MOD1 = 22.1] extrapolates the field along 3D space (R, θ, Z) ion trajectories from the 2D polar map [16]. (a) Construct a 180° two-dimensional map of a median plane field $B_Z(R, \theta)$, proper to simulate the field in a cyclotron as sketched in Fig. 3.1. Use one of the following two methods: either (i) write an independent program, or (ii) use zgoubi and its analytical field model DIPOLE, together with the keyword OPTIONS[CONSTY = ON] [16, cf. INDEX].

Besides: use a uniform mesh (Fig. 3.17) covering from Rmin = 1 to Rmax = 76 cm, with radial increment $\Delta R = 0.5$ cm, azimuthal increment $\Delta\theta = 0.5$ [cm]/ R_0 with R_0 some reference radius (say, 50 cm, in view of subsequent exercises), and constant axial field $B_Z = 5$ kG. The appropriate 6-column formatting of the field map data for TOSCA[MOD.MOD1 = 22.1] to read is the following: $R \cos\theta$, Z, $R \sin\theta$, BY, BZ, BX with θ varying first, R varying second; Z is the vertical direction (normal to the map mesh), $Z \equiv 0$ in the present case. Note that proper functioning of

Fig. 3.17 Principle of a 2D field map in polar coordinates, covering a 180° sector (over the right hand side dee). The mesh nodes $m(R, \theta)$ are distant ΔR radially, $\Delta\theta$ azimuthally. The map is used twice to cover the 360° cyclotron dipole as sketched here, while allowing insertion of an accelerating gap between the two dees

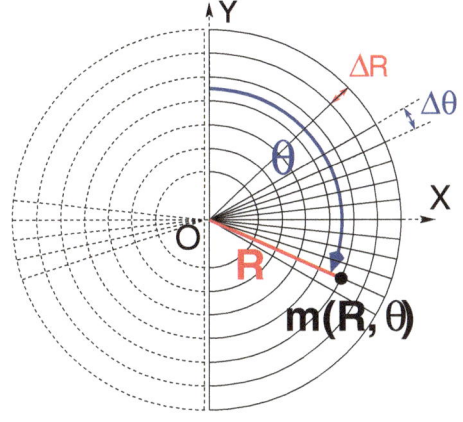

TOSCA requires the field map to begin with the following line of numerical values:
Rmin [cm] ΔR [cm] $\Delta\theta$ [deg] Z [cm]
Produce a graph of the $B_Z(R, \theta)$ field map content.
(b) Ray-trace a few concentric circular mid-plane trajectories centered on the center
of the dipole, ranging in $10 \leq R \leq 80$ cm. Produce a graph of these concentric trajectories in the $(O; X, Y)$ laboratory frame.
Initial coordinates can be defined using OBJET, particle coordinates along trajectories during the stepwise ray-tracing can be logged in zgoubi.plt by setting IL =
2 under TOSCA. In order to find the Larmor radius corresponding to a particular
momentum, the matching procedure FIT can be used. In order to repeat the latter for
a series of different momenta, REBELOTE[IOPT = 1] can be used.
Explain why it is possible to push the ray-tracing beyond the 76 cm radial extent of
the field map.
(c) Compute the orbit radius R and the revolution period T_{rev} as a function of kinetic
energy W or rigidity BR. Produce a graph, including for comparison the theoretical
dependence of T_{rev}.
(d) Check the effect of the density of the mesh (the choice of ΔR and $\Delta\theta$ values,
i.e., the number of nodes $N_\theta \times N_R = (1 + \frac{180°}{\Delta\theta}) \times (1 + \frac{80\,\text{cm}}{\Delta R}))$, on the accuracy of
the trajectory and time-of-flight computation.
(e) Check the effect of the integration step size on the accuracy of the trajectory and time-of-flight computation, by considering a small $\Delta s = 1$ cm and a large
$\Delta s = 10$ cm, at 200 keV and 5 MeV (proton), and comparing with theory.
(f) Consider a periodic orbit, thus its radius R should remain unchanged after stepwise integration of the motion over a turn. However, the size Δs of the numerical
integration step has an effect on the final value of the radius:
For two different cases, 200 keV (a small orbit) and 5 MeV (a larger one), provide a
graph of the dependence of the relative error $\delta R/R$ after one turn, on the integration
step size Δs (consider a series of Δs values in a range $\Delta s : 0.1$ mm $\rightarrow$ 20 cm).
REBELOTE[IOPT = 1] do-loop can be used to repeat the one-turn raytracing with
different Δs.

3.2 Modeling a Cyclotron Dipole: Using an Analytical Field Model
Solution 3.2

This exercise is similar to Exercise 3.1, yet using the analytical modeling DIPOLE,
instead of a field map. DIPOLE provides the Z-parallel median plane field $\mathbf{B}(R, \theta,
Z = 0) \equiv \mathbf{B}_Z(R, \theta, Z = 0)$ at the projected $m(R, \theta, Z = 0)$ ion location (Fig. 3.18),
while $\mathbf{B}(R, \theta, Z)$ at particle location is obtained by extrapolation.
(a) Simulate a 180° sector dipole; DIPOLE requires a reference radius
[16, Eqs. 6.3.19–6.3.21], noted R_0 here; for the sake of consistency with other exercises, it is suggested to take $R_0 = 50$ cm. Take a constant axial field $B_Z = 5$ kG.
Explain the various data that define the field simulation in DIPOLE: geometry, role
of R_0, field and field indices, fringe fields, integration step size, etc.
Produce a graph of $B_Z(R, \theta)$.
(b) Repeat question (b) of Exercise 3.1.
(c) Repeat question (c) of Exercise 3.1.

Fig. 3.18 DIPOLE provides
the value $B_Z(m)$ of the
median plane field at m,
projection of particle
position $M(R, \theta, Z)$ in the
median plane. $\mathbf{B}(R, \theta, Z)$ is
obtained by extrapolation

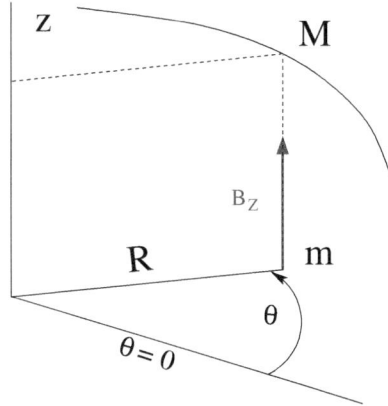

(d) As in question (e) of Exercise 3.1, check the effect of the integration step size on
the accuracy of the trajectory and time-of-flight computation.
Repeat question (f) of Exercise 3.1.
(e) From the two series of results (Exercise 3.1 and the present one), comment on
various pros and cons of the two methods, field map versus analytical field model.

3.3 Resonant Acceleration
Solution 3.3

Based on the earlier exercises, using indifferently a field map (TOSCA) or an analytical model of the field (DIPOLE), introduce a sinusoidal voltage between the two
dees, with peak value 100 kV. Assume that ion motion does not depend on RF phase:
the boost through the gap is the same at all passes, use CAVITE[IOPT = 3] [16,
cf. INDEX] for that. Note that using CAVITE requires prior PARTICUL in order to
specify ion species and data, necessary to compute the energy boost (Eq. 3.2).
(a) Accelerate a proton with initial kinetic energy 20 keV, up to 5 MeV, take harmonic
h = 1. Produce a graph of the accelerated trajectory in the laboratory frame.
(b) Provide a graph of the proton momentum p and total energy E as a function of its
kinetic energy, both from this numerical experiment (ray-tracing data can be stored
using FAISTORE) and from theory, all on the same graph.
(c) Provide a graph of the normalized velocity $\beta = v/c$ as a function of kinetic energy,
both numerical and theoretical, and in the latter case both classical and relativistic.
(d) Provide a graph of the relative change in velocity $\Delta\beta/\beta$ and orbit length $\Delta C/C$
as a function of kinetic energy, both numerical and theoretical. From their evolution,
conclude that the time of flight increases with energy.
(e) Repeat the previous questions, assuming a harmonic h=3 RF frequency.

3.4 Spin Dance
Solution 3.4

Cyclotron modeling in the present exercise can use Exercises 3.1 or 3.2 technique
(i.e., a field map or an analytical field model), indifferently.

(a) Add spin transport, using SPNTRK [16, *cf.* INDEX]. Produce a listing (zgoubi.res) of a simulation, including spin outcomes.
Note: PARTICUL is necessary here, for the spin equation of motion (Eq. 3.30) to be solved [16, Sect. 2]. SPNPRT can be used to have local spin coordinates listed in zgoubi.res (at the manner that FAISCEAU lists local particle coordinates).
(b) Consider proton case, take initial spin longitudinal, compute the spin precession over one revolution, as a function of energy over a range $12\,\text{keV} \rightarrow 5\,\text{MeV}$. Give a graphical comparison with theory.
FAISTORE can be used to store local particle data, which include spin coordinates, in a zgoubi.fai style output file. IL = 2 [16, *cf.* INDEX] (under DIPOLE or TOSCA, whichever modeling is used) can be used to obtain a print out of particle and spin motion data to zgoubi.plt during stepwise integration.
(c) Inject a proton with longitudinal initial spin $\mathbf{S}_i$. Give a graphic of the longitudinal spin component value as a function of azimuthal angle, over a few turns around the ring. Deduce the spin tune from this computation. Repeat for a couple of different energies.
Place both FAISCEAU and SPNPRT commands right after the first dipole sector, and use them to check the spin rotation and its relationship to particle rotation, right after the first passage through that first sector.
(d) Spin dance: the input data file optical sequence here is assumed to model a full turn. Inject an initial spin at an angle from the horizontal plane (this is in order to have a non-zero vertical component), produce a 3-D animation of the spin dance around the ring, over a few turns.
(e) Repeat questions (b–d) for two additional ions: deuteron (much slower spin precession), $^3He^{2+}$ (much faster spin precession).

3.5 Synchronized Spin Torque
Solution 3.5

A synchronized spin kick is superimposed on orbital motion. An input data file for a complete cyclotron is considered as in question Exercise 3.4(d), for instance six $60°$ DIPOLEs, or two $180°$ DIPOLEs.

Insert a local spin rotation of a few degrees around the longitudinal axis, at the end of the optical sequence (i.e., after one orbit around the cyclotron). SPINR can be used for that, rather than a local magnetic field, so to avoid any orbital effect. Track 4 particles on their respective equilibrium orbit, with energies 0.2, 108.412, 118.878 and 160.746 MeV.

Produce a graph of the motion of the vertical spin component S_y along the circular orbit.

Produce a graph of the spin vector motion on a sphere.

3.6 Weak Focusing
Solution 3.6

(a) Consider a $60°$ sector as in earlier exercises (building a field map and using TOSCA as in Exercise 3.1, or using DIPOLE as in Exercise 3.2), construct the sector accounting for a non-zero radial index k in order to introduce axial focusing, say $k = -0.03$, assume a reference radius R_0 for a reference energy of 200 keV (R_0 and B_0 are required in order to define the index k, Eq. 3.11). Ray-trace that 200 keV reference orbit, plot it in the lab frame: make sure it comes out as expected, namely, constant radius, final and initial angles zero.

(b) Using FIT[2], find and plot the radius dependence of orbit rigidity, $BR(R)$, from ray-tracing over a BR range covering 20 keV to 5 MeV; superpose the theoretical curve. REBELOTE[IOPT = 1] can be used to perform the scan.

(c) Produce a graph of the paraxial axial motion of a 1 MeV proton, over a few turns (use IL = 2 under TOSCA, or DIPOLE, to have step by step particle and field data logged in zgoubi.plt). Check the effect of the focusing strength by comparing the trajectories for a few different index values, including close to −1 and close to 0.

(d) Produce a graph of the magnetic field experienced by the ion along these trajectories.

3.7 Loss of Isochronism
Solution 3.7

Compare on a common graphic the revolution period $T_{rev}(R)$ for a field index value $k \approx -0.95, -0.5, -0.03, 0$. The scan method of Exercise 3.6, based on REBELOTE[IOPT=1] preceded by FIT[2], can be referred to.

3.8 Ion Trajectories
Solution 3.8

In this exercise individual ion trajectories are computed. DIPOLE or TOSCA magnetic field modeling can be used, indifferently. Take for instance $B_0 = 5$ kG and for reference R_0 the 200 keV radius. No acceleration here, ions circle around the cyclotron at constant energy.

(a) Produce a graph of the horizontal $x(s)$ and vertical $y(s)$ trajectory coordinates of an ion with rigidity $BR(R_0)$ and paraxial motion, over a few turns around the cyclotron. From the number of turns, give an estimate of the wave numbers. Check the agreement with the expected $\nu_R(k)$, $\nu_y(k)$ values (Eq. 3.18).

(b) Consider now protons in that very cyclotron, far from the reference energy $E(R_0)$, say at 1 MeV and 5 MeV. The wave numbers change with energy: compute k(E) from tracking and check consistency with theory.

(c) Consider a proton, 200 keV energy, plot as a function of s the difference between $x(s)$ from raytracing and its values from Eq. 3.16. Same for y(s) compared to Eq. 3.17. IL = 2 can be used to store in zgoubi.plt the step-by-step particle coordinates across DIPOLE.

(d) Perform a scan of the wave numbers over 200 keV−5 MeV energy interval, computed using OBJET[KOBJ = 5] and MATRIX[IORD = 1, IFOC = 11], or OBJET[KOBJ = 6] and MATRIX[IORD = 2, IFOC = 11], together with REBELOTE[IOPT = 1] to repeat MATRIX for a series of energy values.

3.9 RF Phase at the Accelerating Gap
Solution 3.9

Consider the cyclotron model of exercise 3.6: field index $k = -0.03$ defined at $R_0 = 50$ cm, field $B_0 = 5$ kG on that radius, two dees, double accelerating gap.

Accelerate a proton from 1 to 5 MeV: get the turn-by-turn phase-shift at the gaps; use CAVITE[IOPT=7] to simulate the acceleration. Compare the half-turn $\Delta\phi$ so obtained with the theoretical expectation (Eq. 3.25). Produce similar graphs $B(R)$ and $\Delta W(\phi)$ to Fig. 3.13.

Accelerate over more turns, observe the particle decelerating.

3.10 The Cyclotron Equation
Solution 3.10

The cyclotron model of Exercise 3.3 is considered: two dees, double accelerating gap, uniform field $B = 5$ kG, no field gradient needed here (no vertical motion).

(a) Set up an input data file for the simulation of a proton acceleration from 0.2 to 20 MeV. In particular, assume that $\cos(\phi)$ reaches its maximum value at $W_m = 10$ MeV; find the RF voltage frequency from $d(\cos\phi)/dW = 0$ at W_m.
(b) Give a graph of the energy-phase relationship (Eq. 3.26), for $\phi_i = \frac{3\pi}{4}, \frac{\pi}{2}, \frac{\pi}{4}$, from both simulation and theory.

3.11 Cyclotron Extraction
Solution 3.11

(a) Acceleration of a proton in a uniform field $B = 5$ kG is first considered (field hypotheses as in Exercise 3.3). RF phase is ignored: CAVITE[IOPT = 3] can be used for acceleration. Take a 100 kV gap voltage.
Compute the distance ΔR between turns, as a function of turn number and of energy, over the range $E : 0.02 \to 5$ MeV. Compare graphically with theoretical expectation.
(b) Assume a beam with Gaussian momentum distribution and *rms* momentum spread $\delta p/p = 10^{-3}$. An extraction septum is placed half-way between two successive turns, provide a graph of the percentage of beam loss at extraction, as a function of extraction turn number. COLLIMA can be used for that simulation and for particle counts, it also allows for possible septum thickness.
(c) Repeat (a) and (b) considering a field with index: take for instance $B_0 = 5$ kG and $k = -0.03$ at $R_0 = R(0.2\,\text{MeV}) = 12.924888$ cm.
(d) Investigate the effect of injection conditions (Y_i, T_i) on the modulation of the distance between turns.
Try and confirm numerically that, with slow acceleration, the oscillation is minimized for an initial $|T_i| = |\frac{x_0 \nu_R}{R}|$ (after Ref. [9, p. 133]).

3.12 Acceleration and Extraction of a 6-D Polarized Bunch
Solution 3.12

The cyclotron simulation hypotheses of Exercise 3.10a are considered; account or $k = -0.02$ field index.

Add a short "high energy" extraction line, say 1 m, following REBELOTE in the optical sequence, ending up with a "Beam_Dump" MARKER for instance.

(a) Create a 1,000 ion bunch with the following initial parameters:

– random Gaussian transverse phase space densities, centered on the equilibrium orbit, truncated at 3 sigma, normalized *rms* emittances $\varepsilon_Y = \varepsilon_Z = 1\,\pi\,\mu$m, both emittances matched to the 0.2 MeV orbit optics,
– uniform bunch momentum density $0.2 \times (1 - 10^{-3}) \le p \le 0.2 \times (1 + 10^{-3})$ MeV, matched to the dispersion, namely (Eq. 3.21), $\Delta x = D\frac{\Delta p}{p}$,
– random uniform longitudinal distribution $-0.5 \le s \le 0.5$ mm,

Note: two ways to create this object are, (i) using MCOBJET[KOBJ = 3] which generates a random distribution, or (ii) using OBJET[KOBJ = 3] to read an external particle coordinate file.

Add spin tracking request (SPNTRK), all initial spins normal to the bend plane. Produce a graph of the three initial 2-D phase spaces: (Y, T), (Z, P), (δl, $\delta p/p$), matched to the 200 keV periodic optics.
Provide Y, Z, dp/p, δl and S_Z histograms (HISTO can be used), check the distribution parameters.

(b) Accelerate this polarized bunch to 20 MeV, using the following RF conditions:

– 200 kV peak voltage,
– RF harmonic 1,
– initial RF phase $\phi_i = 3\pi/4$.

Produce a graph of the three phase spaces as observed downstream of the extraction line. Provide the Y, Z, dp/p, δl and S_Z histograms. Compare the distribution parameters with the initial values.
What causes the spins to spread away from vertical?

3.4   Solutions of Exercises of This Chapter

3.1 Modeling a Cyclotron Dipole: Using a Field Map
(a) A field map of a 180° sector of a classical cyclotron magnet.

The first option is retained here: a Fortran program, geneSectorMap.f, given in Table 3.1. constructs the required map of a field distribution $B_Z(R, \theta)$, to be subsequently read and raytraced through using the keyword TOSCA [16, *lookup* INDEX].

Table 3.1 A Fortran program which generates a 180° mid-plane field map. This angle as well as field amplitude can be changed, a field index can be added. This program can be compiled and run, as is. The field map it produces is logged in geneSectorMap.out

```
C geneSectorMap.f program
      implicit double precision (a-h,o-z)
      parameter (pi=4.d0*atan(1.d0), BY=0.d0, BX=0.d0, Z=0.d0)

      open(unit=2,file='geneSectorMap.out')                              ! Field map storage file.

C------------ Hypotheses :
      AT = 180.d0  /180.d0*pi          ! Angular extent of field map. Can be changed 360, 60 deg, etc.).
      BZ=5.d0                                                            ! Field (kG).
      Rmi=1.d0; Rma=76.d0; RM=50.d0  ! cm. Radial extent of field map; reference radius to define mesh.
      dR = 0.5d0 ; NR = NINT((Rma - Rmi)/dR)+1   ! R-distance between nodes in mesh. Number of R-nodes.
C                                      RdA=RM*dA is the distance between two nodes along R=RM arc,
      RdA = 0.5d0  !      given angle increment dA (dA is the "Delta theta" quantity in the main text).
      NX= NINT(RM*AT /RdA) +1 ; RdA= RM*AT / DBLE(NX -1)  ! exact mesh step at RM, corresponding to NX.
      dA = RdA / RM ;A1 = 0.d0 ; A2 = AT                    ! corresponding delta_angle.
C-------------------------------------------------
      write(2,*) Rmi,dR,dA/pi*180.d0,dZ,
     >'     ! Rmi/cm, dR/cm, dA/deg, dZ/cm'
      write(2,*) '# Field map generated using geneSectorMap.f '
      write(2,fmt='(a)') '# AT/rd,  AT/deg, Rmi/cm, Rma/cm, RM/cm,'
     >//' NR, dR/cm, NX, RdA/cm, dA/rd : '
      write(2,fmt='(a,1p,5(e16.8,1x),2(i3,1x,e16.8,1x),e16.8)')
     >'# ',AT, AT/pi*180.d0,Rmi, Rma, RM, NR, dR, NX, RdA, dA
      write(2,*) '# For TOSCA: ',NX,NR,' 1 22.1 1.  !IZ=1 -> 2D ; '
     >//'MOD=22 -> polar map ; .MOD2=.1 -> one map file'
      write(2,*) '#      R*cosA              Z==0,           R*sinA'
     >//'          BY              BZ            BX   ix jr'
      write(2,*) '#       cm                 BZ              cm '
     >//'         cm              kG            cm '
      write(2,*) '#       kG               kG              kG '
      write(2,*) '# '
      do jr = 1, NR
      R = Rmi + dble(jr-1)*dR
      do ix = 1, NX
      A = A1 + dble(ix-1)*dA ; X = R * sin(A) ; Y = R * cos(A)
      write(2,fmt='(1p,6(e16.8),2(1x,i0))') Y,Z,X,BY,BZ,BX,ix,jr
      enddo
      enddo
      stop ' Job complete ! Field map stored in geneSectorMap.out.'
      end
```

Regarding the second option: using the analytical model DIPOLE together with the keyword OPTIONS[CONSTY = ON] to fabricate a field map, examples can be found for instance in the FFAG chapter exercises (Chap. 10).

A polar mesh is retained (Fig. 3.19), rather than Cartesian, consistently with cyclotron magnet symmetry. The program can be compiled (*gfortran-o geneSectorMap geneSectorMap.f* will provide the executable, geneSectorMap) and run, as is. The field map is saved under the name geneSectorMap.out, excerpts of the expected content are given in Table 3.2. That name appears under TOSCA in zgoubi input data file for this simulation (Table 3.3). Figure 3.20 shows the field over the 180° azimuthal extent (using a gnuplot script, bottom of Table 3.2).

Note the following:
(i) the field map azimuthal extent (set at 180° in geneSectorMap) can be changed, for instance to simulate a 60 deg sector instead;
(ii) the field is vertical being the mid-plane field of dipole magnet. The field is taken constant in this exercise, $\forall R$, $\forall \theta$ throughout the map mesh, whereas in upcoming exercises, a *focusing index* will be introduced, which will make $B_Z \equiv B_Z(R)$ an R-dependent quantity (in Chap. 4 which addresses Thomas focusing and the isochronous cyclotron, exercises will further resort to $B_Z \equiv B_Z(R,\theta)$, an R- and θ-dependent quantity).

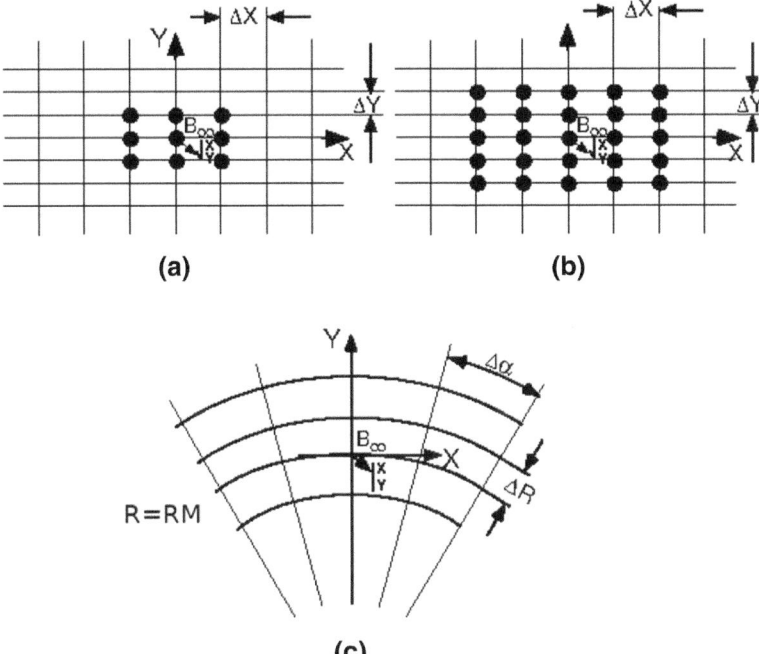

Fig. 3.19 Principle 2-D field map mesh as used by TOSCA, and the (O; X, Y) coordinate system. **A, B**: Cartesian mesh in the (X, Y) plane, case of respectively 9-point and a 25-point interpolation grid; the mesh increments are ΔX and ΔY. **C**: polar mesh and increments ΔR and $\Delta \alpha$ ($\Delta \theta$ in the text), and (O; X, Y) frame moving along a reference arc of radius R_M. The field at particle location is interpolated from its values at the closest 3×3 or 5×5 nodes

This field map can be readily tested using the example of Table 3.3, which raytraces $E_k = 0.12$, 0.2 and 5.52 MeV protons on circular trajectories centered at the center of the field map. Trajectory radii, respectively $R = 10.011$, 12.924 and 67.998 cm (Table 3.3), have been prior determined from

$$\text{Rigidity} \quad B\rho = B_0 \times R \quad \text{and} \quad B\rho = p/c = \sqrt{E_k(E_k + 2M)}/c \qquad (3.34)$$

with $B_0 = 0.5\,\text{T}$ (Table 3.1) and $M = 938.272\,\text{MeV}/c^2$ the proton mass.

The optical sequence for this particle raytracing uses the following keywords:

(i) OBJET to define a (arbitrary) reference rigidity and initial particle coordinates
(ii) TOSCA, to read the field map and raytrace through (and TOSCA's 'IL = 2' flag to store step-by-step particle data into zgoubi.plt)
(iii) FAISCEAU to print out particle coordinates in zgoubi.res execution listing
(iv) SYSTEM to run a gnuplot script (Table 3.3) once raytracing is complete

Table 3.2 First and last few lines of the field map file geneSectorMap.out. The file starts with an 8-line header, the first of which is effectively used by zgoubi (the following 7 are not used) and indicates, in that order: the minimum radius of the map mesh Rmi, the radial increment dR, the azimuthal increment dA, the axial increment dZ (null and not used in the present case of a two-dimensional field map), in units of, respectively, cm, cm, degree, cm. The additional 7 lines provide the user with various indications regarding numerical values used in, or resulting from, the execution of geneSectorMap.f. The first 5 numerical data in line 5 in particular are to be reported in zgoubi input data file under TOSCA keyword. The rest of the file is comprised of 8 columns, the first three give the node coordinates and the next three the field component values at that node, the last two columns are the (azimuthal and radial) node numbers, from (1, 1) to (315, 151) in the present case

```
   1.00      0.500      0.57324840764331209         0.00           |  Rmi/cm, dR/cm, dA/deg, dZ/cm
# Field map generated using geneSectorMap.f
# AT/rd,  AT/deg, Rmi/cm, Rma/cm, RM/cm, NR, dR/cm, NX, RdA/cm, dA/rd :
#  3.14159265E+00  1.800E+02  1.000E+00  7.600E+01  5.000E+01 151  5.000E-01 315  5.00253607E-01  1.00050721E-02
# For TOSCA:       315         151 1 22.1 1.  |IZ=1 -> 2D ; MOD=22 -> polar map ; .MOD2=.1 -> one map file
#
#     R*cosA          Z==0,          R*sinA          BY              BZ              BX     ix jr
#       cm              cm             cm             kG              kG              kG
  1.00000000E+00  0.00000000E+00  0.00000000E+00  0.00000000E+00  5.00000000E+00  0.00000000E+00 1 1
  9.99949950E-01  0.00000000E+00  1.00049052E-02  0.00000000E+00  5.00000000E+00  0.00000000E+00 2 1
  9.99799804E-01  0.00000000E+00  2.00088090E-02  0.00000000E+00  5.00000000E+00  0.00000000E+00 3 1
  9.99549577E-01  0.00000000E+00  3.00107098E-02  0.00000000E+00  5.00000000E+00  0.00000000E+00 4 1
  9.99199295E-01  0.00000000E+00  4.00096065E-02  0.00000000E+00  5.00000000E+00  0.00000000E+00 5 1
  9.99199295E-01  0.00000000E+00  4.00096065E-02  0.00000000E+00  5.00000000E+00  0.00000000E+00 5 1
..........................
 -7.59391464E+01  0.00000000E+00  3.04073010E+00  0.00000000E+00  5.00000000E+00  0.00000000E+00 311 151
 -7.59657679E+01  0.00000000E+00  2.28081394E+00  0.00000000E+00  5.00000000E+00  0.00000000E+00 312 151
 -7.59847851E+01  0.00000000E+00  1.52066948E+00  0.00000000E+00  5.00000000E+00  0.00000000E+00 313 151
 -7.59961962E+01  0.00000000E+00  7.60372797E-01  0.00000000E+00  5.00000000E+00  0.00000000E+00 314 151
 -7.60000000E+01  0.00000000E+00  9.30731567E-15  0.00000000E+00  5.00000000E+00  0.00000000E+00 315 151
```

A gnuplot script to obtain a graph of B(X,Y), Fig. 3.20:

```
# gnuplot_fieldMap.gnu
set key maxcol 1 ; set key t l ; set xtics mirror ; set ytics mirror ; cm2m = 0.01
set xlabel "Y [m]"; set ylabel "X [m]"; set zlabel "B [kG] \n" rotate by 90; set zrange [:5.15]
splot "geneSectorMap.out" u ($1 *cm2m):($3 *cm2m):($5) w l lc rgb "red" notit; pause 1
```

(v) MARKER, to define two particular "LABEL_1" type labels [16, *lookup* INDEX] (#S_halfDipole and #E_halfDipole), to be used with INCLUDE in subsequent exercises.

Three circular trajectories in a dee, resulting from the data file of Table 3.3 are shown in Fig. 3.20. Inspecting zgoubi.res execution listing one finds the D, Y, T, Z, P, S particle coordinates under FAISCEAU, at OBJET (left) and current (right) after a turn in the cyclotron (unchanged, as the trajectory forms a closed orbit):

```
    6  Keyword, label(s) :  FAISCEAU                                                        IPASS= 1
                                        TRACE DU FAISCEAU
                                      (follows element #    5)
                                          2 TRAJECTOIRES
                        OBJET                                              FAISCEAU
         D       Y(cm)    T(mr)   Z(cm)    P(mr)   S(cm)     D-1     Y(cm)    T(mr)   Z(cm)  P(mr)   S(cm)
  o  1  0.7746  10.011   0.000    0.000   0.000  0.0000   -0.2254  10.011   -0.000   0.000  0.000  3.145152E+01  1
  o  1  5.2610  67.998   0.000    0.000   0.000  0.0000    4.2610  67.998   -0.000   0.000  0.000  2.136220E+02  2
```

(b) Concentric trajectories in the median plane.

The optical sequence for this exercise is given in Table 3.4. Compared to the previous sequence (Table 3.3), (i) the TOSCA segment has been replaced by an INCLUDE, for the mere interest of making the input data file for this simulation shorter, and (ii) additional keywords are introduced, including

Table 3.3 Simulation input data file FieldMapSector.inc: it is set to allow a preliminary test regarding the field map geneSectorMap.out (as produced by the Fortran program geneSectorMap, Table 3.1), by computing three circular trajectories centered on the center of the map. This file also defines the INCLUDE segment between the labels (LABEL1 type [16, Sect. 7.7]) #S_halfDipole and #E_halfDipole

```
FieldMapSector.inc
! Uniform field 180 deg sector. FieldMapSector.inc.
'MARKER'   FieldMapSector_S                              ! Just for edition purposes.
'OBJET'
64.62444403717985                    ! Reference Brho ("BORO" in the users' guide) -> 200keV proton.
2
3 1
10.011362 0. 0. 0. 0. 0.7745802 'a'    ! p[MeV/c]= 15.007, Brho[kG.cm]= 50.057, kin-E[MeV]=0.12.
12.924888 0. 0. 0. 0. 1.        'b'                       ! kin-E[MeV]=0.2.
67.997983 0. 0. 0. 0. 5.2610112 'c'    ! p[MeV/c]=101.926, Brho[kG.cm]=339.990, kin-E[MeV]=5.52.
1 1 1
'MARKER'   #S_halfDipole
'TOSCA'
0 2    ! IL=2 to log step-by-step coordinates, spin, etc., to zgoubi.plt (avoid, if CPU time matters).
1. 1. 1. 1.    ! Normalization coefficients, for B, X, Y and Z coordinate values read from the map.
HEADER_8                              ! The field map file starts with an 8-line header.
315 151 1 22.1 1.         ! IZ=1 for 2D map; MOD=22 for polar frame; .MOD2=.1 if only one map file.
geneSectorMap.out
0 0 0 0    ! Possible vertical boundaries within the field map, to start/stop stepwise integration.
2
1.                                  ! Integration step size. Small enough for orbits to close accurately.
2                                                        ! Magnet positioning option.
0. 0. 0. 0.                                              ! Magnet positioning coordinates.
'MARKER'   #E_halfDipole
'FAISCEAU'
'SYSTEM'                     ! This SYSTEM command runs gnuplot, for a graph of the two trajectories.
1
gnuplot <./gnuplot_Zplt.gnu
'MARKER'   FieldMapSector_E                              ! Just for edition purposes.
'END'
```

A gnuplot script to obtain a graph of the orbits, Fig. 3.20:

```
# gnuplot_Zplt.gnu
set key maxcol 1 ; set key t r ; set xtics ; set ytics ; cm2m = 0.01 ; unset colorbox
set xlabel "X_{Lab} [m]" ; set ylabel "Y_{Lab} [m]" ; set size ratio 1 ; set polar
plot for [orbit=1:3] "zgoubi.plt" u ($19==orbit ? $22 :1/0):($10 *cm2m):($19) w l lw 2 lc pal ; pause 1
```

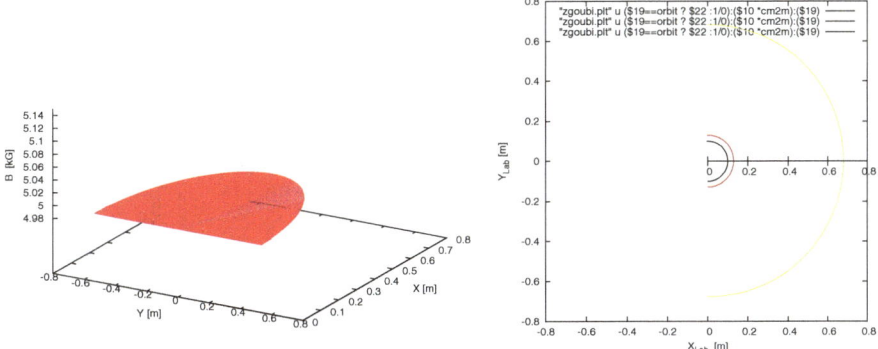

Fig. 3.20 Left: map of a constant magnetic field over a 180 deg sector, 76 cm radial extent. Right: three circular trajectories, at respectively 0.12, 0.2 and 5.52 MeV, computed using that field map

Table 3.4 Simulation input data file: optical sequence to find cyclotron closed orbits at a series of different momenta. An INCLUDE inserts the #S_halfDipole to #E_halfDipole TOSCA segment of the sequence of Table 3.3

```
Uniform field 180 deg. sector. Find orbits.
'MARKER'    FieldMapOrbits_S                                          ! Just for edition purposes.
 'OBJET'
64.62444403717985                               ! Reference Brho ("BORO" in the users' guide) -> 200keV proton.
2
1 1                                                                   ! Just one ion.
12.9248888074 0.  0. 0. 0.   1. 'm'             ! This initial radius yields BR=64.6244440372 kG.cm.
1
'INCLUDE'                                                             ! A half of the cyclotron dipole.
1
FieldMapSector.inc[#S_halfDipole:#E_halfDipole]
'FAISCEAU'
'INCLUDE'                                                             ! A half of the cyclotron dipole.
1
FieldMapSector.inc[#S_halfDipole:#E_halfDipole]
'FIT'
1
2  35  0  6.                        ! Vary momentum, to allow fulfilling the following constraint:
1
3.1 1 2 5 0. 1. 0                 ! request same radius after a half-turn (i.e., after first 180 deg sector,
!                                         this ensures centering of orbit on center of map).
'FAISCEAU' CHECK  ! Allows quick check of particle coordinates, in zgoubi.res: final should = initial.

'REBELOTE'                                                            ! Repeat what precedes,
15  0.1  0 1                                                          ! 15 times.
1
OBJET 30  10:80     ! Prior to each repeat, first change the value of parameter 30 (i.e., Y) in OBJET.
'SYSTEM'
2
gnuplot <./gnuplot_Zplt.gnu
cp gnuplot_Zplt_XYLab.eps gnuplot_Zplt_XYLab_stage1.eps
'MARKER'    FieldMapOrbits_E                                          ! Just for edition purposes.
'END'
```

A gnuplot script to obtain Fig. 3.21:
Note: removing the test '$51==1 ?' on column 51 in zgoubi.plt, would add on the graph the orbit as it is before each FIT.

```
# gnuplot_Zplt.gnu
set key maxcol 1 ; set t r ; set xtics ; set ytics  ; set size ratio 1 ; set polar ; unset colorbox
set xlabel "X_{Lab}  [m] \n" ; set ylabel "Y_{Lab}  [m] \n" ; cm2m = 0.01 ; sector1=4 ; sector2=8 ; pi = 4.*atan(1.)
lmnt1 = 4; lmnt2=8  ### column numer in zgoubi.plt,  $42: NOEL;  $51: FITLST; $49: FIT number
plot for [l=lmnt1/4:lmnt2/4] "zgoubi.plt" u ($42==4*l && $51==1 ? $22 +pi*(l-1):1/0):($10 *cm2m):($49) w p ps .3 lc pal
pause 1
```

– FIT, which finds the circular orbit for a particular momentum,
– FAISCEAU, a means to check local particle coordinates,
– REBELOTE, which repeats the execution of the sequence (REBELOTE sends the execution pointer back to the top of the data file) for a new momentum value which REBELOTE itself defines, prior.

In order to compute and then plot trajectories (Fig. 3.21), zgoubi proceeds as follows: orbit circles for a series of different radii taken in [10, 80] cm are searched, using FIT to find the appropriate momenta. REBELOTE is used to repeat that fitting on a series of different values of R; prior to repeating, REBELOTE modifies the initial particle coordinate Y_0 in OBJET. Stepwise particle data through the dipole field are logged in zgoubi.plt, due to IL = 2 under TOSCA keyword, at the first pass before FIT, and at the last pass following FIT completion. A key point here: a flag, FITLST, recorded in column 51 in zgoubi.plt ([16], Sect. 8.3), is set to 1 at the last pass (the last pass follows the completion of the FIT execution and uses updated FIT variable values).

Fig. 3.21 Circular trajectories in the cyclotron mid-plane, centered on the field map center. The outermost orbit is at R = 80 cm by hypothesis, thus $BR = B_0 \times R = 0.4\,\mathrm{T\,m}$, $E_k = 7.632\,\mathrm{MeV}$. These stepwise (R, θ) data are read from zgoubi.plt, coordinates (Y, X) in zgoubi polar frame nomenclature (Sect. 8.3 [16])

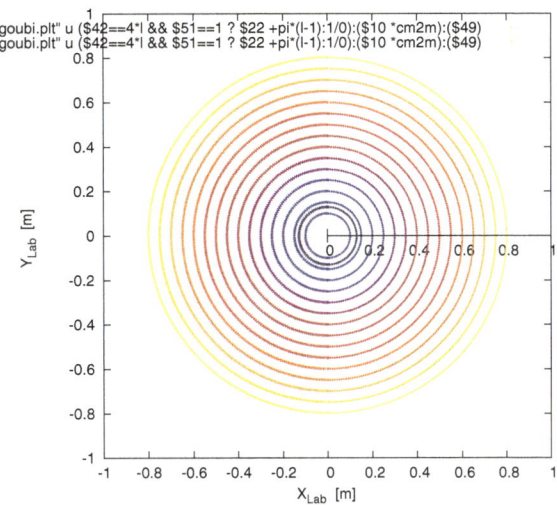

At the bottom of `zgoubi` input data file, a SYSTEM command produces a graph of ion trajectories, by executing a gnuplot script (bottom of Table 3.4). Note the test on FITLST, which allows selecting the last pass following FIT completion. Graphic outcomes are given in Fig. 3.21.

The reason why it is possible to push the raytracing beyond the 76 cm radius field map extent, without loss of accuracy, is that the field is constant. Thus, referring to the polynomial interpolation technique used [16, Sect. 1.4], the extrapolation out of the map will leave the field value unchanged.

(c) Energy and rigidity dependence of orbit radius and time-of-flight.

The orbit radius R and the revolution time T_{rev} as a function of kinetic energy E_k and rigidity BR are obtained by a similar scan to exercise (b). The results are shown in Fig. 3.22.

A slow increase of revolution period with energy can be observed, which is due to the mass increase.

Note that these results are converged for the step size, to high accuracy (see (d)), due to its value taken small enough, namely $\Delta s = 1\,\mathrm{cm}$. This corresponds for instance to 80 steps to complete a revolution for the 120 keV, $R = 12.9\,\mathrm{cm}$ smaller radius trajectory in Fig 3.21.

(d) Numerical convergence: mesh density.

This question concerns the dependence of the numerical convergence of the solution of the differential equation of motion [16, Eq. 1.2.1] upon mesh density.

The program used in (b) to generate a field map (Table 3.1) is modified to construct field maps of $B_Z(R, \theta)$ with various radial and azimuthal mesh densities. Changing these is simply a matter of modifying the quantities dR (radius increment ΔR) and

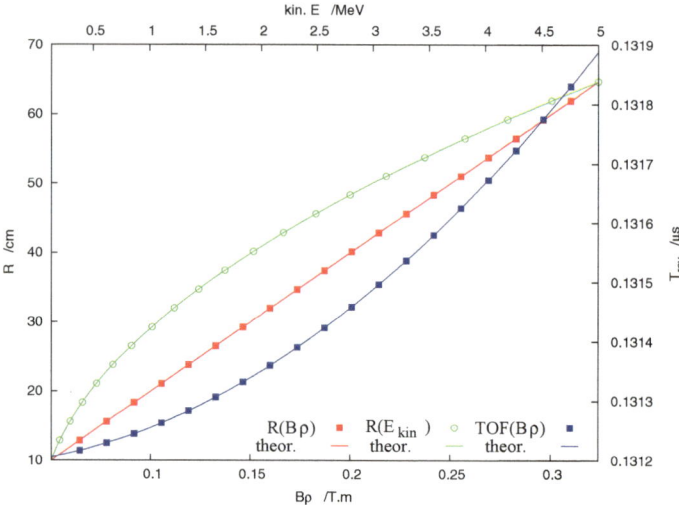

Fig. 3.22 Numerical (markers) and theoretical (solid lines) values of orbit radius, R, and revolution period, T_{rev}, versus kinetic energy (top scale) and rigidity (bottom scale). The mesh density here is $N_\theta \times N_R = 315 \times 151$. The integration step size is $\Delta s = 1$ cm, so ensuring converged results (to $\Delta R/R$ and $\Delta T_{rev}/T_{rev} < 10^{-6}$)

$R \, dA$ (R times the azimuth increment $\Delta\theta$) in the program of Table 3.1. The field maps geneSectorMap.out so generated for various $(dR, R \, dA)$ couples may be saved under different names, and used separately.

Table 3.5 shows the complete, 9 line, TOSCA field map, in the case of a 60° sector covered in $N_\theta \times N_R = \dfrac{60°}{\Delta\theta} \times \dfrac{75\,cm}{\Delta R} = \dfrac{360°}{120°} \times \dfrac{75\,cm}{37.5cm} = 3 \times 3$ nodes. Six sectors are now required to cover the complete cyclotron dipole: zgoubi input data need be changed accordingly, namely stating TOSCA—possibly via an INCLUDE— six times, instead of just twice in the case of a 180° sector.

The result to be expected: with a mesh reduced to as low as $N_\theta \times N_R = 3 \times 3$, compared to $N_\theta \times N_R = 106 \times 151$, radius and time-of-flight should however remain unchanged. This shows in Fig. 3.23 which displays both cases, over a E_k : $0.12 \rightarrow 5$ MeV energy span (assuming protons). The reason for the absence of effect of the mesh density is that the field is constant. As a consequence the field derivatives in the Taylor series based numerical integrator are all zero [16, Sect. 1.2]: only B_Z is left in evaluating the Taylor series, however B_Z is constant. Thus R remains unchanged when pushing the ion by a step Δs, and the cumulated path length— the closed orbit length—and revolution time—path length over velocity—end up unchanged. Note: this will no longer be the case when a radial field index is introduced in order to cause vertical focusing, in subsequent exercises.

Table 3.5 Field map of a 60° constant field sector as read by TOSCA. The field map is complete, with smallest possible $N_\theta \times N_R = 3 \times 3 = 9$ number of nodes. The first line of the header is used by zgoubi (the following 7 are not used), namely, the minimum value of the radius in the map, radius increment, azimuthal increment, and vertical increment (null here, as this is a single, mid-plane map)

```
      1.0       37.50     30.0  0.        !  Rmi/cm, dR/cm, dA/cm, dZ/cm
# Field map generated using geneSectorMap.f
# AT/rd,  AT/deg, Rmi/cm, Rma/cm, RM/cm, NR, dR/cm, NX, RdA/cm, dA/rd :
#   1.04719755E+00  60.    1.    76.    50.    3  37.5  3   26.1799388   0.523598776
# For TOSCA:            3              3 1 22.1 1.  !IZ=1 -> 2D ; MOD=22 -> polar map ; .MOD2=.1 -> one map file
#
#    R*cosA           Z==0,          R*sinA            BY              BZ              BX     ix jr
#      cm              cm              cm              kG              kG              kG
   1.00000000E+00  0.00000000E+00  0.00000000E+00  0.00000000E+00  5.00000000E+00  0.00000000E+00 1 1
   8.66025404E-01  0.00000000E+00  5.00000000E-01  0.00000000E+00  5.00000000E+00  0.00000000E+00 2 1
   5.00000000E-01  0.00000000E+00  8.66025404E-01  0.00000000E+00  5.00000000E+00  0.00000000E+00 3 1
   3.85000000E+01  0.00000000E+00  0.00000000E+00  0.00000000E+00  5.00000000E+00  0.00000000E+00 1 2
   3.33419780E+01  0.00000000E+00  1.92500000E+01  0.00000000E+00  5.00000000E+00  0.00000000E+00 2 2
   1.92500000E+01  0.00000000E+00  3.33419780E+01  0.00000000E+00  5.00000000E+00  0.00000000E+00 3 2
   7.60000000E+01  0.00000000E+00  0.00000000E+00  0.00000000E+00  5.00000000E+00  0.00000000E+00 1 3
   6.58179307E+01  0.00000000E+00  3.80000000E+01  0.00000000E+00  5.00000000E+00  0.00000000E+00 2 3
   3.80000000E+01  0.00000000E+00  6.58179307E+01  0.00000000E+00  5.00000000E+00  0.00000000E+00 3 3
```

Modified TOSCA keyword data, in the case of a 60° sector field map (compared to Tab. 3.3, the sole data line "3 3 1 22.1 1." changes, from "315 151 1 22.1 1." in that earlier 180° sector case):

```
'TOSCA'
0 2          ! IL=2: log step-by-step coordinates, spin, etc., in zgoubi.plt (avoid if CPU time matters).
1. 1. 1. 1.          ! Normalization coefficients, for B, X, Y and Z coordinate values read from the map.
HEADER_8                                    ! The field map file starts with an 8-line header.
3    3    1 22.1 1.          ! IZ=1 for 2D map; MOD=22 for polar frame; .MOD2=.1 if only one map file.
geneSectorMap.out
0 0 0 0      ! Possible vertical boundaries within the field map, to start/stop stepwise integration.
2
1.                       ! Integration step size. Small enough for orbits to close accurately.
2                                                      ! Magnet positioning option.
0. 0. 0. 0.                                            ! Magnet positioning coordinates.
```

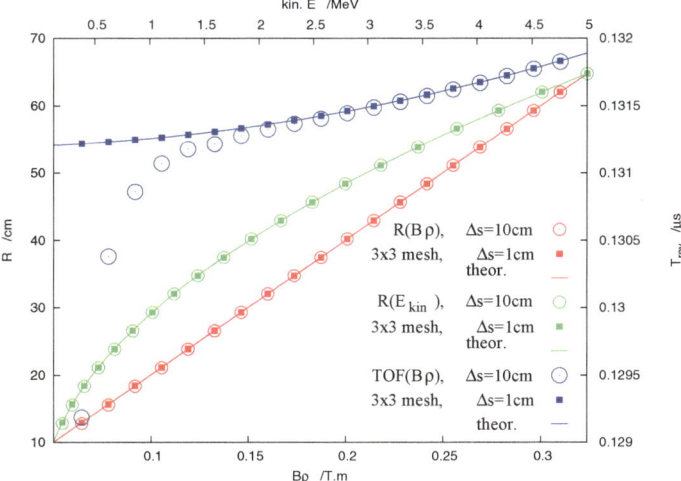

Fig. 3.23 Convergence versus mesh density and step size: a graph of orbit radius R (left axis), and revolution period, T_{rev} (right axis), versus kinetic energy (top scale) and rigidity (bottom scale). Solid markers are for $\Delta s = 1$ cm and $N_\theta \times N_R = 3 \times 3$ node mesh, large empty circles are for $\Delta s = 10$ cm and $N_\theta \times N_R = 106 \times 151$ node mesh. Solid lines are from theory and show convergence in the case 3×3 nodes and $\Delta s = 1$ cm

(e) Numerical convergence: integration step size

Figure 3.23 displays two cases of step sizes, $\Delta s \approx 1\,\mathrm{cm}$ and $\Delta s = 10\,\mathrm{cm}$.

It has been shown (Fig. 3.22) that $\Delta s \approx 1\,\mathrm{cm}$ is small enough that the numerical integration is converged, agreement with theoretical expectation is quite good.

The difference on the value of R, in the case $\Delta s \approx 10\,\mathrm{cm}$, appears to be weak, only noticeable at the scale of the graph for R values small enough that the number of steps over one revolution goes as low as $2\pi R / \Delta s \approx 2\pi \times 14.5/10 \approx 9$. The change in time-of-flight due to the larger step size amounts to a relative 10^{-3}.

Step size is critical in the numerical integration, the reason is that the coefficients of the Taylor series that yield the new position vector $\mathbf{R}(M_1)$ and velocity vector $\mathbf{v}(M_1)$, from an initial location M_0 after a Δs push, are the derivatives of the velocity vector [16, Sect. 1.2] and may take substantial values if $\mathbf{v}(s)$ changes quickly. In such case, taking too large a Δs value makes the high order terms significant and the Taylor series truncation [16, Eq. 1.2.4] is fatal to the accuracy (regardless of a possible additional issue of radius of convergence of the series).

(f) Numerical convergence: $\dfrac{\delta R}{R}(\Delta s)$

Issues faced are the following:

– the increase of $\delta R(\Delta s)/R$ at large Δs has been addressed above;

– a small Δs is liable to cause an increase of $\delta R(\Delta s)/R$, due to computer accuracy: truncation of numerical values at a limited number of digits may cause a Δs push to result in no change in $\mathbf{R}(M_1)$ (position) and $\mathbf{u}(M_1)$ (normed velocity) quantities [16, Eq. 1.2.4].

A detailed answer to the question, including graphs, is left to the reader, the method is the same as in (e).

3.2   Modeling a Cyclotron Dipole: Using an Analytical Field Model

This exercise introduces the analytical modeling of a dipole, using DIPOLE [16, *lookup* INDEX], and compares outcomes to the field map case of Exercise 3.1. The exercise is not entirely solved, however all the material needed for that is provided, and indications are given to complete it.

(a) Analytical modeling.

DIPOLE keyword provides an analytical model of the field to simulate a sector dipole with index, namely [16, *lookup* INDEX]

$$B_Z = \mathcal{F}(\theta) B_0 \left[1 + k \left(\frac{R - R_0}{R_0} \right) + k' \left(\frac{R - R_0}{R_0} \right)^2 + k'' \left(\frac{R - R_0}{R_0} \right)^3 \right] \quad (3.35)$$

R_0 is a reference radius, $B_0 = B_Z(R_0)|_{\mathcal{F} \equiv 1}$ is a reference field value, k is the field index and k', k'' are homogeneous to its first and second derivative with respect to

R (Eq. 3.11). $\mathcal{F}(\theta)$ is an azimuthal form factor, defined by the fringe field model, presumably taking the value 1 in the body of the dipole. In the present case a hard-edge field model is considered, so that

$$\mathcal{F} = \begin{cases} 1 \text{ inside} \\ 0 \text{ outside} \end{cases} \text{the dipole magnet} \qquad (3.36)$$

Setting up the input data list under DIPOLE (Table 3.6) requires close inspection of Fig. 3.24, which details the geometrical parameters such as the full angular opening of the field region that DIPOLE comprises, AT; a reference angle ACN to allow positioning the effective field boundaries at ω^+ and ω^-; field and indices; fringe field regions at $ACN - \omega^+$ (entrance) and $ACN - \omega^-$ (exit); wedge angles, etc.

A 60 deg sector is used here for convenience, it is detailed in Table 3.6 (Table 3.7 provides the definition of a 180 deg sector, for possible comparisons with the present three-sector assembly).

In setting up DIPOLE data the following values have been accounted for:
– $R_0 = 50$ cm, an arbitrary value (consistent with other exercises), more or less half the dipole extent,
– $B_0 = B_Z(R_0) = 5$ kG, as in the previous exercise. Note in passing, $R_0 = 50$ cm thus corresponds to $BR = 0.25$ Tm, $E_k = 2.988575$ MeV proton kinetic energy,
– radial field index $k = 0$ for the time being (constant field at all (R, θ)),
– a hard-edge field model for $\mathcal{F}$ (Eq. 3.36). In that manner for instance, two consecutive 60 deg sectors form a continuous 120 deg sector.

A graph of $B_Z(R, \theta)$ can be produced by computing constant radius orbits, for a series of energies ranging in 0.12–5.52 MeV for instance. DIPOLE[IL = 2] causes logging of step by step particle data in zgoubi.plt, including particle position and magnetic field vector; these data can be read and plotted, to yield similar results to Fig. 3.20.

(b) Concentric trajectories in the median plane.

The optical sequence of Exercise 3.1b (Table 3.4) can be used, by just changing the INCLUDE to account for a 180° DIPOLE (instead of TOSCA), namely

```
'INCLUDE'
1
3* 60degSector.inc[#S_60degSectorUnifB:#E_60degSectorUnifB]
```

wherein 60degSector.inc is the name of the data file of Table 3.6 and
               [#S_60degSectorUnifB:#E_60degSectorUnifB]
is the DIPOLE segment as defined in the latter. Note that the segment represents a 60° DIPOLE, thus it is included 3 times.

The additional keywords in that modified version of the Table 3.4 file include
– FIT, which finds the circular orbit for a particular momentum,
– FAISTORE to print out particle data once FIT is completed,

Table 3.6 Simulation input data file 60degSector.inc: analytical modeling of a dipole magnet, using DIPOLE. That file defines the labels (LABEL1 type [16, Sect. 7.7]) #S_60degSectorUnifB and #E_60degSectorUnifB, for INCLUDEs in subsequent exercises. It also realizes a 60-sample momentum scan of the cyclotron orbits, from 200 keV to 5 MeV, using REBELOTE

Note: this file is available in zgoubi sourceforge repository at
https://sourceforge.net/p/zgoubi/code/HEAD/tree/branches/examples/book/zgoubiMaterial/cyclotron_classical/ProbMdlAnal/

```
60degSector.inc
! Cyclotron, classical. Analytical model of dipole field. File name: 60degSector.inc
'MARKER' ProbMdlAnal_S                              ! Just for edition purposes.
'OBJET'
64.62444403717985                                   ! 200keV proton.
2
1 1                                                 ! Just one ion.
12.9248888074 0. 0. 0. 0. 1. 'm'         ! Closed orbit coordinates for D=p/p_0=1
1                          ! => 200keV proton. Y0=Brho/B=64.624444037[kG.cm]/5[kG].
'PARTICUL'             ! Optioanl - using PARTICUL is a way to get the time-of-flight computed.
PROTON                         ! otherwise, by default zgoubi only requires rigidity.
'FAISCEAU'                                           ! Local particle coordinates.
'MARKER'   #S_60degSectorUnifB             ! Label should not exceed 20 characters.
'DIPOLE'                                   ! Analytical modeling of a dipole magnet.
2          ! IL=2, only purpose is to logged trajectories in zgoubi.plt, for further plotting.
60. 50.                                         ! Sector angle AT; reference radius R0.
30. 5. 0. 0. 0.            ! Reference azimuthal angle ACN; BM=B0 field at RM=R0; indices, N, N', N''.
0. 0.                                           ! EFB 1 is hard-edge,
4 .1455  2.2670 -.6395 1.1558 0. 0. 0.          ! hard-edge only possible with sector magnet.
30. 0. 1.E6 -1.E6 1.E6 1.E6              ! Entrance face placed at omega+=30 deg from ACN.
0. 0.                                           ! EFB 2.
4 .1455  2.2670 -.6395 1.1558 0. 0. 0.
-30. 0. 1.E6 -1.E6 1.E6 1.E6              ! Exit face placed at omega-=-30 deg from ACN.
0. 0.                                           ! EFB 3 (unused).
0 0.      0.      0.      0.   0. 0. 0.
0. 0. 1.E6 -1.E6 1.E6 1.E6 0.
2 10      ! '2' is for 2nd degree interpolation. Could also be '25' (5*5 points grid) or 4 (4th degree).
1.                                              ! Integration step size. Small enough for orbits to close accurately.
2 0. 0. 0. 0.                           ! Magnet positioning RE, TE, RS, TS. Could be instead non-zero, e.g.,
!                                       2 RE=50. 0. RS=50. 0., as long as Yo is amended accordingly in OBJET.
'MARKER'   #E_60degSectorUnifB             ! Label should not exceed 20 characters.
'FAISCEAU'                                           ! Local particle coordinates.
'FIT'                     ! Adjust Yo at OBJET so to get final Y = Y0 -> a circular orbit.
1  nofinal
2 30 0 [12.,65.]                                     ! Variable : Yo.
1 2e-12 199       ! constraint; default penalty would be 1e-10; maximum 199 calls to the function.
3.1 1 2 #End 0. 1. 0                                 ! Constraint: Y_final=Yo.
'FAISTORE'                                 ! Log particle data here, in zgoubi.fai.
zgoubi.fai                                 ! for further plotting (by gnuplot, below).
1
'REBELOTE'                                 ! Momentum scan, 60 samples.
60 0.2 0          1 60 different rigidities; log to video ; take initial coordinates as found in OBJET.
1                                          ! Change parameter(s) as stated next lines.
OBJET 35  1:5.0063899693                   ! Change relative rigity (35) in OBJET; range (0.2 MeV to 5 MeV).
'SYSTEM'
1                                          ! 1 SYSTEM command follows.
/usr/bin/gnuplot < ./gnuplot_TOF.gnu &     ! Launch plot by ./gnuplot_TOF.gnu.
'MARKER'  ProbMdlAnal_E                     ! Just for edition purposes.
'END'
```

A gnuplot script, gnuplot_TOF.gnu, to obtain Fig. 3.25:

```
# gnuplot_TOF.gnu
set xlabel "R [m]"; set ylabel "T_{rev} [{/Symbol m}s]"; set y2label "f_{rev} [MHz]"
set xtics mirror; set ytics nomirror; set y2tics nomirror; set key t l ; set key spacin 1.2
nSector=6; Hz2MHz=1e-6; M=938.272e6; c=2.99792458e8; B=0.5; freqNonRel(x)= Hz2MHz* c**2*B/M/(2.*pi)
set y2range [7.58:7.63] ; set yrange[1/7.63:1/7.58]
plot \
"zgoubi.fai" u 10:($15 *nSector) axes x1y1 w lp pt 5 ps .6 lw 2 linecolor rgb "blue" tit "T_{rev}" ,\
"zgoubi.fai" u 10:(1/($15*nSector)) axes x1y2 w lp pt 6 ps .6 lw 2 linecol rgb "red" tit "f_{rev}" ,\
freqNonRel(x)  axes x1y2 w l lw 2. linecolor rgb "black"  tit "f_{rev},T_{rev} (non rel.)"  ; pause 1
```

– REBELOTE, which repeats the execution of the sequence (REBELOTE sends the execution pointer back to the top of the data file) for a new momentum value which REBELOTE itself defines.

For the rest, follow the same procedure as for Exercise 3.1b. The results are the same, Fig. 3.21.

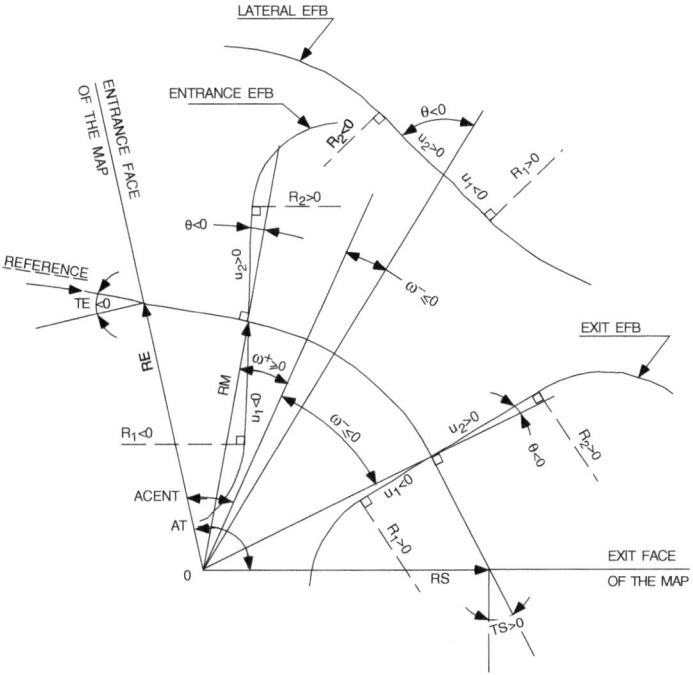

Fig. 3.24 Parameters used to define the geometry of a dipole magnet with index, using DIPOLE. In the text, ACENT is noted ACN [16, Fig. 9]

Table 3.7 A 180° version of a DIPOLE sector, where the foregoing quantities $AT = 60°$, $ACN = \omega^+ = -\omega^- = 30°$ have been changed to $AT = 180°$, $ACN = \omega^+ = -\omega^- = 90°$—a file used under the name 180 degSector.inc in further exercises

Note: this file is available in zgoubi sourceforge repository at
https://sourceforge.net/p/zgoubi/code/HEAD/tree/branches/examples/book/zgoubiMaterial/cyclotron_classical/ProbMdlAnal/

```
! 180degSector.inc
'MARKER'     #S_180degSectorUnifB                        ! Label should not exceed 20 characters.
'DIPOLE'                                                 ! Analytical modeling of a dipole magnet.
2
180. 50.                                      ! Sector angle 180deg; reference radius 50cm.
90.  5.  0.  0.  0.            ! Reference azimuthal angle; Bo field at R0; indices, N, N', N''.
0.  0.                                                   ! EFB 1 is hard-edge,
4  .1455   2.2670  -.6395  1.1558  0.  0.  0.            ! hard-edge only possible with sector magnet.
90. 0.  1.E6  -1.E6  1.E6  1.E6
0.  0.                                                                 ! EFB 2.
4  .1455   2.2670  -.6395  1.1558  0.  0.  0.
-90. 0.  1.E6  -1.E6  1.E6  1.E6
0. 0.                                                                  ! EFB 3.
0 0.      0.      0.      0.    0. 0.  0.
0. 0.  1.E6  -1.E6  1.E6  1.E6 0.
2   10.
0.5                           ! Integration step size. Small enough for orbits to close accurately.
2  0. 0. 0.  0.               ! Magnet positioning RE, TE, RS, TS. Could be isntead non-zero, e.g.,
!                             2 RE=50. 0. RS=50. 0., as long as Yo is amended accordingly in OBJET.
'MARKER'     #E_180degSectorUnifB                        ! Label should not exceed 20 characters.
```

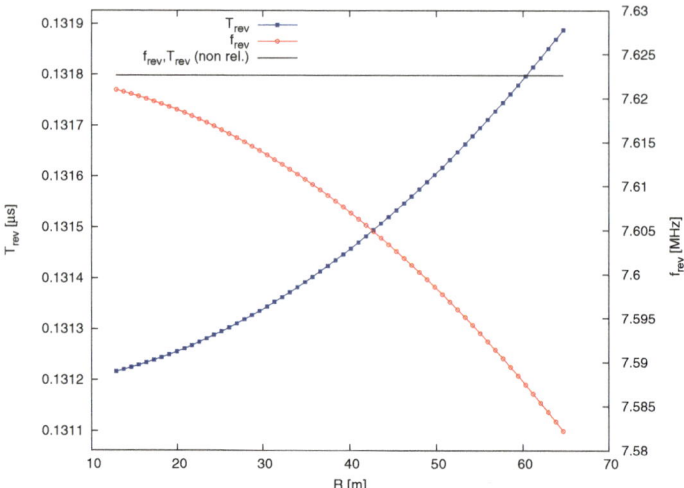

Fig. 3.25 A scan of radius-dependent revolution frequency. An analytical model of a cyclotron dipole is used, featuring uniform field (no radial gradient, at this point)

(c) Energy and rigidity dependence of orbit radius and time-of-flight.

The orbit radius R and the revolution time T_{rev} as a function of kinetic energy E_k and rigidity BR are obtained by a similar scan to exercise (b). The procedure is the same as in Exercise 3.1c. Results are expected to be the same as well (Fig. 3.22).

A comparison of revolution periods can be made using the simulation file of Table 3.6 which happens to be set for a momentum scan and yields Fig. 3.25, to be compared to Fig. 3.22: DIPOLE and TOSCA produce the same results as long as both methods are converged, from the integration step size stand point (small enough), and regarding TOSCA from field map mesh density stand point in addition (dense enough).

(d) Numerical convergence: integration step size; $\dfrac{\delta R}{R}(\Delta s)$.

This question concerns the dependence of the numerical convergence of the solution of the differential equation of motion upon integration step size.

Follow the procedure of Exercise 3.1e: a similar outcome to Fig. 3.23 is expected—ignoring mesh density with the present analytical modeling using DIPOLE.

The $\dfrac{\delta R}{R}$ dependence upon the integration step size Δs is commented in Exercise 3.1e and holds regardless of the field modeling method (field map or analytical model).

(e) Pros and cons.

Using a field map is a convenient way to account for complicated one-, two- or three-dimensional field distributions.

However, using an analytical field model rather, ensures greater accuracy of the integration method.

CPU-time wise, one or the other method may be faster, depending on the problem.

3.3 Resonant Acceleration

The field map and TOSCA [16, *lookup* INDEX] model of a 180° sector is used here (an arbitrary choice, the analytical field modeling DIPOLE would do as well), the configuration is that of Fig. 3.5 with a pair of sectors.

An accelerating gap between the two dees is simulated using CAVITE[IOPT = 3], PARTICUL is added in the sequence in order to specify ion species and data, necessary for CAVITE to operate. Acceleration at the gap does not account for the particle arrival time in the IOPT = 3 option: whatever the later, CAVITE boost will be the same as longitudinal motion is an unnecessary consideration, here).

The input data file for this simulation is given in Table 3.8. It is resorted to INCLUDE, twice in order to create a double-gap sequence, using the field map model of a 180° sector. The INCLUDE inserts the magnet itself, i.e., the #S_halfDipole to #E_halfDipole TOSCA segment of the sequence of Table 3.3. Note: the theoretical field model of Table 3.6, segment #S_60degSectorUnifB to #E_60degSectorUnifB (to be INCLUDEd 3 times, twice), could be used instead: Exercise 3.2 has shown that both methods, field map and analytical field model, deliver the same results.

Particle data are logged in zgoubi.fai at both occurrences of CAVITE, under the effect of FAISTORE[LABEL=cavity], Table 3.8. This is necessary in order to access the evolution of parameters as velocity, time of flight, etc. at each half-turn, given that each half-turn is performed at a different energy

(a) Accelerate a proton.

A proton with initial kinetic energy 20 keV is launched on its closed orbit radius, $R_0 = p/qB = 4.087013$ cm. It accelerates over 25 turns due to the presence to REBELOTE[NPASS = 24], placed at the end of the sequence. The energy range, 20 keV to 5 MeV, and the acceleration rate: 0.1 MeV per cavity, 0.2 MeV per turn, determine the number of turns, $NPASS+1 = (5 - 0.02)/0.2 \approx 25$. The accelerated trajectory spirals out in the fixed magnetic field, it is plotted in Fig. 3.26, reading data from zgoubi.plt.

(b) Momentum and energy.

Proton momentum p and total energy E as a function of kinetic energy, from ray-tracing (turn-by-turn particle data are read from zgoubi.fai, filled up due to FAIS-TORE) are displayed in Fig. 3.27, together with theoretical expectations, namely, $p(E_k) = \sqrt{E_k(E_k + 2M)}$ and $E = E_k + M$.

(c) Velocity.

Proton normalized velocity $\beta = v/c$ as a function of kinetic energy from raytracing is displayed in Fig. 3.27, together with theoretical expectation, namely, $\beta(E_k) = p/(E_k + M)$.

Table 3.8 Simulation input data file: accelerating a proton in a double-dee cyclotron, from 20 keV to 5 MeV, at a rate of 100 kV per gap, independent of RF phase (longitudinal motion is frozen—see question (e) dealing with CAVITE[IOPT = 7] for unfrozen motion). Note that particle data are logged in zgoubi.fai (under the effect of FAISTORE) at both occurrences of CAVITE. The INCLUDE file FieldMapSector.inc is taken from Table 3.3

```
Cyclotron, classical. Acceleration:   20 keV -> 6 MeV.
'MARKER'  ProbAccelGap_S                                      ! Just for edition purposes.
'OBJET'
64.62444403717985                          ! Reference Brho ("BORO" in the users' guide) -> 200keV proton.
2
1 1                                                                    ! Just one ion.
4.087013 0. 0. 0. 0. 0.3162126 'o'         ! D=0.3162126 => Brho[kG.cm]= 20.435064, kin-E[keV]= 20.
1
'PARTICUL'                                       ! Usage of CAVITE requires partical data,
PROTON                                           ! otherwise, by default zgoubi only requires rigidity.
'FAISTORE'                                       ! Store particle data, turn-by-turn.
zgoubi.fai   cavity                  ! Log coordinates at any occurence of LABEL1=cavity, in zgoubi.fai.
1
'INCLUDE'                                        ! Insert a 180 deg sector field map.
1
FieldMapSector.inc[#S_halfDipole:#E_halfDipole]
'FAISCEAU'                                       ! Particle coordinates before RF gap.
'CAVITE'  cavity                                 ! Accelerating gap.
3                                   ! dW = qVsin(phi), independent of time (phi forced to constant).
0. 0.                                                                  ! Unused.
100e3  1.57079632679                            ! Peak voltage 100 kV; RF phase = pi/2.
'INCLUDE'                                        ! Insert a 180 deg sector field map.
1
FieldMapSector.inc[#S_halfDipole:#E_halfDipole]
'FAISCEAU'                                       ! Particle coordinates before RF gap.
'CAVITE'  cavity                                 ! Accelerating gap.
3                                   ! dW = qVsin(phi), independent of time (phi forced to constant).
0. 0.                                                                  ! Unused.
100e3  1.57079632679                            ! Peak voltage 100 kV; RF phase = pi/2.
'REBELOTE'          ! Repeat NPASS=24 times, for a total of 25 turns; K = 99: coordinates at end of
24  0.1 99                          ! previous pass are used as initial coordinates for the next pass.
'FAISCEAU'                                       ! Local particle coordinates logged in zgoubi.res.

'SYSTEM'
2                                                              ! 2 SYSTEM commands follow:
/usr/bin/gnuplot < ./gnuplot_Zplt_XYLab.gnu &                  ! plot trajectories;
/usr/bin/gnuplot < ./gnuplot_awk_Zfai_dTT.gnu &                ! dC/C, dbta/bta, dT/T graph.

'MARKER'  ProbAccelGap_E                                       ! Just for edition purposes.
'END'
```

Two gnuplot scripts, to obtain respectively Fig. 3.26: *and Fig.* 3.28:
The awk command in gnuplot_awk_Zfai_dTT.gnu takes care of a 1-row shift so to subtract next turn data from currant turn ones.

```
# gnuplot_Zplt_XYLab.gnu
set xtics ; set ytics ; set xlabel "X_{Lab} [m]" ; set ylabel "Y_{Lab} [m]"
set size ratio 1 ; set polar ; cm2m = 0.01 ; pi = 4.*atan(1.)
set arrow from 0, 0 to 0, 0.67 nohead lc "red" lw 6; set arrow from 0, -0.75 to 0, 0 nohead lc "blue" lw 6
noel_1=6 ; noel_2=11  # 1st CAVITE is element noel_1; 2nd CAVITE is noel_2. Col. $42 in zgoubi.plt is element numb.
plot for [nl=noel_1:noel_2:5] "zgoubi.plt" u ($42==noel_1? $22:$22+pi ):($10 *cm2m) w p pt 5 ps .2 lc rgb "black"

# gnuplot_awk_Zfai_dTT.gnu
set xtics nomirror; set ytics mirror; set xlabel "E_k [MeV]";
set ylabel "{/Symbol Db}/{/Symbol b},  {/Symbol D}C/C,  {/Symbol D}T_{rev}/T_{rev}"; set logscale y; set yrange [:3]
# zgoubi.fai columns: $25: energy; $14: path length; $23: kinetic E; $29: mass; $15: tim
plot "< awk '/#/ {next;}  { if(prev14>0 && prev25>0) print prev24, ($14 -prev14)/prev14 , prev24} \
{ prev14 = $14; prev24 = $24; prev25=$25 }' < zgoubi.fai" u 1:2 w p pt 5 lc rgb "black" tit "{/Symbol D}C/C" ,\
"< awk '/#/ {next;}  { if(prev14>0 && prev25>0) print prev24, ( -sqrt(prev25**2-$29**2)/prev25 + \
sqrt($25**2-$29**2)/$25 )/(sqrt(prev25**2-$29**2)/prev25) , prev24}   { prev14 = $14; prev24 = $24; prev25=$25 }' \
< zgoubi.fai"  u 1:2 w p pt 6 ps 1.5 lc rgb "red" tit "d{/Symbol b}/{/Symbol b}" ,\
"< awk '/#/ {next;}  { if(prev14>0 && prev25>0) print prev24, ($14 -prev14)/prev14- ( -sqrt(prev25**2-$29**2)/prev25 \
+ sqrt($25**2-$29**2)/$25 )/(sqrt(prev25**2-$29**2)/prev25) , prev24}   { prev14 = $14; prev24 = $24; prev25=$25 }' \
< zgoubi.fai"  u 1:2 w p pt 8 ps 1.5 lc rgb "blue"  tit "{/Symbol D}T/T=dC/C-d{/Symbol b}/{/Symbol b}" ,\
"< awk '/#/ {next;}  { if(prev14>0 && prev15>0) print prev24, ($15-prev15)/prev15 , prev24}   { prev14 = $14; \
prev24 = $24; prev15=$15 }' < zgoubi.fai"  w l lw 2 lc rgb "blue" tit "theor. {/Symbol D}T/T"
```

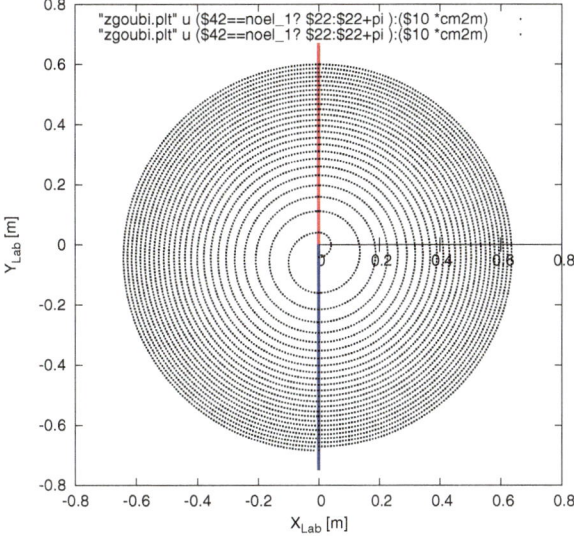

Fig. 3.26 Twenty five turn spiral trajectory of a proton accelerated in a uniform 0.5 T field from 20 keV to 5 MeV at a rate of 200 kV per turn (a 100 kV gap voltage). The vertical thick line materializes the gap, the upper half (red) corresponds to the first occurrence of CAVITE in the sequence (Table 3.8), the lower half (blue) corresponds to the second occurrence of CAVITE

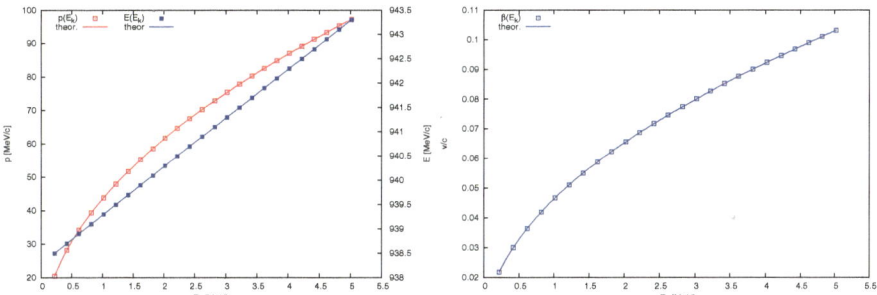

Fig. 3.27 Energy dependence of, left: proton momentum p (left axis) and total energy E (right axis) and of, right: proton normalized velocity $\beta = v/c$. Markers: from raytracing; solid lines: theoretical expectation

(d) Relative velocity, orbit length and time of flight.

The relative increase in velocity is smaller than the relative increase in orbit length as energy increases (this is what Fig. 3.28 shows). Thus the relative variation of the revolution time, Eq. 3.23, is positive; in other words the revolution time increases with energy, the revolution frequency decreases. Raytracing outcomes are displayed in Fig. 3.28, they are obtained using the gnuplot script given in Table 3.8. Note that the path length difference (taken as the difference of homologous quantities in a

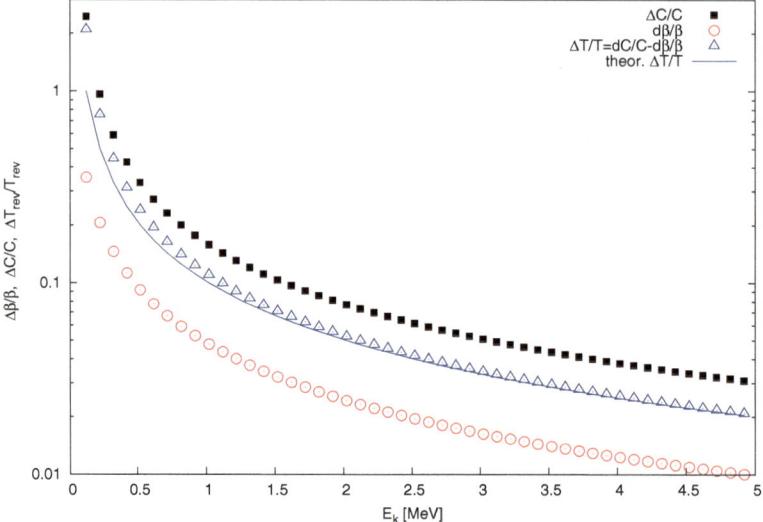

Fig. 3.28 Relative variation of velocity $\Delta\beta/\beta$ (empty circles), circumference $\Delta C/C$ (solid disks) and revolution time $\Delta T/T$ (triangles), as a function of energy, from raytracing. Theoretical expectation for the latter is also displayed (solid line), for comparison

common line) is always between the two CAVITEs (particle data are logged at the two occurrences of CAVITE), crossed successively, which is half a turn. Same for the difference between homologous velocity data on a common line, it corresponds to two successive crossings of CAVITE, i.e., half a turn. The graph includes the theoretical $\delta T_{\text{rev}}/T_{\text{rev}}$ (Eq. 3.23) for comparison with raytracing; some difference appears in the low velocity regime, this may be due to the large $\Delta\beta$ step imparted by the 100 kV acceleration at the gaps.

(e) Harmonic h = 3 RF.

The input data file for this simulation is given in Table 3.9. The RF is on harmonic h = 3 of the revolution frequency. It has been tuned to ensure acceleration up to 3 MeV. The accelerating gap between the two dees is simulated using CAVITE[IOPT = 7]: by contrast with the previous exercise (where CAVITE[IOPT = 3] is used), the RF phase at ion arrival at the gap is now accounted for.

Repeating questions (b–d) is straightforward, changing what needs be changed in Table 3.9 input data file.

3.4 Spin Dance

The DIPOLE analytical field model of Exercise 3.2 (Table 3.6) is used here, as opposed to using a field map and TOSCA, as it allows more straightforward changes in the field, if desired.

Table 3.9 Simulation input data file: accelerating a proton in a double-dee cyclotron, from 20 keV to 5 MeV, using harmonic 3 RF. The INCLUDE file is taken from Table 3.6

```
Cyclotron, classical. Analytical model of dipole field.
'OBJET'
64.62444403717985                                                      ! 200keV proton.
2
1 1                                                                    ! Just one ion.
12.924888 0. 0. 0. 0.  1.  'm'          ! D=1 => 200keV proton. Y0=Brho/B=64.624444037[kG.cm]/5[kG].
1
'PARTICUL'                                    ! This is required for spin motion to be computed,
PROTON                                        ! otherwise, by default zgoubi only requires rigidity.
'INCLUDE'
1                                                                      ! Include a first 180 deg sector.
./180degSector.inc[#S_180degSectorUnifB:#E_180degSectorUnifB]
'CAVITE'
7
0 22862934.0
285e3 -0.5235987755982988
'INCLUDE'
1                                                                      ! Include a second 180 deg sector.
./180degSector.inc[#S_180degSectorUnifB:#E_180degSectorUnifB]
'CAVITE'
7
0 22862934.0                                                           ! RF = 3/T_rev.
285e3 -3.665191429188092                                 ! Peak voltage; synchronous phase.
'REBELOTE'
26 0.4 99                                                              ! 26+1 turn tracking.
'END'
```

(a) Spin transport.

Spin transport is obtained by adding SPNTRK. PARTICUL is necessary in order to get the Thomas-BMT equation of motion solved [16, Sect. 2]. This results in the input data file given in Table 3.10 (excluding FIT and REBELOTE keywords, introduced for the purpose of the following question (b)).

The use of SPNTRK results in the following outcome (an excerpt from zgoubi.res execution listing):

```
 4  Keyword, label(s) : SPNTRK
        Spin  tracking  requested.
                        Particle mass         =      938.2721    MeV/c2
                        Gyromagnetic factor G =      1.792847
                        Initial spin conditions type  1 :
                           All particles have spin parallel to  X  AXIS
        PARAMETRES  DYNAMIQUES  DE  REFERENCE :
                        BORO     =        64.624 kG*cm
                        beta     =      0.02064411
                        gamma    =      1.00021316
                        gamma*G  =      1.7932295094
        POLARISATION  INITIALE  MOYENNE  DU  FAISCEAU  DE        1 PARTICULES :
                        <SX> =      1.000000
                        <SY> =      0.000000
                        <SZ> =      0.000000
                        <S>  =      1.000000
```

Spin coordinates are logged in zgoubi.res execution listing using SPNPRT. Five sample passes around the cyclotron (four iterations by REBELOTE) result in the following outcomes in zgoubi.res, under SPNPRT:

```
      26  Keyword, label(s) : SPNPRT
                  INITIAL                                    FINAL
            SX        SY        SZ       |S|        SX        SY        SZ       |S|      GAMMA
  m  1  1.000000  0.000000  0.000000  1.000000  0.268269  0.963344  0.000000  1.000000   1.0002
  m  1  1.000000  0.000000  0.000000  1.000000  0.268599  0.963252  0.000000  1.000000   1.0002
  m  1  1.000000  0.000000  0.000000  1.000000  0.268949  0.963154  0.000000  1.000000   1.0003
  m  1  1.000000  0.000000  0.000000  1.000000  0.269319  0.963051  0.000000  1.000000   1.0003
  m  1  1.000000  0.000000  0.000000  1.000000  0.269710  0.962942  0.000000  1.000000   1.0003
```

Table 3.10 Simulation input data file: add spin to the cyclotron simulation of Table 3.6. The present input file INCLUDEs six copies of the 60° sector DIPOLE defined therein

```
Cyclotron, classical. Analytical model of dipole field. Spin transport.
'MARKER'  ProbAddSpin_S                                      ! Just for edition purposes.
'OBJET'
64.62444403717985                       ! Reference Brho ("BORO" in the users' guide) -> 200keV proton.
2
1 1                                                          ! Just one ion.
12.9248888074 0. 0. 0. 0.  1.  'm'     ! D=1 => 200keV proton. Y0=Brho/B=64.624444037[kG.cm]/5[kG].
1
'PARTICUL'                                          ! This is required to get the time-of-flight,
PROTON                                              ! otherwise, by default zgoubi only requires rigidity.
'SPNTRK'                                                     ! Request spin tracking.
1                                        ! All spins launched longitudinal (parallel to OX axis).
'INCLUDE'
1
6* ./60degSector.inc[#S_60degSectorUnifB:#E_60degSectorUnifB]         ! 6 * 60 degree sector.
'FAISCEAU'                                          ! Local particle coordinates.
'FIT'                               ! Adjust Yo at OBJET so to get final Y = Y0 -> a circular orbit.
1  nofinal
2 30 0 [12.,65.]                                             ! Variable : Yo.
1  2e-12  199       ! constraint; default penalty would be 1e-10; maximum 199 calls to the function.
3.1 1 2 #End 0. 1. 0                                         ! Constraint: Y_final=Yo.
'FAISCEAU'                               ! Allows checking that Y = Y0 and T = T0 = 0, here.
'SPNPRT'                                             ! Local spin data, logged in zgoubi.res.
'FAISTORE'                                           ! Log particle data here, in zgoubi.fai,
zgoubi.fai                               ! for further plotting of spin coordinates (by gnuplot, below).
1
'REBELOTE'                                           ! Momentum scan, 60 samples.
60 0.2 0            1 60 different rigidities; log to video ; take initial coordinates as found in OBJET.
1                                                    ! Change parameter(s) as stated next lines.
OBJET 35  1:5.0063899693                 ! Change relative rigity (35) in OBJET; range (0.2 MeV to 5 MeV).
'SYSTEM'
1                                                    ! 1 SYSTEM command follows.
/usr/bin/gnuplot < ./gnuplot_Zfai_spin.gnu &
'MARKER'  ProbAddSpin_E                                      ! Just for edition purposes.
 'END'
```

A gnuplot script to obtain Fig. 3.29:
The file zgoubi.1cm is a copy of zgoubi.fai obtained for a $\Delta s = 1$ cm run; zgoubi.fai is for $\Delta s = 0.5$ cm.

```
# gnuplot_Zfai_spin.gnu
set xlabel "G{/Symbol g}"; set ylabel "Spin precession angle  {/Symbol q}_{sp} / 2{/Symbol p}"
set y2label "relative difference num./theor"; set logscale y2
set xtics; set ytics nomirror; set y2tics; am = 938.27208; G = 1.79284735; pi = 4.*atan(1.); set key t c spacin 1.5
plot \
"zgoubi.fai" u ($31*$25/$29):(((4.*pi -atan($21/$20)))/(2.*pi)) w lp pt 4 ps .7 tit "{/Symbol q}_{sp}/2{/Symbol p}" ,\
"zgoubi.1cm" u ($31*$25/$29):(abs((4*pi-atan($21/$20))/pi*180-$31*$25/$29*360.)) axes x1y2 w lp pt 8 ps .7 tit "1 cm",\
"zgoubi.fai" u ($31*$25/$29):(abs((4.*pi -atan($21/$20))/pi*180-$31*$25/$29*360.)) axes x1y2 w lp pt 8 ps .7 tit "5 mm"
```

(b) Spin precession.

Proton case is considered, simulation is performed using Table 3.10 input data file. Initial spin is parallel to the X axis (longitudinal). The particle is raytraced on the circular closed orbit over one revolution, for a particular momentum. Particle data resulting from a FIT (FIT forces orbit closure, by varying the initial Y_0) are logged in zgoubi.fai, by FAISTORE. The computation is repeated using REBELOTE in the very manner that the energy scan was done in Exercise 3.2, over an energy range 12 keV→5 MeV.

Figure 3.29 (obtained using the gnuplot script given in Table 3.10) displays the resulting energy dependence of the spin precession, $\theta_{sp}(E)$, together with its difference to theoretical expected $\theta_{sp}(E) = G\frac{E}{M} \times 2\pi = G\gamma \times 2\pi$ (proton gyromagnetic anomaly $G = 1.792847$).

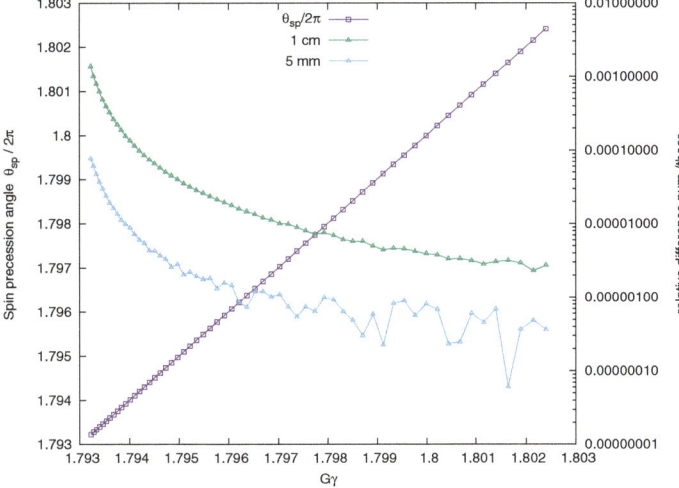

Fig. 3.29 $G\gamma$ dependence of the spin precession angle over a revolution around the cyclotron, in the moving frame (left axis), and relative difference to $G\gamma$ for the two integration step sizes $\Delta s = 0.5$ and 1 cm (right axis), Markers are from raytracing, solid lines are to guide the eye

(c) Spin tune.

Two protons are injected with longitudinal initial spin $\mathbf{S}_i \parallel$ OX axis and respective energies 12 keV and 5.52 MeV, thus the following OBJET (a slight modification to Table 3.10 data):

```
'OBJET'
64.62444403717985                          ! Reference Brho ("BORO" in the users' guide) -> 200keV proton.
2
2 1
12.9248888074 0. 0. 0. 0.  1.   'm'        ! D=1 => 200keV proton. Y0=Brho/B=64.624444037[kG.cm]/5[kG].
67.997983 0. 0. 0. 0. 5.2610112 'o'         ! p[MeV/c]=101.926, Brho[kG.cm]=339.990, kin-E[MeV]=5.52.
1 1
```

FAISCEAU following FIT (Table 3.10) allows to control that momentum and trajectory radius are matched, which means coordinates at OBJET and current coordinates at FAISCEAU are equal. Inspection of zgoubi.res execution listing shows for instance, after 4 turns:

		OBJET						FAISCEAU			
D	Y(cm)	T(mr)	Z(cm)	P(mr)	S(cm)	D-1	Y(cm)	T(mr)	Z(cm)	P(mr)	S(cm)
1.0000	12.925	0.000	0.000	0.000	0.0000	0.0000	12.925	0.000	0.000	0.000	3.248379E+02
5.2610	67.998	0.000	0.000	0.000	0.0000	4.2610	67.998	-0.000	0.000	0.000	1.708976E+03

A graphic of the projection of the spin motion on the longitudinal axis, over a few turns, from the ray tracing, is given in Fig. 3.30, together with the longitudinal component as of the parametric equations of motion

$$\begin{cases} S_X = \hat{S} \cos(G\gamma\theta) \\ S_Y = \hat{S} \sin(G\gamma\theta) \end{cases} \tag{3.37}$$

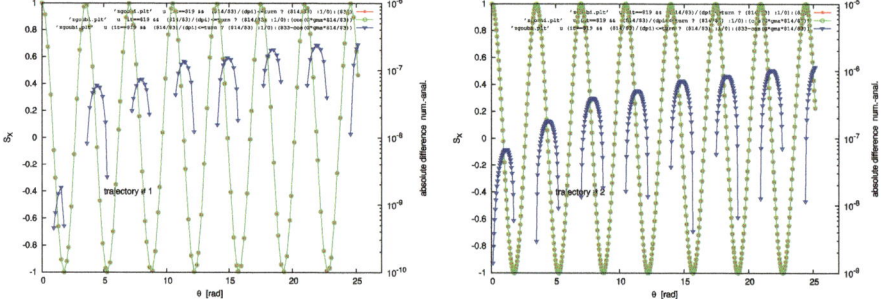

Fig. 3.30 Longitudinal spin component motion (left vertical axis), observed in the moving frame, case of 0.2 MeV energy, R = 12.924888 cm (left graph), and of 5.52 MeV energy, R = 67.998 cm (right graph). Markers are from ray tracing, the solid line is the theoretical expectation (Eq. 3.37). The right vertical axis (triangle markers; solid line is to guide the eye) shows the absolute difference between both. The oscillation is as expected slightly faster at 5.52 MeV: frequencies are in the ratio $\gamma(5.52\,\text{MeV})/\gamma(0.2\,\text{MeV}) = 1.00566$

The motion amplitude is $\hat{S} = \sin\phi$, with ϕ the angle that the spin vector makes with the vertical precession axis. In this simulation **S** is launched parallel to OX, thus $\phi = \pi/2$ and $\hat{S} = 1$.

Now, checking the spin precession:

Placing both FAISCEAU and SPNPRT commands right after the first dipole sector allows checking the spin precession and its relationship to particle rotation, for simplicity right after the first pass through that first sector, as follows. FAISCEAU and SPNPRT (Table 3.10) yield, respectively:

```
                   OBJET                                            FAISCEAU
    D     Y(cm)    T(mr)   Z(cm)   P(mr)    S(cm)   D-1     Y(cm)    T(mr)   Z(cm)   P(mr)    S(cm)
 1.0000   12.925  0.000   0.000   0.000   0.0000  0.0000   12.925   0.000   0.000   0.000   3.248379E+02
 5.2610   67.998  0.000   0.000   0.000   0.0000  4.2610   67.998  -0.000   0.000   0.000   1.708976E+03

             INITIAL                              FINAL                        --- angles ---
          SX       SY       SZ       |S|       SX        SY        SZ      |S|    GAMMA  |Si,Sf|   (Z,Sf)
                                                                                         (deg.)   (deg.)
 m 1  1.000000  0.000000  0.000000  1.000000  -0.302266 -0.953224  0.000000  1.000000  1.0002 -107.594  90.000
 o 1  1.000000  0.000000  0.000000  1.000000  -0.312396 -0.949952  0.000000  1.000000  1.0059 -108.204  90.000
```

SPNPRT tells that,

– case of the first particle, tagged 'm' above; its energy is 200 keV, $\gamma = 1.00021315$,
its spin tune is $\nu_{sp} = G\gamma = 1.793229$

The computed value of the '(S_i, S_f)' angle between initial and final spin vectors is -107.594 (truncated), negative as spin precession has the sign of proton rotation. Theoretical expectation is $G\gamma\alpha = -107.59377$ deg. The resulting spin components are, as above, $S_X = \cos(-107.59377) = -0.302266$ and $S_Y = \sin(-107.59377) = -0.9532235$.

– case of the second particle, tagged 'o'; its energy is 5.52 MeV, $\gamma = 1.00588315$,
its spin tune is $\nu_{sp} = G\gamma = 1.803394$

The computed value of '(S_i, S_f)' is -108.204 (truncated). Theoretical expectation is $G\gamma\alpha = -108.20370$ deg.

Now, accounting for particle rotation in order to get spin coordinates in the laboratory frame:
– the FAISCEAU outcome above shows that, after crossing the 60 deg sector the angles of the two particles have the value $T = 0$, which is expected as they are launched with zero incidence, and as DIPOLE uses a polar coordinate system [16] with particle coordinates computed in the moving (rotating) frame. The latter has also undergone a -60 deg rotation, clockwise, which is therefore the implicit rotation of the particles in the laboratory frame. The spin precession in the laboratory frame results, namely,
– case of the first particle: $(1 + G\gamma)\alpha = -167.59377$ deg.
– case of the second particle: $(1 + G\gamma)\alpha = -168.20370$ deg.

(d) Spin dance.

A 200 keV proton is injected with its initial spin vector at 80 degrees from the vertical axis. The input data file for this simulation is given in Table 3.11, together with a gnuplot script for the animation. The latter plots three things, concurrently:
– the circular trajectory of the particle in the (X,Y) plane: this is the curve at $Z = 0$ in Fig. 3.31, a set of points $\{(R\cos(-X), R\sin(-X), 0)\}$ resulting from the step by step integration. Note that X is counted positive clockwise in zgoubi.fai (consistently with the definition of DIPOLE parameters, Fig. 9 in [16]), hence "–X" the rotation angle;
– the spin vector: its foot is attached to the particle (the previous set of points), whereas its tip is at $\{(S_X \cos(-X) - S_Y \sin(-X), S_X \sin(-X) + S_Y \cos(-X), S_Z\}$, with S_X, S_Y, S_Z the spin vector components in the moving frame as read from zgoubi.fai. S_Z is constant as the precession axis is parallel to the Z axis. The $\begin{pmatrix} \cos(-X) & -\sin(-X) \\ \sin(-X) & \cos(-X) \end{pmatrix}$ rotation applied to the (S_X, S_Y) vector accounts for the transformation from the moving frame to the laboratory frame;
– the cycloidal shape trajectory of the tip of the spin vector (the previous set of points).

A frozen view of that spin dance, over about 2.5 proton revolutions around the ring, is given in Fig. 3.31.

(e) Deuteron

The input data file set up for questions (b–e) can be used *mutatis mutandis*, as follows.

Raytracing a different particle requires changing the reference rigidity, BORO, under OBJET, and changing particle data, under PARTICUL. That reference rigidity is to be determined from the field value in the dipole model (namely, $B_0 = 5$ kG).

Table 3.11 Simulation input data file: spin dance, 20 turns around a uniform field cyclotron. The INCLUDE file 60degSector.inc is taken from Table 3.6

Note: this animation (input data file & gnuplot script) is available in `zgoubi` sourceforge repository at
https://sourceforge.net/p/zgoubi/code/HEAD/tree/branches/exemples/book/zgoubiMaterial/cyclotron_classical/ProbAddSpin/spinDance/

```
Cyclotron, classical. Spin dance.
'MARKER'  ProbAddSpinDance_S                                 ! Just for edition purposes.
'OBJET'
64.62444403717985                          ! Reference Brho ("BORO" in the users' guide) -> 200keV proton.
2
1 1                                                          ! Just one ion.
12.9248888074 0. 0. 0. 0.   1.      'm'    ! D=1 => 200keV proton. Y0=Brho/B=64.624444037[kG.cm]/5[kG].
1
'PARTICUL'                                          ! This is required to get the time-of-flight,
PROTON                                   ! otherwise, by default zgoubi only requires rigidity.
'SPNTRK'                                                     ! Request spin tracking.
4.1                                                          ! All spins are initially
0.984807753012  0. 0.173648177667                            ! at 10 degrees to X axis.
'FAISCEAU'
'INCLUDE'
1
6* ./60degSector.inc[#S_60degSectorUnifB:#E_60degSectorUnifB]        ! 6 * 60 degree sector.
'REBELOTE'                                                   ! Multiturn:
19 0.2  99                                                   19 additional passes.
'SYSTEM'
1
gnuplot < ./gnuplot_Zplt_SDance.gnu
'MARKER'  ProbAddSpinDance_E                                 ! Just for edition purposes.
'END'
```

A gnuplot script to obtain the spin dance in Fig. 3.31. Note a "mag" factor, aimed at artificially increasing the amplitude of the vector tip oscillation in this graphic:

```
set xlabel "X_{Lab}"; set ylabel "Y_{Lab}"; set zlabel "S_Z"; set xtics; set ytics; set ztics  #unset ztics
set zrange  [0:]; set xrange [-25:25]; set yrange [-25:25]; set xyplane 0
dip1=7; dip2=22; dd=3 # positining of 1st and last dipoles in zgoubi.dat sequence, and increment
# magnifies apparent spin tilt    speed up graphic          pi/3              z norm
     mag = 10.          ;        speedUp=1      ;      pi3 = 4.*atan(1.)/3  ;   nz=0.18

# JUST 2D, PROJECTED IN (X,Y) PLANE, FIRST:
set size ratio -1
do for [i=1:239]{ plot \
for [dip=dip1:dip2:dd] "zgoubi.plt" every 1::::speedUp*i u ($19==1 && $42==dip? $10*cos(-$22-pi3*(dip-6.)/3.) :1/0): \
($10*sin(-$22-pi3*(dip-6.)/3.)) w l lw 3 notit ,\
for [dip=dip1:dip2:dd] "zgoubi.plt" every 1::::speedUp*i u ($19==1 && $42==dip? $10*cos(-$22-pi3*(dip-6.)/3.) \
+mag*(cos(-$22-pi3*(dip-6.)/3.)*$33-sin(-$22-pi3*(dip-6.)/3.)*$34) :1/0): \
($10*sin(-$22-pi3*(dip-6.)/3.) +mag*(sin(-$22-pi3*(dip-6.)/3.)*$33+cos(-$22-pi3*(dip-6.)/3.)*$34)) w l notit }
unset size

# 3D, NEXT:
do for [i=1:239]{ splot \
for [dip=dip1:dip2:dd] "zgoubi.plt" every speedUp*i::::speedUp*i u ($19==1&& $42==dip? $10*cos(-$22-pi3*(dip-6)/3):1/0):\
($10*sin(-$22-pi3*(dip-6)/3)):($1*0):(mag*(cos(-$22-pi3*(dip-6)/3)*$33-sin(-$22-pi3*(dip-6)/3)*$34)): \
(mag*(sin(-$22-pi3*(dip-6)/3)*$33+cos(-$22-pi3*(dip-6)/3)*$34))):($35/nz) w vectors notit ,\
for [dip=dip1:dip2:dd] "zgoubi.plt" every 1::::speedUp*i u ($19==1 && $42==dip? $10*cos(-$22-pi3*(dip-6)/3) :1/0): \
($10*sin(-$22-pi3*(dip-6)/3)):($1*0):(($19==1&&$42==dip? $10*cos(-$22-pi3*(dip-6)/3):1/0):($10*sin(-$22-pi3*(dip-6)/3)): \
($1*0) w l lw 3 notit ,\
for [dip=dip1:dip2:dd] "zgoubi.plt" every 1::::speedUp*i u ($19==1 && $42==dip? $10*cos(-$22-pi3*(dip-6)/3)+mag*( \
cos(-$22-pi3*(dip-6)/3)*$33-sin(-$22-pi3*(dip-6)/3)*$34):1/0):($10*sin(-$22-pi3*(dip-6)/3)+mag*(sin(-$22-pi3*(dip-6)/3) \
*$33+cos(-$22-pi3*(dip-6)/3)*$34)):($19==1&&$42==dip? $10*cos(-$22-pi3*(dip-6)/3) +mag*(cos(-$22-pi3*(dip-6)/3)*$33-sin(-$22-pi3*(dip-6)/3)/3 \
*$33 -sin(-$22-pi3*(dip-6)/3)*$34)) :1/0): ($10*sin(-$22-pi3*(dip-6)/3) +mag*(sin(-$22-pi3*(dip-6)/3)*$33+cos(-$22-pi3* \
(dip-6)/3) *$34)):($35/nz) w l lw 3 notit  }
```

Particle data for these two particles are (respectively mass (MeV/c^2), charge (C), G factor):

$$deuteron: \quad 1875.612928 \quad 1.602176487 \times 10^{-19} \quad -0.14301$$

$$^3He^{2+}: \quad 2808.391585 \quad 3.204352974 \times 10^{-19} \quad -4.1841$$

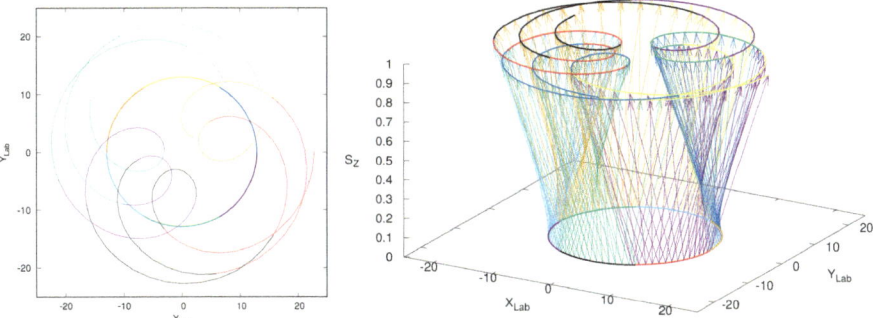

Fig. 3.31 Dance—frozen, here—of the spin of a 200 keV proton over 2.5 turns around the cyclotron. The circle on the left, or bottom closed curve on the right, is the trajectory of the proton. The cycloidal curve represents the motion of the spin vector tip in the moving frame

3.5 Synchronized Spin Torque

The simulation input data file of Exercise 3.4(d) can be used here, with a few addenda or modifications, as follows:

(i) the initial ion coordinate D (rigidity relative to the reference BORO = 64.6244440) under OBJET has to be calculated for the four energies concerned;

(ii) the closed orbit radius at 0.2, 108.412, 118.878 and 160.746 MeV has to be found; calculation is straightforward given that the field considered here is vertical, uniform, namely, B_Z = constant = 5 kG, $\forall R$, so that $R = B\rho / B_Z$; otherwise a FIT procedure can be used to find the orbit radius, given the rigidity, as done already in various exercises [16, lookup "closed orbit"], that could help for instance in the presence of a radial index, or field defects;

(iii) initial spins are set vertical for convenience, but this is not mandatory;

(iv) the multiturn tracking is set to a few tens of turns, in order to allow a few spin precessions;

(v) particle data through DIPOLEs are saved step-by-step all the way in zgoubi.plt by means of IL = 2 (the integration step size is 1 cm (Table 3.6), thus zgoubi.plt may end up bulky);

(vi) turn-by-turn data are saved in zgoubi.fai by means of FAISTORE;

(vii) SPINR is added at the end of the sequence, to impart on spins the requested X-tilt.

This results in the updated simulation input data file given in Table 3.12.

The oscillatory motion of the vertical spin component as the ion orbits around the ring, is displayed in Fig. 3.32. The spin points upward, parallel to the vertical axis at start; SPINR kick is 10 deg in the present case. At $G\gamma = 2$ the spin always finds itself back in the (Y, Z) transverse plane after one proton orbit, this synchronism causes the cumulated spin tilt at SPINR to take the value $N \times 10$ deg (with N the number of orbits). Thus after 18 proton orbits, 36 spin precessions, the spin points downward;

Table 3.12 Simulation input data file: superimposition of a turn-by-turn localized 10 deg X-rotation of the spin (using SPINR[$\phi = 0$, $\mu = 10$]), on top of Thomas-BMT $2\pi G\gamma$ Z-precession. The INCLUDE file 60degSector.inc is taken from Table 3.6

```
Cyclotron, classical. Synchronous spin kick.
'MARKER'  ProbAddSpinTorque_S                                    ! Just for edition purposes.
'OBJET'
64.62444403717985                        ! Reference Brho ("BORO" in the users' guide) -> 200keV proton.
2
4 1
12.9248888074 0. 0. 0. 0.  1.    'm'    ! D=1 => 200keV proton. Y0=Brho/B=64.624444037[kG.cm]/5[kG].
3.0947295453790e2 0. 0. 0. 0.  23.9439548880185 'm'   ! Ggamma=2
3.249214520894 1e2 0. 0. 0. 0.  25.1392067607172 'm'   ! Ggamma=2.02
3.8177333586897e2 0. 0. 0. 0.  29.5378429599586 'm'   ! Ggamma=2.1
1 1 1 1
'PARTICUL'                                          ! This is required for spin motion to be computed,
PROTON                                              ! otherwise, by default zgoubi only requires rigidity.
'SPNTRK'                                                             ! Request spin tracking.
4.1
0. 0. 1.                                             ! Initial spin vector is defined here.

'FAISTORE'
zgoubi.fai
1

'INCLUDE'
1
6* ./60degSector.inc[#S_60degSectorUnifB:#E_60degSectorUnifB]            ! 6 * 60 degree sector.
'FAISCEAU'
'SPINR'
1                                                                   ! Spin rotation,
0. 10.                                              ! about the X-axis, by 10 or 20 dgrees as the case may be.

'REBELOTE'                                                          ! Multiturn ray-tracing.
39 0.2  99
'SYSTEM'
1
gnuplot < ./gnuplot_Zplt_spinTilt.gnu
'MARKER'  ProbAddSpinTorque_E                                    ! Just for edition purposes.
'END'
```

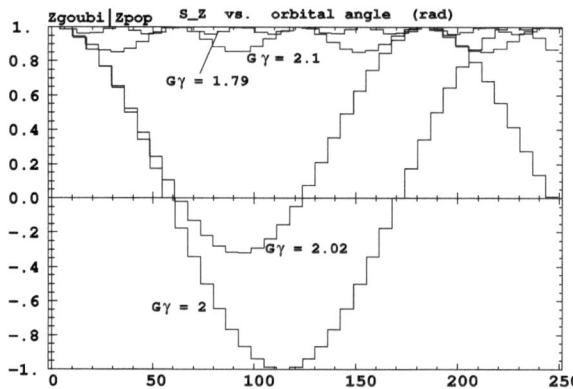

Fig. 3.32 S_Z motion versus orbital angle, while the ion orbits on a circle. S_Z is constant over a turn and then undergoes a discontinuity upon the 10 deg X-tilt, hence the step function. At $G\gamma = 2$ it takes 36 turns, or 226.194 rad, to complete an oscillation. A graph obtained using zpop: menu 7; 1/1 to open zgoubi.plt; 2/[6,23] for S_Z versus θ; 7 to plot

it takes 36 orbits, or 226.194 rad, to complete an oscillation. If $G\gamma$ moves away from an integer, the spin tilts with bounded amplitude, within the limits of a cone.

Additional graphs and details are obtained using the simulation file of Table 3.13. This file simulates spin motion in three different cases, $G\gamma = 1.79322$, $G\gamma = 2$, integer, yielding an integer number of spin precessions over one proton orbit around

Table 3.13 Simulation input data file: a similar simulation to Table 3.12, for different $G\gamma$ values, namely 1.79322, 2 and 2.5. The spin kick at SPINR has been changed to 20 deg. Regarding the use of OBJET[IEX] option: IEX = –9 allows inhibiting the tracking for the particle(s) concerned, all the rest left unchanged; it is necessary here to have at least one particle with IEX = 1, for proper operation of the gnuplot scripts. The INCLUDE file 60degSector.inc is taken from Table 3.6

```
Cyclotron, classical. Synchronized spin kick in a uniform field
'MARKER'  ProbAddSpinSphere_S                                  I Just for edition purposes.
'OBJET'
64.62444403717985                      I Reference Brho ("BORO" in the users' guide) -> 200keV proton.
2
3 1
12.924889  0. 0. 0. 0.    1.        'o'                    ! Ggamma=1.793229 -> 0.200MeV;
309.47295  0. 0. 0. 0. 23.943951797 'i'                    ! Ggamma=2 ->    198.411628MeV;
608.30878  0. 0. 0. 0. 47.064911290 'h'                    ! Ggamma=2.5 -> 370.082556MeV.
1 1 1                    ! For any particle: set to 1 to enable ray-tracing, or to -9 to ignore.
'PARTICUL'                                    ! This is required for spin motion to be computed,
PROTON                                ! otherwise, by default zgoubi only requires rigidity.
'SPNTRK'                                                   ! Request spin tracking.
4.1                                               ! All initial spins taken parallel to Z axis.
0. 0. 1.

'SPNPRT'  PRINT

'INCLUDE'
1
6* ./60degSector.inc[#S_60degSectorUnifB:#E_60degSectorUnifB]        ! 6 * 60 degree sector.
'FAISCEAU'
'SPINR'
1                                                          ! Spin rotation.
0. 20.                                         1 about the X-axis, by 20 degree here.

'REBELOTE'                                I REBELOTE[K=99] for multiturn ray-tracing,
39 0.2 99                                                  ! 39+1 turns total.
'SYSTEM'
3
gnuplot <./gnuplot_Zspnprt_spinOscillation.gnu
gnuplot < ./gnuplot_Zplt_spinTilt.gnu
gnuplot <./gnuplot_Zplt_spinTilt_3D.gnu
'END'
'MARKER'  ProbAddSpinSphere_E                                  I Just for edition purposes.
'END'
```

A gnuplot script to produce spin components versus turn, reading from zgoubi.SPNPRT.Out, Fig. 3.33:

```
# gnuplot_Zspnprt_spinOscillation.gnu
set xlabel "turn"; set ylabel "S_X,   S_Y,   S_Z"; set key b l ; nbtrj=3 # number of trajectories tracked
do for [it=1:nbtrj] {  unset label; set label sprintf("particle %3.5g",it) at 10, 0.8
plot [] [-1:1] \
 'zgoubi.SPNPRT.Out' every nbtrj::(it+2) u ($22):($13) w lp lw .3 pt 4 ps .8 lc rgb "red" ,\
 'zgoubi.SPNPRT.Out' every nbtrj::(it+2) u ($22):($14) w lp lw .3 pt 6 ps .8 lc rgb "blue" ,\
 'zgoubi.SPNPRT.Out' every nbtrj::(it+2) u ($22):($15) w lp lw .3 pt 8 ps .8 lc rgb "black" ; pause .5
set terminal postscript eps blacktext enh
set output sprintf('gnuplot_Zspnprt_spinOsc_trj%i.eps',it); replot; set terminal X11; unset output }
```

A gnuplot script to produce 2D spin motion projection of Fig. 3.33:

```
# gnuplot_Zplt_spinTilt.gnu
set xlabel "S_X"; set ylabel "S_Y"; set size ratio -1; set xrange [-1:1]; set yrange [-1:1]; set key t l
nbtrj=3 # number of trajectories tracked
do for [it=1:nbtrj] {  unset label; set label sprintf("particle %i",it) at -.9, .8
 plot 'zgoubi.plt' u ($19==it? $33 :1/0):($34) w lp lw .3 ps .2 lc rgb "blue"; pause .5
 set terminal postscript eps blacktext color  enh
 set output sprintf('gnuplot_Zplt_SX-SY_trj%i.eps',it); replot; set terminal X11; unset output }
```

A gnuplot script to produce the projection on a sphere of Figs. 3.33:

```
# gnuplot_Zplt_spinTilt_3D.gnu
set xlabel "X"; set ylabel "Y"; set zlabel "Z"; set xrange [-1:1]; set yrange [-1:1]; set zrange [-1:1]
set xyplane 0; set view equal xyz; set view 49, 339; unset colorbox
set urange [-pi/2:pi/2]; set vrange [0:2*pi]; set parametric; R = 1.   # radius of sphere
nbtrj=3 # number of trajectories tracked
do for [it=1:nbtrj] {  unset label; set label sprintf(" particle %i",it) at -1, .9, 1.
 splot  R*cos(u)*cos(v),R*cos(u)*sin(v),R*sin(u) w l lw .2 lc rgb "cyan" notit ,\
 'zgoubi.plt'    u ($19==it? $33 :1/0):($34):($35) w lp lw .2 ps .4 lc palette ; pause .5
 set terminal postscript eps blacktext color  enh
 set output sprintf('gnuplot_Zplt_S3D_trj%i.eps',it); replot; set terminal X11; unset output }
```

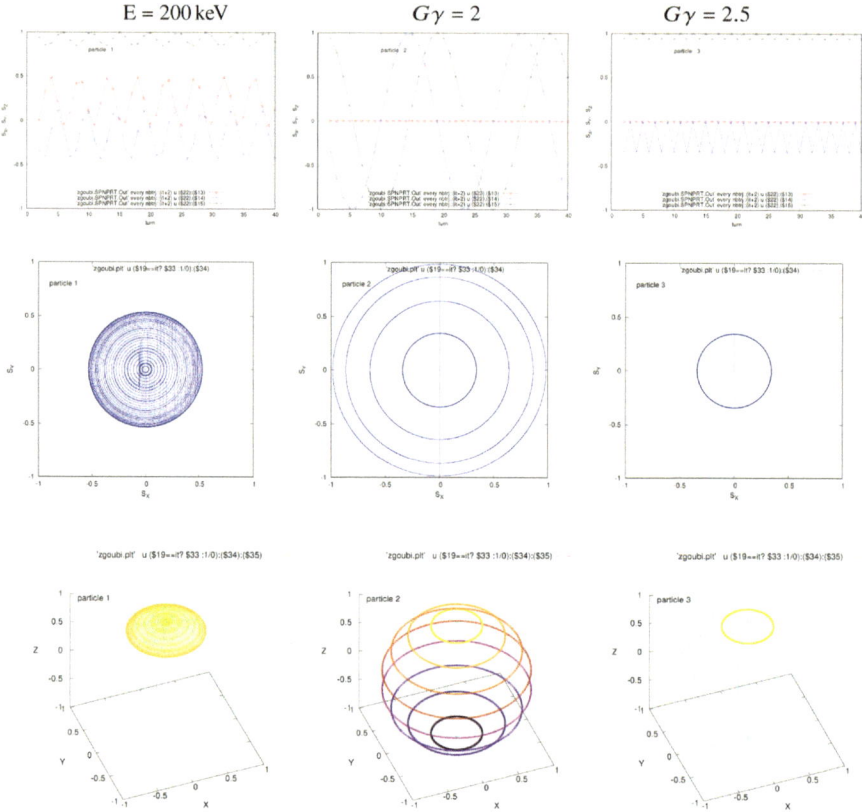

Fig. 3.33 Top row: spin coordinates versus turn; middle row: projection in the median plane (the segment between two consecutive circles materializes the location of the X-kick by SPINR); bottom row: projection on a sphere. $G\gamma = 1.793229$: far from an integer, **S** remains within a cone of reduced aperture. $G\gamma = 2$: the spin vector oscillates between up and down orientations, by 20 deg steps; it takes $180/20 = 9$ orbits for the X-precession at SPINR to flip the spin; $G\gamma = 2.5$: the spin vector finds itself back in the (Y,Z) plane at the location of SPINR, after one orbit and a half-integer number of precessions; it alternates between vertical and 20 deg from vertical, after each orbit around the cyclotron

the cyclotron, and $G\gamma = 2.5$, half-integer, yielding a half-integer number of spin precessions over one proton orbit. Outcomes are given in Fig. 3.33 which shows the spin motion projected on the (X,Y) plane (horizontal), and on a sphere, step-by-step. The spin kick by SPINR is 20 deg in this case. If $G\gamma = 1.793229$, far from an integer, **S**, initially vertical, remains at a bounded angle to the vertical axis, X-kicked from one circle to another, turn after turn; if $G\gamma = 2$ the spin vector flips by $20°$ in the (Y, Z) plane at SPINR, turn after turn; if $G\gamma = 2.5$, half-integer, the spin vector undergoes a half-integer number of precessions over one orbit around the cyclotron, it jumps and alternates between vertical, and the surface of the $20°$ Z-axis cone.

3.6 Weak Focusing

(a) Add a field index.

To the first order in R, in the median plane ($Z = 0$) and noting $R = R_0 + dR$, $B_Z(R_0) = B_0$, $B_Z(R) = B$, the field writes (Sect. 3.2.2) $B(R) = B_0 + dR \left.\frac{\partial B}{\partial R}\right|_{R_0}$. With $k = \frac{R_0}{B_0}\frac{\partial B}{\partial R}$ (Eq. 3.11) this yields

$$B(R) = B_0 + \frac{B_0}{R_0} k\, dR \qquad (3.38)$$

Assume the earlier 200 keV conditions as a reference, thus take
$R_0 = 12.9248888$ cm as the 200 keV radius, whereas $B_0 = B(R_0) = 5$ kG.

Take $k = -0.03$, a slow decrease of the field with R—proper to ensure appropriate vertical focusing with marginal impact on the radial extent of the cyclotron. For instance, with that index value the 5 MeV orbit is at a radius of 75.75467 cm (see OBJET in Table 3.3) (giving $B = 0.3235$ T along the orbit), whereas if k = 0 then $R = 75.75467$ cm is the 6.8463 MeV orbit radius ($B = 0.3788$ T).

The field map is generated using a similar Fortran program to that of Exercise 3.1 (see Table 3.1), *mutatis mutandis*, namely, introducing a reference radius R_0 and field index k. The resulting program is given in Table 3.14, it can be compiled and

Table 3.14 A Fortran program which generates a 60° mid-plane field map with non-zero transverse field k. The field map it produces is logged in geneSectorMapIndex.out

```
C geneSectorMapIndex.f program
      implicit double precision (a-h,o-z)
      parameter (pi=4.d0*atan(1.d0), BY=0.d0, BX=0.d0, Z=0.d0)

      open(unit=2,file='geneSectorMapIndex.out')                    ! Field map storage file.

C------------ Hypotheses :
      AT = 60.d0  /180.d0*pi          ! Angular extent of field map. Can be changed 360, 60 deg, etc.).
      B0 = 5.d0 ;R0 = 12.9248888074d0               ! field at R0 (kG); 200keV radius (cm), B(R0)=B0=5kG.
      ak = -0.03d0                                       ! Field index, defined at R0.
      Rmi=1.d0; Rma=76.d0; RM=50.d0  ! cm. Radial extent of field map; reference radius to define mesh.
      dR = 0.5d0 ; NR = NINT((Rma- Rmi)/dR)+1    ! R-distance between nodes in mesh. Number of R-nodes.
C                                         RdA=RM*dA is the distance between two nodes along R=RM arc,
      RdA = 0.5d0  |      given angle increment dA (dA is the "Delta theta" quantity in the main text).
      NX= NINT(RM*AT /RdA) +1 ; RdA= RM*AT / DBLE(NX -1)  ! exact mesh step at RM, corresponding to NX.
      dA = RdA / RM ; A1 = 0.d0 ; A2 = AT                 ! corresponding delta_angle.
C-------------------------------------------------
      write(2,*) Rmi,dR,dA/pi*180.d0,dZ,
     >'          ! Rmi/cm, dR/cm, dA/deg, dZ/cm'
      write(2,*) '# Field map generated using geneSectorMapIndex.f '
      write(2,fmt='(a)') '# AT/rd,   AT/deg, Rmi/cm, Rma/cm, RM/cm,'
     >//' NR, dR/cm, NX, RdA/cm, dA/rd : '
      write(2,fmt='(a,1p,5(e16.8,1x),2(i3,1x,e16.8,1x),e16.8)')
     >'# ',AT, AT/pi*180.d0,Rmi, Rma, RM, NR, dR, NX, RdA, dA
      write(2,*) '# For TOSCA: ',NX,NR,' 1 22.1 1.  !IZ=1 -> 2D ; '
     >//'MOD=22 -> polar map ; .MOD2=.1 -> one map file'
      write(2,*) '# '
      write(2,*) '#      R*cosA          Z==0,          R*sinA'
     >//'         BY            BZ           BX  ix jr'
      write(2,*) '#        cm             cm            cm '
     >//'        kG            kG           kG '
      do jr = 1, NR
      R = Rmi+ dble(jr-1)*dR
      BZ = B0 + B0/R0 * ak * (R - R0)
      do ix = 1, NX
        A = A1 + dble(ix-1)*dA ; X = R * sin(A) ; Y = R * cos(A)
        write(2,fmt='(1p,6(e16.8),2(1x,i0))') Y,Z,X,BZ,BX,ix,jr
      enddo
      enddo
      stop ' Job complete ! Field map stored in geneSectorMapIndex.out.'
      end
```

Table 3.15 First and last few lines of the field map file geneSectorMapIndex.out. The file starts
with an 8-line header, the first one of which is effectively used by zgoubi, the following 7 are just
comments

```
     1.      0.5      0.5714285714285714Ø      Ø.           ! Rmi/cm, dR/cm, dA/deg, dZ/cm
# Field map generated using geneSectorMapIndex.f
# AT/rd,  AT/deg, Rmi/cm, Rma/cm, RØ/cm, NR, dR/cm, NX, RdA/cm, dA/rd :
#   1.04719755E+00    6.ØE+01   1.ØE+00   7.6ØE+01  1.29248888E+01 151   5.ØE-01 106  4.98665501E-01  9.97331001E-03
# For TOSCA:          106               151  1 22.1 1.  !IZ=1 -> 2D ; MOD=22 -> polar map ; .MOD2=.1 -> one map file
#
#     R*cosA              Z==Ø,            R*sinA            BY              BZ              BX   ix jr
#       cm                 cm               cm               kG              kG              kG
 1.ØØØØØØØØE+00  0.ØØØØØØØØE+00  0.ØØØØØØØØE+00  0.ØØØØØØØØE+00  5.13839448E+00  0.ØØØØØØØØE+00 1 1
 9.99950267E-01  0.ØØØØØØØØE+00  9.97314468E-03  0.ØØØØØØØØE+00  5.13839448E+00  0.ØØØØØØØØE+00 2 1
 9.99801073E-01  0.ØØØØØØØØE+00  1.99452974E-02  0.ØØØØØØØØE+00  5.13839448E+00  0.ØØØØØØØØE+00 3 1
 9.99552432E-01  0.ØØØØØØØØE+00  2.99154662E-02  0.ØØØØØØØØE+00  5.13839448E+00  0.ØØØØØØØØE+00 4 1
 9.99204370E-01  0.ØØØØØØØØE+00  3.98826594E-02  0.ØØØØØØØØE+00  5.13839448E+00  0.ØØØØØØØØE+00 5 1
 .........................
 4.05947602E+01  0.ØØØØØØØØE+00  6.42500229E+01  0.ØØØØØØØØE+00  4.26798081E+00  0.ØØØØØØØØE+00 102 151
 3.99519665E+01  0.ØØØØØØØØE+00  6.46516850E+01  0.ØØØØØØØØE+00  4.26798081E+00  0.ØØØØØØØØE+00 103 151
 3.93051990E+01  0.ØØØØØØØØE+00  6.50469164E+01  0.ØØØØØØØØE+00  4.26798081E+00  0.ØØØØØØØØE+00 104 151
 3.86545219E+01  0.ØØØØØØØØE+00  6.54356779E+01  0.ØØØØØØØØE+00  4.26798081E+00  0.ØØØØØØØØE+00 105 151
 3.80000000E+01  0.ØØØØØØØØE+00  6.58179307E+01  0.ØØØØØØØØE+00  4.26798081E+00  0.ØØØØØØØØE+00 106 151
```

A gnuplot script to obtain Fig. 3.34:

```
# PLOT THE FIELD MAP:
set xtics mirror ; set ytics mirror ; set xlabel "X_{Lab} [m]" ; set ylabel "Y_{Lab} [m]" ; cm2m = 0.01
set zrange [:5.15] ; set view 66, 192 ; unset colorbox
splot "geneSectorMapIndex.out" u ($1 *cm2m):($3 *cm2m):($5) w p lc palette notit ; pause 1

# PLOT THREE TRAJECTORIES
set xtics ; set ytics ; set xlabel "X_{Lab}  [m]" ; set ylabel "Y_{Lab}  [m]" ; cm2m = 0.01 ; set size ratio 1
plot for [trj=1:3] \
  "zgoubi.plt" u ($19==trj ? $10*cm2m*cos($22) :1/0):($10*cm2m*sin($22)) w p pt 7 ps .6 notit ; pause 1
```

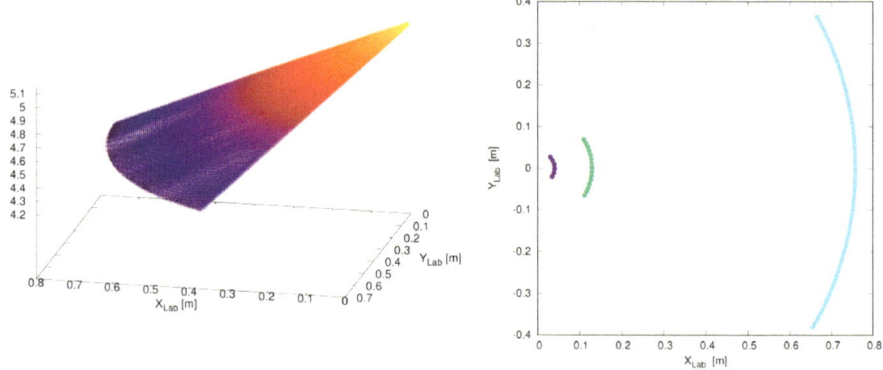

Fig. 3.34 Left: field map of a 60 deg magnetic sector with radial index, 76 cm radial extent. The
field decreases from the center of the ring (at $(X_{\text{Lab}}, Y_{\text{Lab}}) = (0,0)$). Right: three circular arc of
trajectories over a sextant, at respectively from left to right: 0.02 MeV, 0.2 MeV (energy on the
reference radius) and 5 MeV

executed, as is, excerpts of the field data file so obtained are given in Table 3.15, a
graph $B_Z(R,\theta)$ is given in Fig. 3.34. The orbit radius is assessed for three different
energies, and appears to be in accord with theoretical expectation (Fig. 3.34-right).
Comparison with Fig. 3.20-right shows the effect of the negative index on the radial
distribution of the orbits, including a radius about 20% greater in the 5 MeV range.
The input data file to find these trajectories is given in Table 3.16:

Table 3.16 Simulation input data file FieldMapSectorIndex.inc: a file to test trajectories for a field map with radial index. This file also defines the INCLUDE segment between the LABEL_1s #S_60degSectorIndx and #E_60degSectorIndx

```
FieldMapSectorIndex.inc
! Uniform field sector with index. INCLUDE file FieldMapSectorIndex.inc
'MARKER'    FieldMapSectorIdx_S                              ! Just for edition purposes.
'OBJET'
64.62444403717985                        ! Reference Brho ("BORO" in the users' guide) -> 200keV proton.
2
3 1
4.003593 0. 0. 0. 0. 0.3162126 'o'        ! p[MeV/c]= 6.126277, Brho[kG.cm]=20.435, kin-E[MeV]=0.02.
12.92488 0. 0. 0. 0. 1.          'o'       ! Reference ; p[MeV/c]=193.739, Brho[kG.cm]=BORO, kin-E[MeV]=0.2.
75.75467 0. 0. 0. 0. 5.0063900 'o'        ! p[MeV/c]=969.934, Brho[kG.cm]=323.535, kin-E[MeV]=5.
1 1 1
'MARKER'    #S_60degSectorIndx
'TOSCA'
0 2    ! IL=2 to log step-by-step coordinates, spin, etc., in zgoubi.plt (avoid, if CPU time matters).
1. 1. 1. 1.          ! Normalization coefficients, for B, X, Y and Z coordinate values read from the map.
HEADER_8                                  ! The field map file starts with an 8-line header.
106 151 1 22.1 1.            ! IZ=1 for 2D map; MOD=22 for polar frame; .MOD2=.1 if only one map file.
geneSectorMapIndex.out
0 0 0 0       ! Possible vertical boundaries within the field map, to start/stop stepwise integration.
2
1.0                                  ! Integration step size. Small enough for orbits to close accurately.
2                                                            ! Magnet positioning option.
0. 0. 0. 0.                                              ! Magnet positioning coordinates.
'MARKER'    #E_60degSectorIndx
'FIT2'                                    ! This matching procedure finds the closed orbit radius.
3   nofinal
2 30 0 [2.,10.]        ! Variable : Y_0, trajectory 1
2 40 0 [10.,15.]       ! Variable : Y_0, trajectory 2
2 50 0 [50.,80.]       ! Variable : Y_0, trajectory 3
3 1e-20  9999          ! Penalty; max numb of calls to function
3.1 1 2 #End 0. 1. 0   ! Constraint :  Y_final=Y_0, trajectory 1
3.1 2 2 #End 0. 1. 0   ! Constraint :  Y_final=Y_0, trajectory 2
3.1 3 2 #End 0. 1. 0   ! Constraint :  Y_final=Y_0, trajectory 3

!    Carry on with coordinates as found, yet with IL=2 under TOSCA so to log trajectories in zgoubi.plt.
'TOSCA'
0 2    ! IL=2: log step-by-step coordinates, spin, etc., in zgoubi.plt (avoid if CPU time matters).
1. 1. 1. 1.          ! Normalization coefficients, for B, X, Y and Z coordinate values read from the map.
HEADER_8                                  ! The field map file starts with an 8-line header.
106 151 1 22.1 1.            ! IZ=1 for 2D map; MOD=22 for polar frame; .MOD2=.1 if only one map file.
geneSectorMapIndex.out
0 0 0 0       ! Possible vertical boundaries within the field map, to start/stop stepwise integration.
2
1.0                                  ! Integration step size. Small enough for orbits to close accurately.
2                                                            ! Magnet positioning option.
0. 0. 0. 0.                                              ! Magnet positioning coordinates.
'FAISCEAU'                                ! Local particle coordinates logged in zgoubi.res.
'SYSTEM'                                  ! This SYSTEM command runs gnuplot, for a graph of the two trajectories.
1
gnuplot <./gnuplot_Zplt.gnu
'MARKER'    FieldMapSectorIdx_E                             ! Just for edition purposes.
'END'
```

– the file defines an INCLUDE segment, #S_60degSectorIndx to #E_60degSector Indx, used in subsequent exercises;

– the file is set to allow a preliminary test regarding the field map geneSectorMapIndex.out (as produced by the program given in Table 3.14), by computing three circular trajectories centered on the center of the map, at respectively 20 keV, 200 keV (the reference energy for the definition of the gradient index k) and 5 MeV (a large radius);

– note that once the FIT procedure is completed, zgoubi continues in sequence, so raytracing the 3 ions through the field map with, this time, IL set to 2 under TOSCA for stepwise particle data to be logged in zgoubi.plt.

(b) R-dependence of orbit rigidity.

The method is similar to Exercise 3.1(b) (see Table 3.4): FIT finds the closed orbit radius R for a given ion rigidity, and REBELOTE is used to repeat for a series of different momenta, 20 here. The input data file for this exercise is given in Table 3.17, it includes a 21 ion 1-turn raytracing, in sequence with the previous 21-orbit finding.

Table 3.17 Simulation input data file: scan orbits for momentum dependence. Two problems are stacked, executed in sequence: in a first stage FIT finds a closed orbit, whose coordinates are logged in initialRs.fai file when FIT is completed, following what REBELOTE repeats for an additional 20 momenta; in a second stage OBJET grabs the 21-set of ion coordinates from initialRs.fai and these ions are raytraced over 6 sectors, i.e., one full turn. The INCLUDE file FieldMapSectorIndex.inc is taken from Table 3.16

```
Uniform field sector with index. Scan orbits.
'MARKER'   scanSectorIdx_S                               ! Just for edition purposes.
'OBJET'
64.62444403717985                      ! Reference Brho ("BORO" in the users' guide) -> 200keV proton.
2
1 1                                                      ! Just one ion.
4.0039   0. 0. 0. 0. 0.3162126 'o'     ! p[MeV/c]= 6.126277, Brho[kG.cm]=20.435, kin-E[MeV]=0.02.
1
'FAISCEAU'                                               ! Local particle coordinates logged in zgoubi.res.
'INCLUDE'
1
./FieldMapSectorIndex.inc[#S_60degSectorIndx:#E_60degSectorIndx]
'FIT'                                  ! This matching procedure finds the closed orbit radius.
1  nofinal
2 30 0 [3.,80.]        ! Variable : Y_0
1 1e-15  99           ! Penalty; max numb of calls to function
3.1 1 2 #End 0. 0     ! Constraint : Y_final=Y_0
'FAISTORE'
initialRs.fai                                            ! Log coordinates in initialRs.fai.
1
'REBELOTE'               ! A do-loop. Repeat the above, after changing particle rigidity to a new value.
20 0.2  0 1              ! 20 diffrnt rigidities; I/O options; coordinates as from OBJET; changes follow:
1                                 ! Parameter 35 to be changed, in OBJET: relative momentum, namely,
OBJET 35  0.3162126:5.0063900                            ! for energy scan from 0.02 MeV to 5 MeV.

'OBJET'
64.62444403717985                      ! Reference Brho ("BORO" in the users' guide) -> 200keV proton.
3
1 999 1
1 999 1
1. 1. 1. 1. 1. 1. 1. '*'
0. 0. 0. 0. 0. 0. 0.
0
initialRs.fai
'FAISCEAU'                                               ! Local particle coordinates logged in zgoubi.res.
'INCLUDE'
1
6* ./FieldMapSectorIndex.inc[#S_60degSectorIndx:#E_60degSectorIndx]          ! INCLUDE 6 times.
'SYSTEM'
2
gnuplot <./gnuplot_Zplt_orbits.gnu                          ! Plot orbits around the cyclotron.
gnuplot <./gnuplot_Zplt_scanBrho.gnu                               ! Plot R(Brho).
'MARKER'   scanSectorIdx_E                               ! Just for edition purposes.
'END'
```

A gnuplot script to obtain orbits, Fig. 3.35:

```
set xtics ; set ytics ; set xlabel "X_{Lab}  [m]" ; set ylabel "Y_{Lab}  [m]" ; cm2m = 0.01; set polar; set size ratio 1
unset colorbox; pi = 4.*atan(1.); TOSCA1=12; dT=3 # number of 2nd TOSCA & increment in zgoubi.plt listing
plot for [trj=2:21] \
   "zgoubi.plt" u ($19==trj ? $22+($42-TOSCA1)/dT*pi/3 :1/0):($10*cm2m ):($19) w l lw 2 lc palette notit ; pause 1
```

A gnuplot script to obtain Bρ(R), Fig. 3.35:

```
set xtics ; set ytics nomirror ; set y2tics; set xlabel "R  [m]" ; set ylabel "B{/Symbol r} [T m]"
B0=0.5; R0=12.924888e-2; k=-0.03; Brho(x)=  B0* (1.+  (x-R0)/R0* k )*x ; kGcm2Tm=1e-3;  cm2m = 0.01
plot for [trj=2:21] \
   "zgoubi.plt" u ($19==trj? $10*cm2m :1/0):($40*(1.+$2)*kGcm2Tm) w p pt 6 ps 1.2 notit ,\
   Brho(x) axes x1y2 w l lw 2 lc rgb "black" tit "theor." ; pause 1
```

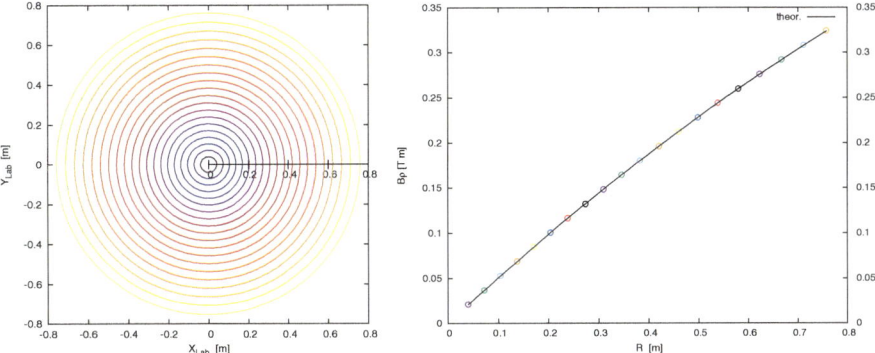

Fig. 3.35 Case of field index k = –0.03. Left: closed orbits at a series of different rigidities. Right: comparison of $B\rho(R)$ from raytracing outcomes (markers) and from theory (solid line, Eq. 3.39)

Table 3.18 Simulation input data file sectorWithIndex.inc: definition of a dipole with index, case of analytical field modeling, namely here k = –0.03 and reference radius $R_0 = 50$ cm. Definition of the [#S_60degSectorWIdx:#E_60degSectorWIdx] segment

```
# sectorWithIndex.inc
'MARKER'    #S_60degSectorWIdx                          ! Label should not exceed 20 characters.
'DIPOLE'                                                ! Analytical modeling of a dipole magnet.
2                       ! IL=2, only purpose is to logged trajectories in zgoubi.plt, for further plotting.
60. 50.                                                 ! Sector angle AT; reference radius.
30. 5. -0.03 0. 0.  ! Reference azimuthal angle ACN; BM field at RM; indices, N (=k=-0.03) at R0=50cm.
0.  0.                                                           ! EFB 1 is hard-edge,
4  .1455   2.2670  -.6395  1.1558  0. 0.  0.            ! hard-edge only possible with sector magnet.
30. 0.  1.E6  -1.E6  1.E6  1.E6
0.  0.                                                  ! EFB 2.
4  .1455   2.2670  -.6395  1.1558  0. 0.  0.
-30. 0.  1.E6  -1.E6  1.E6  1.E6
0. 0.                                                   ! EFB 3 (unused).
0  0.       0.       0.       0.      0 0.  0.
0. 0.  1.E6  -1.E6  1.E6  1.E6 0.
4   10.
0.5                             ! Integration step size. Small enough for orbits to close accurately.
2  0. 0. 0. 0.                                          ! Magnet positioning RE, TE, RS, TS.
'MARKER'    #E_60degSectorWIdx                          ! Label should not exceed 20 characters.
'END'
```

Raytracing outcomes for $k = -0.03$, $R_0 = R(E = 200\,\text{keV}) = 12.924888\,\text{cm}$, $B_0 = B(R_0) = 0.5\,\text{T}$ are given in Fig. 3.35, together with theoretical expectation (with B(R) from Eq. 3.7)

$$\text{Rigidity} \quad BR(R) = B_0 \left(1 + \frac{R - R_0}{R_0} k \right) R \qquad (3.39)$$

(c) Paraxial motion.

A proton with energy 1 MeV is considered, here. DIPOLE [16, *lookup* INDEX] is used rather than a field map, so to allow to freely change the k index value (using TOSCA instead would require computing a new field map when changing k).

The input data for a 60 deg sector are given in Table 3.18, essentially a copy of the uniform dipole field case of Table 3.6 in which the index value $k = -0.03$ has been added (line 3 under DIPOLE). The input data sequence for multiturn trajectory

Table 3.19 Simulation input data file: scan orbits for momentum dependence; the file actually stacks two simulations, executed in sequence; the second simulation uses data produced by the first one, as follows. The first part of the file finds the closed orbits, they depend on the vertical excursion and are not exactly zero, due to the field index; closed orbit coordinates so found are logged in initialRs.fai when FIT is completed. The second part of the file starts at the second occurrence of OBJET which reads initial particle coordinates from initialRs.fai and tracks these particles through a sequence of 120 sector dipoles, i.e., 20 turns. The [#S_60degSectorWIdx:#E_60degSectorWIdx] segment of Table 3.18 is INCLUDEd, here

```
Uniform field sector with index. Scan orbits.
'MARKER'  1MeVVMotion_S                                            ! Just for edition purposes.
!                                     First stage: find closed orbit at 1 MeV, for some k value.
'OBJET'
64.62444403717985                     ! Reference Brho ("BORO" in the users' guide) -> 200keV proton.
1.1
1  1  1  4   1  1
0.  1.  0.  0.1  0.  1.
30.107900 0.  0.  0.   0. 2.2365445724 'm'        ! 1 MeV proton -> Brho/Brho_ref = 2.2365445724.
'INCLUDE'
1
./sectorWithIndex.inc[#S_60degSectorWIdx:#E_60degSectorWIdx]              ! DIPOLE case R0=50cm k=-0.03.
'FIT'                                 ! This matching procedure finds the closed orbit radius.
1   nofinal
2  40  0  .9                          ! Variable : Y_0. Variation can be up to 90%.
1  1e-15  99                          ! Penalty; max numb of calls to function.
3.1 1 2 #End 0. 1. 0                  ! Constraint :  Y_final=Y_0.
'FAISTORE'
initialRs.fai                         ! Log coordinates in initialRs.fai.
1
!                                     Second stage: raytrace the four particles over 20 turns.
'OBJET'
64.62444403717985                     ! Reference Brho ("BORO" in the users' guide) -> 200keV proton.
3
1 999 1
1 999 1
1. 1. 1. 1. 1. 1. 1. '*'
0. 0. 0. 0. 0. 0. 0.
0
initialRs.fai
'FAISCEAU'                            ! Local particle coordinates logged in zgoubi.res.
'INCLUDE'
1
120 * sectorWithIndex.inc[#S_60degSectorWIdx:#E_60degSectorWIdx]    ! INCLUDE 120 sectors (20 turns).
'FAISCEAU'                            ! Local particle coordinates logged in zgoubi.res.
'SYSTEM'
2
gnuplot <./gnuplot_Zplt_1MeVVMotion.gnu
gnuplot <./gnuplot_Zplt_1MeVBField.gnu
'MARKER'  1MeVVMotion_E                                            ! Just for edition purposes.
'END'
```

A gnuplot script to obtain Figs. 3.36, 3.37:

```
# gnuplot_Zplt_1MeVVMotion.gnu
set xtics ; set ytics ; set xlabel "s  [m]" ; set ylabel "Z  [m]" ; cm2m = 0.01; unset colorbox ; set xrange [:36]
plot for [trj=4:1:-1] \
  "zgoubi.plt" u ($19==trj && $42>10? $14*cm2m :1/0):($12*cm2m ):($19) w l lw 2 tit "P[mrad]=0.".(trj-1) ; pause 1
```

A gnuplot script to obtain Fig. 3.38:

```
# gnuplot_Zplt_1MeVBField.gnu
set xtics; set ytics; set xlabel "s  [m]"; set ylabel "Y  [m]"; cm2m = 0.01; unset colorbox
plot for [trj=4:1:-1] \
  "zgoubi.plt" u ($19==trj && $42>10? $14*cm2m :1/0):($10*cm2m ):($19) w l lw 2 tit "P[mrad]=0.".(trj-1) ; pause 1
```

computation around the cyclotron is given in Table 3.19: in a first stage, orbit finding is performed by FIT, for 1 MeV energy; in a subsequent second stage, 4 protons with their initial horizontal coordinates taken on the closed orbit, and differing by their initial vertical take-off angle, are tracked over 120 sectors, i.e., 20 turns around the ring.

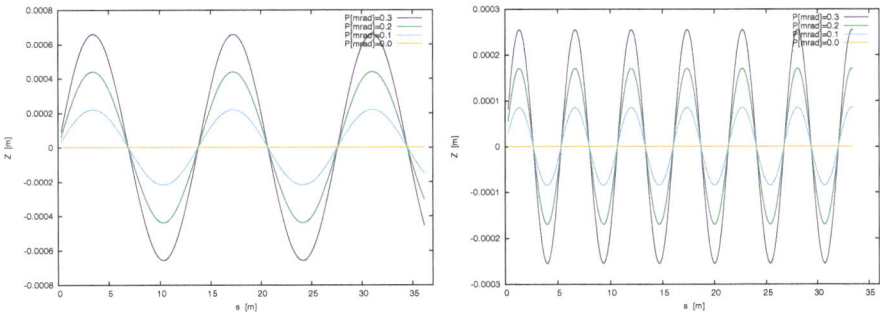

Fig. 3.36 Vertical sine motion over a few turns around the cyclotron, at 1 MeV. Vertical take-off angles are $P_0 = 0$, 0.1, 0.2, 0.3 mrad. Left: k = –0.03, $\nu_Z = \sqrt{0.03} \approx 0.173$ oscillations per turn; right: for k = –0.2, $\nu_Z = \sqrt{0.2} \approx 0.447$ oscillations per turn

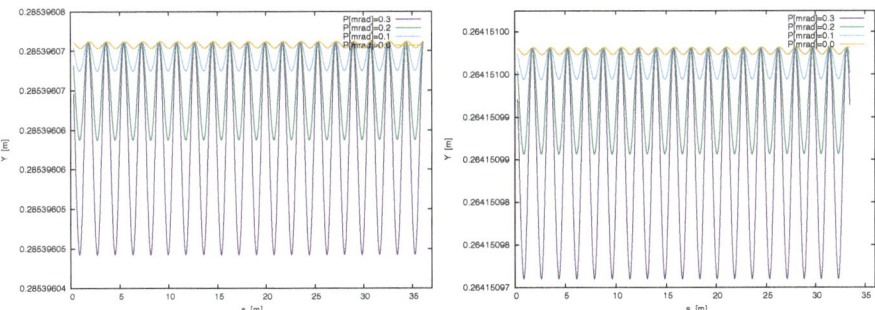

Fig. 3.37 Horizontal motion at 1 MeV, 20 turns around the cyclotron, for vertical take-off angles $P_0 = 0$, 0.1, 0.2, 0.3 mrad. Left: k = –0.03, $\nu_R = \sqrt{1 + 0.03} \approx 1.015$ oscillations per turn; right: for k = –0.2, $\nu_R = \sqrt{1 + 0.2} \approx 1.095$ oscillations per turn

Figure 3.36 displays the vertical sine motion. Stronger index (k closer to –1) results in stronger vertical focusing, hence more oscillations as expected from Eq. 3.18 and smaller motion amplitude as expected from Eq. 3.17. The latter can be written

$$Z(s) = P_0 \frac{R_0}{\sqrt{-k}} \sin \frac{\sqrt{-k}}{R_0}(s - s_0) \quad \text{and} \quad \hat{Z} = P_0 \frac{R_0}{\sqrt{-k}} \tag{3.40}$$

Note that this vertical oscillation results in a modulation of the field along the trajectory (see question (d) of this exercise) which results in a radial oscillation, a second order Y–Z coupling effect (extremely weak), displayed in Fig. 3.37.

(d) Magnetic field.

The magnetic field experienced by 1 MeV protons with four different take-off angles P_0 (Fig. 3.36), along their respective trajectories, case of an index value $k = -0.03$, is displayed in Fig. 3.38. It is essentially constant as expected.

Fig. 3.38 Magnetic field
experienced by 1 MeV
protons with four different
take-off angles P_0
(Fig. 3.36), along their
respective trajectories. Case
$k = -0.03$. The stepwise
structure of these $B_Z(s)$
curves is due to the fact that
field variations are at the
limit of computer truncation
related accuracy

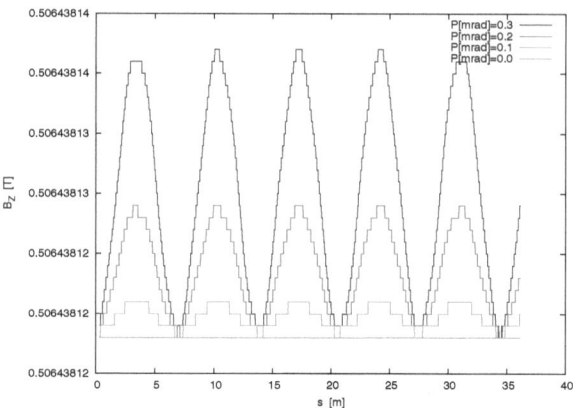

3.7 Loss of Isochronism

In order to scan $T_{rev}(R)$ for different k values, DIPOLE [16, *lookup* INDEX] is used
here, as it allows to easily vary k and subsequently find the closed orbit using FIT.
The method of Exercise 3.6 is employed to obtain a scan. The input data file of
Table 3.17 is a good starting point to do this exercise, changing the INCLUDE to
account for DIPOLE instead of a field map modeling using TOSCA: the proper
INCLUDE formatting can be reproduced from Table 3.19. IL under DIPOLE may
be set at IL = 0 as zgoubi.plt is not used here. Introduce FAISTORE to store local
particle data after FIT (that includes time of flight, the quantity of interest here, which
requires PARTICUL[PROTON] following OBJET).

The new input data file so built for this simulation, is given in Table 3.20.

This input data file is run for four different k values, namely, under DIPOLE (*cf.*
Table 3.18), the line "30. 5. –0.03 0. 0." is successively changed to $\begin{cases} 30.5.\ 0\ 0.\ 0. \\ 30.5.\ -0.5\ 0.\ 0. \\ 30.5.\ -0.95\ 0.\ 0. \end{cases}$.
The corresponding zgoubi.fai files are saved under dedicated copies for plotting, see
"gnuplot script gnuplot_Zfai_scanTrev.gnu" at the bottom of Table 3.20.

The results of these T_{rev} scans are displayed in Fig. 3.39. In the case $k = 0$ the
loss of isochronism is only due to the relativistic change of the mass, a non-zero k
augments the effect. The loss of isochronism is the cause of the ≈ 20 MeV proton
energy limit of the classical cyclotron.

3.8 Ion Trajectories

A `zgoubi` data file is set up for computation of particle trajectories, taking a field value
on reference radius of $B_0(R_0) = 0.5$ T, and reference energy 200 keV (proton). These
hypotheses determine the reference radius value. DIPOLE [16, *lookup* INDEX] is
used (Table 3.21), for its greater flexibility in changing magnet parameters, field and
radial field index amongst other, compared to using TOSCA and a field map.

Table 3.20 Simulation input data file: scan revolution time. The [#S_60degSector WIdx:#E_60degSectorWIdx] segment of Table 3.18 is INCLUDEd, here

```
Uniform field sector with index. Scan orbits.
'MARKER'   isoChroLoss_S                                  ! Just for edition purposes.
'OBJET'
64.62444403717985                      ! Reference Brho ("BORO" in the users' guide) -> 200keV proton.
2
1 1                                                        ! Just one ion.
4.0039   0. 0. 0. 0. 0.3162126 'o'     ! p[MeV/c]= 6.126277, Brho[kG.cm]=20.435, kin-E[MeV]=0.02.
1
'PARTICUL'                             ! Necessary as time of flight computation is needed,
PROTON                                 ! otherwise, by default zgoubi only requires rigidity.
'INCLUDE'
1
./sectorWithIndex.inc[#S_60degSectorWIdx:#E_60degSectorWIdx]         ! DIPOLE case R0=50cm k=-0.03.
'FIT2'                                 ! This matching procedure finds the closed orbit radius.
1  nofinal
2 30 0 [0.5,80.]          ! Variable : Y_0
1 1e-15  99              ! Penalty; max numb of calls to function
3.1 1 2 #End 0. 1. 0      ! Constraint : Y_final=Y_0
'FAISCEAU'                             ! Local particle coordinates logged in zgoubi.res.
'FAISTORE'
zgoubi.fai
1
'REBELOTE'                             ! A do-loop. Repeat the above, after changing particle rigidity to a new value.
20 0.2  0 1              ! 20 diffrnt rigidities; I/O options; coordinates as from OBJET; changes follow:
1                        ! Parameter 35 to be changed, in OBJET: relative momentum, namely,
OBJET 35  0.3162126:5.00639   ! Acceleration to 5MeV. Commented here, for use in subsequent exercises.
! OBJET 35  0.3162126:2.2365445724   ! Substitute to previous, for energy scan from 0.02 MeV to 1 MeV.
'SYSTEM'
1
gnuplot <./gnuplot_Zfai_scanTrev.gnu                                    ! Plot revolution time.
'MARKER'   isoChroLoss_E                                   ! Just for edition purposes.
'END'
```

A gnuplot script to obtain Fig. 3.39:

```
# gnuplot_Zfai_scanTrev.gnu
set xtics ;  set ytics nomirror ; set y2tics; set xlabel "R  [m]" ; set ylabel "T_{rev} [{/Symbol m}s]"
cm2m = 0.01; nSec=6; set y2label "T_{rev}  at  k=0[{/Symbol m}s]" ; set key c r
plot "zgoubi_k0.fai"    u ($10 *cm2m):($15 * nSec) w lp pt 4 ps 1.2 lc rgb "black" tit "k=0"  ,\
     "zgoubi_k0.fai"    u ($10 *cm2m):($15 * nSec) axes x1y2 w lp pt 4 ps 1.2  lc rgb "black" notit ,\
     "zgoubi_k0.03.fai" u ($10 *cm2m):($15 * nSec) w lp pt 7 ps 1.2 tit "k=-03" ,\
     "zgoubi_k0.5.fai"  u ($10 *cm2m):($15 * nSec) w lp pt 8 ps 1.2 tit "k=-5" ,\
     "zgoubi_k0.95.fai" u ($10 *cm2m):($15 * nSec) w lp pt 9 ps 1.2 tit "k=-95" ; pause 1
```

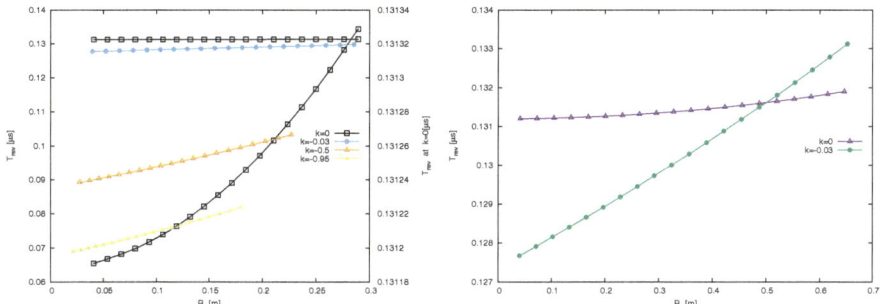

Fig. 3.39 A scan of the revolution time, from 0.02 to 1 MeV, and its dependence on the field index k. The right vertical axis only concerns the case $k = 0$ where the change in revolution time is weak and only due to the mass increase (in $T_{rev} = 2\pi\,\gamma m_0/qB$). The right graph shows, up to 5 MeV, the relatively important contribution of the focusing index, even a weak k = −0.03, compared to the effect of the mass increase (k = 0 curve). Markers are from raytracing, solid lines are from theory

Table 3.21 Input data file 60DegSectorR200.inc: it defines DIPOLE as a sequence segment comprised between the "LABEL_1" type labels [16, Sect. 7.7] #S_60DegSectorR200 and #E_60DegSectorR200. DIPOLE here, has an index $k = -0.03$, reference radius $R_0 \equiv R_0(E_k = 200\,\text{keV}) = 12.924888$ cm and $B_0 = B(R_0) = 0.5$ T. Note that (i) this file can be run on its own: it has been designed to provide the transport MATRIX of that DIPOLE; (ii) for the purpose of some of the exercises, IL = 2 under DIPOLE, optional, causes the printout of particle data in zgoubi.plt, at each integration step (this is at the expense of CPU time, and memory volume)

```
60DegSectorR200.inc
'OBJET'
64.62444403717985                                                    ! 200keV proton.
5
0.01 0.001 0.01 0.001 0. 0.0001
12.9248888074 0.  0.  0. 0. 1.                                       ! 200keV. R=Brho/B=*/.5.
'DIPOLE'   #S_60DegSectorR200               ! Analytical modeling of a dipole magnet.
2        ! IL=2, purpose: log stepwise particle data in zgoubi.plt. Avoid if unused: I/Os take CPU time.
60. 12.924888                                        ! Sector angle AT; reference radius R0.
30.  5. -0.03 0. 0.  ! Reference azimuthal angle ACN; BM=B0 field at RM=R0; indices, N, N', N'' at R0.
0.  0.                                                            ! EFB 1 is hard-edge,
4  .1455   2.2670  -.6395  1.1558  0. 0.  0.             ! hard-edge only possible with sector magnet.
30. 0.   1.E6  -1.E6  1.E6  1.E6
0.  0.                                                               ! EFB 2.
4  .1455   2.2670  -.6395  1.1558  0. 0.  0.
-30. 0.   1.E6  -1.E6  1.E6  1.E6
0. 0.                                                                ! EFB 3 (unused).
0 0.       0.     0.      0.      0. 0.  0.
0. 0.  1.E6  -1.E6  1.E6  1.E6 0.
4   10.
0.5                                      ! Integration step size. Small enough for orbits to close accurately.
2  0. 0. 0. 0.                                         ! Magnet positioning RE, TE, RS, TS.
'FAISCEAU'  #E_60DegSectorR200
'MATRIX'
1 0
'END'
```

(a) Transverse motion.

It first has to be checked that there is consistency between initial orbital radius Y_0 in OBJET at 200 keV proton energy and the value of the reference radius R_0 in DIPOLE (Eq. 3.35). Its theoretical value is $R_0 = BORO/5[kG] = 12.924889$ cm, a closed orbit finding using FIT can be performed, or it can be referred to the solutions of earlier exercises, to check agreement with raytracing outcomes.

(b) Wave numbers at 1 and 5 MeV.

These considerations result in the input data file given in Table 3.22, to compute multiturn trajectories.; note that $R_0 = 12.924889$ cm therein, whereas a value of $R_0 = 50$ cm may be taken instead in other exercises. Field index derivatives k', k'', ... are taken null in the present exercise.

Three particles with paraxial radial and axial motions are raytraced over a few turns. Their starting radius is the closed orbit radius for the respective energies, while a 0.1 mrad take-off angle is imparted to each particle both vertically and horizontally.

The value of the focusing index k_E at an energy E can be expressed in terms of DIPOLE data which are, the index value k at R_0 (Eq. 3.11), reference radius R_0, and field $B_0 = B_Z(R_0)$, namely,

$$k_E = \frac{R_E}{B_E}\frac{\partial B}{\partial R} = \frac{R_0 + \Delta R}{B_0 + \Delta B}\frac{\partial B}{\partial R} \approx k\frac{1 + \Delta R/R_0}{1 + k\Delta R/R_0} \approx k\left[1 + (1-k)\frac{\Delta R}{R_0}\right]$$

Table 3.22 Simulation input data file: raytrace a few turns around the cyclotron, three particles with different momenta, and 0.1 mrad horizontal and vertical take-off angles. The INCLUDE segment is taken from Table 3.21

```
'MARKER'    ProbProjTraj_S
'OBJET'
64.62444403717985                          ! Reference Brho ("BORO" in the users' guide) -> 200keV proton.
2
3 1
12.924888 0.1 0. 0.1 0. 1. 'o'            ! A particle with kin-E=0.2 MeV and 0.1 mrad take-off angles.
30.107898 0.1 0. 0.1 0. 2.2365445 'm'      ! p[MeV/c]=433.306, Brho[kG.cm]=144.535, kin-E[MeV]=1.
75.754671 0.1 0. 0.1 0. 5.0063900 'o'      ! p[MeV/c]=969.934, Brho[kG.cm]=323.535, kin-E[MeV]=5.
1 1 1
'INCLUDE'
1
6* 60DegSectorR200.inc[#S_60DegSectorR200:#E_60DegSectorR200]   ! 6 sectors for an overall 360 deg.
'REBELOTE'
9 0.1 99                                   ! There will be a total of 9+1=10 tunrs.
'SYSTEM'
1
gnuplot < ./gnuplot_Zplt_traj.gnu                         ! Plot the projected Y(s) and Z(s) motions.
'MARKER'    ProbProjTraj_E
'END'
```

A gnuplot script to obtain Fig. 3.41:

```
# gnuplot_Zplt_traj.gnu
set xtics nomirror; set x2tics; set ytics; set xlabel 's  /C_E '; set ylabel 'Y  [cm]'
set palette defined ( 1 "red", 2 "blue", 3 "black" ) ; unset colorbox
array R[3]; R[1]=0.12924888; R[2]=0.301078986; R[3]=0.75754671; pi = 4.*atan(1.); cm2m = 0.01
sector1=3    # number (NOEL) of 1st DIPOLE in \zgoubi.res
# in zgoubi.plt, col. 19: particle number; col. 42: keyword number; col. 14: distance; col. 10: Y ; col. 12: YZ
plot for [sector=1:6] for [trj=1:3] 'zgoubi.plt' u ($19==trj && $42==sector1+2*(sector-1)? $14*cm2m/(2.*pi*R[$19]):1/0) \
:($10*cm2m-R[trj]):($19) w p ps .2 lc palette notit ; pause 1

set ylabel 'Z  [cm]'; plot for [sector=1:6] for [trj=1:3] 'zgoubi.plt' u ($19==trj && $42==sector1+2*(sector-1)? $14*cm2m\
/(2.*pi*R[$19]) :1/0):($12):($19) w p ps .2 lc palette notit ; pause 1
```

with ΔR assumed small, $\partial B/\partial R = k B_E/R_E$ an energy independent quantity, and the index E denoting a quantity taken at the reference energy. The latter property is illustrated in Fig. 3.40, produced using the input data file of Table 3.23.

The resulting radial and axial motions over 10 turns are displayed in Fig. 3.41, which also illustrates, for paraxial motion at some reference energy, the energy dependence of the focusing strength (or wave number) and of the motion amplitude.

An estimate of the wave numbers can be obtained as the inverse of the number of turns per oscillation, namely,

$$\nu_R = \left.\frac{C_E}{\Delta s_M}\right|_E \quad \text{and} \quad \nu_Z = \left.\frac{C_E}{\Delta s_M}\right|_E$$

with Δs_M the measured distance between two consecutive maxima in the sinusoid of concern in Fig. 3.41, C_E the closed orbit length for the energy of concern. Both quantities are obtained from motion records in zgoubi.plt. This yields the values of Table 3.24, where they are compared with the theoretical expectations, namely (Eq. 3.18), $\nu_R = \sqrt{1+k}$ and $\nu_Z = \sqrt{-k}$.

The maximum amplitude of the oscillation is obtained from zgoubi.plt records as well, this yields the results of Table 3.25. For comparison, the theoretical values are (Eqs. 3.16 and 3.17 with respectively $x_0 = 0$, $x_0' = T_0$ and $y_0 = 0$, $y_0' = P_0$) $\hat{Y} = T_0 \frac{R_E}{\sqrt{1+k}}$ and $\hat{Z} = P_0 \frac{R_E}{\sqrt{-k}}$. wherein R_E denotes the closed orbit radius at energy E (for the record: $R_E \equiv R_0$ at energy $E = 200\,\text{keV}$, in the foregoing).

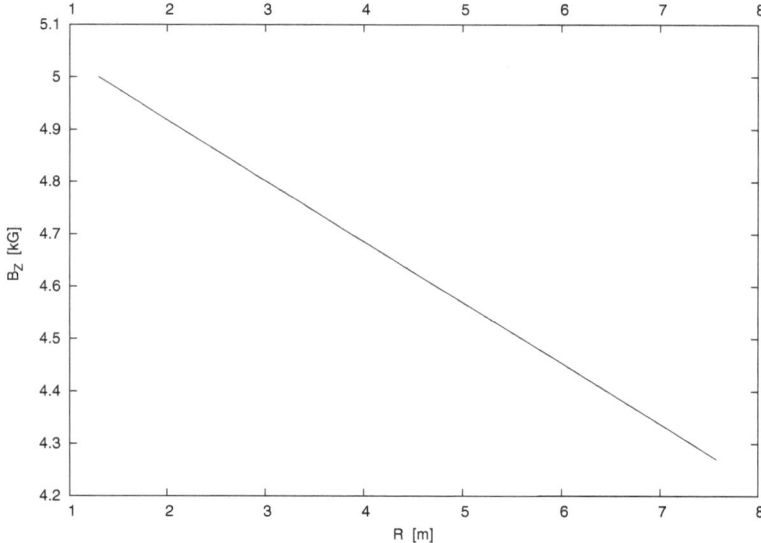

Fig. 3.40 In DIPOLE field model (Eq. 3.35), $\frac{\partial B}{\partial R}$ is constant: this graph shows the linear decrease of the field $B_Z(R)$ (Eq. 3.38), obtained from the raytracing of particles circulating in the median plane on orbits spanning a 0.2–5 MeV energy range

(c) Comparison with theory.

Figure 3.42 shows the difference between numerical and theoretical vertical motion excursion, using an *ad hoc* gnuplot script. An integration step size $\Delta s = 2$ cm is used in the numerical integration.

(d) A scan of energy dependence of wave numbers.

A scan of the wave numbers over 200 keV-5 MeV energy range, computing tunes with MATRIX, is performed using the input data file given in Table 3.26 (essentially a copy of the input data file of Table 3.23, with an INCLUDE accounting for 6 DIPOLEs [16, *lookup* INDEX]).

OBJET[KOBJ = 5] generates 13 particles with paraxial horizontal, vertical and longitudinal sampling, proper to allow the computation of the first order transport coefficients and wave numbers by MATRIX. REBELOTE repeats MATRIX computation for a series of different particle rigidities. It is preceded by FIT which finds the closed orbit. MATRIX includes a PRINT command, which causes the transport coefficients (and various other outcomes of MATRIX computation) to be logged in zgoubi.MATRIX.out. This allows producing the graphic in Fig. 3.43—using the gnuplot script given at the bottom of Table 3.26.

Table 3.23 Simulation input data file for a magnetic field scan. The INCLUDE segment is taken from Table 3.21

```
Field and derivative dB/dR, as a finction of R
'MARKER'    ProbProjTrajB_S
'OBJET'
64.62444403717985                       ! Reference Brho ("BORO" in the users' guide) -> 200keV proton.
2
1 1                                                                ! Just one ion.
12.924888 0.1 0. 0.1 0. 1. 'o'          ! A particle with kin-E=0.2 MeV and 0.1 mrad take-off angles.
1
'INCLUDE'
1                        ! IL=2 is necessary under DIPOLE, for step-by-step log of particle data in zgoubi.plt.
60DegSectorR200.inc[#S_60DegSectorR200:#E_60DegSectorR200]          ! One sector is enough.
'FIT'
1
2 30 0 [12,80]                          ! Vary particle's Y0 at OBJET, to have it match its D (=Brho/BORO).
1 1e-20
3.1 1 2 #End 0. 1. 0                                                ! Consrain Y_final=Y0.
'REBELOTE'
25 0.1 0 1                              ! Scan parameter 35 (relative rigidity, D) in OBJET.
1
OBJET 35 1:5.00639                      ! Scan relative rigidity D from 1 (200 keV) to 5.0063900 (5 MeV).
'SYSTEM'
1
gnuplot < ./gnuplot_Zplt_field.gnu                                 ! Plot B(R), as read fron zgoubi.plt.
'MARKER'    ProbProjTrajB_E
'END'
```

A gnuplot script to obtain Fig. 3.40:

```
# gnuplot_Zplt_field.gnu
set xtics nomirror; set x2tics; set ytics; set xlabel 's /C_E '; set ylabel 'Y [cm]'
set palette defined ( 1 "red", 2 "blue", 3 "black" ) ; unset colorbox
array R[3]; R[1]=0.12924888; R[2]=0.301078986; R[3]=0.75754671; pi = 4.*atan(1.); cm2m = 0.01
sector1=3    # number (NOEL) of 1st DIPOLE in \zgoubi.res (col. 42 in zgoubi.plt)
# in zgoubi.plt, col. 19: particle number; col. 42: keyword number; col. 14: distance; col. 10: Y ; col. 12: YZ
plot   for [i=1:6] for [trj=1:3]
   'zgoubi.plt' u ($19==trj && $42==sector1 +2*(i-1) ? $14*cm2m /(2.*pi*R[$19]) :1/0) \
   :($10*cm2m-R[trj]):($19) w p ps .2 lc palette notit ; pause 1

set ylabel 'Z [cm]' ;
plot   for [i=1:6] for [trj=1:3]
   'zgoubi.plt' u ($19==trj && $42==sector1 +2*(i-1) ? $14*cm2m \
   /(2.*pi*R[$19]) :1/0):($12):($19) w p ps .2 lc palette notit ; pause 1
```

Fig. 3.41 Radial (left) and axial (right) paraxial motion around respectively the 200 keV (smallest amplitude), 1 MeV (intermediate) and 5 MeV (greatest amplitude) closed orbit (the latter is circular, in the median plane, with radius respectively $R_{200\,keV} = 12.924888$ cm, $R_{1\,MeV} = 30.107898$ cm and $R_{5\,MeV} = 75.754671$ cm). The horizontal axis in this graph is s/C_E: path length over closed orbit circumference at energy E, the vertical axis is the motion excursion

Table 3.24 Wave numbers, from numerical raytracing (columns denoted "ray-tr."), from theory, and from discrete Fourier transform ('DFT' cols.) from a multi-turn tracking

E (MeV)	k_E	$\nu_R =$			$\nu_Z =$		
		ray-tr.	$\sqrt{1+k}$	DFT	ray-tr.	$\sqrt{-k}$	DFT
0.2	−0.03	0.98520	0.9849	0.98513	0.17320	0.1732	0.17321
1	−0.07279	0.96187	0.96292	0.96291	0.26980	0.26979	0.26981
5	−0.20586	0.89083	0.89115	0.89115	0.45326	0.45371	0.45371

Table 3.25 Maximum amplitude of the oscillation, from raytracing (columns denoted "ray-tr.") and from theory. R_E is the closed orbit radius for the energy of concern, $T_0 = P_0 = 0.1$ mrad is the trajectory angle at the origin, positions at the origin are zero

E (MeV)	k	$\hat{Y}$		$\hat{Z}$	
		ray-tr. ($\times 10^{-5}$)	$T_0 \frac{R_E}{\sqrt{1+k}}$	ray-tr. ($\times 10^{-5}$)	$P_0 \frac{R_E}{\sqrt{-k}}$
0.2	−0.03	1.3123	1.3125	7.4622	7.4624
1	−0.072787	3.1270	3.1267	1.1160	1.1160
5	−0.20586	8.5010	8.5008	1.6697	1.6697

Fig. 3.42 Vertical excursion of a 1 MeV trajectory over 20 turns (left vertical axis), and difference with theoretical expectation as per Eq. 3.17 (right vertical axis). The plot shows two sinusoidal curves: a segmented one, thicker, from numerical integration, and a thinner one, superimposed, from Eq. 3.17

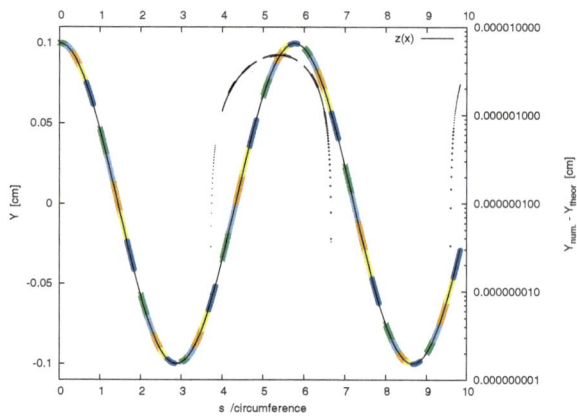

3.9 RF Phase at the Accelerating Gap

According to Sect. 3.2.3 (Fig. 3.13), the RF is taken about half-way of the accelerating range, namely, referring to Fig. 3.39, $T_{rev} = 0.131\,\mu s$ and $f_{rf} = 1/T_{rev} = 7.633$ MHz.

An input data file for this simulation is given in Table 3.27.

In a similar way to the diagrams in Fig. 3.13, the resulting $B(R)$ curve is given in Fig. 3.44, the resulting $\Delta W(\phi)$ curve in Fig. 3.45.

More turns are performed by changing the arguments under REBELOTE in the input data file (Table 3.27), from 42 to 75 in the present case. The resulting energy gain

Table 3.26 Simulation input data file: for this wave number scan, the INCLUDE segment is taken from Table 3.21

```
Field and derivative dB/dR, as a finction of R
'MARKER'    ProbMATRIX_S
'OBJET'
64.62444403717985                        ! Reference Brho ("BORO" in the users' guide) -> 200keV proton.
5                                         ! Define 13 particles for MATRIX computation.
.001 .01 .001 .01 .001 .00001             ! Sampling of the initial coordinates.
12.924888 0. 0. 0. 0. 1.                  ! Reference: p[MeV/c]=193.739, Brho[kG.cm]=BORO, kin-E[MeV]=0.2.
'INCLUDE'
1                     ! IL=2 is necessary under DIPOLE, for step-by-step log of particle data in zgoubi.plt.
6* 60DegSectorR200.inc[#S_60DegSectorR200:#E_60DegSectorR200]          ! Six 60 degree sectors.
'FIT'
1
2 30 0 [12,80]                            ! Vary particle's Y0 at OBJET, to have it match its D (=Brho/BORO).
1 1e-10
3.1 1 2 #End 0. 1. 0                                        ! Consrain Y_final=Y0.
'MATRIX'
1 11 PRINT            ! PRINT: log computation outcome data to zgoubi.MATRIX.out, for further plotting.
'REBELOTE'
25 0.1 0 1                                 ! Scan parameter 35 (particle 1's D) in OBJT.
1
OBJET 35 1:5.00639
'SYSTEM'
1
gnuplot < ./gnuplot_MATRIX_Qxy.gnu
'MARKER'    ProbMATRIX_E
'END'
```

A gnuplot script to obtain Fig. 3.43:

```
# gnuplot_MATRIX_Qxy.gnu
set xlab "kin. E [MeV]";set ylab "{/Symbol n}_x,   ({/Symbol n}_x^2+{/Symbol n}_y^2)^{1/2}";set y2label "{/Symbol n}_y"
set key t 1 maxrow 1; set xtics; set ytics nomirror; set y2tics nomirror
BORO = 64.62444403717985; am = 938.27203e6; c = 2.99792458e8; BrhoRef = BORO *1e-3; eV2MeV = 1e-6
plot  "zgoubi.MATRIX.out" u ((sqrt(($47*BrhoRef*c)**2 + am*am)-am)*eV2MeV):($56) w lp pt 5 lt 1 lw .5 lc rgb "red" \
  tit "{/Symbol n}_x " , \
  "zgoubi.MATRIX.out" u ((sqrt(($47*BrhoRef*c)**2 + am*am)-am)*eV2MeV):($57) axes x1y2 w lp \
  pt 6 lt 3 lw .5 lc rgb "blue" tit "{/Symbol n}_y " ,\
  "zgoubi.MATRIX.out" u ((sqrt(($47*BrhoRef*c)**2 + am*am)-am)*eV2MeV):(sqrt($56**2+$57**2)) \
  w lp pt 7 lt 1 lw .5 lc rgb "black" t "   ({/Symbol n}_x^2+{/Symbol n}_y^2)^{1/2}";pause 1
```

of the proton as a function of RF phase is shown Fig. 3.46. A first graph in Fig. 3.47 shows the evolution of its relative rigidity, namely D-1 as a function of distance, with $D = B\rho(s)/BORO$ and BORO = 64.624444 kG cm the reference rigidity as defined under OBJET; a second graph shows its orbital radius as a function of distance.

3.10 The Cyclotron Equation

Cyclotron model settings of Exercise 3.3 are considered in questions (a) to (c), first: two dees, double accelerating gap, uniform field $B = 0.5$ T. The analytical field modeling DIPOLE [16, *lookup* INDEX] is used.

(a) Simulation data file.

Acceleration is over the energy range $[0.2, 20]$ MeV, the maximum of $\cos(\phi)$ (Fig. 3.14) is placed at $E_k = E_{k,m} = 10$ MeV.

The cyclotron equation (Eq. 3.26) can be written under the form

$$\cos\phi = \cos\phi_i - \frac{\pi}{q\hat{V}}\left[\frac{\omega_{rf}}{2M\omega_{\text{rev}}}(E^2 - E_i^2) - (E - E_i)\right] \qquad (3.41)$$

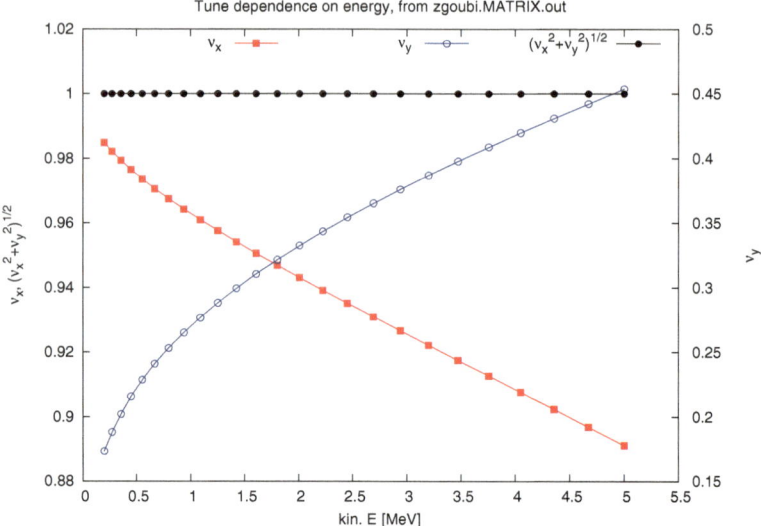

Fig. 3.43 A scan of the energy dependence of the horizontal and vertical wave numbers. Markers are from raytracing, solid lines are from theory (Eq. 3.18). The figure also shows that the raytracing yields $v_R^2 + v_y^2 = 1$, $\forall E$, as expected

where the index i denotes injection parameters, ϕ is the phase of the RF at particle arrival at the accelerating gap, $\hat{V}$ is the peak gap voltage, $E = E_k + M$ is the total energy with M the rest mass. The value of E_k at the maximum of $\cos\phi$ is drawn from $d(\cos\phi)/dE_k = 0$, namely

$$E_{k,m} = \left(\frac{\omega_{\text{rev}}}{\omega_{rf}} - 1\right) M \tag{3.42}$$

Taking $E_{k,m} = 10\,\text{MeV}$ one gets

$$\frac{\omega_{\text{rev}}}{\omega_{rf}} \approx 1 + 0.106578, \qquad f_{rf} \approx 0.989454\omega_{\text{rev}}/2\pi = 7.542209\,\text{MHz}$$

The corresponding input data file is given in Table 3.28. Figure 3.48 shows the case of two particles accelerated at a rate of 400 kV per turn, one resulting from an initial phase at the gap of $\phi_i = \pi/2$ and reaches 20 MeV in about 52 turns, the other resulting from an initial phase $\phi_i = 3\pi/4$ and reaches 20 MeV in about 64 turns. In the latter case, the $\pi/4$ phase shift results from an initial path length offset

$$\delta s = \beta c T_{\text{rf}}/4 = 10.26647\,\text{cm}$$

Table 3.27 Simulation input data file: accelerating a proton to get the evolution of RF phase The [#S_60degSectorWIdx:#E_60degSectorWIdx] segment of Table 3.18 is INCLUDEd, here

```
Cyclotron, classical. Acceleration to 6.02 MeV.
'MARKER' ProbRFPhase_S                              ! Just for edition purposes.
'OBJET'
64.62444403717985                                   ! Reference: 200keV (assuming proton).
2
1 1                                                 ! Just one ion.
12.629892  0.  0.   0.   0.  1. 'm'                 ! Injection energy 200keV, proton.
1
'PARTICUL'                          ! Particle data are necessary as CAVITE is used,
PROTON                              ! otherwise, by default zgoubi only requires rigidity.
'INCLUDE'
1
3* sectorWithIndex.inc[#S_60degSectorWIdx:#E_60degSectorWIdx] ! Three 60 deg sectors. R0=50cm, k=-0.03.
'FAISTORE'                                          ! Log particle coordinates at each turn.
zgoubi.fai
1
'CAVITE'   GAP1
7 PRINT                             ! PRINT: log CAVITE computational dat to zgoubi.CAVITE.out.
0.00     7.63358778626e6           ! f_rf= 1/T_rev, T_rev at about middle of acceleration range.
100e3    -1.57079632679            ! Peak voltage;,  relative phase of 1st cavity.
'INCLUDE'
1
3* sectorWithIndex.inc[#S_60degSectorWIdx:#E_60degSectorWIdx] ! Three 60 deg sectors. R0=50cm, k=-0.03.
'CAVITE'   GAP1
7 PRINT                             ! PRINT: log CAVITE computational dat to zgoubi.CAVITE.out.
0.00     7.63358778626e6           ! f_rf= 1/T_rev, T_rev at about middle of acceleration range.
100e3    +1.57079632679            ! Peak voltage;,  relative phase of 1st cavity.
'FAISCEAU'                                  ! Local particle coordinates logged in zgoubi.res.
'REBELOTE'                  ! K = 99 : coordinates at end of previous pass are used as initial
42  1.1 99                          ! coordinates for the next pass ; idem for spin components.
'SYSTEM'
1                                                   ! 1 SYSTEM command follows.
/usr/bin/gnuplot < ./gnuplot_CAVITE.gnu &           ! Plot phase, as read from zgoubi.CAVITE.out.
'MARKER'  ProbRFPhase_E                              ! Just for edition purposes.
'END'
```

A gnuplot to obtain the accelerated orbit of Fig. 3.45:

```
# gnuplot_CAVITE.gnu
set xlabel "RF phase [rad]" ; set ylabel "{/Symbol D}W [MeV]"; set xtics ;set ytics mirror
plot 'zgoubi.CAVITE.Out' u ($11):($2 - ($6-50)/10000.) w lp notit ; pause 2
```

Fig. 3.44 Radial dependence of the magnetic field over the acceleration range. The field is 0.5 T at a reference radius $R_0 = 0.5$ m, the slope results from the index $k = -0.03$. A graph obtained using zpop: menu 7; 1/1 to open zgoubi.plt; 2/[2,32] for B_Z versus Y; 7 to plot

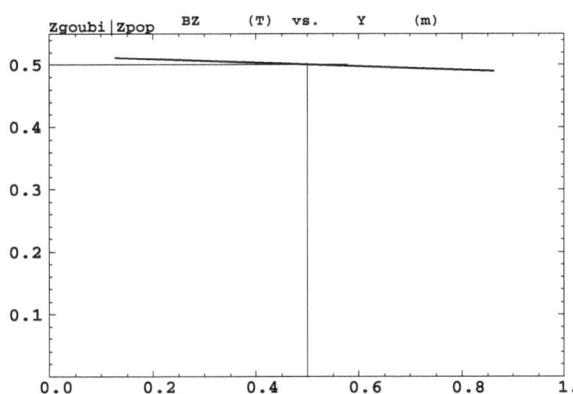

as specified under OBJET ($\beta c = 0.020648c$ is the proton velocity at $E_i = 200$ keV), yielding $\delta\phi = \omega_{rf}\delta s / \beta c = \pi/4$. A third curve in the figure is for to 200 kV voltage and initial phase at gap $\phi = \pi/2$, in that case $\cos(\phi)$ reaches the value of 1 at about 4 MeV, 32 turns, and the particle starts decelerating.

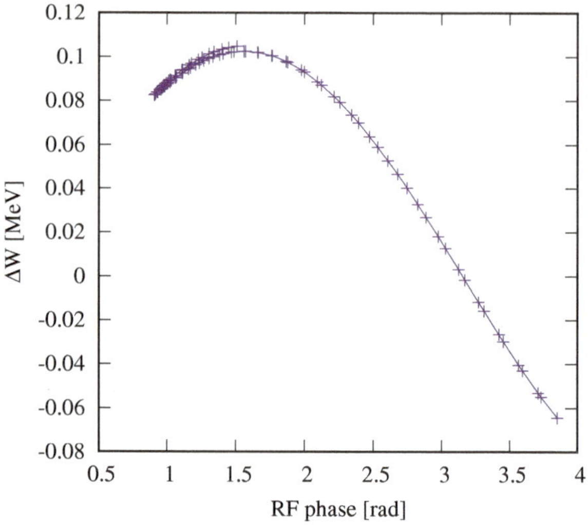

Fig. 3.45 Span in phase of the energy gain $\Delta W = q\hat{V}\sin\phi$ over the acceleration range 200 keV to 5 MeV. The vertical separation of the two $\Delta W(\phi)$ branches on the left ($\Delta\phi < 0$ above and $\Delta\phi > 0$ underneath) is artificial (a "−($6−50)/10000." "trick" in the gnuplot script of Table 3.27), for the sake of clarity—they actually superimpose

Fig. 3.46 Span in phase of the energy gain $\Delta W = q\hat{V}\sin\phi$ over an acceleration and deceleration cycle, starting from 200 keV. The vertical separation of $\Delta W(\phi)$ branches at the left and right ends is artificial

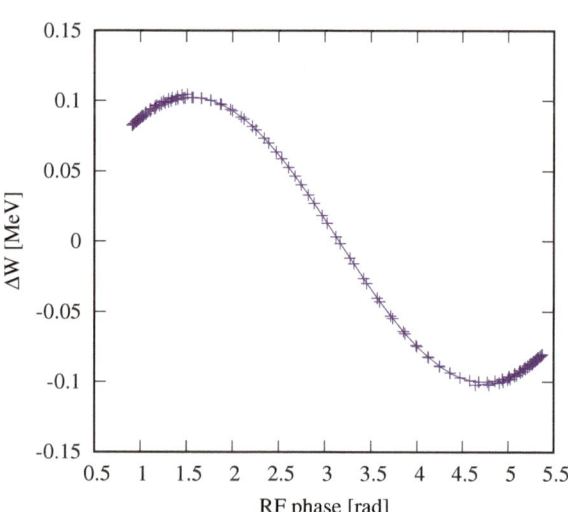

(b) Energy-phase relationship.

A graph of the energy-phase relationship obtained by ray tracing, for $\phi_i = \frac{3\pi}{4}$ and $\frac{\pi}{2}$ at the three different gap voltages $\hat{V} = 100,\ 200$ and 400 kV, is given in Fig. 3.49, together with theoretical expectations (Eq. 3.26).

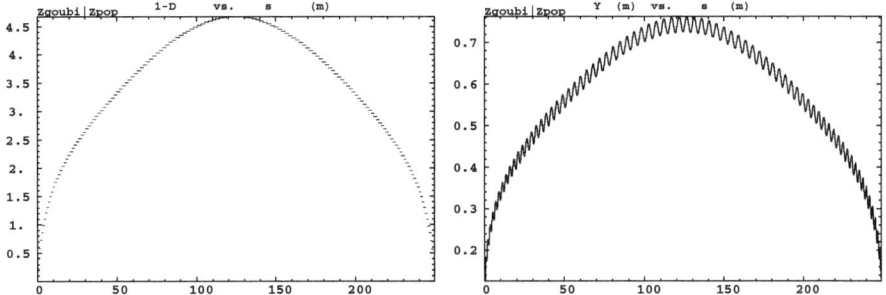

Fig. 3.47 Left: relative rigidity offset of the proton as a function of distance around the ring, accelerating over half the path, and subsequently decelerating back to the initial energy, under the effect of the cumulated phase-shift. Right: increase first and decrease next of the orbital radius as a function of azimuthal distance

3.11 Cyclotron Extraction

(a) Distance between turns.

Simulation input data of Exercise 3.3, Table 3.8, can be referred to as a guidance to build the present simulation file.

A proton is accelerated in 26 turns, in a uniform field $B_0 = 0.5$ T, from 20 keV (rigidity $BORO \times D = 0.064624444 \times 0.3162126 = 0.0204350634608$ Tm, injection radius $Y_0 = BR/B_0 = 4.08701269216$ cm) to 5.02 MeV. The RF phase is ignored thus CAVITE[IOPT = 3] is used, with a 100 kV gap voltage. The input data file for this simulation is given in Table 3.29.

The accelerated orbit and the distance ΔR between turns are displayed in Fig. 3.50. Theoretical expectation (Eq. 3.27) in the case of slow acceleration (typically, the fixed energy closed orbit configuration of Fig. 3.21) is also displayed, for comparison.

(b) Beam losses.

Indications to solve this exercise:
– a beam with Gaussian momentum distribution and *rms* momentum spread $\delta p/p = 10^{-3}$ can be defined using MCOBJET,
– use REBELOTE to accelerate over a given number of turns,
– an extraction septum placed half-way between two successive turns can be simulated using COLLIMA, placed after REBELOTE (the execution pointer will quietly continue beyond REBELOTE do-loop once the latter is completed). COLLIMA counts particles stopped. FAISTORE (or FAISCNL) can be placed after COLLIMA, to log particle data: particles stopped by COLLIMA have their IEX tag set to IEX = –4 [16, *lookup* COLLIMA].

Change the value of NPASS under REBELOTE for a different number of accelerated turns, and COLLIMA positioning data accordingly.

Table 3.28 Simulation input data file: the cyclotron equation (Eq. 3.26). This requires a uniform field, for that the [#S_60degSectorUnifB:#E_60degSectorUnifB] segment of Table 3.6 is INCLUDEd, here. Note the PRINT instruction under CAVITE: it causes a print out of CAVITE computational data in zgoubi.CAVITE.out, during the ray tracing, including RF phase and ion energy which can then be plotted (gnuplot script below, called by SYSTEM, and Fig. 3.48). The second particle under OBJET is launched on the closed orbit, its initial phase at the voltage gap is $\pi/2$. The first and third particles leave with an initial longitudinal shift $\delta s = \mp 10.26647$ cm at OBJET resulting in $\pi/4$ and $3\pi/4$ initial phase at the voltage gap

```
Cyclotron, classical. Acceleration to 6.02 MeV.
'MARKER'  ProbCycloEq_S                                    ! Just for edition purposes.
'OBJET'
64.62444403717985                                          ! Reference: 200keV (assuming proton).
2
3 1                                                        ! A single particle.
12.924888 0. 0. 0 -10.266476 1. '2' ! Path length offset +pi/4, initial phase at gap: phi_rf=pi/2-pi/4.
12.924888 0. 0. 0. 0.           1. '1'           ! Initial phase at gap is phi_rf=pi/2.
12.924888 0. 0. 0. 10.266476 1. '2' ! Path length offset +pi/4, initial phase at gap: phi_rf=pi/2+pi/4.
1 1 1
'PARTICUL'                                            ! Particle data are necessary as CAVITE is used,
PROTON                                                ! otherwise, by default zgoubi only requires rigidity.
'INCLUDE'
1
3 *./60degSector.inc[#S_60degSectorUnifB:#E_60degSectorUnifB]              ! Uniform field, no index.
'CAVITE'   GAP1
7 PRINT                                  ! PRINT: log CAVITE computational dat to zgoubi.CAVITE.out.
0.00  7.54220925334568e6                 ! f_rf= 1/T_rev, T_rev at about middle of acceleration range.
200e3    -1.57079632679                  ! Peak voltage;, relative phase of 1st cavity.
'INCLUDE'
1
3 *./60degSector.inc[#S_60degSectorUnifB:#E_60degSectorUnifB]              ! Uniform field, no index.
'CAVITE'   GAP1
7 PRINT                                  ! PRINT: log CAVITE computational dat to zgoubi.CAVITE.out.
0.00  7.54220925334568e6                 ! f_rf= 1/T_rev, T_rev at about middle of acceleration range.
200e3    +1.57079632679                  ! Peak voltage;, relative phase of 1st cavity.
'FAISTORE'                               ! Log particle coordinates at each turn.
zgoubi.fai
1
'REBELOTE'                     ! K = 99 : coordinates at end of previous pass are used as initial
135 1.1 99                     ! coordinates for the next pass ; idem for spin components.
'SYSTEM'
1                                                      ! 1 SYSTEM command follows.
/usr/bin/gnuplot < ./gnuplot_CAVITE.gnu &    ! Plot Ek versus phase, as read from zgoubi.CAVITE.out.
'MARKER'  ProbCycloEq_E                                 ! Just for edition purposes.
'END'
```

A gnuplot script to obtained Fig. 3.49:

```
# gnuplot_CAVITE.gnu
set xlabel "E_k  [MeV]" ; set ylabel "cos({/Symbol f})"; set xtics; set ytics mirror
pi = 4. * atan(1.); E0 = 938.2720813; qV=400e-3; Ei=0.2; E_km = 10 # locate max of cos(phi) at 10 MeV
omgR = 1. / (1. + E_km/E0); mxTurn=80
plot [0.2:20] [-1.1:1.8]  for [i=2:1:-1] \
'zgoubi.CAVITE.Out' u ($5==i && $6<mxTurn? $10 :1/0):(cos($11)) w p pt i+4 ps .5 notit ,\
cos(pi/2.)  +pi*(1.-omgR *(1.+(x+Ei)/(2*E0))) *(x-Ei)/(.5*qV) w l lw 2 lc rgb "blue" \
tit "V/gap=200kV, {/Symbol f}_0={/Symbol p}/2" ,\
cos(3*pi/4.)+pi*(1.-omgR *(1.+(x+Ei)/(2*E0))) *(x-Ei)/(.5*qV) w l lw 2 lc rgb "red"   \
tit "              {/Symbol f}_0={/Symbol 3p}/4" ,\
```

(c) Change the field index.

The cyclotron model of Table 3.21 is used here, reference field $B_0 = 5$ kG on the 200 keV orbit, and field index k = -0.03. A proton is accelerated over 26 turns, from 20 keV to 5.02 MeV, as in question (a). The 20 keV closed orbit radius (taken as the injection radius) differs from question (a) due to the index k = -0.03, and can be found using a FIT procedure (Table 3.30); it comes out to be $Y_0 = 4.0040586$ cm.

The input data file for this exercise is given in Table 3.31.

The resulting proton trajectory is displayed in Fig. 3.51 (the gnuplot script given in Table 3.31 is used). The greatly different accelerated orbit in this case, compared to the uniform field case in (a) (Fig. 3.50), results from a modulation of the distance between turns, which is an effect of the oscillation motion undergone by the accelerated orbit

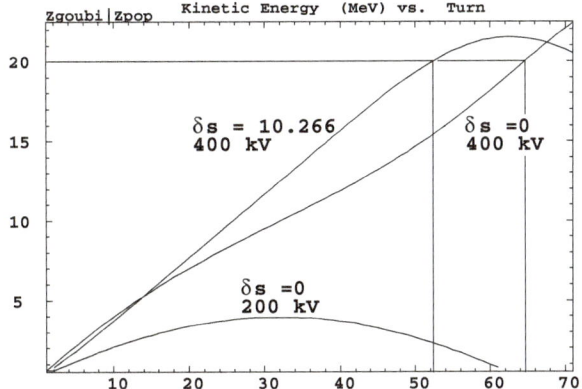

Fig. 3.48 Proton energy versus turn, case of (Table 3.28) voltage 400 kV/turn, two protons with initial phase respectively $\pi/2$ ($\delta s = 0$) and $3\pi/4$ ($\delta s = 10.26647$ cm), which make it up to 20 MeV and beyond. The third case, voltage 200 kV/turn, initial phase $\pi/2$ ($\delta s = 0$), features a maximum energy of 4 MeV and deceleration from there on. A graph obtained using zpop: menu 7; 1/5 to read from zgoubi.fai; 2/[39,2] for Y versus turn

Fig. 3.49 A graph of the energy dependence of the arrival phase at the voltage gap, for a few different values of gap voltage $\hat{V}$ and initial phase ϕ_i. Markers are from raytracing, using the input data file of Table 3.28 repeatedly for the various values of $\hat{V}$ and ϕ_i. Superimposed solid lines are from theory (Eq. 3.26 and Fig. 3.14)

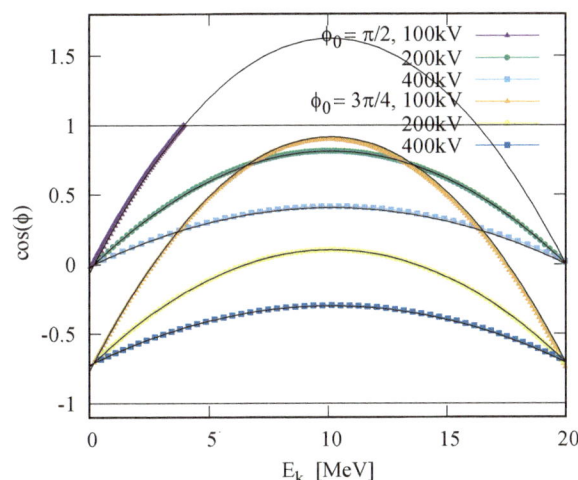

(around the local on-momentum half-circle orbit arc). This effect may be exploited to increase extraction efficiency, by causing such a radial modulation as to maximize turn separation at the location of the septum [17].

(d) Optimize extraction.

The modulation is minimized (or enhanced possibly, at the last turn, for minimized losses at extraction) by optimizing the injection conditions (x_0, x_0').

Table 3.29 Simulation input data file: accelerating a proton to check evolution of $\Delta R / R$, in a dipole field with index. The #S_180degSectorUnifB to #E_180degSectorUnifB segment of Table 3.6 is INCLUDEd

```
Cyclotron extraction. Uniform field.
'MARKER'  ProbdRRUnifB_S                                      ! Just for edition purposes.
 'OBJET'
64.62444403717985                                            ! Reference: 200keV (assuming proton).
2
1 1                                                          ! A single particle.
4.08701 0. 0. 0. 0. 0.3162126 'o'       ! p[MeV/c]= 6.126277, Brho[kG.cm]=20.435, kin-E[MeV]=0.02.
1
! 4.003593 0. 0. 0. 0. 0.3162126 'o'   ! Brho[kG.cm]=20.435, kin-E[MeV]=0.02, case of field with index.
'PARTICUL'                                            ! Particle data are necessary as CAVITE is used,
PROTON                                                ! otherwise, by default zgoubi only requires rigidity.
'INCLUDE'
1
./180degSector.inc[#S_180degSectorUnifB:#E_180degSectorUnifB]    ! one 180 deg sector, uniform field.

'FAISTORE'
zgoubi.fai                                                   ! Log current particle coordinates, in zgoubi.fai.
1

'CAVITE'  cavity                                            ! Accelerating gap.
3                                          ! In this option, dW = qVsin(phi_s), independent of time.
0. 0.
100e3  1.57079632679
'INCLUDE'
1
./180degSector.inc[#S_180degSectorUnifB:#E_180degSectorUnifB]    ! one 180 deg sector, uniform field.

'FAISCEAU'                                                  ! Particle coordinates before gap.
'CAVITE'  cavity                                            ! Accelerating gap.
3                                          ! In this option, dW = qVsin(phi_s), independent of time.
0. 0.
100e3  1.57079632679

'REBELOTE'                        ! K = 99 : coordinates at end of previous pass are used as initial
25  1.1 99                        ! coordinates for the next pass ; idem for spin components.
'FAISCEAU'                                                  ! Local particle coordinates logged in zgoubi.res.
'SYSTEM'
2                                                           ! 2 SYSTEM commands follow.
/usr/bin/gnuplot < ./gnuplot_Zplt_UnifB.gnu &               ! Plot accelerated orbits.
/usr/bin/gnuplot < ./gnuplot_Zfai_dRR.gnu &                 ! Plot delta_R(R).
'MARKER'  ProbdRRUnifB_E                                     ! Just for edition purposes.
'END'
```

A gnuplot script to obtain the accelerated orbit of Fig. 3.50:

```
# gnuplot_Zplt_UnifB.gnu
set xtics ; set ytics ; set xlabel "X_{Lab} [m]" ; set ylabel "Y_{Lab} [m]"
set size ratio 1 ; set polar ; cm2m = 0.01 ; pi = 4.*atan(1.)
set arrow from 0, 0 to 0.7, 0 nohead linecolor "red" lw 6; set arrow from 0, 0 to -0.65, 0 nohead linecolor "blue" lw 6
noel_1=5 ; noel_2=10  # 1st DIPOLE is element $42=noel_1; 4th DIPOLE is $42=noel_2. $42=column number in zgoubi.plt.
plot "zgoubi.plt" u ($42< noel_2? $22   +pi/3.*(($42-noel_1)/2) :1/0):($10 *cm2m) w p pt 5 ps .2 lc rgb "black"  notit ,\
     "zgoubi.plt" u ($42>=noel_2? $22+pi+pi/3.*(($42-noel_2)/2) :1/0):($10 *cm2m) w p pt 5 ps .2 lc rgb "black"  notit; pause 1
```

A gnuplot script to obtain the turn separation curves of Fig. 3.50. *In this script, zgoubi.fai2 is a copy of zgoubi.fai (see exercise3.3) in which the first particle data line (particle data at the first pass) has been removed. This allows drawing the difference ΔR between two successive passes, using the "paste" command (see Tab.* 3.8 *for a similar* 1-*row shift using awk commands):*

```
# gnuplot_Zfai_dRR.gnu
set xtics; set ytics mirror; set key maxrow 2 ; set xlabel "R [cm]" ; set ylabel "{/Symbol D}R [cm]"
set key r c; set logscale y; unset colorbox
plot [8:65] "<paste zgoubi.fai2 zgoubi.fai" u ($10):($10-$63) w p pt 7 ps 1.5 lc rgb "black"  tit "num." ,\
     "zgoubi.fai2" u ($10):($10/2./($38-1)) w l lc rgb "red"  tit "theory" ; pause 1
```

3.12 Acceleration and Extraction of a 6-D Polarized Bunch

This simulation can be set up using material drawn from previous exercises. It is not fully developed here, guidelines are given.

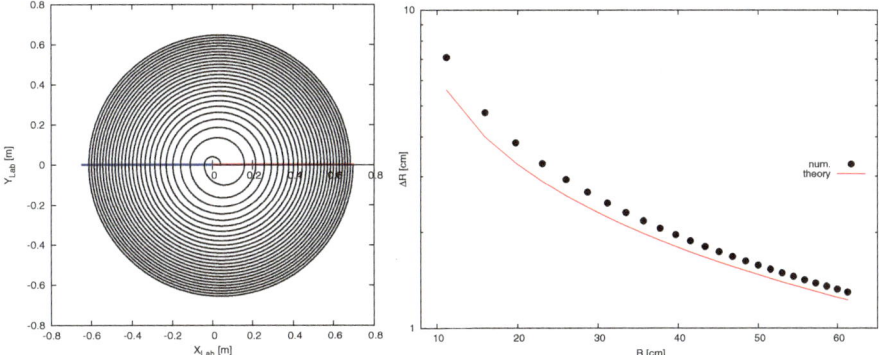

Fig. 3.50 Left: accelerated orbit from 20 keV to 5.02 MeV, at a rate of 200 keV per turn over 26 turns, in a uniform field. The thick horizontal line (colored) figures the accelerating gap. Right: the resulting dependence of orbit separation ΔR on radius, from raytracing (markers) and from theory (solid line); the theoretical curve assumes small dE (adiabatic acceleration, concentric orbits), which is not quite the case here with $\Delta E = 200$ keV/turn

Table 3.30 Simulation input data file: finding the 20 keV injection radius in the presence of a non-zero index k, using FIT The INCLUDE segment is taken from Table 3.21

```
Cyclotron, classical. Find injection radius for k=-0.03
'OBJET'
64.62444403717985                                           ! Reference: 200keV (assuming proton).
2
1 1                                                         ! A single particle.
4.0040586 0. 0. 0. 0. 0.3162126 'o'    ! p[MeV/c]= 6.126277, Brho[kG.cm]=20.435, kin-E[MeV]=0.02.
1
! 4.003593 0. 0. 0. 0. 0.3162126 'o' ! Brho[kG.cm]=20.435, kin-E[MeV]=0.02, case of field with index.
'PARTICUL'
PROTON
'INCLUDE'
1
./60DegSectorR200.inc[#S_60DegSectorR200:#E_60DegSectorR200]      ! One 60 deg sectors with index.

'FIT'
1                                                           ! 1 variable:
1 30 0 1.                    ! variable is Y0 (parameter 30) under OBJET (keyword 1 in the sequence).
2                                                           ! 2 constraints:
3.1 1 2 #End 0. 1. 0                                   ! constraint 1: final Y = Y0.
3    1 3 #End 0. 1. 0                                  ! constraint 2: final T = 0.
'END'
```

The cyclotron simulation hypotheses of Exercise 3.10a are considered, the input data file for this exercise can be built from that of Table 3.28, with a few modifications, namely:
– downstream of REBELOTE, add a 1 meter DRIFT: an embryo of an "high energy line" into which the bunch is steered at extraction;
– that DRIFT is preceded by CHANGREF to center the current reference frame on the final coordinates Y and T of the accelerated orbit; the latter have to be determined by prior raytracing;
– add histograms (to be logged in zgoubi.res) for observation of transverse and longitudinal particle coordinate densities in the bunch at extraction. This uses HISTO, as many times as needed.

Table 3.31 Simulation input data file: accelerating a proton to check evolution of $\Delta R/R$, in a dipole field with index. The [#S_60DegSectorR200:#E_60DegSectorR200] segment of Table 3.21 is INCLUDEd

```
Cyclotron extraction. Field with index.
'MARKER'  ProbdRRIdx_S                                      ! Just for edition purposes.
 'OBJET'
64.62444403717985                                   ! Reference: 200keV (assuming proton).
2
1 1                                                                ! A single particle.
4.0040586 0. 0. 0. 0. 0.3162126 'o'        ! p[MeV/c]= 6.126277, Brho[kG.cm]=20.435, kin-E[MeV]=0.02.
1
! 4.003593 0. 0. 0. 0. 0.3162126 'o'     ! Brho[kG.cm]=20.435, kin-E[MeV]=0.02, case of field with index.
'PARTICUL'                                           ! Particle data are necessary as CAVITE is used,
PROTON                                               ! otherwise, by default zgoubi only requires rigidity.
'INCLUDE'
1
3* ./60DegSectorR200.inc[#S_60DegSectorR200:#E_60DegSectorR200]        ! Three 60 deg sectors with index.

'FAISTORE'
zgoubi.fai                                           ! Log current particle coordinates, in zgoubi.fai.
1

'CAVITE'  cavity                                                   ! Accelerating gap.
3                                         ! In this option, dW = qVsin(phi_s), independent of time.
0. 0.
100e3  1.57079632679
'INCLUDE'
1
3* ./60DegSectorR200.inc[#S_60DegSectorR200:#E_60DegSectorR200]        ! Three 60 deg sectors with index.

'FAISCEAU'                                           ! Particle coordinates before gap.
'CAVITE'  cavity                                                   ! Accelerating gap.
3                                         ! In this option, dW = qVsin(phi_s), independent of time.
0. 0.
100e3  1.57079632679

'REBELOTE'                     ! K = 99 : coordinates at end of previous pass are used as initial
25 1.1 99                      ! coordinates for the next pass ; idem for spin components.
'FAISCEAU'                                           ! Local particle coordinates logged in zgoubi.res.
'SYSTEM'
2                                                                  ! 2 SYSTEM commands follow.
/usr/bin/gnuplot < ./gnuplot_Zplt.gnu &                            ! Plot accelerated orbits.
/usr/bin/gnuplot < ./gnuplot_Zfai_DR.gnu &                         ! Plot delta_R(R).
'MARKER'  ProbdRRIdx_E                                             ! Just for edition purposes.
'END'
```

A gnuplot script to obtain the accelerated orbit of Fig. 3.51:

```
# gnuplot_Zplt.gnu
set xtics ; set ytics ; set xlabel "X_{Lab} [m]" ; set ylabel "Y_{Lab} [m]"
set size ratio 1 ; set polar ; cm2m = 0.01 ; pi = 4.*atan(1.)
set arrow from 0, 0 to  0.8, 0 nohead linecolor "red" lw 6; set arrow from 0, 0 to -0.85, 0 nohead linecolor "blue" lw 6
noel_1=4 ; noel_2=12  # 1st DIPOLE is element $42=noel_1; 4th DIPOLE is $42=noel_2. $42=column number in zgoubi.plt.
plot "zgoubi.plt" u ($42< noel_2? $22   +pi/3.*(($42-noel_1)/2) :1/0):($10 *cm2m) w p pt 5 ps .2 lc rgb "black"  notit ,\
     "zgoubi.plt" u ($42>=noel_2? $22+pi+pi/3.*(($42-noel_2)/2) :1/0):($10 *cm2m) w p pt 5 ps .2 lc rgb "black"  notit; pause 1
```

A gnuplot script to obtain the turn separation curves of Fig. 3.51. *In this script, zgoubi.fai2 is a copy of zgoubi.fai in which the first particle data line (particle data at the first pass) has been removed. This allows drawing the difference ΔR between two successive passes, using the "paste" command - see Tab.* 3.8 *for a similar 1-row shift using awk commands:*

```
# gnuplot_Zfai_DR.gnu
set xtics; set ytics mirror; set key maxrow 2 ; set xlabel "R [cm]" ; set ylabel "{/Symbol D}R [cm]"
set key r c; set logscale y; unset colorbox
plot "<paste zgoubi.fai2 zgoubi.fai" u ($10):($10-$63):($10) w p pt 7 ps 1.5 lw .1 lc palette  notit ; pause 1
```

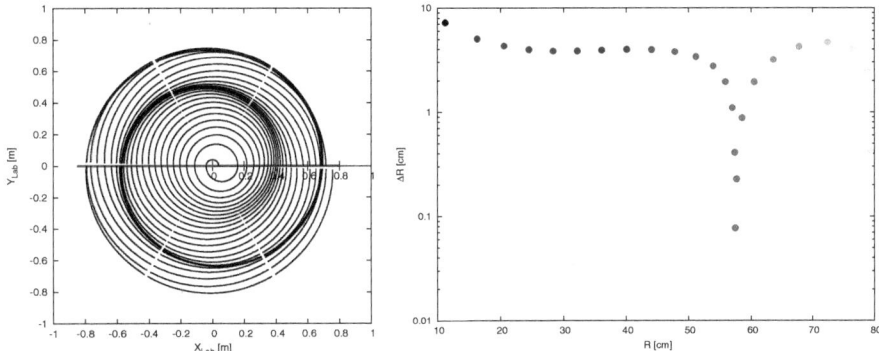

Fig. 3.51 Left: accelerated orbit from 20 keV to 5.02 MeV, at a rate of 200 keV per turn over 26 turns, in a dipole field with index. The thick horizontal line (colored) figures the accelerating gap. Right: the resulting dependence of orbit separation ΔR on radius, observed at the second gap

References

1. A. Sessler, E. Wilson, *Engines of Discovery. A Century of Particle Accelerators* (World Scientific, 2007)
2. E.O. Lawrence, M.S. Livingston, Phys. Rev. **37**, 1707 (1931); Phys. Rev. **38**, 136, (1931); Phys. Rev. **40**, 19 (1932)
3. Credit: Lawrence Berkeley National Laboratory. The Regents of the University of California, Lawrence Berkeley National Laboratory
4. E.O. Lawrence, M.S. Livingston, The production of high speed light ions without the use of high voltages. Phys. Rev. **40**, 19–35 (1932)
5. M.S. Livingston, E.M. McMillan, History of the cyclotron. Phys. Today **12**(10) (1959). https://escholarship.org/uc/item/29c6p35w
6. H.E. Bethe, M.E. Rose, Maximum energy obtainable from cyclotron. Phys. Rev. **52**, 1254 (1937)
7. F.T. Cole, *O Camelot ! A Memoir of the MURA Years* (1994). https://accelconf.web.cern.ch/c01/cyc2001/extra/Cole.pdf
8. L.H. Thomas, the paths of ions in the cyclotron. Phys. Rev. **54**, 580, (1938); M.K. Craddock, AG focusing in the thomas cyclotron of 1938, in *Proceedings of PAC09* (Vancouver, BC, Canada, FR5REP1)
9. T. Stammbach, Introduction to cyclotrons. CERN accelerator school, cyclotrons, linacs and their applications. IBM International Education Centre, La Hulpe, Belgium, 28 April-5 May 1994. Fig. 8; T. Stammbach, Introduction to Cyclotrons. CERN Yellow Report 96-02 (1996), Figure 8, page 15. Copyright/License CERN CC-BY-3.0 https://creativecommons.org/licenses/by/3.0, no change to the material
10. E. Baron et al., The GANIL injector, in *Proceedings of the 7th International Conference on Cyclotrons and their Applications* (Zürich, Switzerland, 1975). http://accelconf.web.cern.ch/c75/papers/b-05.pdf
11. L.B. Cohen, Cyclotrons and synchrocyclotrons, in *Encyclopedia of Physics, Vol. XLIV, Nuclear Instrumentation I*, ed. by S. Flügge (Springer, 1959)
12. J. Le Duff, *Longitudinal Beam Dynamics in Circular Accelerators* (CERN Accelerator School, Jyvaskyla, Finland, 1992)
13. B.W. Montague, Polarized beams in high energy storage rings. Phys. Rep. (Rev. Sect. Phys. Lett.) **113**(1), 1–96 (1984)

14. T. Roser, A. Zelensky, Private communication, BNL, June 2021. Günther Clausnitzer: History of Polarized Ion Source Developments, in *International Workshop on Polarized Ion Sources and Polarized Gas Jets* (KEK, Tsukuba, Japan, 1990). KEK Report 90–15, November 1990, ed. by Y. MORI. https://inis.iaea.org/collection/NCLCollectionStore/_Public/22/051/22051667.pdf
15. F. Méot, Spin dynamics, in *Polarized Beam Dynamics and Instrumentation in Particle Accelerators, USPAS Summer 2021 Spin Class Lectures* (Springer Nature, Open Access, 2023). https://link.springer.com/book/10.1007/978-3-031-16715-7
16. F. Méot, Zgoubi Users' Guide. https://www.osti.gov/biblio/1062013-zgoubi-users-guide. Sourceforge latest version: https://sourceforge.net/p/zgoubi/code/HEAD/tree/trunk/guide/Zgoubi.pdf
17. T. Stammbach, Introduction to Cyclotrons, in *CERN Accelerator School, Cyclotrons, Linacs and Their Applications* (IBM International Education Centre, La Hulpe, Belgium, 1994)

Chapter 4
Relativistic Cyclotron

Abstract This chapter introduces the AVF (azimuthally varying field), isochronous, relativistic cyclotron, and to the theoretical material needed for the simulation exercises. A brief reminder of the historical context is followed by further basic theoretical considerations leaning on the cyclotron concepts introduced in Chap. 3 and including

– Thomas focusing and the AVF cyclotron,
– positive focusing index,
– isochronous optics,
– separated sector cyclotrons,
– spin dynamics in an AVF cyclotron.

Simulation exercises use optical elements and keywords met earlier: the analytical field modeling DIPOLE, TOSCA in case using a field map is preferred, CAVITE to accelerate, SPNTRK to solve spin motion, FAISCEAU, FAISTORE, FIT, etc. The exercises further develop on radial and spiral sector magnets, edge focusing and flutter, isochronous optics, separated sector ring cyclotrons, and their modeling in DIPOLE, DIPOLES and other CYCLOTRON keyword capabilities.

Notations Used in the Text

B; B_0	magnetic field; at a reference radius R_0
$\mathbf{B}$; B_R; B_θ; B_y	field vector; radial, azimuthal and axial components
$BR = p/q$; BR_0	magnetic rigidity; reference rigidity
C; C_0	closed orbit length, $C = 2\pi R$; reference, $C_0 = 2\pi R_0$
E	ion energy, $E = \gamma m_0 c^2$
EFB	effective Field Boundary
$\mathcal{F}$; F	azimuthal field form factor; flutter, $F = \left(\frac{<(\mathcal{F}-<\mathcal{F}>)^2>}{<\mathcal{F}>^2} \right)^{1/2}$
$f_{\text{rev}}, f_{\text{rf}}$	revolution and RF voltage frequencies
h	harmonic number, an integer, $h = f_{\text{rf}}/f_{\text{rev}}$
$k = \frac{R}{B}\frac{dB}{dR}$	geometric index, a global quantity
m; m_0; M	ion mass; rest mass; in units of MeV/c^2

© The Author(s) 2024

F. Méot, *Understanding the Physics of Particle Accelerators*, Particle Acceleration and Detection, https://doi.org/10.1007/978-3-031-59979-8_4

$n = \frac{\rho}{B}\frac{dB}{d\rho}$	focusing index, a local quantity
$\mathbf{p}$; p_0	ion momentum vector; reference momentum
q	ion charge
R; R_0; R_E	average radius of equilibrium orbit; $R = C/2\pi$; $R(p = p_0)$; $R(E)$
$\mathcal{R}$	radial field form factor
RF	Radio-Frequency
s	path variable
T_{rev}, T_{rf}	revolution and accelerating voltage periods
v	ion velocity
$V(t)$; $\hat{V}$	oscillating voltage; its peak value
x', y'	radial and axial coordinates $\left[(*)' = \frac{d(*)}{ds} \right]$
α	trajectory deviation, or momentum compaction
$\beta = v/c$; β_0; β_s	normalized ion velocity; reference; synchronous
$\gamma = E/m_0c^2$	Lorentz relativistic factor
Δp, δp	momentum offset
ε	wedge angle
ϵ_R	strength of a depolarizing resonance
ε_u	Courant-Snyder invariant (u : x, y, l), ...)
ζ	spiral angle of a spiral sector dipole EFB
θ	azimuthal angle
ϕ; ϕ_s	phase of oscillating voltage; synchronous phase

4.1 Introduction

Isochronous cyclotrons are in operation today by the thousands, tens are produced each year. Applications include production of radio-isotopes mostly, proton therapy (Fig. 4.1), high power beams for accelerator-driven systems, secondary particle beam production (Fig. 4.2), and more [1]. The technology and its applications are fostered by cryogeny and high fields which further allow compactness (Fig. 4.1) as well as highest beam rigidities (Fig. 4.3).

At the origin of the evolution of the cyclotron technology, which led to the AVF innovation in the late 1930s, is the energy limitation of the classical cyclotron, at a few tens of MeV/nucleon (Chap. 3). Axial focusing in the latter results from the slow decrease of the guiding field with radius in the wide gap between the electromagnet poles. That negative field index $-1 < k < 0$ (Eqs. 3.11, 3.12) results in both radial and axial periodic stability (Eq. 3.18). Isochronism requires instead the field to increase with radius, i.e. a field index $k > 0$, a consequence of $B(R) \propto \gamma(R)$ (Sect. 4.2). The AVF concept by L.H. Thomas in 1938[1] [5] (Fig. 4.4), solved the problem: AVF entails axial periodic stability as long as the field modulation parameter

---

[1] The very L.H. Thomas of the Thomas-BMT spin motion differential equation, author of the eight years earlier Nature article [7].

Fig. 4.1 COMET protontherapy cyclotron at PSI. A 250 MeV, 500 nA, 4-sector isochronous AVF cyclotron. The spiral poles enhance axial focusing. A 3 m diameter superconducting coil provides the dipole field [2]

$F > \beta\gamma$ (Sect. 4.2.1). The vertical defocusing effect which the radially increasing field causes is compensated by the focusing effect of the AVF. Spiral pole geometry was further introduced in 1954 [8] to increase axial focusing, so allowing greater k and isochronous acceleration to higher energy (Sect. 4.2.2). It took some time, until the late 1950s (see Stammbach's Fig. 3.4), for Thomas' concept to make its way

Fig. 4.2 PSI 590 MeV ring cyclotron delivers a 1.4 MW proton beam. Acceleration takes ∼180 turns; extraction efficiency is >99.99%; overall diameter is 15 m. Beam is used for the production of secondary neutron and muon beams [3]

and lead up to practical realisations[2] [9, 10]. AVF cyclotrons were constructed to accelerate all sorts of ions whereas classical cyclotrons tended to leave the scene (Fig. 3.4). Applications included material science, radiobiology, production of secondary beams, and more. Polarized ion beams became part of the landscape as well from the moment polarized ion sources were made available [11].

The separated sector method was developed in the early 1960s, instances are today's high power PSI 590 MeV spiral sector cyclotron (Fig. 4.2), brought into operation in 1974, and its injector-II, a radial-sector design (Fig. 4.5). Iron-free regions between separated sector dipoles allows room for multiple high-Q RF resonators thus greater turn separation at extraction, for higher efficiency extraction systems and

---

[2] One can read for instance, in 1959s Ref. [10], Cyclotrons and Synchrocyclotrons, regarding engineering aspects, *"Also, no consideration is given to the AVF cyclotron, since none of this type has reached the advanced design stage"*.

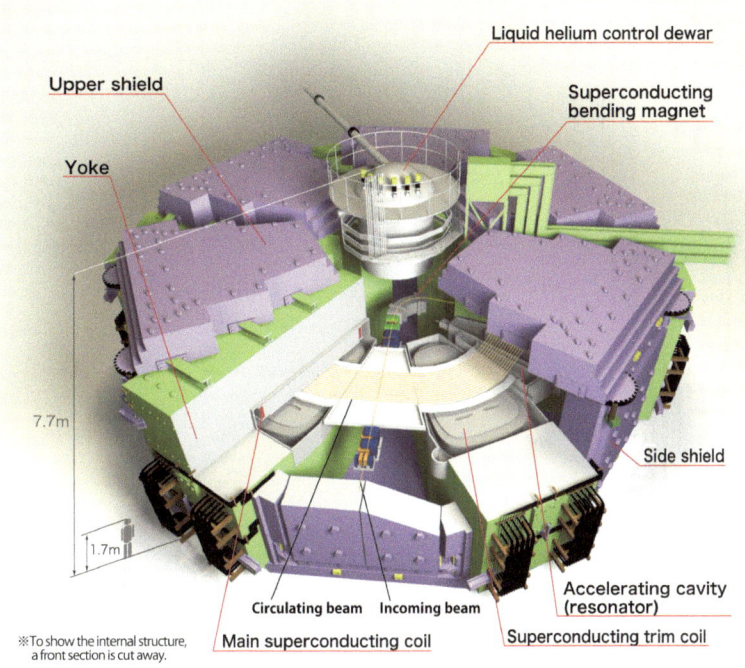

Fig. 4.3 RIKEN K2500, superconducting coil, separated-sector, 8,300 ton ring cyclotron [4]. The dipole field is 3.8 T, rigidity 8 T m, diameter 18.4 m. Beam injection radius is 3.56 m, extraction radius is 5.36 m. The cyclotron is part of a radioactive ion beam accelerator complex

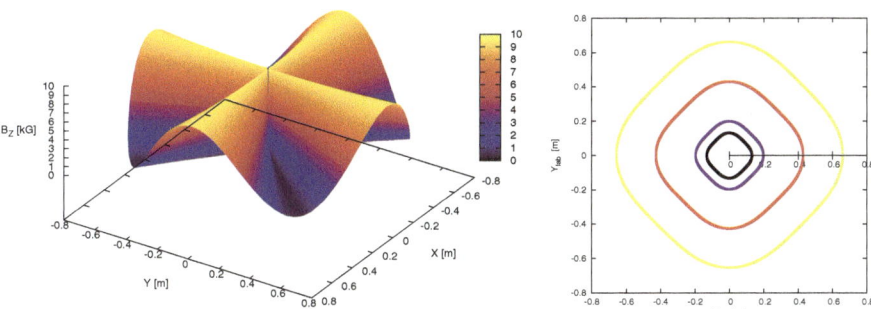

Fig. 4.4 A 4-periodic AVF cyclotron design (after Ref. [5]). Left: mid-plane azimuthally modulated field. Right: closed orbits around the cyclotron feature azimuthally varying curvature, greater on the hills, weaker in the field valleys

Fig. 4.5 PSI injector II, four separated radial sectors, 0.87 MeV injection energy, accelerates protons to 72 MeV in about 100 turns [6]. The drifts include the 50.7 MHz accelerating RF system and a flattop cavity. Injection is from the top, in the central region

thus higher beam current, and for the insertion of beam instrumentation. Cyclotron energy subsequently increased, up to the present days near-GeV range. Cryogeny was introduced in the early 1960s at the Michigan State University superconducting coil K500 cyclotron[3] [12]. Superconducting technology allows higher field and reduction of size, culminating today with RIKEN's K2500 SRC (Fig. 4.3 and Table 4.1).

4.2 Basic Concepts and Formulæ

Mass increase with energy causes loss of synchronism in the classical cyclotron, and the required negative field index (decreasing guiding field with radius) for axial periodic motion stability adds to the effect. Isochronism instead, i.e., constant $\omega_{\text{rev}} = qB/\gamma m_0$, given orbit radius $R = \beta c/\omega_{\text{rev}}$, leads to positive index

$$k = \frac{R}{B}\frac{\partial B}{\partial R} = \frac{\beta}{\gamma}\frac{\partial \gamma}{\partial \beta} = \beta^2 \gamma^2 \qquad (4.1)$$

---

[3] $K = EA/Z^2$, with A the number of mass, Z the number of charge, is a measure of the equivalent proton energy, 500 MeV in this case.

Table 4.1   A comparison between an AVF and a separated sector cyclotron of same energy, 72 MeV, namely, the former AVF injector and the present Injector II of PSI high power cyclotron, after Ref. [11, p. 126]

		AVF	Separated sector
Injection energy	keV	14	870
Extraction energy	MeV	72	72
Beam current	mA	0.2	1.6
Magnet		single dipole	4 sectors
Weight	ton	470	4×180
Dipole gap height	mm	240–450	35
$\langle B \rangle$; B_{max}	T	1.6; 2	0.36; 1.1
RF system		180° dees	2 resonators
Accelerating voltage	kV	2×70	4×250
RF	MHz	50	50
Normalized beam emittance, hor.; vert.	$\pi\,\mu$m	2.4; 1.2	1.2; 1.2
Beam phase width	deg	16–40	12
Energy spread	%	0.3	0.2
Turn separation at extraction	mm	3	18

requiring k to follow the energy increase: the weak focusing condition $-1 < k < 0$ can not be satisfied, transverse periodic stability is lost.

Isochronism requires the revolution period $T_{rev} = 2\pi\gamma m_0/qB$ to be momentum independent; under this condition, differentiating this expression yields the radial field dependence

$$B(R) = \frac{B_0}{\gamma_0}\gamma(R) \tag{4.2}$$

with B_0 and γ_0 some reference conditions,

This led H.A. Bethe and M.E. Rose to conclude, in 1938, *"... it seems useless to build cyclotrons of larger proportions than the existing ones... an accelerating chamber of 37 cm radius will suffice to produce deuterons of 11 MeV energy which is the highest possible..."* [13]. And F.T. Cole to comment, *"If you went to graduate school in the 1940s, this inequality [$-1 < r(dB/dr)/B < 0$] was the end of the discussion of accelerator theory"* [14, Sect. 1.4].

4.2.1   Thomas Focusing

Whereas the classical cyclotron approach assumed revolution symmetry of the field, a 1938 publication stated: *"[...] a variation of the magnetic field with angle, [...] of*

Fig. 4.6 Pole shaping in an AVF cyclotron, an electron model, here [15]. The focusing pattern is FfFfFf, an alternation of strong (hill regions) and weak (valleys) radial focusing [16]

order of magnitude v/c; together with nearly the radial increase of relative amount $\frac{1}{2}v^2/c^2$ of Bethe and Rose; gives stable orbits that are in resonance and not defocused" [5]. In other words, AVF in proper amount (Fig. 4.4) compensates the axial defocusing resulting from the increase of the field with radius (Eq. 4.2). Azimuthal field modulation and radial increase may be obtained by shaping the magnet poles, as illustrated in Fig. 4.6.

Azimuthal Field Modulation, Flutter

A simple approach to the $2\pi/N$-periodic axial symmetry and field modulation may assume a sinusoidal azimuthal form factor

$$\mathcal{F}(\theta) = 1 + f \sin(N\theta) \tag{4.3}$$

This is the case in Fig. 4.4, for instance. The mid-plane field can thus be expressed under the form

$$B(R, \theta) = B_0 \, \mathcal{R}(R) \, \mathcal{F}(\theta) \tag{4.4}$$

with $\mathcal{R}(R)$ the radial dependence. The orbit curvature varies along the $\frac{2\pi}{N}$-periodic orbit, this requires distinguishing between the local focusing index $n = \frac{\rho(s)}{B(s)}\frac{dB}{d\rho}$ and the geometrical index k (Eq. 4.1), a global quantity which determines the wave numbers (Eq. 4.6). A "flutter" factor can be introduced to quantify the effect of the azimuthal modulation of the field on the focusing,

$$F = \left(\frac{< (\mathcal{F} - <\mathcal{F}>)^2 >}{<\mathcal{F}>^2}\right)^{1/2} \overset{\substack{hard \\ edge}}{\longrightarrow} \left(\frac{R}{\rho} - 1\right)^{1/2} \tag{4.5}$$

where $< * > = \oint (*) \, d\theta/2\pi$. If the scalloping of the orbit (i.e., its excursion in the vicinity of R) is of small amplitude, then $R \approx \rho$ and, accounting for the isochronism condition (Eq. 4.1), approximate values of the wave numbers write

$$\nu_R \approx \sqrt{1 + k} \overset{isochr.}{=} \gamma, \qquad \nu_y \approx \sqrt{-k + F^2} \overset{isochr.}{=} \sqrt{-\beta^2 \gamma^2 + F^2} \tag{4.6}$$

Thus the horizontal wave number increases during acceleration, linearly with energy, whereas in the absence of countermeasure the axial wave number would decrease - see Sect. 4.2.2. An additional property is

$$\nu_R^2 + \nu_y^2 = 1 + F^2 \overset{\substack{hard \\ edge}}{\longrightarrow} \frac{R}{\rho} \tag{4.7}$$

The flutter allows designing $-k + F^2 > 0$ (whereas $k > 0$), so ensuring periodic stability of the axial motion. In the hypothesis of a sinusoidal azimuthal field modulation (Eq. 4.3) one has $F = f/\sqrt{2}$ and

$$\nu_y \approx \sqrt{-k + f^2/2}, \qquad \nu_R^2 + \nu_y^2 = 1 + f^2/2 \tag{4.8}$$

Off-Momentum Orbit

The dispersion function $D = \delta x/\delta p/p$ in the revolution symmetry field (Eq. 3.20) has the form $D = \frac{R_0}{1+k}$. Given the isochronous condition $k = \beta^2 \gamma^2$ it can be written

$$D = \frac{R}{\gamma^2} \tag{4.9}$$

An alternate approach consists in considering that, with the isochronism condition $2\pi R/\beta c = T_{\text{rev}}$, a constant, one gets $\frac{dR}{R} = \frac{d\beta}{\beta} = \frac{1}{\gamma^2}\frac{dp}{p}$.

AVF Modeling

A numerical approach to the azimuthal modulation beyond the simple sine modulation of Eq. 4.3, is discussed in Sect. 14.3.3 (Eqs. 14.11, 14.15). It provides a modeling of $\mathcal{F}(\theta)$ over the whole beam excursion area, possibly including an R-dependence,

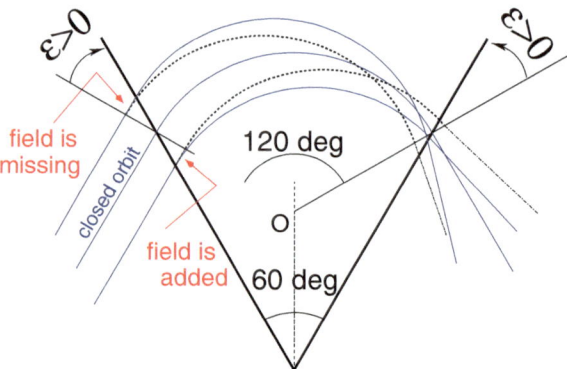

Fig. 4.7 A 120 deg bending of the closed orbit (curvature center at O) is ensured by a 60 deg bending sector. This results in a wedge angle ($\varepsilon > 0$ by convention in this configuration) in the transition regions between valleys and hills, which causes a decrease of the radial focusing (solid incoming trajectories, compared to dotted ones), and axial focusing under the effect of the trajectory angle to the azimuthal field component

$\mathcal{F}(R, \theta)$. The method ensures the continuity of $\mathcal{F}(R, \theta)$ and its derivatives, between neighboring magnetic sectors. It is resorted to in the simulation exercises.

Wedge Focusing

In the entrance and exit regions of a bending sector, closed orbits are at an angle to the iso-field lines, this causes "wedge focusing", an effect sketched in Fig. 4.7: with positive wedge angle ε, case of the AVF configuration, radial focusing decreases whereas the angle of off mid-plane particle velocity vector to the azimuthal component of the field in the wedge region causes axial focusing.

4.2.2 Spiral Sector

Spiral sector geometry was introduced in 1954 in the context of fixed field alternating gradient accelerator (FFAG) studies [8], and found application in cyclotrons (as in PSI's COMET cyclotron, Fig. 4.1). Spiraling the edges (Fig. 4.8) results in stronger axial focusing (Eq. 4.12) compared to a radial sector (Eq. 4.6), it also permits an increase of the wedge angle with radius, so maintaining proper compensation of an increase of $k(R)$ (Eq. 4.1). In a spiral sector bend the wedge angle is positive on one side of the sector, negative on the other side (Fig. 4.8), with a global axial focusing resultant. In a similar approach to the periodic field modulation in a radial sector (Eq. 4.3), a convenient approach to the spiral sector AVF uses azimuthal form factor

$$\mathcal{F}(R, \theta) = 1 + f \sin\left[N\left(\theta - \tan(\zeta(R)) \ln \frac{R}{R_0}\right)\right] \qquad (4.10)$$

Fig. 4.8 Geometrical parameters of a spiral sector dipole. The center of the ring is at O, ζ is the spiral angle (increasing with radius), ε is the wedge angle. In the hard edge field model, a line of constant field inside the sector is an arc of radius R; thus the curvature radius ρ varies along the closed orbit in the dipole

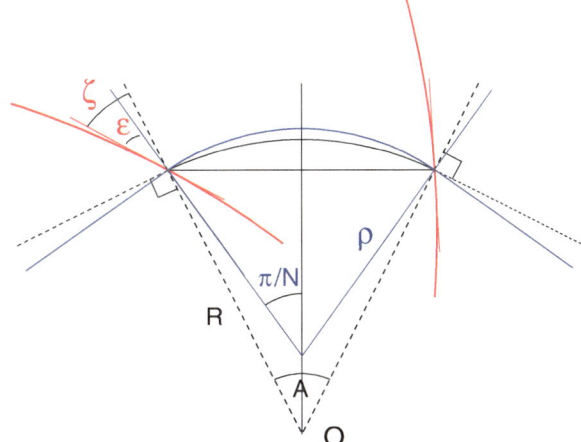

with the spiral angle $\zeta(R)$ an increasing function of radius R, whereas the mid-plane field now writes under the form

$$B(R, \theta) = B_0\, \mathcal{R}(R)\, \mathcal{F}(R, \theta) \tag{4.11}$$

The local magnet edge geometry at R satisfies $r = r_0 \exp(\theta/\tan(\zeta))$, a logarithmic spiral centered at the center of the ring, with ζ the angle between the tangent to the spiral edge and the ring radius (Fig. 4.8). This results in a larger contribution of the flutter term in the axial wave number,

$$\nu_y \approx \sqrt{-k + F^2(1 + 2\tan^2\zeta)} \tag{4.12}$$

As the field index k increases with R to ensure isochronism (Eq. 4.1), the spiral angle follows so to maintain $-k + F^2(1 + 2\tan^2\zeta) > 0$. A limitation here is the maximum spiral angle achievable, obviously $\zeta \to 90\,\text{deg}$.

As an illustration, in TRIUMF cyclotron ζ reaches 72 deg in the 500 MeV region (from zero in the 100 MeV region) whereas $1 + 2\tan^2\zeta$ increases to 20 (from 1 in the 100 MeV region) and compensates a low $F < 0.07$ (down from $F = 0.3$). In PSI 590 MeV cyclotron ζ reaches 35° on the outer radius. Most isochronous cyclotrons of a few tens of MeV use spiral sectors to benefit from the more efficient axial focusing [16].

More can be found in the scaling FFAG chapter (Sect. 10.2.2) regarding the spiral sector, and regarding its numerical simulation.

4.2.3 Isochronism

In the hypothesis of isochronism, the revolution angular frequency satisfies $\omega_{\text{rev}} = c\beta(\gamma)/R(\gamma) = \text{constant}$. An orbital radius $R_\infty = c/\omega_{\text{rev}}$ is reached asymptotically as $\beta = v/c = R/R_\infty \to 1$. In terms of the RF and harmonic number,

$$R_\infty = h\frac{c}{\omega_{\text{rf}}} \tag{4.13}$$

Given $BR_\infty = \gamma m_0 c/q$ and using $\gamma = \left(1 - (R/R_\infty)^2\right)^{-1/2}$, the radial dependence of the field can be expressed in terms of R_∞, namely,

$$B_0 R(R) = \gamma B_0 = \frac{B_0}{\sqrt{1 - (R/R_\infty)^2}} \quad \text{with } B_0 = \frac{m_0 \omega_{\text{rev}}}{q} = \frac{m_0}{q}\frac{\omega_{\text{rf}}}{h} \tag{4.14}$$

and goes to infinity with $R \to R_\infty$. For protons for instance, with $m_0/q = 1.6726 \times 10^{-27}[\text{kg}] / 1.6021 \times 10^{-19}[C] \approx 10^{-8}, BR_\infty[T\,m] = \gamma m_0 c/q \approx 3\gamma$. A typical value for R_∞ can be obtained assuming for instance an upper $\gamma = 1.64$ (600 MeV) in a region of upper field value $B = 1.64\,\text{T}$, yielding $R_\infty \approx 3\,\text{m}$.

Radial Field

From Eq. 4.14 it results that the radial field form factor of Eqs. 4.4, 4.11 can be written

$$R(R) = \left(1 - \left(\frac{R}{R_\infty}\right)^2\right)^{-1/2} \tag{4.15}$$

A possible approach consists in using the Taylor expansion of $R(R)$ (within the limits of radius of convergence of that series), namely

$$R(R) = 1 + \frac{1}{2}\left(\frac{R}{R_\infty}\right)^2 + \frac{3}{8}\left(\frac{R}{R_\infty}\right)^4 + \frac{5}{16}\left(\frac{R}{R_\infty}\right)^6 + \dots \tag{4.16}$$

The coefficients in this polynomial in R/R_∞ are the field index and its derivatives, they can be a starting point for further refinement of the isochronism, including for instance side effects of the azimuthal field form factor $\mathcal{F}(R, \theta)$ (Eqs. 4.3, 4.10).

The radial field index $k(R)$ in the AVF cyclotron is designed to satisfy the condition of isochronism (Eq. 4.1). However, reducing the RF phase slip over the acceleration cycle substantially below $\pm\pi/2$ requires a tolerance below 10^{-5} on field value over the orbit excursion area. This tight constraint requires pole machining, shimming, and other correction coil strategies in order to satisfy Eq. 4.1.

Fast Acceleration

Fixed field and fixed RF allow fast acceleration, the main limitation is in the amount of voltage which can be implemented around the ring. The voltage per turn reaches 4 MV for instance at the PSI 590 MeV ring cyclotron, where bunches are accelerated from 72 MeV to 590 MeV in less than 200 turns.

Harmful resonances may have to be crossed as wave numbers vary during acceleration, including the "Walkinshaw resonance" $\nu_R = 2\nu_y$ as $\nu_R \approx \gamma$ whereas the axial wave number spans $\nu_y \approx 1-\sim1.5$. This coupling resonance may result in an increase of vertical beam size and subsequent particle losses, fast crossing mitigates the effect.

Fast acceleration improves extraction efficiency, as the turn separation dR/dn is proportional to the energy gain per turn (Sect. 4.2.4).

4.2.4 Cyclotron Extraction

The minimum radial distance between the last two turns, where the extraction septum is located, is imposed by beam loss tolerances, which in some cases (high power beams for instance) may be tight, in the 10^{-4} range or less. Space charge in particular matters, as it increases the energy spread, and thus the radial extent of a bunch. In the relativistic cyclotron the separation between two consecutive turns satisfies

$$\Delta R \approx \frac{\gamma}{\gamma + 1} \frac{\Delta E}{E} \frac{R}{\nu_R^2} \tag{4.17}$$

with ΔE the effective acceleration rate per turn. This indicates that greater turn separation at extraction results from increased ring size. As a matter of fact, size is a limitation to intensity in small cyclotrons. It also indicates that extraction efficiency may be increased by moving the radial wave number closer to $\nu_R = 1$.

4.2.5 Resonant Spin Motion

In the quasi-uniform, quasi vertical field $\mathbf{B} \approx \mathbf{B}_y$ of a classical cyclotron dipole, spins quietly perform $G\gamma$ precessions around a vector $\boldsymbol{\omega}_{\mathrm{sp}} \parallel \mathbf{B}$ (Eq. 3.30) as the particle velocity completes a 2π precession around the ring (Sect. 3.2.5) [17].

More is liable to happen in the AVF cyclotron, due to the strong radial field index (Eq. 4.1) and to the azimuthal field modulation (Eqs. 4.3, 4.10): the azimuthal and radial field components B_θ and B_R are non-zero out of the median plane, $\mathbf{B}(R, \theta, y)$ may locally depart from vertical in a substantial manner, and so will the local precession vector $\boldsymbol{\omega}_{\mathrm{sp}}(R, \theta, y)$. The latter varies periodically in addition, as the particle undergoes periodic vertical motion about the median plane. Resonance between spin precession (characterized by spin tune $\nu_{\mathrm{sp}} = G\gamma$, Eq. 3.33) and periodic perturbing field components (characterized by the axial wave number ν_y, Eqs. 4.6, 4.12) occurs if the two motions feature coinciding frequencies. This condition can be expressed under the form

$$\nu_{sp} \pm \nu_y = integer \quad \text{or, equivalently} \quad G\gamma = integer \pm \nu_y \qquad (4.18)$$

The spin precession axis ω_{sp} moves away from the vertical as the spin motion gets closer to resonance (during acceleration as $G\gamma$ varies for instance), to end up in the median plane on the resonance [18, Sect. 3.6].

Consider now an ion bunch, away from any depolarizing resonance. Its polarization is $\langle S_y \rangle$, the average of the projection of the spins on the vertical. If a depolarizing resonance is crossed during acceleration, the initial polarization (far upstream of the resonance; index i) and final polarization (far downstream of the resonance; index f) satisfy the Froissart-Stora law [19],

$$\frac{\langle S_y \rangle_f}{\langle S_y \rangle_i} = 2e^{-\frac{\pi}{2}\frac{|\epsilon_R|^2}{a}} - 1 \qquad (4.19)$$

where $|\epsilon_R|$ is the strength of the resonance: a measure of the strength of the depolarizing fields, its calculation is addressed in a next chapter; a is the resonance crossing speed,

$$a = G\frac{d\gamma}{d\theta} \pm \frac{d\nu_y}{d\theta} \qquad (4.20)$$

The Froissart-Stora formula indicates that, if the resonance is crossed slowly ($a \to 0$), $\langle S_y \rangle_f / \langle S_y \rangle_i \to -1$: spins quietly follow the flipping motion of the precession axis, polarization is flipped and preserved. If the crossing is fast ($a \to \infty$), $\langle S_y \rangle_f / \langle S_y \rangle_i \to 0$, polarization is unaffected. Intermediate crossing speeds cause polarization loss: $|\langle S_y \rangle|$ ends up smaller after the resonance.

4.3 Exercises

Exercises 4.2–4.4 use a field map, designed in Exercise 4.1, to simulate an AVF cyclotron dipole. Note that they can be performed using DIPOLE[S] analytical field model instead, as in Exercise 4.5 (a similar simulation which can be referred to is Exercise 3.2, Classical Cyclotron chapter). As a reminder, regarding the interest of one or the other of the two methods: field maps allow close to real field models (a measured field map for instance, or from a magnet computer code); using an analytical field model allows more flexibility regarding magnet parameters, which can for instance be optimized using a matching procedure.

Note: some of the input data files for these simulations are available in zgoubi sourceforge repository at

[pathTo]/branches/exemples/book/zgoubiMaterial/cyclotron_relativistic/

4.1 Modeling Thomas AVF Cyclotron

Solution: 4.1.

In this exercise a 2D mid-plane field map is built, inspired from Thomas's 1938 article [5]. The method to build the map is that of Exercise 3.1, TOSCA[MOD.MOD1= 22.1] keyword is used to raytrace through and derive the optical parameters of the 4-period AVF cyclotron.

(a) Construct a 360° 2D map of the median plane field $B_Z(R, \theta)$, simulating the field in the 4-period Thomas cyclotron of Fig. 4.4, assuming the following:

– $B_Z(R, \theta) = B_0[\,1 + f \sin(4(\theta - \theta_i))\,]$ (Eq. 4.3), with θ_i some arbitrary origin of the azimuthal angle, to be determined. Hint: depending on θ_i value, the closed orbit may be at an angle to the polar radius, as seen in Fig. 4.4; in that case TOSCA[MOD.MOD1=22.1] would require non-zero in and out positioning angles TE and TS, to be determined and stated using KPOS option [20]; instead, a proper choice of θ_i value allows a simpler TE = TS = 0;
– an average axial field $B_0 = 0.5$ T on the 200 keV radius (the latter, $R_0(B_0)$, is to be determined), $B_Z > 0$ and $0 < f < 1$ modulation.
– an arbitrary field index k—a good idea is to start building and testing the AVF in the case $k = 0$;
– a uniform map mesh in a polar coordinate system (R, θ) as sketched in Fig. 3.17, covering R = 1 to 100 cm; take a radial increment of the mesh $\Delta R = 0.5$ cm, azimuthal increment $\Delta\theta = 0.5$ cm/R_M, with R_M some reference radius, say $R_M = 50$ cm, half way between map boundaries;
– an appropriate 6-column formatting of the field map data for TOSCA to read, as follows:

$$R\cos\theta, \; Z, \; R\sin\theta, \; BY, \; BZ, \; BX$$

with θ varying first, R varying second in that list. Z is the vertical direction (normal to the map mesh), so $Z \equiv 0$ in this 2D mesh.

Provide a graph of $B_Z(R, \theta)$ over the extent of the field map.
(b) Raytrace a few concentric closed trajectories centered on the center of the dipole, ranging in $10 \leq R \leq 80$ cm. Provide a graph of these concentric trajectories in the $(O; X, Y)$ laboratory frame, and a graph of the field along trajectories. Initial coordinates can be defined using OBJET, particle coordinates along trajectories during the stepwise raytracing can be logged in zgoubi.plt by setting IL = 2 under TOSCA.

(c) Check the effect of the integration step size on the accuracy of the trajectory and time-of-flight computation, by considering some Δs values in [0.1,10] cm, and energies in a range from 200 keV to a few tens of MeV (considering protons).
(d) Produce a graph of the energy or radius dependence of wave numbers.

(e) Calculate the numerical value of the axial wave number, ν_y, from the flutter (Eqs. 4.5, 4.6). Comparing with the numerical values, discrepancy is found: repeat (d) for f = 0.1, 0.2, 0.3, 0.6, check the evolution of this discrepancy.

4.2 Designing an Isochronous AVF Cyclotron
Solution: 4.2.

(a) Introduce a radius dependent field index $k(R)$ in the AVF cyclotron designed in Exercise 4.1, proper to ensure R-independent revolution period, in three different cases of modulation: f = 0 (no modulation), f = 0.2 and f = 0.9.

Check this property by computing the revolution period T_{rev} as a function of kinetic energy E_k, or radius R. On a common graph, display both T_{rev} and dT_{rev}/T_{rev} as a function of radius, including for comparison a fourth case: B = constant = 5 kG.

(b) Provide a graph of the energy dependence of wave numbers.

4.3 Acceleration to 200 MeV in an AVF Cyclotron
Solution: 4.3.

In this exercise protons are accelerated to over 100 MeV in an AVF cyclotron: well beyond the about 20 MeV energy reached in the classical cyclotron (see Exercise 3.10).

(a) Produce an acceleration cycle of a proton, from 0.2 to 100 MeV, in the AVF cyclotron designed in Exercise 4.2. Note that a dedicated field map has to be created in order to allow for the higher maximum energy—a 3 m field map outer radius works. Assume proper modulation coefficient f for axial focusing all the way to 300 MeV. Assume a double-dee design, and 400 keV peak voltage in the gap, use CAVITE[IOPT = 7] for acceleration to account for RF phase.

(b) Give a graph of the energy dependence of wave numbers over the acceleration range.

4.4 Thomas-BMT Spin Precession in Thomas Cyclotron
Solution: 4.4.

This exercise uses the field maps and input data file of Exercise 4.3. Dependence of energy boost on RF phase is removed by using CAVITE[IOPT = 3] [20]. Consider helion ions: use PARTICUL[Name = HELION] to define mass, charge and G factor, all quantities needed for the integration of Thomas-BMT differential equation (Eq. 3.30).

(a) By scanning the axial wave number, find the $G\gamma$ value for which the spin motion resonance condition (Eq. 4.18) is satisfied.

(b) Consider a particle with non-zero axial motion, so that it experiences horizontal magnetic field components as it circles around. Track its spin through the resonance, take initial spin vertical $\mathbf{S} \equiv \mathbf{S}_Z$. Provide a graph of S_Z as a function of $G\gamma$ or energy.

(c) Simulate resonance crossing for a series of different vertical motion amplitudes Z_0; produce a graph of these resonance crossings $S_Z(turn)$.

Plot the ratio $S_{y,f}/S_{y,i}(Z_0)$. From a match of this $S_{y,f}/S_{y,i}$ series with Eq. 4.19, show that the resonance strength changes in proportion to the vertical excursion.

(d) Repeat (c) for a series of different resonance crossing speeds instead (Eq. 4.20), leaving Z_0 unchanged.
Show that this $S_{y,f}/S_{y,i}$ series can be matched with Eq. 4.19.

4.5 Isochronism and Edge Focusing in a Separated Sector Cyclotron
Solution: 4.5.

This exercise uses DIPOLE to simulate a 30 deg sector dipole of a 4-period cyclotron, and allow playing with field fall-off extent at dipole EFBs. The configuration of the cyclotron is typically that of PSI 72 MeV injector (Fig. 4.5). DIPOLE allows radial field indices up to the third order $(\partial^3 B_Z/\partial R^3)$ [20, Eq. 6.3.18]. In question (b) however, higher order indices are needed to improve the isochronism, requiring the use of DIPOLES [20, Eqs. 6.3.20, 21].

Take fringe fields into account (see Sect. 14.3.3), with

- $\lambda = 7$ cm the fringe extent (changing λ changes the flutter, Eq. 4.5),
- $C_0 = 0.1455$, $C_1 = 2.2670$, $C_2 = -0.6395$, $C_3 = 1.1558$ and $C_4 = C_5 = 0$, for a realistic field fall-off model.

(a) Assume $k = 0$, here. Produce a model of a period using DIPOLE.
Produce a graph of closed orbits across a period for a few different rigidities (FIT can be used to find them), and a graph of the field along these orbits.

(b) In this question, R-dependence of the mid-plane magnetic field proper to ensuring energy independent revolution period is introduced. Use DIPOLES here, as it allows b_i field indices to higher order, as necessary to reach tight isochronism over the full energy range.
Assume a peak field value $B_0 = 1.1$ T at a radius of 3.5 m in the dipoles. Find the average orbit radius R, and average field B (such that $BR = p/q$), at an energy of 72 MeV.
Determine a series of index values, $b_{i=1,n}$, in the model [20, Eq. 6.3.19]

$$B_Z(R, \theta) = B_0\,\mathcal{F}(R, \theta)\left(1 + b_1\frac{R - R_0}{R_0} + b_2\left(\frac{R - R_0}{R_0}\right)^2 + ...\right) \qquad (4.21)$$

proper to bring the revolution period closest to R-independent, in the energy range 0.9–72 MeV (hint: use a Taylor development of Eq. 4.15 and identify with the R-dependent factors in Eq. 4.21).

(c) Play with the value of λ, concurrently to maintaining isochronism with appropriate b_i values. Check the evolution of radial and axial focusing—OBJET[KOBJ = 5] and MATRIX[IORD = 1, IFOC = 11] or TWISS, or OBJET[KOBJ = 6] and MATRIX[IORD = 2, IFOC = 11], can be used to get the wave numbers.

From raytracing trials, observe that (i) the effect of λ on radial focusing is weak (a second order effect in the particle coordinates); (ii) greater (smaller) λ value results in smaller (greater) flutter and weaker (stronger) axial focusing (a first order effect). Note: the integration step size in DIPOLE[S] has to be consistent with the field fall-off extent (λ value), in order to ensure that the numerical integration is converged.

(d) For some reasonable value of λ (normally, about the height of a magnet gap, say, a few centimeters), compute $F^2 = \frac{\langle (B(\theta) - <B>)^2 \rangle}{<B>^2}$. Check the validity of $v_y = \sqrt{-\beta^2 \gamma^2 + F^2}$ (Eq. 4.6). OBJET[KOBJ = 5] and MATRIX[IORD = 1, IFOC = 11] can be used to compute v_y, or multiturn raytracing and a Fourier analysis.

(e) Check the rule $F^2 \xrightarrow{\text{hard edge}} \frac{R}{\rho} - 1$ (Eq. 4.5), from the field $B(\theta)$ delivered by DIPOLES. Give a theoretical demonstration of that rule.

4.6 A Model of PSI Ring Cyclotron Using CYCLOTRON
Solution: 4.6.

The simulation input data file in Table 4.2 is based on the use of CYCLOTRON, to simulate a period of the eight-sector PSI ring cyclotron and work on the isochronism. That file is the starting point of the present exercise.

(a) With zgoubi users' guide at hand, explain the signification of the data in that simulation input data file.

(b) Compute and plot a few trajectories and field along, across the sector. Provide a graph of field density over the sector.

(c) Compute and plot the radius dependence of the revolution period.

(d) The field indices b_1, b_2, ... are aimed at realizing the isochronism; four, $b_1 - b_4$ are accounted for in (a) and (b), they were drawn from the PSI cyclotron spiral sector magnet field map data. Question (c) proves this small set of indices to result in a poor isochronism of the orbits.

Add higher order indices, until a sufficient number, with proper values, is found that allows FIT to reach a final isochronism improved by an order of magnitude. Provide a revised input data file with updated index series and their values.

4.4 Solutions of Exercises of This Chapter: Relativistic Cyclotron

4.1 Modeling Thomas AVF Cyclotron
(a) A field map of a 360° AVF cyclotron dipole.

A Fortran program, geneAVFMap.f, given in Table 4.3, constructs the required map of a field distribution $B_Z(r, \theta)$, logged under the name geneAVFMap.out for use by TOSCA keyword. A polar mesh is retained (Fig. 3.19), rather than Cartesian, consistently with cyclotron magnet symmetry.

Table 4.2 Simulation input data file: a period of an eight-sector PSI-style cyclotron. The data file is set up for a scan of the periodic orbits, from radius R = 204.1171097 cm to R = 383.7131468 cm, in 15 steps

```
PSI CYCLOTRON

'OBJET'
1249.382414
2
1 1
204.1171097 8.915858372 0.  0.  0.  1.  'o'
1
'PARTICUL'
PROTON

'CYCLOTRON'
2
1   45.0  276.  1.0
0. 0. 0.99212277 51.4590015 0.5 800. -0.476376328 2.27602517e-03 -4.8195589e-06 3.94715806e-09
18.3000E+00  1.  28.  -2.0
8 1.1024358 3.1291507 -3.14287154 3.0858059 -1.43545 0.24047436 0. 0. 0.
11.0  3.5 35.E-3  0.E-4 3.E-8 0. 0. 0.
18.3000E+00  1.  28.  -2.0
8 0.70490173 4.1601305 -4.3309575 3.540416 -1.3472703 0.18261076 0. 0.
-8.5  2.  12.E-3 75.E-6 0.    0. 0. 0.
0. -1
0 0.  0.  0.  0.  0.  0.  0.
0.  0.  0.   0.  0.  0.
2  10.
0.4
2 0.  0. 0.  0.

'FIT2'
2
1 31 0 [-300.,100]
1 35 0 [.1,3.]
2
3.1 1 2 #End 0. 1. 0
3.1 1 3 #End 0. 1. 0
'FAISCEAU'

'FAISTORE'
orbits.fai
1

'REBELOTE'
14 0.2  0 1
1
OBJET 30 221.065356:383.7131468

'SYSTEM'
1
gnuplot <./gnuplot_orbits.gnu
'END'
```

Note the following:

(i) The field map azimuthal extent (set to 360° in geneAVFMap, Table 4.3) can be changed, for instance to simulate a 90 deg sector instead, with a sequence of four simulating the complete ring.

(ii) Assuming mid-plane symmetry, the field in the (O;X,Y) plane is axial. The field is taken radially constant in the first part of this exercise: $k = 0$, thus $\mathcal{R}(R) = 1$ in Eq. 4.11.

(iii) The origin of the azimuthal angle in $B_Z(R, \theta) = B_0[\, 1 + f \sin(4(\theta - \theta_i))\,]$ is taken at $\theta_i = \pi/2$, leading to (Table 4.3)

$$B_Z(R, \theta) = B_0[\, 1 + f \cos(4\theta)\,]$$

With this cosine dependence, and θ covering $0 \rightarrow 2\pi$, the entrance and exit faces of a 360° field map will be a location of maximum field (hill ridge), thus, owing to the $2\pi/4$ cylindrical symmetry of the field, the closed orbit is normal to these entrance

Table 4.3 A Fortran program, geneAVFMap.f, which generates a 360° mid-plane field map. This angle as well as the field amplitude ($B_0 = 5\,$kG at $R_0 = 50\,$cm, here) and its modulation ($f = 0.2$, here) can be changed to any other values, a field index (ak, set to zero here) can be accounted for. The field map produced is logged in geneAVFMap.out, or under different names for the purpose of the exercise, depending upon f, k, or AT values, e.g., geneAVFMap_90deg_f2_k0.out, geneAVFMap_360deg_f9_k0.out...

Note: this file is available in `zgoubi` sourceforge repository at
[pathTo]/branches/exemples/book/zgoubiMaterial/cyclotron_relativistic/ProbThomasAVF/fieldMap/

```
      implicit double precision (a-h,o-z)
      parameter (pi=4.d0*atan(1.d0), BY=0.d0, BX=0.d0, Z=0.d0, dZ=0.d0)
      open(unit=2,file='geneAVFMap.out')
C----------- Hypotheses :
      AT = 360.d0  /180.d0*pi          ! Angular extent of field map. Can be changed 360, 60 deg, etc.).
      f = .2d0                                            ! azimuthal modulation factor.
      B0 = 5.d0 ;R0 = 12.9248888074d0            ! field at R0 (kG); 200keV radius (cm), B(R0)=B0=5kG.
      ak = 0.d0                                          ! Field index, defined at R0.
C
      Rmi=1.d0; Rma=76.d0; RM=50.d0  ! cm. Radial extent of field map.
      dR = 0.5d0 ; NR = NINT((Rma - Rmi)/dR)+1  ! R-distance between nodes in mesh. Number of R-nodes.
C                          RdA=RM*dA is the distance between two nodes along R=RM arc.
      RdA = 1.d0  !       given angle increment dA (dA is the "Delta theta" quantity in the main text).
      NX = NINT(RM*AT /RdA) +1 ; RdA= RM*AT / DBLE(NX -1)  ! exact mesh step at RM, corresponding to NX.
      dA = RdA / RM ; A1 = 0.d0 ; A2 = AT                  ! corresponding delta_angle.
C------------------------------------------------
      write(2,*) Rmi,dR,dA/pi*180.d0,dZ,
     >'   ! Rmi/cm, dR/cm, dA/deg, dZ/cm'
      write(2,*) '# Field map generated using geneAVFMap.f '
      write(2,fmt='(a)') '# AT/rd,  AT/deg, Rmi/cm, Rma/cm, RM/cm,'
     >//'  NR, dR/cm, NX, R*dA/cm, dA/rd : '
      write(2,fmt='(a,1p,5(e16.8,1x),2(i3,1x,e16.8,1x),e16.8)')
     >'# ',AT, AT/pi*180.d0,Rmi, Rma, RM, NR, dR, NX, RdA, dA
      write(2,*) '# For TOSCA: ',NX,NR,' 1 22.1 1.  !IZ=1 -> 2D ; '
     >//'MOD2=22 -> polar map ; .MOD2=.1 -> one map file'
      write(2,*) '# R*cosA (A:0->360), Z==0, R*sinA, BY, BZ, BX '
      write(2,*) '# cm                cm    cm          kG  kG  kG '
      write(2,*) '# '
      do jr = 1, NR
        R = Rmi + dble(jr-1)*dR
        do ix = 1, NX
          A = A1 + dble(ix-1)*dA
          BR = (1.D0 + ak * x/R0)
          BZ = B0 * BR * (1.d0+f*sin(4.d0*A +pi/2.d0))
          X = R * sin(A)  ; Y = R * cos(A)
          write(2,fmt='(1p,6(e16.8),2(1x,i0),3(1x,e16.8))')
     >        Y,Z,X,BY,BZ,BX,ix,jr,A,R,atan(X/Y)
        enddo
      enddo
      stop ' Job complete ! Field map stored in geneAVFMap.out'
      end
```

Top and bottom sections of the field map file geneAVFMap.out with modulation factor f=0.2 and radial index k=0. The file starts with an 8-line header, of which the first line is effectively used by zgoubi, *the following 7 being just comments*:

```
    1.    0.5      1.1464968152866242      0.        ! Rmi/cm, dR/cm, dA/deg, dZ/cm
# Field map generated using geneAVFMap.f
# AT/rd,  AT/deg, Rmi/cm, Rma/cm, RM/cm, NR, dR/cm, NX, R*dA/cm, dA/rd :
#   6.28318531E+00   3.60E+02  1.000E+00   7.6000E+01  5.000E+01 151  5.00E-01 315  1.00050721E+00  2.00101443E-02
# For TOSCA:      315      151 1 22.1 1.  !IZ=1 -> 2D ; MOD=22 -> polar map ; .MOD2=.1 -> one map file
# R*cosA (A:0->360), Z==0, R*sinA, BY, BZ, BX
# cm                cm    cm          kG  kG  kG
#
  1.00000000E+00  0.00000000E+00  0.00000000E+00  0.00000000E+00  5.00000000E+00  0.00000000E+00 1 1
  9.99799804E-01  0.00000000E+00  2.00038090E-02  0.00000000E+00  5.07995514E+00  0.00000000E+00 2 1
  9.99199295E-01  0.00000000E+00  4.00096065E-02  0.00000000E+00  5.15939832E+00  0.00000000E+00 3 1
  9.98198715E-01  0.00000000E+00  5.99943846E-02  0.00000000E+00  5.23782687E+00  0.00000000E+00 4 1
  9.96798463E-01  0.00000000E+00  7.99551413E-02  0.00000000E+00  5.31472063E+00  0.00000000E+00 5 1
  ...............................
  7.57566832E+01  0.00000000E+00 -6.07659074E+00  0.00000000E+00  4.68527937E+00  0.00000000E+00 311 151
  7.58631023E+01  0.00000000E+00 -4.55957323E+00  0.00000000E+00  4.76217913E+00  0.00000000E+00 312 151
  7.59391464E+01  0.00000000E+00 -3.04073010E+00  0.00000000E+00  4.84060168E+00  0.00000000E+00 313 151
  7.59847851E+01  0.00000000E+00 -1.52066948E+00  0.00000000E+00  4.92004486E+00  0.00000000E+00 314 151
  7.60000000E+01  0.00000000E+00 -1.86146313E-14  0.00000000E+00  5.00000000E+00  0.00000000E+00 315 151
```

gnuplot script to obtain Fig. 4.4:

```
set xtics; set ytics; cm2m = 0.01; set xlabel "X [m]"; set ylabel "Y [m]"; set zlabel "B_Z [kG]"; set hidden3d
set view 49, 221 ;   splot "geneAVFMap.out" u ($ 1 *cm2m):($3 *cm2m):($5) w l lc palette notit; pause 1
```

and exit faces, so yielding TE = TS = 0 as KPOS arguments under TOSCA. The same property holds in the case a 90 deg field map is used (θ covering $0 \rightarrow \pi/2$): take entrance and exit faces of the field map along hill ridges, i.e. normal to the closed orbits.

As an indication of expected outcomes of this field map computation, the top and bottom parts of the file generated by geneAVFMap, in the required format for TOSCA[MOD = 22.1] to swallow, are given in Table 4.3. Figure 4.4 displays the midplane field so obtained (it uses the gnuplot script given at the bottom of Table 4.3)

(b) Concentric trajectories.

The input data file to raytrace trajectories with different rigidities (four, here) is given in Table 4.4. The computation is performed in two steps

(1) in a first step, FIT finds the periodic coordinates for a given rigidity; note that for this first step a 90 deg field map is INCLUDed (obtained with $AT = 90$ deg in the Fortran, Table 4.3; and used in subsequent exercises);

(2) upon completion of FIT, a second step computes the closed trajectory over $360°$; note: this double-step is one way to reduce the volume of zgoubi.plt file as it is only written to at the second step, ounce the closed orbit has been found by FIT (note: this can be accomplished in a single step, see Exercise 4.2). This process is repeated for additional rigidities (i.e., additional orbits at different energies) using REBELOTE[IOPT=1].

The following keywords are found in the input data file:

(i) OBJET: define a reference rigidity (arbitrary, taken to be 64.624444 kG cm, here, corresponding to 200 keV protons), and define initial coordinates of a single ion (initial radius Y_0 in the field map frame, other space coordinates zero; relative rigidity D = $B\rho/BORO$),

(ii) TOSCA: read the field map and raytrace; IL = 2 flag causes log of particle data in zgoubi.plt, after each integration step Δs,

(iii) FAISCEAU: log local particle coordinates in zgoubi.res execution listing,

(iv) SYSTEM: run two gnuplot scripts once raytracing is completed, a first one to plot the trajectories, a second one to plot the field along trajectories,

(v) MARKER: define two "LABEL_1" type labels, for use in INCLUDE statements in subsequent exercises.

Four closed orbits resulting from the data file in Table 4.4 (for respective relative rigidities D = 1, 1.5, 3.25, 5, spanning about 70 cm radially) are displayed in Fig. 4.9. Inspecting zgoubi.res one finds the following particle coordinates as logged by "'FAISCEAU' CHECK" (Table 4.4), with initial values (left hand side) equal to final values from the FIT procedure (right hand side), as expected:

```
     8  Keyword, label(s) :  FAISCEAU  CHECK
                                          TRACE DU FAISCEAU
                                          (follows element #      7)
                            OBJET                                          FAISCEAU
          D       Y(cm)    T(mr)   Z(cm)   P(mr)   S(cm)   D-1      Y(cm)    T(mr)  Z(cm)  P(mr)   S(cm)
    m  1  1.0000  13.071   0.000   0.000   0.000   0.0000  0.0000   13.071   0.000  0.000  0.000   2.027537E+01
    m  1  1.5000  19.607   0.000   0.000   0.000   0.0000  0.5000   19.607   0.000  0.000  0.000   3.041305E+01
    m  1  3.2500  42.481   0.000   0.000   0.000   0.0000  2.2500   42.481   0.000  0.000  0.000   6.589494E+01
    m  1  5.0000  65.355   0.000   0.000   0.000   0.0000  4.0000   65.355   0.000  0.000  0.000   1.013768E+02
```

Table 4.4 Simulation input data file FieldMapAVFMag.inc: raytrace a series of ions with different rigidities, spanning 200 keV–5 MeV. This file also defines the optical segment #S_AVFMag_90d to #E_AVFMag_90d, for use in subsequent exercises

Note: this file is available in zgoubi sourceforge repository at
[pathTo]/branches/exemples/book/zgoubiMaterial/cyclotron_relativistic/ProbThomasAVF/concentricTrajectories/

```
FieldMapAVFMag.inc                              | Title line: required at top of file.
!                                               | Additional header lines require a !.
'MARKER'   FieldMapAVFMag_S                               | Just for edition purposes.
'OBJET'
64.62444403717985                   | Reference Brho ("BORO" in the users' guide) -> 200keV proton.
2
1 1
12.9248888 0.  0. 0. 0.  1. 'm'     | Closed orbit coordinates at BR=64.6244440 kG.cm for constant B.
1
'MARKER'   #S_AVFMag_90d
'TOSCA'
0 0                                 | IL=2 to log step-by-step coordinates, spin, etc., in zgoubi.plt.
1. 1. 1. 1.  | Normalization coefficients applied to field, and X, Y Z coordinate values read from map.
HEADER_8                                    | The field map file starts with an 8-line header.
80 151 1 22.1 1.         | IZ=1 for 2D map; MOD=22 for polar frame; .MOD2=.1: only one map file.
geneAVFMap_90deg_f2_k0.out              | Or geneAVFMap_90deg_f9_k0.out for f=0.9 modulation.
0 0 0 0             | Possible boundaries within the field map, to start/stop stepwise integration.
2
.2 ! cm                                                       | Integration step size.
2                                                             | Magnet positioning option.
0. 0. 0. 0.                                                   | Magnet positioning.
'MARKER'   #E_AVFMag_90d
'FAISCEAU'                                                    | Particle coordinates, here.
'FIT'
1
2  30  0 [5.,100.]          | Vary Y0 at OBJET, to allow fulfilling the following two constraint:
2  1e-15
3.1 1 2 5 0. 1. 0                             | request same radius after a period (90 deg);
3   1 3 5 0. 1. 0                             | request orbit angle after a period to be nul.
'FAISCEAU' CHECK   | Allows quick check of particle coordinates, in zgoubi.res: final should = initial.

'TOSCA'
0 2                                 | IL=2 to log step-by-step coordinates, spin, etc., in zgoubi.plt.
1. 1. 1. 1.  | Normalization coefficients applied to field, and X, Y Z coordinate values read from map.
HEADER_8                                    | The field map file starts with an 8-line header.
315 151 1 22.1 1.        | IZ=1 for 2D map; MOD=22 for polar frame; .MOD2=.1: only one map file.
geneAVFMap_360deg_f2_k0.out   | Could also be 4*geneAVFMap_90deg_f2_k0.out (changing IX=315 to IX=90)|
0 0 0 0             | Possible boundaries within the field map, to start/stop stepwise integration.
2
.2 ! cm                                                       | Integration step size.
2                                                             | Magnet positioning option.
0. 0. 0. 0.                                                   | Magnet positioning.
'FAISCEAU'
'REBELOTE'                                                    | Repeat what precedes,
3  0.1  0 1                                                              | 3 times.
1                               | Change the value of parameter 35 in OBJET prior to repeating), i.e.,
OBJET 35   1.5:5                          | the relative rigidity D=1, 1.5, 3.25, 5.
'SYSTEM'
2
gnuplot < ./gnuplot_Zplt_traj.gnu &                  | Plot B(R), as read fron zgoubi.plt.
gnuplot < ./gnuplot_Zplt_field.gnu &                 | Plot B(R), as read fron zgoubi.plt.
'MARKER'   FieldMapAVFMag_E                           | Just for edition purposes.
'END'
```

gnuplot script to obtain Fig. 4.9:

```
# gnuplot_Zplt_traj.gnu
set xtics; set ytics; set xlabel 'X_{lab}  [m]'; set ylabel 'Y_{lab}  [m]'; set size ratio -1; set polar; cm2m=1e-2
plot for [FITnum=1:4] 'zgoubi.plt' u ($49==FITnum? $22:1/0):($10*cm2m):($49) w p pt 4 lc palette notit ; pause 1
```

gnuplot script to obtain Fig. 4.10:

```
# gnuplot_Zplt_field.gnu
set xtics nomirror; set x2tics; set ytics; set xlabel 'angle [rad]'; set ylabel 'B_Z  [kG]'
plot for [FITnum=1:4] 'zgoubi.plt' u ($49==FITnum ? $22 :1/0):($25) w p notit; pause 1
```

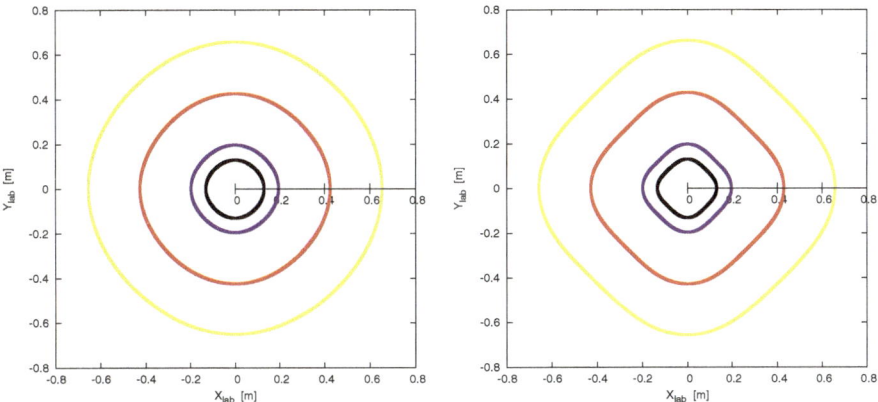

Fig. 4.9 Scalloping closed orbits in the 4-period AVF cyclotron with modulation factor $f = 0.2$ (left) and $f = 0.9$ (right)

The scalloping (orbit oscillation around the reference circle) is small, as can be seen by comparison, below, with the closed orbit radius in the case of constant field. The latter is obtained with a similar computation using a field map generated with $f = 0$; it can also be obtained from $R = p / qB$ with $B = 5$ kG with $p = q \times D \times B\rho_{\text{ref}}$ and $B\rho_{\text{ref}} = 64.6244440$ kG cm the reference rigidity, under OBJET, yielding:

```
      8  Keyword, label(s) :  FAISCEAU   CHECK
                                       TRACE DU FAISCEAU
                                     (follows element #      7)
                      OBJET                                             FAISCEAU
            D      Y(cm)   T(mr)   Z(cm)   P(mr)   S(cm)  D-1     Y(cm)   T(mr)  Z(cm)  P(mr)  S(cm)
  m  1  1.0000  12.925   0.000   0.000   0.000   0.0000 0.0000  12.925 -0.000  0.000  0.000  2.030237E+01
  m  1  1.5000  19.387   0.000   0.000   0.000   0.0000 0.5000  19.387 -0.000  0.000  0.000  3.045355E+01
  m  1  3.2500  42.006   0.000   0.000   0.000   0.0000 2.2500  42.006  0.000  0.000  0.000  6.598270E+01
  m  1  5.0000  64.624   0.000   0.000   0.000   0.0000 4.0000  64.624 -0.000  0.000  0.000  1.015118E+02
```

Figure 4.9 also displays an iteration of this closed orbits computation, yet for the case of a modulation factor f = 0.9 (thus using different field maps, named e.g. geneAVFMap_90deg_f9_k0.out and geneAVFMap_360deg_f9_k0.out, for substitution to the $f = 0.2$ field map names in Table 4.4); the scalloping is increased due to deeper modulation. Inspecting "'FAISCEAU' CHECK" in zgoubi.res execution listing one then finds the following particle coordinates for the 4 different rigidities:

```
      8  Keyword, label(s) :  FAISCEAU   CHECK
                                       TRACE DU FAISCEAU
                                     (follows element #      7)
                      OBJET                                             FAISCEAU
            D      Y(cm)   T(mr)   Z(cm)   P(mr)   S(cm)  D-1     Y(cm)   T(mr)  Z(cm)  P(mr)  S(cm)
  m  1  1.0000  13.165   0.000   0.000   0.000   0.0000 0.0000  13.165  0.000  0.000  0.000  1.978076E+01
  m  1  1.5000  19.747   0.000   0.000   0.000   0.0000 0.5000  19.747  0.000  0.000  0.000  2.967114E+01
  m  1  3.2500  42.786   0.000   0.000   0.000   0.0000 2.2500  42.786  0.000  0.000  0.000  6.428748E+01
  m  1  5.0000  65.825   0.000   0.000   0.000   0.0000 4.0000  65.825  0.000  0.000  0.000  9.890382E+01
```

The magnetic field along these orbits is displayed in Fig. 4.10, it is the same for all four orbits as the field index is zero, here.

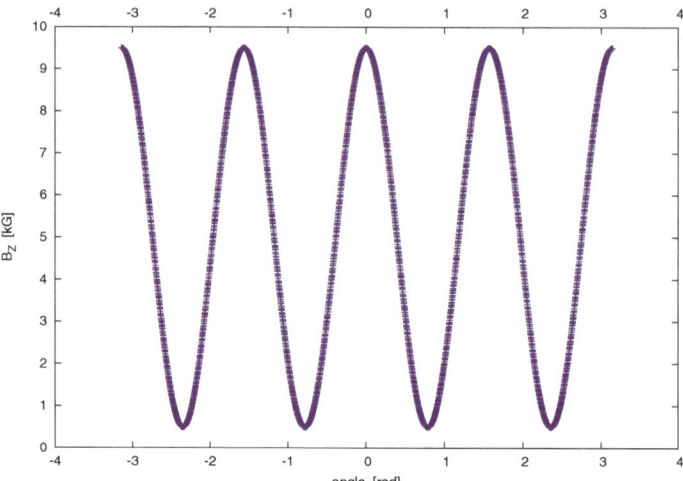

Fig. 4.10 Four-periodic field $B_Z(\theta)$ along the closed orbits, case of a modulation factor $f = 0.9$. The field is the same for all four orbits as the field index is zero. The average value of the field along a closed orbit is $\frac{2}{\pi} \int_{\Delta\theta=\pi/2} B_Z(R, \theta)\, d\theta = 5\,\mathrm{kG}$

(c) Numerical convergence.

Numerical convergence of the stepwise integration is tested using the same input data file as in (b) (Table 4.4, f = 0.2 and k = 0), the integration step size only, Δs, needs be changed, and the resulting change in accuracy translates in a change of closed orbit coordinates as found by FIT. Two values of Δs are tried (in addition to $\Delta s = 0.2\,\mathrm{cm}$ in the previous computations, *cf.* Table 4.4). They yield the following outcomes of FAISCEAU (at its occurrence at the bottom of Table 4.4, prior to REBELOTE), for closed orbits at the four different relative rigidities D = 1, 1.5, 3.25, 5:

◇ Case of $\Delta s = 1\,\mathrm{cm}$

```
         8  Keyword, label(s) :  FAISCEAU    CHECK
                                             TRACE DU FAISCEAU
                                             (follows element #     7)
                               OBJET                                        FAISCEAU
           D      Y(cm)     T(mr)  Z(cm)  P(mr)   S(cm)    D-1      Y(cm)     T(mr)   Z(cm)  P(mr)
 m  1   1.0000   13.099    0.000  0.000  0.000  0.0000   0.0000   13.099   -0.018   0.000  0.000
 m  1   1.5000   19.610    0.000  0.000  0.000  0.0000   0.5000   19.610    0.000   0.000  0.000
 m  1   3.2500   42.481    0.000  0.000  0.000  0.0000   2.2500   42.481    0.000   0.000  0.000
 m  1   5.0000   65.355    0.000  0.000  0.000  0.0000   4.0000   65.355    0.000   0.000  0.000
```

⋄ Case of $\Delta s = 5$ cm

```
         8  Keyword, label(s) :  FAISCEAU  CHECK
                                              TRACE DU FAISCEAU
                                              (follows element #      7)
                              OBJET                                              FAISCEAU
              D       Y(cm)   T(mr)   Z(cm)   P(mr)    S(cm)   D-1      Y(cm)   T(mr)  Z(cm)  P(mr)
     m  1  1.0000   14.177   0.000   0.000   0.000   0.0000  0.0000   14.183  -0.011  0.000  0.000
     m  1  1.5000   20.761   0.000   0.000   0.000   0.0000  0.5000   20.757   0.142  0.000  0.000
     m  1  3.2500   42.624   0.000   0.000   0.000   0.0000  2.2500   42.628   0.000  0.000  0.000
     m  1  5.0000   65.355   0.000   0.000   0.000   0.0000  4.0000   65.354  -0.000  0.000  0.000
```

The change of closed orbit coordinates is substantial for the lowest energy trajectory, smaller circumference $C \approx 2\pi \times 13 \approx 80$ cm, covered in only 16 steps in the case $\Delta s = 5$ cm. Given the strong curvature, the high order derivatives of the field vector take great values so jeopardizing the convergence of the position and velocity vector Taylor series [20, Eq. 1.2.4]. The $\Delta s = 5$ cm case features in addition poor convergence of the FIT procedure, unable to zero the closed orbit angle in the small radius cases, an effect of the field interpolation from a mesh.

(d) Dependence of wave numbers on energy and radius.

A scan of the wave numbers over a relative rigidity interval $D = \frac{B\rho}{BORO} : 1 \rightarrow 5$ is performed using the input data file given in Table 4.5 (BORO is the reference rigidity, under OBJET, D is the sixth coordinate of the reference particle as defined under OBJET[KOBJ = 5]). Wave numbers are computed using MATRIX.

OBJET[KOBJ = 5] generates 13 particles with paraxial radial and axial coordinates, and rigidity sampling, for the computation of transport matrix and wave numbers by MATRIX. REBELOTE repeats this matrix computation sequence, for a series of different rigidities. It is preceded by FIT which finds the closed orbit, this is necessary as, (i) a different rigidity means different orbital radius, (ii) MATRIX computes transport coefficients with respect to particle 1, which requires the latter to be placed on the reference orbit, prior to MATRIX computation.

Inspection of the execution listing zgoubi.res shows the structure of a FIT at the end of the FIT procedure, with the status of the variable (one variable only, here) in a top block, followed by the status of the constraints in a bottom block. Here is an excerpt of the FIT section in zgoubi.res, at the last iteration by REBELOTE (case of relative rigidity D = 5.00639):

```
         6  Keyword, label(s) :  FIT

   STATUS OF VARIABLES  (Iteration # 200 /    199 max.)
   LMNT VAR PARAM  MINIMUM    INITIAL      FINAL      MAXIMUM     STEP     NAME    LBL1
     2   1   30     5.00      65.9       65.908886    100.      0.00      OBJET    -

   STATUS OF CONSTRAINTS (Target penalty =   1.0000E-08)
   TYPE  I   J LMNT#    DESIRED        WEIGHT       REACHED        KI2     NAME    LBL1
     3   1   2    5   0.000000E+00   1.000E+00   1.074504E-04   1.00E+00  MARKER    -
     3   1   3    5   0.000000E+00   1.000E+00   1.537897E-06   2.05E-04  MARKER    -
   Fit reached penalty value   1.1548E-08
```

Details regarding FIT[2] input, algorithms, and outcomes, are found in [20].

Further inspection of the execution listing shows the outcome of a MATRIX command, under the form of two 6×6 blocks, a top one which is the transport matrix $[T_{ij}]$ (see Sect. 14.5.2) from start to end of the optical sequence, and a bottom one, "beam matrix" drawn from the periodicity hypothesis which allows to write (see Sect. 14.5.2) $[T_{ij}] = I \cos(\mu) + J \sin(\mu)$. Here is an excerpt of the MATRIX section

Table 4.5 Simulation input data file: raytrace a set of 13 particles (defined by OBJET[KOBJ = 5]) for a particular reference rigidity, to perform a MATRIX computation. FIT is used to find the closed orbit, prior to MATRIX. Iteration for a series of 35 additional rigidities (relative rigidity D: 1.1→5.00639, in 35 steps) is performed by REBELOTE. This input file INCLUDEs the segment [#S_AVFMag_360d:#E_AVFMag_360d] of file FieldMapAVFMag.inc (Table 4.4)

```
Uniform field sector.
'MARKER'   FieldMapAVFQs_S                                    ! Just for edition purposes.
'OBJET'
64.62444403717985                       ! Reference Brho ("BORO" in the users' guide) -> 200keV proton.
5                                                 ! Define 13 particles for MATRIX computation.
.001 .01 .001 .01 .001 .00001                                ! Sampling of the initial coordinates.
12.9248888 0.  0. 0. 0.  1. 'm'    ! Closed orbit coordinates at BR=64.6244440 kG.cm for constant B.
'INCLUDE'
1
4 *FieldMapAVFMag.inc[#S_AVFMag_90d:#E_AVFMag_90d]
'FIT'
1                                                                              ! One variable.
2  30  0 [5.,100.]                        ! Vary Y0 at OBJET, to allow for the following constraint:
2  1E-8    199     ! Two constraints. Required penalty is 1e-8. Maximum number of iterations is 199.
3.1 1 2 5 0. 1. 0                                           ! request same radius after a period (90 deg);
3    1 3 5 0. 1. 0                            ! request orbital angle after a 1/4-turn to be zero.
'FAISCEAU' CHECK   ! Allows quick check of particle coordinates, in zgoubi.res: final should = initial.
'MATRIX'
1 11 PRINT            ! PRINT: log computation outcome data in zgoubi.MATRIX.out, for further plotting.
'REBELOTE'                                                            ! Repeat what precedes,
35  0.1  0 1                                                          ! 15 times.
1                             ! Change the value of parameter 35 in OBJET, namely, to relative rigidity from
OBJET 35  1.1:5.00639            ! D=1.1 by increment(5.00639-1.1)/35, prior to repeating the sequence.
'SYSTEM'
1
gnuplot < ./gnuplot_MATRIX_Qxy.gnu                                    ! Plot the wave number scan.
'MARKER'   FieldMapAVFQs_E                                    ! Just for edition purposes.
'END'
```

---

gnuplot script to obtain Fig. 4.11:

```
# gnuplot_MATRIX_Qxy.gnu
set xlab "kin. E [MeV]"; set x2lab "R [cm]"; set ylab "{/Symbol n}_R, ({/Symbol n}_R^2+{/Symbol n}_y^2)^{1/2}"
set y2label "{/Symbol n}_y"; set xtics nomirror; set x2tics; set ytics nomirror; set y2tics nomirror
BORO = 64.62444403717985; am = 938.27203e6; c = 2.99792458e8; BrhoRef = BORO *1e-3; eV2MeV = 1e-6
plot "zgoubi.MATRIX.out" u ($59):(1-$56) axes x2y1 w p pt 5 ps 0 notit , \
"zgoubi.MATRIX.out" u ((sqrt(($47*BrhoRef*c)**2 + am*am)-am)*eV2MeV):(1-$56) w lp pt 5 lt 1 lw .5 lc rgb "red" \
tit "{/Symbol n}_R", "zgoubi.MATRIX.out" u ((sqrt(($47*BrhoRef*c)**2+am*am)-am)*eV2MeV):($57) axes x1y2 w lp pt 6 lt 3 \
lw .5 lc rgb "blue" tit "{/Symbol n}_y\n" , "zgoubi.MATRIX.out" u ((sqrt(($47*BrhoRef*c)**2-am*am)-am)*eV2MeV)\
:(sqrt((1-$56)**2+$57**2)) w lp pt 7 lt 1 lw .5 lc rgb "black" t "({/Symbol n}_R^2+{/Symbol n}_y^2)^{1/2} \n"; pause 1
```

in zgoubi.res execution listing, at the last iteration by REBELOTE (relative rigidity D = 5.00639):

```
   8 Keyword, label(s) :  MATRIX
 Reference particle (#    1), path length :   396.12091     cm  relative momentum :    5.00639

          TRANSFER  MATRIX  ORDRE  1  (MKSA units)

     0.945800       0.282906       0.00000       0.00000       0.00000      2.913089E-02
    -0.393398       0.940498       0.00000       0.00000       0.00000      0.232240
     0.00000        0.00000       -0.621747      -0.754564     0.00000      0.00000
     0.00000        0.00000        0.812954      -0.621754     0.00000      0.00000
     0.254138       4.384335E-0    0.00000       0.00000       1.00000      3.80443
     0.00000        0.00000        0.00000       0.00000       0.00000      1.00000

  Beam  matrix  (beta/-alpha/-alpha/gamma)  and  periodic  dispersion  (MKSA units)

     0.848142      -0.007948       0.000000      0.000000      0.000000     0.593089
    -0.007948       1.179123       0.000000      0.000000      0.000000     0.009938
     0.000000       0.000000       0.963418      0.000005      0.000000     0.000000
     0.000000       0.000000       0.000005      1.037971      0.000000     0.000000
     0.000000       0.000000       0.000000      0.000000      0.000000     0.000000
     0.000000       0.000000       0.000000      0.000000      0.000000     0.000000

                          wave numbers
              NU_Y =  0.54102699E-01    NU_Z =  0.64321084
```

The radial wave number ν versus $1 - \nu$ indetermination can be lifted by considering that $k = 0$ so that $\nu_R \approx \sqrt{1 + k} \approx 1$, thus, actually, $\nu_R = 1 - 0.054102699 = 0.945897301$.

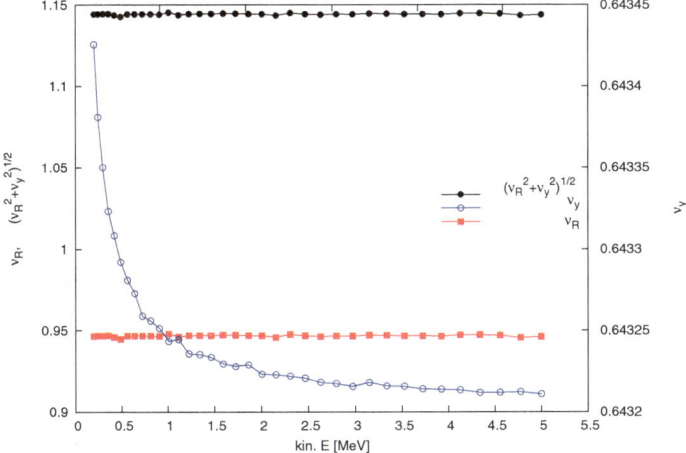

Fig. 4.11 A scan of the wave numbers as a function of proton energy in the cyclotron, with $f = 0.9$ and $k = 0$ here. Fluctuations stem from the use of a field map—performing the scan using DIPOLE analytical field model instead, would yield smooth curves

MATRIX allows a PRINT command (Table 4.5), which causes the transport coefficients to be logged in zgoubi.MATRIX.out as REBELOTE iterates; reading from the latter (gnuplot script given at the bottom of Table 4.5) yields Fig. 4.11. Results appear reasonably close to theoretical approximations $\nu_R \approx \sqrt{1+k} = 1$, $\nu_y \approx F = 0.6364$ and $(\nu_R^2 + \nu_y^2)^{1/2} \approx (1 + F^2)^{1/2} = 1.185$ (Eq. 4.6). The smaller the orbit scalopping (modulation $f \to 0$), the better the agreement (see Table 4.6).

(e) Flutter.

The axial wave number writes (Eq. 4.8) $\nu_y \approx \sqrt{-k + F^2} = F$. The flutter is given by $F = \left(\frac{<(\mathcal{F} - <\mathcal{F}>)^2>}{<\mathcal{F}>^2} \right)^{1/2}$ (Eq. 4.5). The field modulation used here expresses as $\mathcal{F} = 1 + f \cos N\theta$, and $f = 0.9$. From this, one gets

$$< \mathcal{F} > = \frac{2}{\pi} \int_0^{\pi/2} (1 + f \cos N\theta) \, d\theta = 1$$

$$< \mathcal{F}^2 > = \frac{2}{\pi} \int_0^{\pi/2} (1 + f \cos N\theta)^2 \, d\theta = 1 + \frac{f^2}{2}$$

$$F = \left(\frac{< \mathcal{F}^2 > - < \mathcal{F} >^2}{< \mathcal{F} >^2} \right)^{1/2} = \frac{f}{\sqrt{2}} = 0.6364$$

theoretical $\nu_y = F = 0.6364$

Table 4.6 Wave number values in the case k = 0, depending upon the field modulation f, from numerical raytracing ("ray-tr." column) and from Eqs. 4.6, 4.7, namely, $\nu_R = 1$ and $\nu_y = F = f/\sqrt{2}$

	Wave numbers					
	Radial, ν_R		axial, ν_y		$(\nu_R^2 + \nu_y^2)^{1/2}$	
f	ray-tr.	Eq. 4.6	ray-tr.	$f/\sqrt{2}$	ray-tr.	$(1 + f^2/2)^{1/2}$
0.05	0.9999	1	0.0365	0.03535	1.0006	1.0006
0.1	0.9993	1	0.0730	0.0707	1.0020	1.0025
0.2	0.997	1	0.1459	0.1414	1.0076	1.0100
0.34	0.994	1	0.2185	0.2121	1.0177	1.0223
0.6	0.975	1	0.4338	0.4243	1.0671	1.0863
0.9	0.945	1	0.6433	0.6364	1.1432	1.1853

To assess wave numbers for different values of f, a series of field maps is to be computed, as in (a), one for each f value. The outcomes of both numerical integration as in (d), and theoretical calculation as above, for different values of the field modulation factor f, are summarized in Table 4.6. Discrepancy grows with greater modulation, as Eq. 4.6 is a weak-modulation approximation [11, Sect. 3].

4.2 Designing an Isochronous AVF Cyclotron

(a) R-dependent field index.

A field index k(R) proper to ensure R-independent revolution period has to result in (Eq. 4.14)

$$B(R) = \gamma B_0 = \frac{B_0}{\sqrt{1 - (R/R_\infty)^2}} \quad \text{with} \quad B_0 = \frac{M\,\omega_{\text{rev}}}{c^2} = \frac{M}{c^2}\frac{\omega_{\text{rf}}}{h} \quad (4.22)$$

For consistency with similar simulations in the Classical Cyclotron Chap. 3, the following hypotheses are considered:

(i) injection energy $E_{\text{inj}} = 200\,\text{keV}$,

(ii) average radius $R_{\text{inj}} = 0.129248888\,\text{m}$ at that energy,

(iii) average field $B_{\text{inj}} = B(R = R_{\text{inj}}) = 0.5\,\text{T}$.

From this one gets ω_{rev}, the same at all R assuming isochronism, thus in particular

$$\omega_{\text{rev}} = \frac{c^2 B_{\text{inj}}}{M\,\gamma_{\text{inj}}} = 2\pi \times 7.62096882 \times 10^6\,\text{rad/s} \quad \text{wherein} \quad M\,\gamma_{\text{inj}} = M + 200 \times 10^3$$

with $M = 938.27208 \times 10^6\,eV/c^2$, proton rest mass. In this exercise h = 1 is assumed, thus (Eq. 4.13)

$$R_\infty = \frac{c}{\omega_{\text{rf}}} = \frac{2.99792458 \times 10^8}{7.62096882 \times 10^6} = 6.2608118\,\text{m}$$

Using Eq. 4.22 the value $B_0 \equiv B(R = 0)$ results, namely, $B_0 = B_{inj}\sqrt{1 - (R_{inj}/R_\infty)^2}$
$= 4.9989344$, so, finally,

$$B(R) = \frac{B_0}{\sqrt{1 - (R/R_\infty)^2}} = \frac{4.9989344}{\sqrt{1 - (R/6.2608118)^2}}$$

The Fortran program geneAVFMapIsochro.f given in Table 4.7 constructs the map for the $B(R, \theta)$ field distribution. It is derived from the Fortran program of Exercise 4.1 (Table 4.3) by accounting for the isochronism field dependence properties above. In that file, the modulation factor f can be changed, as well as the field index k and the angular extent of the field map, AT. The resulting field distribution over 360 deg is essentially as in Fig. 4.4 as the radial dependence of the field is weak: $B(R) = \gamma B_0$ whereas $\gamma \approx 1$, varying from 1.00000128 to 1.00745 over $R : 10 \rightarrow 76$ cm.

For the purpose of comparisons, four field maps are created and resorted to. Three only differ by the value of the modulation coefficient (Table 4.7): $f = 0$, 0.2, and 0.9, an additional one is a "classical cyclotron" case ("Bcst" index, for constant $B(R, \theta)$). In the latter case in addition
- BR = 1 is substituted to BR = 1/sqrt(1-(R/Rinfty)**2) and
- B0 = T2kG/2 is substituted to B0 = T2kG*Bp2k*sqrt(1-(Rp2k/Rinfty)**2).

In the following these field maps are handled under the following respective names:

geneAVFMap_360deg_f0_isochro.out, geneAVFMap_360deg_f.2_isochro.out,
geneAVFMap_360deg_f.9_isochro.out and geneAVFMap_360deg_Bcst.out.

The input data file to raytrace ion orbits is given in Table 4.9. The FIT procedure finds the closed orbit for the particle defined by OBJET, REBELOTE repeats for a series of additional rigidities in the range 1.1–5×BORO (BORO = 64.624444037 kG cm).

The exercise has been done for modulation factors f = 0, 0.2, or 0.9, and as well for a constant field $B_Z(R, \theta) = 5$ kG, as described above. The latter simulation shows a great difference in the R dependence of the revolution time, compared to the two isochronous cases. The sole cases $f = 0.2$ and $f = 0.9$ simulate an AVF cyclotron, yielding stable axial motion; in the other two cases (f = 0 and constant B) there is no axial focusing, axial motion is unstable.

The resulting sets of closed trajectories are displayed in Fig. 4.12, the R dependence of the revolution period in the four cases is given in Fig. 4.13. The revolution period on the injection orbit for each of the four cases is given in Table 4.8.

(b) Wave numbers.

The energy dependence of wave numbers can be obtained by applying the procedure of Exercise 4.1-d.

4.3 Acceleration to 200 MeV in an AVF Cyclotron
(a) Sufficient modulation has to be considered, for the axial focusing to be efficient up to highest γ (compensating the increase in $k(R)$), namely (Eq. 4.8),

Table 4.7 A Fortran program, geneAVFMapIsochro.f, which generates an $AT = 360°$ mid-plane field map of an isochronous cyclotron. AT as well as the field amplitude ($B_0 = 5$ kG, here) and its modulation ($f = 0.2$, here) can be changed, a field index ($ak = 0$, here) can be accounted for. The field map produced is logged in geneAVFMapIsochro.out, it may be saved under a different name for the purpose of the exercise, depending upon f, k, or AT values

```
C geneAVFMapIsochro.f program
      implicit double precision (a-h,o-z)
      parameter (c=2.99792458d8, am=938.27208d6, T2kG=10.d0)
      parameter (pi=4.d0*atan(1.d0), BY=0.d0, BX=0.d0, Z=0.d0, dZ=0.d0)

      open(unit=2,file='geneAVFMapIsochro.out')
C------------ Hypotheses :
      AT = 360.d0  /180.d0*pi        ! Angular extent of field map. Can be changed 360, 60 deg, etc.).
      f =.9d0                         ! azimuthal modulation factor.
      Bp2k =0.5d0  !For consistency with other exercises, assume a 0.5T average field at 200keV energy.
      Rp2k = 0.129248888074 ; Ep2k = 200.d3    ! Rp2k: 200keV average radius. Reference kinetic energy.
      ah = 1.d0                       ! Harmonic number.
      Rmi=1.d0; Rma=76.d0; RM=50.d0   ! cm. Radial extent of field map; reference radius to define mesh.
      dR = 0.5d0 ; NR = NINT((Rma - Rmi)/dR)+1   ! R-distance between nodes in mesh. Number of R-nodes.
      RdA = 1.d0            ! RdA=RM*dA= distance between two nodes along R=RM arc, dA is angle increment.
      NX= NINT(RM*AT /RdA) +1 ; RdA= RM*AT / DBLE(NX -1)  ! exact mesh step at RM, corresponding to NX.
      dA = RdA / RM ; A1 = 0.d0 ; A2 = AT       ! corresponding delta_angle.
      gma = (Ep2k+am)/am ; omgrv = Bp2k*c**2 / (gma * am)
      omgrf = ah * omgrv ; Rinfty = c/omgrf
      B0 = T2kG * Bp2k * sqrt(1-(Rp2k/Rinfty)**2)                         ! Field at R=Rp2k (kG).
C     B0= T2kG * 0.5d0                                                    ! case B=Cst.
      Rinfty = Rinfty *1.d2                                               ! Convert to cm.
C--------------------------------------------------
      write(2,*) Rmi,dR,dA/pi*180.d0,dZ,
     >'       ! Rmi/cm, dR/cm, dA/deg, dZ/cm'
      write(2,*) '# Field map generated using geneAVFMap.f '
      write(2,fmt='(a)') '# AT/rd,  AT/deg, Rmi/cm, Rma/cm, RM/cm,'
     >//' NR, dR/cm, NX, RdA/cm, dA/rd : '
      write(2,fmt='(a,1p,5(e16.8,1x),2(i3,1x,e16.8,1x),e16.8)')
     >'# ',AT, AT/pi*180.d0,Rmi, Rma, RM, NR, dR, NX, RdA, dA
      write(2,*) '# For TOSCA: ',NX,NR,' 1 22.1 1.  !IZ=1 -> 2D ; '
     >//'MOD=22 -> polar map ; .MOD2=.1 -> one map file'
      write(2,*) '# R*cosA (A:0->360), Z==0, R*sinA, BY, BZ, BX '
      write(2,*) '# cm                    cm        cm        kG kG kG '
      write(2,*) '# '
      do jr = 1, NR
       R = Rmi + dble(jr-1)*dR
        do ix = 1, NX
        A = A1 + dble(ix-1)*dA
        BR = 1.d0 / sqrt(1.d0 - (R/Rinfty)**2)
C       BR = 1.d0 !  case B=Cst
        BZ = B0 * BR * (1.d0+f*sin(4.d0*A +pi/2.d0))
        X = R * sin(A); Y = R * cos(A)
        write(2,fmt='(1p,6(e16.8),2(1x,i0),4(1x,e16.8))')')
     >    Y,Z,X,BY,BZ,BX,ix,jr,A,R,BR,BZ/T2kG*c**2/(BR*am)
        enddo
      enddo
      stop
     >' Job complete ! Field map stored in geneAVFMapIsochro.out.'
      end
```

Top and bottom sections of the field map file geneAVFMapIsochro.out. The file starts with an 8-line header, the first one of which is effectively used by zgoubi, *the following 7 are just comments*

```
# Field map generated using geneAVFMapIsochro.f
# AT/rd,  AT/deg, Rmi/cm, Rma/cm, RM/cm, NR, dR/cm, NX, RdA/cm, dA/rd :
#  6.28318531E+00    3.6E+02  1.0E+00   7.6E+01  5.0E+01 151  5.0E-01 315  1.050721E+00  2.101443E-02
# For TOSCA:       315          151 1 22.1 1.  !IZ=1 -> 2D ; MOD=22 -> polar map ; .MOD2=.1 -> one map file
#
# R*cosA (A:0->360), Z==0, R*sinA, BY, BZ, BX
# cm                  cm        cm      kG kG kG
 1.00000000E+00  0.00000000E+00  0.00000000E+00  0.00000000E+00  9.49798755E+00  0.00000000E+00 1 1
 9.99799804E-01  0.00000000E+00  2.00088090E-02  0.00000000E+00  9.48358368E+00  0.00000000E+00 2 1
 9.99199295E-01  0.00000000E+00  4.00096065E-02  0.00000000E+00  9.44046431E+00  0.00000000E+00 3 1
 9.98198715E-01  0.00000000E+00  5.99943846E-02  0.00000000E+00  9.36890554E+00  0.00000000E+00 4 1
.............................
 7.58631023E+01  0.00000000E+00 -4.55957323E+00  0.00000000E+00  9.43869378E+00  0.00000000E+00 312 151
 7.59391464E+01  0.00000000E+00 -3.04073010E+00  0.00000000E+00  9.51078559E+00  0.00000000E+00 313 151
 7.59847851E+01  0.00000000E+00 -1.52066948E+00  0.00000000E+00  9.55422615E+00  0.00000000E+00 314 151
 7.60000000E+01  0.00000000E+00 -1.86146313E-14  0.00000000E+00  9.56873731E+00  0.00000000E+00 315 151
```

A gnuplot script to obtain a similar field plot to Fig. 4.4 *can be found at the bottom of Tab.* 4.3.

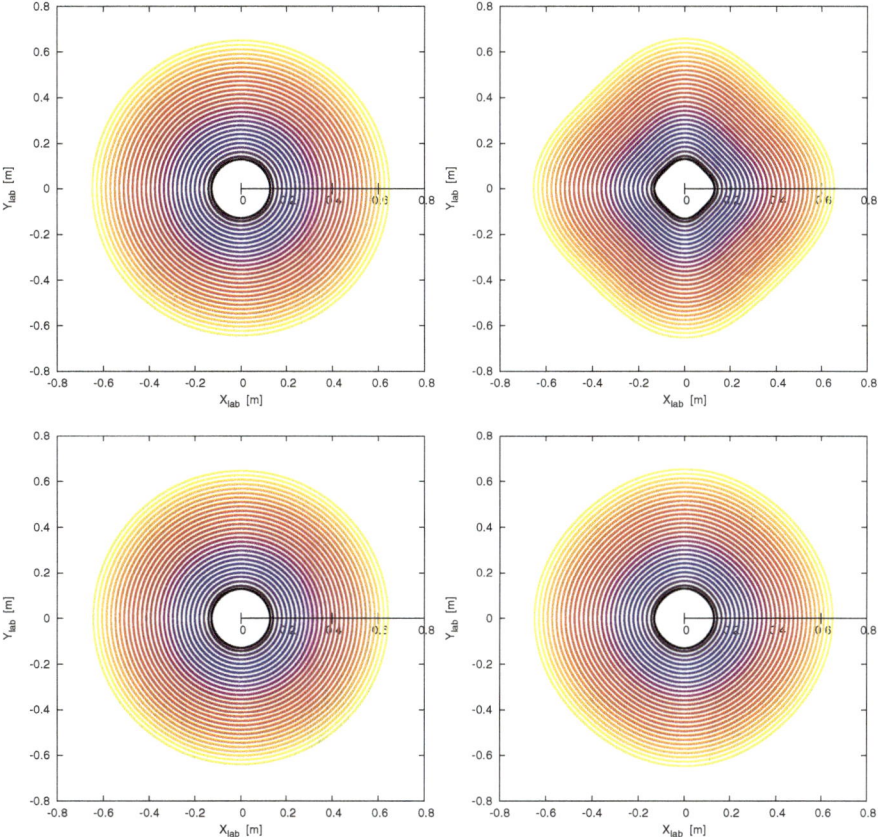

Fig. 4.12 Twenty eight closed orbits in the field of a cyclotron. Top left: constant field $B_Z = 5\,$kG; top right: isochronous $B(R)$ field profile (Eq. 4.15) together with 4-periodic modulation (Eq. 4.3 with N = 4) with f = 0.9; bottom right: same, with f = 0.2; bottom left: same, with f = 0. The f = 0.9 and f = 0.2 cases (right column) satisfy AVF focusing principles, the other two (case of constant B and case f = 0, left column) yield unstable optics due to the absence of axial focusing

$$ f > \beta\gamma\sqrt{2} $$

Assume acceleration of protons, up to over 100 MeV, i.e. $\beta\gamma \gtrsim 0.474$, axial focusing thus requires $f > \beta\gamma\sqrt{2} = 0.67$. A value of f = 0.9 will be taken here.

This results in the 90 deg sector definition given in Table 4.10, which uses a field map with a sufficiently large radial extent, geneAVFMap_90deg_f.9_isochro.out, created using Table 4.7 program. Note that some cyclotron designs feature negative valley field [21] to further increase the flutter (Eq. 4.5) and thus the axial focusing (Eq. 4.6), so potentially allowing higher $k(R)$ and so higher energy (Eq. 4.1).

The voltage gap is simulated using CAVITE[IOPT=7]. Referring to Table 4.8 or Fig. 4.14, the RF frequency has to be around 7.82 MHz, a little tweaking shows that

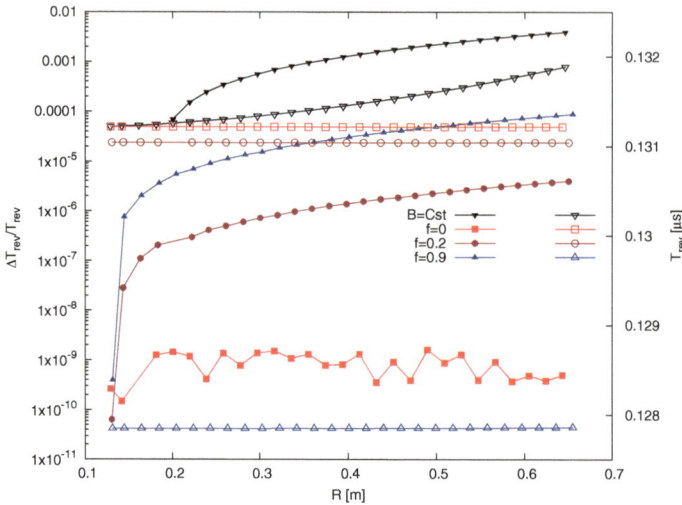

Fig. 4.13 Left vertical scale, solid markers: departure from isochronism in the case of constant field $B_Z = 5$ kG at all (R, θ) (top curve; this is a "classical cyclotron" case) and (from bottom up) of isochronous $B(R)$ (Eq. 4.14) with f = 0, f = 0.2 and f = 0.9 (Eq. 4.3). Right vertical scale, empty markers: revolution time; the "classical cyclotron" case (top curve) features steady increase of revolution time due to mass increase

Table 4.8 Orbit length (C), revolution period (T_{rev}) and revolution frequency ($f_{\text{rev}} = T_{\text{rev}}^{-1}$) at injection, as a function of AVF modulation. Closed orbit length, and thus revolution period, tends to decrease with increasing modulation

f	C (cm)	T_{rev} (μs)	f_{rev} (MHz)
Constant B	81.20948	0.13121691	7.6209688
0	81.20948	0.13121691	7.6209688
0.2	81.1014	0.13104242	7.6311163
0.9	79.12344	0.12784631	7.8218917

$f_{\text{rf}} = 7.7952$ MHz yields efficient use of the RF. A 400 kV peak voltage is applied to the electrode gap. This results in the input data file given in Table 4.11. Acceleration cycles (and deceleration, beyond an RF phase of π) are shown in Figs. 4.15, 4.16.

(b) Energy dependence of wave numbers.

The energy dependence of wave numbers is displayed in Fig. 4.17 in the two modulation cases f = 0.1 and f = 0.9. This simulation has been performed using the input data file of Table 4.12. Two field maps have been generated for that purpose, using the Fortran program in Table 4.7 with proper f values (the latter is the same as used in Exercise 4.2). The argument PRINT under MATRIX causes logging of MATRIX computation outcomes in zgoubi.MATRIX.out, including the wave numbers as plotted in Fig. 4.17.

Table 4.9 Simulation input data file: raytrace a series of closed orbits with different rigidities, spanning 200 keV to 5 MeV

```
Find closed orbits in an [isochronous] 360 degree AVF dipole.
'MARKER'   FieldMapIsochro_S                               ! Just for edition purposes.
'OBJET'
64.62444403717985                    ! Reference Brho ("BORO" in the users' guide) -> 200keV proton.
2
1 1
12.9248888 0.  0. 0. 0.  1. 'm'      ! Closed orbit coordinates at BR=64.6244440 kG.cm for constant B.
1
'PARTICUL'
PROTON
'TOSCA'
0 2                                  ! IL=2 to log step-by-step coordinates, spin, etc., in zgoubi.plt.
1. 1. 1. 1.   ! Normalization coefficients applied to field, and X, Y Z coordinate values read from map.
HEADER_8                             ! The field map file starts with an 8-line header.
315 151 1 22.1 1.           ! IZ=1 for 2D map; MOD=22 for polar frame; .MOD2=.1: only one map file.
geneAVFMap_360deg_f.2_isochro.out    ! Or [...]_360deg_f.9_isochro.out, or [...]_360deg_Bcst.out, etc.
0 0 0 0                   ! Possible boundaries within the field map, to start/stop stepwise integration.
2
.2 ! cm                                                   ! Integration step size.
2                                                         ! Magnet positioning option.
0. 0. 0. 0.                                               ! Magnet positioning.
'FAISCEAU'                                                ! Particle coordinates, here.
'FIT'
1
2  30  0 [5.,100.]                  ! Vary Y0 at OBJET, to allow fulfilling the following constraint:
2  1e-15 200                        ! Penalty 1e-15; a maximum of 200 calls to the function.
3.1 1 2 5 0. .1 0                                        ! request same radius after 360 deg;
3  1 3 5 0. 1. 0                    ! request orbit angle after 360 deg to be zero.

'FAISCEAU' CHECK  ! Allows quick check of particle coordinates, in zgoubi.res: final should = initial.

'FAISTORE'                                     ! Log turn-by-turn particle data in zgoubi.fai.
zgoubi.fai
1

'REBELOTE'                                                ! Repeat what precedes,
27  0.1  0 1                                              ! 27 times.
1
OBJET 35  1.1:5           ! Change the value of parameter 35 (namely, D) in OBJET (prior to repeating).
'SYSTEM'
2
gnuplot < ./gnuplot_Zplt_traj.gnu &
gnuplot < ./gnuplot_Zfai_Trev.gnu &
'MARKER'   FieldMapIsochro_E                               ! Just for edition purposes.
'END'
```

gnuplot script to obtain Fig. 4.12:

```
# gnuplot_Zplt_traj.gnu
set xtics; set ytics; set xlabel 'X_{lab}  [m]'; set ylabel 'Y_{lab}  [m]'; unset colorbox;
set size ratio -1; set polar; cm2m=1e-2 ; nrblt1=27 ; FITlast=1
plot for [FITnb=1:nrblt1] 'zgoubi.plt' u \
($49==FITnb && $51==FITlast ? $22 :1/0):($10*cm2m):($41) w p pt 4 ps .2 lc palette notit ; pause 1
```

gnuplot script to obtain Fig. 4.13. All four cases: B=constant, f=0, 0.2, 0.9, *are plotted together, the respective zgoubi.fai files have been saved under different names for that purpose:*

```
# ./gnuplot_Zfai_Trev_all.gnu
set xtics; set ytics nomirror; set ylabel "{/Symbol D}T_{rev}/T_{rev}"; set y2label "T_{rev} [{/Symbol m}s]"
set xlabel "R [m]"; cm2m=0.01; set key c r maxcol 1; set log y; cm2m=.01; set y2range [.1275:.1325]
# Revolution time on lowest rigidity orbits: 1.6402..., 1.636..., etc. are taken from the respective zgoubi.fai files.
plot \
"zgoubi.Bcst" u ($10*cm2m):(abs($15-1.640561029E-01)/1.640561029E-01 ) w lp pt 11 lc rgb "black"  tit "B=Cst" ,\
"zgoubi.f0"   u ($10*cm2m):(abs($15-1.640211407E-01)/1.640211407E-01 ) w lp pt 5 lc rgb "red"   tit "f=0" ,\
"zgoubi.f2"   u ($10*cm2m):(abs($15-1.638030334E-01)/1.638030334E-01 ) w lp pt 7 lc rgb "brown" tit "f=0.2" ,\
"zgoubi.f9"   u ($10*cm2m):(abs($15-1.598078879E-01)/1.598078879E-01 ) w lp pt 9 lc rgb "blue"  tit "f=0.9" ,\
"zgoubi.Bcst" u ($10*cm2m):($15) axes x1y2 w lp pt 10 ps 1.2 lc rgb "black" notit ,\
"zgoubi.f0"   u ($10*cm2m):($15) axes x1y2 w lp pt 4 ps 1.1 lc rgb "red"   notit ,\
"zgoubi.f2"   u ($10*cm2m):($15) axes x1y2 w lp pt 6 ps 1.1 lc rgb "brown" notit ,\
"zgoubi.f9"   u ($10*cm2m):($15) axes x1y2 w lp pt 8 ps 1.2 lc rgb "blue"  notit ; pause 1
```

Table 4.10 This file provides the simulation of a 90 degree AVF sector, with modulation f = 0.9. It defines an #S_AVFMag_90d_f9 to #E_AVFMag_90d_f9 segment subject to INCLUDE in the input data file of Table 4.11. The END statement is mandatory at the end of an INCLUDE file

```
! fieldMap90deg_f9.inc
'MARKER'    #S_AVFMag_90d_f9
'TOSCA'
0 0                      ! IL=20 to log coordinates, spin, etc., in zgoubi.plt every other 10 integration step.
1. 1. 1. 1.  ! Normalization coefficients applied to field, and X, Y Z coordinate values read from map.
HEADER_8                                          ! The field map file starts with an 8-line header.
80 151 1 22.1 1.            ! IZ=1 for 2D map; MOD=22 for polar frame; .MOD2=.1: only one map file.
geneAVFMap_90deg_f.9_isochro.out                  ! Or [...]_f.1_isochro.out, or [...]_Bcst.out, etc.
0 0 0 0                     ! Possible boundaries within the field map, to start/stop stepwise integration.
2
.2 ! cm                                                         ! Integration step size.
2                                                          ! Magnet positioning option.
0. 0. 0. 0.                                                   ! Magnet positioning.
'MARKER'    #E_AVFMag_90d_f9
'END'
```

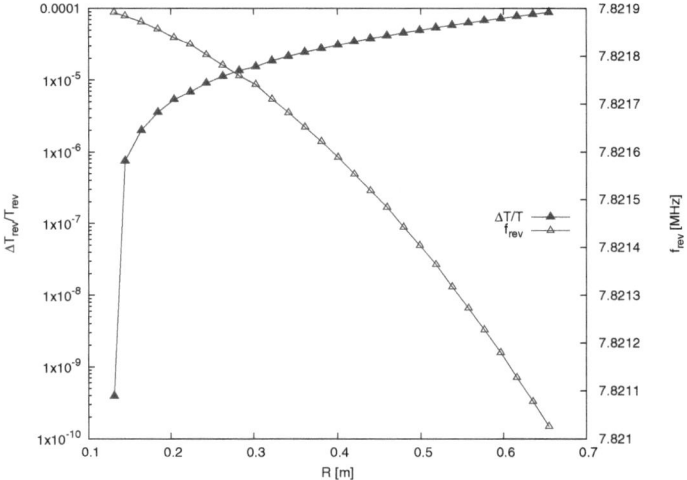

Fig. 4.14 Left vertical scale, solid markers: departure from isochronism as a function of closed orbit radius, case of f = 0.9. Right vertical scale, empty markers: revolution frequency

The theoretical upper limit in energy, for axial stability, is determined by $\beta\gamma < f/\sqrt{2}$, i.e.,

– a theoretical 2.4 MeV for f = 0.1, confirmed in this simulation, and
– a much higher 175 MeV for f = 0.9, whereas this simulation yields 280 MeV (as Eq. 4.6 is a weak modulation approximation).

4.4 Thomas-BMT Spin Precession in Thomas Cyclotron

Simulations use files developed in Exercise 4.3, with *ad hoc* modifications.

Helion is specified using PARTICUL. This determines the value of the gyromagnetic anomaly, as well as mass and charge as they are needed to solve the differential

Table 4.11 Simulation input data file: acceleration gaps (two CAVITE) are added between two 180 deg sectors

```
Uniform field sector. INCLUDE file FieldMapSector.inc.
'MARKER'   FieldMapAVFAccel_S                                    ! Just for edition purposes.
'OBJET'
64.62444403717985                              ! Reference Brho ("BORO" in the users' guide) -> 200keV proton.
2
1 1
13.1650 0. 2. 0. 0. 1. 'm'   ! Closed orbit coord. at BR=64.6244440 kG.cm for f=0.9, isochronous B(R).
1
'MARKER'   AVFAccel400kV_S
'PARTICUL'
PROTON
'FAISTORE'
zgoubi.fai
1
'INCLUDE'
1
2 * fieldMap90deg_f9.inc[#S_AVFMag_90d_f9:#E_AVFMag_90d_f9]                   ! 180 deg sector.
'FAISCEAU'                                               ! Particle coordinates, here.
'CAVITE'   GAP1
7  PRINT                          ! PRINT: log CAVITE computational data in zgoubi.CAVITE.out.
0.00  7.7952e6
400e3    -1.57079632679                         ! Peak voltage;, relative phase of 1st cavity.
'INCLUDE'
1
2 * fieldMap90deg_f9.inc[#S_AVFMag_90d_f9:#E_AVFMag_90d_f9]                   ! 180 deg sector.
'FAISCEAU'                                               ! Particle coordinates, here.
'CAVITE'   GAP1
7  PRINT                          ! PRINT: log CAVITE computational data in zgoubi.CAVITE.out.
0.00  7.7952e6
400e3    +1.57079632679                         ! Peak voltage;, relative phase of 2nd cavity.
'MARKER'   AVFAccel400kV_E
'REBELOTE'                                      ! Repeat what preceds, for a total of 399+1=400 passes.
399 0.1 99
'FAISTORE'
zgoubi.fai
1
'FAISCEAU'
'SYSTEM'
1                                                              ! 1 SYSTEM command follows.
/usr/bin/gnuplot < ./gnuplot_CAVITE.gnu       ! Plot Ek versus phase, as read from zgoubi.CAVITE.out.
'MARKER'   FieldMapAVFAccel_E                            ! Just for edition purposes.
'END'
```

Fig. 4.15 Acceleration followed by deceleration, case of f = 0.9, for three different RF: 7.7942, 7.7952 and 7.7962 MHz

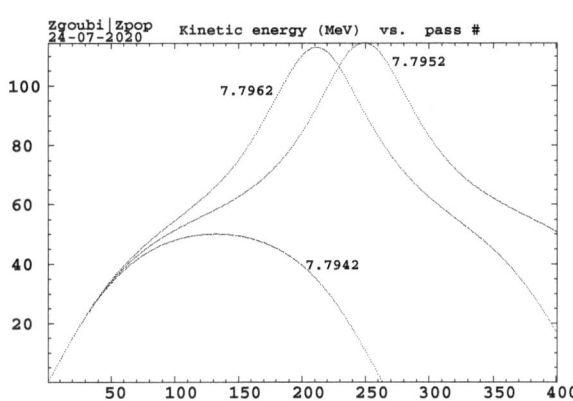

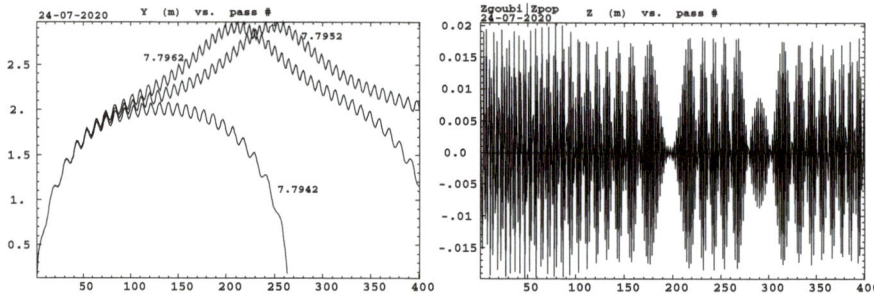

Fig. 4.16 Left: radial excursion during acceleration and deceleration, case of f = 0.9, for the three different RF 7.7942, 7.7952 and 7.7962 MHz. Right: axial excursion, case of f_{rf} = 7.7952 MHz

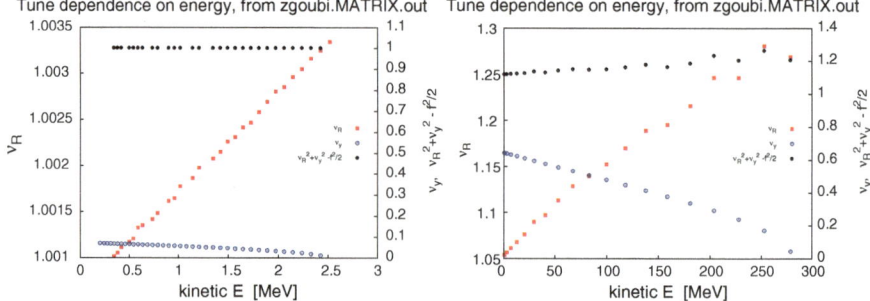

Fig. 4.17 The left and right graphs are for respectively f = 0.1 and f = 0.9 modulation factor. Left vertical scale: radial wave number. Right vertical scale: axial wave number and $v_R^2 + v_y^2 - f^2/2$, the latter expected constant and close to 1 in the small scalloping/weak modulation approximation (Eq. 4.8). The upper limit in energy is determined by v_y decreasing to zero, namely, around 2.4 MeV for f = 0.1, around 280 MeV for f = 0.9

equation of spin motion (Eq. 3.30). PARTICUL results in the following print out in zgoubi.res execution listing:

```
Particle  properties :
HELION
              Mass        =    2808.39      MeV/c2
              Charge      =    3.204353E-19   C
              G factor    =    -4.18415

        Reference  data :
              mag. rigidity (kG.cm)    :    64.624444     =p/q, such that dev.=B*L/rigidity
              mass (MeV/c2)            :    2808.3916
              momentum (MeV/c)         :    38.747842
              energy, total (MeV)      :    2808.6589
              energy, kinetic (MeV)    :    0.26729246
              beta = v/c               :    1.3795851874E-02
              beta*gamma               :    1.3797164913E-02
              G*gamma                  :    -4.184552032
```

Table 4.12 Simulation input data file: energy dependence of wave numbers. The INCLUDE uses the TOSCA segment defined in Table 4.10

```
Scan momentum.
'MARKER'   FieldMapAVFEdep_S                              ! Just for edition purposes.
'OBJET'
64.62444403717985                   ! Reference Brho ("BORO" in the users' guide) -> 200keV proton.
5
.01 .1  .01 .1 .001 .0001
13. 0. 0. 0. 0. 1.
'PARTICUL'
PROTON
'SPNTRK'
4.1
0. 0. 1.
'INCLUDE'
1
4 * fieldMap90deg_f9.inc[#S_AVFMag_90d_f9:#E_AVFMag_90d_f9]          ! 90 deg field extent.
'FAISCEAU'                                               ! Particle coordinates, here.
'FIT'
1
2  30  0  [1.,1e3]
1  1e-10 49
3.1 1 2 7 0. 1. 0
'FAISCEAU'
'MATRIX'
1 11   PRINT
'FAISTORE'
zgoubi.fai
1
'REBELOTE'                                               ! Repeat what precedes,
37  0.1  0 1                                             ! 37 times.
1
OBJET 35  1.1:3.55      ! Change the value of parameter 35 (namely, D) in OBJET (prior to repeating).
'SYSTEM'
1
gnuplot <./gnuplot_MATRIX_Qxy.gnu
'MARKER'   FieldMapAVFEdep_E                              ! Just for edition purposes.
'END'
```

---

gnuplot script to obtain Fig. 4.17:

```
set xlabel "kinetic E  [MeV]"; set ylabel "{/Symbol n}_R,    {/Symbol g}"  # font "roman,24"
set y2label "{/Symbol n}_y,   {/Symbol n}_R^2+{/Symbol n}_y^2 - f^2/2"; set key c r maxcol 1;
set key spacin 1.9; set xtics; set ytics nomirror;  set y2tics nomirror; V2MV=1e-6; BORO = 64.62444403717985;
am=938.27208e6; am2=am*am; BrRef = BORO *1e-3; V2MV = 1e-6; intQx = 1.; c= 2.99792458e8; f = 0.1 # or 0.9
plot \
"zgoubi.MATRIX.out" u ((sqrt(($47*BrRef*c)**2+am2)-am)*V2MV):(intQx+$56) w p pt 5 lc rgb "red" tit "{/Symbol n}_R" ,\
"zgoubi.MATRIX.out" u ((sqrt(($47*BrRef*c)**2+am2)-am)*V2MV):($57) axes x1y2 w p pt 6 lc rgb "blue" tit "{/Symbol n}_y" ,\
"zgoubi.MATRIX.out" u ((sqrt(($47*BrRef*c)**2+am2)-am)*V2MV):((intQx+$56)**2+$57**2-f**2/2.) \
axes x1y2 w p pt 7 lc rgb "black" tit "{/Symbol n}_R^2+{/Symbol n}_y^2 -f^2/2"; pause 1
```

(a) Resonant $G\gamma$.

A preliminary scan of motion wave numbers is performed, using the input file of Table 4.13. This scan shows that, over a kinetic energy range $E_k : 50 \rightarrow 300\,\text{MeV}$, the axial wave number ν_Z decreases from 0.6 to 0.25 about while $G\gamma : -4.25 \rightarrow -4.65$ (Fig. 4.18). It results that at a particular location over that energy range, the relationship

$$G\gamma + \nu_Z = \text{integer} = -4$$

is satisfied, namely here: $G\gamma = -4.4375$.

(b) Helion spin precession.

A spin tracking is launched with the helion ion injected near the $B\rho = 64.624444\,\text{T m}$ closed orbit, namely, $R_{inj} \approx 13\,\text{cm}$ and angle $T_{inj} = 0$, with non-zero axial

Table 4.13 Simulation input data file: a scan of wave numbers, computed using OBJET[KOBJ = 5] and MATRIX, in 74 steps over a relative rigidity range $D : 1 \rightarrow 36$, i.e., helion rigidity $B\rho : 64.624444 \rightarrow 36 \times 64.624444$ T m, energy $E : 0.267292 \rightarrow 2.326479$ MeV. The INCLUDE uses the TOSCA segment defined in Table 4.10. FIT finds particle closed orbit and spin $\mathbf{n}_0$ vector, prior to MATRIX computation

```
MATRIX scan.
'OBJET'
64.6244440                             ! Reference Brho ("BORO" in the users' guide) -> 200keV proton.
5                                      ! KOBJ=5 to define an 11 paticle sample for use by MATRIX.
.001 .01  .001 .01 .001 .0001
13. 0. 0. 0. 0. 1.  ! Initial Y is taken close to its periodic value, FIT will find its precise value),
                                       ! whereas REBELOTE changes it at each of the 73 repeat.
'PARTICUL'
HELION
'INCLUDE'
1
4* ./fieldMap90deg_f9.inc[#S_AVFMag_90d_f9:#E_AVFMag_90d_f9]
'FIT'
1
1 30  0  [1.,1e3]                             ! Vary Y0 (parameter 30) in OBJET (element 1).
2  1e-5 49    ! Periodic orbit constraints apply after a half-turn (i.e., after first 180 deg sector):
3.1 1 2 7 0. 1. 0                             ! particle 1 radius unchenged,
3    1 3 7 0. 1. 0                            ! particle 1 angle T=0.

'FAISCEAU'
'MATRIX'
1 11   PRINT
'REBELOTE'                                    ! Repeat what precedes.
73  0.1  0 1                                  ! 73 times.
1
OBJET 35  1.001:36    ! Change the value of parameter 35 (namely, D) in OBJET (prior to repeating).

'SYSTEM'                  ! SYSTEM is executed in sequence, i.e., when REBELOTE is done.
1
/usr/bin/gnuplot < ./gnuplot_MATRIX_Qxy.gnu
'END'
```

gnuplot script to obtain Fig. 4.18:

```
set xlab "|G{/Symbol g}|"; set x2lab "kinetic E [MeV]"; set ylab "{/Symbol n}_Z, |G{/Symbol g}|-4"; set y2lab "S_Z"
set key c l maxcol 2; set key spacin +1.5; set xtics nomirror; set x2tics; set ytics nomirror; set y2tics
# Particle data:
BORO = 64.62444403717985; q = 2.; amu = 931.4940954e6; am=3.01493224673 * amu; G = 4.1841538; am2=am*am
BrhoRef = BORO *1e-3; eV2MeV = 1e-6; c = 2.99792458e8
# Scales:
Gg1=4.25 ; Gg2=4.65; E1=(Gg1/G-1.)*am/1e6 ; E2=(Gg2/G-1.)*am/1e6; set xrange [Gg1:Gg2] ; set x2range [E1:E2]
plot  \
"zgoubi.fai_spin" u ($25/$29*G):($22) axes x1y2 w lp ps .4 tit "S_Z" ,\
"zgoubi.MATRIX.out_73Qs" u (((sqrt(($47*BrhoRef*c*q)**2 + am2)-am)+am)/am*G):($57) axes x1y1 w p pt 6 lc rgb "blue"  ,\
"zgoubi.MATRIX.out_73Qs" u ((sqrt(($47*BrhoRef*c*q)**2 + am2)-am)/1e6):(((sqrt(($47*BrhoRef*c*q)**2 \
+ am2)-am)+am)/am*G-4.) axes x2y1 w p pt 7 lc rgb "red" tit "|G{/Symbol g}|-4}" ; pause 1
```

amplitude in order to excite the spin resonance, namely, $Z_{\text{inj}} = 2$ cm (vertical take-off angle $P_{\text{inj}} = 0$). The simulation file is given in Table 4.14. Acceleration is over $G\gamma : -4.18 \rightarrow -4.75$, the axial wave number decreases from 0.64312 to 0.23011.

Figure 4.18 displays the vertical spin component, flipping from $+1$ to -1; a close inspection of raytracing outcomes confirms the location of the resonance at $G\gamma_R = -4 - 0.4375$.

(c) Spin resonance crossings. Resonance strength.

This exercise is performed by repeating the simulation of Table 4.14 for a series of different Z_0 values; outcomes are displayed in Fig. 4.19.

As expected (Eq. 4.19) $S_{Z,f}/S_{Z,i}$ tends toward 1 (respectively, toward -1), as the strength of the resonance tends toward zero (respectively, goes $\gg 0$), Fig. 4.20. The

Table 4.14 Simulation input data file: spin tracking through the $G\gamma + \nu_Z = -4$ resonance

```
Track spin through resonance.
'OBJET'
64.62444403717985                        ! Reference Brho ("BORO" in the users' guide) -> 200keV proton.
2
1 1
13.1650 0. 2. 0. 0. 1. 'm'    ! Closed orbit coord. at BR=64.6244440 kG.cm for f=0.9, isochronous B(R).
1
'PARTICUL'
HELION
'SPNTRK'
3
'FAISTORE'
zgoubi.fai
1
'INCLUDE'
1
2* ./fieldMap90deg_f9.inc[#S_AVFMag_90d_f9:#E_AVFMag_90d_f9]
'CAVITE'    GAP1
3
0.00  0.
20e3  +1.57079632679                              ! Peak voltage;, relative phase of 1st cavity.
'INCLUDE'
1
2* ./fieldMap90deg_f9.inc[#S_AVFMag_90d_f9:#E_AVFMag_90d_f9]
'CAVITE'    GAP2
3
0.00  0.
20e3  +1.57079632679                              ! Peak voltage;, relative phase of 2nd cavity.
'REBELOTE'
3999 1.1 99
'SYSTEM'
1                                                 ! 1 SYSTEM command follows.
/usr/bin/gnuplot < ./../gnuplot_MATRIX_Qxy.gnu
'END'
```

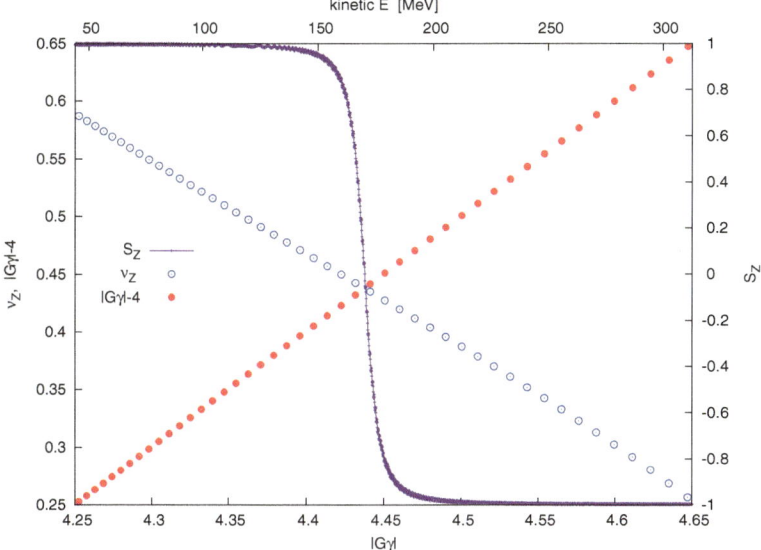

Fig. 4.18 Spin resonance crossing. The graph shows the evolution of the axial wave number ν_Z and of the quantity $|G\gamma| - 4$ (left vertical axis), and of the helion ion spin, initially vertical, $S_Z = 1$ (right vertical axis), as a function of $G\gamma$ (lower horizontal axis) and of energy (upper horizontal axis). ν_Z and $|G\gamma| - 4$ curves cross at $G\gamma = -4.4375$

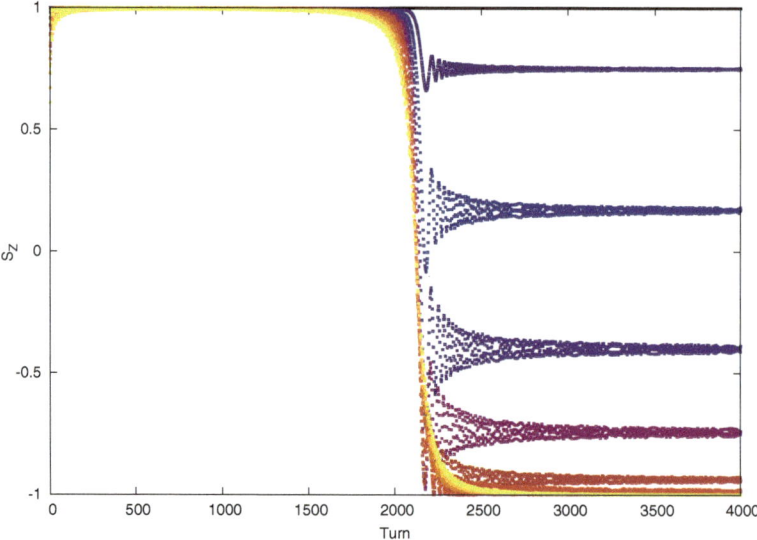

Fig. 4.19 Evolution of S_Z during resonance crossing, for a series of values of the initial axial particle coordinate Z_0. Spin flip occurs at larger Z_0 values

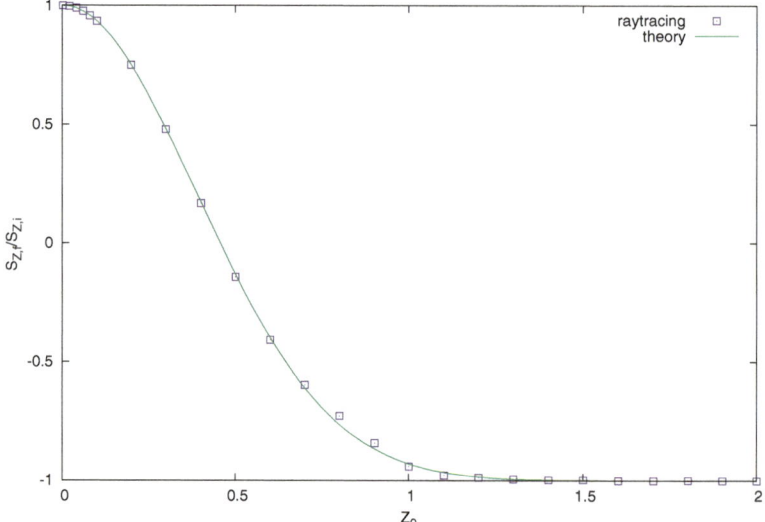

Fig. 4.20 Evolution of $S_{Z,f}/S_{Z,i}$ toward spin flip as the axial motion excursion increases, from raytracing (markers) and from theory (Eq. 4.19 with $|\epsilon_R| \propto Z_0$)

former case corresponds to absence of resonance, i.e., B_Z axial always, as $Z_0 \equiv 0$: the ion motion is in the median plane of the cyclotron dipole. Increasing Z_0 increases the strength of the non-vertical field experienced by the ion as it cycles around the accelerator, and causes spins to undergo greater tilt at traversal of the resonance, toward spin flip with sufficient vertical excursion.

A match of $S_{Z,f}/S_{Z,i}(Z_0)$ to Eq. 4.19 shows that these raytracing outcomes satisfy $|\epsilon_R| \propto Z_0$.

(d) Changing the crossing speed.

The method to answer this question is the same as in (c), repeating the simulation of Table 4.14 for a series of different acceleration rates (i.e. gap voltage, V) in CAVITE, to get $S_{y,f}/S_{y,i}(V)$.

Finally, the relationship to the crossing speed (Eq. 4.20) can be established using the dv_Z/dt data produced in (b) (Fig. 4.18).

4.5 Isochronism and Edge Focusing in a Separated Sector Cyclotron

A separated sector isochoronous cyclotron modeled using DIPOLE.

(a) DIPOLE allows to account for the field fall-off extent at dipole EFBs, which determines the flutter. The input data file for this simulation is given in Table 4.15.

Across the 30 deg sector dipole, a 4-periodic closed orbit undergoes a 90 deg bend, whatever the rigidity. Due to the periodicity and to the field symmetry (the dipole is symmetric with respect to a vertical plane at 15 deg to its EFBs), the closed orbits enter and exit the magnetic sector with angles $TE = TS = 0$.

FIT is used to find the closed orbit at a particular rigidity, the process is repeated (using REBELOTE[IOPT $= 1$]) for a series of different rigidities, in the following way:

– the first constraint under FIT imposes that particle 1 be on a periodic orbit. That constraint is enforced with a weight of 0.1, i.e. greater compared to 1 for the second constraint;
– for that, FIT allows varying B_0, and ends up with the same B_0 always, as expected given $k = 0$. This first constraint is maintained unchanged during the REBELOTE process (which repeats with a different rigidity, yet first changing the relative rigidity D of the second particle - D datum at position 45 in OBJET);
– the second constraint concerns the radial coordinate of closed orbits: it requires that the initial Y coordinate (Y coordinate at OBJET) of particle 1, be equal to its final coordinate (after DIPOLE), a closed orbit condition (Figs. 4.21 and 4.22).

(b) Isochronous B(R).

A similar problem is treated in Exercise 4.6, thus just indications are given here, as to determining a proper radial field law for isochronism.

Table 4.15 Simulation input data file 90degEdgeFocusSector.inc: analytical modeling of a 30 degree magnetic sector of a 4-period separated sector cyclotron. This simulation file includes a search of cyclotron orbits for eight different energies in [0.87, 72] MeV. The LABEL_1s #S_90degCycloSector and #E_90degCycloSector define the dipole segment, for further use in subsequent exercises

```
90degEdgeFocusSector.inc
! Closed orbits and field acroos a 90 degree sector of a 4-period cyclotron
'MARKER'  ProbEdgeFocus_S                                            ! Just for edition purposes.
'OBJET'
1.2493976131130E3         ! Reference Brho, kG.cm ("BORO" in the users' guide) -> case of 72MeV proton.
2
2 1
70. 0. 0. 0. 0. .3 'o'    ! 6.704673 MeV
25.915 0. 0. 0. 0. 0.10789921779517307 'i'    ! Relative rigidity D=0.10789... -> a 0.870 MeV proton.
1 1
'DIPOLE'  #S_90degCycloSector                          ! Analytical field modeling of a dipole magnet.
2    ! IL=2, purpose: log stepwise particle data in zgoubi.plt. Avoid if unused as I/Os take CPU time.
90. 100.                                          ! Sector angle AT; reference radius R0.
45.  12.789066 0. 0. 0.    ! Reference azimuthal angle ACN; BM=B0 field at RM=R0; indices, N, N', N''.
7.  0.                                                ! EFB 1 is  hard-edge.
4 .1455  2.2670  -.6395  1.1558  0. 0.  0.             ! hard-edge only possible with sector magnet.
15. 0.  1.E6  -1.E6  1.E6  1.E6
7.  0.                                                                      ! EFB 2.
4 .1455  2.2670  -.6395  1.1558  0. 0.  0.
-15. 0.  1.E6  -1.E6  1.E6  1.E6
0. 0.                                                  ! EFB 3 (unused).
0 0.     0.     0.      0.      0. 0.  0.
0. 0.  1.E6  -1.E6  1.E6  1.E6 0.
2  10.
1.                        ! Integration step size. The smaller, the more accurately the orbits close.
2 0. 0. 0. 0.                                          ! Magnet positioning RE, TE, RS, TS.
'MARKER'  #E_90degCycloSector
'FAISCEAU' CHECK                                ! Expect intial coordinates = local coordinates, here.
'FIT'
2
3  5 0  1.                             ! Vary field in DIPOLE (constraint is, below: Y_final=Y_OBJET).
2 40 0 [.1,300.]   ! Vary initial coordinate of particle 2 (constraint is, below: Y_final=Y_OBJET).
2
3.1 1 2 #End 0. 0.1 0  ! Constrain particle 1, R=70cm, to being on periodic prbit; great weight (0.1).
3.1 2 2 #End 0. 1.0 0              ! Constrain particle 2 to be on a periodic orbit; weaker weight (1.0).
'REBELOTE'
7 0.1 0 1                           ! IOPT=1 here allows the change of value of parameter 45 in OBJET, below.
1
OBJET 45 .2:1.    ! 7 additional rigidities (follows from REBELOTE[NPASS=7]), from 2.986 to 72 MeV.
'SYSTEM'
2
gnuplot <./gnuplot_Zplt_orbits.gnu                        ! Plot orbits in a quadrant.
gnuplot <./gnuplot_Zplt_field.gnu                         ! Plot field along orbits in a quadrant.
'MARKER'  ProbEdgeFocus_E                                 ! Just for edition purposes.
'END'
```

---

gnuplot script to obtain Fig. 4.21:

```
set xtics; set ytics; set xlabel 'X_{lab}  [m]'; set ylabel 'Y_{lab}  [m]'; cm2m=0.01; set polar
# in zgoubi.plt, col. 19: prticle number; col. 51=1: final past after FIT; col. 22: angle; col. 10: radius
set xrange [0:2.5]; set yrange [0:2.5]; set size ratio -1
plot for [p=1:8] 'zgoubi.plt' u ($19==p && $51==1 ? $22 :1/0):($10 *cm2m) w l notit; pause 1
```

---

gnuplot script to obtain Fig. 4.22:

```
set xtics; set ytics; set xlabel 'X_{lab}  [m]'; set ylabel 'Y_{lab}  [m]'
# in zgoubi.plt, col. 19: prticle number; col. 51=1: final past after FIT; col. 22: angle; col. 25: BZ
plot for [p=1:8] 'zgoubi.plt' u ($19==p && $51==1 ? $22 :1/0):($25) w l notit; pause 1
```

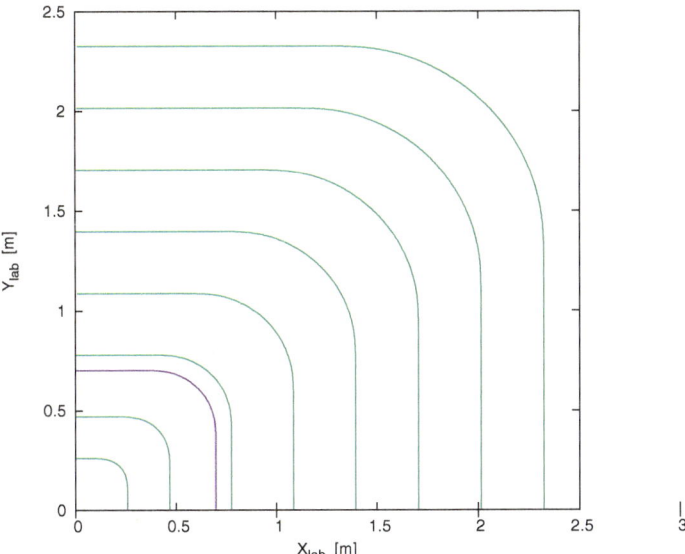

Fig. 4.21 Closed obits across a quadrant, at a few different rigidities, from raytracing

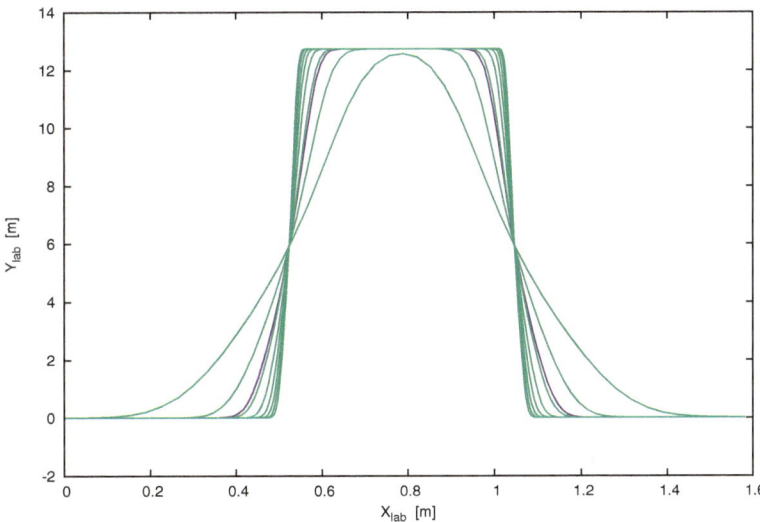

Fig. 4.22 Field along closed obits at different rigidities, over a quadrant, from raytracing

The indices in Eq. 4.21 can be expressed under the form $b_1 = \frac{R_0}{B_0} \frac{\partial B}{\partial R}$, $b_2 = \frac{R_0^2}{2B_0} \frac{\partial^2 B}{\partial R^2}$, $b_3 = \frac{R_0^3}{6B_0} \frac{\partial^3 B}{\partial R^3}$, etc. Expand the $(R - R_0)^i$ terms in Eq. 4.21 and re-organize in increasing powers of R, so writing the radial dependence of the field under the form

$$\mathcal{R}(R) = (1 - b_1 + b_2 - b_3 + b_4 + \ldots) + \frac{R}{R_0}(b_1 - 2b_2 + 3b_3 - 4b_4 + \ldots)$$
$$+ \left(\frac{R}{R_0}\right)^2 (b_2 - 3b_3 + 6b_4 + \ldots) + \left(\frac{R}{R_0}\right)^3 (b_3 - 4b_4 + \ldots) + \ldots \quad (4.23)$$

On the other hand, the Taylor series development of the R-dependent factor of the magnetic field for isochronism, Eq. 4.14, writes

$$\mathcal{R}(R) \approx \frac{1}{\sqrt{1 - (\frac{R}{R_\infty})^2}} = 1 + \frac{(R/R_\infty)^2}{2} + \frac{3(R/R_\infty)^4}{8} + \frac{5(R/R_\infty)^6}{16} + \ldots \quad (4.24)$$

Identify term by term with Eq. 4.23, this yields the indices b_i in terms of powers of $1/R_0$ (R_0 is a known quantity), the very values to be used in defining the field and indices in DIPOLES. Accuracy on isochronism can be improved using FIT[2]: require isochronism (the constraint in FIT[2]) and allow varying the b_i indices in DIPOLES (the variables in FIT[2]) starting from initial values obtained as described above.

(c) Changing field fall-off extent.

Indications:

Changing the fringe field extent λ impacts both the closed orbit landscape and the isochronism. The latter can then be re-optimized by means of FIT, varying the b_i coefficients and constraining, concurrently and over the energy extent of concern, both the orbit periodicity and the isochronism of these orbits. Such a FIT is performed in Exercise 4.6, the same method can be applied here.

(d, e) Flutter, axial wave number.

Indications:

Graphs of $R-$ or $\beta\gamma$-dependence of wave numbers, and relationship to the flutter, are produced in Exercise 4.1, the same techniques can be applied here.

4.6   A Model of PSI Ring Cyclotron Using CYCLOTRON

CYCLOTRON provides a realistic analytical modeling of the field in a radial or spiral sector magnet of a separated sector cyclotron [20, Sect. 6.3 & Part B]. CYCLOTRON keyword belongs in the DIPOLE[S] and FFAG[-SPI] families, with some specificities. The large number of field indices available is one, it simulates pole shaping and allows fine tuning of the isochronism.

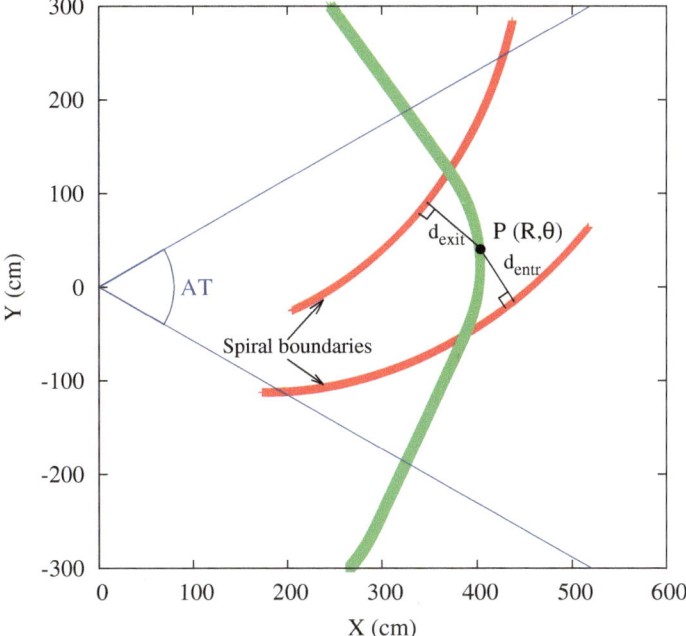

Fig. 4.23 A representation of the EFBs in CYCLOTRON, encompassing a sector of AT angle. The value of the flutter $\mathcal{F}(R, \theta)$ at ion location $P(R, \theta)$ is determined from the distance d to the EFBs. If there are several dipoles within AT, all EFBs are accounted for, in computing the field at $P(R, \theta)$ [23]

(a) CYCLOTRON data list.

A sketch of a PSI cyclotron spiral sector as simulated here, and corresponding to CYCLOTRON input data list of Table 4.2, is given in Fig. 4.23. A commented version, in answer to question (a), is given in Table 4.16. A note on the origin of the data used to simulate a cyclotron sector, in Table 4.2:

(i) The parameters needed in the equation of the spiral effective field boundaries [20, Eq. 6.3.15], and to determine the effective magnetic field length, have been obtained using the magnetic field map of PSI cyclotron [22]. In particular

- at entrance EFB: $\omega = 11$ deg and $\xi\,[deg] = 3.5 + 35.10^{-3}\,r + 3.10^{-8}\,r^3$;
- at exit EFB: $\omega = -8.5$ deg and $\xi\,[deg] = 2 + 12.10^{-3}\,r + 75.10^{-6}\,r^2$.

(ii) The radial field law $\mathcal{R}(R)$ has been obtained by fitting the field fall-off along a series of closed orbits at different radii in the magnetic field map, which yielded the polynomial coefficients b_0 to b_4. A fitting of the 137 MeV closed orbit in particular provided the fringe field coefficients C_0 to C_5.

Table 4.16 Simulation input data file: a period of PSI eight-sector CYCLOTRON model. The data file is set up for a scan of the closed orbits, from radius R = 204.1171097 cm to R = 383.7131468 cm, in 15 steps. Comments have been added, line by line, as a guidance

Note: this file is available in `zgoubi` sourceforge repository at
https://sourceforge.net/p/zgoubi/code/HEAD/tree/branches/exemples/book/zgoubiMaterial/cyclotron_relativistic/ProbPSICyclotron/cyclotron-cell/

```
PSI CYCLOTRON        ! Title. Need one comment line at top of file. More comment lines requires a comment
!                                                              sign, '!', like this one.
'MARKER'  ProbPSICYCLOTRON_S                                  ! Just for edition purposes.
'OBJET'                                          ! Definition of initial particle coordinates.
1249.382414                                                          ! Rigidity [kG cm].
2
1 1
2.67042304E+02 -1.50516664E+01  0.  0.  0.  1.4    'o'
1
'PARTICUL'   ! Type of particle. The only interest here is its allowing computation of time of flight,
PROTON                      ! otherwsie, \zgoubi\ does not need it: it works with the rigidity.

'CYCLOTRON'                            ! Analytical modeling of the field in a separated sector cyclotron.
2                      ! Next line: N, AT, R0 (reference radius), type of sector (radial, spiral, both).
1 45. 276. 1.              ! Next line: ACENT, dR0, FAC, HNORM, K, Rref, field indices b1 to b4.
0. 0. 0.99212277 51.4590015 0.5 800.  -0.476376328 2.27602517e-03 -4.8195589e-06 3.94715806e-09
18.3000E+00  1.  28.  -2.0                          ! lambda=gap, gap's k  g10 g11.
8 1.1024358 3.1291507 -3.14287154 3.0858059 -1.43545 0.24047436 0. 0. 0.  ! NBCOEF, COEFS_C0-7, NORME.
11.0  3.5 35.E-3  0.E-4 3.E-8 0. 0. 0.           ! Entrance EFB: OMEGA, XI0, XI1, XI2, XI3, a,b,c.
18.3000E+00  1.  28.  -2.0                          ! lambda=gap, gap's k  g10 g11.
8 0.70490173 4.1601305 -4.3309575 3.540416 -1.3472703 0.18261076 0. 0. 0. ! NBCOEF, COEFS_C0-5, SHIFT.
-8.5 2.  12.E-3 75.E-6 0.    0. 0. 0.             ! Exit EFB: OMEGA, XI0, XI1, XI2, XI3, a,b,c.
0. -1                                                  ! Lateral EFB, unused.
0 0.  0.  0.  0.  0.  0.                            ! NBCOEF, C0...C5, shift.
0.  0.  0.  0.  0.  0.                              ! omega+, xi, 4 dummies.
2  10.  ! Numerical method for field & derivatives,  flying mesh size is xpas=0.4/10. (KIRD,RESOL).
0.4                                                  ! Integration step size.
2 0.  0.  0.  0.                                    ! magnet positioning.

'FIT2'                                                          ! FIT procedure.
2                                                              ! 2 variables:
2 31 0 [-300.,100]                              ! vary initial angle T0 in OBJET,
2 35 0 [.1,3.]                                  ! vary relative momentum D in OBJET.
2                                                              ! 2 constraints:
3.1 1 2 #End 0. 1. 0                                      ! get Y-Y0=0,
3.1 1 3 #End 0. 1. 0                                      ! get T-T0=0.
'FAISCEAU'

'FAISTORE'                           ! Store coordinates at each pass, in file orbits.fai.
orbits.fai
1

'REBELOTE'                     ! Repeat the complete sequence above, from OBJET, 14 times.
14 0.2  0 1                                            ! IOPT=1 will cause change
1                                      ! of NPRM=1 paameters, as follows:
OBJET 30 281.258209:353.20117  ! in OBJET parameter 30 (Y0) will take 14 values from 281... to 353...

'SYSTEM'                                            ! A "call system". Will execute
1                                                  ! 1 command, as follows:
gnuplot <./gnuplot_orbits.gnu
'MARKER'   ProbPSICYCLOTRON_E                            ! Just for edition purposes.
'END'                          ! End of the sequence. Whatever follows is ignored.
```

gnuplot script to obtain Figs. 4.26 *and* 4.27:

```
# gnuplot_orbits.gnu
set key c t; set xtics; set xlabel "R [cm]"; set ylabel "(T_{rev}-T_{R=314})/T_{R=314}"
colY0=3; T314=1.4737924713529E-02
plot 'orbits.fai' u colY0:(($15-T314)/T314) w lp lt 3 dt 7 lw 2 pt 4 lc rgb "black"; pause 1
#
set xtics nomirror; set xlabel "Y0 [cm]"; set ytics nomirror; set ylabel "Y [cm]"
set x2tics; set x2label "T0 [mrad]"; set y2tics; set y2label "T [mrad]"; colY0=3; colY=10; colT0=4; colT=11
plot 'orbits.fai' u colY0:colY w lp lt 3 dt 7 pt 4 lc 1,'orbits.fai' u colT0:colT axes x2y2 w lp lt 3 pt 4 lc 2
```

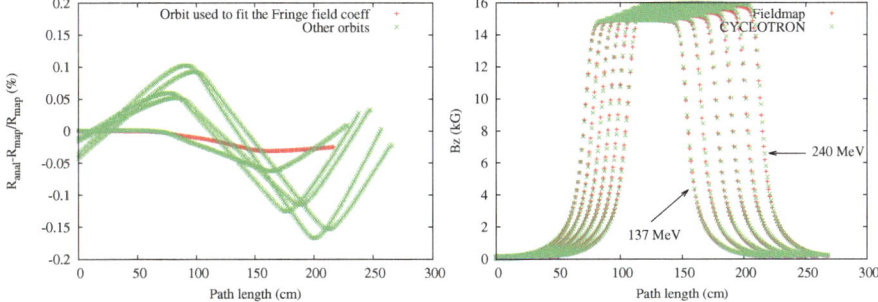

Fig. 4.24 Left: relative difference in radial excursion across a cell, for a few closed orbits (at 137, 156, 176, 196, 218 and 240 MeV), between raytracing in the analytical model CYCLOTRON, and in PSI field map [22]. Right: field profiles along these orbits, using indifferently the analytical model CYCLOTRON, or PSI field map: differences are not noticeable at this scale

(b) Pole and field profiles, closed orbits.

The CYCLOTRON input data file, Table 4.16, can be run for various particle rigidities (change D in the particle coordinates under OBJET[KOBJ=2]), possibly with several particles (OBJET[KOBJ=2,IMAX>1]). IL=2 under CYCLOTRON takes care of storing stepwise particle data in zgoubi.plt, to produce graphs.

Outcomes are illustrated in Fig. 4.24 which compares closed orbit excursions, whether using the analytical CYCLOTRON model, or PSI field map from which it is drawn [22].

Closed orbits at a large number of different rigidities can be raytraced (use OBJET[KOBJ=2,IMAX≫1]), from which a graph of isomagnetic field lines can be produced, by reading from zgoubi.plt. This allows producing Fig. 4.25 in which the EFBs and a series of closed orbits are superposed on a field scale background. Note: the closed orbit for a particular rigidity can be found using FIT, the process can be repeated using REBELOTE[IOPT=1] (as in Table 4.16).

(c) Revolution period.

The Simulation input data file of Table 4.16 performs the scan needed here, comments therein explain the method, which is based on FIT to find the proper rigidity and periodic orbit angle for a particle with periodic radius defined by OBJET, and on REBELOTE[IOPT=1] to repeat the FIT procedure for a new value of the orbit radius, NPASS times.

Figure 4.26 checks the proper completion of the FIT procedure, showing that final orbit radius $R = Y$ (down the magnetic sector) is identical to initial $R = Y_0$ (at OBJET) and final orbit angle T is identical to initial T_0. Note that a global check is provided by the penalty value, an outcome also of the FIT procedure.

Figure 4.27 displays the relative time difference to that of a reference orbit, taken to be orbit number 7 at R = 314 cm in the middle region of the range, bottom of the time of flight parabola.

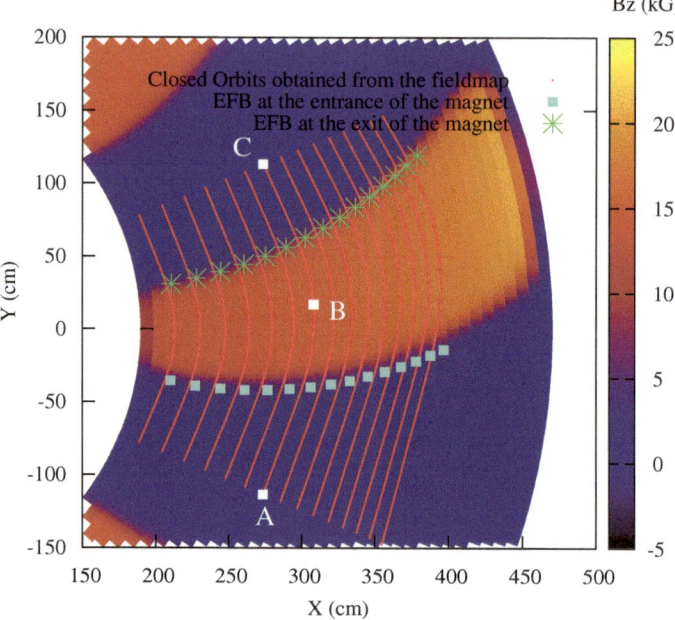

Fig. 4.25 EFBs and field scale, obtained from raytracing using CYCLOTRON or, indifferently, PSI sector field map [22]. B is the location of the maximum field value along the 137 MeV orbit, C is the location of the minimum value in the field valley

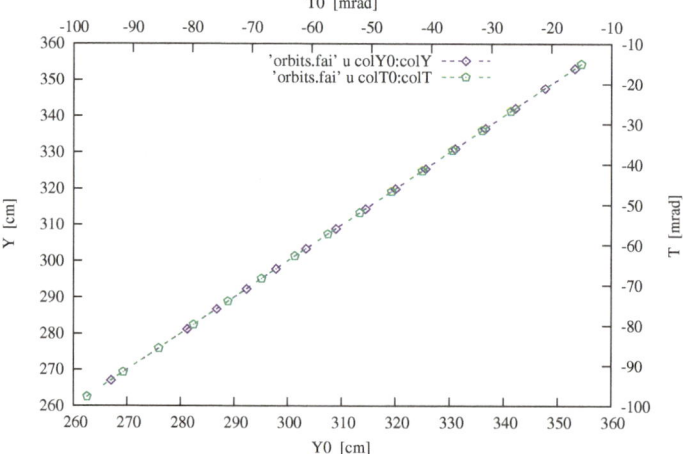

Fig. 4.26 Checking the proper completion of the FIT procedure: final orbit radius $R = Y$ (down the magnetic sector) is identical to initial $R = Y_0$ (at OBJET). The constraint is similar for orbit angles: final orbit angle T identical to initial T_0

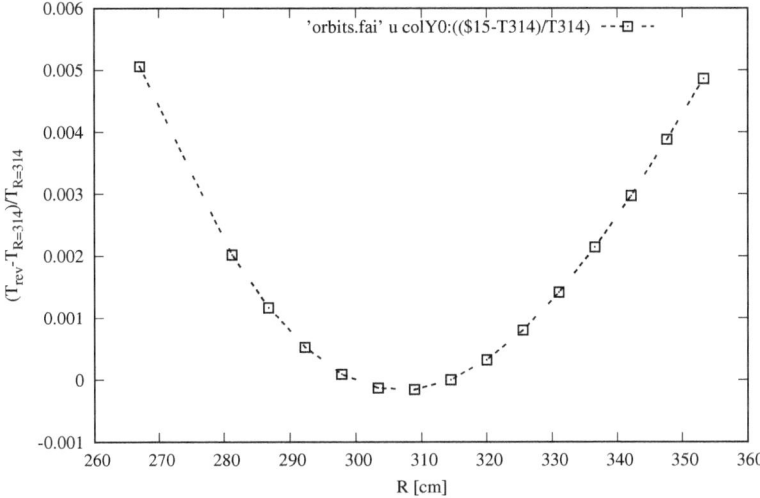

Fig. 4.27 Time of flight difference as a function of closed orbit radius, relative to time of flight $T(R = 314.46264\,\text{cm}) = 0.01473792\,\mu\text{s}$

(d) Improved isochronism.

The number of field indices accounted for in $\mathcal{R}(R)$ (Eq. 4.21) in setting up CYCLOTRON modeling (four only in the case of questions (a) and (b), b_1 to b_4) is increased iteratively, one additional index at a time, and the FIT procedure is re-run each time, until it shows convergence to a series of index values which result in the required degree of isochronism. An additional eight indices allow an improvement of the isochronism by a factor 50. The main field B_0 is also part of the variables, as it allows the orbits to adjust to periodic condition (closed orbits). The simulation data file is given in Table 4.17, including the FIT procedure at the last stage of the iteration on the number of field indices.

Note in the FIT procedure constraints: the time of flight on orbit 4, middle region of the range, bottom of the time of flight parabola, is taken as a reference. It does not act as a constraint in the FIT as its weight is 10^9, compared to 10^{-4} and less for the other 6 orbits. That orbit 4 ends up reaching the revolution time value $T_{\text{ref}} = 0.0147442157\,\mu\text{s}$.

The resulting relative time of flight difference $dT_{\text{rev}}/T_{\text{ref}} = (T_{\text{rev}} - T_{\text{ref}})/T_{\text{ref}}$ as a function of closed orbit radius is displayed in Fig. 4.29, much improved by comparison to the 4 index case (Figs. 4.27 and 4.28).

Table 4.17 Simulation input data file: a period of PSI separated sector cyclotron, using CYCLOTRON for an analytical modeling of the field. The file is setup to FIT 12 field indices, b_1 to b_{12} for improved isochronism. The constant field B_0 is part of FIT variables, to allow for the constraint of orbit periodicity. OBJET[KOBJ = 2, IMAX = 7] creates 7 particles which span the momentum range of interest, via $D : 1.4 \rightarrow 2.1$

Note: this file is available in zgoubi sourceforge repository at

https://sourceforge.net/p/zgoubi/code/HEAD/tree/branches/exemples/book/cyclotron_relativistic/ProbPSICyclotron/adjustIndices

```
'FIT'
27
2 30 0 [200,400]    ! Next 14 variables: initial periodic
2 31 0 [-100,0]     ! radius and angle of the seven orbits.
2 40 0 [200,400]
2 41 0 [-100,0]
2 50 0 [200,400]
2 51 0 [-100,0]
2 60 0 [200,400]
2 61 0 [-100,0]
2 70 0 [200,400]
2 71 0 [-100,0]
2 80 0 [200,400]
2 81 0 [-100,0]
2 90 0 [200,400]
2 91 0 [-100,0]
4 9 0 .2                               ! Vary B0.
4 12 0 1.  ! Next 12 variables: 12 field indices B1-B12.
4 13 0 1.
4 14 0 1.
4 15 0 1.
4 16 0 5.
4 17 0 5.
4 18 0 9.
4 19 0 9.
4 20 0 9.
4 21 0 9.
4 22 0 9.
4 23 0 9.
21 1e-15
3.1 1 2 #End 0. 1. 0      ! First 14 constraints request
3.1 1 3 #End 0. 1. 0            ! periodic radius and angle
3.1 2 2 #End 0. 1. 0            ! for the seven orbits.
3.1 2 3 #End 0. 1. 0
3.1 3 2 #End 0. 1. 0
3.1 3 3 #End 0. 1. 0
3.1 4 2 #End 0. 1. 0
3.1 4 3 #End 0. 1. 0
3.1 5 2 #End 0. 1. 0
3.1 5 3 #End 0. 1. 0
3.1 6 2 #End 0. 1. 0
3.1 6 3 #End 0. 1. 0
3.1 7 2 #End 0. 1. 0
3.1 7 3 #End 0. 1. 0
3.4 1 7 #End 0. .00001 1 4 ! 6 constraints request equal
3.4 2 7 #End 0. .00001 1 4      ! rev. period for all 7
3.4 3 7 #End 0. .00005 1 4           ! particles.
3 4 7 #End 1.4743416128820E-02 1.e9 1 4 ! Reference time
3.4 5 7 #End 0. .0001 1 4        ! is that of particle 4,
3.4 6 7 #End 0. .0001 1 4        ! not actual constraint.
3.4 7 7 #End 0. .00001 1 4
'FAISTORE'   ! Store the 7 particle data ounce FIT done
FITted.fai   ! for a graph, resorting to SYSTEM below.
1
'SYSTEM'
1
gnuplot < gnuplot_Trev.gnu        ! Plot T_rev vs radius.
'MARKER' ProbPSICYCLOTRON_c_E Just for edition purposes.
'END'
```

```
PSI CYCLOTRON. Fit isochronism with 12 indices.
'MARKER'  ProbPSICYCLOTRON_c_S ! Just for edition purposes.
'OBJET'      ! Definition of initial particle coordinates.
1249.382414                         ! Rigidity [kG cm].
2
7 1               ! A set of 7 particles over R=267 to 351 cm).
2.67042304E+02 -1.50516664E+01 0. 0. 0.  1.40000001E+00 'o'
2.81258209E+02 -2.66331145E+01 0. 0. 0.  1.50000001E+00 'o'
2.94829572E+02 -3.87557743E+01 0. 0. 0.  1.60000001E+00 'o'
3.08680463E+02 -5.17104469E+01 0. 0. 0.  1.70778172E+00 'R'
3.19924957E+02 -6.27126203E+01 0. 0. 0.  1.80000001E+00 'o'
3.31557533E+02 -7.45532152E+01 0. 0. 0.  1.90000001E+00 'o'
3.53201171E+02 -9.77647037E+01 0. 0. 0.  2.10037893E+00 'o'
1 1 1 1 1 1 1 ! 7 times 1, for 7 particles (-1 to inhibit).
'PARTICUL'       ! Type of particle, to compute time of flight.
PROTON
'CYCLOTRON'  ! Analytical modeling of the field in a setor.
2
1   45.0 276.  1.0            ! N, AT, R0, type of sector.
0. 0. 0.992122800 51.4311902 0. 0. -4.48715507E-01
2.09658166E-03 -4.52609250E-06 3.9591365E-09 -5.6897260E-14
8.48686076E-17 -2.51326976E-19 4.8763970E-22 -1.5424854E-25
1.75499497E-27 -3.23761721E-29 2.75094168E-32 ! ATTENTION:
!  this line and the previous 2, and the following comment
! line must actually all be on 1 line - no carriage return.
18.3000E+00 1.  28. -2.0   ! lambda=gap, gap's k  g10 g11.
6 1.1024358 3.1291507 -3.1428715 3.085805 -1.4354 0.2404743
11.0  3.5 35.E-3  0.E-4 3.E-8 1. 1. 1.  ! OMEGA,XI0entr, etc
18.3000E+00 1.  28. -2.0   ! lambda=gap, gap's k  g10 g11.
6 0.70490173 4.160130 -4.330957 3.54041 -1.347270 0.1826107
-8.5  2. 12.E-3 75.E-6 0.E-6 1. 1. 1. ! OMEGA,XI0exit, etc.
0. -1                          ! Lateral EFB, unused.
0 0. 0. 0. 0. 0. 0. 0.
0. 0. 0.  0. 0. 0.
2  10.        ! Numerical method for field & derivatives.
1.                           ! Integration step size.
2 0. 0. 0. 0.                      ! Magnet positioning.
```

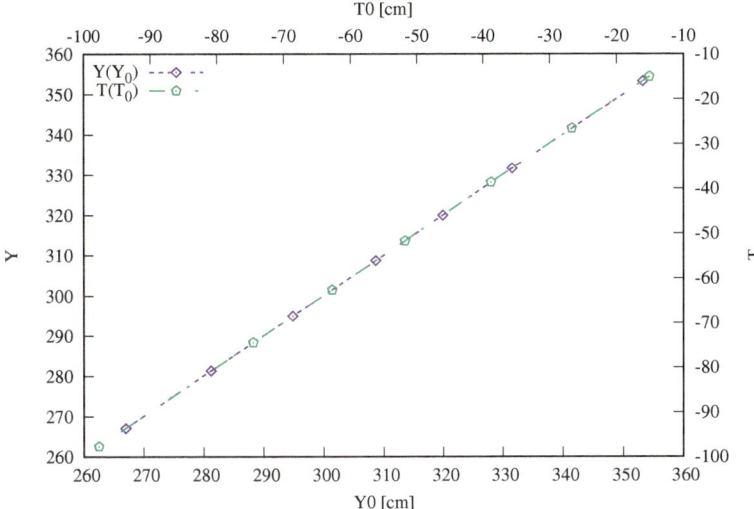

Fig. 4.28 Checking the proper completion of the 12-index FIT procedure: the final orbit radius $R = Y$ (down the magnetic sector) is identical to the initial $R = Y_0$ (at OBJET). Same constraint for orbit angles: the final orbit angle T comes out identical to the initial T_0

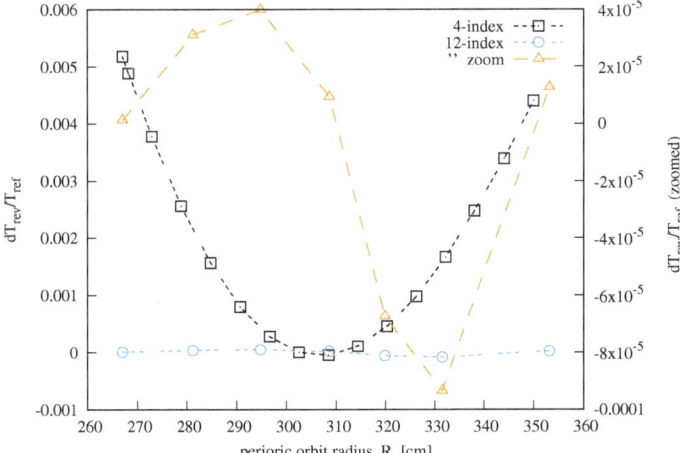

Fig. 4.29 A graph of the improved isochronism with 12 field indices (circles and left vertical axis, and a zoom-in: triangles and right axis; data are read from the file FITted.fai), compared to results obtained in question (**b**) (squares) where 4 indices were used. The isochronism is improved by a factor of $\sim$50

References

1. A link to accelerator conference proceedings and its search tool: https://www.jacow.org/Main/ Proceedings

2. J.M. Schippers, The superconducting cyclotron and beam lines of PSI's new proton therapy facility "PROSCAN", in *17th International Conference on Cyclotrons and their Applications, Tokyo, Japan* (2004). Accessed from 18–22 Oct 2004. http://accelconf.web.cern.ch/c04/data/ CYC2004_papers/20B2.pdf. Figure 4.1: https://accelconf.web.cern.ch/cyclotrons2019/talks/ tub04_talk.pdf Copyrights under license CC-BY-3.0, https://creativecommons.org/licenses/ by/3.0; no change to the material

3. M. Seidel, Production of a 1.3 megawatt proton beam at PSI, in *Proceedings of IPAC'10, Kyoto, Japan, TUYRA03*. http://accelconf.web.cern.ch/IPAC10/talks/tuyra03_talk. pdf. Figure 4.2: https://indico.psi.ch/event/3484/attachments/5948/7494/Cyclotrons_PSI.pdf Copyrights under license CC-BY-3.0, https://creativecommons.org/licenses/by/3.0; no change to the material

4. T. Kawaguchi et al., Design of the sector magnets for the RIKEN superconducting ring cyclotron, in *Proceedings of the 15th International Conference on Cyclotrons and their Applications, Caen, France* (2003). https://accelconf.web.cern.ch/c98/papers/b-14.pdf. Figure 4.3: the picture, and the permission to use it in this publication, have been granted by RIKEN

5. L.H. Thomas, The paths of ions in the cyclotron. Phys. Rev. **54**, 580 (1938)

6. W. Joho, Potential of cyclotrons. https://indico.in2p3.fr/event/115/timetable/?print=1& view=standard

7. L.H. Thomas, Motion of the spinning electron. Nature **117**(2945), 514 (1926)

8. K.R. Symon et al., Fixed-field alternating-gradient particle accelerators. Phys. Rev. **103**, 1837 (1956)

9. A.I. Yavin, The AVF cyclotron. Phys. Today **15**(5), 19–25 (1962). https://doi.org/10.1063/1. 3058175

10. L.B. Cohen, Cyclotrons and synchrocyclotrons, in *Encyclopedia of Physics*, ed. by S. Flügge, Vol. XLIV, Nuclear Instrumentation I. (Springer, 1959)

11. T. Stammbach, Introduction to cyclotrons. CERN accelerator school, cyclotrons, linacs and their applications, IBM International Education Centre, La Hulpe, Belgium (1994). Accessed from 28 April–5 May 1994. Copyright/License CERN CC-BY-3.0. https://creativecommons. org/licenses/by/3.0

12. P.S. Miller et al., The magnetic field of the K500 cyclotron at MSU including trim coils and extraction channels, in *Proceedings of 9th International Conference on Cyclotrons and their Applications, September 1981, Caen, France*. http://accelconf.web.cern.ch/c81/papers/ep-05. pdf

13. H.A. Bethe, M.E. Rose, Phys. Rev. **54**, 588 (1938)

14. F.T. Cole, A memoir of the MURA years (1994). Accessed from April 11 1994. https://epaper. kek.jp/c01/cyc2001/extra/Cole.pdf

15. © The Regents of the University of California, Lawrence Berkeley National Laboratory

16. M.K. Craddock, AG focusing in the Thomas cyclotron of 1938, in *Proceedings of PAC09, Vancouver, BC, Canada, FR5REP1*. http://accelconf.web.cern.ch/PAC2009/papers/fr5rep113. pdf

17. F. Méot, Spin dynamics, in *Polarized Beam Dynamics and Instrumentation in Particle Accelerators, USPAS Summer 2021 Spin Class Lectures*. Springer Nature, Open Access (2023). https:// link.springer.com/book/10.1007/978-3-031-16715-7

18. F. Méot, Spinor methods, in *Polarized Beam Dynamics and Instrumentation in Particle Accelerators, USPAS Summer 2021 Spin Class Lectures*. Springer Nature, Open Access (2023). https://link.springer.com/book/10.1007/978-3-031-16715-7

19. M. Froissart, R. Stora, Depolarisation d'un faisceau de protons polarises dans un synchrotron. Nucl. Instrum. Methods. **7**(3), 297–305 (1960). (June)

20. F. Méot, Zgoubi Users' Guide. https://www.osti.gov/biblio/1062013-zgoubi-users-guide. Sourceforge latest version: https://sourceforge.net/p/zgoubi/code/HEAD/tree/trunk/guide/Zgoubi.pdf
21. P. Mandrillon, Single Stage Cyclotron for an ADS Proceedings of Cyclotrons 2016, Zurich, Switzerland. http://accelconf.web.cern.ch/cyclotrons2016/talks/fra01_talk.pdf
22. M. Haj Tahar, Tutorial - Case Study: PSI Cyclotron, using CYCLOTRON and TOSCA commands in Zgoubi. Zgoubi and OPAL Users Mini-workshop, in FFAG'14 Workshop, BNL (2014). https://www.bnl.gov/ffaworkshop/events/index.php#2014
23. F. Lemuet, F. Méot, Developments in the raytracing code Zgoubi for 6-D multiturn tracking in FFAG rings. NIM A **547**, 638–651 (2005)

Chapter 5
Betatron

Abstract This chapter introduces the betatron fixed orbit cyclic accelerator. It begins with a brief reminder of the historical context, and continues with the Wideröe condition and the principles of fixed orbit acceleration in a betatron. The latter is at the origin of the theory of the "betatron oscillations"—treated in Chaps. 3 and 8. A realistic simulation of a betatron in `zgoubi` would require the simulation of an induction electric field: this is doable from existing dipole models such as DIPOLE[S], and can be seen as an interesting code development exercise. A simpler approach on the other hand only requires two optical elements: DIPOLE and CAVITE. Accounting for synchrotron radiation (SR) energy loss requires SRLOSS. Monte Carlo SR monitoring can use SRPRNT, which logs data in zgoubi.res. SRPRNT[PRINT] in addition logs data in zgoubi.SRPRNT.Out. Electron beam monitoring requires keywords introduced in the previous chapters, such FAISCEAU, FAISTORE. SR monitoring uses SRPRNT. INCLUDE allows simplifying the input data files. Graphs are part of data treatment and simulation outcomes, they are produced using `zpop` or `gnuplot`.

Notations Used in the Text

A; $A_{s,r,y}$	vector potential, **curlA** = **B**; its components
B; $B_{s,r,y}$; $\hat{B}$	magnetic field vector; its components; peak value
B_a	average value of magnetic field circumscribed by the orbit
D	dispersion, $D = \delta x / \delta p/p$
E; $\hat{E}$	electron energy, $E = \gamma m_0 c^2$; peak value
e	elementary charge
E_a	induction accelerating field
E; $E_{s,r,y}$	induction electric field vector; its components
m; m_0; M	electron mass; rest mass; in units of MeV/c^2
$n = -\frac{\rho}{B}\frac{\partial B}{\partial x}$	focusing index
p	electron momentum
R	radius of equilibrium orbit

© The Author(s) 2024   187

F. Méot, *Understanding the Physics of Particle Accelerators*, Particle Acceleration and Detection, https://doi.org/10.1007/978-3-031-59979-8_5

s	path variable
U_s	synchrotron radiation energy loss
x', y'	radial and axial coordinates in the moving frame $\left[(*)' = \frac{d(*)}{ds} \right]$
$\beta = v/c$	normalized velocity
β_u	betatron functions ($u : x, y$)
$\gamma = E/m_0c^2$	Lorentz relativistic factor
δp, Δp	momentum offset
ϵ_c	critical energy of SR, $\epsilon_c = \hbar\omega_c = hc/\lambda_c$
ε_u	Courant–Snyder invariant ($u : x, y$)
ν_u	wave numbers, radial, vertical ($u : x, y$)
ω	angular frequency of the cycling field
ω_c	critical angular frequency of SR, $\omega_c = 3\gamma^3c/2\rho$
ω_{rev}	angular revolution frequency
ω_R, ω_y	betatron frequency, radial, axial
ϕ_a	magnetic flux

5.1 Introduction

The concept of an inductive electric field accelerating electrons, maintained on a constant orbit in a magnetic field, goes back to the early 1920s [1, 2]. That principle of inductive acceleration has been in use over the years in a variety of systems, for instance in synchrotrons for slow extraction [3]; in ion synchrotrons as the acceleration system proper [4]; in induction linacs [5, 6]; in fixed field alternating gradient accelerators (FFAG) [7–9] (Fig. 5.1). The betatron ring is based on this acceleration technique, it is an induction accelerator. The development of the early 1920s concept, toward success in 1940, was fostered by the need for high energy electrons and for X-rays. The earlier classical cyclotron was of no help, it only works for under-relativistic particles, which for electrons means keVs. And five more years would be needed for phase focusing to revolutionize cyclic acceleration.

Betatron acceleration on a constant orbit requires Widerøe's condition, established by the latter in the late 1920s [1]. Transverse confinement in addition requires focusing. Operation of a betatron ring, bringing these two prerequisite together, was first achieved in 1940, with the acceleration of electrons to 2.3 MeV, on the rising slope of a several 100 Hz magnetic field impulse [10, 11] (Fig. 5.2). This was the first cyclic accelerator, and the first electron ring. The energy reach of the technology culminated with a 315 MeV betatron at the University of Chicago, in the late 1940s, with an orbital radius of 1.2 m for a guiding field strength close to 1 T, requiring a 350 ton magnet. At that energy, synchrotron radiation starts causing significant energy loss, which can be somewhat mitigated with additional RF power, yet will eventually jeopardize Widerøe's condition for a fixed orbit. This was a limitation for higher energy. The betatron on the other hand is inefficient to accelerate ions from

Fig. 5.1 Induction acceleration in (left) the MURA 180 keV spiral sector FFAG electron model, circa 1957 [7], and in (right) the Ion Beta 2.5 MeV spiral sector injector of KURRI 150 MeV ADS FFAG facility, circa 2005 [9]. In both cases, the induction system appears as a large "betatron core" (red tori, in Ion Beta) which surrounds an arc of the ring

Fig. 5.2 A quite special betatron: first to be operated, in 1940 [10, 11], it was also used for the first demonstration of a synchrotron [12, 13]. In 1946, in UK, equipped with an RF gap, it accelerated electrons from 4 MeV where they were brought in betatron mode, up to 8 MeV using resonant acceleration on a fixed orbit

low energy, as the energy gain which the acceleration pulse allows is small (ions at $v \ll c$ perform much less revolutions than electrons at $v \approx c$, during the pulse); let alone the large magnet that this would require.

From a historical standpoint the betatron has this particularity that it was used to demonstrate the concept of longitudinal phase stability in 1946 [12] (preceding shortly the synchro-cyclotron operation of a 37-in classical cyclotron [14]). This synchrotron mode of operation of a betatron pioneered the principle of resonant, cyclic acceleration on a fixed orbit. Within a few years however, the latter took over for the acceleration of electrons, to potentially far higher energy, with smaller magnets confined along the electron orbit.

Betatrons when they were first developed and produced were destined for nuclear physics, as X-ray generators, medical and other applications. Linacs, much more compact and lighter, easy to include in a gantry system, allowing much higher dose, have taken over in radiotherapy and in a number of light and compact X-ray source applications. Betatrons nowadays, in the 3–10 MeV range, are mostly designed for use as portable, low dose-rate X-ray sources; typical uses include material radiography, cargo inspection [15].

5.2 Basic Concepts and Formulæ

Referring to the schematic in Fig. 5.3: the current in the coils is pulsed (an impulse wave-form at a repetition rate of tens to hundreds Hz) so producing a time-varying magnetic field $B(t)$ which induces an accelerating azimuthal electromotive force along the orbit, while maintaining the electron beam on that constant radius orbit during acceleration. The vacuum chamber is made from non-conducting material (glass for instance), or includes non-conducting gaps. Keeping the beam on a closed orbit around the ring requires the Widerøe condition: the magnetic field along the orbit must be half the average magnetic field which it circumscribes, a rule established in 1927 [2]. As in the classical cyclotron (Chap. 3), and in the still to come early synchrotron (Chap. 8), a tapered gap in the region of the vacuum chamber provides the necessary weak focusing for stable transverse "betatron oscillations" [16] around the closed orbit, over the hundreds of thousands of turns that the acceleration to top energy requires.

The electron energy E, field B and radius R of the fixed orbit are related by $BR = p/e = \beta E / ec$, where $\beta \to 1$ as the electron accelerates. It results that the maximum energy which can be reached satisfies

$$\hat{E}/e = \hat{B}Rc$$

Fig. 5.3 Betatron ring, principle. The equilibrium orbit is at radius R where the field value is one-half of the average field encompassed by the orbit. The induction electric field, resulting from varying $B(t)$, is tangent to the orbit. The magnetic field in the region of the vacuum chamber is shaped to provide weak focusing

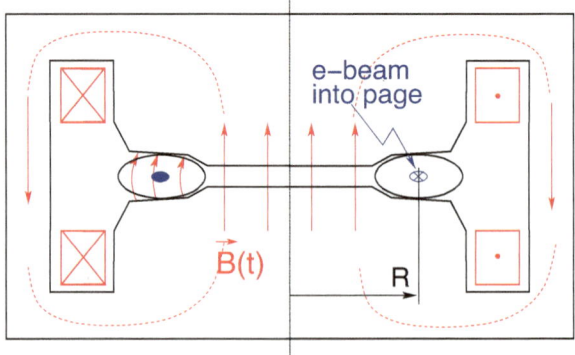

An energy $\hat{E} \approx 300\,\text{MeV}$ can be reached for instance with an orbit radius $R \approx 1\,\text{m}$ and a guiding field cycled up to $\hat{B} \approx 1\,\text{T}$.

5.2.1 Betatron Condition. Acceleration

The inductive electric field $\mathbf{E}$ resulting from the varying magnetic field $\mathbf{B} = \mathbf{curlA}$ satisfies $\mathbf{E} = -\partial \mathbf{A}/\partial t$. The magnetic vector potential $\mathbf{A}$ has the symmetry of the current, so $A_r = A_y = 0$ and $\mathbf{A} \equiv A_s \mathbf{s}$. Thus the magnetic field components satisfy

$$B_s = 0, \quad B_r = -\frac{\partial A}{\partial y}, \quad B_y = \frac{1}{r}\frac{\partial r A}{\partial r}$$

From the expression of B_y it results that the magnetic flux through the area circumscribed by the orbit of radius R is

$$\phi_a(t) = 2\pi \int_0^R B_y(t)\, r\, dr = 2\pi R A(t) \tag{5.1}$$

Note $B_a \equiv \overline{B_y} = \phi_a/\pi R^2$ the average value of the vertical field component through the orbital circle. From $\mathbf{E}_s = -\partial \mathbf{A}_s/\partial t$ one gets the accelerating field under the form

$$E_s(t) = \frac{R}{2}\frac{d B_a(t)}{dt} \tag{5.2}$$

With $dp/dt = e E_s$, and on the other hand $dp/dt = d(e B R)/dt$, by integration over $0 \to t$ one gets the betatron condition

$$B(t) = \frac{B_a(t)}{2} \tag{5.3}$$

The guiding field $B(t)$ along the orbit is one-half of the average field through the area defined by the orbit. The actual shape of the magnetic field pulse does not matter, as long as it induces an accelerating *emf*. Field cycling can for instance take the form

$$B(t) = \frac{1}{2}\hat{B}\,(1 - \cos \omega t) \tag{5.4}$$

such that $0 \leq B(t) \leq \hat{B}$. With acceleration taking place over a quarter of a period of $B(t)$, the duty cycle can get close to 25%. With a pulse over $[-\hat{B}, \hat{B}]$, two beams can be accelerated both ways, on the positive slope of $B(t)$ for one, the negative slope for the other. If they hit a common internal target, two X-ray pulses are emitted, in opposite directions, at twice the repetition rate $\omega/2\pi$.

The gain in energy during acceleration is the same for all electrons, regardless of their transverse excursions and azimuthal location along the orbit. The beam spreads around the betatron during the acceleration cycle, as there is no longitudinal focusing.[1]

5.2.2 Transverse Motion

The theory of the betatron established the stability of the eponymous oscillations which electrons undergo under the effect of a radial field index, as they circle in the vicinity of the closed orbit during acceleration [16]. Elements of theory are introduced in Chap. 3, Classical Cyclotron, and Chap. 8, Weak Focusing Synchrotron. They can be referred to for more details.

The properties of betatron oscillations are derived assuming an "adiabatic change of the magnetic field", i.e., the magnetic field on the orbit changes slowly compared to the betatron oscillation. Let's figure the respective characteristic times of betatron oscillation and field impulse. The frequencies of the transverse oscillations in a structure with cylindrical symmetry and radial field index $0 < n = -\frac{\rho_0}{B_0}\frac{\partial B_y}{\partial x} < 1$ write (Eq. 3.18)

$$\omega_R/\omega_{\text{rev}} = \nu_R = \sqrt{1-n} \quad \text{and} \quad \omega_y/\omega_{\text{rev}} = \nu_y = \sqrt{n} \tag{5.5}$$

They are commensurate with the revolution period as, in the betatron, $\nu_R \approx \nu_y \approx 0.5$. Take for instance $R = 1$ m, then $T_{\text{rev}} = 2\pi R/c = 21$ ns. On the other hand, assume 60 Hz cycling, meaning magnetic field and momentum ramps of several milliseconds. This validates the adiabaticity hypothesis, time-varying corrective terms associated with $\mathbf{B}$ and $\mathbf{p}$ in the equations of transverse motion can be neglected. In this hypothesis the differential equations of motion (cf. section "Betatron Motion", Eq. 8.10) have for solution the harmonic oscillations (cf. section "Betatron Motion", Eq. 8.21)

$$\begin{cases} u(s) \approx \sqrt{\beta_u(s)\varepsilon_u/\pi} \, \cos\left(\nu_u \frac{s}{R} + \phi\right) \\ u'(s) \approx -\sqrt{\dfrac{\varepsilon_u/\pi}{\beta_u(s)}} \, \sin\left(\nu_u \frac{s}{R} + \phi\right) + \alpha_u(s) \, \cos\left(\nu_u \frac{s}{R} + \phi\right) \end{cases} \tag{5.6}$$

with $u(s)$ standing for $x(s) = \delta R(s)$ (distance to the reference radius R in the moving frame), or for $y(s)$, and other notations as defined in section "Betatron Motion".

Transverse motion properties include the following [16]:

– on-momentum electrons, whose momentum satisfies $p = eBR$, undergo betatron oscillations around the reference orbit of radius R;

---

[1] Longitudinal focusing is a property of synchrotron acceleration, see Chap. 7.

– electrons with a momentum offset δp undergo betatron oscillations around a reference orbit of radius $R + D\delta p/p$, with $D = R/(1 - n)$ the dispersion, a constant (Eq. 3.20);
– the amplitude of these oscillations damps in proportion to $1/\sqrt{p}$ under the effect of acceleration, $\beta\gamma\varepsilon_u$ is a constant of the motion (Eq. 8.31).

Beam injection is based on the property that the injected orbit undergoes betatron oscillations which damp during acceleration, thus allowing part of the injected beam to miss the injector on successive turns.

5.2.3 Synchroton Radiation

The topic is introduced in this chapter, as the effect of SR on beam dynamics, i.e., orbit spiraling, was first experienced in a betatron. This theoretical material is resorted to in the exercises.

Given the energy reached by electrons in a betatron, SR was to be expected. The emission of radiation by a charged particle had been established about half a century earlier [18]. SR was resorted to, rightly or not in the early times, in operating machines as they were reaching meaningful energy, as a possible explanation of some undesired beam dynamics effects observed. As a matter of fact, the deleterious effect of SR on the operation of betatrons was pointed out at the time [19]. Measurements of properties of the radiation were undertaken, that was the beginning of a long story, still underway...

Some key properties of SR are summarized below, with particular insight in the stochastic process and the resulting energy loss, as it is the basis of its simulation in Zgoubi. More is addressed in Chap. 9.

Stochastic Emission of Photons

A detailed theory of SR and its properties can be found in [20]. Energy loss by synchrotron radiation is comprised of three random processes, namely [21],

– the emission of $k \geq 0$ photons, which abides by the Poisson distribution

$$p(k) = \frac{\Lambda^k}{k!}e^{-\Lambda} \tag{5.7}$$

with

$$\Lambda = <k> = <k^2> = \frac{5er_0}{2\hbar\sqrt{3}}B\rho\Delta\theta \tag{5.8}$$

the average number of photons radiated over a trajectory arc $\Delta\theta$. In this expression, $r_0 = e^2 / 4\pi\varepsilon_0 m_0 c^2$ is the classical radius of the electron ($r_0 = 2.818 \cdot 10^{-15}$ m), e its charge, $m_0 c^2$ its rest energy, $\varepsilon_0 = 1 / 36\pi \times 10^9$, $\hbar = h/2\pi$ with h the Planck constant;

– the energy ϵ of the emitted photon(s), which abides by the probability law

$$P(\epsilon/\epsilon_c) = \frac{3}{5\pi} \int_0^{\epsilon/\epsilon_c} \frac{d\epsilon}{\epsilon_c} \int_{\epsilon/\epsilon_c}^{\infty} K_{5/3}(x)dx \qquad (5.9)$$

with $K_{5/3}$ the modified Bessel function and

$$\epsilon_c = \frac{3\hbar\gamma^3 c}{2\rho} \qquad \left(\text{or, } \lambda_c = \frac{2\rho}{3\gamma^3}\right) \qquad (5.10)$$

the critical energy (critical wavelength) of the radiation;

– the photon emission angle ξ with respect to electron momentum vector, a stochastic quantity which causes scattering of the latter. A cylindrical-symmetric Gaussian distribution may be assumed for simplicity,

$$p(\xi) = \exp(-\frac{\xi^2}{2\sigma_\xi^2}) \qquad (5.11)$$

For simplicity as well the *rms* $\sigma_\xi \approx 1/\gamma$ can be considered independent of photon energy. The scattering angle of the momentum vector is quite small anyway, and usually ignored.

Energy Loss

The average energy loss by a ultra-relativistic electron ($\beta = v/c \approx 1$) over an arc of trajectory $\Delta\theta$ in a uniform field B writes [20]

$$\overline{\Delta E} = \frac{2}{3} r_0 E_0 \gamma^4 \frac{\Delta\theta}{\rho} = \frac{2}{3} r_0 e c \gamma^3 B \Delta\theta \qquad (5.12)$$

The stochastic nature of photon emission causes an energy spread which averages to

$$\sigma_{\Delta E/E} = \frac{\sqrt{110\sqrt{3}\hbar c / \pi\varepsilon_0}}{24 E_0/e} \gamma^{5/2} \frac{\sqrt{\Delta\theta}}{\rho} \qquad (5.13)$$

Over a revolution ($\Delta\theta = 2\pi$ in Eq. 5.12), the energy loss writes

$$U_s \text{ [MeV/turn]} = C_\gamma \frac{E_s^4 \text{ [GeV]}}{\rho_{[m]}}, \qquad C_\gamma = \frac{4\pi}{3} \frac{r_0}{(m_0 c^2/e)^3} \qquad (5.14)$$

For electrons, $C_\gamma = 8.85852 \times 10^{-5} \text{ m/GeV}^3$.

5.3 Exercises

5.1 Develop a Betatron Magnet in Zgoubi
Solution 5.1

The subroutine which governs the functioning of DIPOLE, dipi.f, can be used as a template: it can be copy-pasted under a different name, inddip.f for instance, and modified as needed. The subroutine inddip.f, in addition to computing the guiding field, will have to allow for a pulsed field, computation of the induction electric field it entails, and its effect on electron coordinates and momentum.

A keyword such as BETADIPOL can be created (BETATRON is already used! to simulate a betatron core for slow extraction). A new keyword needs to be added in LSTKEY.H, and zgoubi.f has to be updated to account for it.

Test this BETADIPOL with the next exercises.

5.2 A 315 MeV Betatron
Solution 5.2

(a) Build an input data file for the Chicago 315 MeV betatron magnet, according to the parameter list of Table 5.1.

If you did not do Exercise 5.1, then proceed in the following way:

Split DIPOLE into $N = 72$ magnetic sectors to simulate the 360° dipole. No acceleration in this preliminary step.

Check the closed orbit, and the effect of the integration step size.

Check periodic stability and tunes, using TWISS.

Table 5.1 Parameter table (after [17])

Top energy	MeV	315
Injection energy	keV	135
Orbit radius R	(m)	1.22
Maximum guide field value	kG	9.2
Repetition rate	Hz	6
$\dot{B}$	T/s	To be determined

(b) Simulate an acceleration cycle, assuming a linear $B(t)$ ramp, for an electron launched on the closed orbit for simplicity.

Interleave the split DIPOLE with CAVITE[IOPT = 3] to simulate acceleration. Note that this will assume longitudinal E_s, with no dependence on transverse coordinates.

Use FAISTORE to store turn-by-turn electron data. Assume no synchrotron radiation in this preliminary step. Check the accelerated orbit.

(c) Simulate the previous acceleration cycle for 2 electrons featuring a paraxial horizontal excursion for one, a paraxial vertical excursion for the other. Check the transverse damping of the betatron oscillations.

5.3 Acceleration with Radiation Loss in the 315 MeV Betatron
Solution 5.3

Referring to [17], SR caused a 9% energy loss in the 315 MeV betatron (it was compensated with a voltage impulse from a separate system). In this exercise we check consistency of the energy loss in the betatron model of Exercise 5.2, with that number. The SR on/off switch is provided by SRLOSS [22, INDEX]. SRPRNT[PRINT] can be introduced to log SR data in zgoubi.res and, an effect of the PRINT command, additional details in zgoubi.SRPRNT.Out.

In order to perform a convergence test on the Monte Carlo SR loss, a 10,000 electron bunch is launched, with 315 MeV energy, for a single turn along the closed orbit.

DIPOLE field has to be set for 315 MeV; a possibility for that if using data files from the previous exercise where DIPOLE is set for 135 keV, and update the field using the global command SCALING.

(a) Get the energy loss per turn U_s, and some of the radiated photon properties (critical energy, etc.). Check these outcomes against theoretical expectations, sections "Stochastic Emission of Photons" and "Energy Loss".

(b) Simulate an acceleration cycle from 0.135 keV to 315 MeV energy region, for an electron launched on the fixed closed orbit. Check the orbit spiraling. Check the energy dependence of the synchrotron radiation energy loss. Check the aforementioned 9% experimental outcome.

5.4 Solutions of Exercises of This Chapter: Betatron

5.1 Develop a Betatron Magnet in Zgoubi
This code development exercise and its benchmarking are left for the reader to complete.

DIPOLE Fortran source subroutine can be used as a template, namely [pathTo]/ zgoubi-code/zgoubi/dipi.f. This routine includes the assignment of the necessary dipole parameters for DIPOLE simulation, as specified in zgoubi.dat. It includes in addition an ENTRY, 'ENTRY DIPF(...)', which is called (from chamc.f) during stepwise raytracing and provides the magnetic field at particle location.

5.2 A 315 MeV Betatron

(a, b) A 315 MeV betatron input data file. Acceleration on the closed orbit.

Acceleration up to 320 MeV is simulated actually, to have some insight a little beyond 315 MeV. Take kinetic energy (Table 5.1): $E_k : 0.135 \rightarrow 320$ MeV, thus $B\rho$: $0.00131829454 \rightarrow 1.069108254786$ T m and guiding field $B(t) : 0.0010805693 \rightarrow 0.876318241628$ T (consistent with a maximum 9.2 kG, Table 5.1). Take repetition rate 6 Hz, assume symmetric saw-tooth like $B_a(t)$ excitation, hence a ramp up in $1/12$ s. It results $\dot{B} = (0.876318241628 - 0.0010805693)/(1/12) = 10.50285$ T/s.

From Eqs. 5.2 and 5.3 one gets $E_s = R\dot{B} = 12.81$ V/m. Energy gain over a turn is $W/e = \oint E_s ds = 2\pi R E_s = 2\pi R^2 \dot{B} = 98.211$ eV/turn. With a ring comprised of 72 induction modules of 5° angle each, this means $98.211/72 = 1.364188$ eV/module. Thus use CAVITE[IOPT $= 3, \hat{V} = 1.364188$] for the induction module.

The number of turns from 0.135 to 320 MeV is $(320 - 0.135) \times 10^6/98.211 = 3.2565 \times 10^6$. Thus use REBELOTE[NPASS 3.2565 $\times 10^6$, IOPT $= 99$] for multi-turn raytracing.

The top file in Table 5.2 defines a 5° induction module. The bottom file in Table 5.2 is set to create a sequence of 72 induction sectors and request acceleration over an induction cycle.

Tracking outcomes:
• Checking closed orbit and effect of step size.

A 1-turn tracking using the input data files of Table 5.2, with acceleration off (temporarily set CAVITE[IOPT $= 0$]), results in the following at the bottom of zgoubi.res, with initial coordinates at OBJET on the left hand side, the final coordinates after a turn on the right hand side:

```
  220  Keyword, label(s) :  FAISCEAU
                                   TRACE DU FAISCEAU
                                (follows element #    219)
                                     1 TRAJECTOIRES
                          OBJET                                                 FAISCEAU
       D      Y(cm)    T(mr)    Z(cm)    P(mr)    S(cm)    D-1    Y(cm)    T(mr)    Z(cm)   P(mr)   S(cm)
  o 1 1.0000 122.000   0.000    0.000    0.000   0.0000  0.0000 122.000  -0.000    0.000   0.000  7.665486E+02   1
               Time of flight (mus) :  4.17943566E-02 mass (MeV/c2) :  0.510999
```

The file [b_]zgoubi.fai has the coordinates to greater accuracy, in particular, final radius after one turn around the ring $Y = 1.2200000000000026$ m, and angle $T = -2.2846308178614549 \times 10^{-12}$ rad, essentially identical to the starting values 122 cm and 0 rad respectively. An integration step size near 1 cm ensures closed orbit closure with such accuracy.

After a 3.25656×10^6 turn acceleration cycle (energy increase shown in Fig. 5.4) the electron coordinates in zgoubi.res (bottom of the file) appear to be:

```
  220  Keyword, label(s) :  FAISCEAU
                                   TRACE DU FAISCEAU
                                (follows element #    219)
                                     1 TRAJECTOIRES
                          OBJET                                                 FAISCEAU
       D      Y(cm)    T(mr)    Z(cm)    P(mr)    S(cm)    D-1    Y(cm)    T(mr)    Z(cm)   P(mr)   S(cm)
  o 1 1.0000 122.000   0.000    0.000    0.000   0.0000 809.9784 122.000  -0.001    0.000   0.000  2.496323E+09   1
               Time of flight (mus) :  83333.546    mass (MeV/c2) :  0.510999
```

Table 5.2 Top input data file: simulation of a 5° "induction sector" module. Bottom input data file: INCLUDEs this module, and tracks electrons (actually, positrons, here, for convenience) from 135 keV to 320 MeV

- 5-degree sector and longitudinal boost (CAVITE):

```
Betatron. 5 degree induction sector
'OBJET'
1.3182945462093764                                                      ! 135 keV electron.
2
1 1
122. 0. 0. 0. 0. 1. 'o'                          ! Closed orbit coordinates for D= Brho/BORO= 1.
1
'PARTICUL'                                        ! This is required to get the time-of-flight,
POSITRON                                          ! and to accelerate.
'FAISCEAU'                                        ! Local particle coordinates.
'MARKER'    #S_Scaling                            ! Label should not exceed 20 characters.
'SCALING'
1 1
DIPOLE
-1                                    ! Causes field increase in DIPOLE, in correlation to particle
1.                                    ! rigidity increase by CAVITE.
1
'MARKER'    #S_5degSector                         ! Label should not exceed 20 characters.
'DIPOLE'                                          ! Analytical modeling of a dipole magnet.
0                 ! IL=2, only purpose is to logged trajectories in zgoubi.plt, for further plotting.
5. 122.                                           ! Sector angle AT; reference radius RM.
2.5  0.010805693 -0.1 0. 0.          ! Reference azimuthal angle ACN; BM field at RM; indices, N, N', N''.
0. 0.                                             ! EFB 1 is hard-edge.
4  .1455   2.2670  -.6395  1.1558  0. 0. 0.       ! hard-edge only possible with sector magnet.
2.5 0.  1.E6  -1.E6  1.E6  1.E6                    ! Entrance face placed at omega+=2.5deg from ACN.
0. 0.                                             ! EFB 2.
4  .1455   2.2670  -.6395  1.1558  0. 0. 0.
-2.5 0.  1.E6  -1.E6  1.E6  1.E6                   ! Exit face placed at omega-=-2.5deg from ACN.
0. 0.                                             ! EFB 3 (unused).
0 0.    0.    0.    0.    0. 0. 0.
0. 0.   1.E6  -1.E6  1.E6  1.E6 0.
2 10    ! '2' is for 2nd degree interpolation. Could also be '25' (5*5 points grid) or 4 (4th degree).
0.96                                    ! Integration step size. Small enough for orbit  to close accurately.
2 0. 0. 0. 0.                                     ! Magnet positioning RE, TE, RS, TS.
'FAISCEAU'
'CAVITE'                                          ! Accelerating gap.
3                               ! dW = qVsin(phi), independent of time (phi forced to constant).
0. 0.                                             ! Unused.
1.3641881785 1.5709632679            ! Acceleration is per 5 deg sector, 72 as much per turn ~ 1.332 Volt.
'MARKER'    #E_CAVITE                             ! Label should not exceed 20 characters.
'END'
```

- Complete betatron dipole, including induction boost:

```
Betatron. R=1.22 m.  E:  0.135 -> 320 MeV.
! Brho: 1.3182945462E-3 -> 1.069108254786 Tm -> B: 1.0805693E-3 -> 0.876318241628 T.
! B-dot ~ 10.257 T/s
'OBJET'
1.3182945462093764                                                      ! 135 keV electron.
2
1 1
122. 0. 0. 0. 0. 1. 'o'                          ! Closed orbit coordinates for D= Brho/BORO= 1.
1
'PARTICUL'                                        ! This is required to get the time-of-flight,
POSITRON                                          ! and to accelerate.
'FAISCEAU'                                        ! Local particle coordinates.
'FAISTORE'                            ! Log particle data at LABEL1='turn', in b_zgoubi.fai, for further
b_zgoubi.fai turn         ! plotting (by gnuplot, below); "b_" indicates binary formatting (faster I/O).
11111                                   ! Only save turn 1, every other 11111 turn, and last turn.
'OPTIONS'                             ! Inhibit print out to zgoubi.plt (save on CPU time and file volume).
1 2
WRITE OFF                             ! Inhibit writes to zgoubi.res (save on CPU time and file volume).
.plt 0                                ! Inhibit print out to zgoubi.plt (save on CPU time and file volume).

'INCLUDE'
1
72 * 5dInductionSector.inc[#S_Scaling:#E_CAVITE]
'MARKER' turn             ! Local particle coordinates printed out in b_zgoubi.fai, as per FAISTORE.

'REBELOTE'                                                              ! Multiturn.
3256566 0.5  99
'OPTIONS'                                         ! Restore print out to zgoubi.res.
1 1
WRITE ON
'FAISCEAU'                                        ! Local particle coordinates.
'END'
```

Fig. 5.4 Kinetic energy versus turn in the betatron, from 0.135 to 320 MeV in 3256566 turns. A graph obtained using zpop: menu 7; 1/2 to open b_zgoubi.fai; 2/[39,20] for E versus turn; 7 to plot

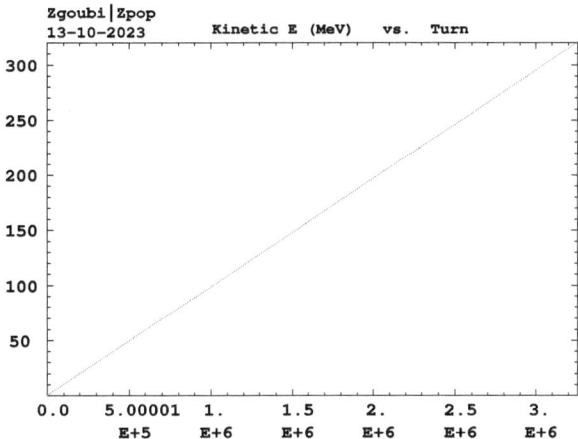

These outcomes confirm the stability of the closed orbit, and yields the expected distance $N \times C = 3.25656 \times 10^6 \times 2\pi R = 24.9632 \times 10^6$ m, as well as the time of flight, $N \times C/c = 0.0833$ s.

- TWISS computation is produced, by running this input data file:

```
Betatron. a TWISS computation.
'OBJET'
1.3182945462093764                                           ! 135 keV electron.
5
.001 .01 .001 .01 .0 .0001                          ! 13-particle set sampling.
122. 0. 0. 0. 0. 1.                    ! Closed orbit coordinates for D= Brho/BORO= 1.

'INCLUDE'
1
72 * 5dInductionSector.inc[#S_Scaling:#E_CAVITE]

'TWISS'
2 1. 1.
'END'
```

The outcome is the following (bottom of zgoubi.res):

```
Reference, before change of frame (particle #  1  - Y,T,Z)  :
  1.22000000E+02  -1.41929654E-10   0.00000000E+00   0.00000000E+00

TWISS parameters,  periodicity of   1  is  assumed
                    -  COUPLED  -
   Beam  matrix  (beta/-alpha/-alpha/gamma) and  periodic  dispersion  (MKSA units)

      1.285993     0.000000     0.000000     0.000000     0.000000     1.355556
      0.000000     0.777609     0.000000     0.000000     0.000000     0.000000
      0.000000     0.000000     3.857979     0.000000     0.000000    -0.000000
      0.000000     0.000000     0.000000     0.259203     0.000000     0.000000
      0.000000     0.000000     0.000000     0.000000     0.000000     0.000000
      0.000000     0.000000     0.000000     0.000000     0.000000     0.000000

                    Betatron  tunes  (Q1 Q2 modes)
          NU_Y =  0.94868330        NU_Z =  0.31622777

                      Momentum compaction :
                  dL/L / dp/p =   1.1111111

                      Transition gamma  =  9.48683297E-01

                      Chromaticities :
          dNu_y / dp/p = -6.44168625E-02        dNu_z / dp/p =  0.19325030
```

Fig. 5.5 Transverse
excursion (markers) and
$(Y_0 - 1.22) \times p_0/p$ (solid
line) versus energy. A graph
obtained using zpop: menu
7; 1/2 to open [b_]zgoubi.fai;
2/[20, 2] for Y versus E; 7 to
plot; $(Y_0 - 1.22) \times p_0/p$
graph is superposed using
option 20

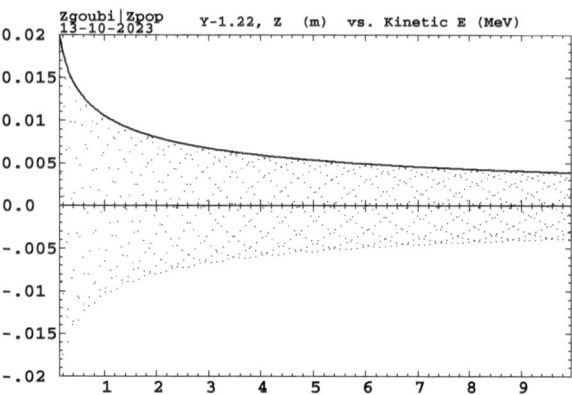

These results show that the closed orbit is at R = 1.22 m as expected, and that the
lattice is periodically stable.

(c) Transverse damping.

Simulation of the acceleration cycle for 2 electrons featuring a paraxial horizontal
excursion for one, a paraxial vertical excursion for the other uses the bottom file of
Table 5.2 with the following OBJET[KOBJ = 2, IMAX = 3]:

```
'OBJET'
1.3182945462093764                                      ! 135 keV electron.
2
3 1
122.  0. 0.  0. 0. 1. 'o'        ! Closed orbit coordinates for D= Brho/BORO= 1.
124.  0. 0.  0. 0. 1. 'o'                               ! dY0 = 2 cm.
122.  0. 2.  0. 0. 1. 'o'                               ! Z0 = 2 cm.
1 1 1
```

Results obtained are displayed in Fig. 5.5. The damping amounts to

$$\sqrt{\frac{p_{\text{initial}}}{p_{\text{final}}}} = \sqrt{\frac{\beta\gamma_{\text{initial}}}{\beta\gamma_{\text{final}}}} = \frac{1}{\sqrt{26.46}} = 0.194$$

thus Y and Z motions both damp from $Y_0 - 122 = Z_0 = 2$ cm at 135 keV to
$\hat{Y} - 122 \approx \hat{Z} \approx 0.39$ cm at 320 MeV.

A different way to check the convergence is to launch a few particles on an
ellipse, and check that $\varepsilon_{Y,\text{initial}}/\varepsilon_{Y,\text{final}} = p_{\text{initial}}/p_{\text{final}}$. In that aim, change
OBJET[KOBJ = 2] to the following OBJET[KOBJ = 8]:

```
'OBJET'
1.3182945462093764                                      ! 135 keV electron.
8
30 1 1
122.e-2  0. 0.  0. 0. 1. 'o'       ! Closed orbit coordinates for D= Brho/BORO= 1.
0. 1.285993 1e-6               ! Definition of the invariant, and its value, horizontal,
0. 3.857979 1e-6                                              !vertical.
0. 1. 0.                                              ! longitudinal.
```

The values of the optical functions in this OBJET are taken from the previous TWISS.

The input data file has two FAISCEAU (Table 5.2). The first one right after the object definition by OBJET, the second FAISCEAU at the end of the tracking. In the resulting listing zgoubi.res the first FAISCEAU provides the following concentration ellipse data (see Sect. 14.5.1):

```
--------------- Concentration ellipses :
  surface/pi         alpha        beta        <X>          <XP>         numb. of prtcls  ratio    space
                                                                        in ellips, out
  5.0000E-07 [m.rad]  1.5176E-16  1.2860E+00  1.220000E+00  2.222614E-19     30      30   1.000    (Y,T)
  {\tiny
```

which indicate that the initial horizontal ellipse formed by the 30 electrons is centered at $R = 1.22$ m, its surface is twice the concentration ellipse surface, i.e., $2 \times 5.0000E-07 = 10^{-6}$ m rad, as stated in OBJET. In zgoubi.res the second FAISCEAU (bottom of the file) provides the following concentration ellipse data:

```
--------------- Concentration ellipses :
  surface/pi         alpha        beta        <X>          <XP>         numb. of prtcls  ratio    space
                                                                        in ellips, out
  1.9065E-08 [m.rad]  -1.7993E-06  1.2861E+00  1.220019E+00  -2.823702E-06   30      30   1.000    (Y,T)
  {\tiny
```

Note that this final concentration ellipse is found offset by 20 μm at 1.22002 m (from the launch position at $R = 1.22$ m), a numerical effect with various possible causes such as the need for more electrons to better define the positioning of the matching ellipse, an investigation left to the reader. The surface of the ellipse is $2 \times 1.9065E-08$ m rad.

Horizontal phase space ellipses are displayed in Fig. 5.6, showing that $\varepsilon_{Y,\text{final}}/\varepsilon_{Y,\text{initial}} = (2 \times 1.9066 \times 10^{-8})_{[mm.mrad]}/10^{-6}_{[mm.mrad]} = 0.0381$, close to the expected $1/26.456 = 0.0378$. Both are expected much closer actually. Same for ellipse centering: from the Min-Max Horizontal data at the foot of the graphs, ellipse center from 30 particles is loosely estimated at $(1.21945 + 1.22171)/2 = 1.22058$ m (left), quite different from the statement OBJET[$Y0 = 122_{[cm]}$], and at $(1.21980 + 1.22024)/2 = 1.22002$ m (right). Zgoubi raytracing accuracy does allow better precision; this requires more electrons so to properly define the invariant they lie on, and the parameters of the latter from concentration ellipses.

Checking convergence of the numerical integration for the induction element size (5° sector or less) and the step size: changing the modular "induction dipole" from a 5° sector to 10 times less, 0.5°, and the step size accordingly, does not change these results.

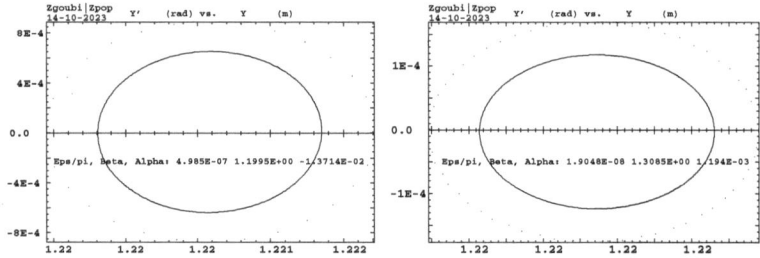

Fig. 5.6 Horizontal phase space positions of a few tens of electrons distributed on an initial ellipse invariant, and the matching concentration ellipse. Left: 135 keV; right: 320 MeV. The invariant value is twice the ellipse surface, thus 10^{-6} and $2 \times 1.9066 \times 10^{-8}$, respectively. A graph obtained using zpop: menu 7; 1/2 to open b_zgoubi.fai; 2/[2, 3] for T versus Y; 7 to plot

5.3 Acceleration with Radiation Loss in the 315 MeV Betatron

(a) SR loss over a turn.

The input data file to track 10,000 electrons at 320 MeV, over a turn with SR loss, is given in Table 5.3

It can be checked in the execution listing zgoubi.res, under DIPOLE data list, that DIPOLE field is scaled to 320 MeV (from its value set for 135 keV):

```
Field has been * by scaling factor      798.32335697
```

SRPRNT delivers statistical data computed from step-by-step Monte Carlo SR. The results are the following (bottom of zgoubi.res file):

```
* Monte Carlo S.R. statistics, from beginning of structure,
     10000 particles,  a total of      8640000 integration steps :
  Average energy loss per particle per pass :        0.7252923    keV.
  Critical energy of photons (average) :             5.7107747E-02 keV
  Average energy of radiated photon :                1.7725422E-02 keV

 rms energy of radiated photons :            3.2685214E-02 keV
 Smallest, BIGEST photon :                   1.2536E-14    9.8729E+03   keV
 Number of photons radiated - Total :                 409182.0
                            - per particle per pass :  40.91820
                            - per particle, per step : 4.7359028E-02
```

It is an interesting exercise, left to the reader, to compare these step-by-step Monte Carlo SR outcomes to theoretical expectations. For instance:

Table 5.3 Input data file to track 10,000 electrons over a turn, with SR loss. SCALING[SCL = 1.3182945462093764 * 798.32335697] is used to scale DIPOLE field (set for 135 keV) to the current reference rigidity OBJET[BORO = 1.3182945462093764 * 798.32335697] which corresponds to a 320 MeV electron

```
Betatron. R=1.22 m.  SR loss at 320 MeV (single turn).
'OBJET'
1.3182945462093764 * 798.32335697                        ! Rigidity in kG cm: 320 MeV electron.
1
10000 1 1 1 1 1
0. 0. 0. 0. 0. 0.                                        ! Coordinate sampling.
122. 0. 0. 0. 0. 1.                     ! Closed orbit coordinates for D= Brho/BORO= 1.
'PARTICUL'                              ! This is required to get the time-of-flight,
POSITRON                                                 ! and to accelerate.
'FAISCEAU'                                   ! Local particle coordinates.

'SRLOSS'
1
DIPOLE
0 123456

'MARKER'      #S_Scaling                           ! Label should not exceed 20 characters.
'SCALING'
1 1
DIPOLE
-1
798.32335697                                    ! Scale DIPOLE field to 320 MeV.
1

'INCLUDE'
1
72 * 5dInductionSector.inc[#S_5degSector:#E_5degSector] ! Only grabs DIPOLE (not SCALING, notCAVITE).

'SRPRNT'                                    ! Synchrotron radiation statistical data.
'FAISCEAU'                                   ! Local particle coordinates.
'END'
```

– average energy loss per particle per pass: 0.7252923 keV, with theory's 0.7148976 keV (Eq. 5.14);
– critical energy of photons (average): 5.7107747E-02 keV, with theory's 0.0568316 keV (Eq. 5.14).

Note that similar cases of SR simulations may be found in various examples in zgoubi sourceforge repository, https://sourceforge.net/p/zgoubi/code/HEAD/tree/branches/exemples/ folder.

(b) An acceleration cycle up to 315 MeV region, including SR.

Building the simulation file starts from Table 5.2. Add SRLOSS as in Table 5.3, and add REBELOTE[NPASS = 3256566] for a complete acceleration cycle. Change OBJET to a single particle. Make sure INCLUDE grabs the "induction" acceleration simulation by CAVITE, together with the 5° DIPOLE, i.e., the segment [#S_Scaling:#E_CAVITE]. The resulting file is as follows:

```
Betatron. R=1.22 m.     E:  0.135 -> 320 MeV.
! Brho: 1.3182945462E-3 -> 1.069108254786 Tm -> B: 1.0805693E-3 -> 0.876318241628 T.
! B-dot ~ 10.257 T/s

'OBJET'
1.3182945462093764                                          ! 320 MeV electron.
1                                          ! Number of electrons to be tracked is given here.
1 1 1 1 1                                                      ! Sampling.
0. 0. 0. 0. 0. 0.                                             ! Coordinate sampling.
122. 0. 0.0 0. 0. 1.                          ! Closed orbit coordinates for D=p/p_0=1.
'PARTICUL'                                    ! This is required to get the time-of-flight,
POSITRON                                                  ! and to accelerate.
'FAISCEAU'                                          ! Local particle coordinates.
'FAISTORE'                                    ! Log particle data at LABEL1='turn', in zgoubi.fai,
zgoubi.fai turn                                   ! for further plotting (by gnuplot, below).
1                                             ! Only save turn 1, every other 11111 turn, and last turn.

'SRLOSS'
0    PRINT
BEND
0 123456

'OPTIONS'                            ! Inhibit print out to zgoubi.plt (save on CPU time and file volume).
1 2
WRITE OFF                            ! Inhibit writes to zgoubi.res (save on CPU time and file volume).
.plt 2                               ! Inhibit any print out to zgoubi.plt (save on CPU time and file volume).

'INCLUDE'
1
5dInductionSector.inc[#S_Scaling:#E_CAVITE]              ! Just one 5 degreee sector is included.

'MARKER' turn                        ! Local particle coordinates printed out in b_zgoubi.fai, as per FAISTORE.

'REBELOTE'                                                    ! Multiturn.
234472752 0.5  99        ! A cycle to 315 MeV takes 3256000 turns.  234472752 = 3256000 * 72 turns.

'OPTIONS'                                                 ! Restore print out to zgoubi.res.
1 1
WRITE ON

'SRPRNT'
'FAISCEAU'                                                 ! Local particle coordinates.
'END'
```

The simulation is straightforward, refer to the preliminary SR loss simulations above and to [17] for the interpretation of the tracking outcomes.

References

1. J.P. Slepian, 'X-Ray Tube', US Patent No 1,645,304, Filed April 1, 1922, published Oct. 11, 1927. See Fig. 2 in F. Scarlat, E. Badita, E. Stancu, A. Scarisoreanu, Basic principles of conventional and laser driven therapy accelerators. Adv. Med. Imaging Health Inf. **2019**(1), 1–23 (2019). https://kosmospublishers.com/basic-principles-of-conventional-and-laser-driven-therapy-accelerators/
2. R. Widerøe, A new principle for generation of high voltages. Thesis, Aachen, October 29, 1927. Archiv für Elektrotechnik **21**, 387–406 (1928)
3. P.-A. Chamouard, Saturne 2 : 20 years for physics (Sect. 7: The extracted beams of Saturne 2), in *The 20 Years of the Synchrotron Saturne 2*, ed. by A. Boudard, P.-A. Chamouard. (World Scientific, 2000)
4. K. Takayama, KEK digital accelerator and its beam commissioning, in *Talk slides, IPAC 2011*, September 4–9, 2011, San Sebastian. https://accelconf.web.cern.ch/IPAC2011/talks/weoba02_talk.pdf
5. S. Nath, Linear induction accelerators at the Los Alamos National Laboratory DARHT facility, in *TH304 Proceedings of Linear Accelerator Conference LINAC2010*, Tsukuba, Japan. https://accelconf.web.cern.ch/LINAC2010/papers/th304.pdf
6. M.A. Green, S. Yu, Superconducting magnets for induction phase-rotation in a Neutrino Factory. Tech. Note LBNL-48445; SCMAG-749. https://www.osti.gov/servlets/purl/795332

7. The induction system is apparent in the photos of the MURA accelerators, Figs. 1, 2, 3 and 9, in K.R. Symon, MURA Days, in *Proceedings of the 2003 Particle Accelerator Conference*. https://accelconf.web.cern.ch/p03/PAPERS/WOPA003.PDF Fig. 5.1: Copyrights under license CC-BY-3.0, https://creativecommons.org/licenses/by/3.0; no change to the material

8. S. Boucher, et al., The Radiatron: a high average current betatron for industrial and security applications, in *TUPP150 Proceedings of EPAC08*, Genoa, Italy. https://accelconf.web.cern.ch/e08/papers/tupp150.pdf

9. K. Okabe, et al., Development of H- injection of proton-FFAG at kurri, in *THPEB009 Proceedings of IPAC'10*, Kyoto, Japan. https://accelconf.web.cern.ch/IPAC10/papers/thpeb009.pdf Fig. 5.1: Copyrights under license CC-BY-3.0, https://creativecommons.org/licenses/by/3.0; the photo has been trimmed to mostly leave the 2.5 MeV injector

10. A. Sessler, E. Wilson, *A Century of Particle Accelerators* (World Scientific, 2007)

11. D.W. Kerst, The acceleration of electrons by magnetic induction. Phys. Rev. **60**, 47–53 (1941)

12. F.K. Goward, D.E. Barnes, Experimental 8 MeV synchrotron for electron acceleration. Nature **158**, 413 (1946)

13. E.J.N. Wilson, Fifty years of synchrotrons, in *Proceedings of EPAC96*. https://accelconf.web.cern.ch/e96/PAPERS/ORALS/FRX04A.PDF Fig. 5.2 : Copyrights under license CC-BY-3.0, https://creativecommons.org/licenses/by/3.0

14. D. Bohm, L. Foldy: Theory of the synchrocyclotron. Phys. Rev. **72**, 649–661 (1947). (Demonstration of phase stability using Berkeley 37-inch and 184-inch cyclotrons) https://journals.aps.org/pr/abstract/10.1103/PhysRev.72.649

15. V.A. Fomichev, Mobile accelerator based on ironless pulsed betatron for dynamic objects radiographing, in *IPAC2019*, Melbourne, Australia 019-THP. https://accelconf.web.cern.ch/ipac2019/papers/thpmp026.pdf

16. D.W. Kerst, R. Serber, Electronic orbits in the induction accelerator. Phys. Rev. **60**, 53–58 (1941)

17. D.W. Kerst et al., Operation of a 300 MeV betatron. Phys. Rev. **78**, 297–1 (1950)

18. A. Liénard, Champ électrique et magnétique produit par une charge concentrée en un point et animée d'un mouvement quelconque. L'Éclairage Électrique. **16**, 5 (1898)

19. D. Iwanenko, I. Pomeranchuk, On the maximal energy attainable in a Betatron. Phys. Rev. **65**, 343–1 (1944)

20. A. Hofmann, *The Physics of Synchrotron Radiation*. Cambridge Monographs on Particle Physics, Nuclear Physics and Cosmology (20) (Cambridge University Press, 2004)

21. F. Méot, Simulation of radiation damping in rings, using stepwise ray-tracing methods, JINST 10 T06006 (2015). https://doi.org/10.1088/1748-0221/10/06/T06006. http://iopscience.iop.org/1748-0221/10/06/T06006

22. F. Méot: Zgoubi Users' Guide. https://www.osti.gov/biblio/1062013-zgoubi-users-guide. Sourceforge latest version: https://sourceforge.net/p/zgoubi/code/HEAD/tree/trunk/guide/Zgoubi.pdf

Chapter 6
Microtron

Abstract This chapter introduces the microtron, and to the theoretical material needed for the simulation exercises. It begins with a brief reminder of the historical context, and continues with the beam optics and acceleration techniques that the microtron method leans on, relying in that on basic charged particle optics and acceleration concepts introduced in the previous chapters. It further addresses the following aspects:

- spiraling accelerated orbits tangenting at the accelerating gap,
- beam recirculation through an accelerating system, via return arcs,
- methods for periodic motion stability,
- harmonic number jump acceleration.

The simulation of a classical microtron only requires one optical element: DIPOLE, and CAVITE for acceleration. Simulation of a racetrack microtron adds DRIFT to simulate drift sections, possibly BEND for small steering dipoles along the recirculation paths, and a series of CAVITE to simulate a linac section. Particle monitoring requires keywords introduced in the previous chapters, such as FAISCEAU, FAISTORE, PICKUPS, and some others. Beam path and optics optimization use FIT[2]. SYSTEM is used to, mostly, resort to gnuplot so as to end simulations with some nice graphs (orbits, fields, or else) obtained by reading data from output files such as zgoubi.fai (resulting from the use of FAISTORE), zgoubi.plt (resulting from IL = 2), or other zgoubi.*.out files resulting from a PRINT command.

Notations Used in the Text

B	magnetic field
$B\rho = p/e$	magnetic rigidity
C_n	length of nth orbit
E; E_0; E_i; E_n	electron energy; at rest; kinetic at injection; on nth orbit
e	elementary charge
f_{rev}, f_{rf}	revolution and RF voltage frequencies

© The Author(s) 2024
F. Méot, *Understanding the Physics of Particle Accelerators*, Particle Acceleration and Detection, https://doi.org/10.1007/978-3-031-59979-8_6

h	harmonic number, an integer, $h = f_{\text{rf}}/f_{\text{rev}}$
$k = \frac{R}{B}\frac{dB}{dR}$	radial field index
$l;\ m;\ n$	$(C_n - C_{n-1})/\lambda_{rf}$; C_1/λ_{rf}; orbit number
m_0	electron rest mass
p	electron momentum
R	orbit radius
RF	radio-Frequency
s	path variable
$T_{\text{rev}}, T_{\text{rf}}$	revolution and accelerating voltage periods
v	electron velocity
$V(t);\ \hat{V}$	oscillating voltage; its peak value
x', y'	radial and axial coordinates in the moving frame $\left[(*)' = \frac{d(*)}{ds} \right]$
$\beta = v/c$	normalized electron velocity
$\gamma = E/m_0 c^2$	Lorentz relativistic factor
ΔE	energy gain in accelerating cavity or linac
$\Delta p, \delta p$	momentum offset
ε_u	Courant–Snyder invariant ($u\ :\ x,\ y$)
λ_{rf}	RF wavelength

6.1   Introduction

Although a similar geometry and uniform fixed field structure to the classical cyclotron (Fig. 6.1) and based as well on resonant acceleration using a fixed frequency oscillating voltage [1–3], it was not until 1944, more than a decade after Lawrence's cyclotron, that the original classical microtron concept, a ultra-relativistic electron beam accelerator, appeared in the literature [1]. A first specimen was built in Canada four years later [4]. Figure 6.1 shows an early principle schematic of a classical microtron, with typical parameters as given in Table 6.1.

The concept evolved into the racetrack microtron (RTM), and a 4-sector RTM, using AVF focusing, was brought to operation a decade later [5]. A technology still topical today, in specific energy ranges and applications [6]. During this period theoretical studies and developments addressed injection efficiency, resonant acceleration and phase focusing, transverse focusing, modes of operation, etc. [7]. A typical RTM schematic is shown in Fig. 6.2: return straights ensure recirculation of the bunches through a linac section, via a pair of 180° dipoles. The latter feature the necessary weak index focusing for vertical stability. Today, MAMI (Mainz Microtron) is the highest energy microtron installation, a four-RTM cascade delivering 100 µA CW polarized electron beam up to 1.6 GeV [8]. The RTM is an early stage of the present-day recirculating linear accelerator (RLA), yet with its recirculating arcs staked horizontally rather than vertically. In its "double-sided" design (cf. MAMI-C, 43 recirculations in 15 MeV steps) the RTM may be seen as akin to CEBAF-style two-linac RLA.

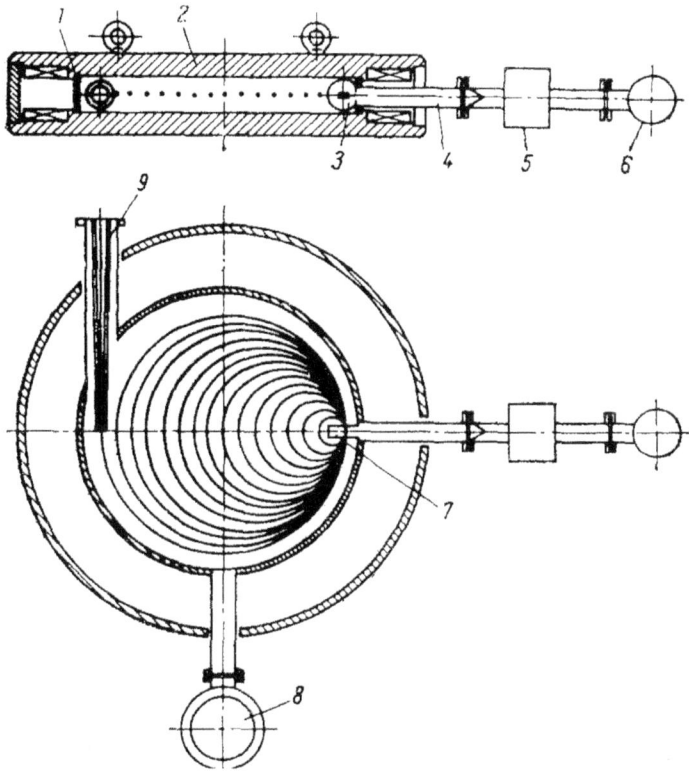

Fig. 6.1 Schematic of a classical microtron [2]. **1** Vacuum chamber; **2** magnet; **3** accelerating resonator; **4** waveguide; **5** ferrite; **6** magnetron; **7** electron emitter; **8** high-vacuum pump; **9** extraction channel

Table 6.1 Typical parameters of an early classical microtron [2]

Energy	$10-18$ MeV
Magnetic field	$1-2$ kG
Magnet pole diameter	75 cm
RF	3 GHz
Cavity voltage	≈ 511 keV
Repetition rate	400 Hz
Average current	$50\,\mu$A
Bunch emittances, H, V	$2 \times 1.5, 4 \times 15$ mm×mrad
Bunch length	$\gtrsim \lambda_{rf}/20$

Microtrons are fixed-field, fixed RF bunch-train accelerators. They provide fast acceleration, in just a few recirculations through a cavity or a short linac, thus essentially preserving the properties of the bunch out of the source, namely short bunches

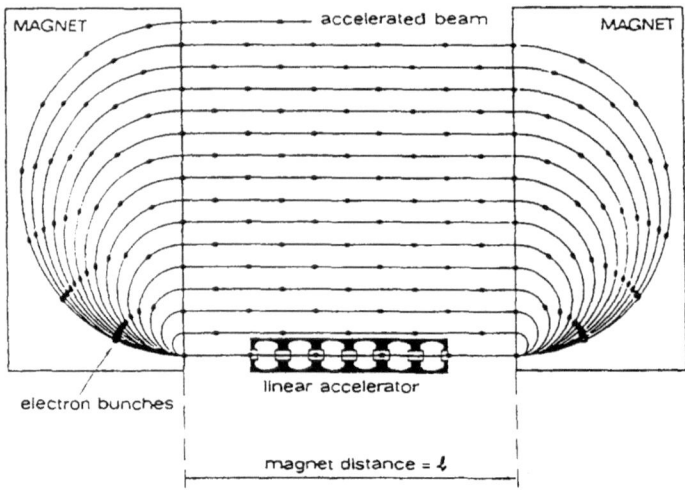

Fig. 6.2 A principle schematic of a racetrack microtron [3]. Room is allowed for a high energy boost linac section

with small transverse emittance and small momentum spread. The isochronism constraint requires accelerated bunches to be ultrarelativistic from the first recirculation, thus the method is of interest for lepton beams. The method is also of interest for the acceleration of polarized electron beams: an RTM microtron is essentially a long beam line, with accelerating sections, and so exempt from adverse effects of resonant depolarization as met in cyclic accelerators (see Sects. 4.2.5, 8.2.4, 9.2.7), thus electron bunch polarization from the source is preserved.

The classical microtron allows a few tens of MeV electron energy range, using an X-band (9.4 GHz), C-band (5.9 GHz), or S-band (2.8 GHz) high-gradient RF system. There are various designs of racetrack microtrons, they use short S-band or L-band (1.3 GHz) standing-wave linacs. Repetition rate is pulsed linacs' tens of Hz range, duty cycle up to several %, current up to 100 μA. Racetrack microtrons may use high fields, up to 2–3 T with normal conducting bends, up to 7–8 T with cryogenic magnets, this makes them compact electron recirculators [9].

The small 6D emittance electron bunches it delivers make the microtron an appropriate option in a number of applications, such as injector for synchrotron light source [9–11], free electron lasers (Fig. 6.3); industrial electron beams; γ-ray sources [13, 14]; radiation therapy; photonuclear production of isotopes, etc.

Fig. 6.3 Classical microtron injector (white arrow; the black arrow points to its RF cavity magnetron), at Kaeri far infrared free electron laser, a 2 m undulator (yellow arrow) down a short beam line (red arrow) [12]

6.2 Basic Concepts and Formulæ

The basic principles of the microtron are as follows:

- a fixed frequency accelerating RF system,
- bunch acceleration based on harmonic number jump,
- a fixed magnetic field to recirculate the bunches through the RF system, using

  - in the case of the classical microtron (Fig. 6.1): a 360° uniform field dipole with radial index $-1 < k < 0$;
  - in the case of the RTM (Fig. 6.2):

    two or more [5] split sectors, possibly featuring a field index,
    return straights to recirculate the beam,
    the recirculation path length an integer multiple of the RF wavelength.

6.2.1 Classical Microtron

Referring to Fig. 6.1, and assuming $\beta \approx 1$, in order to maintain isochronism between revolution time and oscillating voltage the length of any return orbit is a multiple of the RF wavelength λ_{rf}. As the energy $E = m_0 c^2$ is increased m increases, thus the revolution frequency decreases, a known property of the classical cyclotron (Chap. 3).

The length of the first orbit is

$$C_1 = m\lambda_{rf}, \qquad m \geq 2, \text{ integer} \tag{6.1}$$

with the constraint that it must clear the cavity. In passing, from this constraint an order of magnitude of the RF can be figured out: assume $C_1 \approx 30\,\text{cm}$, thus $f_{rf} = c/\lambda_{rf} \approx 2c/C_1 = O(10^9)$, in the GHz range.

Assume an orbit length increment of $l\lambda_{rf}$, l an integer (note that $l = 1$ minimizes the overall orbit excursion), thus the length of the nth orbit is

$$C_n = [m + (n-1)l]\,\lambda_{rf} \tag{6.2}$$

From $BR = p/e$ with $p = \beta E/c$, and assuming $\beta \approx 1$ as a necessary condition for isochronism of all orbits, one gets the energy increment by the cavity,

$$\Delta E = ceB\Delta R = \frac{ceB}{2\pi}l\lambda_{rf} = \frac{c^2eB}{2\pi}\Delta T_{rev} \tag{6.3}$$

with

$$\Delta T_{rev} = \Delta C/c = l\lambda_{rf}/c \tag{6.4}$$

the revolution time increment. The energy of the nth orbit comes out to be

$$E_n = \frac{ceB}{2\pi}[m + (n-1)l]\,\lambda_{rf} = \Delta E\left(\frac{m}{l} + n - 1\right) \tag{6.5}$$

Combining the expression for the energy at the first turn

$$E_1 = E_0 + E_i + \Delta E \tag{6.6}$$

with ΔE from Eq. 6.5 yields the energy gain at the accelerating gap,

$$\Delta E = \frac{E_0 + E_i}{m/l - 1} \tag{6.7}$$

The dependence of the field strength on m and l results, namely

$$B = \frac{2\pi}{ce\,\lambda_{rf}}\frac{\Delta E}{l} = \frac{2\pi}{ce\lambda_{rf}}\frac{E_0 + E_i}{m - l} \tag{6.8}$$

The guiding field B is maximized, and so the magnet size is minimized, for $m - l = 1$, which works for $m = 2$ (length of first orbit is 2 wavelengths) and $l = 1$ (orbit length increment is 1 wavelength). With $m \gtrsim 2$ and $l < m$, Eq. 6.7 indicates that the energy increment at the accelerating gap in order to preserve the synchronism is of the order of magnitude of the rest mass of the particle. A constraint which precludes considering this method for the acceleration of ions.

Weak focusing, as in the classical cyclotron (Sect. 3.2.2), ensures transverse motion stability:

- geometrical focusing maintains horizontal beam confinement in the vicinity of the closed orbit,
- a small field index (Eq. 3.11) ensures vertical focusing, with minor decrease of geometrical focusing, and marginal effect on the synchronism and microtron Eqs. 6.1–6.8.

6.2.2 Racetrack Microtron

In the racetrack microtron the recirculating dipole is split into two halves (Fig. 6.2). This allows room for a longer linac and greater energy gain, for efficient electron injection systems, and for beam instrumentation such as orbit correctors and beam position monitors (BPM). The linac occupies a straight section between the two magnets, whereas a series of return straight sections connect the trajectory arcs of increasing radius/energy.

As in the classical microtron, isochronous acceleration requires that the time of flight of any return orbit be a multiple of the RF period $T_{rf} = 2\pi/\lambda_{rf}$. Thus, as long as $\beta \approx 1$, the time difference between two orbits (Eq. 6.4)

$$\Delta T_{rev} = \Delta C/c = l\lambda_{rf}/c$$

still holds. With an added drift length L between the two 180° bends, the energy gain at the linac writes [15]

$$\Delta E = \frac{E_0 + E_i}{m/l - 1 - 2L/l\lambda_{rf}} \tag{6.9}$$

which reduces to Eq. 6.7 if $L = 0$. The dependence of the field strength on m and l is unchanged compared to the classical microtron, namely (Eq. 6.8),

$$B = \frac{2\pi}{ce\,\lambda_{rf}} \frac{\Delta E}{l} \tag{6.10}$$

The previous two equations show that the bending field and the linac boost must be adapted to the injection energy in order to satisfy the isochronism condition. Obviously, the top energy E_n after n passes in the linac, injection energy E_i and linac boost ΔE satisfy

$$E_n = E_i + n \times \Delta E \tag{6.11}$$

with in addition E_i and ΔE linked by Eq. 6.9.

The stable RF phase interval satisfies

$$\tan(\Delta\phi) = \frac{2}{\pi l} \tag{6.12}$$

A phase slip may result from $\beta < 1$ at the first turn, a few degrees expectedly, and should be well within the stable phase interval. Additional considerations on energy variation tolerances, and longitudinal acceptance, can be found for instance in [15] which addresses the design and properties of a variable energy RTM.

Orbit correction is paramount. It is needed to ensure proper beam steering over many recirculations, and in particular beam alignment on the linac axis. Techniques for that include horizontal and vertical steerers placed along the return straights, and BPMs. Tight orbit control may allow to relax on the dipole field homogeneity and on their positioning constraint.

Various methods may be resorted to regarding vertical focusing and transverse stability, depending on the general design of the racetrack. They may include

- wedge focusing [5] (see Sect. 14.4.1),
- active field clamps at the main dipole edges,
- a field index in the main dipoles, as in the classical microtron,
- AVF focusing [16] in the RTM sectors (see Sect. 4.2.1), as in the relativistic cyclotron,
- additional focusing in the return straights.

6.2.3 Synchrotron Radiation

Synchrotron radiation (SR) matters in high energy microtrons, in a similar way that it matters in beam lines [17] and recirculation linacs as CEBAF [18].

Effects of SR include energy loss and radial and longitudinal emittance growth. SR was first observed in a betatron (visually), and its effects on orbit as well, above 300 MeV where it required compensation by an ad hoc RF system. For this reason it is introduced in the Betatron chapter (Chap. 5), which can be referred to. The orbit spiraling which energy loss by SR causes in a high γ RTM (cf. section "Energy Loss", Eq. 5.14) requires compensation measures.

Emittance growth upon SR matters in high γ rings, whose future possibly includes the muon collider [19] and other FCC lepton and hadron collider rings [20]. It is introduced in that context, in the Strong Focusing Synchrotron chapter, Sect. 9.2.5, which can be referred to. As a matter of fact, in a GeV range microtron SR induced emittance growth matters as well, in relation with linac pipe aperture for instance.

6.3 Exercises

Note: Some of the input data files for these simulations are available in `zgoubi` sourceforge repository at

[pathTo]/branches/exemples/book/zgoubiMaterial/microtron/

6.1 Build a Classical Microtron
Solution 6.1

A 10-pass classical microtron simulation is worked out in this exercise, using S-band RF. Machine parameters are checked first, by raytracing. A bunch is then tracked, over an acceleration cycle.

(a) Build an input data file for a classical microtron (following the principle sketch of Fig. 6.1) with parameters taken from Table 6.2 (after Ref. [14]). DIPOLE can be used for the guiding field simulation (exercises in cyclotron Chaps. 3 and 4 can be referred to, field simulations are similar). Be sure to introduce a weak field index in order to ensure vertical motion stability. CAVITE[IOPT = 3] can be used as a zero-length accelerating system.

Validate the input data file by tracking an on-momentum electron, on the reference orbit, from injection to extraction energy. From the raytracing outcomes, check

– the expected circumference of the nth orbit, C_n (Eq. 6.2),
– the energy E_n of the nth orbit (Eq. 6.5).

(b) Track a 2×10^3-electron bunch over a complete acceleration cycle. Take initial bunch transverse emittances $\varepsilon_x = \varepsilon_y = 0.1\,\pi\,\mu$m, and momentum spread $\delta p/p = 10^{-3}$.

Produce a graph of a few trajectories in the laboratory frame.
Produce graphs of the final transverse and longitudinal phase spaces.

Table 6.2 Parameters of a 9.5 MeV classical microtron, after Ref. [14]. Injection energy E_i is at cavity entrance

Injection energy (E_i)	0.409 MeV
Extraction energy (E_x)	9.6 MeV
Number of orbits (n)	10
Energy gain in accel. gap (ΔE)	To be determined
RF (f_{rf})	2.8 GHz
Length of first orbit (C_1)	$2 \times \lambda_{rf}$
Length of nth orbit (C_n)	$(n+1)\lambda_{rf}$

6.2 Build a 100 MeV Racetrack Microtron
Solution 6.2

The microtron schematic of Fig. 6.2 is considered in this exercise. Take the following parameters [23, Sect. 3.1]: 5 MeV linac, 3 GHz RF, injection energy 50 keV, spacing between the half-dipoles 1 m, final energy 100 MeV in 20 linac passes.

DIPOLE can be used for the sectors of the 180° bends. Assume hard-edge for simplicity. BEND is convenient to simulate correction dipoles in the return straights, one at each end for instance, if necessary. DRIFT is used for the field-free spaces. A series of 6 CAVITE elements can be used to simulate the linac section.

(a) Assemble the RTM: the two double-sector 180° bends, drift sections with correction dipoles, linac section. Check the geometry by producing a graph of the accelerated orbit. No need to set up all 20 passes, 3–4 passes are enough to establish the effectiveness of the simulation and assess the various parameter adjustments to be performed to make it work.

(b) Check the transverse stability of the optics.

(c) Accelerate a 6D bunch. Check the evolution of its parameters: energy, transverse and longitudinal phase spaces.

6.4 Solutions of Exercises of This Chapter: Microtron

6.1 Build a Classical Microtron
(a) A classical microtron, input data file.
Based on the microtron parameters given in Table 6.2, the simulation input file of Table 6.3 results. In particular,

– a single particle is launched under OBJET; its launch radius is taken in (Fig. 6.4)

$$0 < Y_0 < C_1/\pi = 2\lambda_{\rm rf}/\pi \implies 0 < Y_0 < 6.8\,{\rm cm}$$

a necessary condition for DIPOLE to function, as per its geometrical definition [21, Fig. 9];
– the integration step size in DIPOLE is taken substantially less than $C_1 = 2\lambda_{\rm rf} = 21.4$ cm, to ensure accurate numerical integration of the equations of motion;
– the energy gain under CAVITE is (Eq. 6.7 with $l = m - l = 1$) $\Delta E = E_0 + E_i \lesssim 2E_i$;
– REBELOTE[IPASS = 9] ensures a 10-turn acceleration cycle;
– in addition: a weak filed index is added (Eq. 3.11), to ensure vertical motion stability. The value is taken from the Classical Cyclotron Chap. 3, Exercise 3.6, namely,

$$k = \frac{R_0}{B_0}\frac{\partial B_y}{\partial R}\bigg|_{R=R_0, y=0} = -0.03$$

Table 6.3 Simulation input data file microtron_one360degDipole.dat. An analytical modeling of a dipole magnet field, using DIPOLE. That file defines the labels (LABEL1 type [21, Sect. 7.7]) #S_piMicrotronSector and #E_piMicrotronSector for INCLUDEs in subsequent questions. It also accelerates a single particle launched in the vicinity of the reference orbit

Note: this file, and the gnuplot scripts files it resorts to, are available in zgoubi sourceforge repository at

https://sourceforge.net/p/zgoubi/code/HEAD/tree/branches/exemples/book/zgoubiMaterial/microtron/classical/

```
File name: microtron_360degDipole.dat.
'MARKER'  ProbMicroClassic_S                                   ! Just for edition purposes.
'OBJET'
2.551881                                                        ! Brho_ref: 409 keV electron.
2
1 1                      ! Initial radius Y0<C1/pi=2*lambda_rf/pi=2*10.70687/pi=6.8cm; a small
6. 0. .01 0. 0. 1. 'o'   ! vertical amplitude is added (0.1mm), to check the vertical focusing.
1
'PARTICUL'                                                     ! This is required for acceleration,
POSITRON                                          ! otherwise \zgoubi\ only requires rigidity.

'CAVITE'                                             ! Electrons enter accelerating gap, here.
3                       ! For this preliminary check, synchrotron motion is not necessary, IOPT=3 is fine.
0. 0.
919999  1.57079632679                                          ! Peak voltage; synchronous phase.

'MARKER'   #S_piMicrotronSector                     ! Label should not exceed 20 characters.
'DIPOLE'                                             ! Analytical definition of a dipole field.
2
180. 40.                                                        ! Sector angle; reference radius.
90. 1.800876 -0.03 0. 0.             ! Reference azimuthal angle; field; indices, N, N', N''.
0.  0.                                                          ! EFB 1 is  hard-edge,
4  .1455   2.2670  -.6395  1.1558  0. 0. 0.        ! hard-edge only possible with sector magnet.
90. 0.  1.E6  -1.E6  1.E6  1.E6
0.  0.                                                          ! EFB 2 is hard-edge.
4  .1455   2.2670  -.6395  1.1558  0. 0. 0.
-90. 0.  1.E6  -1.E6  1.E6  1.E6
0. 0.                                                          ! EFB 3 unused.
0 0.    0.    0.      0.      0. 0. 0.
0. 0.  1.E6  -1.E6  1.E6  1.E6 0.
2   10.
0.2                                                  ! Integration step size (cm). Has to be << C_1.
2 0. 0. 0. 0.                                                   ! Magnet positioning RE, TE, RS, TS.
'MARKER'    #E_piMicrotronSector                    ! Label should not exceed 20 characters.
'FAISCEAU'                                                      ! Local particle coordinates.
'DIPOLE'                                             ! Analytical definition of a dipole field.
2
180. 40.                                                        ! Sector angle; reference radius.
90. 1.800876 -0.03 0. 0.             ! Reference azimuthal angle; field; indices, N, N', N''.
0.  0.                                                          ! EFB 1 is hard-edge,
4  .1455   2.2670  -.6395  1.1558  0. 0. 0.        ! hard-edge only possible with sector magnet.
90. 0.  1.E6  -1.E6  1.E6  1.E6
0.  0.                                                          ! EFB 2 is hard-edge.
4  .1455   2.2670  -.6395  1.1558  0. 0. 0.
-90. 0.  1.E6  -1.E6  1.E6  1.E6
0. 0.                                                          ! EFB 3 unused.
0 0.    0.    0.      0.      0. 0. 0.
0. 0.  1.E6  -1.E6  1.E6  1.E6 0.
2   10.
0.2                                                  ! Integration step size. Has to be << R_1 ~cm.
2 0. 0. 0. 0.                                                   ! Magnet positioning RE, TE, RS, TS.

'FAISTORE'                                                      ! Log turn-by-turn particle data.
zgoubi.fai
1
'REBELOTE'
9   0.1 99                                         ! 9 more passes, for a total of 10 revolutions.

'FAISCEAU'

'SYSTEM'
4                                                               ! Four graphs follow:
gnuplot <./gnuplot_Zplt_XY-Lab_polar.gnu                        ! accelerated orbit;
gnuplot <./gnuplot_Zfai_s-vs-pass.gnu          ! cumulated distance and energy versus turn;
gnuplot <./gnuplot_Zplt_sB.gnu                          ! field versus path distance s;
gnuplot <./gnuplot_Zplt_sYZ.gnu                                 ! vertical motion Z(s).

'MARKER'  ProbMicroClassic_E                                    ! Just for edition purposes.
'END'
```

Fig. 6.4 Accelerated orbit
from 0.4 to 9.5 MeV, in a
microtron operated classical
cyclotron magnet

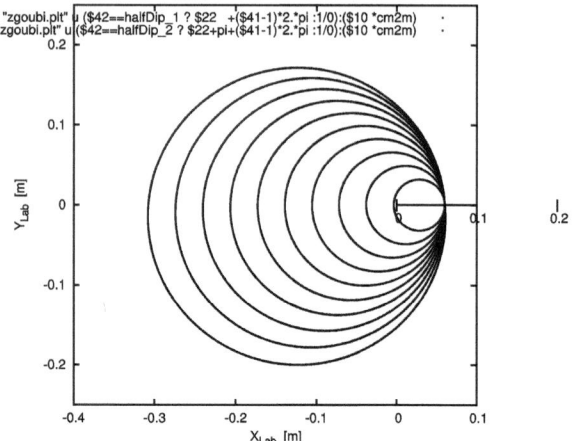

This is small enough a value that the general microtron Eqs. 6.1–6.8 still hold,
with marginal perturbation. Note that k is defined with respect to the center of the
dipole, as a consequence the magnetic field over a revolution is not homogeneous;
– a marginal vertical motion is added ($Z_0 = 0.1$ mm), for monitoring the vertical
focusing. It is small enough that the trajectory of the particle only marginally
departs from the reference accelerated orbit in the median plane. That trajectory
can thus still be used to check raytracing outcomes against theory, Eqs. 6.1–6.8.

The input data file (Table 6.3) is set for a 10-turn tracking of a single particle
launched near the reference orbit. The resulting accelerated orbit, from injection to
extraction energy, is displayed in Fig. 6.4. The dependence of orbit circumference and
energy on turn number is displayed in Fig. 6.5, with comparison to theory. Figures 6.6
and 6.7 show the field experienced along, and coordinates of the accelerated orbit.

(b) Acceleration of a 6D bunch.

The simulation file is given in Table 6.4. One or the other of the available variables,
or several: RF system parameters (synchronous phase and voltage under CAVITE);
the initial bunch conditions (injection phase via coordinate s_0 under MCOBJET,
injection energy); the dipole field, need adjustment to maximize the transmission.

This optimization exercise is left to the reader. A possibility is to develop, or use
existing algorithms, for beam steering and for the longitudinal motion. A brute force
method can be used as well, consisting in scanning the variables, with the objective of
maximizing the number of particles transmitted, with proper final energy. Zgoubi's
FIT procedure can also resorted to, it allows such constraints as the average value of
one or the other of the final coordinates of the particles (FIT[IC = 3]) [21], including
their momentum. More constraints, adapted to this particular type of problem, can
be added in the zgoubi source subroutine concerned, ff.f [22].

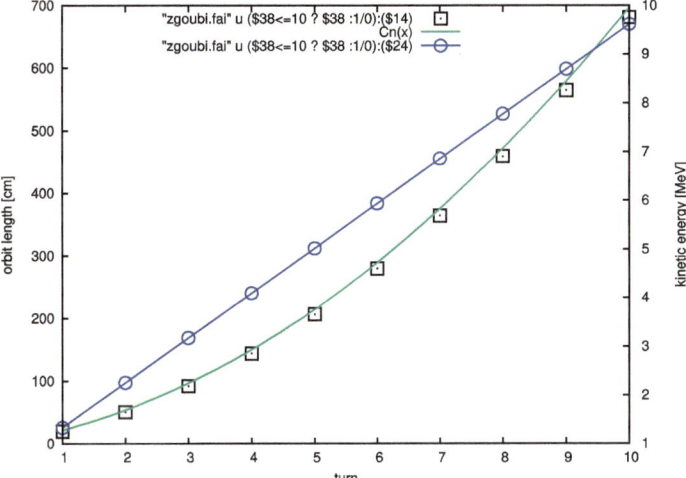

Fig. 6.5 Dependence of orbit circumference (square markers) and energy (circles) on turn number, from raytracing, data read from zgoubi.fai [21, Sect. 8.2]. Solid lines: theory, Eqs. 6.2 and 6.5, respectively. Note: orbit length comes closer to theory if the field index k is set to zero

Fig. 6.6 The magnetic field varies along the accelerated orbit under the effect of the transverse field index, however only weakly, so preserving near-synchronism with the RF voltage

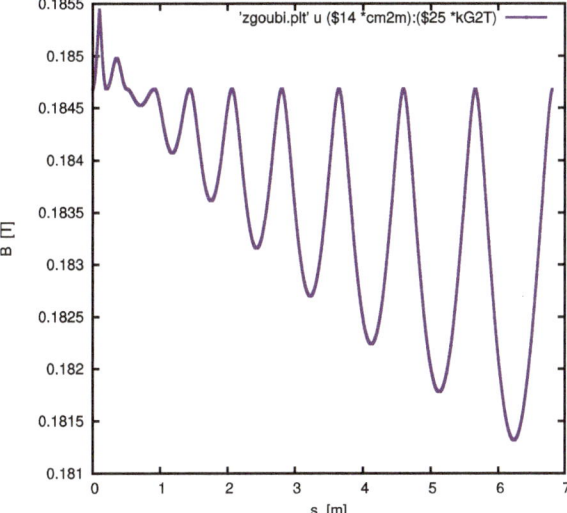

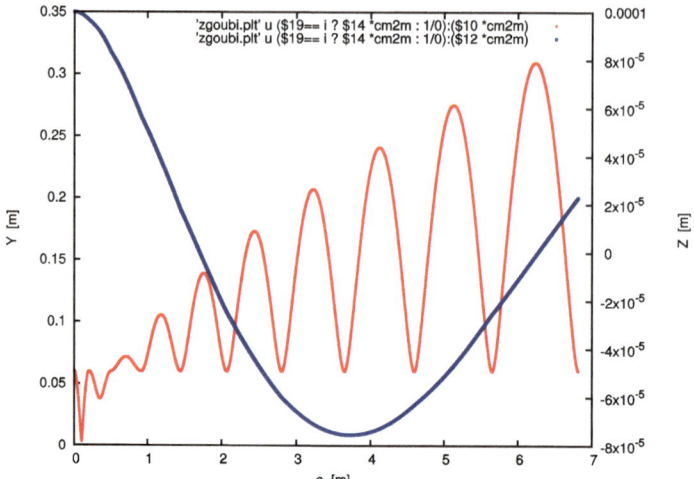

Fig. 6.7 Horizontal (red, fast oscillation) and vertical (blue, slow oscillation) coordinates of the accelerated particle, over the 10 revolutions

Table 6.4 Simulation input data file for 6D bunch acceleration. CAVITE[IOPT = 2] is used to accelerate, accounting for RF phase motion

```
Microtron, classical. Bunch acceleration
'MCOBJET'
2.551881                                              ! Brho_ref: 409 keV electron.
3
2000
2 2 2 2 2 2
3.e-2 0. 0. 0. 0.00815 1.
0. 1. 1e-9    3
0. 1. 1e-9    3
0. 1. 1e-9    1
123456 234567 345678
'PARTICUL'                                            ! This is required for acceleration,
POSITRON                                     ! otherwise \zgoubi\ only requires rigidity.

'OPTIONS'
1 1
.plt 0                                       ! Inhibit any print out to zgoubi.plt (saves on CPU time).

'CAVITE'
2                                                     ! Phase motion is accounted for.
0.107068735  1.                          ! Orbit length, first pass: determines f_rf=2.8 GHz.
920200  1.56976                                       ! Peak voltage; synchronous phase.

'INCLUDE'
1
2* microtron_360degDipole.dat[#S_piMicrotronSector:#E_piMicrotronSector]

'FAISTORE'                                             ! Log turn-by-turn particle data.
zgoubi.fai
1

'REBELOTE'
9  0.1 99                                    ! 9 more passes, for a total of 10 revolutions.

'FAISCEAU'
'END'
```

6.2 Build a 100 MeV Racetrack Microtron

Setting up this simulation, and checking it, is left to the reader. The input data file of the previous exercise, Table 6.3, can be used as a starting point for a split DIPOLE in the recirculation path.

Guidance in setting up the optical parameters for this RTM can be found in [23, Sect. 3.1].

It is a good idea for such recirculator simulation from start to end in a single job, to resort to GOTO keyword. GOTO in this context is used to switch the beam to proper sub-systems (i.e., return straights, 180° dipoles, linac). In this case it may be required to end the job using FINISH (END may not work in this context). AUTOREF also is useful to recenter the beam when presented at these sub-systems.

Guidance in setting up the input data file using these keywords can be found in an existing complete recirculating linac simulation, an energy recovery linac (ERL), 12 passes up 11 passes down, in https://sourceforge.net/p/zgoubi/code/HEAD/tree/branches/exemples/didacticExercises/LR-eRHIC/ folder. This ERL simulation is based on the keywords

- REBELOTE[IOPT = 1] in combination with GOTO[PASS#] and GOTO [GOBACK] to switch the beam to proper subsystems (return loops, spreader, combiner and linac),
- REBELOTE[IOPT = 1, LMNT = DRIFT[DltaPhase]] to flip the RF phase by pi at entrance to the linac, for energy recovery from pass number 14 on,
- abundant use of INCLUDE so to allow separate data files for the recirculating channel(s), spreaders, combiners and linac—which has the merit additional of simplifying the main input data file,
- FINISH, which ends the job, in lieu of END which cannot be used in this context.

References

1. V.I. Veksler: Proc. USSR Acad. Sci. **43**, 346 (1944); J. Phys. USSR **9**, 153 (1945)
2. S.P. Kapitsa, The microtron and areas of its application. Translated from Atomnaya Énergiya **18**(3), 203–209 (1965)
3. P. Lidbjörk, The microtron, in *CERN Accelerator School, Fifth General Accelerator Physics Course, Proceedings*, University of Jyväskylä, Finland, 7–18 September 1992, vol. 2. Report number: CERN-94-01, CERN-YELLOW-94-01. http://cds.cern.ch/record/235242/files/CERN-94-01-V2.pdf. Figure 7.18: Copyrights under license CC-BY-3.0, https://creativecommons.org/licenses/by/3.0; no change to the material
4. W.J. Henderson, H. Le Caine, R. Montalbetti, A magnetic resonance accelerator for electrons. Nature **162**, 699–700 (1948). https://doi.org/10.1038/162699a0
5. E. Brannen, H. Froelich, Preliminary operation of a four-sector racetrack microtron. J. App. Phys. **32**, 1179 (1961)
6. C. Hori, et al., Optical design of AVF weak-focusing accelerator, in *TUP036, Proceedings of the 22nd International Conference on Cyclotrons and Their Applications*, Cape Town, South Africa (2019), pp. 242–244. https://accelconf.web.cern.ch/cyclotrons2019/papers/tup036.pdf

7. A.P. Grinberg, The microtron. Soviet Physics Uspekhi **4**(6), 857–879 (1962). https://doi.org/10.1070/PU1962v004n06ABEH003391

8. M. Dehn, K. Aulenbacher, R. Heine, et al., The MAMI-C accelerator. Eur. Phys. J. Spec. Top. **198**, 19 (2011). https://doi.org/10.1140/epjst/e2011-01481-4. https://www.blogs.uni-mainz.de/fb08-nuclear-physics/accelerators-mami-mesa/the-mainz-microtron/

9. T. Hori, Ten years of compact synchrotron light source AURORA, in *WEP55, Proceedings of the 1999 Particle Accelerator Conference*, New York (1999). https://accelconf.web.cern.ch/p99/PAPERS/WEP55.PDF

10. Nadji, et al., Status of SESAME project. Proceedings of PAC09, Vancouver, BC, Canada WE5RFP022. https://accelconf.web.cern.ch/PAC2009/papers/we5rfp022.pdf

11. W.H.C. Theuws, et al., The 75 MeV racetrack microtron Eindhoven, in *Proceedings of the Linac 96 Conference*. https://accelconf.web.cern.ch/l96/PAPERS/MOP18.PDF

12. Y.U. Jeong, Compact terahertz free-electron laser as a users facility, in *Proceedings of APAC 2004*, Gyeongju, Korea. Fig. 6.3: Copyrights under license CC-BY-3.0, https://creativecommons.org/licenses/by/3.0; no change to the material

13. R. Hajima, et al., Compact gamma-ray source for non-destructive detection of nuclear material in cargo. THPS098, in *Proceedings of IPAC2011*, San Sebastián, Spain. https://accelconf.web.cern.ch/IPAC2011/papers/thps098.pdf

14. R.J. Abrams, et al., Compact, microtron-based gamma source, in *THPMR052, Proceedings of IPAC2016*, Busan, Korea. https://accelconf.web.cern.ch/ipac2016/papers/thpmr052.pdf

15. W.H.C. Theuws, et al., Continuous electron-energy variation of the Eindhoven racetrack microtron, in *17th IEEE Particle Accelerator Conference (PAC 97)*. https://accelconf.web.cern.ch/pac97/papers/pdf/7W020.PDF

16. H.R. Frœlich, J.J. Manca, Performance of a multicavity racetrack microtron. IEEE Trans. Nucl. Sci. (Proceedings of PAC75) **NS-22**(3) (1975). https://accelconf.web.cern.ch/p75/PDF/PAC1975_1758.PDF

17. G. Leleux, et al., Synhrotron radiation perturbations in long beam lines, in *Proceeding of the PAC 1991 Accelerator Conference*, May 6–9, 1991 San Francisco, California, USA. https://accelconf.web.cern.ch/p91/PDF/PAC1991_0517.PDF

18. D.R. Douglas, et al., Control of synchrotron radiation effects during recirculation, in *IPAC2015*, Richmond, VA, USA. https://accelconf.web.cern.ch/IPAC2015/papers/tupma035.pdf

19. B.J. King, Further studies on the prospects for many-TeV muon colliders, in *Proceedings of the PAC 2001 Accelerator Conference*, 18–22 Jun 2001, Chicago, IL, USA. https://accelconf.web.cern.ch/p01/PAPERS/RPPH314.PDF

20. M. Benedikt, F. Zimmermann, Status of the future circular collider study, in *Proceedings of RuPAC2016*, St. Petersburg, Russia. https://accelconf.web.cern.ch/rupac2016/papers/tuymh01.pdf

21. F. Méot, Zgoubi users' guide. https://www.osti.gov/biblio/1062013-zgoubi-users-guide. Sourceforge latest version: https://sourceforge.net/p/zgoubi/code/HEAD/tree/trunk/guide/Zgoubi.pdf

22. https://sourceforge.net/p/zgoubi/code/HEAD/tree/trunk/zgoubi/ff.f

23. P. Lidbjörk, The microtron, in *CERN Accelerator School, Fifth General Accelerator Physics Course, Proceedings*, University of Jyväskylä, Finland, 7–18 September 1992, vol. 2. Report number: CERN-94-01, CERN-YELLOW-94-01. http://cds.cern.ch/record/235242/files/CERN-94-01-V2.pdf

Chapter 7
Synchrocyclotron

Abstract This chapter introduces the concept of phase focusing by synchronous acceleration, and the synchrocyclotron which confirmed the principle. Synchrocyclotron style of acceleration in a fixed field alternating gradient accelerator (FFAG) is also addressed. The theoretical material needed for the simulation exercises is essentially that of the Weak Focusing Synchrotron, Chap. 8, regarding phase stability, and that of the Classical Cyclotron, Chap. 3, or FFAG optics, Chap. 10, regarding transverse stability. The chapter begins with a brief reminder of the historical context, and continues with the theoretical material which the synchrocyclotron optics and acceleration techniques lean on. The simulation of a synchrocyclotron is achieved using just three keywords: DIPOLE for the magnet, and CAVITE and SCALING for acceleration. FFAG dipoles have their specific keywords, FFAG and FFAG-SPI (Chap. 10). Particle monitoring uses FAISCEAU, FAISTORE, and some others. Optics matching and optimization, and the design of RF programs as well, use FIT[2]. INCLUDE is resorted to, although there is no obligation, in order mostly to simplify the input data files. SYSTEM calls to gnuplot scripts allow ending simulations with various graphs; gnuplot reads data from output files such as zgoubi.fai (produced by FAISTORE), zgoubi.plt (resulting from IL = 2) and from files zgoubi.*.out resulting from a PRINT command.

Notations Used in the Text

B	magnetic field value
$B\rho = p/q$; $B\rho_0$	particle rigidity; reference rigidity
C; C_0	orbit length
E; E_s	particle energy, $E = \gamma m_0 c^2$; synchronous energy
f_{rev}, $f_{\text{rf}} = h\, f_{\text{rev}}$	revolution and RF voltage frequencies
h	RF harmonic number, $h = f_{\text{rf}}/f_{\text{rev}}$
m; m_0; M	particle mass; rest mass; mass in units of MeV/c^2
$k = \frac{R}{B}\frac{dB}{dR}$	radial field index
$\mathbf{p}$; p; p_0	momentum vector; its modulus; reference
q	particle charge

© The Author(s) 2024
F. Méot, *Understanding the Physics of Particle Accelerators*, Particle Acceleration and Detection, https://doi.org/10.1007/978-3-031-59979-8_7

R	average orbit radius, $C = 2\pi R$
$R,\ \theta,\ y$	particle coordinates, radial, azimuthal, axial
α	momentum compaction, or trajectory deviation
$\beta = v/c;\ \beta_0;\ \beta_s$	normalized particle velocity; reference; synchronous
$\gamma = E/m_0c^2;\ \gamma_{tr}$	Lorentz relativistic factor; transition γ
$\nu_{R,\theta}$	wave numbers, radial and axial
$\phi;\ \phi_s$	particle phase at voltage gap; synchronous phase

7.1 Introduction

The synchrocyclotron (SC) accelerator is an outcome of the 1945 concept of phase focusing synchronous acceleration [1–3]. Demonstration of the latter successfully used a small classical cyclotron [4] (following closely a proof-of-principle in fixed closed orbit regime, using a betatron [5]).

Synchronous acceleration opened the way to the highest energies: this is the acceleration method in today's high energy colliders. Acceleration techniques at the time had intrinsic energy limitations: electrostatic generators around a few MeV due to insulation break-down at high electric field; the classical cyclotron, in the few MeV ion energy range (a few keV in the case of electrons) due to the loss of isochronism resulting from relativistic increase in mass (Chap. 3); the bulky isochronous ion cyclotron in the GeV range in relation with extraction efficiency (Chap. 4); the betatron due to the loss of the Widerøe condition resulting from synchrotron radiation (Chap. 5).

Phase focusing in a SC requires varying the frequency of the accelerating voltage, for it to follow the increase of the revolution period (Eq. 3.3) as the accelerated bunch spirals outward. A consequence is that the acceleration has to be cycled, at a rate determined by the time it takes to bring a bunch from injection to extraction energy. The repetition rate of the RF frequency cycling is typically of the order of $10^2 - 10^3$ Hz, determined by orbit size, energy, and RF voltage. This is orders of magnitude below cyclotron CW regime (with a bunch repetition rate of typically tens of MHz) as is the delivered average current.

The aforementioned successful demonstration using a 37-inch cyclotron resulted in the modification of Berkeley 184-inch cyclotron (Fig. 3.3), under commissioning at the time, into a synchrocyclotron which was brought into operation in 1946, and allowed producing 200 MeV proton beams and 400 MeV alpha-particle beams.

The highest beam energy from a frequency modulated cyclotron was achieved in Gatchina (Leningrad, Russia) in the late 1960s (Fig. 7.1), producing 1 GeV proton bunches at a 40–60 Hz repetition rate. A 600 MeV SC was CERN's first accelerator, providing beams from 1957, for particle and nuclear physics; only leaving particle physics to the CERN PS in 1964; supplying short-lived beam to ISOLDE from 1967 until its shut down in 1990. Its parameters, typical of a high energy SC, are given in Table 7.1; note the small peak accelerating voltage, more than an order of magnitude what an isochronous cyclotron requires. The SC technology is still topical today,

Fig. 7.1 Gatchina 1 GeV synchrocyclotron [6]

Table 7.1 Parameters of the CERN 600 MeV SC

Magnet pole diameter	5 m
Magnetic field	1.9 T
Peak voltage	25 kV
RF sweep	29→ 16.5 MHz
Repetition rate	50 Hz
Average current	1 μA

in particular in protontherapy application as the use of superconducting magnet technology allows for compact devices (Fig. 7.2).

A typical history line in that respect is Orsay synchrocyclotron (Fig. 7.3, parameters in Table 7.2): since 1991 it was one of the two protontherapy accelerators in hospital environment in France [9] (with MEDICYC in Nice [10]). The 157 MeV, 450 Hz repetition rate SC delivered a first beam in 1957, a typical nuclear physics research installation; the facility was shut-down in 1975 for evolution to 200 MeV; in 1993 the installation was converted to a hadrontherapy hospital, IC-CPO (Institut Curie-Centre de Protontherapie d'Orsay).

In a general manner the synchrocyclotron method can be understood as applying the phase focusing technique in a fixed-field ring accelerator. This is one way FFAGs are operated (they may also use induction acceleration [11, 12]—Chap. 10). By contrast with the classical cyclotron, FFAGs use high gradient radial or spiral sector dipoles and feature strong focusing optics (Chap. 10).

Fig. 7.2 The 230 MeV superconducting S2C2 [7], a 1 kHz repetition rate compact synchrocyclotron for protontherapy. Parameters: RF frequency 60–90 MHz; magnetic field 5 T; overall diameter 2.5 m; weight 50 Ton

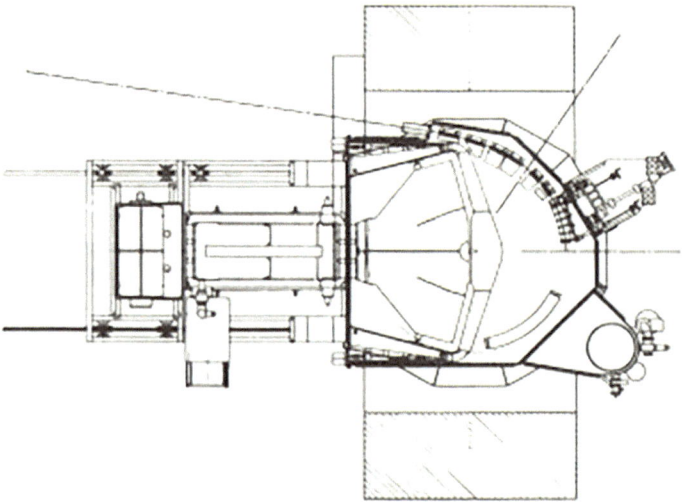

Fig. 7.3 Layout of Orsay 200 MeV synchrocyclotron [8], a 450 Hz repetition rate machine using a rotating condenser for RF cycling

Table 7.2 Parameters of the Orsay 200 MeV SC, first a nuclear physics instrument, then converted to hadrontherapy

Magnet pole diameter	2.4 m
Magnetic field	1.6 T
Peak voltage	25 kV
RF sweep	25 → 20 MHz
Repetition rate	450 Hz
Average current	3 μA

7.2 Basic Concepts and Formulæ

The classical cyclotron offered the opportunity of implementing the concept of phase stability, using existing technology. This further allowed a leap in ion energy, up to GeV energy range [6].

In the small classical cyclotron used for the demonstration, the oscillating electric voltage was applied between a dee and a flat electrode facing it. The voltage can be low, in the kVolt range, an easier technology compared to hundreds of kVolts required by the isochronism condition in a cyclotron. Many more turns are thus needed, of the order of 10^5 compared to a few 100 in a cyclotron, however a large number of turns no longer matters thanks to the phase focusing.

A drawback of synchronous acceleration in a cyclotron is that the RF system, thus bunch delivery, has to be cycled, due to the time of flight variation (increasing with energy). Only particles which maintain correct RF phase at the accelerating gaps, within a few degrees, are held in a bunch. The magnetic field is fixed, though (by contrast with pulsed synchrotrons which also require cycling the magnets) thus allowing a large repetition rate, nevertheless $4-5$ orders of magnitudes below a cyclotron CW regime, to the detriment of the average current.

In FFAGs a synchronous RF system is comprised of modular cavities, providing one or more accelerating gaps, similar to the RF technology found in synchrotrons. Drift sections between the dipoles provide the space for inserting these cavities (Chap. 10).

7.2.1 Phase Stability

The two accelerator lattice species, weak and strong focusing, differ by the energy range of their transition gamma, $\gamma_{tr} = 1/\sqrt{\alpha}$ (γ_{tr} is defined in Sect. 8.2.2), a property of the lattice which determines two different phase focusing and thus acceleration regimes: either below transition, or above transition.

Weak focusing results in $\gamma_{tr} \approx \nu_R$ (Eq. 8.34), while due to revolution symmetry $\nu_R < 1$ (Eq. 3.19). Thus in a classical cyclotron $\gamma_{tr} < 1 < \gamma$, regardless of γ, acceleration is above transition always, which in a practical manner means on the negative slope of the accelerating wave. This is sketched in Fig. 7.4: a particle with slightly greater energy than the synchronous particle takes more time to go around the ring (Eq. 3.3) (path length increase is larger than velocity increase), it tends to arrive later at the RF gap (at $\phi > \phi_s$), thus experiences smaller voltage which tends to speed it up. A particle with a lower energy is faster and arrives at the gap earlier, $\phi < \phi_s$, it experiences greater voltage which tends to slow it down. In both cases the non-synchronous particle is pulled towards the synchronous phase, this results in an overall stable oscillatory motion around ϕ_s, the particles stay bunched in the vicinity of the synchronous phase.

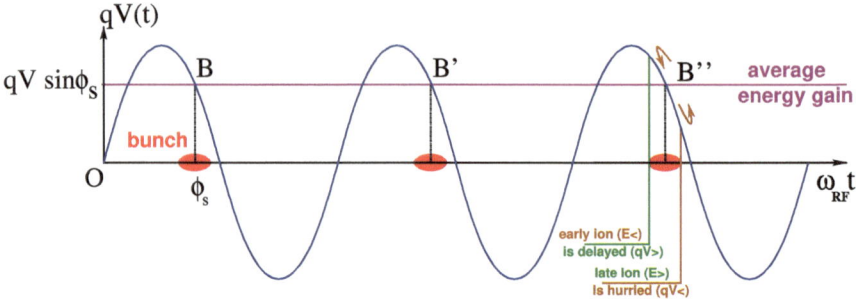

Fig. 7.4 A sketch of the mechanism of phase stability in a synchrocyclotron. Stability occurs for particles falling in the vicinity of a constant synchronous phase ϕ_s, turn after turn, at locations B, B', B" ...

In an FFAG $\gamma_{tr} \approx \nu_R \approx \sqrt{1+k} > 1$ (cf. Sect. 10.2, Eqs. 10.13 and 10.18), with k the field index. Generally $\gamma < \gamma_{tr}$: particle acceleration is below transition (cf. Sect. 8.2.2, Fig. 8.15).

The phase focusing technique results in synchrotron oscillations, the particle motion in the longitudinal (momentum, RF phase) phase space. This is addressed in Sect. 8.2.2, Weak Focusing Synchrotron chapter.

7.2.2 Transverse Stability

The classical cyclotron features weak vertical focusing based on a small $dB/dR < 0$, and geometrical horizontal focusing. This ensures periodic stability, the technique is addressed in Sect. 3.2.2.

FFAGs instead are based on strong focusing optics, using high transverse gradient combined function dipoles (typically, a dB/dR gradient of several T/m, as in strong focusing synchrotrons—Sect 9.2.1). FFAG lattices use radial sector magnets, yielding "Alternating Gradient" optics, the method is addressed in Sect. 10.2. They also use spiral sector dipoles, the method is addressed in Sects. 4.2.1 and 10.2.2.

JINR phasotron, a 680 MeV, $\approx$ 3 m radius, proton synchrocyclotron has a similar structure, of radially increasing average magnetic field, and spiral azimuthal field variation for periodic stability [13].

7.3 Exercises

7.1 Operate a Cyclotron Dipole in Synchrocyclotron Mode
Solution 7.1

Using a dipole magnet simulation taken from the classical cyclotron exercises (Sect. 3.3), add SCALING[IOPT = −2] and CAVITE[IOPT = 6] to simulate an acceleration cycle, in the synchrocyclotron mode.

7.2 Operate an FFAG in Synchrocyclotron Mode
Solution 7.2

Using a lattice simulation taken from the scaling FFAG exercises (radial, Sect. 10.3.1, or spiral, Sect. 10.3.2), add SCALING[IOPT = −2] and CAVITE[IOPT = 6] to simulate an acceleration cycle, in the synchrocyclotron mode.

7.4 Solutions of Exercises of This Chapter: Synchrocyclotron

7.1 Operate a Cyclotron Dipole in Synchrocyclotron Mode
The problem requires a classical cyclotron dipole model. This can be based, indifferently, on

– TOSCA keyword [14] and a field map: the method is devised in Exercise 3.6, with solution 3.6. Zgoubi input data file given in Table 3.16, "FieldMapSectorIndex.inc", which resorts to a field map of a 180° sector dipole with index, can be used;
– or DIPOLE keyword [14] for an analytical field model: the method is devised in Exercise 3.6 as well (3.6 (c)). The simulation of a sector DIPOLE with index given in Table 3.18, "sectorWithIndex.inc", can replace TOSCA in the input data file of Table 3.16.

Acceleration
Synchrotron acceleration then needs to be installed. CAVITE[IOPT = 6] is used, the option IOPT = 6 allows reading the RF voltage law (frequency, voltage, etc.) from a ancillary file zgoubi.freqLaw.In, see below. A similar problem is solved in the case of synchrocyclotron operation of a radial FFAG lattice and can be referred to, Exercise 10.5. The same procedure is repeated here, it comprises three steps, as follows.

(i) Find a set of closed orbits in the acceleration range of concern, 20 keV to 6 MeV about, and their revolution period (a few tens of closed orbits is fine, it does not need to be turn-by-turn, zgoubi will interpolate from zgoubi.freqLaw.In content). This can be performed using FIT, the input data file for that is given in Table 7.3, orbits

Table 7.3 Simulation input data file orbit_20to6000keV_FIT.dat. It finds 30 cyclotron orbits, evenly spaced from 200 keV to 6 MeV, using REBELOTE and FIT. This file also defines the LABEL1s #S_halfDipole_SC and #S_halfDipole_SC for use in subsequent data files

```
! File orbit_20to6000keV_FIT.dat
Find closed orbits for RFprogram.f to build zgoubi.freqLaw.In
'MARKER'   orbit_20to6000keV_S                                          ! Just for edition purposes.
'OBJET'
64.62444403717985                        ! Reference Brho ("BORO" in the users' guide) -> 200keV proton.
2
1 1                                                                     ! Just one ion.
4.087013 0. 0. 0. 0. 0.3162126 'o'       ! D=0.3162126 => Brho[kG.cm]= 20.435064, kin-E[keV]= 20.
1
'PARTICUL'                                              ! Usage of CAVITE requires partical data,
PROTON                                          ! otherwise, by default zgoubi only requires rigidity.
'FAISCEAU'                                          ! Local particle coordinates logged in zgoubi.res.

'MARKER'    #S_halfDipole_SC
'TOSCA'
0 2    ! IL=2 to log step-by-step coordinates, spin, etc., to zgoubi.plt (avoid, if CPU time matters).
1. 1. 1. 1.       ! Normalization coefficients, for B, X, Y and Z coordinate values read from the map.
HEADER_8                                           ! The field map file starts with an 8-line header.
315 151 1 22.1 1.          ! IZ=1 for 2D map; MOD=22 for polar frame; .MOD2=.1 if only one map file.
geneSectorMap.out
0 0 0 0     ! Possible vertical boundaries within the field map, to start/stop stepwise integration.
2
0.1 ! cm                            ! Integration step size. Set to 0.1 due to small radius orbits.
2                                                                       ! Magnet positioning option.
0. 0. 0. 0.                                                             ! Magnet positioning.
'MARKER'    #E_halfDipole_SC
'FAISCEAU'                                          ! Local particle coordinates logged in zgoubi.res.

'FIT'
1
2 30 0 [4,100]
1
3.1 1 2 #End 0. 1. 0                                 ! Request same radius after 180 deg orbital angle.

'FAISTORE'                                              ! Store particle data, turn-by-turn.
orbits.fai
1

'REBELOTE'
30 0.2 0 1
1
OBJET 35  0.31621261:5.4856827652488418                                 ! 20keV to 6MeV.

'SYSTEM'
2   ! gnuplot files found in sourceforge, in [pathTo]/zgoubi-code//toolbox/gnuplotFiles/gnuplot_Zplt/.
gnuplot <./gnuplot_Zplt_XYLab.gnu        ! Plot trajectories in lab coordinates, from zgoubi.plt records.
gnuplot <./gnuplot_Zfai_TOFvsBrho.gnu              ! Plot TOF vs. Brho, from orbits.fai records.

'MARKER'   orbit_20to6000keV_E                                          ! Just for edition purposes.
'END'
```

and time of flights produced by its execution are displayed in Fig. 7.5. The output file of interest here is orbits.fai, it is needed in step (ii).

(ii) Run an interface program, essentially a read-write procedure: read from orbits.fai, write in zgoubi.freqLaw.In with the proper formatting (Table 7.4).

(iii) Build the appropriate `zgoubi` input data file for acceleration (Table 7.5). This requires the following.

Either one of the aforementioned TOSCA modeling or DIPOLE modeling files can be used as a starting point, *mutatis mutandis*, as follows:

– PARTICUL[PROTON] is necessary as CAVITE is used: it allows converting energy change in rigidity change (zgoubi pushes particles using rigidity),

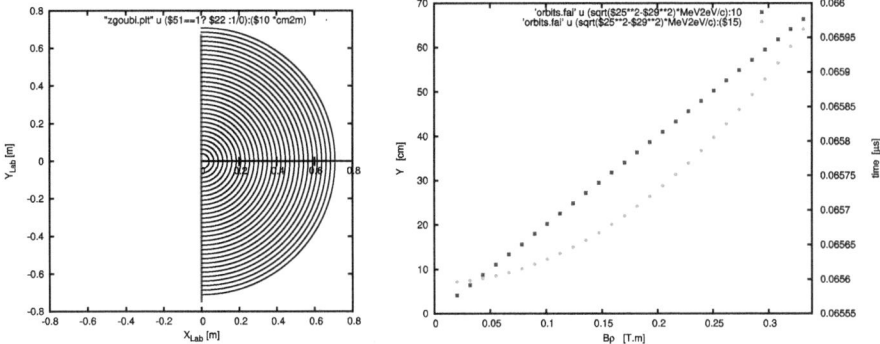

Fig. 7.5 Left: thirty constant-energy closed orbits across a half-dipole, obtained by running the input data file of Table 7.3. Stepwise integration data are read from zgoubi.plt, filled up upon IL = 2 under TOSCA. Right, right vertical axis (circles): time of flight along these half-circle orbits, versus rigidity; left vertical axis (squares): orbital radius

Table 7.4 The content of zgoubi.freqLaw.In (top and bottom parts), as read by zgoubi when using CAVITE[IOPT = 6]. Zgoubi actually only uses the turn number, column 1 (and will interpolate as needed), and the revolution time which is the cumulated time-of-flight across the cells, column 4

turn#	cell time of flight, tauCell	cumulated phase (unused)	cumulated time of flight, oclock	kinetic energy Ekin / MeV
1.00000000E+00	6.55958718E-02	0.00000000E+00	6.55958718E-02	2.00000091E-02
3.44520453E+00	6.55978931E-02	0.00000000E+00	1.31193765E-01	4.89040996E-02
5.89040905E+00	6.56008024E-02	0.00000000E+00	1.96794567E-01	9.05178122E-02
8.33561357E+00	6.56046001E-02	0.00000000E+00	2.62399167E-01	1.44839474E-01
....				
6.21301131E+01	6.59122230E-02	0.00000000E+00	1.70845545E+00	4.54511888E+00
6.45753177E+01	6.59363393E-02	0.00000000E+00	1.77439179E+00	4.89008111E+00
6.70205222E+01	6.59613305E-02	0.00000000E+00	1.84035312E+00	5.24755827E+00
6.94657267E+01	6.59871956E-02	0.00000000E+00	1.90634032E+00	5.61753613E+00
7.19109312E+01	6.60139336E-02	0.00000000E+00	1.97235425E+00	6.00000000E+00

- SCALING[IOPT = −2] provides the RF program to CAVITE (by reading it from an external file, zgoubi.freqLaw.In). The RF program here simply provides the turn dependence of revolution time (Table 7.4),
- CAVITE[IOPT = 6] boosts the particle(s) at each pass, following that pre-defined RF program,
- REBELOTE sends the execution pointer back to the top of the input data file, for multiturn tracking. A $2 \times 20\,\text{kV}$ acceleration rate per turn may be obtained from peak voltage $\hat{V} = 40\,\text{kV}$ and synchronous phase $\phi_s = 30°$ at the accelerating gap. This determines the number of passes for a $0.02 \rightarrow 6\,\text{MeV}$ cycle, namely, $(6 - 0.02)/0.02 = 300$, or 150 turns in the dipole,
- FAISTORE stores turn-by-turn particle data (to some user defined file, e.g., zgoubi.fai).

Outcomes of these synchrocyclotron acceleration simulations are displayed in Fig. 7.6.

Table 7.5 Simulation input data file for synchronous acceleration in a cyclotron dipole, from 20 keV to ∼ 6 MeV, in 150 turns.

```
Synchronous acceleration in a classical cyclotron
'MARKER'   accel_20to6000keV_S                                      ! Just for edition purposes.
'OBJET'
64.62444403717985                             ! Reference Brho ("BORO" in the users' guide) -> 200keV proton.
2
2 1                                                                            ! Just one ion.
4.087013    0. 0. 0. 0. 0.3162126 'o'         ! D=0.3162126 => Brho[kG.cm]= 20.435064, kin-E[keV]= 20.
4.087013  100. 0. 0. 0. 0.3162126 'o'         ! D=0.3162126 => Brho[kG.cm]= 20.435064, kin-E[keV]= 20.
1 1
'PARTICUL'                                                   ! Usage of CAVITE requires partical data,
PROTON                                          ! otherwise, by default zgoubi only requires rigidity.

'OPTIONS'
1 1
.plt 20    ! Global control of IL: applies to all optical elements (here, sets its value under TOSCA).

'SCALING'                                                     ! SCALING is used to control CAVITE.
1   1
CAVITE
-2     ! Option -2 causes read of RF law for CAVITE from external file (default is zgoubi.freqLaw.In).
1                                                               ! unused with option -2.
1                                                               ! unused with option -2.

'FAISCEAU'                                                   ! Local particle coordinates logged in zgoubi.res.
'FAISTORE'                                                   ! Store particle data, turn-by-turn.
zgoubi.fai   AftCAV         ! Log coordinates at any occurence of LABEL1=AftCAV, in zgoubi.fai.
1
'INCLUDE'
1
orbit_20to6000keV_FIT.dat[#S_halfDipole_SC:#E_halfDipole_SC]
'MARKER'   #S_halfDipole

'CAVITE'   cavity                                                        ! Accelerating gap.
6 PRINT                                                ! RF law is read from zgoubi.freqLaw.In.
0. 0.                                                                      ! Unused.
40e3   2.61799387799                                  ! Peak voltage 40 kV; RF phase = 150deg.
'MARKER'   AftCAV                                       ! Storage by FAISTORE is effective here.

'REBELOTE'
300 0.2 99                                      ! 300 passes in a half-dipole and a voltage gap.
'FAISCEAU'                                                     ! Particle coordinates here.

'SYSTEM'
2                                                                    ! 2 SYSTEM command follow:
/usr/bin/gnuplot < ./gnuplot_Zplt_XYLab.gnu &
/usr/bin/gnuplot < ./gnuplot_Zfai_YvsPass.gnu &
'MARKER'   accel_20to6000keV_E                                           ! Just for edition purposes.
'END'
```

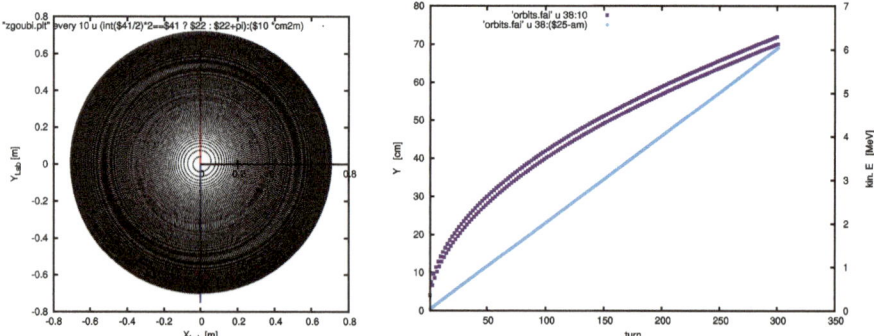

Fig. 7.6 Left: spiral trajectory of the synchronous particle over 150 turns from 0.02 to 6 MeV about, obtained by running the input data file of Table 7.5. Stepwise integration data are read from zgoubi.plt, filled up upon IL = 2 under TOSCA. Right, right vertical axis (circles): increasing energy, half-turn by half-turn; left vertical axis (squares): increasing orbital radius

7.2 Operate an FFAG in Synchrocyclotron Mode

Synchronous acceleration is generally used in FFAGs—in addition to induction acceleration, for instance in the MURA FFAGs [11] and in the ion-Beta at the KURRI Institute [12].

The present problem is treated as part of the FFAG Chapter exercises, Sects. 10.3.1 and 10.3.2. It requires an FFAG ring lattice, radial or spiral, indifferently.
In the former case:

– a radial FFAG lattice is devised in Exercise 10.1, solutions are found 10.1;
– synchrotron acceleration in that lattice is devised in Exercise 10.5, solutions are found 10.5.

Regarding the second type of lattice, spiral sector, similar synchronous acceleration simulations are performed in Exercise 10.13 (solutions 10.12). They resort to the analytical field modeling FFAG-SPI, and to CAVITE[IOPT = 6] as well for synchronous acceleration.

References

1. V. Veksler, A new method of acceleration of relativistic particles. J. Phys. USSR **9**, 153–158 (1945)
2. E.M. McMillan, The synchrotron. Phys. Rev. **68**, 143–144 (1945)
3. L. Jones, F. Mills, A. Sessler et al., *Innovation Was Not Enough* (World Scientific, 2010)
4. J.R. Richardson, et al., Frequency modulated cyclotron. Phys. Rev. **69**, 669 (1946); J.R. Richardson, et al., Development of the frequency modulated cyclotron. Phys. Rev. **73**, 424 (1948). https://journals.aps.org/pr/abstract/10.1103/PhysRev.73.424 University of California, Berkeley, California, Lawrence Radiation Laboratory: The 184-inch synchrocyclotron. https://www.gutenberg.org/files/33397/33397-h/33397-h.htm#THE_184-INCH_SYNCHROCYCLOTRON
5. F.K. Goward, D.E. Barnes, Experimental 8 MeV synchrotron for electron acceleration. Nature **158**, 413 (1946)
6. O.A. Shcherbakov, et al., Spallation neutron source at the 1 GeV synchrocyclotron of PNPI, in *Proceedings of the RuPAC-2016*, Peterhof, St. Petersburg, 21–25 November, 2016. https://accelconf.web.cern.ch/rupac2016/talks/wezmh01_talk.pdf Fig. 7.1: Copyrights under license CC-BY-3.0, https://creativecommons.org/licenses/by/3.0; no change to the material
7. S. Henrotin, et al., Commissioning and testing of the first IBA S2C2, in *TUP07 Proceedings of Cyclotrons2016*, Zurich, Switzerland. https://accelconf.web.cern.ch/cyclotrons2016/papers/tup07.pdf Fig. 7.2: Copyrights under license CC-BY-3.0, https://creativecommons.org/licenses/by/3.0; no change to the material
8. A. Laisné, et al., The Orsay 200 MeV Synchrocyclotron. https://accelconf.web.cern.ch/c78/papers/a-09.pdf Fig. 7.3: Copyrights under license CC-BY-3.0, https://creativecommons.org/licenses/by/3.0; no change to the material
9. S. Meyroneinc, et al., Beam quality for protontherapy at C.P.O, in *Proceedings of the 15th International Conference on Cyclotrons and their Applications*, Caen, France (1998). https://accelconf.web.cern.ch/c98/papers/a-04.pdf
10. P. Mandrillon, et al., Commissioning and implementation of the MEDICYC CYCLOTRON PROGRAMME, in *Proceedings of the Twelfth International Conference on Cyclotrons and their Applications*, Berlin, Germany. https://accelconf.web.cern.ch/c89/papers/e-05.pdf

11. F.T. Cole, O Camelot, a Memoir of the MURA Years. Cyclotron Conference, East Lansing, USA, May 13–17 (2001). https://accelconf.web.cern.ch/accelconf/c01/cyc2001/extra/Cole.pdf
12. M. Tanigaki, et al., Construction of FFAG accelerators in KURRI for ADS study, Proceedings of the EPAC 2004 Accelerator Conference, pp. 2676–2678 (2004). http://accelconf.web.cern.ch/accelconf/e04/PAPERS/THPLT078.PDF
13. L.M. Onischenko, JINR Phasotron, in *Proceedings of the Pac 1987 Conference.* https://accelconf.web.cern.ch/p87/PDF/PAC1987_0878.pdf
14. F. Méot, Zgoubi Users' Guide. https://www.osti.gov/biblio/1062013-zgoubi-users-guide. Sourceforge latest version: https://sourceforge.net/p/zgoubi/code/HEAD/tree/trunk/guide/Zgoubi.pdf

Chapter 8
Weak Focusing Synchrotron

Abstract This chapter introduces the weak focusing synchrotron, and the theoretical material needed for the simulation exercises. It begins with a brief reminder of the historical context, and continues with the beam optics and acceleration techniques that the weak focusing synchrotron principle and methods lean on, relying on basic charged particle optics and acceleration concepts introduced in the previous chapters. It further addresses the following aspects:

– fixed closed orbit,
– periodic structure,
– periodic motion stability,
– optical functions,
– synchrotron motion,
– depolarizing resonances.

The simulation of a weak focusing synchrotron lattice only requires two optical elements: DIPOLE or BEND to simulate combined function dipoles, and DRIFT to simulate straight sections. A third element, CAVITE, is required for acceleration. Computation of synchrotron radiation (SR) Poynting and spectral brightness uses zpop. Particle monitoring requires keywords introduced in the previous chapters, including FAISCEAU, FAISTORE, PICKUPS, and some others. Spin motion computation and monitoring resort to SPNTRK, SPNPRT and FAISTORE. Optics matching and optimization use FIT[2]. INCLUDE is used, mostly here in order to shorten the input data files. SYSTEM is used to, mostly, resort to gnuplot so as to end simulations with some specific graphs (orbits, fields, or else) obtained by reading data from output files such as zgoubi.fai (resulting from the use of FAISTORE), zgoubi.plt (resulting from IL = 2), or other zgoubi.*.out files resulting from a PRINT command.

© The Author(s) 2024   237
F. Méot, *Understanding the Physics of Particle Accelerators*, Particle Acceleration
and Detection, https://doi.org/10.1007/978-3-031-59979-8_8

Notations Used in the Text

B; $\mathbf{B}$; $B_{x,y,s}$	field; field vector; its components in the moving frame
$B\rho = p/q$; $B\rho_0$	particle rigidity; reference rigidity
C; C_0	orbit length, $C = 2\pi R + \left[\begin{smallmatrix}\text{straight}\\\text{sections}\end{smallmatrix}\right.$; reference, $C_0 = C(p = p_0)$
$\mathbf{E}$; E_σ, E_π	SR electric field impulse; its parallel and normal components
E; E_s	particle energy, $E = \gamma m_0 c^2$; synchronous energy
EFB	Effective Field Boundary
f_{rev}, $f_{\text{rf}} = h\, f_{\text{rev}}$	revolution and RF voltage frequencies
G	gyromagnetic anomaly, $G = 1.792847$ for proton
h	RF harmonic number, $h = f_{\text{rf}}/f_{\text{rev}}$
m; m_0; M	particle mass; rest mass; mass in units of MeV/c^2
$n = -\frac{\rho}{B}\frac{\partial B}{\partial x}$	focusing index
$\mathbf{n_0}$	stable spin precession direction
$\mathbf{p}$; p; p_0	momentum vector; its modulus; reference
$\mathbf{P} = \mathbf{E} \times \mathbf{B}$	SR Poynting vector
P_i, P_f	beam polarization, initial, final
q	particle charge
R	average orbit radius, $R = C/2\pi$
s	path variable
v	particle velocity
$V(t)$; $\hat{V}$	oscillating voltage; its peak value
x', y'	horizontal and vertical coordinates in the moving frame
α	momentum compaction; or trajectory deviation; or depolarizing resonance crossing speed
$\beta = v/c$; β_0; β_s	normalized particle velocity; reference; synchronous
β_u	betatron functions ($u : x, y$)
$\gamma = E/m_0 c^2$	Lorentz relativistic factor
δp, Δp	momentum offset
ϵ_c	critical energy of SR, $\epsilon_c = \hbar\omega_c = hc/\lambda_c$
ε	wedge angle
ε_u	Courant-Snyder invariant; or beam emittance ($u : x, y, l$)
ϵ_R	strength of a depolarizing resonance
μ_u	betatron phase advance per period, $\mu_u = \int_{\text{period}} \frac{ds}{\beta_u(s)}$ ($u : x, y$)
ν_u	wave numbers, horizontal, vertical, synchrotron ($u : x, y, l$)
ρ; ρ_0	curvature radius; reference
σ	beam matrix
ϕ; ϕ_s	particle phase at voltage gap; synchronous phase
φ	spin angle to the vertical axis
ω	angular frequency
ω_c	critical angular frequency of SR, $\omega_c = 3\gamma^3 c/2\rho$

8.1 Introduction

The synchrotron is an outcome of the mid-1940s phase focusing resonant acceleration concept [1, 2]. Phase focusing, or synchronous, acceleration with slow variation of the magnetic field to maintain the beam on a constant orbit and constant RF phase was demonstrated with the acceleration of electrons from 4 to 8 MeV, in an existing betatron, using fixed RF, in 1946 [3]. This proof-of-principle was closely followed by the construction and operation, at GEC, of a 70 MeV synchrotron (weak focusing... no other choice at the time). The latter happened to be the opportunity for the first observation of visible SR, a serendipity resulting from the fact that the vacuum chamber was made of glass [4]. Observations included color of the radiation changing from blue to yellow when energy was decreased to 40 MeV[1] [5]—more in the Poynting simulation Exercise 8.3. Measurements of properties of the radiation were undertaken at the time, whereas SR acquired a status of a beam monitoring tool, that was the beginning of a long story, still underway...

Transverse beam confinement in the weak focusing synchrotron version of the synchrotron, over the thousands of turns needed for acceleration to top energy, was based on the technique known at the time, inherited from cyclotron and betatron: weak focusing,

Phase focusing states that stability of longitudinal motion (longitudinal focusing), is obtained if the particles in a bunch arrive at the accelerating gap in the vicinity of a proper phase of the oscillating voltage, the synchronous phase, such that the bunch stays together during acceleration. Synchrotrons operate in general in a non-isochronous regime: the revolution period changes with energy. As a consequence the RF, $f_{rf} = h f_{rev}$, has to change continuously from injection to top energy in order to maintain an accelerated bunch on the synchronous phase. The reference orbit in a synchrotron is maintained at constant radius by ramping the guiding field in the main dipoles in synchronism with the acceleration, as in the betatron [6].

The synchrotron concept increased the energy reach of particle accelerators at the time. It led to the construction of a series of proton rings with increasing energy [8]: 1 GeV at Birmingham (1953), 3.3 GeV at the Cosmotron (Brookhaven National Laboratory, 1953–1969), 6.2 GeV at the Bevatron (Berkeley, 1954–1993), 10 GeV at the Synchro-Phasotron (JINR, Dubna, 1957–2003), and a few others in the late 1950s. Weak focusing magnets are quite bulky, creating a practical limit to further increase in energy.[2] This issue was overcome with the strong focusing method, devised in the early 1950s (Chap. 9). The general layout of these first weak focusing synchrotrons included straight sections (often 4, Figs. 8.1 and 8.2), to allow for the insertion of injection and extraction systems, accelerating cavities, orbit correction and beam monitoring equipment (Fig. 8.3).

---

[1] At 70 MeV with a bending radius of say 0.5 m, the critical wavelength $\lambda_c = 4\pi\rho/3\gamma^3$ (Sect. 5.2.3) falls in the visible range.

[2] The story has it that it was possible to ride a bicycle in the vacuum chamber of Dubna's Synchro-Phasotron.

Fig. 8.1 SATURNE 1 at
Saclay [7], a 3 GeV,
4-period, 68.9 m
circumference, weak
focusing synchrotron,
constructed in 1956–1958.
The injection line is seen in
the foreground. Injection is
from a 3.6 MeV Van de
Graaff (not visible)

Fig. 8.2 A slice of the
SATURNE 1 dipole [7]. The
slight gap tapering,
increasing outward,
determines the weak index
condition $0 < n < 1$

The weak focusing synchrotron was used in fixed-target nuclear and particle physics, material science, medicine, industry, etc. Remarkably, it was a landmark (if not the starting point[3]) of the history of collider rings, the AdA *Anello di Accumulazione*, which demonstrated long term beam storage (and the Touschek effect), and

---

[3] The third electron model built by the MURA group, a 50 MeV fixed field alternating gradient (FFAG) ring, started in 1961, was operated in collider mode with two counter-rotating electron beams [12, 13].

Fig. 8.3 Loma Linda University medical synchrotron [9], during commissioning in 1989 at the Fermilab National Accelerator Laboratory where it was designed

produced the first e+e- collisions in the early 1960s, was a weak focusing synchrotron, a 250 MeV ring based on a $n = 0.55$ gradient dipole [14].

Polarized beams

Synchrotrons allowed the acceleration of polarized beams to high energy.[4] The possibility was considered from the early times at Argonne ZGS (Zero-Gradient Synchrotron), a 12 GeV weak focusing synchrotron operated over 1964–1979 [16] (Fig. 8.4). ZGS accelerated polarized proton beams to 17.5 GeV/c with *appreciable polarization* [17]. Polarization preservation techniques included harmonic orbit correction and fast betatron tune jumps at the strongest depolarizing resonances [18] (cf. Sect. 8.2.4, Fig. 8.19). Experiments were performed to assess the possibility of polarization transmission in strong focusing synchrotrons, and potential polarization lifetime in colliders [19]. Acceleration of polarized deuteron was achieved in the late 1970s [20].

The weak focusing synchrotron is still topical today, due for a large part to its relative simplicity, with low energy beam application where relatively low current is not a concern, such as in the hadrontherapy (Fig. 8.3) [10, 11]. It only requires a single type of a simple weak gradient dipole, a single power supply, a single accelerating gap. It has an advantage of beam manipulation flexibility, when needed, compared to (synchro-)cyclotrons.

---

[4] Polarized proton and deuteron beams had been accelerated in electrostatic columns (Sect. 2.1), and soon after in cyclotrons, when polarized beam sources were made available.

Fig. 8.4 The ZGS at Argonne during construction [15]. A 12 GeV, 8-dipole, 4-period, 172 m circumference, wedge focusing synchrotron. The two persons inside and outside the ring, in the background, give an idea of the size of the magnets

8.2 Basic Concepts and Formulæ

The synchrotron is based on two key principles. First, a slowly varying magnetic field maintains a constant orbit during acceleration,

$$B(t)\,\rho = p(t)/q, \quad \rho = constant, \tag{8.1}$$

with $p(t)$ the particle momentum and ρ the bending radius in the dipoles. Second, longitudinal phase stability enables synchronous acceleration. In a regime where velocity change with energy cannot be ignored (non-ultrarelativistic particles), the latter requires a modulation of the accelerating voltage frequency to satisfy

$$f_{\mathrm{rf}}(t) = h f_{\mathrm{rev}}(t) \qquad \text{with h an integer} \tag{8.2}$$

Synchronism between accelerating voltage oscillations and particle revolution keeps the bunch on a synchronous phase. Synchronous acceleration is technologically simpler in the case of electrons above a few MeV, because frequency modulation is unnecessary. For instance, from $v/c = 0.9987$ at 10 MeV to $v/c \to 1$ the relative change in revolution frequency amounts to $\delta f_{\mathrm{rev}}/f_{\mathrm{rev}} = \delta\beta/\beta < 0.0013$.

Varying field and RF on the one hand, fixed orbit in addition, are major evolutions compared to cyclotron, where instead, field and RF are fixed, and the accelerated orbit spirals out. A fixed orbit reduces the radial extent of individual guiding magnets, allowing a structure comprised of a circular string of dipoles. A synchrocyclotron instead uses a single, massive dipole (the volume of iron increases more than quadratically with bunch rigidity) with a wide radial extent allowing for a span of the field integral over $\oint B_{\text{injection}} \, dl = \frac{2\pi p_{\min}}{q} - \oint B_{\text{extraction}} \, dl = \frac{2\pi p_{\max}}{q}$.

Either a weak index ($-1 < k < 0$, Sect. 3.2.2) and/or wedge focusing (cf. Sect. 14.4.1) are used in weak focusing synchrotrons. Transverse stability was based solely on the latter at Argonne ZGS. Weak focusing in the ZGS resulted in weak depolarizing resonances, an advantage in that matter [19].

The synchrotron is a pulsed accelerator due to the necessary ramping of the field in order to maintain a constant orbit. The acceleration is cycled, from injection to top energy, repeatedly. The cycling repetition rate depends on the type of power supply. If the ramping uses a constant electromotive force, then

$$B(t) \propto (1 - e^{-\frac{t}{\tau}}) = 1 - \left[1 - \left(\frac{t}{\tau} \right) + \left(\frac{t}{\tau} \right)^2 - \ldots \right] \approx \frac{t}{\tau} \quad (8.3)$$

essentially linear. $\dot{B} = dB/dt$ does not exceed a few Tesla/second, the repetition rate of the acceleration cycle is of the order of a Hertz. If instead the magnet winding is part of a resonant circuit then the field oscillates from an injection threshold to a maximum value, $B(t) : B_0 \to B_0 + \hat{B}$, as in the betatron. In this case the repetition rate can be up to a few tens of Hertz. In both cases anyway B imposes its law and the other quantities, RF frequency in particular, follow.

For comparison: in a synchrocyclotron the field is constant, thus acceleration can be cycled as fast as the swing of the voltage frequency allows (hundreds of Hz are common practice). A conservative 10 kV per turn requires of the order of 10,000 turns for a proton to reach 100 MeV, with velocity $0.046 < v/c < 0.43$ from 1 to 100 MeV. Take $v \approx c$ for simplicity, and a circumference of a few meters, the acceleration thus takes $\approx 10^4 \times C/c \approx$ ms, potentially allowing a repetition rate in the kHz range, more than an order of magnitude beyond the reach of a rapid-cycling pulsed synchrotron.

8.2.1 Periodic Stability

This section introduces various ingredients concerning transverse focusing and the conditions for periodic stability. It builds on material introduced in Chap. 3, Classical Cyclotron.

Closed Orbit

The closed orbit is fixed, as in the betatron, and maintained during acceleration by ensuring that the relationship of Eq. 8.1 is satisfied. In a perfect ring, the closed orbit

Fig. 8.5 A 4-fold symmetric
structure with four drift
spaces of length $2l$. Orbit
length on reference
momentum p_0 is
$C = 2\pi\rho_0 + 8l$. (O; s, x, y) is
the moving frame, along the
reference orbit. The orbit for
momentum $p = p_0 + \Delta p$
($\Delta p < 0$, here) is at constant
distance $\Delta x = D_x \frac{\Delta p}{p_0}$ from
the reference orbit

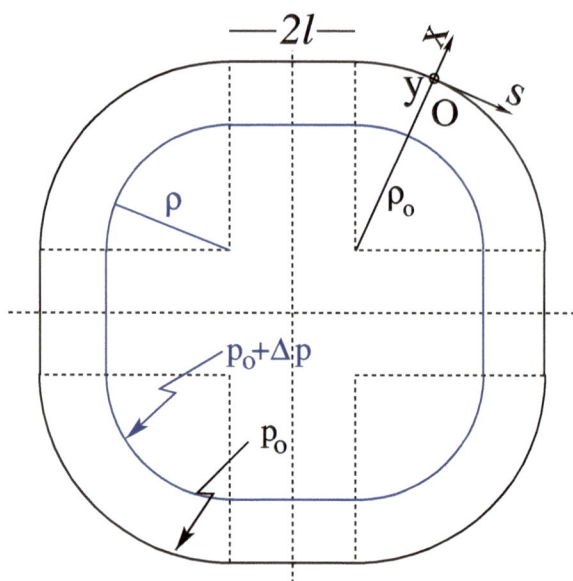

is along an arc in the bending magnets and straight along the drifts, Fig. 8.5. Particle
motion is defined in the Serret-Frénet frame (O; s, x, y), Fig. 3.8.

Transverse Focusing

Radial motion stability around a reference closed orbit in an axially symmetric dipole
field requires a field index (Sect. 3.2.2),

$$n = -\frac{\rho_0}{B_0}\frac{\partial B_y}{\partial x}\bigg|_{x=0,\ y=0} \tag{8.4}$$

This quantity, evaluated on the reference arc in the dipoles, satisfies the weak focusing
condition (Eq. 3.12 with $n = -k$)

$$0 < n < 1 \tag{8.5}$$

This condition can be obtained with a tapered gap (as in SATURNE 1 dipole, Fig. 8.2)
resulting in both radial and axial focusing (Figs. 8.6 and 8.7). Note the sign convention
here, opposite to that used for the cyclotron (Eq. 3.11). This condition holds regardless
of the presence or not of drifts. Adding drifts brings to defining two radii, namely,

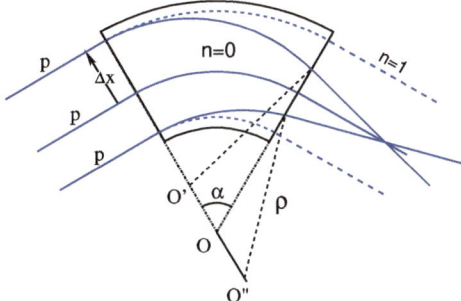

Fig. 8.6 Geometrical focusing: in a sector dipole with focusing index $n = 0$, parallel incoming rays of equal momenta experience the same curvature radius ρ, so their trajectories converge as outer trajectories have a longer path in the field. An index value n $= 1$ cancels that effect: parallel incoming rays exit parallel

Fig. 8.7 Axial motion stability requires proper shaping of field lines: B_y has to decrease with radius. The Laplace force pulls a positive charge (located at I) with velocity pointing out of the page, toward the median plane. Increasing the field gradient (n closer to 1, gap opening up faster) increases the focusing

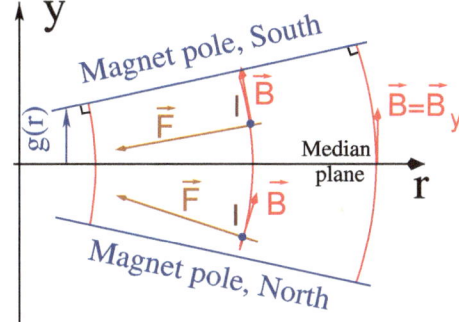

(i) the magnet curvature radius ρ_0,

(ii) an average radius $R = C/2\pi = \rho_0 + Nl/\pi$ (with C the length of the reference closed orbit, N the number of drifts and $2l$ their length) (Fig. 8.5) which can also be written

$$R = \rho_0(1 + k), \qquad k = \frac{Nl}{\pi\rho_0} \tag{8.6}$$

Adding drift spaces decreases the average focusing around the ring.

Geometrical focusing

The limit $n \to 1$ of the transverse motion stability domain corresponds to a cancellation of the geometrical focusing (Fig. 8.6): in a constant field dipole (radial field index n $= 0$) the longer (respectively shorter) path in the magnetic field for parallel trajectories entering the magnet at greater (respectively smaller) radius result in convergence. This effect is cancelled (i.e., the bend angle is the same whatever the entrance radius) if the curvature center is made independent of the entrance radius:

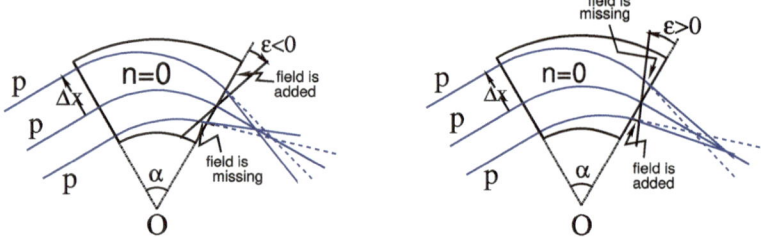

Fig. 8.8 Left: a focusing wedge ($\varepsilon < 0$); opening the sector increases horizontal focusing and decreases vertical focusing. Right: a defocusing wedge ($\varepsilon > 0$), closing the sector, has the reverse effect. This is the origin of the focusing in the ZGS zero-gradient dipoles

$OO' = 0$, $O''O = 0$. This occurs if trajectories at an outer (inner) radius experience a smaller (greater) field such as to satisfy $BL = B\rho = C^{st}$. Differentiating $B\rho = C^{st}$ gives $\frac{\Delta B}{B} + \frac{\Delta \rho}{\rho} = 0$, with $\Delta \rho = \Delta x$, so yielding $n = -\frac{\rho_0}{B_0}\frac{\Delta B}{\Delta x} = 1$. The focal distance associated with the curvature is (Eq. 3.13 with $R = \rho_0$) $f = \frac{\rho_0^2}{\mathcal{L}}$.

Wedge Focusing

Entrance and exit wedge angles may be used to ensure transverse focusing, Fig. 8.8: opening the magnetic sector increases the horizontal focusing (and decreases the vertical focusing); closing the magnetic sector has the reverse effect (cf. Sect. 14.4.1).

At the wedge the trajectory undergoes a deviation proportional to the distance to the optical axis, amounting to

$$\Delta x' = \frac{\tan \varepsilon}{\rho_0} \Delta x, \quad \Delta y' = -\frac{\tan(\varepsilon - \psi)}{\rho_0} \Delta y \tag{8.7}$$

The angle ψ is a correction for the fringe field extent (Eq. 14.20); the effect is of the first order on the vertical focusing, and second order horizontally.

Profiling the magnet gap in order to adjust the focal distance complicates the magnet; a parallel gap, $n = 0$, makes it simpler, for that reason edge focusing may be preferred. The method benefited the acceleration of polarized beams in the ZGS, as radial field components (which are responsible for depolarization), met at the EFBs of the eight main dipoles, were therefore weak [17]. Preserving beam polarization at high energy required tight control of the tunes, achieved at the 0.01 level by means of pole face winding added at the ends of the dipoles [21, 22].

Drawbacks of the weak focusing method include interdependence of radial and axial focusing, see *Working point* Section, below.

Betatron Motion

The first order differential equations of motion in the moving frame (Fig. 8.5) derive from the Lorentz equation

$$\frac{d m \mathbf{v}}{d t} = q \mathbf{v} \times \mathbf{B} \;\Rightarrow\; m \frac{d}{d t} \left\{ \begin{matrix} \frac{ds}{dt} \mathbf{s} \\ \frac{dx}{dt} \mathbf{x} \\ \frac{dy}{dt} \mathbf{y} \end{matrix} \right\} = q \left\{ \begin{matrix} \left(\frac{dx}{dt} B_y - \frac{dy}{dt} B_x \right) \mathbf{s} \\ -\frac{ds}{dt} B_y \mathbf{x} \\ \frac{ds}{dt} B_x \mathbf{y} \end{matrix} \right\} \tag{8.8}$$

Motion in a weak index dipole field is solved in Sect. 3.2.2, Classical Cyclotron chapter: in Eq. 3.7 substitute ρ to R, $n = -\frac{\rho_0}{B_0} \frac{\partial B_y}{\partial x}$ to $-k$ (Eq. 3.11), and evaluate on the reference orbit. Taylor expansions of the transverse field components in the moving frame lead to

$$B_y(\rho)|_{y=0} = B_0(1 - n\frac{x}{\rho_0}) + O(x^2)$$
$$B_x(0 + y) = -n\frac{B_0}{\rho_0} y + O(y^3) \tag{8.9}$$

Assume transverse stability: $0 < n < 1$. In the approximation $ds \approx vdt$ (Eq. 3.14) Eqs. 8.8, and 8.9 lead to the differential equations of motion

$$\frac{d^2x}{ds^2} + \frac{1-n}{\rho_0^2} x = 0, \quad \frac{d^2y}{ds^2} + \frac{n}{\rho_0^2} y = 0 \tag{8.10}$$

In an periodic structure comprised of gradient dipoles, wedges and drift spaces, the differential equation of motion takes the general form of Hill's equation, namely (with u standing for x or y),

$$\left\{ \begin{matrix} \dfrac{d^2u}{ds^2} + K_u(s)u = 0 \\ K_u(s + S) = K_u(s) \end{matrix} \right. \quad \text{with} \quad \left\{ \begin{matrix} \text{in dipoles :} \begin{cases} K_x = \frac{1-n}{\rho_0^2} \\ K_y = \frac{n}{\rho_0^2} \end{cases} \\ \text{at a wedge at } s = s_0 : K_{\substack{x \\ y}} = \frac{\pm \tan \varepsilon}{\rho_0} \delta(s - s_0) \\ \text{in drift spaces :} \frac{1}{\rho_0} = 0, \; K_x = K_y = 0 \end{matrix} \right.$$

$$\tag{8.11}$$

Here $K_u(s)$ is periodic, $S = 2\pi R/N$ ($S = C/4$ for instance in a 4-period ring, Figs. 8.1 and 8.5).

The solution of Eqs. 8.11 is not as straightforward as in the cyclotron where a constant K_u around the ring (Eq. 3.15) results in a sinusoidal motion (Eq. 3.17). A sinusoidal motion, with adding drifts, however remains a reasonable approximation, see below, *Weak focusing approximation*.

Floquet established [23] that the two independent solutions of Hill's second order differential equation with periodic coefficient have the form [24]

$$
\begin{cases}
u_1(s) = \sqrt{\beta_u(s)}\, e^{i \int_0^s \frac{ds}{\beta_u(s)}} \\
du_1(s)/ds = \dfrac{i - \alpha_u(s)}{\beta_u(s)}\, u_1(s)
\end{cases}
\quad \text{and} \quad
\begin{cases}
u_2(s) = u_1^*(s) \\
du_2(s)/ds = du_1^*(s)/ds
\end{cases}
\tag{8.12}
$$

where $\beta_u(s)$ and $\alpha_u(s) = -\beta_u'(s)/2$ are periodic functions, from what it results that

$$
u_{\frac{1}{2}}(s + S) = u_{\frac{1}{2}}(s)\, e^{\pm i \int_{s_0}^s \frac{ds}{\beta_u(s)}}
\tag{8.13}
$$

where $\int_{s_0}^s \frac{ds}{\beta_u(s)}$ is the betatron phase advance at s, from the origin s_0. A real solution of Hill's equation is the linear combination $A\, u_1(s) + A^*\, u_2^*(s)$. With $A = \frac{1}{2}\sqrt{\varepsilon_u/\pi}\, e^{i\phi}$ following conventional notations, ϕ the phase of the motion at the origin $s = s_0$, the general solution of Eq. 8.11 is

$$
\begin{cases}
u(s) = \sqrt{\beta_u(s)\varepsilon_u/\pi}\, \cos\left(\int_{s_0}^s \frac{ds}{\beta_u} + \phi\right) \\
u'(s) = -\sqrt{\dfrac{\varepsilon_u/\pi}{\beta_u(s)}}\, \sin\left(\int_{s_0}^s \frac{ds}{\beta_u} + \phi\right) + \alpha_u(s)\, \cos\left(\int_{s_0}^s \frac{ds}{\beta_u} + \phi\right)
\end{cases}
\tag{8.14}
$$

The Courant-Snyder invariant of the motion is

$$
\frac{\varepsilon_u}{\pi} = \frac{1}{\beta_u(s)}\left[u^2 + \left(\alpha_u(s)u + \beta_u(s)u'\right)^2\right]
\tag{8.15}
$$

At a given azimuth s of the periodic structure the observed turn-by-turn motion lies on that ellipse (Fig. 8.9). The form and orientation of the ellipse feature a weak dependence on the observation azimuth s, via the respective local values of $\alpha_u(s)$ (small at all s) and $\beta_u(s)$ (weakly modulated), and its area ε_u is an invariant. Equation 8.14 taken for $\alpha_u(s) = 0$ (an observation azimuth s where the ellipse is upright) shows that motion along the ellipse is clockwise. Note that in the coordinate system $(u, (\alpha_u(s)u + \beta_u(s)u'))$ the particle moves on a circle of radius ε_u/π.

The phase advance over a turn (from one position to the next on the ellipse, Fig. 8.9) in an N-periodic ring yields the wave number

$$
\nu_u = \frac{1}{2\pi} \int_{s_0}^{s_0+NS} \frac{ds}{\beta_u(s)} = \frac{N}{2\pi} \int_{\text{period}} \frac{ds}{\beta_u(s)} = \frac{N\mu_u}{2\pi}
\tag{8.16}
$$

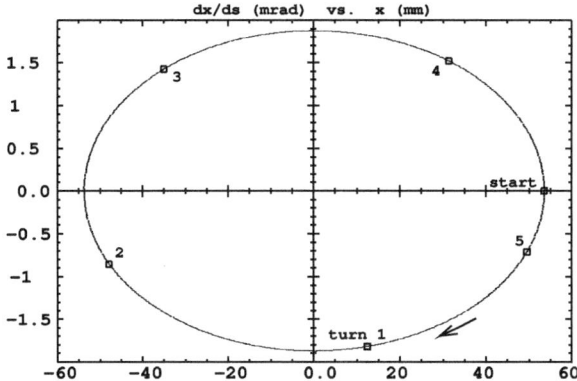

Fig. 8.9 A thousand passes in a ZGS 43 m cell, observed at the center of the long drift where $\alpha_x(s) = 0$, materialize the upright horizontal Courant-Snyder invariant. The first five passes are marked, motion goes clockwise with a cell phase advance of $0.21 \times 2\pi$. The aspect ratio of the ellipse only weakly depends on s, its area ($\varepsilon_x = 100\pi\ \mu$ m here) is an invariant of the motion

Weak focusing approximation

In a cylindrically symmetric structure the sinusoidal motion is the exact solution of the first order differential equations of motion (Eqs. 3.16 and 3.17, Classical Cyclotron chapter), the coefficients $K_x = (1 - n)/\rho_0^2$ and $K_y = n/\rho_0^2$ are independent of s. Adding drift spaces results in Hill's differential equation with periodic coefficient $K(s + S) = K(s)$ (Eq. 8.11), with solution a pseudo harmonic motion (Eq. 8.14). Due to the weak focusing the beam envelope is only weakly modulated (cf. below), thus also is $\beta_u(s)$. In practice the modulation of $\beta_u(s)$ does not exceed a few percent, justifying the introduction of the average value $\overline{\beta}_u$ to approximate the phase advance by

$$\int_0^s \frac{ds}{\beta_u(s)} \approx \frac{s}{\overline{\beta}_u} = \nu_u \frac{s}{R} \tag{8.17}$$

The right equality is obtained by applying this approximation to the phase advance per period, namely

$$\mu_u = \int_{s_0}^{s_0+S} \frac{ds}{\beta_u(s)} \approx \frac{S}{\overline{\beta}_u} \tag{8.18}$$

and introducing the wave number of the N-period optical structure (Eq. 8.16) so that

$$\overline{\beta}_u = \frac{R}{\nu_u} \tag{8.19}$$

the wavelength of the betatron oscillation. With $k \ll 1$ and using Eq. 8.23,

$$\overline{\beta_x} = \frac{\rho_0(1 + k/2)}{\sqrt{1 - n}}, \quad \overline{\beta_y} = \frac{\rho_0(1 + k/2)}{\sqrt{n}} \tag{8.20}$$

Substituting $\nu_u \frac{s}{R}$ to $\int \frac{ds}{\beta_u(s)}$ in Eq. 8.14 yields the approximate solution

$$\begin{cases} u(s) \approx \sqrt{\beta_u(s)\varepsilon_u/\pi} \, \cos\left(\nu_u \frac{s}{R} + \phi\right) \\ u'(s) \approx -\sqrt{\frac{\varepsilon_u/\pi}{\beta_u(s)}} \, \sin\left(\nu_u \frac{s}{R} + \phi\right) + \alpha_u(s) \, \cos\left(\nu_u \frac{s}{R} + \phi\right) \end{cases} \tag{8.21}$$

Beam envelopes

The beam envelope $\hat{u}(s)$ (with u standing for x or y) is determined by a particle on the maximum invariant ε_u/π. It is given at all s by

$$\hat{u}(s) = \pm\sqrt{\beta_u(s)\frac{\varepsilon_u}{\pi}} \tag{8.22}$$

As $\beta_u(s)$ is S-periodic, so also is the envelope, $\hat{u}(s + S) = \hat{u}(s)$. In a cell with symmetries, the beam envelopes feature the same symmetries, as shown in Fig. 8.10

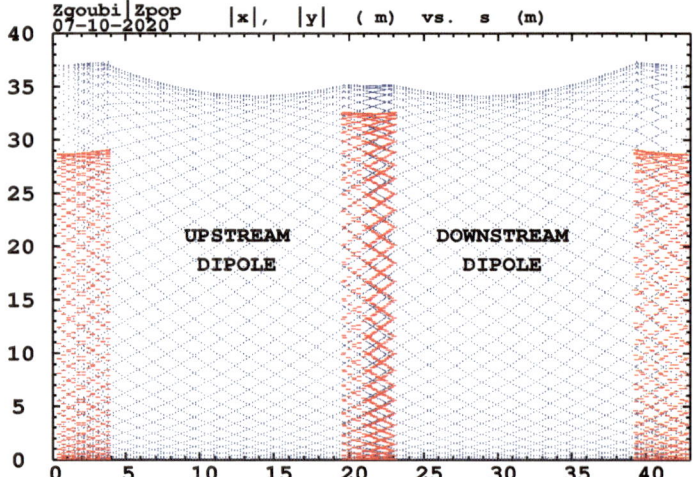

Fig. 8.10 Multiturn particle excursion (absolute values, $|x(s)|$ and $|y(s)|$) along the ZGS 2-dipole 43 m cell. The motion extrema ($[\beta_u(s)\varepsilon_u/\pi]^{1/2}$, Eq. 8.22) tangent the envelops, respectively horizontal (red, across the dipoles), and vertical (blue). Envelops are only weakly modulated. They feature the symmetry of the cell

for the ZGS: a symmetry with respect to the center of the cell. Envelope extrema are at azimuth s of $\beta_u(s)$ extrema, i.e. where $d\hat{u}(s)/ds \propto \beta_u'(s) = 0$ or $\alpha_u = 0$ as $\beta_u' = -2\alpha_u$.

Working point

The "working point" of the synchrotron is the wave number pair (v_x, v_y) at which the accelerator is operated, it fully characterizes the focusing. In a structure with cylindrical symmetry (such as the classical cyclotron) $v_x = \sqrt{1-n}$ and $v_y = \sqrt{n}$ (Eq. 3.18) so that $v_x^2 + v_y^2 = 1$: when the radial field index n is changed the working point stays on a circle of radius 1 in the stability diagram (or "tune diagram", Fig. 8.11). If drift spaces are added, from Eqs. 8.19 and 8.20, with $1 + \frac{k}{2} \approx \sqrt{R/\rho_0}$ (Eq. 8.6), it comes

$$v_x \approx \sqrt{(1-n)\frac{R}{\rho_0}}, \quad v_y \approx \sqrt{n\frac{R}{\rho_0}}, \quad v_x^2 + v_y^2 \approx \frac{R}{\rho_0} \tag{8.23}$$

Thus the working point is located on a circle of radius $\sqrt{R/\rho_0} > 1$ (Fig. 8.11), tunes can not exceed the limits

$$0 < v_{x,y} \lesssim \sqrt{R/\rho_0}$$

Horizontal and vertical focusing are not independent (Eq. 8.11): if v_x increases then v_y decreases and vice versa. This is a lack of flexibility which strong focusing overcomes by providing two knobs allowing separate adjustment.

Fig. 8.11 Location of the working point in the tune diagram. (A) field with revolution symmetry: (v_x, v_y) is on a circle of radius 1; (B) sector field with index $0 < n < 1$ and drift spaces: (v_x, v_y) is on a circle of radius $\sqrt{R/\rho_0}$; (C) strong focusing, AG index $|n| \gg 1$ or separated function: v_x and v_y are large, set independently

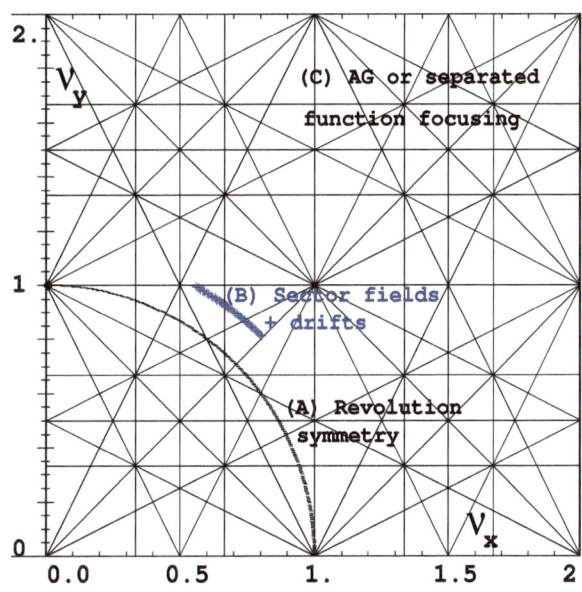

Fig. 8.12 In a sector dipole
with radial index $n \neq 0$,
closed orbits follow arcs of
constant field. A closed orbit
at $p_0 + \Delta p$ follows an arc of
radius $\rho_0 + \Delta \rho$,
$\Delta \rho = \Delta p / (1 - n) q B_0$

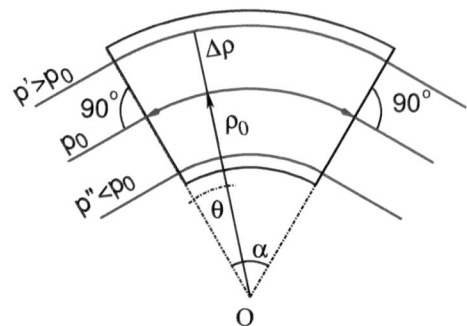

Off-momentum orbits; periodic dispersion

In the linear approximation in $\Delta p / p_0$, a momentum offset $\Delta p = p - p_0$ changes
mv to $mv(1 + \Delta p / p_0)$ in Eq. 8.8. This changes the horizontal equation of motion
(Eq. 8.10) to

$$\frac{d^2 x}{ds^2} + K_x x = \frac{1}{\rho_0} \frac{\Delta p}{p_0}, \quad \text{or} \quad \frac{d^2 x}{ds^2} + K_x \left(x - \frac{1}{\rho_0 K_x} \frac{\Delta p}{p_0} \right) = 0 \qquad (8.24)$$

A change of variable $x - \frac{1}{K_x \rho_0} \frac{\Delta p}{p_0} \to x$ (with $1/\rho_0 K_x = \rho_0/(1 - n)$) restores the
unperturbed equation of motion; thus orbits of different momenta $p = p_0 + \Delta p$ are
separated by

$$\Delta x = \frac{\rho_0}{1 - n} \frac{\Delta p}{p_0} \qquad (8.25)$$

from the reference orbit (Fig. 8.12). Introducing the geometrical radius $R = (1 + k)\rho_0$ (Eq. 8.6) to account for the added drifts, this yields the dispersion function

$$D_x = \frac{\Delta x}{\Delta p / p_0} \equiv \frac{\Delta R}{\Delta p / p_0} = \frac{R}{(1 - n)(1 + k)} = \frac{\rho_0}{1 - n}, \quad \text{constant, positive}$$
$$(8.26)$$

D_x is the chromatic dispersion of the orbits, an s-independent quantity: in a structure
with axial symmetry, comprising drift sections (Fig. 8.5) or not (classical and AVF
cyclotrons for instance), the ratio $\Delta x / \Delta p / p_0$ is independent of the azimuth s, the
distance of a chromatic orbit to the reference orbit is constant around the ring.

Given that $n < 1$,

– higher momentum orbits, $p > p_0$, have a greater radius,
– lower momentum orbits, $p < p_0$, have a smaller radius.

The horizontal motion of an off-momentum particle is a superposition of the
betatron motion (solution of Hill's Eq. 8.21 taken for $u = x$) and of a particular
solution of the inhomogeneous equation (Eq. 8.24), namely

$$x(s) = \sqrt{\beta_u(s)\varepsilon_u/\pi} \, \cos\left(\nu_u \frac{s}{R} + \phi\right) + \frac{p_0}{1-n} \frac{\Delta p}{p_0} \qquad (8.27)$$

The vertical motion is unchanged.

Chromatic orbit length

In an axially symmetric structure the difference in closed orbit length $\Delta C = 2\pi\,\Delta R$ resulting from the difference in momentum comes from the dipoles, as all orbits are parallel in the drifts (Fig. 8.5). Hence, from Eq. 8.26, the relative closed orbit lengthening factor, or momentum compaction, is

$$\alpha = \frac{\Delta C}{C} \bigg/ \frac{\Delta p}{p_0} \equiv \frac{\Delta R}{R} \bigg/ \frac{\Delta p}{p_0} = \frac{1}{(1-n)(1+k)} \approx \frac{1}{\nu_x^2} \qquad (8.28)$$

with $k = Nl/\pi\rho_0$ (Eq. 8.6). Note that the relationship $\alpha \approx 1/\nu_x^2$ between momentum compaction and horizontal wave number established for a revolution symmetry structure (Eq. 3.22) still holds when adding drifts.

8.2.2 Acceleration

The field B in a synchrotron is varied during acceleration (a function performed by the magnet power supply) concurrently with the variation of the bunch momentum p (a function performed by the accelerating cavity) in such a way that the beam stays on the design orbit. Given the energies involved, the magnet supply imposes its law $B(t)$ (Fig. 8.13), and the cavity follows the best it can. The accelerating voltage $\hat{V}(t)\sin\omega_{rf}t$ is maintained in synchronism with the revolution motion by ensuring, as well as possible,

$$\omega_{rf} = h\omega_{rev} = h\frac{c}{R} \frac{B(t)}{\sqrt{\left(\frac{m_0 c}{q\rho}\right)^2 + B^2(t)}}$$

Typically, for a $C = 2\pi R \approx 70\,\text{m}$ circumference ring,[5] accelerating from $\beta = v/c \approx 0.09$ at injection (3.6 MeV protons) to $\beta \approx 1$ at top energy (3 GeV), the revolution period $T_{rev} = C/\beta c$ and frequency $\omega_{rev}/2\pi = 1/T_{rev}$ span

$$\begin{cases} T_{rev} : \ 2.6\,\mu\text{s} \ \to \ 0.24\,\mu\text{s} \\ f_{rev} : \ 380\,\text{kHz} \to 4.2\,\text{MHz} \end{cases}$$

---

[5] Case of the SATURNE 1 weak focusing synchrotron (Fig. 8.1), cf. Exercise 8.1, Table 8.1

Energy gain

The variation of the particle energy over one turn amounts to the work of the force $F = dp/dt = q\rho dB/dt$ on the charge at the cavity, namely

$$\Delta W = F \cdot 2\pi R = 2\pi R q \rho \dot{B} \tag{8.29}$$

In a slow-cycling synchrotron $\dot{B}$ is usually constant over most of the acceleration cycle (Eq. 8.3), and so is ΔW. At SATURNE 1, for instance

$$\frac{\Delta W}{q} = 2\pi R \rho \dot{B} = 68.9_{[m]} \times 8.42_{[m]} \times 1.8_{[T/s]} = 1044 \, \text{volts/turn}$$

The field ramp lasts

$$\Delta t = (B_{\max} - B_{\min})/\dot{B} \approx B_{\max}/\dot{B} = 0.8 \, \text{s}$$

The number of turns to the top energy ($W_{\max} \approx 3 \, \text{GeV}$) is

$$N = \frac{W_{\max}}{\Delta W} = \frac{3 \, 10^9 \, \text{eV}}{1044 \, \text{eV/turn}} \approx 3 \times 10^6 \, \text{turns}$$

The dependence of particle mass on field writes

$$m(t) = \gamma(t) m_0 = \frac{q\rho}{c} \sqrt{\left(\frac{m_0 c}{q\rho}\right)^2 + B^2(t)}$$

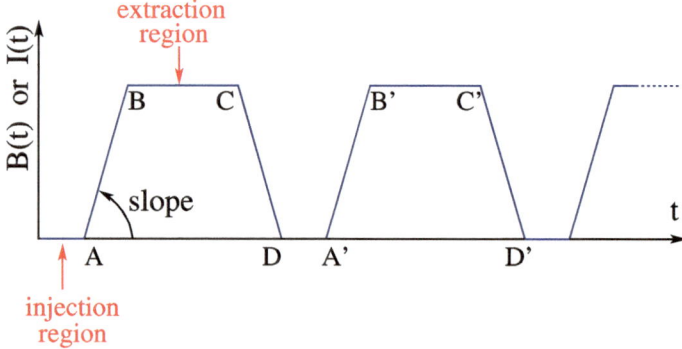

Fig. 8.13 Cycling $B(t)$ in a pulsed synchrotron. Ignoring saturation, $B(t)$ during the ramp is proportional to the magnet power supply current $I(t)$. Beam injection occurs at low field, in the region of A, while extraction occurs at top energy on the high field plateau. (AB): field ramp up (acceleration); (BC): flat top; (CD): field ramp down; (DA'): thermal relaxation. (AA'): repetition period; (1/AA'): repetition rate; *slope*: ramp velocity $\dot{B} = dB/dt$ (T/s)

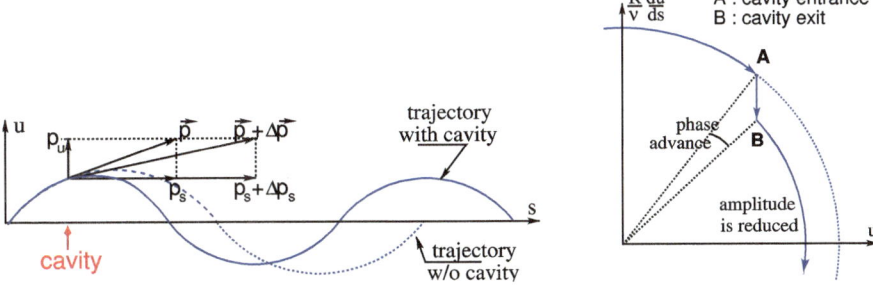

Fig. 8.14 Adiabatic damping of betatron oscillations from $u' = p_u/p_s$ to $u'_2 = p_u/(p_s + \Delta p_s)$ at the accelerating cavity. In transverse phase space the particle motion invariant ε_u decreases, as a result of $\Delta\left(\frac{du}{ds}\right)$

Adiabatic damping of the betatron oscillations

Particle momentum increases at the accelerating gap, resulting in a decrease of the amplitude of betatron oscillations (an increase if deceleration). The mechanism is sketched in Fig. 8.14 (the solution of the equations of motion is addressed in Sect. 10.2.3). The slope at the cavity is

$$\text{before the cavity:} \quad \frac{du}{ds} = \frac{m\frac{du}{dt}}{m\frac{ds}{dt}} = \frac{p_u}{p_s}, \quad \text{after:} \quad \left.\frac{du}{ds}\right|_2 = \frac{m\frac{du}{dt}}{m\frac{ds}{dt}}\bigg|_2 = \frac{p_{u,2}}{p_{s,2}}$$

with u standing for x or y. As the kick in momentum is longitudinal, $dp_u/dt = 0$ thus $p_{u,2} = p_u$ and the increase in momentum is purely longitudinal, $p_{s,2} = p_s + \Delta p_s$. Thus

$$\left.\frac{du}{ds}\right|_2 = \frac{p_u}{p_s + \Delta p_s} \approx \frac{p_u}{p_s}(1 - \frac{\Delta p_s}{p_s})$$

and as a consequence the slope du/ds varies across the cavity,

$$\Delta\left(\frac{du}{ds}\right) = \left.\frac{du}{ds}\right|_2 - \frac{du}{ds} = -\frac{du}{ds}\frac{\Delta p_s}{p_s}, \quad \text{proportional to the slope}$$

If $\Delta p/p > 0$ (acceleration) then the slope decreases. This variation has two consequences on the betatron oscillation (Fig. 8.14):
- a change of the betatron phase,
- a modification of the betatron amplitude.

Coordinate transport

At the cavity

$$\begin{cases} u_2 = u \\ u'_2 \approx \frac{p_u}{p_s}(1 - \frac{dp}{p}) = u'(1 - \frac{dp}{p}) \end{cases}$$

In matrix form,

$$\begin{pmatrix} u_2 \\ u_2' \end{pmatrix} = [C] \begin{pmatrix} u \\ u' \end{pmatrix} \quad \text{with} \quad [C] = \begin{bmatrix} 1 & 0 \\ 0 & 1 - \frac{dp}{p} \end{bmatrix} \tag{8.30}$$

Since $det[C] = 1 - \frac{dp}{p} \neq 1$ the system is non-conservative and the area of the beam ellipse in phase space is not conserved. Assume one cavity in the ring and note $[T] \times [C]$ the one-turn coordinate transport matrix with origin at entrance of the cavity. Its determinant is

$$det[T] \times det[C] = det[C] = 1 - \frac{dp}{p}$$

The variation of the transverse ellipse area satisfies $\varepsilon_u = (1 - \frac{dp}{p})\varepsilon_0$ or, with $d\varepsilon_u = \varepsilon_u - \varepsilon_0$, $\frac{d\varepsilon_u}{\varepsilon_u} = -\frac{dp}{p}$, The solution is

$$p\,\varepsilon_u = \text{constant}, \quad \text{or} \quad \beta\gamma\varepsilon_u = \text{constant} \tag{8.31}$$

Over N turns the coordinate transport matrix is $[T_N] = ([T][C])^N$, thus the ellipse area changes by a factor

$$det[C]^N = (1 - \frac{dp}{p})^N \approx 1 - N\frac{dp}{p}$$

Phase stability

The motion of a particle in the longitudinal phase space $(\phi, \delta p/p)$ is stabilized in the vicinity of a synchronous phase, ϕ_s, by the mechanism of phase stability, or longitudinal focusing (Fig. 8.15). It requires

   (i) the presence of an RF cavity with frequency locked on the revolution time,

   (ii) bunch centroid to be positioned either on the rising slope of the oscillating voltage (low energy regime), or on the falling slope (high energy regime).

   The synchronous particle follows the reference closed orbit, its velocity satisfies $v(t) = \frac{qB\rho(t)}{m}$. At each turn it reaches the accelerating gap when the oscillating voltage is at the synchronous phase ϕ_s, and undergoes an energy gain

$$\Delta W = q\hat{V}\sin\phi_s$$

The condition $|\sin\phi_s| < 1$ imposes a lower limit to the cavity voltage for acceleration to happen. According to Eq. 8.29,

$$\hat{V} > 2\pi R\rho\dot{B}$$

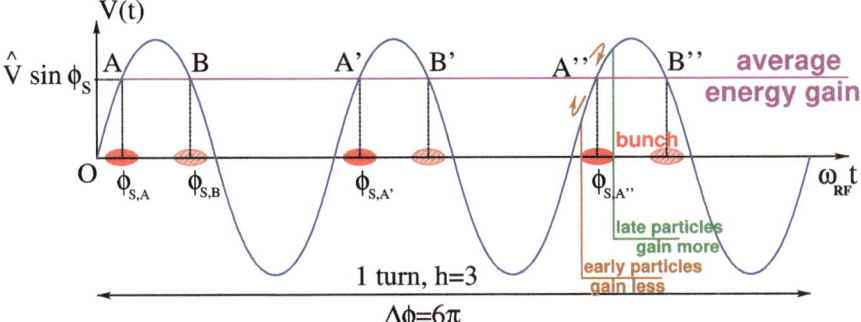

Fig. 8.15 A sketch of the mechanism of phase stability, $h = 3$ in this example. Below transition phase stability occurs for a synchronous phase taken at either one of A, A', A" arrival times at the gap. Beyond transition the stable phase is at either one of B, B', B' locations

Referring to Fig. 8.15, the synchronous phase can be placed on the left (A, A', A" ... series) or on the right (B, B', B" ... series) of the oscillating voltage crest. One and only one of these two possibilities, and which one depending upon the optical lattice and on particle energy, ensures that particles in a bunch remain grouped in the vicinity of the synchronous particle.

The transition is between these two time-of-flight regimes. Consider a particle with higher energy compared to the synchronous particle:

– if the increase in path length around the ring is faster than the increase in velocity (case of classical cyclotron and synchrocyclotron; and of high energy electron synchrotron, where velocity essentially does not change), a revolution around the ring takes more time, the particle arrives at the accelerating gap late ($\phi(t) > \phi_s$); in order for it to be pulled toward bunch center (i.e., take less time around the ring) it has to lower its energy increase; this is the B series, above transition;

– if the velocity increase is faster than the path length increase (case in general of synchrotrons at low energy), revolution around the ring takes less time, the particle arrives at the accelerating gap early ($\phi(t) < \phi_s$); in order for it to be pulled toward bunch center (i.e., take more time around the ring) it has to lower its energy increase; this is the A series, below transition.

Transition energy

The transition between the two time-of-flight regimes occurs when $\dfrac{dT_{\text{rev}}}{T_{\text{rev}}} = 0$. With $T = 2\pi/\omega = C/v$, this can be written

$$\frac{d\omega_{\text{rev}}}{\omega_{\text{rev}}} = -\frac{dT_{\text{rev}}}{T_{\text{rev}}} = \frac{dv}{v} - \frac{dC}{C}$$

With $\frac{dv}{v} = \frac{1}{\gamma^2}\frac{dp}{p}$ and momentum compaction $\alpha = \dfrac{dC}{C} / \frac{dp}{p}$, (Eq. 8.28), it becomes

$$\frac{d\omega_{\text{rev}}}{\omega_{\text{rev}}} = -\frac{dT_{\text{rev}}}{T_{\text{rev}}} = \left(\frac{1}{\gamma^2} - \alpha\right)\frac{dp}{p} = \eta\frac{dp}{p} \qquad (8.32)$$

which introduces the phase slip factor

$$\eta = \overbrace{\frac{1}{\gamma^2}}^{\text{kinematics}} - \underbrace{\alpha}_{\text{lattice}} = \frac{1}{\gamma^2} - \frac{1}{\gamma_{\text{tr}}^2} \qquad (8.33)$$

The "transition gamma", γ_{tr}, is a property of the lattice.

In a weak focusing lattice, after Eq. 8.28 and classical cyclotron's Eq. 3.22,

$$\gamma_{\text{tr}} = 1/\sqrt{\alpha} \approx \nu_x \qquad (8.34)$$

Thus the phase stability regime is

$$\begin{aligned} &\text{below transition, i.e. } \phi_s < \pi/2, \quad \text{if} \quad \gamma < \nu_x \\ &\text{above transition, i.e. } \phi_s > \pi/2, \quad \text{if} \quad \gamma > \nu_x \end{aligned} \qquad (8.35)$$

In a weak focusing synchrotron the horizontal tune $\nu_x = \sqrt{(1-n)R/\rho_0}$ (Eq. 8.23) may be $\gtrsim 1$, and subsequently $\gamma_{\text{tr}} > 1$ is a possibility, γ_{tr} may have to be crossed during acceleration. There is no transition gamma if $\nu_x < 1$. At SATURNE 1 for instance, with $\nu_x \approx 0.7$ (Table 8.1) and $\gamma_{\text{tr}} < 1$. So, ramping in energy did not require crossing transition-gamma.[6]

8.2.3 Synchrotron Radiation Poynting

Visible SR was first observed in the GEC 70 MeV weak focusing synchrotron [4]. So, the bases of SR theory may opportunistically be recalled here [26, 27]. This theoretical material serves the purpose of the exercises in addition. The topic is further explored in Sect. 9.2.6, which addresses some aspects of the use of visible SR for high energy electron or proton beam imaging.

In addressing low energy SR, the Poynting vector

$$\mathbf{P} = \mathbf{E} \times \mathbf{B}$$

---

[6] Transition-γ crossing (Sect. 8.2.2) is a common longitudinal phase space beam manipulation during acceleration in strong focusing synchrotrons. It requires an RF phase jump [25].

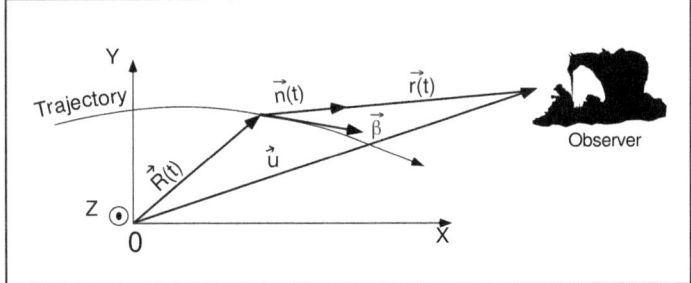

Fig. 8.16 The frame and vectors entering in the definition of the electric field radiated by the accelerated particle (Eq. 8.36). `Zgoubi` notations are used here (Fig. 1.3): (X, Y): horizontal plane; Z: vertical axis; $\mathbf{R}(t)$ is the particle position in the laboratory frame (O, X, Y, Z). Besides, $\mathbf{u}$ is the position of the observer; $\mathbf{r}(t) = \mathbf{u} - \mathbf{R}(t)$ is the position of the particle with respect to the observer; $\mathbf{n}(t) = \mathbf{r}(t)/|\mathbf{r}(t)|$ is the (normalized) direction of observation; $\boldsymbol{\beta} = (1/c)d\mathbf{R}/dt$ is the normalized velocity vector of the particle

is the relevant quantity [27, 28]. The electromagnetic field is given by the Liénard-Wiechert equations in the long distance approximation

$$\mathbf{E}(\mathbf{n}, \tau) = \frac{q}{4\pi\varepsilon_0 c} \frac{\mathbf{n}(t) \times [(\mathbf{n}(t) - \boldsymbol{\beta}(t)) \times d\boldsymbol{\beta}/dt]}{r(t)\,(1 - \mathbf{n}(t) \cdot \boldsymbol{\beta}(t))^3} , \qquad \mathbf{B} = \frac{1}{c}\mathbf{n} \times \mathbf{E} \qquad (8.36)$$

where $\mathbf{n} = \mathbf{r}/r$ is the direction of observation (Fig. 8.16), $\boldsymbol{\beta} = \mathbf{v}/c$, $\dot{\boldsymbol{\beta}} = d\boldsymbol{\beta}/dt$, t is the "retarded time", at which the particle emitted the radiation, τ is the observer time, a little later. Namely, when at position $\mathbf{r}(t)$ with respect to the observer, the particle emits a signal which reaches the observer at time

$$\tau = t + r(t)/c \qquad (8.37)$$

Electric impulse, From Raytracing [29, Sect. 3.2.1]

The vectors $\mathbf{n}$, $\boldsymbol{\beta}$, $\dot{\boldsymbol{\beta}}$ are sub-products of the stepwise integration of particle motion. They are used to compute $\mathbf{E}(\mathbf{n}, \tau)$ in `zpop` (its subroutine sref.f, actually).

The electric field impulse $\mathbf{E}(\mathbf{n}, \tau)$ is decomposed into two polarization components $E_\sigma(\mathbf{n}, \tau)$ and $E_\pi(\mathbf{n}, \tau)$, respectively parallel and normal to the bend plane (Fig. 8.17).

As an example, results for GEC 70 MeV synchrotron are given in Fig. 8.17: the electric field impulses have been derived from the electron trajectory obtained by stepwise raytracing, using Eqs. 8.36 and 8.37. The spectral brightness follows, Eqs. 8.38 and 8.39.

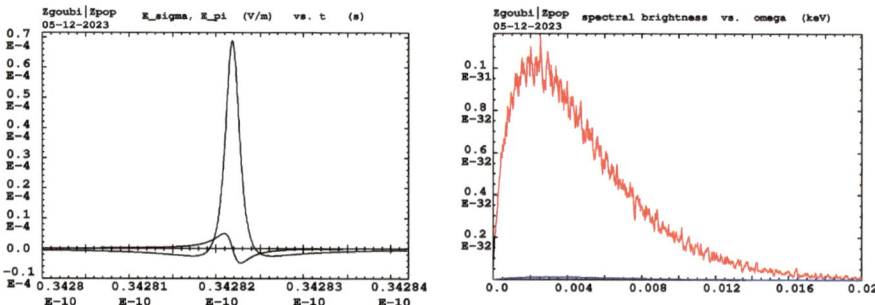

Fig. 8.17 Left: typical shape of $E_{\sigma,\pi}(\tau)$ impulses as observed in the direction $\phi = 0$, $\psi \approx 0.1/\gamma$, in GEC 70 MeV synchrotron (by comparison, at $\phi = \psi = 0$, $E_\sigma(\tau)$) is marginally different, whereas $E_\pi(\tau) \equiv 0$). Right: spectral brightness, peaking near $\hbar\omega_c = 2\gamma^3 c/2\rho \approx 2.7$ eV ($\lambda_c = 0.47\,\mu$m); at such small $\psi \approx 0.1/\gamma$, the π component of the radiation (blue curve) is quite weak compared to the σ component (red curve)

Spectral brightness [29, Sect. 3.2.2]

The respective Fourier transforms of the electric field impulse components $E_\sigma(\mathbf{n}, \tau)$ and $E_\pi(\mathbf{n}, \tau)$, namely

$$\tilde{E}_{\sigma,\pi}(\phi, \psi, \omega) = \int E_{\sigma,\pi}(\phi, \psi, t)e^{-i\omega\tau}d\tau \, / \, \sqrt{2\pi} \tag{8.38}$$

provide the spectral angular brightness (Fig. 8.17)

$$\frac{\partial^3 P_{\sigma,\pi}}{\partial\phi\partial\psi\partial\omega} = 2\epsilon_0 cr^2 \left| \tilde{E}_{\sigma,\pi}(\phi, \psi, \omega) \right|^2 \tag{8.39}$$

with ϕ and ψ the angles of $\mathbf{n}$ to respectively the (X, Z) and (X, Y) planes. Zpop computes $\tilde{E}_{\sigma,\pi}(\phi, \psi, \omega)$ and $\frac{\partial^3 P_{\sigma,\pi}}{\partial\phi\partial\psi\partial\omega}$ (its subroutine srdw.f, actually).

Electric impulse, Analytical [27] [29, Sect. 3.2]

The following theoretical reminders are resorted to in the exercises, for instance for comparison to numerical outcomes from the computation of the electric impulse $\mathbf{E}(\mathbf{n}, \tau)$ (Eq. 8.36) from numerical integration. Referring to Fig. 8.16, the observer direction and velocity vectors write

$$\mathbf{n} = (\cos\psi\cos\phi, \cos\psi\sin\phi, \sin\psi), \quad \boldsymbol{\beta} = \beta(\cos\omega_0 t, \sin\omega_0 t, 0) \tag{8.40}$$

The observer time τ and, to order $1/\gamma^2$, its differential element are obtained from the particle time t (the numerical integration time) using

$$\tau = \frac{1+\gamma^2\psi^2}{2\gamma^2}t + \frac{\omega_0^2}{6}t^3, \quad \frac{d\tau}{dt} = 1 - \mathbf{n}(t)\cdot\boldsymbol{\beta}(t) \simeq \frac{1+\gamma^2\psi^2}{2\gamma^2} + \frac{1}{2}(\omega_0 t - \phi)^2 \tag{8.41}$$

with $\omega_0 \approx c/\rho$ and ρ the local curvature radius. The origin of observer time is at $\phi = \psi = 0$, i.e. in the direction tangent to particle trajectory. The radiated electric impulses result, namely

$$E_\sigma(t) = \frac{q\omega_0\gamma^4}{\pi\epsilon_0 cr} \frac{(1+\gamma^2\psi^2) - \gamma^2(\omega_0 t - \phi)^2}{\left(1+\gamma^2\psi^2 + \gamma^2(\omega_0 t - \phi)^2\right)^3} \mathrm{rect}\left(\frac{t}{2T}\right)$$

$$E_\pi(t) = \frac{q\omega_0\gamma^4}{\pi\epsilon_0 cr} \frac{-2\gamma\psi\gamma(\omega_0 t - \phi)}{\left(1+\gamma^2\psi^2 + \gamma^2(\omega_0 t - \phi)^2\right)^3} \mathrm{rect}(\frac{t}{2T}) \qquad (8.42)$$

where $\mathrm{rect}(x) = 1$ if $-\frac{1}{2} < x < \frac{1}{2}$, zero otherwise, defines the boundary of the numerical integration, namely over a particle deviation angle $\alpha = \pm\omega_0 T$. The impulse components $E_{\sigma,\pi}(t)$ in particle time have similar shapes to $E_{\sigma,\pi}(\tau)$ at the observer, Eq. 8.43 (Fig. 8.17), they differ in width as the latter is squeezed according to $\tau(t)$ contraction (Eq. 8.41)—a squeeze resulting from a double Doppler shift from electron trajectory arc in the lab, to electron frame, and back to observed radiation in the lab, $\Delta\tau \approx \Delta t/\gamma^2$. The cubic dependence $\tau(t)$ (Eq. 8.41) has an analytical solution; substitution of that solution $t(\tau)$ in Eq. 8.42 yields analytical expressions for the field impulses in observer time,

$$E_\sigma(\phi, \psi, \tau) = \frac{q\omega_0\gamma^4}{\pi\epsilon_0 cr} \frac{1 - 4\,\mathrm{hsin}^2[\frac{1}{3}\mathrm{Ahsin}\,u(\phi, \psi, \tau)]}{(1+\gamma^2\psi^2)^2(1 + 4\,\mathrm{hsin}^2[\frac{1}{3}\mathrm{Ahsin}\,u(\phi, \psi, \tau)])^3} \mathrm{rect}[\frac{\tau}{2\Gamma(\phi, \psi)}] \quad (8.43)$$

$$E_\pi(\phi, \psi, \tau) = \frac{q\omega_0\gamma^4}{\pi\epsilon_0 cr} \frac{4\gamma\psi\,\mathrm{hsin}[\frac{1}{3}\mathrm{Ahsin}\,u(\phi, \psi, \tau)]}{(1+\gamma^2\psi^2)^{5/2}(1 + 4\,\mathrm{hsin}^2[\frac{1}{3}\mathrm{Ahsin}\,u(\phi, \psi, \tau)])^3} \mathrm{rect}[\frac{\tau}{2\Gamma(\phi, \psi)}]$$

where $\Gamma(\phi, \psi)$ is the observation direction dependent signal duration and $u = \frac{1}{2}\frac{\gamma\phi}{\sqrt{1+\gamma^2\psi^2}}\left(3 + \frac{\gamma^2\phi^2}{1+\gamma^2\psi^2}\right) - 2\frac{\omega_c}{(1+\gamma^2\psi^2)^{3/2}}\tau$. The critical frequency ω_c partitions the power spectrum in two equal parts, $\int_0^{\omega_c} \frac{dP}{d\omega} = \int_{\omega_c}^\infty \frac{dP}{d\omega}$. Equation 8.43 can be used to check numerical integration outcomes. Refer to [26, 27, Sect. 4.4] for additional details.

The typical width of the impulse in observer time is the familiar

$$\Delta\tau_c = \pm\frac{1}{\omega_c}\left(1+\gamma^2\psi^2\right)^{3/2} \quad \text{with} \quad \omega_c = \frac{3\gamma^3 c}{2\rho} \text{ the critical frequency} \quad (8.44)$$

Accounting for the γ^2 double Doppler shift contraction, $\Delta\tau_c$ thus corresponds to a trajectory arc length $l_c \approx c\Delta\tau_c \times \gamma^2 \approx \rho/\gamma$. From the point of view of the observer, the SR power, integrated over the all spectrum, is mostly contained in an *rms* opening angle [26, Sect. 5.5.2] (Fig. 8.18)

$$\Delta\phi_{c,\mathrm{rms}} = 0.83\frac{1}{\gamma} \quad (8.45)$$

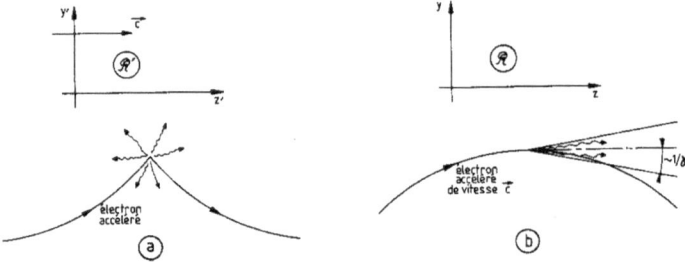

Fig. 8.18 Electron trajectory in the lab frame $\mathcal{R}$ (right) and in $\mathcal{R}'$ traveling parallel to, and at, its velocity (left). The radiation, spanning a $\pm pi/2$ angle in $\mathcal{R}'$, is confined in a forward cone of opening $\pm\Delta\phi_c \approx \pm 1/\gamma$ in $\mathcal{R}$

For a given observation frequency ω, the *rms* opening angle is a function of frequency and satisfies [26, Sect. 5.5.1]

$$\Delta\phi_c(\omega) \approx 0.72 \frac{1}{\gamma} \left(\frac{\omega_c}{\omega}\right)^{1/3} \tag{8.46}$$

Lower radiation energy has wider opening.

8.2.4 Depolarizing Resonances

The field index is zero in the ZGS, transverse focusing is ensured by wedge angles at the ends of the eight dipoles, the only locations where non-zero horizontal field components are present. The latter are weak and as a consequence so also are depolarizing resonances: *"As we can see from the table, the transition probability [from spin state $\psi_{1/2}$ to spin state $\psi_{-1/2}$] is reasonably small up to $\gamma = 7.1$"* [17], i.e. proton $G\gamma = 12.73$, $p = 6.6$ GeV/c. The table referred to stipulates a transition probability $P_{\frac{1}{2},-\frac{1}{2}} < 0.042$, whereas resonances beyond that energy range feature $P_{\frac{1}{2},-\frac{1}{2}} > 0.36$. Beam depolarization up to 6 GeV/c, under the effect of these resonances, is illustrated in Fig. 8.19.

In a synchrotron using gradient dipoles, particles experience radial fields $B_x(y) = -n\frac{B_0}{\rho_0} y$ as they undergo vertical betatron oscillations [17, 30, 31]. As n is small these radial field components are weak, and so is their effect on spin motion.

In a P-periodic ring, the vertical betatron motion excites "systematic intrinsic" spin resonances, located at

$$G\gamma_R = k P \pm \nu_y, \quad k \in \mathbb{N}$$

If the P periodicity of the optics is lost (due to an optical defect), all resonances, systematic and non-systematic, $G\gamma_R = $ integer $\pm \nu_y$ are excited. In the ZGS for

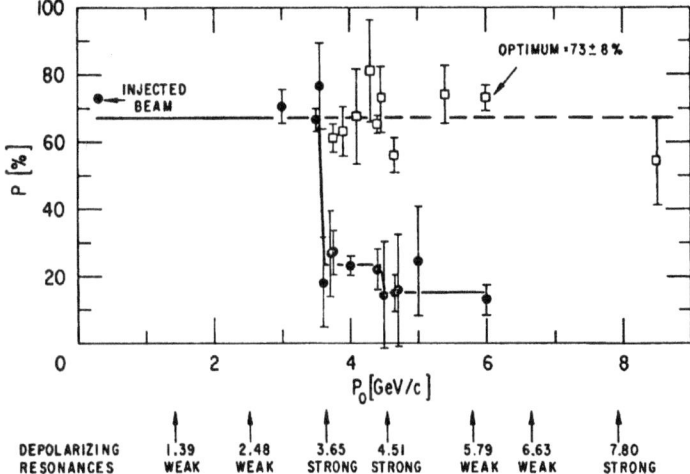

Fig. 8.19 Polarization loss at the ZGS [33] through the strong intrinsic resonances $G\gamma_R = 7.2$ ($p = 3.65\,\text{GeV/c}$) and 8.8 ($4.51\,\text{GeV/c}$) (black circles). A vertical tune jump method preserves polarization (empty circles)

instance, $\nu_y \approx 0.8$ (Table 8.2), the ring is $P = 4$-periodic, thus $G\gamma_R = 4k \pm 0.8$. Strongest intrinsic resonances are located at

$$G\gamma_R = k\,m\,P \pm \nu_y$$

with m the number of cells per superperiod [32, Sect. 3.II]. In the ZGS, with $m = 2$ the strongest resonances occur at (Fig. 8.19)

$$G\gamma_R = 2 \times 4k \pm 0.8 = 7.2\ (3.65\,\text{GeV/c});\ \ 8.8\ (4.51\,\text{GeV/c});\ \ 15.2\ (7.9\,\text{GeV/c});\ \ldots$$

In the presence of vertical orbit defects, non-zero transverse fields are experienced along the closed orbit, they excite "imperfection", *aka* "integer", depolarizing resonances, located at

$$G\gamma_R = k, \qquad k \in \mathbb{N}$$

In the case that the periodicity of the orbit is that of the lattice, P, the sole imperfection resonances, located at $G\gamma_R = kP$, are excited. The strongest imperfection resonances are located at [32, Sect. 3.II]

$$G\gamma_R = k\,m\,P$$

Spin precession axis. Resonance width

Consider the spin vector

$$\mathbf{S}(\theta) = (S_\eta, S_\xi, S_y)$$

of a particle, in the laboratory frame, with θ the orbital angle around the accelerator. Introduce the projection $s(\theta)$ of $\mathbf{S}$ in the bend plane

$$s(\theta) = S_\eta(\theta) + jS_\xi(\theta) \qquad (\text{and } S_y^2 = 1 - s^2) \qquad (8.47)$$

In the case of a stationary solution of the spin motion, viz. stationary spin precession axis around the ring (Fig. 8.21) [31, Sect. 3.6.1], s satisfies [31] (Fig. 8.20)

$$s^2 = \cfrac{1}{1 + \cfrac{\Delta^2}{|\epsilon_R|^2}} \qquad (8.48)$$

with $\Delta = G\gamma - G\gamma_R$ the distance to the resonance; thus the resonance width appears to be a measure of its strength. The quantity of interest is the angle, ϕ, of the spin precession direction to the vertical axis.

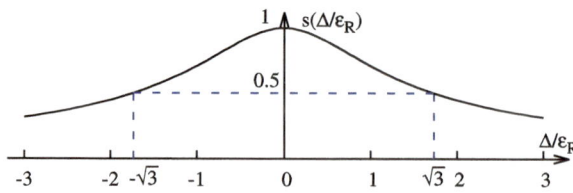

Fig. 8.20 A graph of $s(\Delta) = \sqrt{1 - S_y^2(\Delta)}$. $s = 1$ on the resonance ($\Delta = 0$), the spin vector lies in the bend plane. $s = 1/2$ at distance $\Delta = \pm\sqrt{3}\epsilon_R$ from $G\gamma_R$, the spin vector is at $30°$ to the y axis

Fig. 8.21 Near an integer resonance, at any azimuth θ around the ring spins $\mathbf{S}(m)$ (m is the turn number, $\mathbf{S}(m)$ started vertical, here) precess at frequency $\omega = \sqrt{\Delta^2 + |\epsilon_R|^2}$ around a stationary axis $\mathbf{n}_0(\theta)$, whose orientation varies along the ring. $\mathbf{n}_0$ is aligned along $\overline{\mathbf{S}}$, average of $\mathbf{S}(m)$ over turns

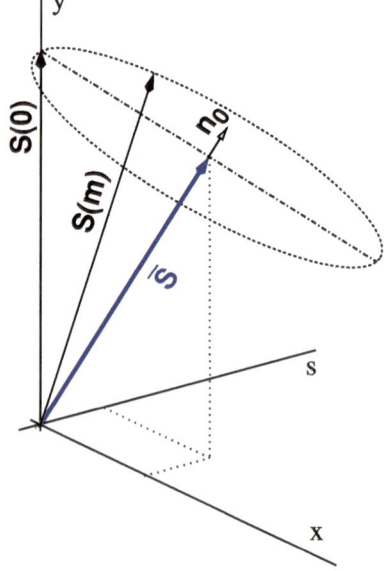

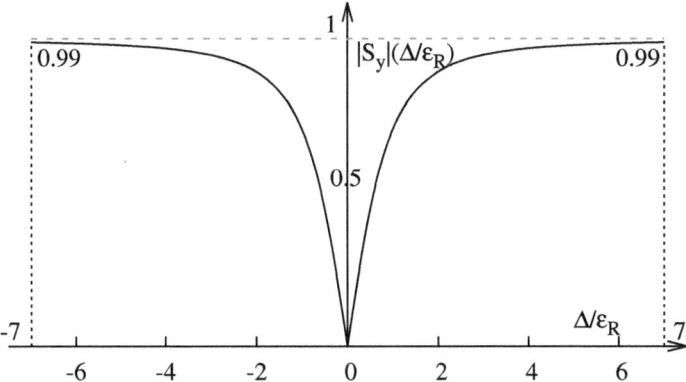

Fig. 8.22 Dependence of polarization on the distance to the resonance. For instance $|S_y| = 0.99$, 1% depolarization at $\Delta = \pm 7|\epsilon_R|$. $S_y = 0$, full depolarization on the resonance ($\Delta = 0$), the precession axis lies in the bend plane

It is given by (Fig. 8.22)

$$\cos^2 \phi(\Delta) \equiv S_y^2(\Delta) = 1 - s^2 = \frac{\Delta^2/|\epsilon_R|^2}{1 + \Delta^2/|\epsilon_R|^2} \tag{8.49}$$

On the resonance, with $\Delta = 0$, the spin precession axis lies in the bend plane: $\phi = \pm \pi/2$. A depolarization by 1% ($|S_y| = 0.99$) corresponds to a distance to the resonance $\Delta = 7|\epsilon_R|$, spin precession axis at an angle $\phi = \mathrm{acos}(0.99) = 8°$ from the vertical.

Conversely,

$$\frac{\Delta^2}{|\epsilon_R|^2} = \frac{S_y^2}{1 - S_y^2} \tag{8.50}$$

The precession axis is common to all spins, while S_y is a measure of the polarization along the vertical axis,

$$S_y = \frac{N^+ - N^-}{N^+ + N^-}$$

where N^+ and N^- denote the number of particles in spin states $\frac{1}{2}$ and $-\frac{1}{2}$ respectively.

Things complicate a little in the vicinity of an intrinsic resonance [31, Sect. 3.6.2], the precession axis is not stationary, spins precess around it while it precesses itself around the vertical, Fig. 8.23.

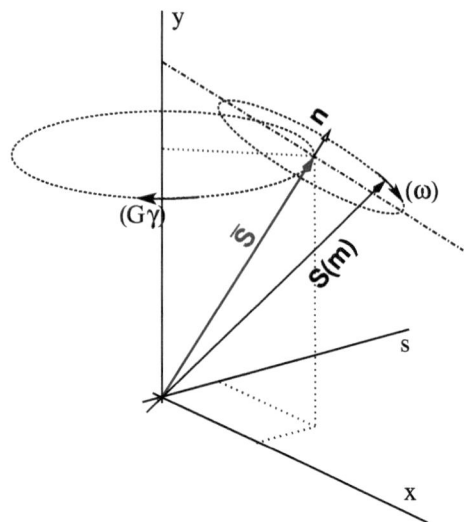

Fig. 8.23 Near an intrinsic resonance, spins $\mathbf{S}(m)$ precess at frequency ω around an axis $\mathbf{n}$, which itself precesses around the vertical axis at frequency $G\gamma$

Resonance crossing

Crossing an isolated resonance (Figs. 8.19 and 8.24) polarization is affected according to the Froissart-Stora law [34], [31, Sect. 2.3.6],

$$\frac{P_f}{P_i} = 2e^{-\frac{\pi}{2}\frac{|\epsilon_R|^2}{\alpha}} - 1 \tag{8.51}$$

from a value P_i upstream to an asymptotic value P_f downstream of the resonance. In this expression

$$\alpha = G\frac{d\gamma}{d\theta} = \frac{1}{2\pi}\frac{\Delta E}{M} \tag{8.52}$$

is the crossing speed for an energy gain ΔE per turn.

Spin motion through weak resonances

Depolarizing resonances are weak up to several GeV in a weak focusing synchrotron because the radial and/or longitudinal fields are weak. Thus assume $S_{y,f} \approx S_{y,i}$, with $S_{y,f}$ and $S_{y,i}$ the asymptotic vertical spin component values respectively upstream and downstream of the resonance. With the origin of the orbital angle taken at the resonance (Fig. 8.24), and introducing the Fresnel integrals [31]

$$C(x) = \int_0^x \cos\left(\frac{\pi}{2}t^2\right) dt, \qquad S(x) = \int_0^x \sin\left(\frac{\pi}{2}t^2\right) dt$$

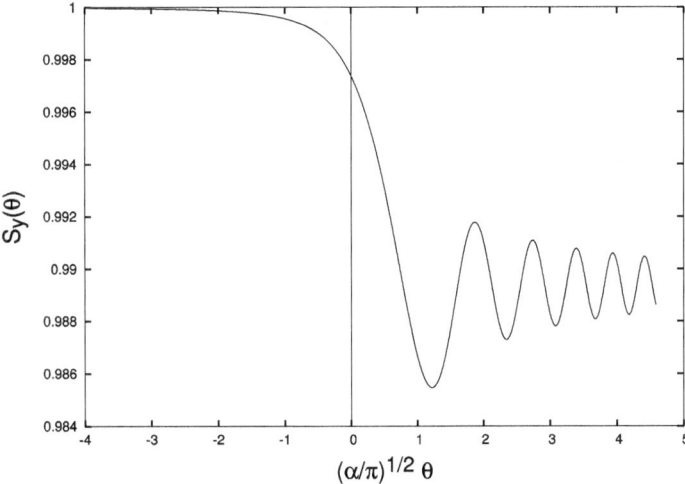

Fig. 8.24 Vertical component of spin motion $S_y(\theta)$ through a weak depolarizing resonance (Eq. 8.53). The vertical line is at the location of the resonance, which coincides with the origin of the orbital angle

the polarization satisfies

$$\text{if } \theta < 0 : \left(\frac{S_y(\theta)}{S_{y,i}}\right)^2 = 1 - \frac{\pi|\epsilon_R|^2}{\alpha}\left\{\left[\tfrac{1}{2} - C\left(-\theta\sqrt{\frac{\alpha}{\pi}}\right)\right]^2 + \left[\tfrac{1}{2} - S\left(-\theta\sqrt{\frac{\alpha}{\pi}}\right)\right]^2\right\}$$

$$\text{if } \theta > 0 : \left(\frac{S_y(\theta)}{S_{y,i}}\right)^2 = 1 - \frac{\pi|\epsilon_R|^2}{\alpha}\left\{\left[\tfrac{1}{2} + C\left(\theta\sqrt{\frac{\alpha}{\pi}}\right)\right]^2 + \left[\tfrac{1}{2} + S\left(\theta\sqrt{\frac{\alpha}{\pi}}\right)\right]^2\right\}$$

(8.53)

In the asymptotic limit,

$$\frac{S_y(\theta)}{S_{y,i}} \xrightarrow{\theta \longrightarrow \infty} 1 - \frac{\pi}{\alpha}|\epsilon_R|^2 \tag{8.54}$$

which agrees with a Taylor development of Froissart-Stora formula, Eq. 8.51, to first order in $|\epsilon_R|^2/\alpha$. This approximation holds in the limit that higher order terms can be neglected.

8.3 Exercises

8.1 Construct SATURNE 1 Synchrotron. Spin Resonances
Solution 8.1.

In this exercise, the weak focusing 3 GeV synchrotron SATURNE 1 (Fig. 8.1) is modeled. Spin resonances in a weak dipole gradient lattice are observed.

(a) Construct a model of SATURNE 1 90° cell dipole in the hard-edge model, using DIPOLE. Use the parameters given in Table 8.1, and Fig. 8.25 as a guidance. For beam monitoring purposes, split the dipole in two 45° halves. It is judicious to take RM = 841.93 cm in DIPOLE, as this is the reference radius for the definition of the radial index. Take an integration step size in centimeter range—small enough to ensure numerical convergence, as large as doable for faster multiturn raytracing.

Validate the model by producing the 6×6 transport matrix of the cell dipole (MATRIX[IFOC=0] can be used for that, with OBJET[KOBJ=5] to define a proper set of paraxial initial coordinates) and checking against theory (Sect. 14.1, Eq. 14.6).

(b) Construct a model of SATURNE 1 cell, with origin at the center of the drift. Find the closed orbit, that particular trajectory which has all its coordinates zero in the drifts: use DIPOLE[KPOS] to cancel the closed orbit coordinates at DIPOLE ends. While there, check the expected value of the dispersion (Eq. 8.26) and of the momentum compaction (Eq. 8.28), from the raytracing of a chromatic closed orbit—i.e., the orbit of an off-momentum particle. Plot these two orbits (on- and off-momentum), over a complete turn around the ring, on a common graph.

Table 8.1 Parameters of SATURNE 1 weak focusing synchrotron [35]. ρ_0 denotes the reference bending radius in the dipole; the reference orbit, field index, wave numbers, etc., are taken along that radius

Orbit length, C	cm	6890
Average radius, $R = C/2\pi$	cm	1096.58
Drift length, $2l$	cm	400
Magnetic radius, ρ_0	cm	841.93
$R/\rho_0 = 1 + k$		1.30246
Field index n, nominal		0.6
Wave numbers ν_x, ν_y, nominal		0.72, 0.89
Stability limit		$0.5 < n < 0.757$
Injection energy (proton)	MeV	3.6
Field at injection	kG	0.326
Top energy	GeV	2.94
Field at top energy, B_{max}	kG	14.9
$\dot{B}$	kG/s	18
Synchronous energy gain	keV/turn	1.160
RF harmonic		2

Fig. 8.25 A schematic layout of SATURNE 1, a $2\pi/4$ axial symmetry structure, comprised of 4 radial field index 90° dipoles and 4 drift spaces. The cell in the simulation exercises is taken as a $\pi/2$ quadrant: half-drift/90°-dipole/half-drift

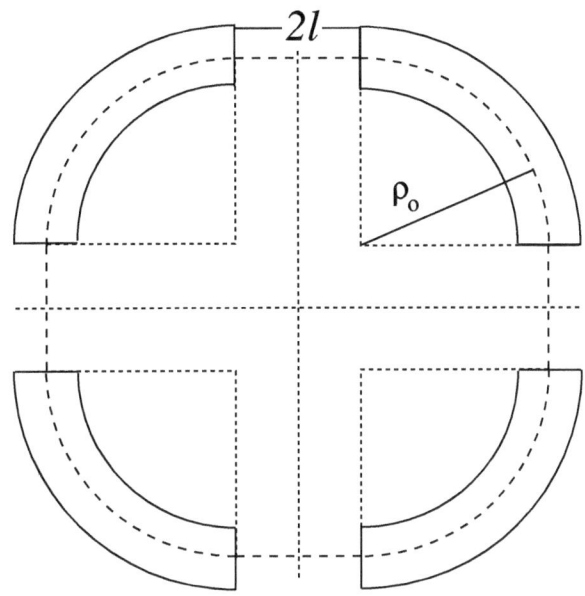

Compute the cell periodic optical functions and tunes, using either OBJET[KOBJ=5] and MATRIX[IORD=1,IFOC=11], or TWISS, or OBJET[KOBJ=6] and MATRIX[IORD=2,IFOC=11]; check their values against theory. Check consistency with previous dispersion function and momentum compaction outcomes.

Move the origin of the lattice at a different azimuth s along the cell: verify that, while the transport matrix depends on the origin, its trace does not.

Produce a graph of the optical functions (betatron functions and dispersion) along the cell. Check the expected average values of the betatron functions (Eq. 8.20).

Produce a scan of the tunes over the field index range $0.5 \leq n \leq 0.757$. REBELOTE[IOPT=1] can be used to repeatedly change n over that range. Superimpose the theoretical curves $\nu_x(n)$, $\nu_y(n)$.

(c) Justify considering the betatron oscillation as sinusoidal, namely,

$$y(\theta) = A \cos(\nu_y \theta + \phi)$$

wherein $\theta = s/R$, $R = \oint ds/2\pi$.

(d) Launch a few particles evenly distributed on a common paraxial horizontal Courant-Snyder invariant, vertical motion taken null (OBJET[KOBJ=8] can be used), for a single pass through the cell. Store particle data along the cell in zgoubi.plt, using DIPOLE[IL=2] and DRIFT[split,N=20,IL=2]. Use these to generate a graph of the beam envelopes.

Using Eq. 8.22 compare with the results obtained in (b). Find the minimum and maximum values of the betatron functions, and their azimuth $s(min[\beta_x])$, $s(max[\beta_x])$. Check the latter against theory.

Repeat for the vertical motion, taking $\varepsilon_x = 0$, ε_y paraxial.

Repeat, using, instead of several particles on a common invariant, a single particle traced over a few tens of turns.

(e) Produce an acceleration cycle from 3.6 MeV to 3 GeV, for a few particles launched on a common $10^{-4} \pi m$ initial invariant in each plane. Ignore synchrotron motion (CAVITE[IOPT=3] can be used in that case). Take a peak voltage $\hat{V} = 200 \, kV$ (for faster raytracing—unrealistic though, as it would result in prohibitive $\dot{B}$ (Eq. 8.29)) and synchronous phase $\phi_s = 150°$ (justify $\phi_s > \pi/2$).

Check the betatron damping over the acceleration range: compare with theory (Eq. 8.31).

How close to symplectic the numerical integration is (it is by definition *not* symplectic, being a truncated Taylor series method [36, Eq. 1.2.4]), depends on the integration step size, and on the size of the flying mesh in the DIPOLE[IORDRE,Resol] method [36, Fig. 20]; check a possible departure of the betatron damping from theory as a function of these parameters.

Produce a graph of the horizontal and vertical wave number values over the acceleration cycle.

(f) Some spin motion, now. Adding SPNTRK at the beginning of the sequence used in (e) will ensure spin tracking.

Based on the input data file worked out for question (d), simulate the acceleration of a single particle, through the intrinsic resonance $G\gamma_R = 4 - \nu_y$, from a distance of a few times the resonance strength upstream (this requires determining BORO value under OBJET) to a distance of a few times the resonance strength downstream of the resonance, at an acceleration rate of 10 kV/turn.

OBJET[KOBJ=8] can be used to allow to easily define an initial invariant value.

Start with spin vertical. On a common graph, plot $S_y(turn)$ for a few different values of the vertical betatron invariant (the horizontal invariant value does not matter—explain that statement, it can be taken zero). Derive the resonance strength from this tracking, check against theory.

Repeat, for different crossing speeds.

Push the tracking beyond $G\gamma = 2 \times 4 + \nu_y$: verify that the sole systematic resonances $G\gamma = integer \times P \pm \nu_y$ are excited—with $P = 4$ the periodicity of the ring.

Break the 4-periodicity of the lattice by perturbing the index in one of the 4 dipoles (say, by 10%), verify that all resonances $G\gamma = integer \pm \nu_y$ are now excited.

(g) Consider a case of weak resonance crossing, single particle (i.e., a case where $P_f/P_i \approx 1$, taken from (f); crossing speed may be increased, or particle invariant decreased if needed), show that it satisfies Eq. 8.53. Match its turn-by-turn tracking data to Eq. 8.53 so to get the vertical betatron tune ν_y, the location of the resonance $G\gamma_R$, and its strength.

(h) Stationary spin motion (i.e. at fixed energy) is considered in this question. Track a few particles with distances from the resonance $\Delta = G\gamma - G\gamma_R = G\gamma - (4 - \nu_y)$ evenly spanning the interval $\Delta \in [0, 7 \times \epsilon_R]$.

Produce on a common graph the spin motion $S_y(turn)$ for these particles, as observed at some azimuth along the ring.

Produce a graph of the average over turns, $\langle S_y \rangle|_{turn}(\Delta)$ (as in Fig. 8.22). Produce the vertical betatron tune ν_y, the location of the resonance $G\gamma_R$, and its strength ϵ_R, obtained from a match of $\langle S_y \rangle|_{turn}(\Delta)$ to (Eq. 8.49)

$$\langle S_y \rangle(\Delta) = \frac{|\Delta|}{\sqrt{|\epsilon_R|^2 + \Delta^2}}$$

(i) Track a 200-particle 6-D bunch, with Gaussian transverse densities $\varepsilon_{x,y}$ a few μm, and Gaussian $\delta p/p$ with $\sigma_{\delta p/p} = 10^{-4}$. Produce a graph of the average value of the vertical spin component S_y over a 200 particle set, as a function of $G\gamma$, across the $G\gamma_R = 4 - \nu_y$ resonance. Indicate on that graph the location of the resonant $G\gamma_R$ values.

Perform this resonance crossing for five different values of the particle invariant: $\varepsilon_y/\pi = 2, 10, 20, 40, 200\,\mu$m. Compute P_f/P_i in each case, check the dependence on ε_y against theory.

Compute the resonance strength, ε_y, from this tracking.

Re-do this crossing simulation for a different crossing speed (take for instance $\hat{V} = 10$ kV) and a couple of vertical invariant values, compute P_f/P_i so obtained. Check the crossing speed dependence of P_f/P_i against theory.

8.2 Construct the ZGS Synchrotron. Spin Resonances
Solution 8.2.

In this exercise, the ZGS 12 GeV synchrotron is modeled. Spin resonances in a zero-gradient, wedge focusing synchrotron are addressed.

A photo taken in the ZGS tunnel is given in Fig. 8.4; a schematic layout of the ring is shown in Fig. 8.26, and a sketch of the double dipole cell in Fig. 8.27. Table 8.2 details the parameters of the synchrotron resorted to in these simulations.

(a) Construct a model of ZGS 45° cell dipole in the hard-edge model, using DIPOLE. Use the parameters given in Table 8.2, and Figs. 8.26 and 8.27 as a guidance. For beam monitoring purposes, split the dipole in two 22.5° halves. Take the closed orbit radius as the reference RM = 2076 cm in DIPOLE: it will be assumed that the orbit is the same at all energies.[7] Take an integration step size in centimeter range—small enough to ensure numerical convergence, as large as doable for fast multiturn raytracing.

Validate the model by producing the 6×6 transport matrices of both dipoles (MATRIX[IFOC=0] can be used for that, with OBJET[KOBJ=5] to define a proper set of paraxial initial coordinates) and checking against theory (Sect. 14.1, Eq. 14.6).

---

[7] Note that in reality the reference orbit in ZGS moved outward during acceleration [37].

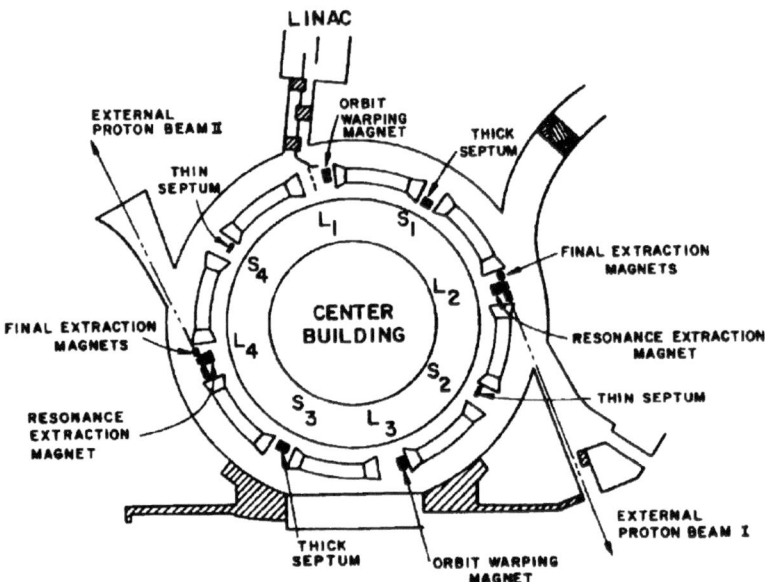

Fig. 8.26 A schematic layout of the ZGS [33], a π-periodic structure, comprised of 8 zero-index dipoles, 4 long and 4 short straight sections

Fig. 8.27 A sketch of ZGS cell layout. In defining the entrance and exit faces (EFBs) of the magnet, beam goes from left to right. Wedge angles at the long straight sections (ε_1) and at the short straight sections (ε_2) are different

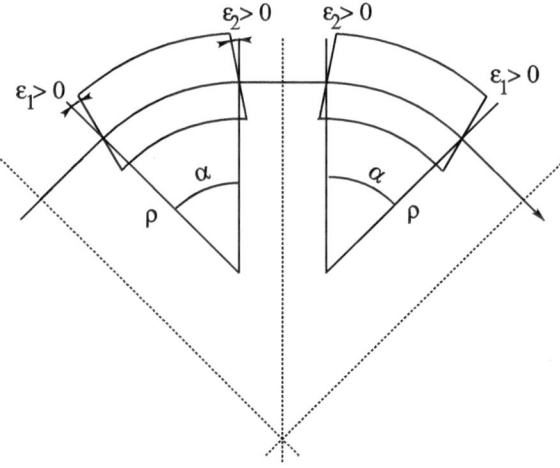

Table 8.2 Parameters of the ZGS weak focusing synchrotron after Refs. [37, 38] [33, pp. 288–294, p. 716] (2nd column, when they are known) and in the present simplified model and numerical simulations (3rd column). Note that the actual orbit moves during ZGS acceleration cycle, tunes change as well—this is not taken into account in the present modeling, for simplicity

		From Refs. [37, 38]	Simplified model
Injection energy	MeV	50	
Top energy	GeV	12.5	
$G\gamma$ span		1.888387–25.67781	
Length of central orbit	m	171.8	170.90457
Length of straight sections, total	m	41.45	40.44
Lattice			
Wave numbers ν_x; ν_y		0.82; 0.79	0.849; 0.771
Max. β_x; β_y	m		32.5; 37.1
Magnet			
Length	m	16.3	16.30486 (magnetic)
Magnetic radius	m	21.716	20.76
Field min.; max.	kG	0.482; 21.5	0.4986; 21.54
Field index		0	
Yoke angular extent	deg	43.02590	45
Wedge angle	deg	≈10	13 and 8
RF			
Rev. frequency	MHz	0.55–1.75	0.551–1.751
RF harmonic h=$\omega_{\mathrm{rf}}/\omega_{\mathrm{rev}}$		8	
Peak voltage	kV	20	200
B-dot, nominal/max.	T/s	2.15/2.6	
Energy gain, nominal/max.	keV/turn	8.3/10	100
Synchronous phase, nominal	deg	150	
Beam			
ε_x; ε_y (at injection)	$\pi\,\mu$m	25; 150	
Momentum spread, rms		3×10^{-4}	
Polarization at injection	%	>75	100

Add fringe fields in DIPOLE[λ,$C_0 - C_5$], the rest of the exercise will use that model. Take fringe field extent and coefficient values

$$\lambda = 60 \, \text{cm}, \; C_0 = 0.1455, \; C_1 = 2.2670, \; C_2 = -0.6395, \; C_3 = 1.1558, \; C_4 = C_5 = 0$$
$$(8.55)$$

($C_0 - C_5$ determine the shape of the field fall-off, they have been computed from a typical measured field profile $B(s)$).

(b) Construct a model of ZGS cell accounting for dipole fringe fields, with origin at the center of the long drift. In doing so, use DIPOLE[KPOS] to cancel the closed orbit coordinates at DIPOLE ends.

Compute the periodic optical functions at cell ends, and cell tunes, using MATRIX[IORD=1,IFOC=11] (or OBJET[KOBJ=6] and MATRIX[IORD=2, IFOC=11]); check their values against theory.

Move the origin at the location (azimuth s along the cell) of the betatron functions extrema: verify that, while the transport matrix depends on the origin, its trace does not. Verify that the local betatron function extrema, and the dispersion function, have the expected values.

Produce a graph of the optical functions (betatron functions and dispersion) along the cell.

(c) Additional verifications regarding the model.

Produce a graph of the field B(s)

– along the on-momentum closed orbit, and along off-momentum chromatic closed orbits, across a cell;

– along orbits at large horizontal excursion;

– along orbits at large vertical excursion.

For all these cases, verify qualitatively, from the graphs, that $B(s)$ appears as expected.

(d) Justify considering the betatron oscillation as sinusoidal, namely,

$$y(\theta) = A \, \cos(\nu_y \theta + \phi)$$

wherein $\theta = s/R$, $R = \oint ds/2\pi$.

(e) Produce an acceleration cycle from 50 MeV to $\sim$17 GeV about, for a few particles launched on a common $10^{-5} \, \pi$ m vertical initial invariant, with small horizontal invariant. Ignore synchrotron motion (CAVITE[IOPT=3] can be used in that case). Take a peak voltage $\hat{V} = 200$ kV (this is unrealistic but yields 10 times faster computing than the actual $\hat{V} = 20$ kV, Table 8.2) and synchronous phase $\phi_s = 150°$ (justify $\phi_s > \pi/2$). Add spin, using SPNTRK, in view of the next question, (f).

Check the accuracy of the betatron damping over the acceleration range, compared to theory. How close to symplectic the numerical integration is (it is by definition *not* symplectic [29, Eq. 1.2.4]), depends on the integration step size, and on the size of the flying mesh in the DIPOLE method [36, Fig. 20]; check a possible departure of the betatron damping from theory as a function of these parameters.

Produce a graph of the evolution of the horizontal and vertical wave numbers during the acceleration cycle.

(f) Using the raytracing material developed in (e): produce a graph of the vertical spin component of a few particles, and the average value over the 200 particle bunch, as a function of $G\gamma$. Indicate on that graph the location of the resonant $G\gamma_R$ values.

(g) Based on the simulation file used in (f), simulate the acceleration of a single particle, through one particular intrinsic resonance, from a few thousand turns upstream to a few thousand turns downstream.

Perform this resonance crossing for different values of the particle invariant. Determine the dependence of final/initial vertical spin component value, on the invariant value; check against theory.

Re-do this crossing simulation for a different crossing speed. Check the crossing speed dependence of final/initial vertical spin component so obtained, against theory.

(h) Introduce a vertical orbit defect in the ZGS ring.

Find the closed orbit.

Accelerate a particle launched on that closed orbit, from 50 MeV to $\sim$17 GeV about, produce a graph of the vertical spin component.

Select one particular resonance, reproduce the two methods of (g) to check the location of the resonance at $G\gamma_R$ = integer, and to find its strength.

8.3 Visible SR from GEC 70 MeV Synchrotron
Solution 8.3.

Produce the electric field impulse radiated by a 70 MeV electron in GEC synchrotron, as observed at a vertical elevation $\psi \neq 0$ in a plane tangent to the orbit. MULTIPOL can be used to simulate a trajectory arc of a few $1/\gamma$ deviation. Set IL=2 to log stepwise particle coordinates in zgoubi.plt.

Produce the spectral brightness of the radiation in that direction.

Zpop menu 8/16 can be used for these two questions. An alternative is to program the equations of concern (Eq. 8.36 et seqs.) in an interface, in python for instance.

The 70 MeV orbit radius in GEC synchrotron is 29.2 cm.

8.4 Solutions of Exercises of This Chapter: Weak Focusing Synchrotron

8.1 Construct SATURNE 1 Synchrotron. Spin Resonances
Figure 8.1 displays a photo of SATURNE 1 synchrotron. A schematic layout of the ring and 90° cell is given in Fig. 8.25. This figure and Table 8.1 will be referred to in building SATURNE 1 ring in the following.

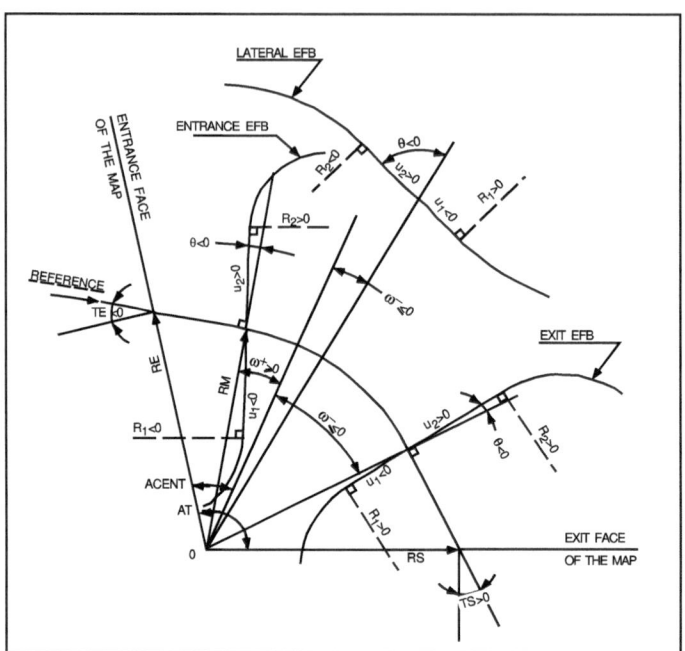

Fig. 8.28 A representation of the data that define a dipole magnet, using DIPOLE [36]

(a) A model of SATURNE 1 synchrotron.

DIPOLE is used to simulate the 90° cell dipole, data are set for a hard-edge model in this exercise (for a DIPOLE model including fringe fields, refer to the ZGS case, Exercise 8.2).

Some guidance regarding DIPOLE data, referring to Fig. 8.28:

- DIPOLE is defined in a cylindrical coordinate system.
- AT is given the value of the bending sector extent: $AT = 90°$. The dipole EFBs coincide with DIPOLE entrance and exit boundaries.
- RM is given the curvature radius value, $RM = B\rho/B = 0.274426548\,[\mathrm{T\,m}]/0.03259493\,[\mathrm{T}] = 8.4193\,\mathrm{m}$, as it fits the geometry of the optical axis around the ring. The field value matches the reference rigidity under OBJET, these are the injection energy values, 3.6 MeV, proton.
- ACENT=45° is the reference azimuth, for the positioning of the entrance and exit EFBs. It is taken half-way of the AT range, an arbitrary choice.
- KPOS=2 allows cancelling the coordinates of particle 1 (considered here as the reference trajectory, coinciding with the optical axis around the ring) at entrance and exit of DIPOLE:

  – The entrance and exit radii in and out of the AT sector for a particle on the closed orbit (i.e., a particle traveling along the design optical axis) are $RE = RS = RM$.

– The angle TE identifies with the closed orbit angle at the entrance EFB: TE=0, the closed orbit is normal to the EFB. Same for TS at exit EFB.

Simulation of a 90° sector in the hard edge model is given in Table 8.3; note that the sector has been split in two 45° halves, this is in order to allow a possible insertion of a beam monitor, so requiring $AT = 45°$, $\omega^+ = -\omega^- = 22.5°$. FAISCEAU located next to DIPOLE indicates that a trajectory entering DIPOLE at radius $R = RM$, normally to the EFB (thus, $Y_0 = 0$ and $T_0 = 0$ in OBJET) exits with $Y = 0$ and $T = 0$. Data validation at this stage can be performed by comparing DIPOLE's transport matrix computed with MATRIX (Table 8.4), and theoretical expectations (Sect. 14.1, Eq. 14.6):

$$
\begin{array}{c}
\alpha=\pi/2, \\
\rho=8.4193 \\
n=0.6 \\
[T_{ij}] \\
= \\
(Eq.\ 14.6)
\end{array}
\begin{pmatrix}
0.545794 & 11.15444 & 0 & 0 & 0 & 9.560222 \\
0.062944 & 0.545794 & 0 & 0 & 0 & 1.324865 \\
0 & 0 & 0.346711 & 10.19506 & 0 & 0 \\
0 & 0 & -0.086295 & 0.346711 & 0 & 0 \\
1.324865 & 9.560222 & 0 & 0 & 1 & 5.17640 \\
0 & 0 & 0 & 0 & 0 & 1
\end{pmatrix}
\tag{8.56}
$$

Introducing fringe fields

SATURNE ring simulations which follow use the hard edge model, however, it might be desired at this point to add DIPOLE fringe fields. The following changes are needed in that case:

• The bending sector is 90°, however the field region extent AT has to encompass the fringe fields, at both ends of the 90° sector. A 5° extension is taken (namely, $ACENT - \omega^+ = AT - ACENT + \omega^- = 5°$), for a total $AT = 100°$ which allows $RM \times \tan(ACENT - \omega^+) \approx 74$ cm; this large extension ensures absence of truncation of the fringe fields at the AT sector boundaries, over the all radial excursion of the beam.
• ACENT $= 50°$ is the reference azimuth (an arbitrary value; taken half-way of the AT range for convenience), for the positioning of the entrance and exit EFBs.
• The entrance radius in the AT sector is $RE = RM/\cos(ACENT - \omega^+) = RM/\cos(5°)$, with $\omega^+ = 45°$ the positioning of the entrance EFB with respect to ACENT. And similarly for the positioning of the exit reference frame, $RS = RM/\cos(AT - (ACENT - \omega^-)) = RM/\cos(5°)$ with $\omega^- = -45°$ the positioning of the exit EFB. Note that $\omega^+ - \omega^- = 90°$, the value of the sector angle.
• The entrance angle TE identifies with the angular increase of the sector: TE=5°. And similarly for the positioning of exit frame, 5° downstream of the exit EFB, thus TS $= 5°$.
• Negative drifts with equal lengths
$$RM \times \tan(ACENT - \omega^+) = RM \times \tan(AT - (ACENT - \omega^-)) = 0.7366545469\ \text{cm}$$
need to be added upstream and downstream of DIPOLE, to account for the optical axis additional length over the 5° angular extent.

Table 8.3 Simulation input data file: SATURNE 1 90° DIPOLE is split into a pair of adjacent 45° sectors in the hard edge model. FAISTORE or (here) FAISCEAU is inserted between the latter two, for beam monitoring. The reference optical axis has equal entrance (RE) and exit (RS) positions, and null angles (TE and TS), it coincides with the arc of radius $R = RM$ inside the sector. This input data file is named SatI_DIP.inc and defines the cell sequence segment S_SatI_DIP to E_SatI_DIP, for INCLUDE statements in subsequent exercises. The present file equipped with OBJET[KOBJ=5] and MATRIX[IFOC=0] computes the dipole transport matrix

```
File name: SatI_DIP.inc
! SATURNE 1. Hard edge dipole model. Transport matrix.
'MARKER'   SatI_DIP.inc_S                              ! Just for edition purposes.
'OBJET'
0.274426548e3                                          ! Reference Brho: 3.6 MeV proton.
5                                      ! Create a 13 particle set, proper for MATRIX computation.
.001 .01 .001 .01 .001 .0001                           ! Coordinate sampling.
0. 0. 0. 0. 0. 1.        ! Reference trajectory: all initial coordinates nul, relative rigidity D=1.
1

'MARKER' S_SatI_DIP      ! Cell dipole begins here. A marker used for INCLUDEs in subsequent exercises.

'DIPOLE'  upstream_half                                ! Analytical modeling of a dipole magnet.
0                        ! set IL=2 here, to log trajectory coordinates in zgoubi.plt, at integration steps.
45. 841.93               ! Field region angle=90; reference radius set to curvature radius value.
22.5 0.3259493638 -0.6 0. 0.    ! Reference angle ACENT set to AT/2; Bo field at RM; indices, all zero.
.0 0.                                                  ! EFB 1, hard-edged.
4  .1455   2.2670  -.6395  1.1558  0. 0. 0.            ! Enge coefficients.
22.5 0.  1.E6  -1.E6  1.E6  1.E6          ! Angle to ACENT; face angle; face is straight.
.0 0.                                                  ! EFB 2, hard-edged.
4  .1455   2.2670  -.6395  1.1558  0. 0. 0.
-22.5 0.  1.E6  -1.E6  1.E6  1.E6
0. 0.                                                  ! EFB 3. Unused.
0 0.      0.      0.      0.      0. 0. 0.
0. 0.  1.E6  -1.E6  1.E6  1.E6 0.
2 1                                ! Degree of interpolation polynomial; flying grid sizing.
2.                                 ! Integration step size. It can be large in uniform field.
2 841.93 0. 841.93 0.                              ! Positioning of entrance and exit frames.
'MARKER' half-dipole !.plt            ! Uncomment LABEL_2='.plt' (may go with IL=2 under DIPOLE) to
                                                        ! log particle data in zgoubi.plt.
'FAISCEAU'       ! Provides local coordinates, and ellipse parameters, at center of SATURNE 1 dipole.
'DIPOLE'  downstream_half                              ! Analytical modeling of a dipole magnet.
0                        ! set IL=2 here, to log trajectory coordinates in zgoubi.plt, at integration steps.
45. 841.93               ! Field region angle=90; reference radius set to curvature radius value.
22.5 0.3259493638 -0.6 0. 0.    ! Reference angle ACENT set to AT/2; Bo field at RM; indices, all zero.
.0 0.                                                  ! EFB 1, hard-edged.
4  .1455   2.2670  -.6395  1.1558  0. 0. 0.            ! Enge coefficients.
22.5 0.  1.E6  -1.E6  1.E6  1.E6          ! Angle to ACENT; face angle; face is straight.
.0 0.                                                  ! EFB 2, hard-edged.
4  .1455   2.2670  -.6395  1.1558  0. 0. 0.
-22.5 0.  1.E6  -1.E6  1.E6  1.E6
0. 0.                                                  ! EFB 3. Unused.
0 0.      0.      0.      0.      0. 0. 0.
0. 0.  1.E6  -1.E6  1.E6  1.E6 0.
2 1                                ! Degree of interpolation polynomial; flying grid sizing.
2.                                 ! Integration step size. It can be large in uniform field.
2 841.93 0. 841.93 0.                              ! Positioning of entrance and exit frames.

'MARKER' E_SatI_DIP      ! Cell dipole ends here. A marker used for INCLUDEs in subsequent exercises.

'FAISCEAU'                                             ! Local particle coordinates.
'MATRIX'                                ! Compute transport matrix, from trajectory coordinates.
1 0

'MARKER'   SatI_DIP.inc_E                              ! Just for edition purposes.
'END'
```

(b) SATURNE 1 cell.

A cell with origin in the middle of the drift is given Table 8.5, it INCLUDEs the split dipole, with a pair of 2 m half-drifts added at both ends (Fig. 8.25).

Closed orbit; chromatic closed orbit

The on-momentum closed orbit is set to zero along the drifts (see coordinates in Table 8.4: $Y_0 = Y = 0$, $T_0 = T = 0$) thanks to DIPOLE[KPOS=2, $RE = RS = RM$, $TE = TS = 0$] (Table 8.3).

Table 8.4 Outcomes of the DIPOLE simulation of Table 8.3

An excerpt from zgoubi.res execution listing. Coordinates of the first particle (considered here as the reference trajectory) and its path length under FAISCEAU, at OBJET on the left hand side below, locally on the right hand side:

```
    3  Keyword, label(s)  :  FAISCEAU
                                          TRACE DU FAISCEAU
                                          (follows element #      2)

                        OBJET                                        FAISCEAU
          D      Y(cm)   T(mr)    Z(cm)   P(mr)   S(cm)    D-1     Y(cm)   T(mr)   Z(cm)   P(mr)   S(cm)
   0  1   1.000   0.00    0.00    0.00     0.00    0.000   0.000   0.00    0.00    0.00    0.000   1.322501E+03
```

Transport matrix of SATURNE 1 90 degree sector bend, in the hard edge model, in two different cases of integration step size, namely, 4 cm and 1 m (an excerpt of MATRIX computation, from zgoubi.res execution listing). It can be checked against expectations, Eq. 8.56. The reference trajectory (first one) exits better aligned (i.e. reference coordinates, before change of frame for MATRIX computation, are closer to zero):

- *Case of* 4 cm *step size*:

```
    4  Keyword, label(s)  :  MATRIX

  Reference, before change of frame (particle #  1  - D-1,Y,T,Z,s,time)  :
  0.00000000E+00    4.53054326E-07   6.27843350E-07   0.00000000E+00   0.00000000E+00   1.32250055E+03   4.41138700E-02

              TRANSFER  MATRIX  ORDRE   1  (MKSA units)
         0.545795         11.1544         0.00000         0.00000         0.00000         9.56022
        -6.294423E-02      0.545795        0.00000         0.00000         0.00000         1.32487
         0.00000          0.00000          0.346711       10.1951         0.00000         0.00000
         0.00000          0.00000         -8.629576E-02    0.346711        0.00000         0.00000
         1.32487          9.56022          0.00000         0.00000         1.00000         5.17640
         0.00000          0.00000          0.00000         0.00000         0.00000         1.00000

       DetY-1 =        0.0000000278,    DetZ-1 =        0.0000000045
```

- *Case of* 1 m *step size*:

```
    4  Keyword, label(s)  :  MATRIX

  Reference, before change of frame (particle #  1  - D-1,Y,T,Z,s,time)  :
  0.00000000E+00   -7.54923113E-03  -1.08904867E-02   0.00000000E+00   0.00000000E+00   1.32249873E+03   4.41138091E-02

              TRANSFER  MATRIX  ORDRE   1  (MKSA units)
         0.545757         11.1567         0.00000         0.00000         0.00000         9.56154
        -6.295274E-02      0.546125        0.00000         0.00000         0.00000         1.32517
         0.00000          0.00000          0.346697       10.1954         0.00000         0.00000
         0.00000          0.00000         -8.629900E-02    0.346750        0.00000         0.00000
         1.32486          9.56148          0.00000         0.00000         1.00000         5.17692
         0.00000          0.00000          0.00000         0.00000         0.00000         1.00000

       DetY-1 =        0.0003978566,    DetZ-1 =        0.0000685588
```

The radial coordinate of an off-momentum chromatic orbit can be estimated from the dispersion, Eq. 8.26, namely,

$$Y_\delta = \frac{\rho_0}{1-n}\frac{\delta p}{p} = \frac{841.93}{1-(-0.6)}10^{-4} \approx 0.21048\,\text{cm}$$

whereas the orbit angle is zero, around the ring (on- and off-momentum closed orbits are parallel to the optical axis).

Besides,

– computation of an accurate value of Y_δ is performed adding FIT at the end of the cell;

Table 8.5 Simulation input data file: SATURNE 1 cell, assembled by INCLUDE-ing DIPOLE taken from Table 8.3 together with two half-drifts. This input data file is named SatI_cell.inc and defines the SATURNE 1 cell sequence segment S_SatI_cell to E_SatI_cell, for INCLUDE statements in subsequent exercises

```
File name: SatI_cell.inc.
| SATURNE 1, one cell of the 4-period ring.
'MARKER'   SatICellMATRIX_S                                  | Just for edition purposes.
'OBJET'
0.274426548e3                                                | Reference Brho: 3.6 MeV proton.
5                                       | Create a 13 particle set, proper for MATRIX computation.
.001 .01 .001 .01 .001 .0001                                 | Coordinate sampling.
0. 0. 0. 0. 0. 1.           | Reference trajectory: all initial coordinates nul, relative rigidity D=1.

'MARKER'  S_SatI_cell
'DRIFT' half_drift
200.
'INCLUDE'
1
./SatI_DIP.inc[S_SatI_DIP:E_SatI_DIP]
'DRIFT' half_drift
200.
'MARKER'  E_SatI_cell
'FAISCEAU'                                                   | Local particle coordinates.
'TWISS'   | Produce transport matrix, beam matrix, and periodic optical functions along the sequence.
2 1. 1.
'MARKER'   SatICellMATRIX_E                                  | Just for edition purposes.
'END'
```

– in order to raytrace three particles, respectively on-momentum and at $\delta p/p = \pm 10^{-4}$, OBJET[KOBJ=2] is used;

– in order to raytrace around the ring, for the purpose of plotting the closed orbit coordinates, a 4-cell sequence follows the FIT procedure.

This results in the input data file given in Table 8.6. Running this file produces the following coordinates as per the FIT procedure (an excerpt from zgoubi.res execution listing):

```
STATUS OF VARIABLES  (Iteration #    4 /    999 max.)
LMNT VAR PARAM  MINIMUM   INITIAL        FINAL      MAXIMUM     STEP       NAME   LBL1
  2   1    30   0.168     0.211       0.21056000    0.253    1.040E-05  OBJET    -
  2   2    40   0.00      0.00        0.0000000     0.00     1.040E-05  OBJET    -
  2   3    50  -0.253    -0.210      -0.21040403   -0.168    1.040E-05  OBJET    -
STATUS OF CONSTRAINTS (Target penalty =   1.0000E-10)
TYPE  I   J LMNT#   DESIRED        WEIGHT       REACHED       KI2     NAME   LBL1
  3   1   2   12  0.000000E+00  1.000E+00  1.466978E-06  6.70E-01 MARKER  E_SatI_cell
  3   2   2   12  0.000000E+00  1.000E+00  6.028957E-07  1.13E-01 MARKER  E_SatI_cell
  3   3   2   12  0.000000E+00  1.000E+00  8.357183E-07  2.17E-01 MARKER  E_SatI_cell
Fit reached penalty value   3.2139E-12
```

The local coordinates Y, T and initial coordinates Y_0, T_0 (as defined under OBJET) are identical to better than 5 μm, 0.5 μrad, respectively, confirming the periodicity of these chromatic trajectories. Orbit coordinates around the ring are displayed in Fig. 8.29.

Table 8.6 Simulation input data file: first find the three periodic orbits at $\delta p/p = 0, \pm 10^{-4}$, through a cell; once FIT is done, complete a 4-cell turn

```
SatI_Orbits.INC.dat: SATURNE 1, on-momentum and chromatic orbits.
'MARKER'    SatI_Orbits_S                                      ! Just for edition purposes.
'OBJET'
0.274426548e3                                                  ! Reference Brho: 3.6 MeV proton.
2                                                              ! Create particles individually.
3 1                                                            ! Three particles.
+.210560 0. 0. 0. 0. 1.0001 'p' ! Chromatic orbit coordinates Y0 and T0 for D=1.001 relative rigidity.
0. 0. 0. 0. 0. 1. 'o'                                          ! On-momentum orbit.
-.210404 0. 0. 0. 0. 0.9999 'm' ! Chromatic orbit coordinates Y0 and T0 for D=0.999 relative rigidity.
1 1 1

'INCLUDE'
1
./SatI_cell.inc[S_SatI_cell:E_SatI_cell]

'FIT'
3
2 30 0 .2                                                      ! Vary Y_0(particle 1) under OBJET.
2 40 0 .2                                                      ! Vary Y_0(particle 2) under OBJET.
2 50 0 .2                                                      ! Vary Y_0(particle 3) under OBJET.
3
3.1 1 2 #End 0. 1. 0                                           ! Constrain Y(particle 1)=Y_0(particle 1).
3.1 2 2 #End 0. 1. 0                                           ! Constrain Y(particle 2)=Y_0(particle 2).
3.1 3 2 #End 0. 1. 0                                           ! Constrain Y(particle 3)=Y_0(particle 3).

!        When FIT is done converging on the constraints, execution quietly carries on with the periodic
!              coordinates , raytracing through 4 cells to complete a turn around the ring.
'INCLUDE'
1
4 * ./SatI_cell.inc[S_SatI_cell:E_SatI_cell]

'SYSTEM'
1
gnuplot < gnuplot_Zplt_traj.gnu                                ! Plot the orbit radial coordiante.
'MARKER'   SatI_Orbits_E                                       ! Just for edition purposes.
'END'
```

A gnuplot script (excerpt) to obtain a graph of particle coordinates, from zgoubi.plt (as in Fig. 8.29):

```
# gnuplot_Zplt_traj.gnu
traj1 = 1 ; traj2 = 3
plot \
for [i=traj1:traj2]  'zgoubi.plt' u ($19== i ? $14 *cm2m : 1/0):($10 *cm2m):($19) w p ps .4 lc palette
```

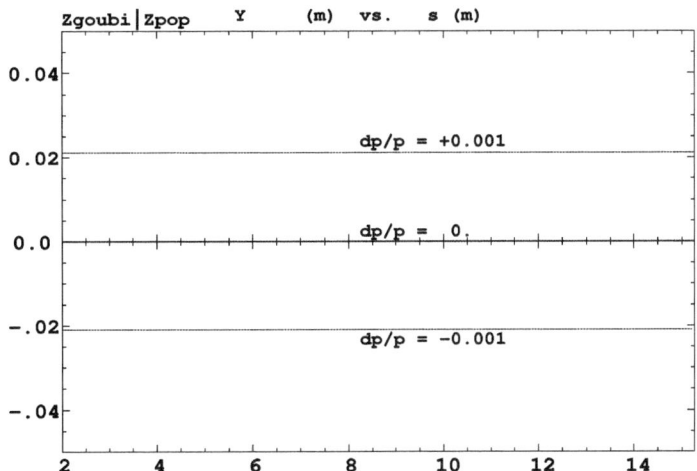

Fig. 8.29 Radial coordinate of the orbits around the ring, on-momentum, and for $dp/p = \pm 10^{-3}$. A graph obtained using zpop: menu 7; 1/1 to open zgoubi.plt; 2/[6,2] for Y versus distance s; 7 to plot. A gnuplot script for a similar graph is given in Table 8.6

Table 8.7 Results obtained running the simulation input data file of Table 8.5, SATURNE 1 cell—an excerpt from zgoubi.res execution listing

```
    14  Keyword, label(s) :  TWISS

Reference, before change of frame (particle #  1  - D-1,Y,T,Z,s,time)  :
 0.00000000E+00   6.02895730E-07   6.54169939E-07   0.00000000E+00   0.00000000E+00   1.72250055E+03   6.57784696E-01

       Beam  matrix  (beta/-alpha/-alpha/gamma) and  periodic  dispersion  (MKSA units)
          14.418595     0.000000     0.000000     0.000000     0.000000    21.048250
           0.000000     0.069355     0.000000     0.000000     0.000000     0.000000
           0.000000     0.000000    11.411041     0.000000     0.000000    -0.000000
           0.000000     0.000000     0.000000     0.087634     0.000000     0.000000
           0.000000     0.000000     0.000000     0.000000     0.000000     0.000000
           0.000000     0.000000     0.000000     0.000000     0.000000     0.000000

                        Betatron  tunes  (Q1 Q2 modes)
                NU_Y =  0.18103144        NU_Z =  0.22214599

                    dL/L / dp/p =   1.9194487

                         Transition gamma  =  7.21791469E-01

                         Chromaticities :
                dNu_y / dp/p = -0.60221729              dNu_z / dp/p =  0.38005442
```

Table 8.8 The header part of zgoubi.TWISS.out listing resulting from the SATURNE 1 cell simulation of Table 8.5. The ring wave numbers are 4 times the present cell values Q1, Q2. Optical functions (betatron function and derivative, orbit, phase advance, etc.) along the optical sequence follow this header in zgoubi.TWISS.out

```
 @ LENGTH      %le     17.22500552
 @ ALFA        %le      1.919448707
 @ ORBIT5      %le                  -0
 @ GAMMATR     %le      0.7217914685
 @ Q1          %le      0.18103144E+00  1  [frac., int.]
 @ Q2          %le      0.22214599E+00  1  [frac., int.]
 @ DQ1         %le     -0.6022172911
 @ DQ2         %le      0.3800544183
 @ DXMAX       %le      2.10586311E+01    2.10482503E+01  @ DXMIN
 @ DYMAX       %le      0.00000000E+00    0.00000000E+00  @ DYMIN
 @ XCOMAX      %le      2.10528899E-01    0.00000000E+00  @ XCOMIN
 @ YCOMAX      %le      0.00000000E+00    0.00000000E+00  @ YCOMIN
 @ BETXMAX     %le      1.57006971E+01    1.44132839E+01  @ BETXMIN
 @ BETYMAX     %le      1.30884296E+01    1.14110171E+01  @ BETYMIN
 @ XCORMS      %le      6.05227342E-04
 @ YCORMS      %le      0.    not computed
 @ DXRMS       %le      2.98427468E-03
 @ DYRMS       %le      0.00000000E+00
```

Lattice parameters

The TWISS command down the sequence (Table 8.5) produces the periodic beam matrix results shown in Table 8.7; MATRIX[IFOC=11] would, as well. It also produces a zgoubi.TWISS.out file which details the optical functions along the sequence (at the downstream end of the optical elements). The header of that file details the optical parameters of the structure (Table 8.8).

Moving the origin of the cell

The origin of the sequence can be moved by placing both drifts at an end of DIPOLE. It can also be taken in the middle of DIPOLE, as the latter has been split. An input data sequence (with INCLUDEs expanded) is provided at the top of the execution listing zgoubi.res, it can be used to copy-paste pieces around. It can then be checked that betatron tunes, chromaticities, momentum compaction (Table 8.7) do not change, whereas the beam matrix does.

Optical functions along the cell

They are computed by transporting the beam matrix, from the origin. A Fortran program available in `zgoubi` sourceforge package toolbox, betaFromPlt [36], performs this computation in the following way: OBJET[KOBJ=5.1] provides the initial beta function values (determined in the previous question); IL=2 under DIPOLE logs stepwise particle data in zgoubi.plt; 'split 10 2' added under DRIFT does it, too [36, DRIFT]. The program betaFromPlt computes the transport matrix T_{step_i} from the origin of the sequence (at OBJET) to the considered step$_i$ along the sequence, using particle coordinates read in zgoubi.plt—a similar computation to what MATRIX does [36, MATRIX Sect.]. The beam matrix $\sigma = \begin{bmatrix} \beta & -\alpha \\ -\alpha & \gamma \end{bmatrix}$ is then transported, from the origin to step$_i$, using (Eq. 14.46)

$$\sigma_{\text{step}_i} = T_{\text{step}_i} \, \sigma_{\text{origin}} \, \tilde{T}_{\text{step}_i}$$

The result is displayed in Fig. 8.30.

Tune scan

A simulation is given in Table 8.9, derived from Table 8.5: MATRIX[IFOC=11] has been substituted to TWISS, a REBELOTE do loop repeatedly changes n. A graph of the scan is given in Fig. 8.31, a few values are detailed in Table 8.10.

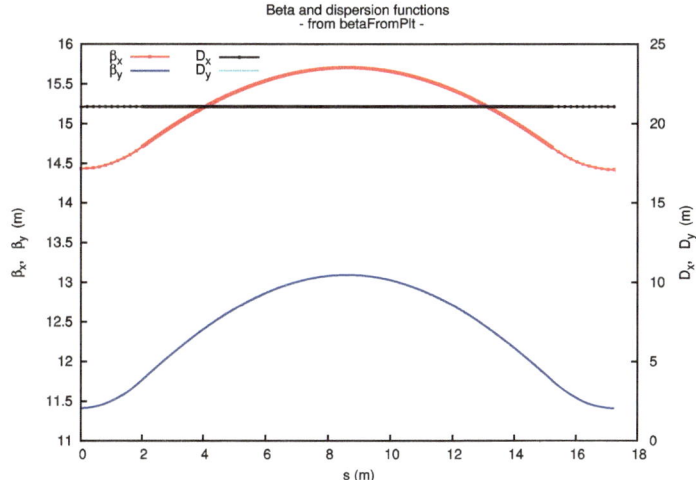

Fig. 8.30 Optical functions along SATURNE 1 cell. They are obtained from the transport of the beta functions, from the origin (at OBJET), using transport matrices computed from step-by-step particle coordinates stored in zgoubi.plt

Table 8.9 Simulation input data file: tune scan, using REBELOTE to repeatedly change DIPOLE field index *n*. Beam matrix and wave numbers are computed by MATRIX, from the coordinates of the 13 particle sample generated by OBJET[KOBJ=5]

```
SATURNE 1, tune scan.
'MARKER'   SatI_Qscan_S                                    ! Just for edition purposes.
'OBJET'
0.274426548e3                                              ! Reference Brho: 3.6 MeV proton.
5                                          ! Create a 13 particle set, proper for MATRIX computation.
.001 .01 .001 .01 .001 .0001                               ! Coordinate sampling.
0. 0. 0. 0. 0. 1.        ! Reference trajectory: all initial coordinates nul, relative rigidity D=1.
1

'MARKER'   S_SatI_cell
'DRIFT' half_drift
200.
'INCLUDE'
1
./SatI_DIP.inc[S_SatI_DIP:E_SatI_DIP]
'DRIFT' half_drift
200.
'MARKER'   E_SatI_cell
'FAISCEAU'                                                 ! Local particle coordinates.
'MATRIX'
1 11  PRINT   ! Comoute a 10+4 period transport matrix, and tunes. Save outcomes to zgoubi.MATRIX.out.

'REBELOTE'                   ! A do loop: repeat the section above commencing at the top of the file,
10  1.1  0 1                                               ! 10 times.
1
DIPOLE 6  -0.757:-0.5   ! Change the value of parameter 30 (namely, n) in DIPOLE (prior to repeating).
                                                           ! in any DIPOLE in the sequence.
'SYSTEM'
1
gnuplot <./gnuplot_MATRIX_Qxy.gnu                          ! Plot tunes vs index.
'MARKER'   SatI_Qscan_E                                    ! Just for edition purposes.
'END'
```

gnuplot script to obtain Fig. 8.31:

```
# ./gnuplot_MATRIX_Qxy.gnu
set xlabel "index n";set ylabel "{/Symbol n}_x,      ({/Symbol n}_x^2+{/Symbol n}_y^2)^{1/2}"
set y2label "{/Symbol n}_y"; set xtics; set ytics nomirror; set y2tics nomirror; ncell=4
set key t l; set key maxrow 2; set yrange [:1.3]; set y2range [:1.06]
n1 = -0.757; dn=(.757-.5)/10.; R=10.9658; rho=8.4193
plot \
"zgoubi.MATRIX.out" u (n1+($61-1)*dn): \
($61>1? $56 *ncell :1/0) w p pt 5 lt 1 lw .5 lc rgb "red"           tit "{/Symbol n}_x " ,\
"zgoubi.MATRIX.out" u (n1+($61-1)*dn):($61>1? sqrt((1+(n1+($61-1)*dn))*R/rho)): \
1/0) w l lt 1  lc rgb "red" tit "theor. " ,\
"zgoubi.MATRIX.out" u (n1+($61-1)*dn): \
($61>1? $57 *ncell :1/0) axes x1y2 w p pt 6 lt 3 lw .5 lc rgb "blue"   tit "{/Symbol n}_y " ,\
"zgoubi.MATRIX.out" u (n1+($61-1)*dn): \
($61>1? sqrt((-(n1+($61-1)*dn))*R/rho):1/0) axes x2y2 w l lt 3  lc rgb "blue" tit "theor. " ,\
"zgoubi.MATRIX.out" u (n1+($61-1)*dn): \
($61>1? sqrt($56**2+$57**2) *ncell :1/0) w p pt 7 lt 1 lc rgb "black" tit "({/Symbol n}_x^2+{/Symbol n}_y^2)^{1/2}" ,\
"zgoubi.MATRIX.out" u (n1+($61-1)*dn):($61>1? sqrt(R/rho):1/0)  w l lt 1  lc rgb "black" tit "theor. "
pause 1
```

(c) Sinusoidal approximation of the betatron motion.

The approximation

$$y(\theta) = A \, \cos(\nu_Z \theta + \phi)$$

is checked here considering the vertical motion (considering the horizontal motion leads to similar conclusions). The value of the various parameters in that expression are determined as follows:

– the particle raytraced for comparison is launched with an initial excursion $Z_0(\theta = 0) = 5$ cm. At the launch point (middle of the drift) the beam ellipse is upright (see below), whereas phase space motion is clockwise, thus take

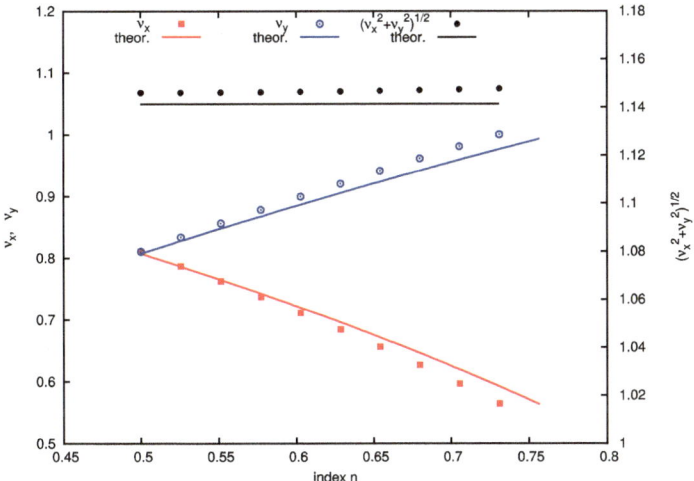

Fig. 8.31 A scan of the wave numbers, and of $\sqrt{\nu_Y^2 + \nu_Z^2} \approx \sqrt{R/\rho_0} = 1.141$, in SATURNE 1 for $0.5 \leq n \leq 0.757$. Solid curves are from theoretical approximations (Eq. 8.23), markers are from numerical simulations

Table 8.10 Dependence of wave numbers on index n, from numerical raytracing (columns denoted "ray-tr.") and from theory

n	ν_Y		ν_Z	
	ray-tr.	$\sqrt{(1-n)\frac{R}{\rho_0}}$	ray-tr.	$\sqrt{n\frac{R}{\rho_0}}$
0.5	0.810353	0.806987	0.810353	0.806987
0.6	0.724125	0.721791	0.888583	0.884010
0.7	0.626561	0.625089	0.960806	0.954840
0.757	0.563635	0.562580	0.999804	0.992955

$$A = 5\,\text{cm} \quad \text{and} \quad \phi = 0$$

– the vertical betatron tune of the 4-cell ring is (Table 8.8)

$$\nu_Z = 4 \times 0.222146 = 0.888284$$

$-\theta = s/R$ and $R = \oint ds/2\pi$ with (Table 8.8)

$$2\pi R = \text{circumference} = 2\pi \times 10.9658 = 68.9\,\text{m}$$

Consistency with sinusoidal approximation is shown in Fig. 8.32.

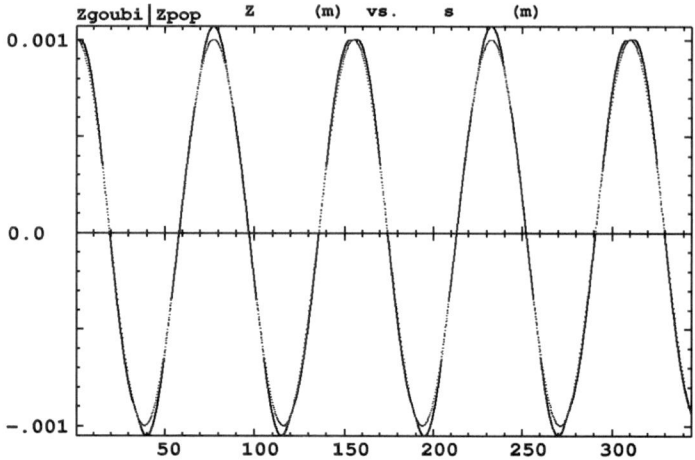

Fig. 8.32 Vertical betatron motion, five turns around SATURNE 1 ring, from raytracing (modulated oscillation), and sine approximation, superimposed

(d) Beam envelopes, periodic ellipses.

A few particles are launched through the cell with initial coordinates taken on a common invariant (horizontal and/or vertical), using OBJET[KOBJ=8]. The input data file is given in Table 8.11. The initial ellipse parameters (under OBJET) are the periodic values $\alpha_Y = \alpha_Z = 0$, $\beta_Y = 14.426\,\text{m}$, $\beta_Z = 11.411\,\text{m}$, found in zgoubi.TWISS.out (Table 8.8). The envelopes so generated, and the quantities $u^2(s)/\varepsilon_u/\pi$ (Eq. 8.22), are displayed in Fig. 8.33. The extremum extremorum of $u^2(s)/\varepsilon_u/\pi$ comes out to be, respectively, $\hat{\beta}_Y = 15.7\,\text{m}$ and $\hat{\beta}_Z = 13.08\,\text{m}$, consistent with earlier derivations (BETXMAX and BETYMAX values in Table 8.8 and Fig. 8.30).

This raytracing also provides the coordinates of the particles on their common upright invariant (Fig. 8.34)

$$u^2/\beta_u + \beta_u u'^2 = \varepsilon_u/\pi$$

at start and at the end of the cell ($\varepsilon_u/\pi = 10^{-4}$, here). This allows checking that the initial ellipse parameters (under OBJET, Table 8.11) are effectively periodic values, and that the raytracing went correctly, namely by observing that the initial and final ellipses do superimpose.

(e) An acceleration cycle. Symplecticity checks.

Eleven particles are launched for a 30,000 turn acceleration at a rate of

$$\Delta W = q\hat{V}\sin\phi_s = 200 \times \sin 150^0 = 100\,\text{keV/turn}$$

Table 8.11 Simulation input data file: raytrace 60 particles across SATURNE 1 cell to generate beam envelopes. Store particle data in zgoubi.plt, along DIPOLEs and split DRIFTs. The INCLUDE file and segment are defined in Table 8.5

```
SATURNE 1 envelopes.
'MARKER'  SatI_envelopes_S                                ! Just for edition purposes.
'OBJET'
0.274426548e3                                             ! Reference Brho: 3.6 MeV proton.
8                              ! Create a set of 60 particles evenly distributed on the same invariant;
1 60 1    ! case of 60 particles on a vertical invariant; use 60 1 1 instead for horizontal invariant.
0. 0. 0. 0. 0. 1.
0.   14.426 1e-4
0.   11.411 1e-4
0. 1. 0.

'FAISTORE'                                   ! This logs the coordinates of the particle to zgoubi.fai,
zgoubi.fai  S_SatI_cell  E_SatI_cell                   ! at the two LABEL1s as indicated.
1

'MARKER'  S_SatI_cell                                    ! SATURNE 1 cell begins here.
'DRIFT' half_Drift                             ! Option 'split' devides the drift in 10 pieces,
200. split 10  2                            ! 'IL=2' causes log of particle data to zgoubi.plt.

'INCLUDE'
1
./SatI_DIP.inc[S_SatI_DIP:E_SatI_DIP]

'DRIFT' half_Drift                             ! Option 'split' devides the drift in 10 pieces,
200. split 10  2                            ! 'IL=2' causes log of particle data to zgoubi.plt.

'MARKER'  E_SatI_cell                                    ! SATURNE 1 cell ends here.
'FAISCEAU'
'MARKER'   SatI_envelopes_E                              ! Just for edition purposes.
'END'
```

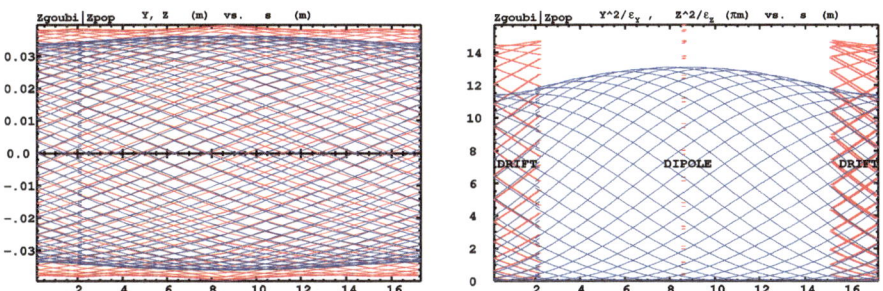

Fig. 8.33 Left: horizontal and vertical envelopes as generated by plotting the coordinates Y(s) (greater excursion, red, along the drifts and dipole) or Z(s) (smaller excursion, blue) across the SATURNE 1 cell, of 60 particles evenly distributed on a common $10^{-4}\,\pi$ m invariant, either horizontal or vertical (while the other invariant is zero). Right: a plot of $Y^2(s)/\varepsilon_Y/\pi$ and $Z^2(s)/\varepsilon_Z/\pi$; their extrema identify with $\beta_Y(s)$ and $\beta_Z(s)$, respectively. Graphs obtained using zpop: menu 7; 1/1 to open zgoubi.plt; 2/[6,2] (or [6,4]) for Y versus s (or Z versus s); 7 to plot; option 3/14 to raise Y (or Z) to the square and normalize to $\varepsilon_{Y,Z}/\pi$

(E : 3.6 MeV $\rightarrow$ 3.0036 GeV), all evenly distributed on the same initial vertical invariant

$$Z^2/\beta_Z + \beta_Z Z'^2 = \varepsilon_Z/\pi \tag{8.57}$$

with $\varepsilon_Z/\pi = 10^{-4}$ m, or, normalized, $\beta\gamma\varepsilon_Z/\pi = 0.08768 \times 10^{-4}$ m.

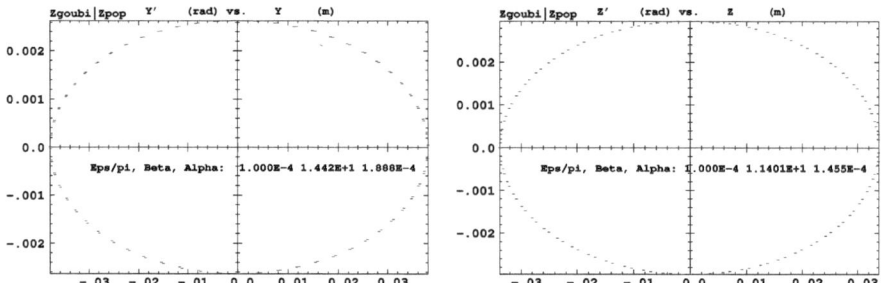

Fig. 8.34 Sixty particles evenly distributed on a common periodic invariant (either $\varepsilon_Y = 10^{-4}\pi\,\mathrm{m}$ and $\varepsilon_Z = 0$, left graph, or the reverse, right graph) have been tracked through the cell. Initial and final phase space coordinates are displayed in these graphs: the initial and final ellipses superimpose. Optical function values given in the figures result from an *rms* match, of indifferently the initial or final coordinates; they do agree with TWISS outcomes (Table 8.8). A graph obtained using zpop: menu 7; 1/5 to open zgoubi.fai; 2/[2,3] (or [4,5]) for T versus Y (or P vs. Z); 7 to plot

The simulation file is given in Table 8.12. CAVITE[IOPT=3] is used, it provides an RF phase independent boost

$$\Delta W = q\hat{V}\sin\phi_s$$

as including synchrotron motion is not necessary here, even better, anticipating on spin question (f) this ensures constant depolarizing resonance crossing speed, so precluding any possibility of multiple crossing (it can be referred to [40] regarding that effect).

Betatron damping

Figure 8.35 shows the damped vertical motion of the individual particles, over the acceleration range, together with the initial and final distributions of the 11 particles on elliptical invariants. Departure from the matching ellipse at the end of the acceleration cycle, 3 GeV (Eq. 8.57 with $\varepsilon_Z/\pi = 1.0745 \times 10^{-6}\,\mathrm{m}$), is marginal.

Degree of non-symplecticity of the numerical integration

The degree of non-symplecticity as a function of integration step size is illustrated in Fig. 8.36. The initial motion is taken paraxial, vertical motion is considered as it resorts to off-mid plane Taylor expansion of fields [36, DIPOLE Sect.], a stringent test as the latter is expected to deteriorate further the non-symplecticity inherent to the Lorentz equation integration method (a truncated Taylor series method [36, Eq. 1.2.4]).

Evolution of the wave numbers

The Fortran tool tunesFromFai_iterate can be used to computes tunes as a function of turn number or energy, it reads turn-by-turn particle data from zgoubi.fai and computes a discrete Fourier transform over so many turns (a few tens, 100 here

Table 8.12 Simulation input data file: track 11 particles launched on the same vertical invariant. The INCLUDE adds the SATURNE 1 cell four times, the latter is defined in Table 8.5

```
SATURNE 1 ring. Polarization landscape.
'MARKER'  SatIPolarLand_S                             ! Just for edition purposes.
'OBJET'
0.274426548e3                                         ! Reference Brho: 3.6 MeV proton.
8                      ! Create a set of 60 particles evenly distributed on the same invariant;
1 11 1    ! case of 11 particles on a vertical invariant; use 11 1 1 instead for horizontal invariant.
0. 0. 0. 0. 0. 1.
0.  14.426 1e-4                      ! Periodic optical functions and invariant value, horizontal and
0.  11.411 1e-4                                                                      ! vertical.
0. 1. 0.                                              ! No momentum spread.

!'MCOBJET'                                             ! Commented.
!1.03527036749193e3                         ! Reference Brho: 50 MeV proton.
!3                         ! Create a 13 particle set, proper for MATRIX computation.
!200
!2 2 2 2 2 2
!0. 0. 0. 0. 0. 1.
!0. 14.426 25e-6 3                            ! Periodic alpha_Y, beta_Y, and invariant value;
!0. 11.411 10e-6 3                            ! Periodic alpha_Z, beta_Z, and invariant value.
!0. 1. 1.e-8 3
!123456 234567 345678

'PARTICUL'
PROTON                ! Necessary data in order to allow  (i) spin trackingand, and  (ii) acceleration.
'SPNTRK'                                              ! Switch on spin tracking,
3                                                     ! all initial spins vertical.
'FAISCEAU'
'FAISTORE'
b_polarLand.fai                           ! Log particle data in b_polarLand.fai, turn-by-turn; "b_" imposes
7                                         ! binary write, which results in faster i/o.

'SCALING'
1 1
DIPOLE
-1                                        ! Causes field increase in DIPOLE, in correlation to particle
1.                                        ! rigidity increase by CAVITE.
1

! 4 cells follow.
'INCLUDE'
1
4* ./SatI_cell.inc[S_SatI_cell:E_SatI_cell]

'CAVITE'
3                             ! With IOPT=3, synchronous phase below of 30deg or 150deg is equivalent.
0 0
200e3  0.523598775598                        ! Acceleration rate is 200*0.5=100keV/turn.
! 20e3  0.523598775598                   ! Commented: an acceleration rate of 20*0.5=10keV/turn.

'REBELOTE'
30000 0.2 99                         ! Case of 100 keV/turn: ~30,000 turns from 3.6 MeV to 3 GeV.
! 300000 0.3 99          ! Commented: case of 10 keV/turn: ~300,000 turns from 3.6 MeV to 3 GeV.

'FAISCEAU'
'MARKER'  SatIPolarLand_E                             ! Just for edition purposes.
'SPNPRT'

'END'
```

for instance), every so many turns (300, here) [41]. Typical results are displayed in Fig. 8.37, tunes have the expected values: $\nu_Y = 0.7241$, $\nu_Z = 0.8885$. An acceleration rate of 100 keV/turn has been taken (namely, $\hat{V} = 200$ kV and still $\phi_s = 150°$), to save on computing time. SCALING with option NTIM=−1 causes the magnet field to strictly follow the momentum boost by CAVITE.

(f) Crossing an isolated intrinsic depolarizing resonance.

The simulation uses the input data file of Table 8.12, with the following changes:

- Under OBJET:
  - 1st line, change the reference rigidity BORO for an initial $G\gamma \approx 2.95$, upstream of $G\gamma_R = 4 - \nu_Z \approx 3.1$;
  - 3rd line, request a single particle ("1 1 1", in lieu of 11, "1 11 1");

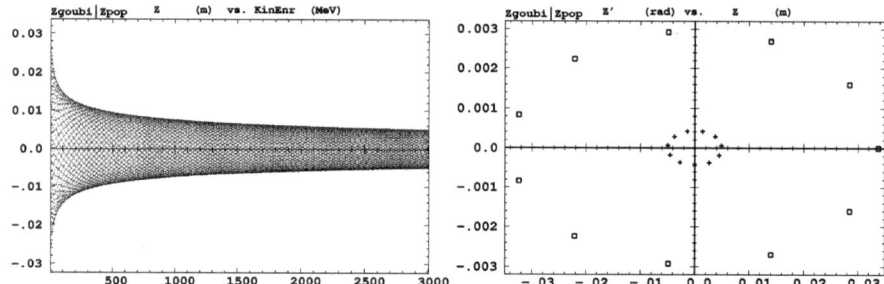

Fig. 8.35 Left: damped vertical motion, from 3.6 MeV to 3.004 GeV in 30,000 turns. Right: the initial coordinates of the 11 particles (squares) are taken on a common invariant $\varepsilon_Z(0) = 10^{-4}\,\pi$m (at 3.6 MeV, $\beta\gamma = 0.0877$, thus $\beta\gamma\varepsilon_Z(0) = 8.77 \times 10^{-6}\,\pi$m); the final coordinates after 30,000 turns (crosses) appear to still be (with negligible departure) on a common invariant, of value $\varepsilon_Z(final) = 2.149 \times 10^{-6}\,\pi$m, or $\beta\gamma\varepsilon_Z(final) = 8.77 \times 10^{-6}\,\pi$m (at 3.004 GeV, $\beta\gamma = 4.08045$), equal to the initial value $\beta\gamma\varepsilon_Z(0)$

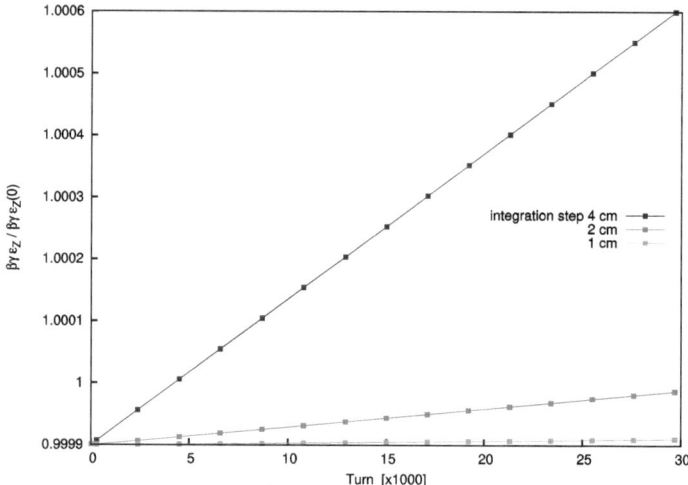

Fig. 8.36 Turn-by-turn evolution of the normalized invariant, $\beta\gamma\varepsilon_Z(turn)/\beta\gamma\varepsilon_Z(0)$ (initial $\varepsilon_Z(0)$ taken paraxial), for integration step sizes 1, 2 and 4 cm

- 6th line, set the invariant ε_Z/π to the desired value, ε_Y/π value is indifferent. Resulting OBJET:

```
'OBJET'
4.08807740024e3                      ! Reference Brho -> G*gamma=2.949312341 -> 605.22655 MeV proton.
8                                    ! Create a (set of) particle(s) on a given invariant.
1 1 1                                                                  ! case of 1 particle.
0. 0. 0. 0. 0. 1.
0. 1. 0.                             ! Horizontal invariant taken zero.
0. 11.411 1e-4                       ! Periodic alpha_Z, beta_Z, and invariant value.
0. 1. 0.                                                         ! No momentum spread.
```

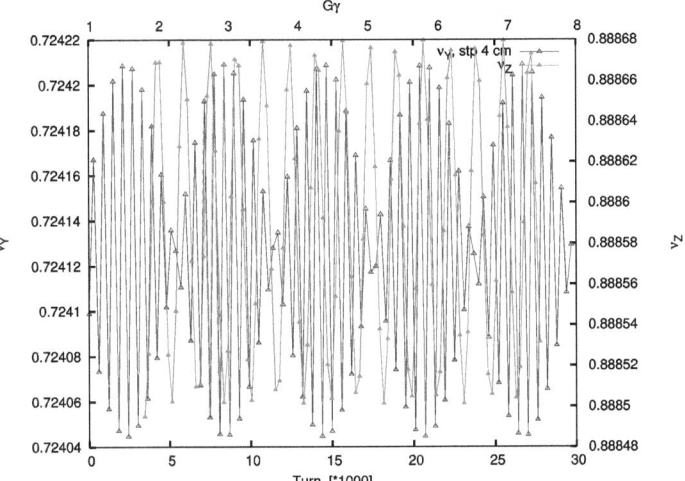

Fig. 8.37 Horizontal fractional ring tune (left vertical axis), $\nu_Y \approx 0.7241$, and vertical ring tune (right vertical axis), $\nu_Z \approx 0.8885$, as a function of turn number, over 30,000 turns ($E : 0.0036 \rightarrow$ 3 GeV at a rate of 100 keV/turn)

- change the field value under DIPOLE (via SCALING for instance) consistently with the new BORO value, so to maintain a curvature radius $\rho_0 = BORO/B = 8.4193$ m (Table 8.1),
- under CAVITE, set the peak voltage to the required value,
- under REBELOTE, set the number of turns to an appropriate value: a total of 15,000, of which 8,000 about upstream of the resonance, is convenient for an acceleration rate of 10 keV/turn.

Changing the particle invariant value

Particle spin motion through the isolated resonance for seven different invariant values, $\varepsilon_Z/\pi = 1, 2, 10, 20, 40, 80, 200\,\mu$m, observed at the beginning of the optical sequence (FAISTORE[b_polarLand.fai] location, Table 8.12), is displayed in Fig. 8.38.

The intrinsic resonance strength satisfies $|\epsilon_R|^2 = A\,\varepsilon_Z$, with A a factor which characterizes the lattice (see Sect.9.2.7, Eq. 9.53). On the other hand, from the Froissart-Stora formula (Eq. 8.51) one gets

$$|\epsilon_R|^2 = \frac{2\alpha}{\pi} \ln\left(\frac{2}{1+S_{Z,f}/S_{Z,i}}\right) \xrightarrow{S_{Z,f}\approx S_{Z,i}} \frac{\alpha}{\pi}\left(1 - \frac{S_{Z,f}}{S_{Z,i}}\right) \qquad (8.58)$$

with α, crossing speed (Eq. 8.52), a constant. Thus one expects to find $\frac{1}{\varepsilon_Z} \ln\left(\frac{2}{1+S_{Z,f}/S_{Z,i}}\right)$ constant.

Calculation of the resonance strength from P_f/P_i tracking outcomes, using Eq. 8.58, requires the value of the crossing speed, which is

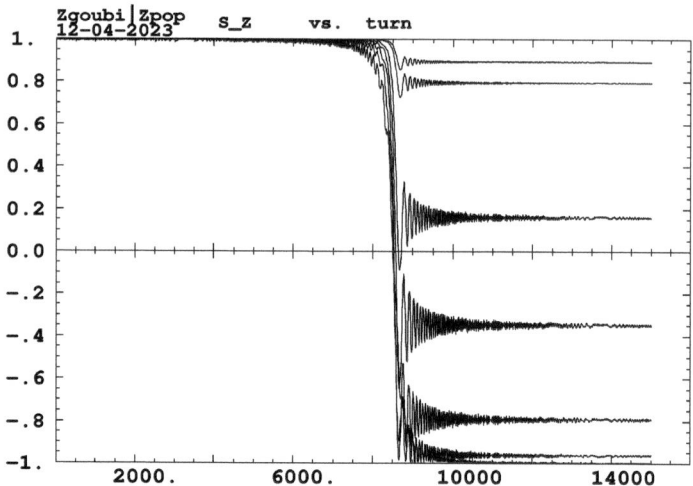

Fig. 8.38 Turn-by-turn spin motion through the isolated resonance $G\gamma_R = 4 - \nu_Z$, for 7 different values of the particle invariant from (top to bottom) $1\,\mu$m to $200\,\mu$m where full spin flip occurs. A graph obtained using `zpop`: menu 7; 1/8 to open b_polarLand.fai; 2/[39,23] for S_Z versus turn; 7 to plot

Table 8.13 Relationship between the invariant value ε_Z/π and the quantity $\ln\left(\frac{2}{1+S_{Z,f}/S_{Z,i}}\right) \propto$ $|\epsilon_R|^2$ (Eq. 8.58). $\hat{V} = 20\,$kV, here, crossing speed $\alpha = 1.696 \times 10^{-6}$ (Eq. 8.59). $S_{Z,i} = 1$ always, and $S_{Z,f}$ (col. 2) is a rough estimate from Fig. 8.38. The rightmost column gives the resulting ratio $|\epsilon_R|^2/\varepsilon_Z/\pi$, essentially constant

ε_Z/π (μm)	$\frac{S_{Z,f}}{S_{Z,i}} \equiv S_{Z,f}$	$\ln\frac{2}{1+S_{Z,f}}$	$\dfrac{|\epsilon_R|^2}{\varepsilon_Z/\pi}$ $(\times 10^{-8})$
1	0.89	0.024568	2.652645
2	0.795	0.046965	2.535451
10	0.17	0.232844	2.514034
20	−0.35	0.488116	2.635115
40	−0.78	0.958607	2.587537
80	−0.975	1.903089	2.568474

$$\alpha = \frac{1}{2\pi}\frac{\Delta E}{M} = \frac{1}{2\pi}\frac{20 \times 10^3 \times \sin 30° \ [\text{eV/turn}]}{938.27208 \times 10^6 [\text{eV}]} = 1.696 \times 10^{-6} \qquad (8.59)$$

Table 8.13, rightmost column, displays the ratio $|\epsilon_R|^2/\varepsilon_Z/\pi$ so obtained, essentially constant as expected.

Table 8.14 Relationship between the acceleration rate $\Delta E \propto \hat{V}$ and the quantity $\ln\left(\frac{2}{1+S_{Z,f}/S_{Z,i}}\right)$. Normalized to ε_Z/π, their product (rightmost column) appears to be essentially constant, as expected

ε_Z/π (μm)	$\hat{V}$ (kV)	$\frac{S_{Z,f}}{S_{Z,i}} \equiv S_{Z,f}$	$\ln\frac{2}{1+S_{Z,f}}$	$\frac{\hat{V}}{\varepsilon_Z/\pi} \times \ln\frac{2}{1+S_{Z,f}}$
1	10	+0.79	0.048	0.482
10	10	−0.33	0.475	0.475
20	10	−0.78	0.959	0.479
1	20	+0.89	0.025	0.49
2	20	+0.795	0.047	0.47

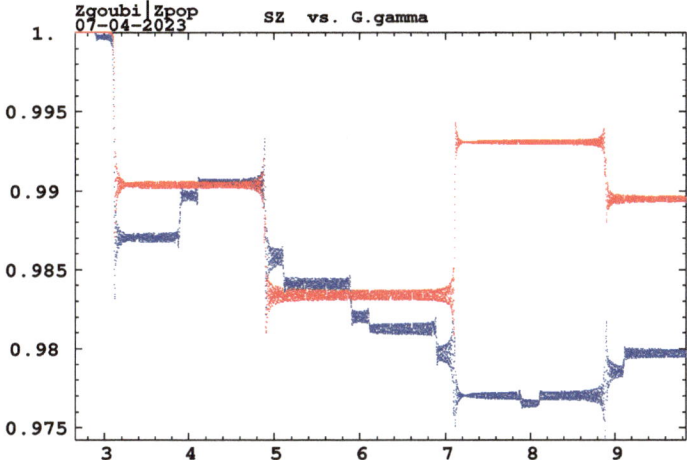

Fig. 8.39 Resonance crossing in SATURNE 1, a turn-by-turn record of $S_Z(G\gamma)$. Case of systematic resonances $G\gamma = 4k \pm \nu_Z$ in a 4-period lattice (red), and of random resonances $G\gamma = k \pm \nu_Z$ in a 1-periodic perturbed optics lattice (blue). A graph obtained using zpop: menu 7; 1/8 to open b_polarLand.fai; 2/[59,23] for S_Z versus $G\gamma$; 7 to plot

Changing the crossing speed

The crossing speed is reduced by a factor of 2, using $\hat{V} = 10\,\text{kV}$, and accordingly the number of turns is doubled, to 30,000, the only modifications to the input data simulation file used in the previous question. Tracking results, Table 8.14, show that $\frac{\hat{V}}{\varepsilon_Z/\pi} \times \ln\left(\frac{2}{1+S_{Z,f}/S_{Z,i}}\right)$ is constant, as expected.

Systematic resonances, random resonances

A single-particle tracking is pushed beyond $G\gamma = 8 + \nu_Z \approx 8.89$, 40,000 turns at a rate of $100\,\text{kV/turn}$. The resulting $S_Z(G\gamma)$, Fig. 8.39, shows that in a 4-periodic lattice the sole systematic resonances are excited, whereas all resonances are excited if the 4-periodicity is broken—here, by changing the index to $n = -0.66$ in one DIPOLE, the periodicity is 1.

(g) Spin motion across a weak depolarizing resonance.

The goal is to check numerical outcomes against the Fresnel integral model (Eq. 8.53). A weak resonance is obtained using small amplitude vertical motion and fast crossing.

A single particle is raytraced, in the following conditions:

– resonance to be crossed: $G\gamma_R = 4 - \nu_y \approx 3.1115$,

– acceleration: peak voltage $\hat{V} = 100\,\text{kV}$, synchronous phase $\phi_s = 30°$,

– particle invariant $\varepsilon_Z/\pi = 10^{-6}\,\text{m}$.

The initial rigidity is taken a few hundred turns upstream of the resonance, namely, $B\rho_{\text{ref}} = 4.0880774\,\text{T m}$, $605.226550\,\text{MeV}$, $G\gamma = 2.94931241$, a distance to $G\gamma_R$ of $4 - \nu_Z - 2.949312415 \approx 0.16223$. Tracking extends a few thousand turns beyond $G\gamma_R$ so that S_Z reaches its asymptotic value, from which the resonance strength $|\epsilon_R|$ can be calculated, using Eq. 8.58.

The simulation file is given in Table 8.15. Note the new setting of the SCALING factor SCL: DIPOLE field was set for a curvature radius $\rho_0 = 8.4193\,\text{m}$, given a reference rigidity $B\rho_{\text{ref}} \equiv BORO = 0.274426548\,\text{Tm}$ (Table 8.3). However the reference rigidity is now changed to $B\rho_{\text{ref}} = 4.0880774\,\text{Tm}$, thus maintaining ρ_0 requires scaling the field in DIPOLE by $4.0880774/0.274426548 = 14.8968$ at turn 1: this is the new factor, $SCL = 14.8968$, under SCALING (Table 8.15). Option NT$=-1$ under SCALING ensures that the scaling factor will automatically follow, turn-by-turn, the rigidity boost by CAVITE so preserving constant curvature radius $\rho_0 = 8.4193\,\text{m}$.

The resulting turn-by-turn spin motion is displayed in Fig. 8.40. The Fresnel integral model (Eq. 8.53) has been superimposed. Parameters in the latter are as follows:

– crossing speed $\alpha = \dfrac{1}{2\pi}\dfrac{\Delta E}{M} = \dfrac{1}{2\pi}\dfrac{10^5 \times \sin 30°\,[\text{eV/turn}]}{938.27208 \times 10^6[\text{eV}]} = 8.4812 \times 10^{-6}$,

– asymptotic $S_{Z,f} = 0.999780$, whereas initial $S_{Z,i} = 1$, thus (Eq. 8.58)

$$|\epsilon_R|^2 = 5.939 \times 10^{-10}$$

– orbital angle origin set at the location of $G\gamma_R$, which is turn 1699.

(h) Stationary spin motion near a resonance.

The simulation input data file of Table 8.15 can be used for these fixed energy trials, with some changes, as follows:

– OBJET[KOBJ=1] is used as it allows to define a set of particles with sampled momentum offset, namely:

```
'OBJET'
4.4393621786553803e3        ! BORO taken as close to resonant G.gamma as prior knowledge of nu_Z allows.
1                                                  ! Create a set of particles.
1  1  1  1  1    41                                          ! 41 particles sampling a
0. 0. 0. 0. 0.  .00001              ! momentum offset, in -20*1e-5< D-1 < 20*1e-5.
0. 0. 3. 0. 0.  1.                              ! All particles have initial Z=3cm.
```

Table 8.15 Simulation input data file: track a particle launched on a vertical invariant $\varepsilon_y/\pi = 10^{-6}$ m, with horizontal motion indifferent, taken zero here. The INCLUDE adds the SATURNE 1 cell four times, the latter is defined in Table 8.5

```
SATURNE 1 ring. Crossing Ggamma=4-nu_Z, weak resonance case (small vertical invariant)
'MARKER'   SatIWeakXing_S                          ! Just for edition purposes.
'OBJET'
4.08807740024e3                                    ! Reference Brho: 605226550 MeV proton.
8                                       ! Create a (set of) particle(s) on a given invariant.
1  1  1                                            ! create a single particle.
0. 0. 0. 0. 0. 1.
0.   14.426 0                                      ! Horizontal invariant is null.
0. 11.411  1e-6                     ! Periodic alpha_Z, beta_Z, and invariant value.
0. 1. 0.                                           ! No momentum spread.
'PARTICUL'
PROTON             ! Necessary data in order to allow (i) spin trackingand, and (ii) acceleration.
'SPNTRK'                                           ! Switch on spin tracking,
3                                                  ! nitial spin vertical.
'FAISCEAU'
'FAISTORE'
xing4-Qy.fai                             ! Log particle data in xing.fai, turn-by-turn.
1

'SCALING'
1 1
DIPOLE
-1                      ! Causes field increase in DIPOLE to follow rigidity increase by CAVITE.
14.8968                                            ! Relative rigidities at turn 1.
1

! 4 cells follow.
'INCLUDE'
1
4* ./SatI_cell.inc[S_SatI_cell:E_SatI_cell]

'CAVITE'
3
0 0
200e3  0.523598775598                              ! Acceleration rate is 200*0.5=100keV/turn.

'REBELOTE'
3999 0.3 99                                        ! A total of 3999+1=4000 turns.

'FAISCEAU'
'MARKER'   SatIWeakXing_E                          ! Just for edition purposes.
'SPNPRT'

'END'
```

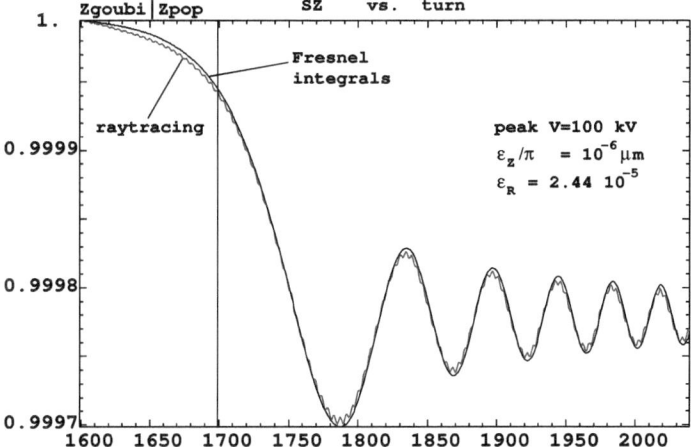

Fig. 8.40 Turn-by-turn spin motion through the isolated resonance $G\gamma_R = 4 - \nu_Z$, case of weak resonance strength. Modulated curve (blue): from raytracing. Smooth curve (black): Fresnel integral model

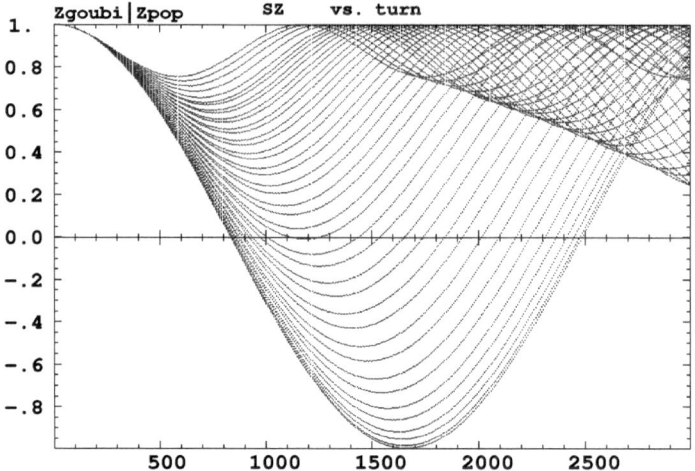

Fig. 8.41 Turn-by-turn value of the vertical component of spins precessing at fixed energy in SATURNE 1, observed at the beginning of the sequence, where spins start vertical ($S_Z = 1$). The greater (respectively smaller) the distance to the resonance, the closer the precession axis is to the vertical (resp., to the bend plane), and the greater (resp. smaller) the oscillation frequency $\omega = \sqrt{\Delta^2 + |\epsilon_R|^2}$

– with BORO changed, closer to $G\gamma_R = 4 - \nu_y \approx 3.1115$, DIPOLE field needs to be set to 5.27284 kG,

– a number of turns REBELOTE[IPASS≈a few thousand] results in at least half an oscillation of $S_Z(turn)$ (the precession frequency increases with the distance to the resonance, with a minimum of $\omega = |\epsilon_R|$ on the resonance [43, Fig. 3.4]), which is convenient for determining $\langle S_Z \rangle$.

Figure 8.41 displays the turn-by-turn evolution of the vertical component of the spins as they precess around the eigenvector **n** [43, Sect. 3.6.2, Fig. 3.3].

A quick, and accurate enough, approximation to the vertical component of the precession axis is $\langle S_Z \rangle|_{\text{period}} = \frac{1}{2}\{min\,[S_Z(\theta)] + max\,[S_Z(\theta)]\}$, it yields the $\langle S_Z \rangle\,(\Delta)$ graph of Fig. 8.42.

A match of the $\langle S_Z \rangle$ values by (Eq. 8.49)

$$|S_Z(\Delta)| = \frac{|\Delta|}{\sqrt{\Delta^2 + |\epsilon_R|^2}}$$

given $G\gamma_R = 4 - \nu_Z$, yield vertical tune and resonance strength values, respectively,

$$\nu_Z = 0.88845 \quad \text{and} \quad |\epsilon_R| = 2.77 \times 10^{-4}$$

The vertical tune ν_Z is fairly consistent with earlier results, and $|\epsilon_R| = 2.77 \times 10^{-4}$ for $\varepsilon_Z/\pi = 79 \times 10^{-6}$ m also is with $|\epsilon_R| = 2.44 \times 10^{-5}$ for $\varepsilon_Z/\pi = 10^{-6}$ m in the previous question (h). The difference deserves further inspection, this is left to the reader.

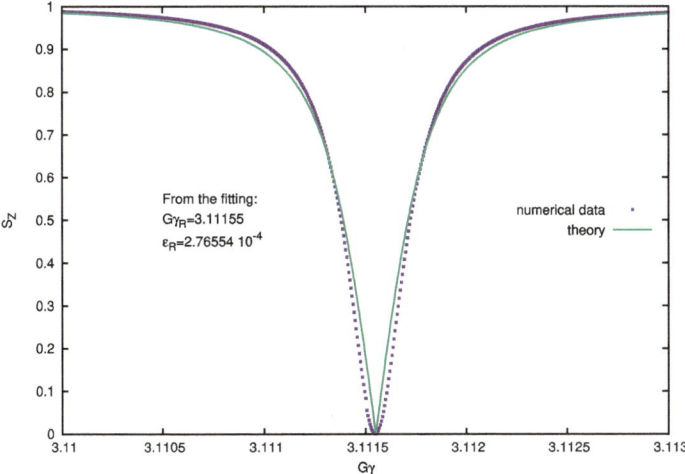

Fig. 8.42 Vertical component (absolute value) of the spin precession axis as a function of $G\gamma$, in the vicinity of the resonance. Markers are from tracking (656 particles), solid curve and numerical values of $G\gamma_R$ and v_Z are from a fitting using Eq. 8.49

(i) Bunch depolarization.
Spin depolarizing resonances in SATURNE 1 are located at (Figs. 8.43 and 8.44)

$$G\gamma_R = k \pm v_Z = k \pm 0.888284 \quad \equiv 4 - 0.888284, \ 4 + 0.888284, \ 8 - 0.888284$$

where v_Z has been taken from Table 8.8, or from Fig. 8.37. $G\gamma_R$ is bounded by $G\gamma(3\,\text{GeV}) = 7.525238 < 8 + v_Z$

The simulation data file to track through these resonances is the same as in question (e), Table 8.12, except for the following:
  – substitute MCOBJET (to be uncommented) to OBJET (to be commented),
  – under CAVITE substitute a peak voltage $V = 20\,\text{kV}$ to $V = 200\,\text{kV}$,
  – under REBELOTE, request a 300,000 turn cycle rather than 30,000.
MCOBJET creates a 200 particle bunch with Gaussian transverse and longitudinal densities, with the following *rms* values at 3.6 MeV:

$$\varepsilon_Y/\pi = 25\,\mu\text{m}, \quad \varepsilon_Z/\pi = 10\,\mu\text{m}, \quad \frac{dp}{p} = 10^{-4}$$

CAVITE accelerates that bunch from 3.6 MeV to 3 GeV at a rate of $q\hat{V}\sin(\phi_s) = 10\,\text{keV/turn}$ ($\hat{V} = 20\,kV$, $\phi_s = 150°$), in 300,000 turns.

Figure 8.43 shows sample S_Z spin components of a few particles taken among the 200 tracked. Figure 8.44 displays $\langle S_Z \rangle$, the average polarization of the bunch (gnuplot script given in Table 8.16).

The strength of any one of the three resonances crossed can be computed, from the upstream and downstream bunch polarization averaged over the 200 particles,

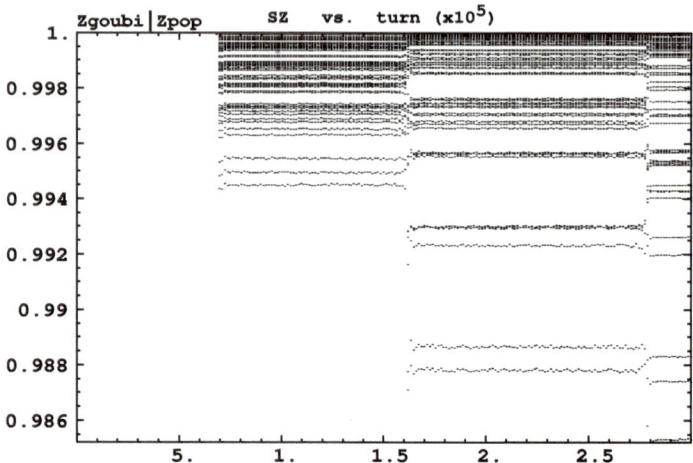

Fig. 8.43 Vertical spin component of a few particles accelerated from 3.6 MeV to 3 GeV. A graph obtained using `zpop`: menu 7; 1/2 to open b_zgoubi.fai; 2/[39,23] for S_Z versus turn; 7 to plot

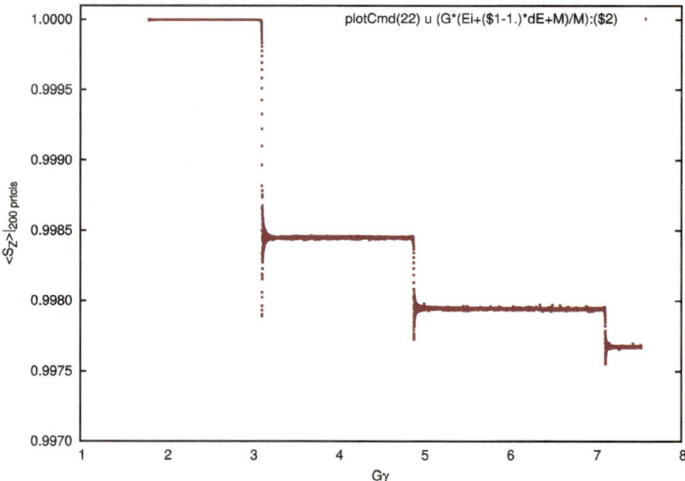

Fig. 8.44 Average vertical spin component of a 200 particle bunch, accelerated from 3.6 MeV to 3 GeV

using Eq. 8.58. Dependence upon the vertical emittance of the bunch can be performed repeating this tracking simulation, with a different vertical emittance (under MCOBJET).

Checking dependence upon crossing speed of the depolarizing effect of the resonances can be performed by repeating this tracking simulation with a different accelerating rate $\hat{V}\,\sin(\phi_s)$.

Table 8.16 A gnuplot script to plot the average vertical spin component of the 200 particle set, along the acceleration ramp (Fig. 8.44). The average is prior computed by an awk script, which reads the necessary data from zgoubi.fai.

```
# gnuplot_avrgFromFai.gnu
fName = 'zgoubi.fai'; plotCmd(col_num)=sprintf('< gawk -f average.awk -v col_num=%d %s', col_num, fName)
set xtics; set ytics; set xlabel "G{/Symbol g}"; set ylabel "<S_Z>|_{200 prtcls}"
set format y '%0.4f'; set grid; set xr [:]; set yr [.997:1.0001]
Qy=0.888248;
do for [intgr=1:2] {  set arrow nohead from 4*intgr-Qy, 0.997 to 4*intgr-Qy, 1.0001 lw 1 dt 2
                      set arrow nohead from 4*intgr+Qy, 0.997 to 4*intgr+Qy, 1.0001 lw 1 dt 2  }
M=938.27208; Ei = 3.6; G = 1.79284735; Qy = 0.888284; dE = 0.01  # MeV/turn
plot  plotCmd(22) u (G*(Ei+($1-1.)*dE+M)/M):($2) w p pt 5 ps .4 lc rgb 'dark-red'; pause 1
```

average.awk script to compute $\langle S_Z \rangle$ [42]:

```
function average(x, data){
    n = 0;mean = 0;
    val_min = 0;val_max = 0;
    for(val in data){
        n += 1;
        delta = val - mean;
        mean += delta/n;
        val_min = (n == 1)?val:((val < val_min)?val:val_min);
        val_max = (n == 1)?val:((val > val_max)?val:val_max);
    }
    if(n > 0){
        print x, mean, val_min, val_max;
    }}
{    curr = $38;
    yval = $(col_num);

    if(NR==1 || prev != curr){
        average(prev, data);
        delete data;
        prev = curr;    }
    data[yval] = 1;   }
END{
    average(curr, data);   }
```

8.2 Construct the ZGS Synchrotron. Spin Resonances

(a) A model of ZGS synchrotron.

DIPOLE is used to simulate both cell dipoles. It is necessary to have Fig. 8.28 at hand (in addition to the users' guide), when filling up the data list under DIPOLE. Some comments regarding these data:

- DIPOLE field is defined in a cylindrical coordinate system.
- The bending sector is $45°$, this is also the field region extent angle AT in the preliminary hard-edge model.
- When accounting for fringe fields, the angular extent AT has to encompass the fringe fields, at both ends of the $45°$ sector: an extra $5°$ takes care of that, for a total $AT = 55°$, which ensures absence of truncation of the fringe fields at the AT sector boundaries, over the all radial excursion of the beam.
- RM is given the curvature radius value, $RM = B\rho/B = 1.035270_{[T\,m]}/0.04986851_{[T]} = 20.76\,\text{m}$, this makes magnet positioning and closed orbit checks easier (see below).
- The field and reference rigidity are for injection energy, $50\,\text{MeV}$, an arbitrary choice.
- ACENT$=27.5°$ is the reference azimuth for the positioning of the entrance and exit EFBs. It is taken in the middle of the AT range, an arbitrary choice.

- The entrance radius in the AT sector is $RE = RM/\cos(AT - \omega^+) = RM/\cos(5°)$, with $\omega^+ = 22.5°$ the positioning of the entrance EFB with respect to ACENT (Fig. 8.28). And similarly for the positioning of the exit reference frame, $RS = RM/\cos(AT - (ACENT - \omega^-)) = RM/\cos(5°)$ with $\omega^- = -22.5°$ the positioning of the exit EFB. Note that $\omega^+ - \omega^- = 45°$, the value of the bend angle.
- The entrance angle TE identifies with the extension to the 45° sector, namely, TE=-5°. And similarly for the positioning of exit frame, 5° downstream of the exit EFB, TS=5°.

In order to build the cell, and in the first place the two cell dipoles (they are mirror symmetric, thus build one, the other follows), it is a good idea to proceed by steps:

(i) first build a 45° sector in the hard edge model (Table 8.17). Outcomes of FAISCEAU located next to DIPOLE indicate that a trajectory entering DIPOLE at radius $R = RM$, normal to the EFB (thus, $Y_0 = 0$ and $T_0 = 0$ in OBJET), does exit with $Y = 0$ and $T = 0$. Data validation at this stage can be performed by comparing DIPOLE's transport matrix computed with MATRIX, and the theoretical expectation (after Eq. 14.6):

$$
T = \begin{pmatrix}
\cos\alpha & \rho\sin\alpha & 0 & 0 & 0 & \rho(1-\cos\alpha) \\
-\frac{1}{\rho}\sin\alpha & \cos\alpha & 0 & 0 & 0 & \sin\alpha \\
0 & 0 & 1 & \rho\alpha & 0 & 0 \\
0 & 0 & 0 & 1 & 0 & 0 \\
\sin\alpha & 0 & 0 & 0 & 1 & \rho(\alpha-\sin\alpha) \\
0 & 0 & 0 & 0 & 0 & 1
\end{pmatrix}
\underset{\substack{\alpha=\pi/4,\\ \rho=20.76}}{=}
\begin{pmatrix}
0.7071 & 14.6795 & 0 & 0 & 0 & 6.0804 \\
-0.03406 & 0.7071 & 0 & 0 & 0 & 0.7071 \\
0 & 0 & 1 & 16.3048 & 0 & 0 \\
0 & 0 & 0 & 1 & 0 & 0 \\
0.7071 & 0 & 0 & 0 & 1 & 1.6253 \\
0 & 0 & 0 & 0 & 0 & 1
\end{pmatrix}
$$

Table 8.17 Simulation input data file: a 45° sector bend in the hard edge model. The reference trajectory has equal entrance and exit positions, and opposite sign angles. It coincides with the arc $R = RM$. MATRIX computes the transport matrix of the dipole (bottom of this Table), for comparison with the fringe field model

```
ZGS. Hard edge dipole model. Transport matrix.
'OBJET'
1.03527036749193e3                                      ! Reference Brho: 50 MeV proton.
5                                       ! Create a 13 particle set, proper for MATRIX computation.
.001 .01 .001 .01 .001 .0001                                      ! Coordinate sampling.
0. 0. 0. 0. 0. 1.        ! Reference trajectory: all initial coordinates nul, relative rigidity D=1.
1
'DIPOLE'                                            ! Analytical modeling of a dipole magnet.
20                    ! IL=2 here, to log trajectory coordinates in zgoubi.plt, at integration steps.
45. 2076.                   ! Field region angle=45; reference radius set to curvature radius value.
22.5 0.4986851481175 0. 0. 0.  ! Reference angle ACENT set to AT/2; Bo field at RM; indices, all zero.
.0 0.                                                          ! EFB 1, hard-edged.
4  .1455    2.2670 `-.6395  1.1558  0. 0.  0.                ! Enge coefficients.
22.5 0.  1.E6  -1.E6  1.E6  1.E6             ! Angle to ACENT; face angle; face is straight.
.0 0.                                                          ! EFB 2, hard-edged.
4  .1455    2.2670  -.6395  1.1558  0. 0.  0.
-22.5 0.  1.E6  -1.E6  1.E6  1.E6
0. 0.                                                        ! EFB 3. Unused.
0  0.       0.        0.        0.       0. 0.  0.
0. 0.  1.E6  -1.E6  1.E6  1.E6 0.
2 1                             ! Degree of interpolation polynomial; flying grid sizing.
200.                            ! Integration step size. It can be large in uniform field.
2 2076. 0. 2076. 0.                             ! Positioning of entrance and exit frames.
! reference frames.
'FAISCEAU'                                             ! Local particle coordinates.
'MATRIX'                                    ! Compute transport matrix, from trajectory coordinates.
1 0
'END'
```

Table 8.18 Outcomes of the simulation file of Table 8.17

An excerpt from zgoubi.res execution listing. Coordinates of the first particle (the reference trajectory) and its path length under FAISCEAU, at OBJET on the left hand side below, locally on the right hand side:

```
3  Keyword, label(s) :  FAISCEAU
                                   TRACE DU FAISCEAU
                                   (follows element #      2)
                                        13 TRAJECTOIRES
                          OBJET                                                FAISCEAU
        D     Y(cm)    T(mr)   Z(cm)    P(mr)    S(cm)    D-1    Y(cm)    T(mr)   Z(cm)    P(mr)    S(cm)
  0  1  1.0000   0.000    0.000   0.000    0.000    0.0000   0.0000  -0.000   -0.000   0.000    0.000   1.630487E+03
```

Transport matrix of a 45 degree sector, hard edge model, two difference cases of integration step size, namely, 4 cm and 2 m (an excerpt of MATRIX computation, from zgoubi.res execution listing). It can be checked against expectations, Eq. 8.60. The reference trajectory (first one) exits better aligned (reference coordinates, before change of frame for MATRIX computation, are closer to zero):

- Case of 4 cm step size:

```
    4  Keyword, label(s) :  MATRIX

  Reference, before change of frame (particle #  1  - D-1,Y,T,Z,s,time)  :
  0.00000000E+00  -3.25144356E-10  -4.13789229E-10   0.00000000E+00   0.00000000E+00   1.63048659E+03   5.43871783E-02

              TRANSFER  MATRIX  ORDRE  1  (MKSA units)
     0.707107      14.6795       0.00000       0.00000       0.00000       6.08046
    -3.406102E-02   0.707107     0.00000       0.00000       0.00000       0.707107
     0.00000       0.00000       1.00000      16.3049        0.00000       0.00000
     0.00000       0.00000       7.285552E-16  1.00000       0.00000       0.00000
     0.707107      6.08046       0.00000       0.00000       1.00000       1.62533
     0.00000       0.00000       0.00000       0.00000       0.00000       1.00000

     DetY-1 =     0.0000000025,   DetZ-1 =     0.0000000002
  First order symplectic conditions (expected values = 0) :
     2.5100E-09   2.3381E-10     0.000        0.000        0.000        0.000
```

- Case of 2 m step size:

```
    4  Keyword, label(s) :  MATRIX

  Reference, before change of frame (particle #  1  - D-1,Y,T,Z,s,time)  :
  0.00000000E+00  -2.01277929E-03  -2.51514609E-03   0.00000000E+00   0.00000000E+00   1.63048722E+03   5.43871994E-02

              TRANSFER  MATRIX  ORDRE  1  (MKSA units)
     0.707105      14.6795       0.00000       0.00000       0.00000       6.08056
    -3.406102E-02   0.707108     0.00000       0.00000       0.00000       0.707120
     0.00000       0.00000       1.00000      16.3051        0.00000       0.00000
     0.00000       0.00000       1.457135E-17  1.00003       0.00000       0.00000
     0.707109      6.08048       0.00000       0.00000       1.00000       1.62531
     0.00000       0.00000       0.00000       0.00000       0.00000       1.00000

     DetY-1 =    -0.0000010903,   DetZ-1 =     0.0000286273
     R12=0 at    -20.76    m,     R34=0 at   -16.30    m
  First order symplectic conditions (expected values = 0) :
    -1.0903E-06   2.8627E-05     0.000        0.000        0.000        0.000
```

MATRIX computation outcomes from raytracing can be found for comparison in Table 8.18.

(ii) next, add fringe fields, including the $5°$ extensions that add to AT (Table 8.19). Negative drifts with length $RM \tan(5°) = 181.62646548$ cm have been added at both ends, so to recover the actual $45°$ sector opening. A FIT procedure finds the field value necessary for recovering the exact orbit deviation, as the latter is perturbed when introducing fringe fields. Again, FAISCEAU allows checking the correctness of DIPOLE data: exit coordinates come out to be $Y = 0$ and $T = 0$; however the path across the dipole is changed under the effect of the fringe fields, thus its length: s =

Table 8.19 Simulation input data file: ZGS 45° sector bend, with entrance and exit EFBs wedge angles and fringe fields. The reference trajectory has equal entrance and exit position, and opposite sign angles. It runs closely to the arc $R = RM$, not strictly coinciding with the latter due to the fringe fields. MATRIX computes the transport matrix of the dipole, for comparison with the hard edge model. Negative drifts with length $RM \tan(5°) = 181.62646548$ cm are added to recover the hard edge path length

```
ZGS. Simplfied model. Find centered orbit in DIPOLE.
'OBJET'
1.03527036749193e3                                      ! Reference Brho: 50 MeV proton.
5                                    ! Create a 13 particle set, proper for MATRIX computation.
.001 .01 .001 .01 .001 .0001                                       ! Coordinate sampling.
0. 0. 0. 0. 0. 1.           ! Reference trajectory: all initial coordinates nul, relative rigidity D=1.
1
'DRIFT'
-181.62646548
'DIPOLE'                                                ! Analytical modeling of a dipole magnet.
0                     ! IL=2 here, to log trajectory coordinates in zgoubi.plt, at integration steps.
55. 2076.                            ! Field region angle=45; reference radius set to curvature radius value.
27.5 0.49860858 0. 0. 0.        ! Reference angle ACENT set to AT/2; Bo field at RM; indices, all zero.
60.  0.                                                 ! EFB 1 with fringe field extent.
4  .1455    2.2670 -.6395  1.1558  0. 0.  0.            ! Enge coefficients.
22.5 13.   1.E6 -1.E6  1.E6  1.E6    ! EFB angle to ACENT; 13 deg EFB tilt angle; EFB is straight.
60.  0.                                                 ! EFB 2 with fringe field extent.
4  .1455    2.2670 -.6395  1.1558  0. 0.  0.
-22.5 -8.  1.E6 -1.E6  1.E6  1.E6    ! EFB angle to ACENT; -8 deg EFB tilt angle; EFB is straight.
0. 0.                                                   ! EFB 3. Unused.
0 0.      0.      0.      0. 0. 0.
0. 0.  1.E6 -1.E6  1.E6  1.E6 0.
2 1                          ! Degree of interpolation polynomial; flying grid sizing, proper for accuracy.
4.0                                                     ! Integration step size.
2 2084.5090 -0.087266462599717 2084.5090 0.087266462599717      ! Positioning of entrance and exit.
'DRIFT'
-181.62646548
'FIT'
2
3 5 0 .1                                                ! Vary DIPOLE field.
3 64 3.66  .1
2  1e-15  999
3   1 2 #End  0. 1. 0                     ! Request nul trajcory position  at exit of DIPOLE.
3   1 3 #End  0. 1. 0                     ! Request nul trajcory angle  at exit of DIPOLE.

'FAISCEAU'                                              ! Local particle coordinates.
'MATRIX'                                  ! Compute transport matrix, from trajectory coordinates.
1 0
'END'
```

Transport matrix of ZGS 45 degree sector with EFB wedge angles and fringe fields (an excerpt of MATRIX computation, from zgoubi.res execution listing). It can be checked against matrix transport expectations. The "first order symplectic conditions" are small, which is an indication of accurate numerical integration of the trajectories across DIPOLE:

```
      7  Keyword, label(s) :  MATRIX

   Reference, before change of frame (particle #  1  - D-1,Y,T,Z,s,time) :
     0.00000000E+00  -2.19331903E-08  -2.24434360E-08   0.00000000E+00   0.00000000E+00   1.63080750E+03   6.65146963E-02

               TRANSFER  MATRIX  ORDRE  1  (MKSA units)
        0.870365        14.6806         0.00000         0.00000         0.00000         6.08068
      -2.030224E-02     0.806503        0.00000         0.00000         0.00000         0.748209
        0.00000         0.00000         0.827040        16.3143         0.00000         0.00000
        0.00000         0.00000        -1.580329E-02    0.897394        0.00000         0.00000
        0.774666        6.08004         0.00000         0.00000         1.00000         1.63006
        0.00000         0.00000         0.00000         0.00000         0.00000         1.00000

        DetY-1 =      -0.0000003451,    DetZ-1 =      0.0000000379
   First order symplectic conditions (expected values = 0) :
        -3.4507E-07     3.7861E-08     0.000          0.000         0.000          0.000
```

1630.459 cm is slightly different, compared to the hard edge case (an arc of radius radius $RM = 2076$ cm and length 1630.487 cm)

(iii) next, add the EFB angles: the sector is closing (wedge angles $\varepsilon_1 > 0$ and $\varepsilon_2 > 0$ by convention) thus the EFB tilt angle θ under DIPOLE is positive at entrance,

negative at exit (Fig. 8.28). In order to reach proper wave number values (this is addressed below), the wedge angles are taken to be $\varepsilon_1 = 13°$ and $\varepsilon_2 = 8°$. These considerations result in the following:
– the entrance (respectively exit) EFB of the upstream dipole of the cell is tilted with respect to the reference orbit by an angle $\theta = +13°$ (resp. $\theta = -8°$) (Fig. 8.27),
– the entrance (resp. exit) EFB of the downstream dipole is tilted with respect to the reference orbit by an angle $\theta = +8°$ (resp. $\theta = -13°$).
This final step requires again re-adjusting the radial positioning of the dipole (RE and RS, entrance and exit radius respectively), and field. In that aim the FIT procedure in Table 8.19 is added a variable: the RE and RS radii, coupled, and a constraint: the reference orbit has zero radial excursion at exit of the dipole. This FIT results in re-adjusted magnetic field and RE, RS positioning, with the respective values

$$B_0 = 0.49860858 \, kG \quad \text{and} \quad RE = RS = 2084.5090 \, cm$$

These are the values used in the ZGS cell simulation in Table 8.20,

(iv) and, finally, assemble this dipole and its mirror symmetric, in a cell (Fig. 8.27 and Table 8.20). The mirror symmetric is obtained by just permuting the entrance and exit wedge angles. The cell includes a half long-drift at each end, and a short drift between the dipoles. The three have been taken equal for simplification, 3.37 m long.

Lattice parameters

The TWISS command down the sequence (Table 8.20) produces the periodic beam matrix results shown in Table 8.21. It also produces a zgoubi.TWISS.out file which details the optical functions along the sequence (at the downstream end of the optical elements). The header of that file details the optical parameters of the structure (Table 8.22).

(b) Betatron functions of the ZGS cell.

Among the various ways to produce the betatron functions along the sequence (and throughout the DIPOLEs), here are two possibilities, based on the storage of particle coordinates in zgoubi.plt during stepwise raytracing:

1. a direct way consists in using OBJET[KOBJ=5] and transport the 13-particle set so obtained across the sequence; then, betaFromPlt from zgoubi toolbox [39] can be used to compute the transport matrix from the origin, step by step along the sequence, from particle coordinate values logged in zgoubi.plt during the stepwise integration;
2. an indirect way consists in launching a few particles on a common invariant (horizontal and/or vertical) and subsequently plot the s-dependent quantities $\hat{Y}^2(s)/\varepsilon_Y$ and/or $\hat{Z}^2(s)/\varepsilon_Z$. The maximum value of the latter, a function of the distance s, is the betatron function along the sequence, $\beta_{Y,Z}(s)$.

The second method is used here (this is an arbitrary choice. Exercises may be found in the various Chapters, that use the first method and may be referred to).

Table 8.20 Simulation input data file: ZGS cell simplified model, obtained by assembling DIPOLE taken from Table 8.19 and its mirror symmetric, and adding drift spaces. This input data file defines the ZGS cell sequence segment S_ZGS_cell to E_ZGS_cell. It also defines the dipole segments S_ZGS-DIP_UP to E_ZGS-DIP_UP (first dipole of the cell) and S_ZGS-DIP_DW to E_ZGS-DIP_DW (second dipole of the cell). In further INCLUDE statements, this file is used under the name ZGS_cell.inc

```
File ZGS_cell.inc
! A period of the 4-period ZGS.
'MARKER'    ZGSCellMATRIX_S                                     ! Just for edition purposes.
'OBJET'
1.03527036749193e3                                             ! Reference Brho: 50 MeV proton.
5                                             ! Create a 13 particle set, proper for MATRIX computation.
.001 .01 .001 .01 .001 .0001                                          ! Coordinate sampling.
0. 0. 0. 0. 0. 1.      ! Reference trajectory: all initial coordinates nul, relative rigidity D=1.
1

'MARKER'  S_ZGS_cell                                            ! ZGS cell begins here.
'DRIFT' half_longDrift
337.

'MARKER'  S_ZGS-DIP_UP                                          ! 1st dipole of cell begins here.
'DRIFT'
-181.62646548
'DIPOLE'  DIP_UP                                   ! Analytical modeling of a dipole magnet.
2                   ! IL=2 here, to log trajectory coordinates in zgoubi.plt, at integration steps.
55. 2076.                         ! Field region angle=45; reference radius set to curvature radius value.
27.5 0.49860858 0. 0. 0.        ! Reference angle ACENT set to AT/2; Bo field at RM; indices, all zero.
60. 0.                                               ! EFB 1 with fringe field extent.
4 .1455   2.2670  -.6395  1.1558  0. 0.  0.                           ! Enge coefficients.
22.5 13.  1.E6  -1.E6  1.E6  1.E6          ! EFB angle to ACENT; EFB tilt angle; EFB is straight.
60. 0.                                               ! EFB 2 with fringe field extent.
4 .1455   2.2670  -.6395  1.1558  0. 0.  0.
-22.5 -8.  1.E6  -1.E6  1.E6  1.E6          ! EFB angle to ACENT; EFB tilt angle; EFB is straight.
0. 0.                                               ! EFB 3. Unused.
0 0.       0.      0.       0.       0. 0.  0.
0. 0.  1.E6  -1.E6  1.E6  1.E6 0.
2 1                 ! Degree of interpolation polynomial; flying grid sizing is step, proper for accuracy.
2.0                                                  ! Integration step size.
2 2084.5090 -0.087266462599717 2084.5090 0.087266462599717     ! Positioning of entrance and exit.
'DRIFT'
-181.62646548
'MARKER' E_ZGS-DIP_UP                                           ! 1st dipole of cell ends here.

'DRIFT' shortDrift
337.

'MARKER' S_ZGS-DIP_DW                                          ! 2nd dipole of cell begins here.
'DRIFT'
-181.62646548
'DIPOLE'  DIP_DW                                   ! Analytical modeling of a dipole magnet.
2                   ! IL=2 here, to log trajectory coordinates in zgoubi.plt, at integration steps.
55. 2076.                         ! Field region angle=45; reference radius set to curvature radius value.
27.5 0.49860858 0. 0. 0.        ! Reference angle ACENT set to AT/2; Bo field at RM; indices, all zero.
60. 0.                                               ! EFB 1 with fringe field extent.
4 .1455   2.2670  -.6395  1.1558  0. 0.  0.                           ! Enge coefficients.
22.5 8.  1.E6  -1.E6  1.E6  1.E6          ! EFB angle to ACENT; EFB tilt angle; EFB is straight.
60. 0.                                               ! EFB 2 with fringe field extent.
4 .1455   2.2670  -.6395  1.1558  0. 0.  0.
-22.5 -13.  1.E6  -1.E6  1.E6  1.E6          ! EFB angle to ACENT; EFB tilt angle; EFB is straight.
0. 0.                                               ! EFB 3. Unused.
0 0.       0.      0.       0.       0. 0.  0.
0. 0.  1.E6  -1.E6  1.E6  1.E6 0.
2 1                 ! Degree of interpolation polynomial; flying grid sizing is step, proper for accuracy.
2.0                                                  ! Integration step size.
2 2084.5090 -0.087266462599717 2084.5090 0.087266462599717     ! Positioning of entrance and exit.
'DRIFT'
-181.62646548
'MARKER' E_ZGS-DIP_DW                                           ! 2nd dipole of cell ends here.

'DRIFT' half_longDrift
337.
'MARKER'  E_ZGS_cell                                            ! ZGS cell ends here.

'FAISCEAU'                                                      ! Local particle coordinates.
'TWISS'   ! Produce transport matrix, beam matrix, and periodic optical functions along the sequence.
2 1. 1.
'MARKER'    ZGSCellMATRIX_E                                     ! Just for edition purposes.
'END'
```

Table 8.21 Results obtained running the simulation input data file of Table 8.20, ZGS cell—an excerpt from zgoubi.res execution listing

```
13  Keyword, label(s) :  TWISS

Reference, before change of frame (particle #  1  - D-1,Y,T,Z,s,time)  :
 0.00000000E+00  -1.59240732E-05  -9.81570020E-07  0.00000000E+00  0.00000000E+00   4.27261430E+03   4.53811009E-01


   Beam  matrix  (beta/-alpha/-alpha/gamma) and  periodic  dispersion  (MKSA units)
      28.633680       0.000002       0.000000       0.000000       0.000000      36.853463
       0.000002       0.034924       0.000000       0.000000       0.000000       0.000003
       0.000000       0.000000      37.008846       0.000001       0.000000      -0.000000
       0.000000       0.000000       0.000001       0.027021       0.000000       0.000000
       0.000000       0.000000       0.000000       0.000000       0.000000       0.000000
       0.000000       0.000000       0.000000       0.000000       0.000000       0.000000

                    Betatron  tunes  (Q1 Q2 modes)
          NU_Y =  0.21235913        NU_Z =  0.19286706

                       Momentum compaction :
                    dL/L / dp/p =   1.4126935


                       Transition gamma  =  8.41348710E-01

                       Chromaticities :
          dNu_y / dp/p =  4.70986585E-02        dNu_z / dp/p =  4.45745634E-02
```

Table 8.22 An excerpt of zgoubi.TWISS.out file resulting from ZGS cell simulation of Table 8.20. Note that the ring (4-period) wave numbers are 4 times the cell values Q1, Q2 below. Optical functions (betatron function and derivative, orbit, phase advance, etc.) along the optical sequence are listed as part of zgoubi.TWISS.out following the header. The top and bottom parts of that listing are given below

```
@ LENGTH      %le   42.72614305
@ ALFA        %le   1.412693458
@ ORBIT5      %le                 -0
@ GAMMATR     %le   0.8413487096
@ Q1          %le   0.2123591260
@ Q2          %le   0.1928670550
@ DQ1         %le   0.4709865847E-01
@ DQ2         %le   0.4457456345E-01
@ DXMAX       %le   3.81566835E+01   @ DXMIN   %le   3.68534544E+01
@ DYMAX       %le   0.00000000E+00   @ DYMIN   %le   0.00000000E+00
@ XCOMAX      %le   3.68530296E-01   @ XCOMIN  %le  -1.59240732E-07
@ YCOMAX      %le   0.00000000E+00   @ YCOMIN  %le   0.00000000E+00
@ BETXMAX     %le   3.25272034E+01   @ BETXMIN %le   2.86307346E+01
@ BETYMAX     %le   3.73198843E+01   @ BETYMIN %le   3.50936471E+01
@ XCORMS      %le   8.67153286E-04
@ YCORMS      %le   0.
@ DXRMS       %le   6.22665688E-01
@ DYRMS       %le   0.00000000E+00
```

Top and bottom four lines (truncated) of zgoubi.TWISS.out optical functions listing, including the periodic β_x, β_y (β_Y, β_Z in zgoubi notations) and D_x (η_Y in zgoubi notations) values at cell ends:

```
# alfx         btx           alfy           bty          alfl        btl           Dx            Dxp
-2.2668087e-6  2.8636996e+1 -1.0203802e-6  3.7013045e+1 0.0000000e+0 0.0000000e+0  3.6856253e+1  1.2733594e-4  etc.
-2.2589191e-6  2.8636995e+1 -9.9937511e-7  3.7013042e+1 0.0000000e+0 0.0000000e+0  3.6853463e+1  2.6012859e-6  etc.
-2.2589191e-6  2.8636995e+1 -9.9937511e-7  3.7013042e+1 0.0000000e+0 0.0000000e+0  3.6853463e+1  2.6012859e-6  etc.
-1.1768220e-1  2.9033592e+1 -9.1049986e-2  3.7319884e+1 0.0000000e+0 0.0000000e+0  3.6853472e+1  2.6012859e-6  etc.
.............................
 1.1775697e-1  2.9027748e+1  9.1140989e-2  3.7313084e+1 0.0000000e+0 0.0000000e+0  3.6853454e+1  2.6012859e-6  etc.
 1.1775697e-1  2.9027748e+1  9.1140989e-2  3.7313084e+1 0.0000000e+0 0.0000000e+0  3.6853454e+1  2.6012859e-6  etc.
 5.1297527e-5  2.8630735e+1  7.3912348e-5  3.7005690e+1 0.0000000e+0 0.0000000e+0  3.6853463e+1  2.6012859e-6  etc.
 5.1297527e-5  2.8630735e+1  7.3912348e-5  3.7005690e+1 0.0000000e+0 0.0000000e+0  3.6853463e+1  2.6012859e-6  etc.
```

Table 8.23 Simulation input data file: raytrace 60 particles across ZGS cell to generate beam envelopes. Store particle data in zgoubi.plt, along DIPOLEs and split DRIFTs. The INCLUDE file and segments are defined in Table 8.20

```
ZGS envelopes.
'OBJET'
1.03527036749193e3                                              ! Reference Brho: 50 MeV proton.
8                                     ! Create a set of 60 particles evenly distributed on the same invariant;
1 60 1    ! case of 60 particles on a vertical invariant; use 60 1 1 instead for horizontal invariant.
0. 0. 0. 0. 0. 1.
0.  2.8637014E+1 0e-4
0.  3.7012633E+1 1e-4
0. 1. 0.

'FAISTORE'                                        ! This logs the coordinates of the particle to zgoubi.fai,
zgoubi.fai  S_ZGS_cell  E_ZGS_cell                              ! at the two LABEL1s as indicated.
1

'MARKER' S_ZGS_cell
'DRIFT' half_longDrift                                ! Option 'split' devides the drift in 10 pieces,
337. split 10  2                                     ! 'IL=2' causes log of particle data to zgoubi.plt.

'INCLUDE'
1
./ZGS_cell.inc[S_ZGS-DIP_UP:E_ZGS-DIP_UP]

'DRIFT' shortDrift
337.  split 10  2

'INCLUDE'
1
./ZGS_cell.inc[S_ZGS-DIP_DW:E_ZGS-DIP_DW]

'DRIFT' half_longDrift
337.  split 10  2
'MARKER' E_ZGS_cell

'FAISCEAU'
'END'
```

The input data file to derive the betatron function following method (2) above is given in Table 8.23. The initial ellipse parameters (under OBJET) are the periodic values, namely, $\alpha_Y = \alpha_Z = 0$, $\beta_Y = 28.63$ m, $\beta_Z = 37.01$ m, they are a subproduct of the TWISS procedure performed in (a), to be found in zgoubi.TWISS.out (Table 8.22). The resulting envelopes and their squared value are shown in Fig. 8.45. Note that this raytracing also provides the coordinates of the 60 particles on their common upright invariant

$$x^2/\beta_x + \beta_x x'^2 = \varepsilon_x/\pi$$

at start and at the end of the cell (with x standing for either Y or Z, and $\varepsilon_{Y,Z}/\pi = 10^{-4}$, here). This allows checking that the initial ellipse parameters (under OBJET, Table 8.23) are effectively periodic values, and that the raytracing went correctly, namely by observing that the initial and final ellipses do superimpose (Fig. 8.46).

Dispersion function

Raytracing off-momentum particles on their chromatic closed orbit provides the periodic dispersion function. In order to do so, the input data file of Table 8.23 can be used, it just requires changing OBJET to the following:

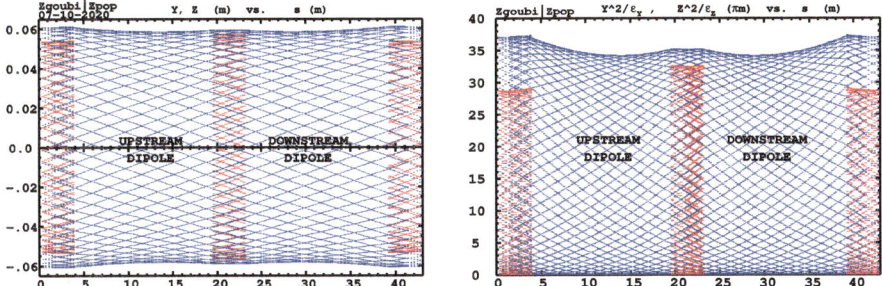

Fig. 8.45 Left: horizontal and vertical envelopes as generated by plotting the coordinates Y(s) (thick lines, red, along the drifts only) or Z(s) (thin lines, blue) across the ZGS cell, of 60 particles evenly distributed on a common $10^{-4}\,\pi$ m invariant, either horizontal or vertical (while the other invariant is zero). Right: a plot of $Y^2(s)/\varepsilon_Y$ and $Z^2(s)/\varepsilon_Z$: the extrema identify with $\beta_Y(s)$ and $\beta_Z(s)$, respectively. The extrema extremorum values are $\hat{\beta}_Y = 32.5$ m and $\hat{\beta}_Z = 37.1$ m, respectively. Graphs obtained using zpop: menu 7; 1/1 to open zgoubi.plt; 2/[6,2] (or [6,4]) for Y versus s (or Z versus s); 7 to plot; option 3/14 to raise Y (or Z) to the square and normalize to $\varepsilon_{Y,Z}/\pi$

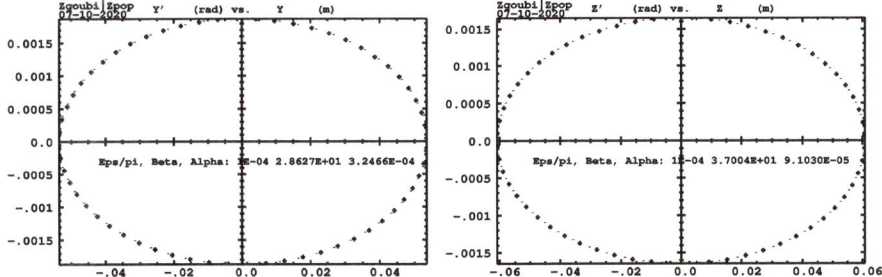

Fig. 8.46 Sixty particles evenly distributed on a common periodic invariant (of value either $\varepsilon_Y = 10^{-4}\pi$ m and $\varepsilon_Z = 0$, left graph, or the reverse, right graph) have been tracked from start to end of the cell. These periodic invariants are defined assuming the periodic ellipse parameters determined from prior TWISS, given in Table 8.22; values resulting from an *rms* match of the coordinates are given in the figure, and do agree with those TWISS data. The figure shows the good superposition of the start and end invariants (the start and end *rms* match ellipse parameters show negligible difference), which confirms the correct value of the periodic ellipse parameters, namely, left graph: horizontal phase space at start (crosses) and end (dots) of the cell; right graph: vertical phase space at start (crosses) and end (dots) of the cell

```
'OBJET'
1.03527036749193e3                          ! Reference Brho: 50 MeV proton.
2                                           ! Create particles individually'
3 1                                         ! three particles.
+36.85e-1 0. 0. 0. 0. 1.001 'p' ! Chromatic orbit coordinates Y0 and T0 for D=1.001 relative rigidity.
0. 0. 0. 0. 0. 1. 'o'                        ! On-momentum orbit.
-36.85e-1 0. 0. 0. 0. 0.999 'm' ! Chromatic orbit coordinates Y0 and T0 for D=0.999 relative rigidity.
1 1 1
```

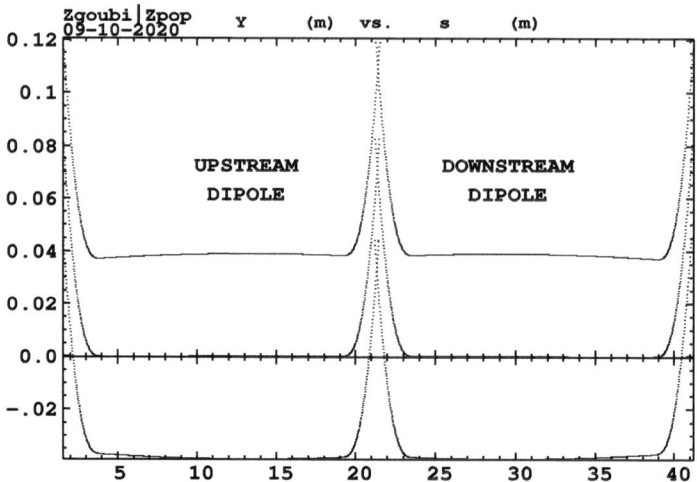

Fig. 8.47 A graph of the radial excursion, within DIPOLE range (namely, $AT = 55°$ extent, Table 8.20), of an on-momentum particle (its radial position in the dipole body is $R_0 \approx 20.7628$ m, corresponding to Y=0 in this graph) and two particles at respectively $dp/p = \pm 10^{-3}$. The diverging parts at DIPOLE ends are in the 5° fringe field regions. A graph obtained using zpop: menu 7; 1/1 to open zgoubi.plt; 2/[6,2] for Y versus distance; 7 to plot

The position and angle of the chromatic particles, which are offset by $\Delta p/p = \pm 10^{-3}$, are drawn from the value of the periodic dispersion $\eta_Y = 36.85$ m and its derivative $\eta'_Y \approx 0$ (Table 8.22), namely, $Y_0 = \eta_Y \, \Delta p/p = \pm 3.685$ cm and $T_0 = \eta_Y \, \Delta p/p = 0$.

Running Table 8.23 simulation with this new OBJET produces the following coordinates at FAISCEAU, located at the end of the sequence (an excerpt from zgoubi.res execution listing):

```
   18  Keyword, label(s) :  FAISCEAU
                                      TRACE DU FAISCEAU
                                    (follows element #      17)
                                       3 TRAJECTOIRES
                          OBJET                                             FAISCEAU
           D       Y(cm)   T(mr)  Z(cm)  P(mr)   S(cm)       D-1     Y(cm)   T(mr)  Z(cm)  P(mr)   S(cm)
     p  1  1.0010   3.685   0.000  0.000  0.000  0.0000    0.0010    3.685   0.000  0.000  0.000  4.278650E+03
     o  1  1.0000   0.000   0.000  0.000  0.000  0.0000    0.0000    0.000   0.000  0.000  0.000  4.272614E+03
     m  1  0.9990  -3.685   0.000  0.000  0.000  0.0000   -0.0010   -3.685  -0.000  0.000  0.000  4.266579E+03
```

The local coordinates Y, T (under FAISCEAU, right hand side) are equal to the initial coordinates Y_0, T_0 (under OBJET, left hand side), to better than 5 μm, 0.5 μrad accuracy respectively (zgoubi.fai can be consulted for double precision on these values), so confirming the periodicity of these chromatic trajectories. Figure 8.47 shows the particle trajectories through the two DIPOLEs. A difference between the on- and off-momentum trajectories yields as expected a quasi-constant $\eta_Y \approx 36.8$ m whereas $\eta'_Y \approx 0$.

Fig. 8.48 Dispersion function along ZGS cell, obtained by orbit difference. The discontinuities are artifacts, they are located in the overlapping regions between the optical sequence DIPOLEs and DRIFTs (Table 8.23)

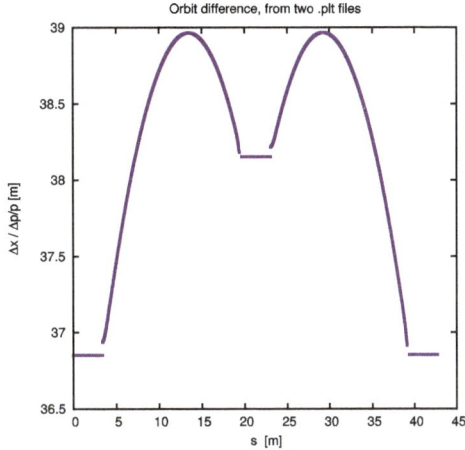

Orbit difference

The method can be used to compute the dispersion function. This requires tracking particles with $\pm dp/p$ momentum offset. A gnuplot script can compute and plot the orbit difference, and normalize to dp/p; the result is the periodic dispersion, displayed in Fig. 8.48.

(c) Some verifications regarding the model.

The field along large excursion orbits can be logged in zgoubi.plt, using option IL=2 (or 20, or 200, etc. for printout every 10, or 100, etc. integration step) under DIPOLE.

The simulation file of Table 8.23 is used to raytrace five particles, with OBJET changed to the following:

```
'OBJET'
1.03527036749193e3                              ! Reference Brho: 50 MeV proton.
2                                               ! Create particles individually,
5 1                                             ! five particles.
+36.85e-1 0. 0. 0. 0. 1.01  'p'    ! Chromatic orbit coordinates for D=1.01 relative rigidity.
0. 0. 0. 0. 0. 1.  '0'                      ! On-momentum closed orbit.
-36.85e-1 0. 0. 0. 0. 0.99  'm'    ! Chromatic orbit coordinates for D=0.99 relative rigidity.
0. 0. 5. 0. 0.  1.   'm'                  ! Initial vertial excursion is Z0= 5 cm off-mid-plane.
0. 0. 20. 0. 0.  1.  'm'                 ! Initial vertial excursion is Z0=20 cm off-mid-plane.
1 1 1 1 1
```

Apart from the on-momentum particle (2nd in the list) this OBJET defines two particles on $\Delta p/p = \pm 1\%$ chromatic orbit (1st and 3rd in the list), this is an excursion of a few tens of centimeters, large as requested, as $\Delta x \approx 38 \times dp/p$. OBJET also defines 2 particles launched into the cell at respectively $Z_0 = 5\,\text{cm}$ and $Z_0 = 20\,\text{cm}$.

The magnetic field as a function of the azimuthal angle in DIPOLE frame, along these trajectories across the upstream DIPOLE of the cell, is shown in Fig. 8.49. The field curves for the first four trajectories essentially superimpose except for the fringe field regions (Fig. 8.49), due to the wedge angles. This behaves as expected. Detailed

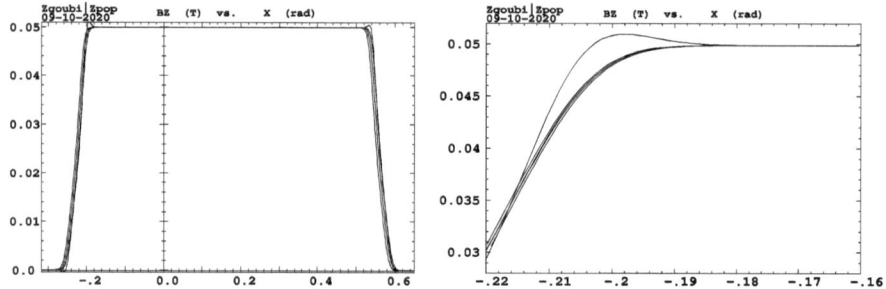

Fig. 8.49 Magnetic field along 5 different trajectories across the upstream DIPOLE, including four large horizontal and vertical excursion cases, and a zoom in on the entrance fringe field region

inspection is possible, from particle coordinate and field data in zgoubi.plt—this is out of the scope of the present question.

The field along the 5th particle trajectory features overshoots (Fig. 8.49), this is due to the large vertical excursion ($Z \approx 20\,$cm in the entrance fringe field region). It looks reasonable, however it may be an artifact in the case that the high order derivatives of the field in that region are large, resulting from the truncated Taylor series method used for off mid-plane field extrapolation [36, Sect. 1.3.3].

(d) Sinusoidal approximation of the betatron motion.

The approximation

$$y(\theta) = A\,\cos(\nu_Z\theta + \phi)$$

is checked here considering the vertical motion (considering the horizontal motion leads to similar conclusions). The value of the various parameters in that expression are determined as follows:

– the particle raytraced for comparison is launched with an initial excursion $Z_0(\theta = 0) = 5\,$cm (4th particle in OBJET, above). At the launch point (middle of the long drift) the beam ellipse is upright (see below), whereas phase space motion is clockwise, thus take

$$A = 5\,\text{cm} \quad \text{and} \quad \phi = \pi/2$$

– the vertical betatron tune of the 4-cell ring is (Table 8.22)

$$\nu_Z = 4 \times 0.192869 = 0.77147$$

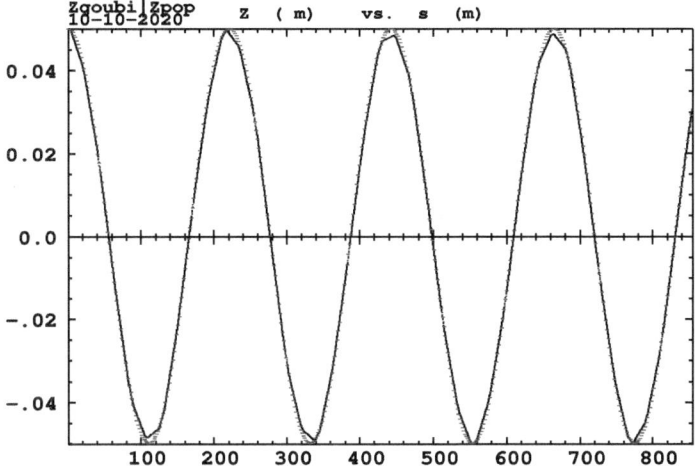

Fig. 8.50 Vertical betatron motion, five turns around the ZGS ring, from raytracing (continuous line), and sine approximation, superimposed (dashed line)

$-\theta = s/R$ and $R = \oint ds/2\pi$ with (Table 8.22)

$$2\pi R = \text{circumference} = 4 \times 42.72614331 = 170.90457 \, \text{m}$$

Consistency with sinusoidal approximation is shown in Fig. 8.50.

(e) An acceleration cycle. Symplecticity checks.

Eleven particles are launched for 65,000 turn tracking at a rate of

$$\Delta W = q\hat{V} \sin \phi_s = 400 \times \sin 150^0 = 200 \, \text{keV/turn}$$

$(E : 0.05 \rightarrow 13.05 \, \text{GeV})$, all evenly distributed on the same initial vertical invariant

$$Z^2/\beta_Z + \beta_Z Z'^2 = \varepsilon_Z/\pi \tag{8.60}$$

with $\varepsilon_Z/\pi = 10^{-4}$ m, or, normalized, $\beta\gamma\varepsilon_Z/\pi = 0.33078 \times 10^{-4}$ m.

The simulation file is given in Table 8.24. CAVITE[IOPT=3] is used, it provides an RF phase independent boost

$$\Delta W = q\hat{V} \sin \phi_s$$

as including synchrotron motion is not necessary here, even better, anticipating on spin question (f) this ensures constant depolarizing resonance crossing speed, so precluding any possibility of multiple crossing (it can be referred to [40] regarding that effect).

Table 8.24 Simulation input data file: track 11 particles launched on the same vertical invariant, with zero horizontal invariant. The INCLUDE adds the ZGS cell four times, the latter is defined in Table 8.20. An MCOBJET is commented, it is used in a subsequent spin tracking exercise

```
ZGS ring. Polarization landscape.
'MARKER'   ZGSPolarLand_S                                        ! Just for edition purposes.
'OBJET'
1.03527036749193e3                                               ! Reference Brho: 50 MeV proton.
8                                                    ! Create a 13 particle set, proper for MATRIX computation.
1 11 1    ! Define 9 particles, all with ~0 horiz. invariant, evenly spread on same vertical invariant.
0. 0. 0. 0. 0. 1. 'o'    ! Reference trajectory: all initial coordinates nul, relative rigidity D=1.
0. 28.63 0. ! Horiz. invariant taken zero. Nominal would be 0.14mu_m norm. i.e. 4.6e-8 non-normalized.
0. 37.01 150e-6    ! epsilon_Z/pi = beta.gamma * epsilon_norm, latter =0.05e-6 m, beta.gamma=0.3398.
0. 1. 0. 0.                                                      ! All paricles are on-momentum.

! 'MCOBJET'                                                      ! Commented.
!1.03527036749193e3                                              ! Reference Brho: 50 MeV proton.
!3                                                               ! Create random coordinates.
!200
!2 2 2 2 2 2
!0. 0. 0. 0. 0. 1.
!0. 28.63 25e-6 3                                 ! Periodic alpha_Y, beta_Y, and invariant value;
!0. 37.01 10e-6 3                                 ! Periodic alpha_Z, beta_Z, and invariant value.
!0. 1. 1.e-8 3
!123456 234567 345678

'PARTICUL'
PROTON              ! Necessary data in order to allow  (i) spin trackingand, and  (ii) acceleration.
'SPNTRK'                                                         ! Switch on spin tracking,
3                                                                ! all initial spins vertical.
'FAISCEAU'
'FAISTORE'
b_polarLand.fai                      ! Log particle data in b_polarLand.fai, turn-by-turn; "b_" imposes
7                                                   ! binary write, which results in faster i/o.

'SCALING'
1 1
DIPOLE
-1                                   ! Causes field increase in DIPOLE, in correlation to particle
1.                                                   ! rigidity increase by CAVITE.
1

! 4 cells follow.
'INCLUDE'
1
4* ./ZGS_cell.inc[S_ZGS_cell:E_ZGS_cell]

'CAVITE'
3                    ! With IOPT=3, synchronous phase below of 30deg or 150deg is equivalent.
0 0
400e3  0.523598775598                              ! Acceleration rate is 400*0.5=200keV/turn.

'REBELOTE'
87000 0.3 99

'FAISCEAU'
'MARKER'   ZGSPolarLand_E                                        ! Just for edition purposes.
'SPNPRT'

'END'
```

Betatron damping

Figure 8.51 shows the damped vertical motion of the individual particles, over the acceleration range, together with the initial and final distributions of the 11 particles on elliptical invariants. Departure from the matching ellipse at the end of the acceleration cycle, 13 GeV (Eq. 8.60 with $\varepsilon_Z/\pi = 2.2244 \times 10^{-7}$ m), is marginal.

Degree of non-symplecticity of the numerical integration

The degree of non-symplecticity as a function of integration step size is illustrated in Fig. 8.52. The initial motion is taken paraxial, vertical motion is considered as it resorts to off-mid plane Taylor expansion of fields [36, DIPOLE Sect.], a stringent test as the latter is expected to deteriorate further the non-symplecticity inherent to the Lorentz equation integration method (a truncated Taylor series method [36, Eq. 1.2.4]).

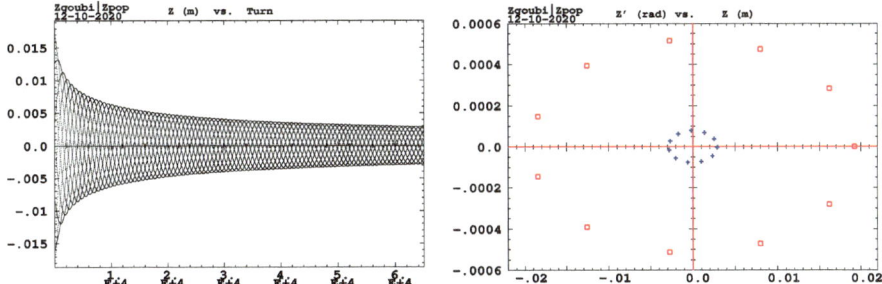

Fig. 8.51 Left: damped vertical motion, from 50 MeV to 13.05 GeV, 65,001 turns. Right: the initial coordinates of the 11 particles (squares) are taken on a common invariant $\varepsilon_Z(0) = 10^{-5}\,\pi\mathrm{m}$ (at 50 MeV, $\beta\gamma = 0.33078$, thus $\beta\gamma\varepsilon_Z(0) = 0.33078 \times 10^{-5}\,\pi\,\mathrm{m}$); the final coordinates after 65,000 turns (crosses) appear to still be (with negligible departure) on a common invariant of value $\varepsilon_Z(final) = 2.2244 \times 10^{-7}\,\pi\,\mathrm{m}$, or $\beta\gamma\varepsilon_Z(final) = 0.33076 \times 10^{-5}\,\pi\,\mathrm{m}$ (at 13 GeV, $\beta\gamma = 14.869842$), equal to the initial value

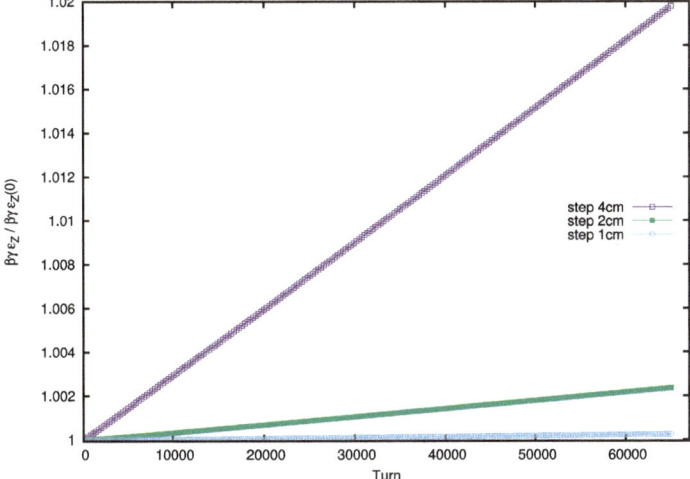

Fig. 8.52 Turn-by-turn evolution of the normalized invariant, $\beta\gamma\varepsilon_Z(turn)/\beta\gamma\varepsilon_Z(0)$ (initial $\varepsilon_Z(0)$ taken paraxial), for integration step sizes 1, 2 and 4 cm

Evolution of the wave numbers

The Fortran tool tunesFromFai_iterate can be used to computes tunes as a function of turn number or energy, it reads turn-by-turn particle data from zgoubi.fai and computes a discrete Fourier transform over so many turns (a few tens, for instance), every so many turns [41]. Typical results are displayed in Fig. 8.53, tunes have the expected values: $\nu_Y = 0.849$, $\nu_Z = 0.771$. An acceleration rate of 200 keV/turn has been taken (namely, $\hat{V} = 400$ kV and still $\phi_s = 150^0$), to save on computing time. Note that turn-by-turn raytracing allows determining the tune value at all γ along the acceleration cycle (and thus for instance the γ values at which the resonance

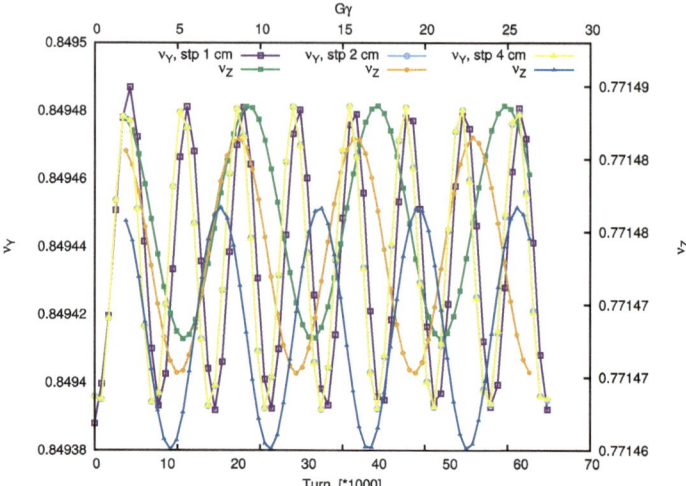

Fig. 8.53 Horizontal ring tune (left vertical axis), $\nu_Y \approx 0.8494$, and vertical ring tune (right vertical axis), $\nu_Z \approx 0.77147$, as a function of turn number, over 65,000 turns ($E : 0.05 \rightarrow 13\,\text{GeV}$ at a rate of 200 keV/turn). The graph displays results for integration step sizes 1, 2 and 4 cm, essentially converged

occurs, see (f)). In these simulations anyway the horizontal and vertical tunes are essentially constant over the all cycle: it is determined by the wedge angle, which will not charge as long as the reference orbit is not changed. The latter holds here, as SCALING[NTIM=-1] causes the magnet field to strictly follow the momentum boost by CAVITE.

(f) Crossing an isolated intrinsic depolarizing resonance.

The simulation uses the input data file of Table 8.24, with the following changes:

- Under OBJET:
  - 1st line, change the reference rigidity $BORO$ to the proper value, a few thousand turns upstream of the resonance to be crossed;
  - 3rd line, request a single particle ("1 1 1", in lieu of "1 11 1");
  - 6th line, set the invariant ε_Z/π to the desired value, ε_Y/π value is indifferent;

- change the field value under DIPOLE consistently with the new BORO value, so to maintain the expected curvature radius $\rho_0 = BORO/B = 20.76\,\text{m}$ (Table 8.2);
- under CAVITE, provide the desired peak voltage $\hat{V}$;
- under REBELOTE, set the number of turns: a few thousands of turns upstream and downstream of the resonance.

Similar simulations are performed in questions (f)–(i) of Exercise 8.1. Please refer to the solutions of these SATURNE 1 simulations.

(g) Study of an imperfection depolarizing resonance.

The simulation data files of question (f) can be used here, *mutatis mutandis*, and the methodology in (f) can be followed.

Similar simulations are proposed in questions (f)–(i) of Exercise 8.1, as well as in the "Strong Focusing Synchrotron" Chapter exercises. Please refer to the solutions of these exercises.

(h) Spin tracking. Bunch polarization.

Spin depolarizing resonances in the ZGS are located at

$$G\gamma_R = kP \pm \nu_Z \quad = 4 - \nu_Z, \; 4 + \nu_Z, \; 8 - \nu_Z, \; 8 + \nu_Z, \; 12 - \nu_Z, \; \text{etc.}$$

with P = 4 the superperiodicity of the ring, and $\nu_Z = 0.77147$ taken from Table 8.22, or from Fig. 8.53. $G\gamma_R$ is bounded, in the present simulation, by $G\gamma(17.4\,GeV) = 35.0 < 9P - \nu_Z$. Resonances are expected to be stronger at $G\gamma_R = 2 \times 4k \pm \nu_Z = 8 - \nu_Z, \; 8 + \nu_Z, \; 16 - \nu_Z$, etc., with the additional factor 2 the number of cells per superperiod [32, Sect. 3.II].

The simulation data file to track through these resonances is the same as in question (e), Table 8.24, except for the substitution of MCOBJET (to be uncommented) to OBJET (to be commented). MCOBJET creates a 200 particle bunch with Gaussian transverse and longitudinal densities, with the following *rms* values at 50 MeV:

$$\varepsilon_Y/\pi = 25\,\mu\text{m}, \quad \varepsilon_Z/\pi = 10\,\mu\text{m}, \quad \frac{dp}{p} = 10^{-4}$$

which are presumably close to ZGS polarized proton runs [33]. CAVITE accelerates that bunch from 50 MeV to 17.4 GeV about, at a rate of $q\hat{V}\sin(\phi_s) = 200\,\text{keV/turn}$ ($\hat{V} = 400\,kV$, $\phi_s = 30°$), in 87,000 turns about.

Figure 8.54 shows sample S_Z spin components of a few particles taken among the 200 tracked. Figure 8.55 displays $\langle S_Z \rangle$, the vertical polarization component of the bunch (gnuplot script given in Table 8.16).

8.3 Visible SR from GEC 70 MeV Synchrotron

The input data file to simulate an electron trajectory in GEC dipole is given in Table 8.25.

The critical energy is $\omega_c/2\pi = 3\gamma^3 c/(4\pi\rho) = 2.6635\,\text{eV}$, a critical wavelength of $\lambda_c = 0.4655\,\mu\text{m}$, in the visible range. The GEC glass vacuum chamber was transparent, this allowed the observation of this synchrotron light.

The *rms* angular aperture of the radiation is $\sim 1/\gamma = 7.247\,\text{rad}$. By taking a 22 cm long magnet, the cone captured by observing tangentially to the arc extends over $\Delta\phi = 22_{[cm]}/(\rho/\gamma) \approx \pm 52$ times the SR angular aperture $1/\gamma$, well beyond the peak region of the electric impulse. Thus this truncature only concerns quasi-zero tail regions of the impulse (Fig. 8.17), its Fourier transform can be considered accurate enough.

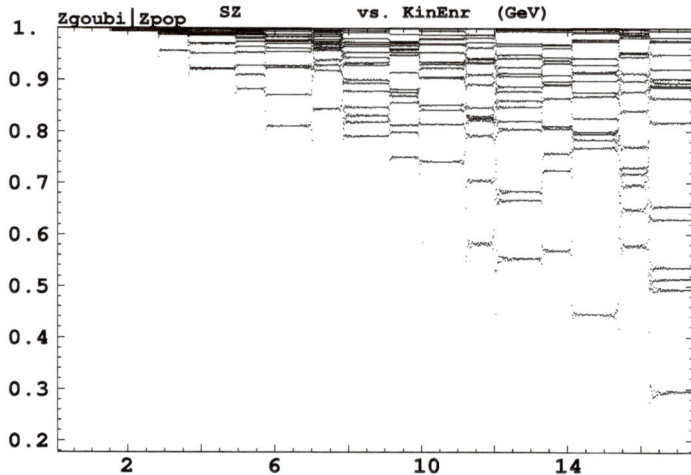

Fig. 8.54 Individual vertical spin component of 20 particles accelerated in ZGS from 50 MeV to 17.4 GeV, at a rate of 200 keV/turn. A graph obtained using zpop: menu 7; 1/2 to open b_zgoubi.fai; 2/[20,23] for S_Z versus energy; 7 to plot

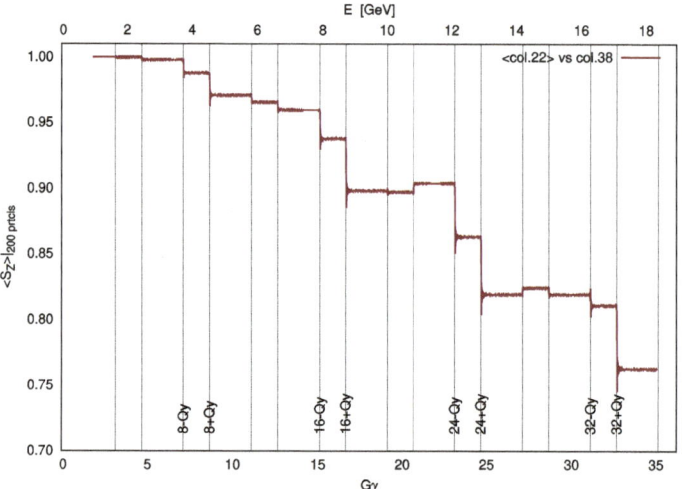

Fig. 8.55 Average vertical component of the polarization vector of a 200 particle bunch, accelerated from 50 MeV to 17.4 GeV. The vertical lines materialize the locations $G\gamma_R = 4k \pm \nu_Z$ of the depolarizing resonances. Resonances are strongest at $G\gamma_R = 8k \pm \nu_Z$ (as labeled)

Table 8.25 Simulation input data file: computation of the 29.2 cm radius trajectory of a 70 MeV electron in a dipole field

```
Electron trajectory on a +/-50gamma arc of GEC 70 MeV synchrotron.
'OBJET'
0.235193199210899 *1e3                         Rigidity(kG.cm) - 70 MeV electrons.
2
1  1                                                      ! Defines a single electron.
0. 385.06672 0. 0. 0. 1. '0'             ! Incidence of electron, half the dipole bend angle.
1
'MULTIPOL'
2
22.  10.  8.0545616168116  0. 0. 0. 0. 0. 0. 0. 0.           ! Length(cm), unused here, field (kG).
0. 0.   0. 0. 0.  .0 .0   0. 0. 0. 0.                        ! Entrance fringe field (hard edge),
6  .1122 6.2671 -1.4982 3.5882 -2.1209 1.723                 !   Enge coefficients.
0. 0.   0. 0. 0.  0. 0.   0. 0. 0. 0.                        ! Exit fringe field.
6  .1122 6.2671 -1.4982 3.5882 -2.1209 1.723
0. 0. 0. 0. 0. 0. 0. 0. 0. 0.
#200|9999|200                                               ! 1000 integration steps in body.
1 0. 0. 0.
'FAISCEAU'
'END'
```

The two components of the electric field impulse in the direction $(\phi, \psi) = (0, 0.1/\gamma)$, and their spectral brightness, are displayed in Fig. 8.17. The σ component peaks near $\hbar\omega_c = 2.6635$ eV, as expected.

References

1. V.I. Veksler, A new method of accelerating relativistic particles. Comptes-Rendus de l'Académie des Sciences de l'URSS **43**(8), 329–331 (1944)
2. E.M. McMillan, The synchrotron. Phys. Rev. **68**, 143–144 (1945)
3. F.K. Goward, D.E. Barnes, Experimental 8 MeV synchrotron for electron acceleration. Nature **158**, 413 (1946)
4. H.C. Pollock, The discovery of synchrotron radiation. Am. J. Phys. (1983)
5. F.R. Elder et al., Radiation from electrons in a synchrotron. Phys. Rev. **71**, 829–Published 1 June 1947
6. D.W. Kerst, The acceleration of electrons by magnetic induction. Phys. Rev. **60**, 47–53 (1941)
7. SATURNE 1 photos: credit CEA Saclay. Archives historiques du CEA. Copyright CEA/Service de documentation
8. A. Sessler, E. Wilson, *Engines of Discovery*. A Century of Particle Accelerators (World Scientific, 2007)
9. Fig. 8.3 : Credit Reider Hahn, Fermilab
10. K. Endo et al., Compact proton and carbon ion synchrotrons for radiation therapy. MOPRI087, in *Proceedings of EPAC 2002*, Paris, France, pp. 2733–2735. https://accelconf.web.cern.ch/e02/PAPERS/MOPRI087.pdf
11. V.A. Vostrikov, et al., Novel approach to design of the compact proton synchrotron magnetic lattice. tupsa17, 26th Russian Particle Accelerator Conference RUPAC2018, Protvino, Russia (2018). https://accelconf.web.cern.ch/rupac2018/papers/tupsa17.pdf
12. K.R. Symon, MURA Days, in *Proceedings of the 2003 Particle Accelerator Conference*. https://accelconf.web.cern.ch/p03/PAPERS/WOPA003.PDF
13. T. Ohkawa, Two-beam fixed field alternating gradient accelerator. Rev. sci. Instrum. **29**, 108–17, 1
14. Bernardini, C., AdA: the first electron-positron collider. Phys. Perspect. **6** (2004). 156–183 1422-6944/04/020156-28. https://link.springer.com/article/10.1007/s00016-003-0202-y
15. Image by Argonne National Laboratory, *Comm* (L.A, Martinez, ANL, Apr. 2023)

16. D. Cohen, Feasibility of accelerating polarized protons with the argonne ZGS. Rev. Sci. Instrum. **33**, 161 (1962). https://doi.org/10.1063/1.1746524
17. L.G. Ratner, T.K. Khoe, Acceleration of polarized protons in the zero gradient synchrotron, in *Procs. PAC 1973 Conference, Washington* (1973). http://accelconf.web.cern.ch/p73/PDF/PAC1973_0217.PDF
18. J. Bywatwr, T. Khoe et al., A pulsed quadrupole system for preventing depolarization. IEEE Trans. Nucl. Sci. **20**(3) (1973)
19. Y. Cho et als., Effects of depolarizing resonances on a circulating beam of polarized protons during or storage in a synchrotron. IEEE Trans. Nucl. Sci. NS-**24**(3) (1977)
20. E.F. Parker, High energy polarized deuterons at the argonne national laboratory zero gradient synchrotron. IEEE Trans. Nucl. Sci. NS-**26**(3), 3200–3202 (1979)
21. D.E. Suddeth et als., Pole face winding equipment for eddy current correction at the zero gradient synchrotron, in *Procs. PAC 1973 Conference, Washington* (1973). http://accelconf.web.cern.ch/p73/PDF/PAC1973_0397.PDF
22. A.V. Rauchas, A.J. Wright, Betatron tune profile control in the zero gradient synchrotron (ZGS) using the main magnet pole face windings, in *Procs. PAC1977 Conference, IEEE Transactions on Nuclear Science*, NS-**24**(3) (1977)
23. G. Floquet, Sur les équations différentielles linéaires à coefficients périodiques. Annales scientifiques de l'E.N.S. 2e série, tome 12, pp. 47–88 (1883). http://www.numdam.org/item?id=ASENS_1883_2_12__47_0
24. G. Leleux, *Accélérateurs Circulaires* Lecture Notes (INSTN, CEA Saclay, 1978)
25. T. Risselada, Transition gamma jump schemes, in *Proceedings of Jyv askyl a CERN Accelerator School*, 7–18 Sept. 1992. Yellow Report CERN 94-01
26. A. Hofmann, The physics of synchrotron radiation. Cambridge monographs on particle physics, in *Nuclear Physics and Cosmology* (20) (Cambridge University Press, 2004)
27. F. Méot, A theory of low frequency far-field synchrotron radiation. Part. Accel. **62**, 215–239 (1999)
28. F. Méot, L. Ponce, N. Ponthieu, Low frequency interference between short SR sources. PRST-AB **4**, 062801 (2001)
29. F. Méot, Zgoubi Users' Guide. https://www.osti.gov/biblio/1062013-zgoubi-users-guide. Sourceforge latest version. https://sourceforge.net/p/zgoubi/code/HEAD/tree/trunk/guide/Zgoubi.pdf
30. G. Leleux, Traversée des résonances de dépolarisation. Rapport Interne LNS/GT-91-15, SATURNE, Groupe Théorie, CEA Saclay (février, 1991)
31. F. Méot, Spin dynamics. In: Polarized beam dynamics and instrumentation in particle accelerators, in *USPAS Summer 2021 Spin Class Lectures* (Springer Nature, Open Acess, 2023). https://link.springer.com/book/10.1007/978-3-031-16715-7
32. S.Y. Lee, *Spin Dynamics and Snakes in Synchrotrons* (World Scientific, 1997)
33. T.K. Khoe et al., The high energy polarized beam at the ZGS, in *Procs. IXth International Conference on High Energy Accelerators, Dubna*, pp. 288–294 (1974). Figure 8.19: Copyrights under license CC-BY-3.0. https://creativecommons.org/licenses/by/3.0; no change to the material
34. M. Froissart, R. Stora, Dépolarisation d'un faisceau de protons polarisés dans un synchrotron. Nucl. Inst. Meth. **7**, 297 (1960)
35. H. Bruck, P. Debraine, R. Levy-Mandel, J. Lutz, I. Podliasky, F. Prevot, J. Taieb, S.D. Winter, R. Maillet, Caractéristiques principales du Synchrotron à Protons de Saclay et résultats obtenus lors de la mise en route, rapport CEA no. 93, CEN-Saclay (1958)
36. F. Méot, Zgoubi Users' Guide. https://www.osti.gov/biblio/1062013-zgoubi-users-guide Sourceforge latest version. https://sourceforge.net/p/zgoubi/code/HEAD/tree/trunk/guide/Zgoubi.pdf
37. M.H. Foss et al., The argonne ZGS magnet. IEEE **1965**, 377–382 (1965)
38. L.A. Klaisner et al., IEEE **1965**, 133–137 (1965)
39. A post-processing tool to transport betatron functions step-by-step, using raytracing data stored in zgoubi.plt. https://sourceforge.net/p/zgoubi/code/HEAD/tree/trunk/toolbox/betaFromPlt/

40. T. Aniel et al., Polarized particles at SATURNE. Journal de Physique, Colloque C2, supplément au n02, Tome 46, février, pp. C2–499 (1985). https://hal.archives-ouvertes.fr/jpa-00224582
41. The Fortran tunesFrmFai_iterate.f, together with a README and an example of its use, can be found at https://sourceforge.net/p/zgoubi/code/HEAD/tree/trunk/toolbox/tunesFromFai/
42. https://stackoverflow.com/questions/42677017/plot-average-of-nth-rows-in-gnuplot
43. F. Méot, Spinor methods, in *Polarized Beam Dynamics and Instrumentation in Particle Accelerators*. USPAS Summer 2021 Spin Class Lectures (Springer Nature, Open Access, 2023). https://link.springer.com/book/10.1007/978-3-031-16715-7

Chapter 9
Strong Focusing Synchrotron

Abstract This chapter introduces the strong focusing alternating gradient (AG) and separated function synchrotrons. It provides the theoretical material which the simulation exercises lean on. The chapter begins with a brief reminder of the historical context, and continues with beam optics, chromaticity, acceleration, resonances and resonant extraction, dynamical effects of synchrotron radiation (SR), the electromagnetic SR impulse, and depolarizing resonances. This resorts to basic charged particle optics, acceleration, and dynamics in magnetic fields introduced in the previous chapters. The simulation of a strong focusing AG synchrotron requires just two optical elements from `zgoubi` library: DIPOLE or MULTIPOL to simulate a combined function dipole, and DRIFT to simulate straight sections. Main dipoles in a separated function synchrotron can use BEND. It requires in addition quadrupoles, simulated using QUADRUPO or MULTIPOL. The latter can simulate higher order lenses, which can otherwise resort to SEXTUPOL, OCTUPOLE, etc. Acceleration uses CAVITE. Accounting for synchrotron radiation (SR) energy loss requires SRLOSS. Monte Carlo SR monitoring can use SRPRNT, which logs data in zgoubi.res. SRPRNT[PRINT] in addition logs data in zgoubi.SRPRNT.Out. Computation of synchrotron radiation (SR) Poynting and spectral brightness uses `zpop`. Particle monitoring requires keywords introduced in the previous Chapters, including FAISCEAU, FAISTORE, possibly PICKUPS, and some others. Spin motion computation and monitoring resort to SPNTRK, SPNPRT, FAISTORE. Optics matching and optimization use FIT[2]. INCLUDE is used, mostly here in order to simplify the input data files. SYSTEM is used to, mostly, resort to gnuplot so as to end simulations with some specific graphs. Data for the latter are read from output files filled up during the execution of the code, such as zgoubi.fai (resulting from the use of FAISTORE), zgoubi.plt (resulting from IL=2), or other zgoubi.*.out files resulting from a PRINT command. Stepwise particle data logged in zgoubi.plt are used by the interface zpop to compute the electric field impulse of SR and subsequent spectral angular energy density of the radiation.

© The Author(s) 2024

F. Méot, *Understanding the Physics of Particle Accelerators*, Particle Acceleration
and Detection, https://doi.org/10.1007/978-3-031-59979-8_9

Notations used in the Text

$\mathbf{B}$; $B_{x,y,s}$; B	Field vector; its components in the moving frame; its modulus
$B\rho = p/q$; $B\rho_0$	Particle rigidity; reference rigidity
C; C_0	Orbit length; $C = 2\pi R + \left[\begin{array}{c}\text{straight}\\\text{sections}\end{array}\right]$; reference, $C_0 = C$ $(p = p_0)$
$\mathbf{E}$; E_σ, E_π	SR electric field impulse; its parallel and normal components
E; E_s	Particle energy, $E = \gamma m_0 c^2$; synchronous energy
EFB	Effective Field Boundary
f_{rev}, $f_{\text{rf}} = h\, f_{\text{rev}}$	Revolution and RF voltage frequencies
G	Gyromagnetic anomaly, $G = 1.792847$ for proton
G; $K = G/B\rho$	Quadrupole gradient; focusing strength
h	RF harmonic number
m; m_0; M	Particle mass; rest mass; in units of MeV/c^2
$n = -\frac{\rho}{B}\frac{\partial B}{\partial x}$	Focusing index
$\mathbf{n}_0$	Stable spin precession direction
$\mathbf{P} = \mathbf{E} \times \mathbf{B}$	SR Poynting vector
P_i, P_f	Beam polarization, initial, final
$\mathbf{p}$; p; p_0	Momentum vector; its modulus; reference
q	Particle charge
r; R	Orbital radius; average radius, $R = C/2\pi$
S	Periodicity of the lattice
s	Path variable
U_s	SR energy loss
$\mathbf{v}$; v	Particle velocity vector; its modulus
$V(t)$; $\hat{V}$	Oscillating voltage; its peak value
x, x$'$, y, y$'$, 1, $\frac{dp}{p}$	Particle coordinates in the moving frame, $[(*)' = d(*)/ds]$
α	Momentum compaction; or trajectory deviation; or depolarizing resonance crossing speed
$\beta = v/c$; β_0; β_s	Normalized particle velocity; reference; synchronous
β_u	Betatron functions (u : x, y, Y, Z)
$\gamma = E/m_0 c^2$	Lorentz relativistic factor
γ_{tr}	Transition γ, $\gamma_{\text{tr}} = 1/\sqrt{\alpha}$
δp, Δp	Momentum offset
ϵ_c	Critical energy of SR, $\epsilon_c = \hbar\omega_c = hc/\lambda_c$
ε	Wedge angle
ε_u/π	Courant-Snyder invariant; emittance//π (u : x, y, l)
ϵ_R	Strength of a depolarizing resonance
η	Phase slip factor, $\eta = \frac{1}{\gamma^2} - \alpha$
μ_u	Betatron phase advance per period, $\mu_u = \int_{\text{period}} \frac{ds}{\beta_u(s)}$ (u : x, y)
ν_u	Wave numbers, horizontal, vertical, synchrotron (u : x, y, l)
ρ; ρ_0	Curvature radius; reference
σ	Beam matrix

ϕ; ϕ_s	Particle phase at voltage gap; synchronous phase
φ_u	Betatron phase advance, $\varphi_u = \int ds/\beta_u$ (u : $x, y, Y, or Z$)
φ	Spin angle to the vertical axis
ω_c	Critical angular frequency of SR, $\omega_c = 3\gamma^3 c/2\rho$
ω_s; Ω_s	$2\pi f_{rev}$; synchrotron frequency

9.1 Introduction

In the very manner that the 1930s–1940s cyclotron, betatron, microtron, weak focusing synchrotron, which are still in use today, have since essentially not changed in their concepts and design principles, today the gap profile, yoke and current coil geometry of combined function alternating-gradient (AG) dipoles remain essentially as patented in 1950 (Fig. 9.1) [1].

In 1952, in the context of studies concerning the Cosmotron, strong focusing was devised at the Brookhaven National Laboratory (BNL): "*Strong focusing forces result from the alternation of large positive and negative n-values in successive sectors of the magnetic guide field in a synchrotron. This sequence of alternately converging and diverging magnetic lenses [...] leads to significant reductions in oscillation amplitude*" [3]. It led to the construction of the first two high-energy AG proton synchrotrons (PS), in the 30 GeV range, in the late 1950s: the CERN PS, and the AGS at BNL (Fig. 9.2). Both remain major pieces, 60 years later, of the respective injection chains of the two largest colliders in operation, the LHC and RHIC. Early works at BNL provided theoretical formalism, still at work today, for the analysis of beam dynamics in synchrotrons [4].

Separated function focusing, whereby beam guiding is ensured by uniform field dipoles while focusing is ensured separately by quadrupoles (Fig. 9.3), followed from the development of the latter (Fig. 9.4), a spin-off of the strong index technology [7].

The dramatic reduction of transverse beam size by strong focusing allows guiding and focusing magnets with small aperture, from lowest energies: medical syn-

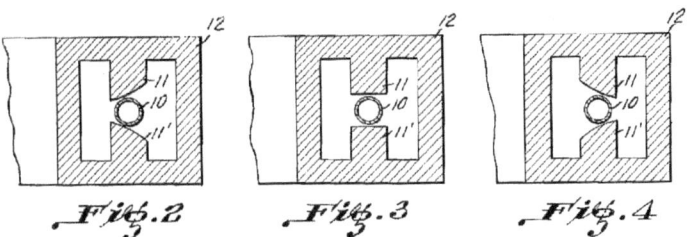

Fig. 9.1 Bending magnet pole profiles for a *focusing system for ions and electrons* [1]. Assuming curvature center to the left, the left (respectively right) profile is defocusing (resp. focusing), the middle profile has zero index

Fig. 9.2 Top: the AGS combined function main dipole. The hyperbolic profile poles are visible, partly hidden by the field coils. Bottom: the 809 m circumference AGS synchrotron, comprised of 240 such dipoles [2]

chrotrons in the 100 MeV range for instance, to highest ones: hundreds of GeV to multi-TeV range particle physics and nuclear physics colliders (Fig. 9.5). Beams in all these machines are essentially confined in a sub-centimeter or sub-millimeter scale transverse space. A synchrotron is a string of dipole and multipole magnets through which runs a vacuum pipe of a few centimeters diameter (hadron rings) or a few millimeters (electrons). The size of the ring is essentially determined by its circumference, proportional to the magnetic rigidity. This revolutionized the race to high energies, from the prior few GeV weak focusing synchrotrons and their huge magnets, to todays 7 TeV, 27 km long LHC and with further plans for 100 TeV,

Fig. 9.3 SATURNE 2 strong focusing 3 GeV synchrotron at Saclay [5], successor in the late 1970s of SATURNE 1 weak focusing synchrotron (Fig. 8.1). It was the first strong focusing synchrotron to accelerate polarized ion beams

Fig. 9.4 A quadrupole magnet at LBL in 1957, used for beam lines at the 184-inch cyclotron. An early specimen here, obviously, being a spin-off of the early 1950s concept of strong focusing [6]

Fig. 9.5 In RHIC tunnel at the Brookhaven National Laboratory [2]. The two rings of the 255 GeV polarized proton beams and heavy ion collider run parallel over 3.8 km, and intersect at two experiments, STAR and SPHENIX

Fig. 9.6 The ion rapid cycling medical synchrotron (iRCMS) [9], an ion beam RCS for the treatment of cancer tumors

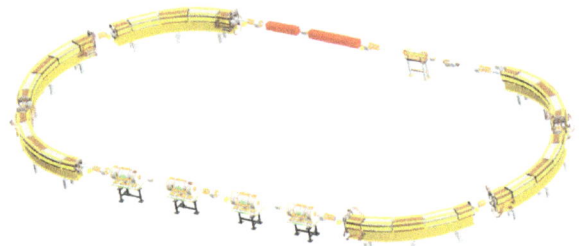

100 km circumference colliders [8]. Strong focusing fostered the development of high energy synchrotron light sources around the world, with high brightness synchrotron radiation (SR) from UV to gamma rays produced in electron storage rings in up to multi-GeV energy range.

AG focusing is still resorted to today, for instance in the hadrontherapy application (Fig. 9.6), light source lattice [10], and other high energy collider design [11], as it has the merit of compactness. On the other hand, the flexibility of separated function optics made it more popular: it allows to introduce modular functions in complex ring designs such as dispersion suppression sections, low-beta or insertion device sections, long straights, et cetera. Low-emittance, high-brightness light source lattices have complicated focusing further, by introducing longitudinal field gradient bending systems to minimize equilibrium beam emittance [12].

Due to the necessary ramping of the field in order to maintain a constant orbit, synchrotron accelerators are pulsed, storage rings in some cases as well, high energy colliders in particular to bring beams to highest store energy. The acceleration is cycled and the accelerating voltage frequency as well in ion accelerators, from injection to top energy. If the ramping uses a constant electromotive force, then (Eq. 8.3)

$$B(t) \approx \frac{t}{\tau} \tag{9.1}$$

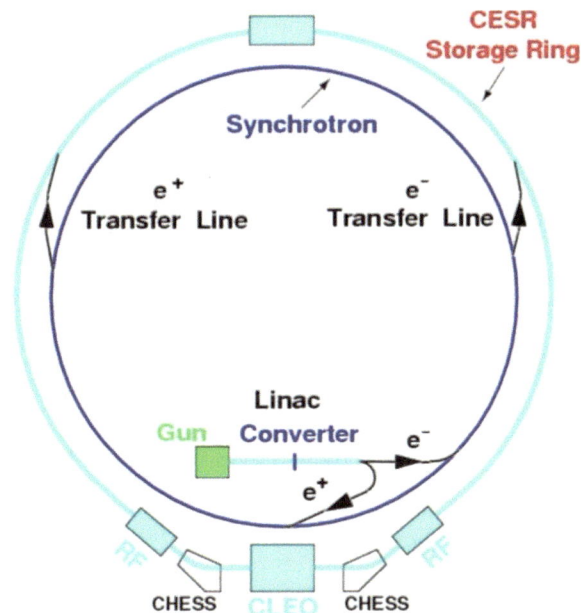

Fig. 9.7 Cornell rapid cycling synchrotron, 5 GeV injector of CESR storage ring [13]

$\dot{B} = dB/dt$ does not exceed a few Tesla/second, thus the repetition rate of the acceleration cycle is of the order of a Hertz. If instead the magnet winding is part of a resonant circuit then the field oscillates,

$$B(t) = B_0 + \frac{\hat{B}}{2}(1 - \cos \omega t) \tag{9.2}$$

so that, in the interval of half a voltage repetition period (i.e., $t : 0 \rightarrow \pi/\omega$) the field increases from an injection threshold value to a maximum value at highest rigidity, $B(t) : B_0 \rightarrow B_0 + \hat{B}$. The latter determines the highest achievable energy: $\hat{E} = pc/\beta = q\hat{B}\rho c/\beta$. The repetition rate with resonant magnet cycling can reach a few tens of Hertz, a technique known as a rapid-cycling synchrotron (RCS). In both cases anyway B imposes its law and other parameters, comprising the acceleration cycle, the RF frequency in particular, will follow B(t).

Instances of RCS rings include Cornell 12 GeV, 60 Hz electron AG synchrotron [14] (Fig. 9.7), commissioned in 1967 with a 7 GeV beam, a world record at the time, and still in operation half a century later as the injector of Cornell 5 GeV storage ring (CESR/CHESS) [15]; Fermilab 8 GeV, 60 Hz Booster, which provides protons for the production of neutrino beams; the 30 GeV 500 kW proton beam J-PARC facility in Japan. Rapid cycling is also considered in ion-therapy applications (Fig. 9.6).

To conclude on these preliminaries, lets mention the giants among accelerator facilities which nuclear (NP) and particle (HEP) physics research laboratories are: so far, strong focusing synchrotrons happen to be the building blocks from which

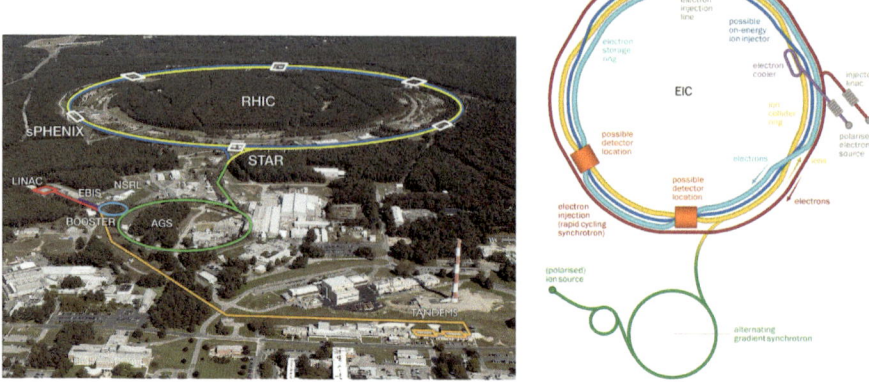

Fig. 9.8 RHIC complex at the Brookhaven National Laboratory (left) [2], a cascade of 4 strong focusing ion synchrotrons: the AGS and its Booster, and the 3.8 km circumference intersecting RHIC rings, in motion towards the EIC project (right) [16] which will add 2 electron synchrotrons: an 18 GeV storage ring and its RCS injector

they are constructed. This is so at the CERN LHC complex. This is apparent also in Fig. 9.8 which shows RHIC heavy ion collider complex, and its planned evolution, the Electron-Ion Collider [17][1]. The next colliders could be linacs, it was at SLAC with the SLC [18], it was the plan with such projects as TESLA [19], the NLC [20]. The interest of NP and HEP will decide on the research tools: more large synchrotron rings for a muon collider [21], an FCC-ee, -hh and other -eh [8], or high gradient linacs for the ILC [22] or for ReLic e^+e^- collider [23]. Or new acceleration methods and technologies?

9.2 Basic Concepts and Formulæ

Alternating gradient focusing is sketched in Fig. 9.9. An order of magnitude of the focusing index can be estimated from the fields met in these structures: say a maximum B~1 Tesla in the dipole gap, same at pole tip in quadrupoles ~10 cm off axis. The latter results in $\frac{\Delta B}{\Delta x} \sim 10$ T/m, the former in meters to tens of meters dipole curvature radius. All in all, in absolute value,

$$n = -\frac{\rho}{B}\frac{\partial B}{\partial x} \sim \frac{10^{0\sim2}_{[m]}}{1_{[T]}} \times 10_{[T/m]} \sim 10^{1\sim3} \quad \gg 1 \tag{9.3}$$

much greater than in a weak focusing structure, characterized by $0 < n < 1$.

---

[1] Beam polarization studies have been using zgoubi in all five EIC synchrotrons.

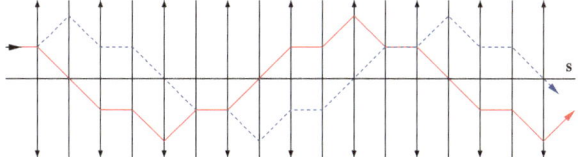

Fig. 9.9 Horizontally focusing lenses (field index $n \gg 0$, the solid red trajectory) are vertically defocusing ($n \ll 0$, the dashed blue trajectory), and vice versa. This imposes alternating gradients in order for a sequence to be globally focusing, for both planes

9.2.1 Components of the Strong Focusing Optics

Combined Function (AG) Optics

This is, typically, the BNL AGS and CERN PS optics, using dipoles that ensure both beam guiding and focusing (Fig. 9.2). Separate quadrupole and multipole lenses have later been introduced as they provide knobs for the adjustment of optical functions and other parameters. AG optics is still topical in modern designs, as in the iRCMS whose six 60° arcs are comprised of a sequence of five focusing and defocusing combined function dipoles [9], Fig. 9.6.

Field

Referring to normal conducting magnet technology, a hyperbolic pole profile (Fig. 9.1) is an equipotential (a line of constant scalar potential V) of equation

$$V_{\text{pole}} = A\,xy$$

at the origin of a magnetic field $\mathbf{B} = \mathbf{grad}\,V$, everywhere perpendicular to the equipotential. A combined function dipole with mid-plane geometrical symmetry is defined by materializing two equipotentials, at $\pm V_{\text{pole}}$ (Fig. 9.10). This results in a vertical field component $B_y = \partial V/\partial y = Ax$, and therefore a radial field index

$$n = -\frac{\rho}{B_y}\frac{\partial B_y}{\partial x}\bigg|_{y=0} = \frac{\rho}{B_y}A$$

A is a constant, typically up to $\sim$10 T/m, cf. Eq. 9.3. The pole profile opens up either inward (toward the center of curvature, a horizontally focusing dipole, vertically defocusing) or outward (a vertically focusing dipole, horizontally defocusing), Fig. 9.11.

In a bent AG dipole a line of constant field is an arc of a circle; the field guides the reference particle along the arc in the median plane. The mid-plane field can be expressed under the form

Fig. 9.10 Symmetric
materialization of pole
profiles, at $\pm V$. Nothing
would preclude materializing
poles at V_1 and $-V_2$
potentials, with the same
resulting field between the
poles

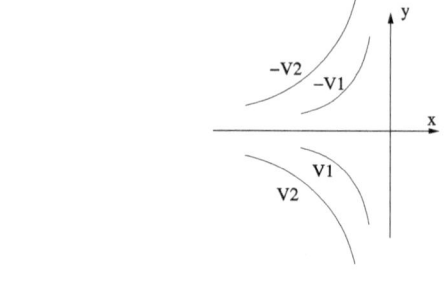

Fig. 9.11 Beam focusing in
combined function dipoles.
The center of curvature is to
the left. The pole profile
follows an equipotential
$V = Axy$. Top: the pole
profile opens up towards the
center of curvature → the
dipole is horizontally
converging (vertically
diverging: current I comes
out of the page, force **F**
results from field **B**).
Bottom: pole profile closing
toward the center of
curvature → the dipole is
horizontally diverging,
vertically converging

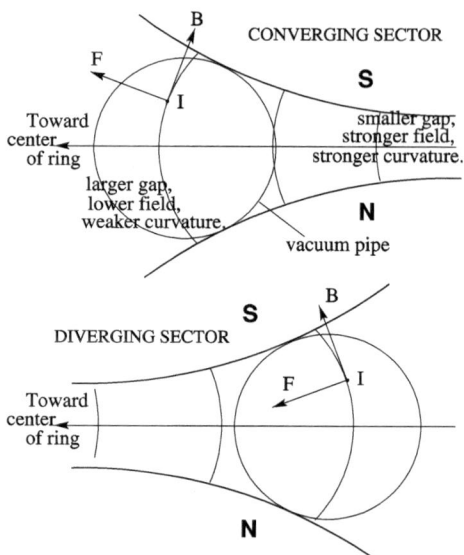

$$B_y(r, \theta) = \mathcal{G}(r, \theta)\, B_0 \left(1 + n\, \frac{r - r_0}{r_0} + n'\left(\frac{r - r_0}{r_0}\right)^2 + n''\left(\frac{r - r_0}{r_0}\right)^3 + \ldots \right)$$
(9.4)

with r_0 the reference (normally the orbit) radius. Higher order indices, sextupole n',
octupole n'', . . ., may be residual effects from fabrication tolerances, magnetic saturation, deformation of yoke with years, etc., or included by design, with significant
value.

In a straight AG dipole, a line of constant field is a straight line; an instance
is the AGS main magnet (Fig. 9.2). Another instance is the Fermilab recycler arcs
permanent magnet dipole, which includes quadrupole and sextupole components [24,
25]. The modeling of the field in a straight combined function dipole can be derived
from the scalar potential of Eq. 9.5.

Separated Function Optics

In a separated function lattice quadrupole lenses ensure the essential of the focusing, main bends have zero index. In smaller rings though, geometrical focusing in bending magnets may be significant (Sect. 8.2.1, Fig. 8.6). Wedge angles in addition may be introduced and contribute horizontal and vertical focusing/defocusing (Fig. 8.8).

Higher order multipole lenses are used for the compensation of adverse effects: coupling, aberrations, space charge, impedance, etc., and for beam manipulations: controlling the coupling, resonant extraction, etc.

The field in a multipole of order n ($n = 1$, 2, 3, etc.: dipole, quadrupole, sextupole, etc.) derives, via $\mathbf{B} = \mathbf{grad}V$, from the Laplace potential [26]

$$V_n = (n!)^2 \left\{ \sum_{q=0}^{\infty} (-1)^q \alpha_{n,0}^{(2q)}(s) \frac{(x^2 + y^2)^q}{4^q q!(n+q)!} \right\} \left\{ \sum_{m=0}^{n} \frac{x^{n-m} y^m}{m!(n-m)!} \sin m \frac{\pi}{2} \right\} \quad (9.5)$$

where $\alpha_{n,0}^{(2q)}(s) = d^{2q} \alpha_{n,0}(s)/ds^{2q}$ accounts for the s-dependence of the potential. Technologies for multipoles and combined multipoles include pole profiling, permanent magnets [24, 27], superconducting $\cos n\theta$ winding as in RHIC and LHC colliders, and variants.

In a hard-edge field model the $\sum_{q=0}^{\infty}$ series is reduced to the $q = 0$ term, with the following outcomes [28, 29].

Quadrupole

The equipotential (the pole profile) is an equilateral hyperbola, of equation $Gxy =$ constant in an upright quadrupole (left figure below), and $G(x^2 - y^2) =$ constant in a $\pi/4$ skew quadrupole (right). The resulting field writes

$$B_x = \frac{\partial V}{\partial x} = Gy$$

$$B_y = \frac{\partial V}{\partial y} = Gx$$

$$B_x = Gx$$

$$B_y = -Gy$$

Upright quadrupoles are used for focusing, skew quadrupoles are used to compensate, or introduce, transverse coupling. The focusing strength

$$K = \frac{1}{L} \frac{\int G(s) \, ds}{p/q} \quad (9.6)$$

is momentum-dependent.

Sextupole

The equipotential satisfies $H(3x^2y - y^3) = $ constant in an upright sextupole (left), $H(x^3 - 3xy^2) = $ constant in a $\pi/6$ skew sextupole (right), with resulting field

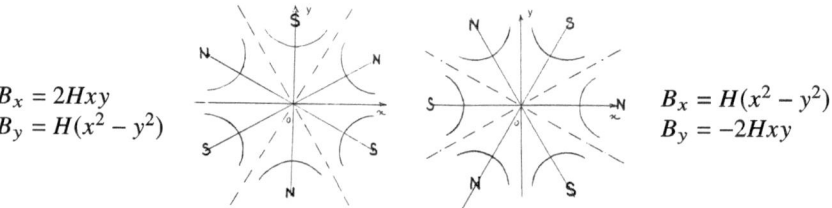

$B_x = 2Hxy$
$B_y = H(x^2 - y^2)$

$B_x = H(x^2 - y^2)$
$B_y = -2Hxy$

Upright sextupoles introduce a vertical field component $B_y \propto x^2$, they are used to correct optical aberrations, to modify the momentum dependence of the wave numbers ν_x, ν_y, and in beam manipulations such as resonant extraction. Skew sextupoles introduce a radial field component $B_x \propto y^2$, they are used to correct optical aberrations.

Octupole

The equipotential pole profile satisfies $O(x^3y - xy^3) = $ constant in an upright octupole (left), $O(x^4 - 6x^2y^2 + y^4) = $ constant in a $\pi/8$ skew octupole (right), yielding the field

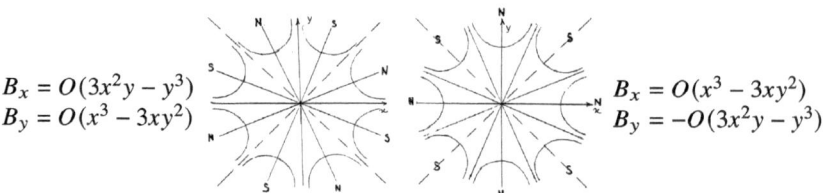

$B_x = O(3x^2y - y^3)$
$B_y = O(x^3 - 3xy^2)$

$B_x = O(x^3 - 3xy^2)$
$B_y = -O(3x^2y - y^3)$

Upright octupoles are used to introduce a vertical field component $B_y \propto x^3$; skew octupoles introduce a vertical field component $B_y \propto y^3$. Octupoles are used to correct aberrations, or to modify the amplitude dependence of wave numbers.

9.2.2 Transverse Motion

The transverse motion of a particle in the S-periodic lattice of a cyclic accelerator, at design momentum p_0 and with curvature radius ρ_0, satisfies Hill's equations[2]

---

[2] Acceleration, or deceleration, adds a velocity term, betatron damping results. This is addressed in "Betatron damping", Sect. 10.2.3, where it accounts in addition for a non-constant varying orbital radius.

$$\frac{d^2x}{ds^2} + K_x(s)x = \frac{1}{\rho_0}\frac{\Delta p}{p_0}, \qquad \frac{d^2y}{ds^2} + K_y(s)y = 0 \tag{9.7}$$

where $K_x(s)$, $K_y(s)$ have the periodicity of the lattice ($K_{\substack{x\\y}}(s+S) = K_{\substack{x\\y}}(s)$), and depend locally on the nature of the optical elements, in the following way.

Case of

$$-\text{dipole}:\quad\begin{cases} K_x = \dfrac{1-n}{\rho_0^2} \\[2mm] K_y = \dfrac{n}{\rho_0^2} \end{cases} \qquad \left(n = -\frac{\rho_0}{B_0}\frac{\partial B_y}{\partial x}\right) \tag{9.8}$$

$$-\text{a wedge at } s = s_w:\quad \begin{cases} K_{\substack{x\\y}} = \pm\dfrac{\tan\varepsilon}{\rho_0}\delta(s-s_w) \end{cases} \left(\text{with } \varepsilon \lessgtr 0 \text{ if } \begin{matrix}\text{focusing}\\\text{defocusing}\end{matrix}\right)$$

$$-\text{quadrupole}: K_{\substack{x\\y}} = \frac{\pm G}{B\rho};\; \frac{1}{\rho_0} = 0 \quad \left(\text{gradient } G = \frac{\text{field at pole tip}}{\text{radius at pole tip}}\right)$$

$$-\text{drift space}: K_x = K_y = 0;\; \frac{1}{\rho_0} = 0$$

By contrast with the betatron and weak focusing technologies, strong focusing with its independent focusing ($G > 0$) and defocusing ($G < 0$) gradient families allows separate adjustment of the horizontal and vertical focusing strengths, and wave numbers as a consequence.

The on-momentum ($p = p_0$) closed orbit coincides with the reference axis of the optical elements. The betatron motion for an on-momentum particle satisfies Eq. 9.7 with $\Delta p = 0$. Solving the latter (see section "Betatron Motion") requires introducing two independent solutions $u_{\frac{1}{2}}(s)$ (Eq. 8.12), the linear combination of which yields the pseudo harmonic motion (Eq. 8.14)

$$\left| \begin{aligned} u(s) &= \sqrt{\beta_u(s)\varepsilon_u/\pi}\,\cos\left(\int\frac{ds}{\beta_u(s)} + \varphi_u\right) \\[3mm] u'(s) &= -\sqrt{\frac{\varepsilon_u/\pi}{\beta_u(s)}}\,\sin\left(\int\frac{ds}{\beta_u(s)} + \varphi_u\right) + \alpha(s)\cos\left(\int\frac{ds}{\beta_u(s)} + \varphi_u\right) \end{aligned}\right. \tag{9.9}$$

The motion satisfies the Courant-Snyder invariant, namely (Fig. 9.12)

$$\gamma_u(s)u^2 + 2\alpha_u(s)uu' + \beta_u(s)u'^2 = \frac{\varepsilon_u}{\pi} \tag{9.10}$$

i.e., the surface of the phase space ellipse is a constant of the motion. Its form and orientation (Fig. 9.12) change along the period as a consequence of the strong

Fig. 9.12 Courant-Snyder
invariant and turn-by-turn
harmonic motion along the
invariant, observed at some
azimuth s. The aspect ratio
of the ellipse depends on the
observation azimuth s but its
area ε_u is invariant

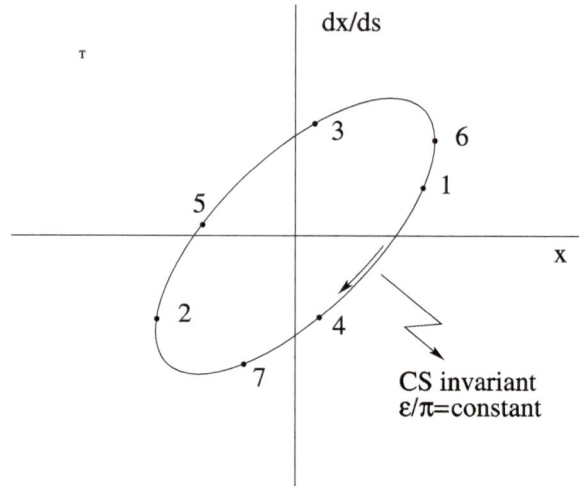

modulation of the betatron functions (Fig. 9.13), far more than in a weak focusing lattice which features weak betatron modulation: $\alpha_u(s) \approx 0$ and $\beta_u(s) \approx$ constant (Figs. 8.9 and 8.10).

Beam envelopes are given by the extrema,

$$\hat{x}_{\mathrm{env}}(s) = \pm\sqrt{\beta_x(s)\frac{\varepsilon_x}{\pi}}, \qquad \hat{y}_{\mathrm{env}}(s) = \pm\sqrt{\beta_y(s)\frac{\varepsilon_y}{\pi}} \tag{9.11}$$

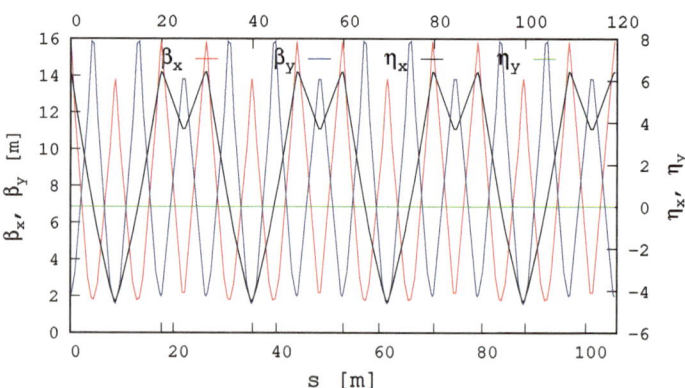

Fig. 9.13 Optical functions around SATURNE 2 synchrotron, a 4-period FODO cell lattice

Phase Space Motion

Write the two independent solutions $u_{\frac{1}{2}}(s)$ (Eq. 8.12) under the form

$$u_1(s) = \underbrace{F(s)}_{S-\text{periodic}} \times \underbrace{e^{i\mu\frac{s}{S}}}_{\frac{2\pi S}{\mu}-\text{periodic}} \quad \text{and} \quad u_2(s) = u_1^*(s) = F^*(s)\, e^{-i\mu\frac{s}{S}} \qquad (9.12)$$

where

$$F(s) = \sqrt{\beta_u(s)}\, e^{i\left(\int_0^s \frac{ds}{\beta_u(s)} - \mu\frac{s}{S}\right)} \qquad (9.13)$$

Introduce

$$\psi_u(s) = \int_0^s \frac{ds}{\beta_u(s)} - \mu\frac{s}{S} \qquad (9.14)$$

so that $F(s) = \sqrt{\beta_u(s)}\, e^{i\psi_u(s)}$. Equation 9.9 thus takes the form

$$\left|\begin{array}{l} u(s) = \overbrace{\sqrt{\beta_u(s)\varepsilon_u/\pi}}^{S-\text{periodic}}\ \cos\overbrace{\left[\nu\frac{s}{R} + \underbrace{\psi_u(s)}_{S-\text{per.}} + \varphi_u\right]}^{\frac{2\pi S}{\mu}-\text{periodic}} \\[2em] u'(s) = -\sqrt{\frac{\varepsilon_u/\pi}{\beta_u(s)}}\ \sin\left[\nu\frac{s}{R} + \psi_u(s) + \varphi_u\right] + \alpha(s)\ \cos\left[\nu\frac{s}{R} + \psi_u(s) + \varphi_u\right] \end{array}\right. \qquad (9.15)$$

where $\nu = \dfrac{N\mu}{2\pi}$. Thus, as the betatron function $\beta_u(s)$ and phase $\psi_u(s)$ are S-periodic, the turn-by-turn motion observed at a given azimuth s (i.e., $u(s)$, $u(s + S)$, $u(s + 2S)$, ...) is sinusoidal and its frequency is $\nu = N\mu/2\pi$. Successive particle positions $(u(s), u'(s))$ in phase space lie on the Courant-Snyder invariant (Eq. 9.10). The working point (ν_x, ν_y) fully characterizes the first order optical setting of the lattice.

Off-Momentum Motion

The motion of an off-momentum particle satisfies the inhomogeneous Hill's horizontal differential Eq. 9.7. The chromatic closed orbit

$$x_{\text{ch}}(s) = D_x(s)\frac{\delta p}{p} \qquad (9.16)$$

is a particular solution of the equation, its periodicity is that of the cell.

By contrast with a weak focusing lattice where chromatic closed orbits are parallel (Eq. 8.26), in a strong focusing lattice they are distorted (Fig. 9.13), their excursion depends on the distribution along the cell of (i) the dispersive elements which are the dipoles, and (ii) the focusing.

The horizontal motion of an off-momentum particle is a superposition of the particular solution (Eq. 9.16) and of the betatron motion, solution of the homogeneous Hill's equation (Eq. 9.15), namely

$$x(s) = x_\beta(s) + x_{\mathrm{ch}}(s) = \sqrt{\beta_x(s)\frac{\varepsilon_x}{\pi}} \cos\left(\int \frac{ds}{\beta_x} + \varphi_x\right) + D_x(s)\frac{\delta p}{p_0} \qquad (9.17)$$

whereas the vertical motion is unchanged (Eq. 9.15 taken for $u(s) \equiv y(s)$).

Chromaticity

The focusing strength of combined function dipoles and quadrupoles is a decreasing function of particle rigidity $B\rho = p/q$ (Eq. 9.8). In a ring this affects the horizontal and vertical wave numbers, an effect quantified as the chromaticity, $\xi_{x,y}$. To the first order in $\delta p/p$, this writes

$$\delta\nu_{x,y} = \xi_{x,y}\frac{\delta p}{p} \qquad (9.18)$$

A linear lattice has a natural chromaticity. Over a distance $\mathcal{L}$ it is given by

$$\xi_{x,y} = \frac{-1}{4\pi}\int_{\mathcal{L}}\beta_{x,y}(s)K_{x,y}(s)ds \qquad (9.19)$$

Use a circular integral, $\oint$ in the case of a ring. The natural chromaticity is a negative quantity: focusing decreases with increasing momentum.

One consequence of the chromaticity is that beam momentum spread $\delta p/p$ results in a tune spread $\delta\nu_{x,y} = \xi_{x,y} \times \delta p/p$, a beam occupies an extended area in the tune diagram. For this reason in particular, the chromaticity is usually corrected. This is realized by placing sextupoles in dispersive sections, at least two families: a family of horizontal lenses (strength H_x) located at large β_x and a family of vertical lenses (strength H_y) located at large β_y.

The effect leaned on is the following:

– betatron motion $x_\beta(s)$ of particles with momentum $p_0 + \Delta p$ is around an off-centered, chromatic closed orbit $x_{\mathrm{ch}}(s)$ (Eq. 9.16);
– introducing a sextupole results in a local gradient as $B_y \propto (x_{\mathrm{ch}} + x_\beta)^2 = x_{\mathrm{ch}}^2 + 2x_{\mathrm{ch}}x_\beta + x_\beta^2$, namely, $\left.\frac{\partial B_y}{\partial x}\right|_{x=x_{\mathrm{ch}}} = 2x_{\mathrm{ch}} = 2D_x\frac{\Delta p}{p}$. This results in a focusing force proportional to $\delta p/p$. Sextupoles contribute to chromaticity (or its compensation) following

$$\xi_{x,y} = \frac{1}{4\pi}\int H_{x,y}(s)\beta_{x,y}(s)D_x(s)ds \qquad (9.20)$$

9.2.3 Resonances

Consider the excitation of transverse beam motion by a generator of frequency Ω located at some azimuth along the ring [29]. The action of the excitation $S \times \sin \Omega t$ on the oscillating motion $u(t)$ can be written under the form

$$\frac{d^2 u}{dt^2} + \omega^2 u = S \sin \Omega t \tag{9.21}$$

Assume harmonic motion for simplicity (as in a weak focusing lattice). Take generator amplitude $S = $ constant, the solution (superposition of the solution of the homogeneous differential equation and of a particular solution of the inhomogeneous differential equation) writes

$$u(t) = U \cos(\omega t + \varphi_u) + \frac{S}{\omega^2 - \Omega^2} \sin \Omega t \tag{9.22}$$

If betatron motion and excitation are in synchronism, i.e. on the resonance, $\omega = \Omega$, a particular solution of Eq. 9.21 is

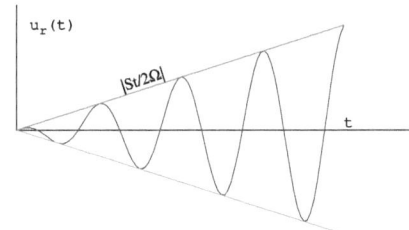

$$u_r(t) = -\frac{S\,t}{2\Omega} \cos \Omega t$$

the amplitude of the oscillatory motion grows rapidly with time, at a rate $|St/2\Omega|$.

Assume the amplitude S to be T'-periodic instead, angular frequency $\omega' = 2\pi/T'$, take its Fourier expansion

$$S(t) = \sum_{p=0}^{\infty} a_p \cos(p\omega' t + \varphi_p)$$

the equation of motion thus writes

$$\frac{d^2 u}{dt^2} + \omega^2 u = \sum_{p=0}^{\infty} a_p \cos(p\omega' t + \varphi_p) \sin \Omega t =$$
$$\sum_{p=0}^{\infty} \frac{a_p}{2} \Big[\sin[(\Omega - p\omega')t + \varphi_p] + \sin[(\Omega + p\omega')t + \varphi_p] \Big] \tag{9.23}$$

Resonance may occur at generator frequencies $\Omega = \omega \pm p\omega'$, the strength depends on the amplitude a_p of the excitation harmonics. A generator at some point in the lattice excites all harmonics with equal amplitudes a_p. In the case of an extended excitation source, low harmonics only matter.

Sextupole and Octupole Resonances

The horizontal motion in the presence of sextupoles $(B_y(\theta)|_{y=0} = S(\theta)x^2)$ satisfies

$$\frac{d^2x}{d\theta^2} + \nu_x^2 x = S(\theta)x^2 \tag{9.24}$$

Assume weak perturbation of the motion, so that $x(\theta) \approx \hat{x}\cos(\nu_x\theta + \varphi_x)$, the solution for unperturbed motion. Assume also $S(\theta)$ 2π-periodic. Substitute its Fourier series expansion $S(\theta) = \sum_{p=0}^{\infty} a_p \cos(p\theta + \varphi_p)$ in Eq. 9.24, develop to get

$$\frac{d^2x}{d\theta^2} + \nu_x^2 x = \frac{\hat{x}^2}{2}\left[\sum_{p=0}^{\infty} a_p \cos(p\theta + \varphi_p) \right.$$

$$\left. + \frac{1}{2}\sum_{p=0}^{\infty} a_p\left[\cos[(p - 2\nu_x)\theta + \varphi_p - 2\varphi_x] + \cos[(p + 2\nu_x)\theta + \varphi_p + 2\varphi_x]\right]\right] \tag{9.25}$$

Thus resonance may occur at the betatron frequency families $\nu_x = \pm p$, $\nu_x = \pm(p - 2\nu_x)$, and $\nu_x = \pm(p + 2\nu_x)$, i.e.,

$$\begin{bmatrix} \nu_x = p \\ 3\nu_x = p \end{bmatrix}$$

In the case of a single sextupole in the ring, all the harmonics p are excited with the same amplitude a_p.

An octupole introduces a field component $B_y(\theta)|_{y=0} = O(\theta)x^3$. A similar development yields

$$\begin{bmatrix} \nu_x = p \\ 2\nu_x = p \\ 4\nu_x = p \end{bmatrix}$$

Resonances in a general manner occur at betatron frequencies satisfying

$$m\nu_x + n\nu_y = \text{integer}$$

In this coupling regime one has

$$\frac{\varepsilon_x}{m} - \frac{\varepsilon_y}{n} = \text{constant}, \quad \text{an invariant of the motion} \tag{9.26}$$

From this it results that,

– if m and n have opposite signs the resonance causes energy exchange between the horizontal and vertical motions: $\frac{\varepsilon_x}{|m|} + \frac{\varepsilon_y}{|n|} = \text{constant}$, an increase of ε_x correlates with a decrease of ε_y and vice-versa. In the presence of linear coupling for instance, $\nu_x - \nu_y =$ integer, $\varepsilon_x + \varepsilon_y =$ constant. An increase in motion amplitude anyway may cause particle loss, an issue in cyclotrons where the Walkinshaw resonance $\nu_x = 2\nu_y$ causes vertical beam loss due to the increase of ε_y;

– if m and n have the same sign the resonance is liable to induce motion instability: $\frac{\varepsilon_x}{m} - \frac{\varepsilon_y}{n} =$ constant, ε_x and ε_y may both increase with no limit.

Resonant Extraction

Resonant extraction is based on the effect of a non-linear force on a dynamical system. A linear regime, under the effect of linear forces, satisfies Eq. 9.7. If $x(s)$ is a stable solution, so is $\lambda x(s)$ (λ a proportionality constant). Introducing a non-linear force modifies the equation of motion, into for instance

$$\diamond \quad \frac{d^2x}{ds^2} + K_x(s)x = S(s)x^2: \text{ sextupole perturbation,}$$

$$\diamond \quad \frac{d^2x}{ds^2} + K_x(s)x = O(s)x^3: \text{ octupole perturbation,}$$

If $x(s)$ is a stable solution, it may no longer be the case for $\lambda x(s)$. If $x(s)$ is small enough the motion, subject to linear and non-linear forces, is quasi-linear and stable. However, increasing the motion amplitude will at some point result in unstable motion. In the (x, x') phase space, the stable regime is bounded by a separatrix. Outside the latter the motion is essentially unstable, or liable to reach amplitudes beyond transverse acceptance of the accelerator (Fig. 9.14).

Fig. 9.14 Horizontal motion near a 3rd integer resonance. Within the triangle separatrix the motion is stable. Outside the triangle, motion reaches large amplitudes. An electrostatic septum extracts particles which jump to the right of the septum (into the extraction channel) during their motion

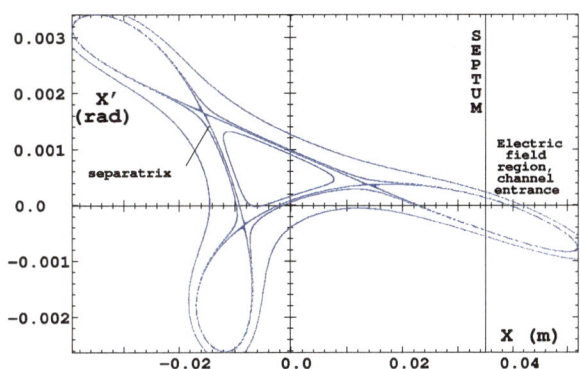

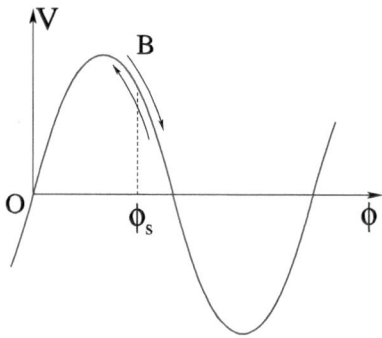

Fig. 9.15 In the presence of
RF, particles oscillate in the
vicinity of the synchronous
phase. Above transition, in
this schematic

9.2.4 Acceleration. Synchrotron Motion

Particle motion in longitudinal phase space (phase, momentum) and its stability
are determined by the lattice and by the acceleration parameters, as introduced in
Sect. 8.2.2. They include the

- RF $f_{\mathrm{rf}} = h f_{\mathrm{rev}}$,
- voltage $V(t) = \hat{V} \int \sin \omega_{\mathrm{rf}} dt$,
- synchronous phase ϕ_s (phase of the particle in synchronism with the RF oscillation), which increases by $2\pi h$ per turn,
- transition $\gamma_{\mathrm{tr}} = 1/\sqrt{\alpha}$ (Fig. 8.15).

In the case of weakly modulated betatron functions (weak focusing lattice; AG
lattice to some extent), $\alpha \approx 1/\nu_x^2$ so that

$$\gamma_{\mathrm{tr}} \approx \nu_x$$

This is the case of SATURNE 1: a weak focusing lattice (see Chap. 8 and simulation
exercises there) operated above transition as $\gamma_{\mathrm{tr}} = \nu_x \approx 0.6$. In the AGS at BNL the
working point is $\nu_x \approx 8.7$ whereas $\gamma_{\mathrm{tr}} = 8.4 \approx \nu_x$; transition is crossed as proton
beams are accelerated from $\gamma \approx 3$ to $\gamma \approx 25$. Instead, SATURNE 2 strong focusing
lattice was operated at negative α, $\eta = \frac{1}{\gamma^2} - \alpha$ does not cancel, γ_{tr} is pure imaginary.

The energy gain per turn at the cavity is

$$\Delta W = 2\pi R\, q\rho \dot{B} = q\hat{V} \sin \phi_s$$

ΔW is imposed by the field law in order to ensure that at all time the synchronous
particle momentum satisfies

$$p_s(t) = q B(t)\rho$$

Phase Stability

Particles with phase and momentum offsets $(\Delta\phi, \Delta p/p) = (\phi - \phi_s, (p - p_s)/p_s)$ in the vicinity of the synchronous particle at (ϕ_s, p_s) undergo periodic longitudinal oscillations (Fig. 9.15). The longitudinal motion satisfies the differential equations

$$\frac{d\Delta\phi}{dt} = h\eta\omega_s \frac{\Delta p}{p}, \qquad \frac{d(\Delta p/p)}{dt} = \frac{e\hat{V}\omega_s}{2\pi\beta_s^2 E_s}[\sin\phi - \sin\phi_s] \qquad (9.27)$$

If peak amplitudes are small the differential Eq. 9.27 yield

$$\frac{d^2\Delta\phi}{dt^2} + \Omega_s^2\Delta\phi = 0 \qquad (9.28)$$

the motion is sinusoidal, with a synchrotron angular frequency

$$\Omega_s = \frac{c}{R}\sqrt{\frac{\eta h q \hat{V}\cos\phi_s}{2\pi E_s}} \qquad (9.29)$$

The synchrotron tune, number of synchrotron oscillations per revolution, writes

$$\nu_s = \frac{\Omega_s}{\omega_{\mathrm{rev}}} = \frac{1}{\beta_s}\sqrt{\frac{\eta h q \hat{V}\cos\phi_s}{2\pi E_s}} \qquad (9.30)$$

Synchrotron oscillations are slow compared to betatron oscillations, typically $\nu_s \sim \nu_{x,y}/10^{2\sim3}$. Motion stability requires $\Omega_s^2 > 0$, or

$$\eta\cos\phi_s > 0$$

Longitudinal motion in $(\phi, \dot{\phi}/\Omega_s)$ phase space is on a circle. The extent in phase and energy, or momentum, of the small amplitude oscillations satisfy

$$\widehat{\Delta\phi} = \frac{h\eta E_s}{p_s R\Omega_s}\frac{\widehat{\Delta E}}{E_s} = \frac{h\eta E_s}{p_s R\Omega_s}\beta_s^2\frac{\widehat{\Delta p}}{p} \qquad (9.31)$$

The bunch length is

$$L_{\mathrm{bunch}} = \frac{R}{h}\widehat{\Delta\phi} \qquad (9.32)$$

Separatrix

If peak amplitudes are large the oscillations are non-linear and, assuming slow acceleration

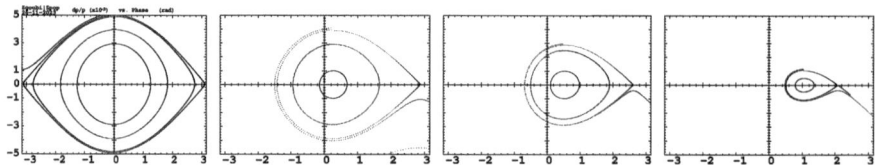

Fig. 9.16 Longitudinal motion separatrix in $(\phi, dp/p)$ phase space, and some stable as well as unbounded motions. Case of SATURNE 2 at injection energy, 50 MeV. From left to right: case of $\phi_s = 0$ (stationary bucket), $\phi_s = 15$, 30, and 60°. Small motions are centered on ϕ_s, their synchrotron tunes satisfy Eq. 9.30. The momentum acceptance (height of the separatrix) satisfies Eq. 9.36, with respectively $\pm\frac{\widehat{\Delta p}}{p} \approx 0.00496$, 0.00392, 0.00290 and 0.00107

$$\frac{d^2 \Delta\phi}{dt^2} + \Omega_s^2 \frac{\sin\phi - \sin\phi_s}{\cos\phi_s} = 0 \tag{9.33}$$

A first integral of this equation is the equation of the separatrix (Fig. 9.16)

$$\frac{\dot\phi^2}{2} - \Omega_s^2 \frac{\cos\phi + \phi\sin\phi_s}{\cos\phi_s} = -\Omega_s^2 \frac{\cos(\pi - \phi_s) + (\pi - \phi_s)\sin\phi_s}{\cos\phi_s} \tag{9.34}$$

This defines two locations where $\dot\phi$ changes sign, i.e. $\dot\phi = 0$, namely,

(i) $\phi_1 = \pi - \phi_s$,
(ii) ϕ_2 such that $\cos\phi_2 + \phi_2\sin\phi_s = \cos(\pi - \phi_s) + (\pi - \phi_s)\sin\phi_s$.

The motion is stable, oscillatory, within the domain $\phi \in [\phi_1, \phi_2]$, the "bucket", and unbounded beyond. The bucket height is obtained for $\phi = \phi_s$, namely, from Eq. 9.34

$$\frac{\dot\phi_{\max}}{\Omega_s} = \sqrt{2\left[2 - (\pi - 2\phi_s)\tan\phi_s\right]} \tag{9.35}$$

Expressed in momentum,

$$\pm\frac{\widehat{\Delta p}}{p} = \pm\frac{1}{\beta_s}\sqrt{\frac{q\hat{V}}{\pi h \eta E_s}\left[2\cos\phi_s - (\pi - 2\phi_s)\sin\phi_s\right]} \tag{9.36}$$

Its dependence on ϕ_s is represented in Fig. 9.17. Stationary bucket mode, i.e. $\sin\phi_s = 0$, has greatest acceptance. The latter decreases in accelerated bucket mode as $\phi_s \rightarrow \pi/2$ (Fig. 9.16).

Adiabatic Damping of Synchrotron Oscillations

The equation of motion, Eq. 9.33, assumes a slow acceleration rate, $dT_{\text{rev}}/dt \ll 1$, such that $p_s(t)$, η, possibly $\hat{V}$, and thus Ω_s change slowly during synchrotron

Fig. 9.17 Dependence of the momentum extent of the bucket (normalized to $\frac{1}{\beta_s}\sqrt{\frac{q\hat{V}}{\pi h \eta E_s}}$) on the synchronous phase ϕ_s. It takes its value in $\sqrt{2} \rightarrow 0$ for $\sin \phi_s : 0 \rightarrow 1$

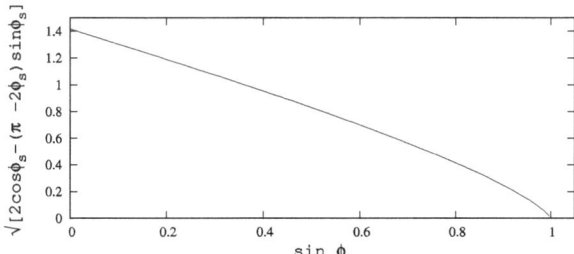

oscillations and therefore can be considered constant. The extreme phase and momentum excursions during acceleration satisfy

$$\widehat{\Delta\phi} \propto \left(\frac{\eta}{R^2 \gamma \hat{V} \cos \phi_s} \right)^{1/4}$$

$$\frac{\widehat{\Delta p}}{p} \propto \frac{1}{\beta_s} \left(\frac{\hat{V} \cos \phi_s}{\eta \gamma^3 R^2} \right)^{1/4} \tag{9.37}$$

In the case of acceleration on a fixed orbit (constant radius R),

$$\widehat{\Delta\phi} \times \widehat{\Delta p} = \text{constant} \tag{9.38}$$

Adiabatic Damping of the Betatron Oscillations

The mechanism is described in Sect. 8.2.2 (Fig. 8.14), the equations of motion are addressed in Sect. 10.2.3. In the case of an adiabatic change of momentum $p = \beta\gamma m_0 c$ (a slow change compared to the betatron motion oscillation frequency) the transverse motion damping satisfies

$$p\,\varepsilon_u = \text{constant}, \quad \text{or} \quad \beta\gamma\varepsilon_u = \text{constant} \tag{9.39}$$

Coordinate damping satisfies (Eq. 10.22 with orbit radius $R = \text{constant}$)

$$x, y \propto 1/\sqrt{p}, \quad x', y' \propto 1/\sqrt{p}. \tag{9.40}$$

9.2.5 Synchrotron Radiation, Dynamical Effects

Emittance growth upon SR matters in high γ rings, electron rings so far, muon collider possibly in the future [30] and other FCC lepton and hadron collider [8].

The stochastic nature of SR and the energy loss it results in, have been introduced in Chap. 5. Dynamical effects in a synchrotron ring are further addressed here [31, 32].

Motion Invariants

In the absence of perturbation by synchrotron radiation, particle motion satisfies the Courant-Snyder (Eq. 9.41) and longitudinal (Eq. 9.42) phase-space invariants

$$\varepsilon_u = \gamma_u(s)u^2 + 2\alpha_u(s)uu' + \beta_u(s)u'^2 \quad (u = x \text{ or } y) \tag{9.41}$$

$$\varepsilon_l = \frac{\alpha E_s}{2\Omega_s} \left(\frac{\widehat{\delta E}}{E_s} \right)^2 \tag{9.42}$$

Under the effect of stochastic SR, individual invariants can in general not be determined, averages over particle ensembles are considered instead (noted $\overline{(*)}$ in the following), they evolve according to

$$\frac{d\bar{\varepsilon}_u}{dt} = -\frac{\bar{\varepsilon}_u}{\tau_u} + C_u \tag{9.43}$$

towards a stationary solution

$$\varepsilon_{u,eq} = C_u \tau_u \tag{9.44}$$

where C_u is a constant at fixed energy (storage ring), with characteristic time

$$\tau_u = \frac{T_{rev} E_s}{U_s J_u} \tag{9.45}$$

$J_{n=x,y,l}$ are the partition numbers (lattice properties), respectively horizontal, vertical, longitudinal,

$$J_x = 1 - \mathcal{D}, \quad J_y = 1, \quad J_l = 2 + \mathcal{D} \tag{9.46}$$

where

$$\mathcal{D} = \frac{\overline{D_x(1 - 2n)/\rho^3}}{\overline{1/\rho^2}}$$

In this expression, $\overline{(*)} = \frac{1}{2\pi R} \int_{\text{dipoles}} (*) ds$, n is the field index—case of combined function dipoles, D_x is the dispersion function, The partition numbers satisfy the Robinson theorem

$$J_x + J_y + J_l = 4 \tag{9.47}$$

Table 9.1 Common expressions for the energy loss per turn, U_s (E-loss), for the damping times and equilibrium emittances, in the hypothesis of an isomagnetic lattice. Their scaling with γ is given in the 2nd row

	E-loss	$\varepsilon_{l,eq}$	σ_l	τ_l	$\varepsilon_{x,eq}$[a]	τ_x	$\varepsilon_{y,eq}$	τ_y
Scaling:	γ^4	$\gamma^{3/2}$	$1/\gamma^{1/2}$	$1/\gamma^3$	γ^2	$1/\gamma^3$		$1/\gamma^3$
	$C_\gamma \dfrac{E_s^4}{\rho}$	$\dfrac{\alpha E_s}{\Omega_s}\dfrac{C_q\gamma^2}{J_l\,\rho}$	$\dfrac{\alpha c}{\Omega_s}\sigma_{\frac{\Delta E}{E}}$	$\dfrac{T_{rev}E_s}{U_s J_l}$	$\dfrac{C_q\gamma^2}{J_x\rho}\overline{\mathcal{H}}$	$\dfrac{T_{rev}E_s}{U_s J_x}$	$\ll \varepsilon_x$	$\dfrac{T_{rev}E_s}{U_s J_y}$

[a] $\overline{\mathcal{H}} = \dfrac{1}{L_{\mathrm{dip}}}\int_{\mathrm{dip}}\dfrac{ds}{\beta_x}\left[D_x^2 + (\alpha_x D_x + \beta_x D_x')^2\right]$, integral over the dipoles

Common expressions for the calculation of the energy loss and equilibrium quantities, in the hypothesis of an isomagnetic lattice, are recalled in Table 9.1.

Vertical emittance results from coupling, always present in a ring, due for instance to a loss of median plane symmetry, or to fringe fields, or excited on purpose to control the vertical emittance as in light sources. Given the coupling factor κ—normally <0.1, the vertical and horizontal emittances satisfy

$$\epsilon_y = \kappa\epsilon_x, \qquad \epsilon_x + \epsilon_y = \epsilon_0 \tag{9.48}$$

where ϵ_0 is the equilibrium horizontal emittance in the absence of coupling (Table 9.1).

The basic considerations above hold for a defect-free planar ring. Things can be (as usual) more complicated, for instance in the presence of vertical dispersion.

Field Scaling

Particle stiffness decrease upon SR loss causes these to experience increased field strength ($1/\rho$ in dipoles, $G/B\rho$ in quadrupoles, etc.). In the case of beam lines (which may include high energy ERLs [11]), this effect may be taken care of by scaling the magnetic fields to the theoretical average energy loss (Eq. 5.12), namely

$$\Delta E_{scaling} = \sum_{bends}\frac{2}{3}r_0 e c\gamma^3 B\Delta\theta \tag{9.49}$$

In a storage ring the energy lost by SR is restored by the RF system, bends and lenses are operated at constant field. In pulsed regime such as in a booster injector, bends and lenses are operated at constant strength during acceleration.

9.2.6 Visible Synchrotron Radiation. Interference

Visible SR was first observed at the GEC 70 MeV. For this reason it has been introduced in the Weak Focusing Synchrotron chapter, Sect. 8.2.3. The SR spectrum at that energy peaks—has its critical frequency—in the visible region. The matter

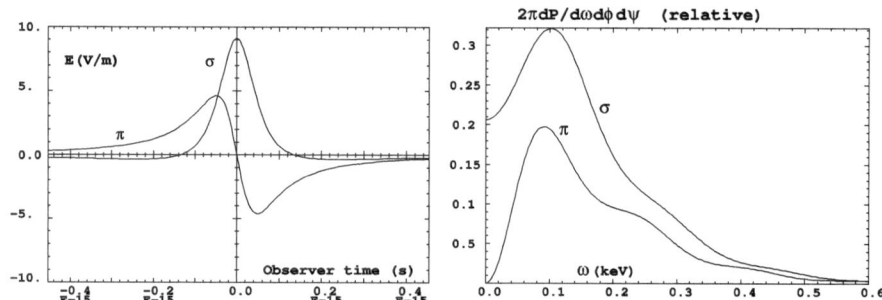

Fig. 9.18 Left: typical shape of the $E_\sigma(\tau)$ and $E_\pi(t)$ electric impulse components of the Poynting vector, emitted by a 2.5 GeV electron on a $\rho = 53.6$ m circular trajectory in a $l = 20$ cm-long dipole, as observed in the laboratory. $E_{\sigma,\pi}(\tau)$ are obtained from the stepwise integration of electron motion through the magnet, which provides the ingredients to compute Eq. 8.36, accounting for the retarded time $t = \tau - r(t)/c$ (Eq. 8.37). Right: the spectral brightness of the σ component of the radiation allows comfortable beam diagnostics conditions in the visible range ($\hbar\omega \sim 0.5$ eV)

is developed further in the present chapter, in regard with the use of visible SR for beam diagnostics in electron and high energy proton rings [31, 33].

An example of the use of visible SR from a proton beam is found at the CERN SPS, where edge radiation was used at 270 GeV for beam imaging [34]. At that energy in the SPS, the critical frequency (the peak brightness) is in the infrared region. Undulator radiation, more intense, was used down to 200 GeV [35], in the p − p̄ collider era (1980s). Another example is the LHC synchrotron light profile monitor, a major beam monitoring tool at injection energy, 450 GeV [36], [37, Appendix C].

An example of the use of visible SR from a high energy electron beam is found at the former LEP, where it was produced in a dedicated 4-dipole miniwiggler. The critical frequency in a high energy electron ring is way above the visible range. In such case, visible SR can be dealt with in terms of low-frequency SR [38], a method which can be extended to the analytical treatment of SR interference [37]. The underlying theoretical material is recalled here. It is resorted to in the exercises, to cross check Poynting computation from raytracing (using Eq. 8.36).

Low Frequency SR

A typical electric field impulse from a LEP miniwiggler dipole, and the resulting spectral brightness, as observed in the laboratory, are displayed in Fig. 9.18. The LEP 4-dipole miniwiggler was subject to visible light interference from 4 coherent sources, the effect is illustrated in Fig. 9.19.

A doublet of LEP miniwiggler dipoles, in both cases of same sign and opposite sign dipoles, is the object of numerical simulations in Exercise 9.6. It is on the other hand treated theoretically in [37, Sect. 3.1]. The latter provides all necessary

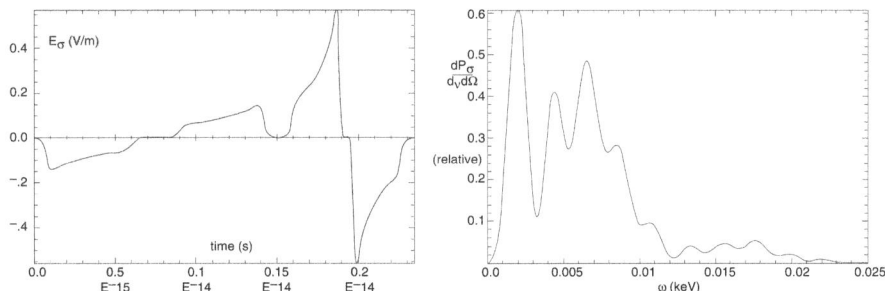

Fig. 9.19 An interferencial spectrum, case of LEP 4-dipole miniwiggler [39]. By contrast with the single dipole case (Fig. 9.18), the spectral brightness of the σ component cancels in the low energy end of the spectrum

material for cross checks of numerical outcomes from the stepwise integration of electron motion.

9.2.7 Polarization, Resonances

In a weak focusing optics lattice, radial field components experienced by a particle in the course of its vertical betatron motion are small, which results in weak depolarizing resonances (Sect. 8.2.4). By contrast, strong focusing field gradients in the combined function dipoles and/or focusing lenses of strong focusing optics results in strong radial field components and therefore strong depolarizing resonances.

Spin precession and resonant spin motion in the magnetic components of a cyclic accelerator have been introduced in Sects. 3.2.5 and 4.2.5. The general conditions for depolarizing resonance to occur have been introduced in Sect. 8.2.4. In a strong focusing synchrotron they essentially result from the radial field components in the focusing magnets and their strength is determined by the lattice optics, as follows.

Strength of Imperfection Resonances

Imperfection, or integer, depolarizing resonances are driven by a non-vanishing vertical closed orbit $y_{co}(\theta)$ which causes spins to experience periodic radial fields in focusing magnets, dipoles in combined function lattices and quadrupoles in separated function lattices, namely,

$$B_x(\theta) = G\, y(\theta) = K(\theta) \times B_0\rho_0 \times y_{co}(\theta) \tag{9.50}$$

with θ the orbital angle and $B_0\rho_0$ the lattice rigidity. Resonance occurs if the spin undergoes an integer number of precessions over a turn: it then experiences

1-turn-periodic torques, which cause it to move away from the stable $\mathbf{n}_0$ direction as field perturbations along the closed orbit add up coherently. Thus resonances occur at integer values

$$G\gamma_n = n$$

A Fourier development of these perturbative fields yields the strength of the $G\gamma_n$ harmonic [40, Sect. 2.3.5.1]

$$\epsilon_n^{imp} = (1 + G\gamma)\frac{R}{2\pi} \oint K(\theta)\, y_{co}(\theta)\, e^{-jG\gamma(\theta - \alpha)}\, e^{jn\theta}\, d\theta$$

In the thin-lens approximation, near the resonance where $G\gamma - n \to 0$, this simplifies into a series over the quadrupole fields,

$$\epsilon_n^{imp} = \frac{1 + G\gamma_n}{2\pi} \sum_{Qpoles} [\cos G\gamma_n\, \alpha_i + \sin G\gamma_n\, \alpha_i]\, (KL)_i\, y_{co}(\theta_i) \tag{9.51}$$

with θ_i the quadrupole location, $(KL)_i$ the integrated strength (slice the dipoles as necessary in an AG lattice for this series to converge) and α_i the cumulated orbit deviation.

Orbit harmonics near the betatron tune ($n = G\gamma_n \approx \nu_y$) excite strong resonances. Imperfection resonance strength is further amplified in P-superperiodic rings, with m-cell superperiods, if the betatron tune $\nu_y \approx$ integer $\times\, m \times P$ [41, Chap. 3-I].

Strength of Intrinsic Resonances

Intrinsic depolarizing resonances are driven by betatron motion, which causes spins to experience strong radial field components in quadrupoles, namely

$$B_x(\theta) = G\, y(\theta) = K(\theta) \times B_0\rho_0 \times y_\beta(\theta) \tag{9.52}$$

The effect of resonances on spin depends upon betatron amplitude and phase, their effect on beam polarization depends on beam emittance. Longitudinal fields from dipole ends are usually weak by comparison and ignored. The location of intrinsic resonances depends on betatron tune, it is given in an M-periodic structure by

$$G\gamma_n = nM \pm \nu_y$$

A Fourier development of the perturbative fields yields the two families of strengths [40, Sect. 2.3.5.2]

$$\epsilon_n^{intr\pm} = \frac{\lambda_x \rho_0}{4\pi} \int_0^{2\pi} K(\theta)\sqrt{\beta_y(\theta)\frac{\varepsilon_y}{\pi}}\, e^{\pm j\left(\int_0^{s(\theta)} \frac{ds}{\beta_y} - \nu_y\theta\right)}\, e^{-jG\gamma(\theta - \alpha(\theta))}\, e^{jn\theta}\, d\theta$$

In the thin-lens approximation, near the resonance where $G\gamma \pm \nu_y - n \to 0$, this simplifies into a series over the quadrupole fields,

$$\left\{ \begin{array}{c} \mathcal{R}e(\epsilon_n^{\text{intr}\pm}) + \\ j \, \mathcal{I}m(\epsilon_n^{\text{intr}\pm}) \end{array} \right\} = \frac{1 + G\gamma_n}{4\pi} \sum_{\text{Qpoles}} \left\{ \begin{array}{c} \cos(G\gamma_n\alpha_i \pm \varphi_i) + \\ j \, \sin(G\gamma_n\alpha_i \pm \varphi_i) \end{array} \right\} (KL)_i \sqrt{\beta_{y,i} \frac{\varepsilon_y}{\pi}} \tag{9.53}$$

Spin Diffusion

Spin diffusion stems from the stochastic emission of photons in magnetic fields (Sect. 5.2.3). A change δ in the energy offset ΔE of a particle, due to the emission of a photon, causes a change $\partial \mathbf{n}/\partial\delta$ of the local spin precession direction. In dispersive sections it also causes a change in the horizontal invariant, $\partial\epsilon_x/\partial\delta$, and in vertical invariant as well, $\partial\epsilon_y/\partial\delta$ in the presence of vertical dispersion, which in turn result in perturbations $\partial \mathbf{n}/\partial\epsilon_{x,y}$.

As far as numerical integration is concerned, spin diffusion is a sub-product of the stepwise integration of Thomas-BMT equation (Sect. 3.2.5), and of the simulation of stochastic emission of photons (Sect. 5.2.3). It is at work in Cornell RCS simulation, Exercise 9.4.

9.3 Exercises

In complement to the present exercises, a tutorial on depolarizing resonances in a strong focusing synchrotron can be found in [40, Chap. 14]. Proton, helion and electron beams are considered, using the lattice of the AGS Booster at BNL. The simulations explore methods for preservation of polarization, including tune-jump quadrupoles, a solenoid, Siberian snakes, spin rotators in the case of electrons, including synchrotron radiation and effects on polarization life time.

Note: input data files for these simulations are available in zgoubi sourceforge repository at https://sourceforge.net/p/zgoubi/code/HEAD/tree/branches/exemples/book/zgoubiMaterial/synchrotron_strongFocusing/.

9.1 Construct SATURNE 2 Synchrotron

Solution 9.1.

Over the years 1978–1997 the 3 GeV synchrotron SATURNE 2 at Saclay (Figs. 9.3 and 9.20) delivered polarized proton beams, and polarized deuteron and ^{6}Li beams up to 1.1 GeV/nucleon, for intermediate energy nuclear physics research, including meson production [42, 43, 45]. The separated function synchrotron was designed *ab initio* for the acceleration of polarized ion beams [44], and the first strong focusing synchrotron to do so—ZGS, first to accelerate polarized beams, protons and deuterons, was a weak focusing synchrotron (Chap. 8).

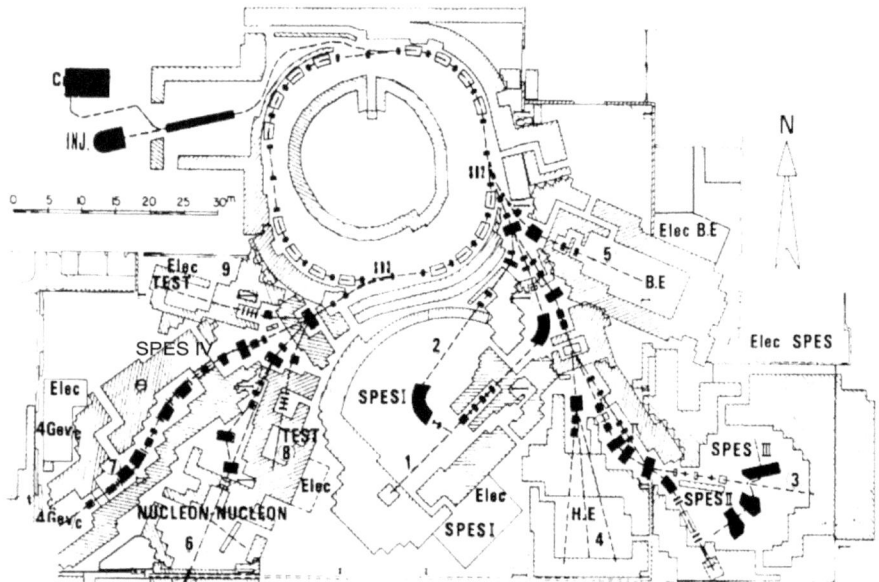

Fig. 9.20 SATURNE 2 synchrotron and its experimental areas, including mass spectrometers SPES I to SPES IV, a typical nuclear physics accelerator facility. The polarized ion sources Dioné and Hypérion are at the top left, followed by a 20 MeV linac. In the early 1980s a synchrotron booster, MIMAS, was added for higher polarized ion performance

SATURNE 2 is a FODO lattice with missing dipole. Its parameters are given in Table 9.2.

(a) Simulate the main dipole using BEND. Dipole fringe fields matter in this small ring, take them into account assuming $\lambda = 8$ cm extent and the following Enge coefficient values (Eq. 14.11, Sect. 14.3.3):

$$C_0 = 0.2401, \quad C_1 = 1.8639, \quad C_2 = -0.5572, \quad C_3 = 0.3904, \quad C_4 = C_5 = 0$$

Produce the transport matrix of the dipole, check against theory. Compare with the matrix of the hard edge model.

Produce a graph of the field across the dipole, in the median plane and at 5 cm vertical distance. OPTIONS[CONSTY=ON] can be used to force a particle to constant Y and Z.

Simulate the F and D quadrupoles, using respectively QUADRUPOLE and MULTIPOL. Compare matrices with theory.

Construct the cell. Produce machine parameters (tunes, chromaticities), check against data, Table 9.2.

Construct the 4-cell ring. Produce a graph of the optical functions. Produce the beam matrix.

Table 9.2 Parameters of SATURNE 2 separated function FODO lattice. ρ_0 is the radius of the reference orbit in the main dipole

Orbit length, C	m	105.5556
Average radius, $R = C/2\pi$	m	16.8
Straight sections, length:		
– Short	m	0.716256
– Long	m	3.92148
Dipole:		
– Bend angle, α	deg	22.5
– Magnetic radius, ρ_0	m	6.3381
– Wedge angle, ε	deg	2.45
Quadrupole:		
– Gradient range	T/m	0.5–10.56
– Magnetic length F/D	m	0.46723/0.486273
Wave numbers, typical, ν_x; ν_y		3.64; 3.60
Chromaticities, ξ_x; ξ_y		Negative, a few units
Momentum compaction α		0.015
Injection energy (proton)	MeV	20
Top energy	GeV	3
$\dot{B}$	T/s	4.2
Synchronous energy gain	keV/turn	1.160
RF harmonic		3

(b) Accelerate a bunch comprised of a few tens of particles with Gaussian density distributions (it can be defined using MCOBJET), from injection to top energy, 50 MeV to 3 GeV. Use harmonic 3 RF frequency, take a (unrealistic, for a reduced number of turns) peak RF voltage $\hat{V} = 1$ MV, and synchronous phase $\phi_s = 30°$.

Produce a graph of Y, Z and dp/p versus turn. Check the transverse damping against theory.

(c) Determine the momentum acceptance of the ring at 50 MeV, with $\hat{V} = 10$ kV peak voltage, in the following four cases: stationary bucket (synchronous phase $\phi_s = 0$) and accelerated buckets with $\phi_s = 15$, 30, and 60°.

Reproduce the longitudinal phase space graphs displayed in Fig. 9.16.

9.2 Non-Linear Motion in SATURNE 2

Solution 9.2.

(a) Simulate horizontal particle motion near a third integer resonance. Provide a graph of the transverse phase space.

(b) Simulate horizontal particle motion near a quarter integer resonance. Provide a graph of the transverse phase space.

9.3 SVD Orbit Correction

Solution 9.3.

Using SATURNE 2 ring, inject dipole defects and use SVDOC to find the corrected orbit.

It can be done in the following way:

– place a horizontal pickup (HPU), a dipole defect (HDEF, using a thin-lens MUL-TIPOL, length e.g. 1e-3 cm) and a dipole corrector (HKIC, using a thin-lens MUL-TIPOL) in the middle of the QF quadrupole of the FODO cells,
– in a similar manner, place a VPU, a VDEF and a VKIC just upstream of the FODO cell QD,
– excite V and H closed orbits by injecting random defects in HDEF and VDEF, using ERRORS.

Use SVDOC to find the orbit correction.

Provide a graph of the orbit at the PUs, before and after correction.

In the previous setting, there is 24 defects (12 H and 12 V) and 24 correctors (12 H and 12 V). Repeat for 24 defects and only 12 correctors per plane.

9.4 Cornell Electron RCS. Radiative Energy Loss

Solution 9.4.

Note: details regarding these simulations and their solutions can be found in the Tech. Note EIC/57;BNL-114452-2017-IR [46].

The goal in this exercise is to simulate Cornell RCS lattice and accelerate beam, first without synchrotron radiation, then taking it into account. In a fourth step electron spin is added and polarization transmission through the acceleration cycle assessed.

(a) Details of the RCS geometry and lattice can be found in Ref. [14], however a simplified 6-superperiodic version of the ring is considered here, with six identical long straights and six identical arcs. The RCS parameters are given in Table 9.3. The input data files are given in

– Tables 9.4 and 9.5: definition of the focusing and defocusing bends, and of the focusing and defocusing doublets;
– Table 9.6: definition of a FODO cell;
– Table 9.7: definition of a supercell;
– Table 9.8: definition of the 6-supercell ring.

Produce the optical parameters of the ring. A TWISS command can be used for that. Produce graphs of the closed orbit and optical functions around the ring.

(b) Raytrace a few tens of particles over 2300 turns around the ring, from 320 MeV to 8 GeV about, ignoring radiative energy loss. Assume normalized emittances $\varepsilon_x = \varepsilon_y = 25\,\pi\mu$m, Gaussian densities, initial rms $\delta p/p = 5 \times 10^{-3}$. Use CAVITE[IOPT=3] for acceleration. Produce a graph of the three phase spaces produce graphs of transverse and longitudinal excursions versus turn number, check damping again expectations.

Table 9.3 Cornell RCS parameters in the present simplified lattice simulation

Top energy	GeV	7
Injection energy	MeV	320
Circumference, simplified 6-supercell case	m	786.947
Bunch		
$\varepsilon_x, \varepsilon_y$ at injection	$\pi\mu m$	25
Bunch length	mm	6
dE/E at injection		5×10^{-3}
Combined function lattice		$48\times$FFDD
Nb of F and D cell dipoles		192
ρ_F, ρ_D	m	$\approx$95, 92
Field at 7 GeV	T	0.25
Max. β_x, β_y	m	29, 26
ν_x, ν_y, natural		9.62, 13.82
ξ_x, ξ_y, natural		$-13, -16$
RF, synch. radiation		
Repetition rate	Hz	up to 60
Acceleration rate	MV/turn	3
E-loss per turn at 5, 10 GeV	MeV	0.6, 9
τ_x ($\approx \frac{2.5}{E^3}$) at 5, 10 GeV	ms	16, 2

(c) Re-do (b) with synchrotron radiation energy loss, following SR loss theoretical material introduced in the "Betatron" Chap. 5. Use SRLOSS for radiation, and CAVITE[IOPT=11, Facility=CornellSynch, $U_{00} = 9.48145321 \times 10^{-6}$] for acceleration. Check equilibrium emittances.

(d) Produce a graph of the average bunch polarization over the acceleration cycle in (c), starting with all spins up at injection energy. Check against the resonance spectrum over $a\gamma$: $0.7 \rightarrow 18$.

9.5 Coupling in a Light Source Storage Ring

Solution 9.5.

In this exercise, it is proposed to reproduce SR damping simulations, in a case of coupled light source lattice, detailed in JINST article [48].

Simulation of radiation damping in rings, using stepwise ray-tracing methods (the original (1990s) ESRF lattice is concerned—today's ESRF lattice is completely different, minimal emittance, un-isomagnetic).

An input data file for the early ESRF lattice can be found at https://sourceforge.net/p/zgoubi/code/HEAD/tree/branches/exemples/SRDamping/ESRFRing/coupled. It accounts for $\kappa = 0.58$ optical coupling, by a single skew quadrupole placed at the begining of the lattice.

Reproduce the numerical results for this coupled case, as detailed in Sect. 5 of that JINST article [48].

Table 9.4 Simulation input data files for the focusing (left) and defocusing (right) combined function dipoles. They define the segments, respectively, F_BEND_S:F_BEND_E and D_BEND_S:D_BEND_E, for use by INCLUDE commands in further input data files. These files can be run as is: FIT will center the closed orbit across the magnet, accounting for the field scaling by the *ad hoc* coefficient under SCALING

```
RCS focusing combined function dipole              RCS defocusing combined function dipole
| File: F_BEND.inc                                 ! File: D_BEND.inc
'OBJET'                                            'OBJET'
1. *1e3                                            1. *1e3
5                                                  5
.001 .001 .001 .001 0. .0001                       .001 .001 .001 .001 0. .0001
0. 0. 0. 0. 0. 1.                                  0. 0. 0. 0. 0. 1.

'SCALING'                                          'SCALING'
1  1                                               1  1
MULTIPOL F_BEND                                    MULTIPOL D_BEND
-1                                                 -1
0.98523998                                         1.1078694
1                                                  1

'MARKER'  F_BEND_S                                 'MARKER'  D_BEND_S

'MULTIPOL'   F_BEND                                'MULTIPOL'   D_BEND
0  .Dip                                            0  .Dip
320.2700  10. 0.1021746 0.0435214 0. 0. 0. 0. 0. 0. 0.   320.0150 10. 0.1022560 -0.0437325 0. 0. 0. 0. 0. 0. 0.
0. 0. 10.00  4.0  0.800 0.00 0.00 0.00 0.00 0. 0. 0. 0.   0. 0. 10.00  4.0  0.800 0.00 0.00 0.00 0.00 0. 0. 0. 0.
4  .1455   2.2670  -.6395  1.1558  0. 0.  0.       4  .1455   2.2670  -.6395  1.1558  0. 0.  0.
0. 0. 10.00  4.0  0.800 0.00 0.00 0.00 0.00 0. 0. 0. 0.   0. 0. 10.00  4.0  0.800 0.00 0.00 0.00 0.00 0. 0. 0. 0.
4  .1455   2.2670  -.6395  1.1558  0. 0.  0.       4  .1455   2.2670  -.6395  1.1558  0. 0.  0.
0. 0. 0. 0. 0. 0. 0. 0. 0. 0.                      0. 0. 0. 0. 0. 0. 0. 0. 0. 0.
#30|320|30        ! YCE offset found by FIT        #30|320|30          ! YCE offset found by FIT
3  0.0000000000E+00 0.52818473  -1.6362461735E-02   3  0.0000000000E+00  -1.4110319  -1.6362461735E-02

'MARKER'  F_BEND_E                                 'MARKER'  D_BEND_E

'FIT'                                              'FIT'
1                                                  1
4 65 0 [-4.,4.]                                    4 65 0 [-2.,2.]
2                                                  2
3.1 1 2 #End 0. 1. 0                                3.1 1 2 #End 0. 1. 0
3.1 1 3 #End 0. 1. 0                                3.1 1 3 #End 0. 1. 0

'END'                                              'END'
```

Table 9.5 Definition of focusing (left) and defocusing (right) doublets, for use by further INCLUDE commands

```
! File: BF2.inc                        ! File: BD2.inc

'MARKER' BF2_S                         'MARKER' BD2_S

'DRIFT'                                'DRIFT'
23.999061                              24.126561
'INCLUDE'                              'INCLUDE'
1                                      1
F_BEND.inc[F_BEND_S:F_BEND_E]          D_BEND.inc[D_BEND_S:D_BEND_E]
'DRIFT'                                'DRIFT'
23.999061                              24.126561
'DRIFT'                                'DRIFT'
23.999061                              24.126561
'INCLUDE'                              'INCLUDE'
1                                      1
F_BEND.inc[F_BEND_S:F_BEND_E]          D_BEND.inc[D_BEND_S:D_BEND_E]
'DRIFT'                                'DRIFT'
23.999061                              24.126561

'MARKER' BF2_E                         'MARKER' BD2_E

'END'                                  'END'
```

Table 9.6 Simulation input data file for a FODO cell

```
! File: FD.inc

'MARKER'   FD_S
'INCLUDE'
1
BF2.inc[BF2_S:BF2_E]
'INCLUDE'
1
BD2.inc[BD2_S:BD2_E]
'MARKER'   FD_E
'END'
```

9.6 SR Electric Impulse and Interference in a Miniwiggler

Solution 9.6.

In this exercise, the electric field component of synchrotron radiation in short dipoles is produced. An interferential spectrum is prodcued from a pair of dipoles. This exercise is based on the LEP miniwigller configuration [37].

(a) Produce the input data file for the simulation of an electron trajectory in one of the LEP miniwiggler dipoles schemed in Fig. 9.21. Dipole length is $L = 52.602$ cm, bend angle 0.8 mrad. Electron energy is $E = 45$ GeV. Produce the electric field impulse observed at long distance in the direction $\phi = \psi = 0$. Produce its spectrum.

Check the various quantities: duration of the electric field impulse, critical frequency of the spectrum, etc.

(b) Consider the dipole pair of Fig. 9.21. Take distance between dipoles $d = 23.098$ m. Produce the electric field impulse observed at long distance in the direction $\phi = \psi = 0$. Produce its spectrum.

Check the various quantites: duration of the electric field impulse, critical frequency of the spectrum.

Repeat, in the direction $\phi = 0$, $\psi = 0.2$ mrad.

(c) Repeat (b), for the dipole pair disposed as in Fig. 9.21 [37, Sect. A].

(d) Repeat (c) for the configuration of Fig. 9.22, a case of edge radiation interference [37, Sect. B].

9.7 Depolarizing Resonances in SATURNE 2

Solution 9.7.

Unexpectedly as it is not a systematic resonance, $G\gamma = 7 - \nu_y$ was found to be harmful to beam polarization. Produce a crossing of that resonance, for a few particles with different momenta, and vertical invariant $\varepsilon_Z \approx 10\pi$ μm. Take peak voltage 6 kV and synchronous phase $\phi_s = 0.2363176$ rad.

The input data file given in Table 9.14, an outcome of Exercise 9.15, can be used as a starting point for this simulation.

Table 9.7 Simulation input data file for a supercell

```
File : superCell.inc
'OBJET'
1. *1e3                    ! Rigidity is 1 T m.
5
.001 .001 .001 .001 0. .0001
0. 0. 0. 0. 0. 1.

'MARKER'  superCell_S

'INCLUDE'
1
F_BEND.inc[F_BEND_S:F_BEND_E]
'DRIFT'
40.988209
'DRIFT'
40.988209
'INCLUDE'
1
F_BEND.inc[F_BEND_S:F_BEND_E]
'DRIFT'
15.600113
'DRIFT'
15.600113
'INCLUDE'
1
D_BEND.inc[D_BEND_S:D_BEND_E]
'DRIFT'
24.062811
'DRIFT'
24.062811
'INCLUDE'
1
D_BEND.inc[D_BEND_S:D_BEND_E]
'DRIFT'
15.600113
'DRIFT'
15.600113
'INCLUDE'
1
F_BEND.inc[F_BEND_S:F_BEND_E]
'DRIFT'
40.988209
'DRIFT'
40.988209
'INCLUDE'
1
F_BEND.inc[F_BEND_S:F_BEND_E]
'DRIFT'
15.600113
'INCLUDE'
1
D_BEND.inc[D_BEND_S:D_BEND_E]
'DRIFT'
15.600113
'DRIFT'
15.600113
'INCLUDE'
1
D_BEND.inc[D_BEND_S:D_BEND_E]
'DRIFT'
40.988209
'DRIFT'
40.988209
'INCLUDE'
1
F_BEND.inc[F_BEND_S:F_BEND_E]
'DRIFT'
15.600113
'DRIFT'
15.600113
'INCLUDE'
1
F_BEND.inc[F_BEND_S:F_BEND_E]
'DRIFT'
24.062811
'INCLUDE'
1
BD2.inc[BD2_S:BD2_E]
```

```
'DRIFT'
24.062811
'MULTIPOL'
0 .Dip
44.6375  10. 0.1022198 -0.0437325 0. 0.0 0.0 0.0 0.0 0.0 0.0 0.0 0.0
0. 0. 10.00  4.0  0.800 0.00 0.00 0.00 0.00 0. 0. 0. 0.
4  .1455   2.2670  -.6395  1.1558  0. 0.  0.
0. 0. 10.00  4.0  0.800 0.00 0.00 0.00 0.00 0. 0. 0. 0.
4  .1455   2.2670  -.6395  1.1558  0. 0.  0.
0. 0. 0. 0. 0. 0. 0. 0. 0. 0.
#30|45|30    Dip  B129VA
3   0.0000000000E+00  1.77777778E-02  -2.2814180400E-03
'MULTIPOL'
0 .Dip
275.505  10. 0.1022165  0.0842350 0. 0.0 0.0 0.0 0.0 0.0 0.0 0.0 0.0
0. 0. 10.00  4.0  0.800 0.00 0.00 0.00 0.00 0. 0. 0. 0.
4  .1455   2.2670  -.6395  1.1558  0. 0.  0.
0. 0. 10.00  4.0  0.800 0.00 0.00 0.00 0.00 0. 0. 0. 0.
4  .1455   2.2670  -.6395  1.1558  0. 0.  0.
0. 0. 0. 0. 0. 0. 0. 0. 0. 0.
#30|276|30    Dip  B129HB
3   0.0000000000E+00   0.65358025  -1.4081043695E-02
'DRIFT'
24.062811

'DRIFT'
60.800000
'DRIFT'
244.000000
'DRIFT'
244.000000
'DRIFT'
60.800000

'DRIFT'
24.062811
'MULTIPOL'
0 .Dip
275.505  10. 0.1022165 -0.0844460 0. 0.0 0.0 0.0 0.0 0.0 0.0 0.0 0.0
0. 0. 10.00  4.0  0.800 0.00 0.00 0.00 0.00 0. 0. 0. 0.
4  .1455   2.2670  -.6395  1.1558  0. 0.  0.
0. 0. 10.00  4.0  0.800 0.00 0.00 0.00 0.00 0. 0. 0. 0.
4  .1455   2.2670  -.6395  1.1558  0. 0.  0.
0. 0. 0. 0. 0. 0. 0. 0. 0. 0.
#30|276|30    Dip  B128VA
3   0.0000000000E+00   0.6397805  -1.4081043695E-02
'MULTIPOL'
0 .Dip
44.6375  10. 0.1022198  0.0435214 0. 0.0 0.0 0.0 0.0 0.0 0.0 0.0 0.0
0. 0. 10.00  4.0  0.800 0.00 0.00 0.00 0.00 0. 0. 0. 0.
4  .1455   2.2670  -.6395  1.1558  0. 0.  0.
0. 0. 10.00  4.0  0.800 0.00 0.00 0.00 0.00 0. 0. 0. 0.
4  .1455   2.2670  -.6395  1.1558  0. 0.  0.
0. 0. 0. 0. 0. 0. 0. 0. 0. 0.
#30|45|30    Dip  B128HB
3   0.0000000000E+00  1.77777778E-02  -2.2814180400E-03
'DRIFT'
24.062811

'INCLUDE'
1
5 * FD.inc[FD_S:FD_E]
'DRIFT'
7.073665    ! -24.126561 + 2*15.600113

'MARKER'  superCell_E

'TWISS'
2  1. 1.

'SYSTEM'
1
gnuplot <./gnuplot_TWISS.gnu

'END'
```

Table 9.8 Simulation input data file for Cornell RCS ring

```
                                        'INCLUDE'
                                        1
                                        6 * superCell.inc[superCell_S:superCell_E]

File: ring.INC.dat. Cornell RCS ring    'OPTIONS'
'OBJET'                                 1 1
1. *1e3                                 WRITE ON
5
.001 .001 .001 .001 0. .0001            !'TWISS'        ! Uncomment to get a TWISS and graphs.
0. 0. 0. 0. 0. 1.                       !2  1. 1.
                                        !'SYSTEM'
'OPTIONS'                               !1
1 1                                     !gnuplot <./gnuplot_TWISS.gnu
WRITE OFF                               !'END'

'SCALING'                               'FIT2' ! Set SCALING coefficients for requested tunes.
1  3                                    2
MULTIPOL                                3  8 0 .2
-1                                      3 12 0 .2
1.                                      2
1                                       0.1 7 0 #End 0.62 1. 0
MULTIPOL F_BEND                         0.1 8 0 #End 0.82 1. 0
-1
0.99292280                              !'MATRIX'
1                                       !1 11
MULTIPOL D_BEND                         'TWISS'
-1                                      2 1. 1.
1.1294084                               'END'
1
```

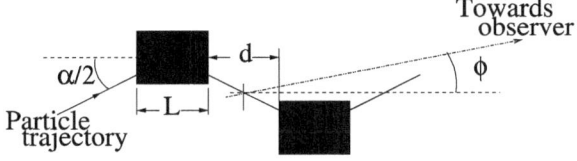

Fig. 9.21 Synchrotron radiation electric field impulse from a pair of dipoles is observed in the direction (ϕ, ψ), with ϕ the bend plane angle as shown, and ψ the angle to the bend plane. This schematic defines the observation direction $\phi = 0$

Fig. 9.22 Both dipoles have same sign. This schematic defines the observation direction $\phi = 0$

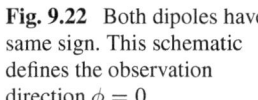
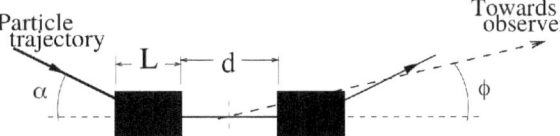

9.8 Ion and Electron Polarization. Preservation of Polarization

More simulations regarding

– spin polarized ions and special devices and methods for the preservation of polarization during acceleration, including tune jump, partial and full Siberian snakes, etc.,
– electron spin diffusion in a storage ring and its suppression, spin matching, polarization lifetime, etc.,
    can be found, with complete solutions, in the USPAS Summer 2021 Spin Class Lectures, "Polarized Beam Dynamics and Instrumentation in Particle Accelerators" [47, Chap. 14].

Table 9.9 Simulation input data file saturneBEND.inc: computes the transport matrix of SAT-URNE 2 dipole. It does so via parameter adjustment resorting to FIT, by imposing symmetric positioning and null in and out reference orbit coordinates. This file defines the SATURNE 2 sequence segment satBEND_S to satBEND_E for use in subsequent INCLUDE statements

```
File name: saturneBEND.inc
! SATURNE bending magnet.
'MARKER'   SatII_BEND.inc_S                              ! Just for edition purposes.
'OBJET'
1000.                                                                    ! 1 T m rigidity.
5                                   ! Allows computation of transport matrix from 13 rays.
.001 .01 .001 .01 .0 .0001
0. 0. 0. 0.  0. 1.                         ! Initial coordinates of the reference particle (particle #1 of 13).

'OPTIONS'
1 1
.plt 2                              ! Set IL=2 in optical elements (log stepwise data in zgoubi.plt).

'MARKER' satBEND_S
'BEND'
0
247.30039  0.  1.57776             ! Straight length of the magnet: 2*rho*sin(alpha/2); unused; field.
30. 8. .04276056667
4 .2401  1.8639  -.5572  .3904 0. 0.
30. 8. .04276056667
4 .2401  1.8639  -.5572  .3904 0. 0.
#60|220|60   bend      ! Integration step sizes: 60 steps in fringe field regions, 220 in body.
3 0. -2.1866472E-02 0.    ! Option KPOS=3 computes change of frame such that magnet gets half-bend
!                           in and out tilt (CHANGREF upstream and downstream could be used instead).
'MARKER' satBEND_E

'FIT2'            ! Find the symmetric orbit arc, by allowing a small YCE offset of the dipole.
1                                                                  ! 1 variable:
5 72 0 [-1,1]                      ! YCE offset (parameter 72) of BEND (element #4 in sequence).
2  1e-15                 ! 2 constraints to ensure symmetrically positionned trajectory in BEND:
3.1 1 2 #End 0. 1. 0                                          ! Dinal Y=Y0 (= 0).
3.1 1 3 #End 0. 1. 0                                          ! Final T=T0 (= 0).

'MATRIX'
1 0                                      ! Compute transport matrix, from OBJET down to here.
'MARKER'    SatII_BEND.inc_E                               ! Just for edition purposes.
'END'
```

9.4   Solutions of Exercises of This Chapter: Strong Focusing Synchrotron

9.1 Construct SATURNE 2 Synchrotron

(a) A model of SATURNE 2 synchrotron.

● Simulation of the main dipole, using BEND, is given in Table 9.9, the simulation includes finding the reference orbit using FIT, and delivering the transport matrix. The theoretical transport matrix, Eq. 14.7, can be used to check the latter.

Looking up the execution listing zgoubi.res, one finds parameters used for the stepwise raytracing through BEND (an excerpt):

```
                      4 Keyword, label(s) :   BEND

                          Length    =   2.473004E+02 cm
                          Arc length =   2.488966E+02 cm
                          Deviation  =   2.250000E+01 deg.,     3.926991E-01 rad

                          Field = 1.5777600E+00 kG  (i.e.,    1.5777600E+00 * SCAL)
                          Reference curvature radius (Brho/B) =   6.3380996E+02 cm
                          Skew  angle =   0.000000E+00 rad

                          Entrance face
                          DX =    20.000      LAMBDA =      8.000
                          Wedge  angle =  0.042761 RD
                          Exit     face
DX =     20.000     LAMBDA =      8.000
Wedge  angle =  0.042761 RD
Fringe  field  coefficients :
   0.24010  1.86390 -0.55720  0.39040  0.00000  0.00000
```

• The transport matrix of BEND is found under MATRIX, bottom of zgoubi.res execution listing:

```
     0.940314        2.42532         0.00000         0.00000         0.00000         0.482408
    -4.777393E-02    0.940252        0.00000         0.00000         0.00000         0.385925
     0.00000         0.00000         0.985387        2.48909         0.00000         0.00000
     0.00000         0.00000        -1.165918E-02    0.985379        0.00000         0.00000
     0.385938        0.482407        0.00000         0.00000         1.00000         6.371162E-02
     0.00000         0.00000         0.00000         0.00000         0.00000         1.00000
```

By comparison, from Eq. 14.7 one gets reasonable agreement, for instance

$$T_{11} = \frac{\cos(\alpha - \varepsilon)}{\cos \varepsilon} = 0.9389152, \quad T_{12} = \rho \sin \alpha = 2.42548586, \quad T_{26} = \frac{\sin(\alpha - \varepsilon) + \sin \varepsilon}{\cos \varepsilon} = 0.3856742$$

$$T_{33} = 1 - \alpha \tan \varepsilon = 0.984570, \quad T_{43} = -\frac{\tan \varepsilon}{\rho}(2 - \tan \varepsilon) = -0.012302$$

Note that the wedge angle ε in T_{33} and T_{43}, which are the hard edge model coefficients, should actually be corrected for the extent of the fringe field ($\lambda = 8\,\mathrm{cm}$), Eqs. 14.19, 14.20: the present agreement shows that this effect is marginal.

• The magnetic field across the bend is produced using the input data file of Table 9.10, which resorts to OPTIONS[CONSTY ON]. It is displayed in Fig. 9.23. BEND is defined in a Cartesian frame, its axis is a straight line, thus OPTIONS [CONSTY], which maintains a constant distance to that axis, results in straight trajectories. Using DIPOLE instead would result in circular trajectories, at constant distance from its RM-radius reference arc, as DIPOLE is defined in a polar frame.

• SATURNE cell is built from the half-arc (Table 9.11) followed by a FODO cell (Table 9.12). Simulation of a cell follows, Table 9.13, the ring is a 4-cell assembly, Table 9.14.

The TWISS command performed to determine the optical functions along the ring (Table 9.14) has the virtue of logging in zgoubi.res execution listing the transport matrices of individual optical elements as the propagation of the optical functions along the sequence proceeds. Looking up zgoubi.res one finds for the focusing quadrupole for instance:

```
              TRANSFER  MATRIX  OF  LAST  ELEMENT  (MKSA units)

     0.917793        0.454355        1.517751E-16    3.462630E-17    0.00000         7.608860E-07
    -0.346989        0.917793       -1.555146E-16   -1.398444E-16    0.00000         3.196050E-06
     1.738006E-16    3.728060E-17    1.08452         0.480321        0.00000         2.368020E-12
     2.539950E-17    5.274520E-19    0.366819        1.08452         0.00000         9.062457E-12
     3.872045E-06    9.241877E-07    3.708880E-16    5.131748E-15    1.00000        -1.383900E-07
     0.00000         0.00000         0.00000         0.00000         0.00000         1.00000
```

Table 9.10 Simulation input data file to produce the magnetic field across BEND, on a straight axis and 5 cm off-mid plane. Another possibility for the same result would be, rather than using OPTIONS[CONSTY], to launch the two particles with a large relative rigidity D, they would therefore go straight

```
SaturneBEND_field.INC.dat.
! Field in SATURNE bending magnet.
'OBJET'
1000.                                                        ! 1 T m rigidity.
2
2 1
0. 0. 0. 0.  0. 1. 'o'       ! D=1 here. Could be set to 1e10 for a similar effect to OPTIONS[CONSTY].
0. 0. 5. 0.  0. 1. 'Z'
1 1

'OPTIONS'
1 2
.plt 2                                  ! Set IL=2 in optical elements (log stepwise data in zgoubi.plt).
CONSTY  ON                              ! Force particles to maintain their initial Y_0 and Z_0 coodinates.

'MARKER' satBEND_S
'BEND'
0
247.30039  0.  1.57776                  ! Straight length of the magnet: 2*rho*sin(alpha/2); unused; field.
20. 8. .04276056667
4 .2401 1.8639 -.5572 .3904 0. 0.
20. 8. .04276056667
4 .2401 1.8639 -.5572 .3904 0. 0.
#60|220|60    bend                      ! Integration step sizes: 60 steps in fringe field regions, 220 in body.
1 0. 0. 0.
'MARKER' satBEND_E

'FAISCEAU'
'END'
```

Fig. 9.23 Magnetic field across the bend, in the mid-plane and 5 cm off. In the latter case, varying field causes overshoots in the fall-off regions. A graph obtained using zpop: menu 7; 1/1 to open zgoubi.plt; 2/[6, 32] for B_Z versus s; 7 to plot

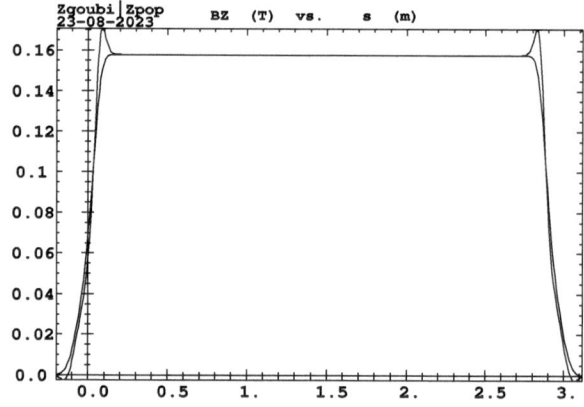

Agreement can be observed with transport coefficient values obtained from Eq. 14.27 taken with $L = 46.723$ cm, $K = 0.763695$ m^{-1}, namely,

$$T_{11} = \cos L\sqrt{K} = 0.9177929, \; T_{21} = -\sqrt{K}\sin L\sqrt{K} - 0.3469888$$
$$T_{33} = \cosh L\sqrt{K} = 1.084523, \; T_{21} = \sqrt{K}\sinh L\sqrt{K} = 0.3668189$$

Similar agreement is observed for the defocusing quadrupole.

The optical functions of the ring are computed using the input data file of Table 9.13, they are displayed in Fig. 9.24.

Beam matrix and optical parameters (tunes, momentum compaction, ...) are found in zgoubi.res execution listing resulting from the ring simulation (Table 9.14),

```
Beam matrix (beta/-alpha/-alpha/gamma) and periodic dispersion (MKSA units)

     15.152044    2.789149    0.000000    0.000000    0.000000    6.384097
      2.789149    0.579417    0.000000    0.000000    0.000000    0.680738
      0.000000    0.000000    2.074406   -0.398720    0.000000   -0.000000
      0.000000    0.000000   -0.398720    0.558703    0.000000    0.000000
      0.000000    0.000000    0.000000    0.000000    0.000000    0.000000
      0.000000    0.000000    0.000000    0.000000    0.000000    0.000000

       Betatron tunes (Q1 Q2 modes)                Momentum compaction :
     NU_Y = 0.63825247      NU_Z = 0.60416649       dL/L / dp/p = 1.50488289E-02

     Transition gamma  = 8.15170861E+00
```

namely:

```
Chromaticities :    dNu_y / dp/p = -4.1965255      dNu_z / dp/p = -5.1988031
```

Table 9.11 Simulation input data file: half SATURNE 2 arc, from center of QF to center of QF; computation of its transport matrix. This input data file saturneHalfArc.inc defines the arc sequence segment sat_HalfArc_S to sat_HalfArc_E for use in subsequent INCLUDE statements. PUH, HDEF, HKIC as well as PUV, VDEF, VKIC markers and thin lenses are not used here, they are provisions for a subsequent SVD orbit correction simulation using SVDOC

```
File name: saturneHalfArc.inc
! SATURNE 2 half arc.
'MARKER'            SatII_HalfArc.inc_S  ! Just for edition purposes.
'OBJET'
1000.
5
.001 .01 .001 .01 .0 .0001
0. 0. 0. 0. 0. 1.

'MARKER' sat_HalfArc_S
'MARKER'      PUH       ! Provision for further SVDOC simulations.
'MULTIPOL'    HDEF      ! Provision for further SVDOC simulations.
0
1e-3 10. 0.   0. 0. 0. 0. 0. 0. 0. 0.
0. 0. 0. 0. 0. 0. 0. 0.
6 .1122 6.2671 -1.4982 3.5882 -2.1209 1.723
0. 0. 0. 0. 0. 0. 0. 0.
6 .1122 6.2671 -1.4982 3.5882 -2.1209 1.723
0. 0. 0. 0. 0.   0. 0. 0. 0.
#10|3|10    ! Quad
1 0. 0. 0.
'MULTIPOL'    HKIC      ! Provision for further SVDOC simulations.
0
1e-3 10. 0.   0. 0. 0. 0. 0. 0. 0. 0.
0. 0. 0. 0. 0. 0. 0. 0.
6 .1122 6.2671 -1.4982 3.5882 -2.1209 1.723
0. 0. 0. 0. 0. 0. 0. 0.
6 .1122 6.2671 -1.4982 3.5882 -2.1209 1.723
0. 0. 0. 0. 0.   0. 0. 0.
#10|3|10    ! Quad
1 0. 0. 0.
'QUADRUPO'   Half_QF
0              ! A trick here: with pole tip radius 10cm, the field value
23.3615 10. .763695         ! in kG identifies with gradient in T/m.
0.  0.
6 .1122 6.2671 -1.4982 3.5882 -2.1209 1.723
0.  0.
6 .1122 6.2671 -1.4982 3.5882 -2.1209 1.723
1.
1 0. 0. 0.
'ESL'
71.6256

'INCLUDE'
1
saturneBEND.inc[satBEND_S:satBEND_E]

'ESL'
71.6256
'MARKER'   PUV                        ! For SVDOC simulations.
```

```
'MULTIPOL'          VDEF
0
1e-3 10. 0.   0. 0. 0. 0. 0. 0. 0. 0.
0. 0. 0. 0. 0. 0. 0. 0. 0.
6 .1122 6.2671 -1.4982 3.5882 -2.1209 1.723
0. 0. 0. 0. 0. 0. 0. 0. 0. 0.
6 .1122 6.2671 -1.4982 3.5882 -2.1209 1.723
1.57079632679 0. 0. 0. 0. 0.   0. 0. 0. 0.
#10|3|10    ! Quad
1 0. 0. 0.
'MULTIPOL'          VKIC
0
1e-3 10. 0.   0. 0. 0. 0. 0. 0. 0. 0.
0. 0. 0. 0. 0. 0. 0. 0. 0.
6 .1122 6.2671 -1.4982 3.5882 -2.1209 1.723
0. 0. 0. 0. 0. 0. 0. 0.
6 .1122 6.2671 -1.4982 3.5882 -2.1209 1.723
1.57079632679 0. 0. 0. 0. 0.   0. 0. 0.
#10|3|10    ! Quad
1 0. 0. 0.
'MULTIPOL'          QD
0
48.6273 10. 0. -.765533 0. 0. 0. 0. 0. 0. 0. 0.
0. 0. 0. 0. 0. 0.   0. 0. 0. .0 .0
6 .1122 6.2671 -1.4982 3.5882 -2.1209 1.723
0. 0. 0. 0. 0. 0. 0. 0. 0. 0.
6 .1122 6.2671 -1.4982 3.5882 -2.1209 1.723
0. 0. 0. 0. 0. 0. 0. 0. 0.
1.
1 0. 0. 0.
'ESL'
71.6256

'INCLUDE'
1
saturneBEND.inc[satBEND_S:satBEND_E]

'ESL'
71.6256
'QUADRUPO'
0
23.3615 10. .763695
0.  0.
6 .1122 6.2671 -1.4982 3.5882 -2.1209 1.723
0.  0.
6 .1122 6.2671 -1.4982 3.5882 -2.1209 1.723
1.
1 0. 0. 0.
'MARKER' sat_HalfArc_E

'FAISCEAU'
'MATRIX'
1 0

'MARKER'    SatII_HalfArc.inc_E
'END'
```

Table 9.12 Simulation input data file: SATURNE 2 FODO cell, from center of QF to center of QF; computation of its transport matrix. This input data file saturneFODO.inc defines the FODO sequence segment sat_HalfArc_S to sat_HalfArc_E for use in subsequent INCLUDE statements. PUH, HDEF, HKIC as well as PUV, VDEF, VKIC markers and thin lenses are not used here, they are provisions for a subsequent SVD orbit correction simulation using SVDOC

```
                                                              'MULTIPOL'        VDEF
                                                              0
                                                              1e-3  10. 0.    0. 0. 0. 0. 0. 0. 0. 0.
File name: saturneFODO.inc                                    0.  0. 0.  0. 0.  0. 0.  0. 0.  0. 0.
! SATURNE 2 FODO.                                             6  .1122 6.2671 -1.4982 3.5882 -2.1209 1.723
'MARKER'   SatII_FODO.inc_S          ! Just for edition purposes.   0.  0. 0. 0. 0.  0. 0.   0. 0. 0. 0.
'OBJET'                                                       6  .1122 6.2671 -1.4982 3.5882 -2.1209 1.723
1000.                                    ! 1 T m rigidity.    1.57079632679 0. 0. 0. 0. 0.   0. 0. 0. 0.
5            ! Allows computation of transport matrix from 13 rays.   #10|3|10   ! Quad
.001 .01 .001 .01 .0 .0001                                    1 0. 0. 0.
0. 0. 0. 0. 0. 1.                                             'MULTIPOL'        VKIC
                                                              0
'MARKER'  sat_FODO_S                                          1e-3  10. 0.    0. 0. 0. 0. 0. 0. 0. 0.
'MARKER'    PUH                                               0.  0. 0.  0. 0.  0. 0.  0. 0.  0. 0.
'MULTIPOL'        HDEF                                        6  .1122 6.2671 -1.4982 3.5882 -2.1209 1.723
0                                                             0.  0. 0. 0. 0.  0. 0.   0. 0. 0. 0.
1e-3  10. 0.    0. 0. 0. 0. 0. 0. 0. 0.                       6  .1122 6.2671 -1.4982 3.5882 -2.1209 1.723
0.  0. 0.  0. 0.  0. 0.  0. 0.  0. 0.                         1.57079632679 0. 0. 0. 0. 0.   0. 0. 0. 0.
6  .1122 6.2671 -1.4982 3.5882 -2.1209 1.723                 #10|3|10   ! Quad
0.  0. 0. 0. 0.  0. 0.   0. 0. 0. 0.                          1 0. 0. 0.
6  .1122 6.2671 -1.4982 3.5882 -2.1209 1.723                 'MULTIPOL'        QD
0. 0. 0. 0. 0. 0.   0. 0. 0. 0.                               0
#10|3|10    ! Quad                                            48.6273 10. 0.  -.765533 0. 0. 0. 0. 0. 0. 0. 0.
1 0. 0. 0.                                                    0. 0. 0. 0. 0.  0. 0. 0. 0. 0. .0 .0
'MULTIPOL'        HKIC                                        6  .1122 6.2671 -1.4982 3.5882 -2.1209 1.723
0                                                             0. 0. 0. 0. 0.  0. 0. 0. 0. 0. 0.
1e-3  10. 0.    0. 0. 0. 0. 0. 0. 0. 0.                       6  .1122 6.2671 -1.4982 3.5882 -2.1209 1.723
0.  0. 0.  0. 0.  0. 0.  0. 0.  0. 0.                         0. 0. 0. 0. 0. 0. 0. 0. 0. 0. 0.
6  .1122 6.2671 -1.4982 3.5882 -2.1209 1.723                 1.
0.  0. 0. 0. 0.  0. 0.   0. 0.                                1 0. 0. 0.
6  .1122 6.2671 -1.4982 3.5882 -2.1209 1.723                 'ESL'        LSS_dw
0. 0. 0. 0. 0. 0.   0. 0. 0. 0.                               392.148
#10|3|10    ! Quad                                            'QUADRUPO'   Half_QF
1 0. 0. 0.                                                    0
'QUADRUPO'   Half_QF                                          23.3615 10. .763695
0                                                             0.  0.
23.3615 10. .763695                    ! Half a QF quad.      6  .1122 6.2671 -1.4982 3.5882 -2.1209 1.723
0.  0.                      ! Hard-edge entrance field model. 0.  0.
6  .1122 6.2671 -1.4982 3.5882 -2.1209 1.723                 6  .1122 6.2671 -1.4982 3.5882 -2.1209 1.723
0.  0.                         ! Hard-edge exit field model.  1.
6  .1122 6.2671 -1.4982 3.5882 -2.1209 1.723                 1 0. 0. 0.
1.                               ! Integration step size.     'MARKER'  sat_FODO_E
1 0. 0. 0.
'ESL'        LSS_up                                           'FAISCEAU'
392.148                                                       'MATRIX'
'MARKER'    PUV                                               1 0     ! Compute transport matrix, from OBJET.

                                                              'MARKER'   SatII_FODO.inc_E
                                                              'END'
```

(b) Six-D bunch acceleration. Betatron damping.

The input data file for this simulation is given in Table 9.15.

A major change is BORO value, set to 1.03527036749193 T m, i.e. a kinetic energy of 50 MeV. SCALING is introduced to scale the field in all magnetic elements to that value of the rigidity. SCALING[NTIM = −1] will cause fields to further scale to the increase of the reference rigidity by CAVITE.

Periodic beam parameters under MCOBJET are taken from the beam matrix resulting from the TWISS run in question (a). CAVITE is programmed with harmonic $h = 3$ and synchronous phase $30°$.

Various tracking results are displayed in Fig. 9.25. During acceleration, betatron oscillation amplitudes damp as $1/\sqrt{\beta\gamma}$ (Eq. 9.40). With all parameters maintained constant except for the energy, $\widehat{\Delta p/p}$ damps as $1/\beta\gamma^{1/4}$ (Eq. 9.37).

Table 9.13 Simulation input data file: SATURNE 2 cell, comprised of two half-arcs and a FODO cell; computation of its optical functions using TWISS. The gnuplot script gnuplot_TWISS.gnu is copied from the [pathTo]/branches/zgoubi-code/toolbox/gnuplotFiles/gnuplot_Zfai/ library which is part of `zgoubi` sourceforge package

```
File name: saturneCell.inc
! SATURNE 2 Cell.
'MARKER'   SatII_CELL.inc_S  ! Just for edition purposes.
'OBJET'
1000.
5
.001 .01 .001 .01 .0 .0001
0. 0. 0. 0. 0. 1.

'MARKER' satCell_S
'INCLUDE'
1
2* saturneHalfArc.inc[sat_HalfArc_S:sat_HalfArc_E]
'INCLUDE'
1
saturneFODO.inc[sat_FODO_S:sat_FODO_E]
'MARKER' satCell_E

'TWISS'
2 1. 1.

'SYSTEM'
1                       ! Produces graphs of the orbits and
gnuplot <./gnuplot_TWISS.gnu        ! optical functions.

'MARKER'   SatII_CELL.inc_E  ! Just for edition purposes.
'END'
```

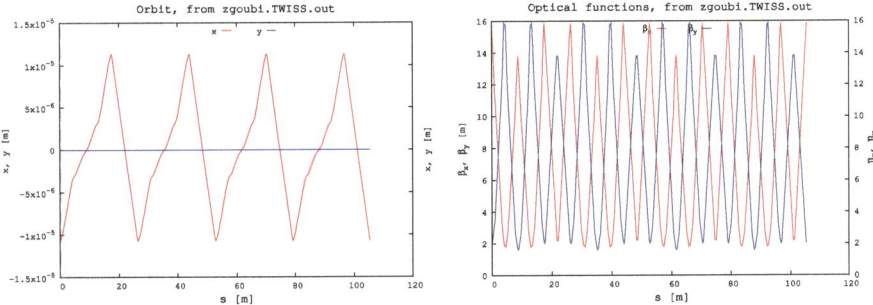

Fig. 9.24 Quasi-zero closed orbit, and the optical functions in SATURNE 2, from a TWISS command (Table 9.14)

(c) Momentum acceptance. Longitudinal separatrix.

The input data file of Table 9.15 can be used here, *mutatis mutandis*, as follows. Longitudinal phase space motion is computed

- over a few thousand turns: use REBELOTE[IPASS=few thousand],
- for a few particles in the vicinity of the separatrix: use OBJET[KOBJ=2] to define individual particles. Particles are launched with null longitudinal offset (OBJET[$S_0 = 0$]), dp/p near momentum acceptance. For instance, for the stationary bucket case:

Table 9.14 Simulation input data file: SATURNE 2 ring, comprised of 4 cells; computation of its optical functions using TWISS; a SYSTEM call to a gnuplot script gets them plotted

```
File name: saturneRing.inc
! SATURNE 2 Ring.
'MARKER'  SatII_RING.inc_S  ! Just for edition purposes.
'OBJET'
1000.
5  ! Define a 13-particle sample for MATRIX computation.
.001 .01 .001 .01 .0 .0001       ! Coordinate sampling.
0. 0. 0. 0. 0. 1.                ! Reference particle (#1).

'INCLUDE'
1
4 * saturneCell.inc[satCell_S:satCell_E]

'FIT2'          ! Find initial coordinates (at OBJET) of
4                               ! the closed orbit.
2 30  0 [-1.,1.]
2 31  0 [-10.,10.]
2 32  0 [-1.,1.]
2 33  0 [-10.,10.]
4  1e-15
3.1 1 2 #End 0. 1. 0
3.1 1 3 #End 0. 1. 0
3.1 1 4 #End 0. 1. 0
3.1 1 5 #End 0. 1. 0

'TWISS'
2 1. 1.

'SYSTEM'
1
gnuplot <./gnuplot_TWISS.gnu

'MARKER'  SatII_RING.inc_E  ! Just for edition purposes.
'END'
```

```
'OBJET'
1.03527036749193 * 1e3                          ! Reference Brho: 50 MeV proton.
2                                               ! Create 4 initial conditions.
4 1
0. 0. 0. 0. 0. 1.0005 'o'
0. 0. 0. 0. 0. 1.001077 'o'
0. 0. 0. 0. 0. 1.00108 'o'
0. 0. 0. 0. 0. 1.00115 'o'
1 1 1 1 1
```

– the synchronous phase in CAVITE has to be set to proper value.

Tracking hypotheses are recapped in Table 9.16 outcomes are displayed in Fig. 9.16 (Sect. 9.2.4). The paraxial synchrotron tune and bucket height, Table 9.17, can be checked against Eqs. 9.30, 9.36, respectively.

9.2 Non-Linear Motion in SATURNE 2

(a, b) Third- or fourth-integer phase space portraits.

It is necessary to run the lattice near 3rd integer horizontal tune, $\nu_Y \approx 3 + \frac{2}{3}$, taken near the nominal tune in order to avoid excessive perturbation of the optics. (respectively 4th integer horizontal tune, $\nu_Y \approx 3 + \frac{3}{4}$.) The vertical tune is left unchanged.

A preliminary FIT will provide the required SCALING coefficients for QUADRUPO (the focusing quadrupole family) and MULTIPOL (defocusing quadrupole family). The file for that is given in Table 9.18. FIT outcomes, found in the execution listing zgoubi.res, are as follows:

– case of $\nu_Y = 3 + \frac{2}{3}$:

Table 9.15 Simulation input data file: acceleration from 50 MeV to 3 GeV in SATURNE 2 ring. To save on CPU-time and on voluminous log file zgoubi.plt, OPTIONS[.plt 0] makes sure that any $IL \neq 0$ value in the different magnets (Tables 9.9, 9.11 and 9.12) is overriden with $IL = 0$. This simulation, with 30 particles, takes about 6 min with a 3 GHz clock CPU. The gnuplot script gnuplot_Zfai_YZD.vs.turn_multiplot.gnu is copied from the [pathTo]/branches/zgoubi-code/toolbox/gnuplotFiles/gnuplot_Zfai/ library which is part of zgoubi sourceforge package Note: this file, and all INCLUDE files it resorts to, are available in zgoubi sourceforge repository at https://sourceforge.net/p/zgoubi/code/HEAD/tree/branches/exemples/book/zgoubiMaterial/synchrotron_strongFocusing/SaturneII/accelerationCycle

```
File name: Saturne_accel.INC.dat
! Acceleration in SATURNE 2 Ring.
'MARKER'  SatII_accel.INC.dat_S                        ! Just for edition purposes.
'MCOBJET'
1.03527036749193 * 1e3                                 ! Reference Brho: 50 MeV proton.
3                            ! Create a 13 particle set, proper for MATRIX computation.
30
2 2 2 2 2 2                                            ! Gaussian densities, horiz., vertic. and long.
0. 0. 0. 0. 0. 1.
-0.062696 15.827646 1e-6 1    ! Periodic alpha_Y, beta_Y, and emittance truncated at 1-sigma;
-0.082216 2.001473 1e-6 1     ! periodic alpha_Z, beta_Z, and emittance truncated at 1-sigma;
0. 1. 1.e-8 1                           ! Gaussian dp/p, rms 1e-4, truncated at 1-sigma.
123456 234567 345678
'PARTICUL'
PROTON      ! Necessary data in order to allow acceleration (CAVITE needs to know mass and charge).

'OPTIONS'
1 1
.plt 0                        ! Inhibits possible logging to zgoubi.plt by any optical element.

'FAISTORE'
zgoubi.fai
7                                       ! Log every 7 turn - to save on CPU time and file volume.

'SCALING'                  ! Scale field in the following 3 optical elements, during acceleration.
1 3                       ! The field in these magnets must be set for a rigidity of 1 Tm, which is the
QUADRUPO                                 ! case in their respective INCLUDEd files.
-1      ! This option '-1' causes the field to be ramped in proportion of rigidity increase by CAVITE.
1.03527036749193                               ! Initial value of the field scaling factor.
1
MULTIPOL
-1
1.03527036749193
1
BEND
-1
1.03527036749193
1

'INCLUDE'
1
4 * saturneCell.inc[satCell_S:satCell_E]

'CAVITE'  ! IOPT=2 accounts for RF pahse at particle arrival (use IOPT=3 for constant energy-kick).
2         ! PRINT added here would log RF data to zgoubi.cavite.Out (for post-treatment, plot, debug ...).
105.555614 3.                ! Length of closed orbit (taken from former TWISS run); harmonic number.
1.e6  0.523598775598                                ! Peak voltage; synchronous phase.

'REBELOTE'
6000 0.3 99

'FAISCEAU'

'SYSTEM'
1
gnuplot < ./gnuplot_Zfai_YZD.vs.turn_multiplot.gnu              ! Produce Y, Z, D vs. turn graphs.

'MARKER'  SatII_accel.INC.dat_E                        ! Just for edition purposes.
'END'
```

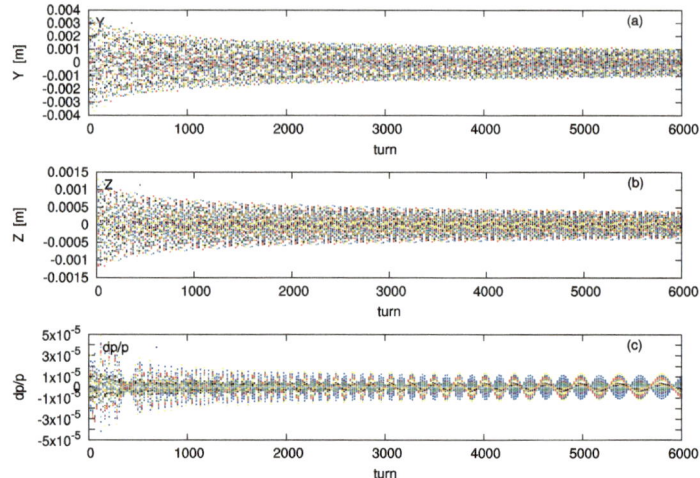

Fig. 9.25 Acceleration over $E : 50\,\mathrm{MeV} \to 3\,\mathrm{GeV}$ in SATURNE 2: damped Y, Z motions, and $D - 1 = \delta p/p$, as a function of turn. Turn-by-turn data read from zgoubi.fai

Table 9.16 Stationary bucket tracking: working hypotheses

E_s (MeV)	Proton mass + 50
$B\rho$ (T m)	1.035
h	3
V (kV)	10
ϕ_s (deg)	0, 15, 30 or 60
alpha	0.01505
eta	0.8863
T rev. (μs)	0.35
RF (MHz)	8.5

Table 9.17 Accelerating bucket: theoretical expectations (left) and tracking outcomes (right). The numerical value of ν_s is obtained as the inverse of the number of turns necessary to close the phase space ellipse

		Theoretical	Numerical
Synch. tune ν_s ($\times 10^{-4}$) (Eq. 9.30)		7.5779 (Eq. 9.30)	7.5758 (Eq. 9.30)
$\dfrac{\widehat{\Delta p}}{p}$ ($\times 10^{-3}$)	$\phi_s = 0$	4.9562	5.05
(Eq. 9.36)	$\phi_s = 15$	3.9249	3.97
	$\phi_s = 30$	2.9002	2.94
	$\phi_s = 60$	1.0693	1.10

Table 9.18 Simulation input data file: find focusing and defocusing quadrupole family settings, for horizontal tune value $\nu_Y = 3 + \frac{2}{3}$ (or $\nu_Y = 3 + \frac{3}{4}$). The INCLUDE file saturneCell.inc is defined in Exercise 9.1-a

```
File name: Saturne_FITnuY.INC.dat
'MARKER'   SatII_FITnuY_S                      ! Just for edition purposes.
'OBJET'
1. * 1e3
5
0.001 0.01 0.001 0.01 0.001 0.0001
0. 0. 0. 0. 0. 1.

'OPTIONS'
1 1
.plt 0                         ! Inhibits possible logging to zgoubi.plt by any optical element.

'SCALING'                      ! Scale field in the following 3 optical elements.
1 3
QUADRUPO                                        ! Focusing family.
-1     ! This option '-1' causes the field to be ramped in proportion of rigidity increase by CAVITE.
1.0056333  ! (case b, nuY=0.75: 1.0220177)     ! FIT variable #1.
1
MULTIPOL                                        ! Defocusing family.
-1
1.0006548  ! (case b, nuY=0.75: 1.0025472)     ! FIT variable #2.
1
BEND
-1
1.
1

'INCLUDE'
1
4 * saturneCell.inc[satCell_S:satCell_E]

'FIT'
2
4 4 0 .2                                        ! 2 variables:
4 8 0 .2                                        ! vary QUADRUPO scaling coefficient,
2                                               ! vary MULTIPOL scaling coefficient.
0.1 7 0 #End 0.6666 1. 0                        ! 2 constraints:
0.1 8 0 #End 0.60416 1. 0                       ! horizontal (fractional) tune (or 0.75, case b),
                                                ! vertical (fractional) tune, unchanged.
'MARKER'   SatII_FITnuY_E                        ! Just for edition purposes.
'END'
```

```
                        FIT variables and constraints in good order, FIT will proceed.
         STATUS OF VARIABLES (Iteration #   1 /   999 max.)
         LMNT VAR PARAM  MINIMUM   INITIAL      FINAL     MAXIMUM    STEP      NAME    LBL1   LBL2
           4   1     4   0.805      1.01      1.0056333     1.21   1.341E-03  SCALING    -     -
           4   2     8   0.801      1.00      1.0006548     1.20   1.334E-03  SCALING    -     -

         STATUS OF CONSTRAINTS (Target penalty =   1.0000E-01)
         TYPE  I    J LMNT#    DESIRED      WEIGHT      REACHED      KI2    NAME    LBL1     LBL2
           0   7    0   160  6.666000E-01  1.000E+00  6.665979E-01  3.90E-01 MARKER  satCell_E  -
           0   8    0   160  6.041600E-01  1.000E+00  6.041574E-01  6.10E-01 MARKER  satCell_E  -
         Fit reached penalty value  1.1302E-11
```

– case of $\nu_Y = 3 + \frac{3}{4}$:

```
                        FIT variables and constraints in good order, FIT will proceed.
         STATUS OF VARIABLES (Iteration #  11 /   999 max.)
         LMNT VAR PARAM  MINIMUM   INITIAL      FINAL     MAXIMUM    STEP      NAME    LBL1   LBL2
           4   1     4   0.805      1.02      1.0220177     1.21   6.131E-07  SCALING    -     -
           4   2     8   0.801      1.00      1.0025472     1.20   6.101E-07  SCALING    -     -

         STATUS OF CONSTRAINTS (Target penalty =   1.0000E-10)
         TYPE  I    J LMNT#    DESIRED      WEIGHT      REACHED      KI2    NAME    LBL1     LBL2
           0   7    0   160  7.500000E-01  1.000E+00  7.500061E-01  8.10E-01 MARKER  satCell_E  -
           0   8    0   160  6.041600E-01  1.000E+00  6.041629E-01  1.90E-01 MARKER  satCell_E  -
         Fit reached penalty value  4.5228E-11
```

Finally, non-linear horizontal phase space portraits, at fixed energy, are generated using the input data file given in Table 9.19, essentially a copy of the file in Table 9.15 with the following modifications:

– use OBJET[KOBJ=2] to define individual particles with diverse invariant values. The value of the rigidity does not matter;
– modify the SCALING factors for QUADRUPO and MULTIPOL families, so to set the horizontal tune to a proper value, near 3rd- or 4th-integer;

Table 9.19 Simulation input data file: generate a non-linear phase space, at fixed energy. The input data file of Table 9.15 is used, with a few modifications. In particular, the QUADRUPO family and MULTIPOL family SCALING coefficients are updated according to the previous FIT outcomes

```
File name: SATII_3rdInt.INC.dat   (or SATII_4thInt.INC.dat)
! Acceleration in SATURNE 2 Ring.
'MARKER'    SatII_3rdInt_S                                     ! Just for edition purposes.
'OBJET'
1. * 1e3
2
5 1
.1  0. 0. 0. 0. 1. 'o'
.5  0. 0. 0. 0. 1. 'a'
1.  0. 0. 0. 0. 1. 'b'
2.  0. 0. 0. 0. 1. 'c'
2.  2. 0. 0. 0. 1. 'e'
1 1 1 1 1                             ! '-1' will inhibit tracking of the particle(s) concerned.

'OPTIONS'
1 1
.plt 0                               ! Inhibits possible logging to zgoubi.plt by any optical element.

'FAISTORE'
zgoubi.fai
7                                    ! For smaller volume zgoubi.fai: log particle data every 7 passes.

'SCALING'                 ! Scale field in the following 3 optical elements, during acceleration.
1  3                      ! The field in these magnets must be set for a rigidity of 1 Tm, which is the
QUADRUPO                                        ! case in their respective INCLUDEd files.
-1     ! This option '-1' causes the field to be ramped in proportion of rigidity increase by CAVITE.
1. *1.0056                                          ! Case b, nuY=0.75:  * 1.0220177.
1
MULTIPOL
-1
1. *1.0006548                                       ! Case b, nuY=0.75:  * 1.0025472.
1
BEND
-1
1.
1

'MULTIPOL'                 ! Non-linear thin lens, with sextupole or octupole component.
0
1e-2  10.  0. 0. 1.e1 0. 0.  0. 0. 0. 0. 0.         ! Case b, nuY=0.75:  1e4 octupole component.
0. 0. 0. 0. 0.  0. 0. 0. 0.  .0 .0
6  .1122 6.2671 -1.4982 3.5882 -2.1209 1.723
0. 0. 0. 0. 0.  0. 0. 0. 0.  0. 0.
6  .1122 6.2671 -1.4982 3.5882 -2.1209 1.723
0. 0. 0. 0. 0. 0. 0. 0. 0. 0.
#40|4|40
1 0. 0. 0.

'INCLUDE'
1
4 * saturneCell.inc[satCell_S:satCell_E]

'REBELOTE'
9999 0.3 99

'FAISCEAU'

'SYSTEM'
1
gnuplot < ./gnuplot_Zfai_YZD.vs.turn_multiplot.gnu              ! Produce Y, Z, D vs. turn graphs.

'MARKER'    SatII_3rdInt_E                           ! Just for edition purposes.
'END'
```

- remove CAVITE;
- add a thin lens to excite the resonance. MULTIPOL can be used, with a sextupole component to excite the 3rd integer resonance motion, near $\nu_Y = 3 + \frac{2}{3}$ (or an octupole component to excite the 4th integer resonance motion, near $\nu_Y = 3 + \frac{3}{4}$).

The resulting phase space portraits are displayed in Fig. 9.26.

9.3 SVD Orbit Correction

An input data file for orbit correction in SATURNE 2, using SVDOC, is given in Table 9.20. Its data are commented for clarification of the *modus operandi*. This file

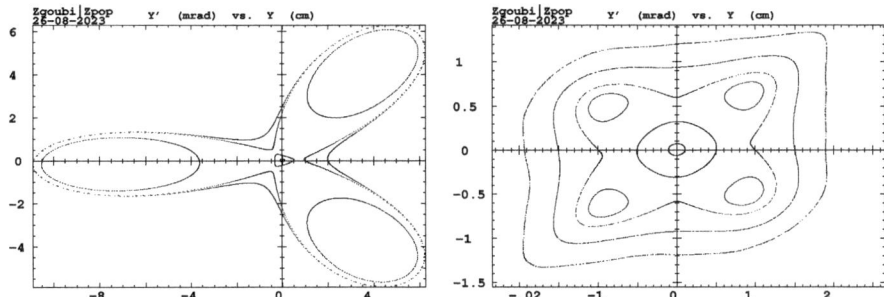

Fig. 9.26 Third-integer and fourth-integer resonance phase space portraits, about 10^4 turns. Graphs obtained using zpop: menu 7; 1/5 to open zgoubi.fai; 2/[2,3] for T versus Y; 7 to plot

INCLUDEs SATURNE ring sequence taken from Exercise 9.1, thus it has a 12 PUH and a 12 PUV family; a 12 HKIC and a 12 VKIC family, thin lenses used to correct the orbit; a 12 HDEF and a 12 VDEF family, thin lenses used to excite H and V orbits. The latter are located at respectively focusing and defocusing quadrupoles.

Closed orbit coordinates are found by FIT2, their values at the pick-ups, PUH and PUV, both before and after correction, are logged in the output file zgoubi. SVDOrbits.out, which is one of the files resulting from SVDOC execution. This file also contains the corrector settings. The A matrix is logged in zgoubi.SVD_ Amatrix.out.

The outcomes of two different SVDOC runs, involving 24 PUs and 24 correctors for one, 24 PUs and 12 correctors for the other, are displayed in Fig. 9.27.

Note: running an SVDOC problem also produces a copy of the original file, zgoubi.SVD.out.dat, updated with corrector values as found by SVDOC (in a similar way that FIT produces a copy of the original problem in zgoubi.FIT.out.dat, updated with variable values found by FIT). This file zgoubi.SVD.out.dat can be run as is and will produce the same results as in Fig. 9.27 (make sure it includes a FAISTORE[zgoubi.fai PU*] instruction, to save particle data at all PUH and PUV), or FAISTORE[zgoubi.fai all] instruction, to save particle data at all optical elements).

9.4 Cornell Electron RCS. Radiative Energy Loss

(a) Cornell RCS optics.

The input file in Table 9.8 is run to produce lattice parameters. However, prior to doing so it is necessary to set the respective quadrupole components of the focusing and defocusing combined function dipole families (BF and BD) to their expected nominal value, yielding $\nu_Y \approx 9.62$, $\nu_Z \approx 13.82$ (Table 9.3).

This requires running iteratively, a couple of times,

- the input files in Table 9.4, which computes the orbit across BF and BD so to zero it in and out,
- the input file in Table 9.8 which, using FIT, computes the BF and BD families field SCALING coefficients proper to yield the expected tune values.

Table 9.20 Simulation input data file: SVD orbit correction in SATURNE 2 ring, using SVDOC. saturneCell.inc file is used (Table 9.13), which includes (see Tables 9.11 and 9.12) 12 PUH and 12 PUV; 12 HKIC and 12VKIC (correctors); 12 HDEF and 12 VDEF (excite orbit defects). Comment for instance every other corrector for a 24 PUs × 12 correctors system

```
File name: saturneRing.inc
! SATURNE 2 Ring.
'MARKER'  SatII_RING_SVDOC_S                              ! Just for edition purposes.
'OBJET'
1000.
2
1 1                           ! A single particle. Its closed orbit coordiantes will be found by FIT.
0. 0. 0. 0. 0. 1. 'o'
1
'FAISCEAU'
'OPTIONS'
1 1
WRITE ON
'FAISTORE'
zgoubi.fai  none ! all
1
'ERRORS'
0  2 123456   PRINT
MULTIPOL{HDEF} 1 BP A U 0. 1.e5  3  ! Huge field coefficient due to H/VDEF and H/VKIC being thin lens
MULTIPOL{VDEF} 1 BP A U 0. 1.e5  3                        ! of length 1e-5 cm.

'INCLUDE'
1
4 * saturneCell.inc[satCell_S:satCell_E]
'FAISCEAU'  atFIT

'FIT2'                                        ! Find the closed orbit, at each SVDOC loops.
4              ! final (NOT nofinal) is mandatory, that causes store of actual c.o. in zgoubi.SVD.out.
2 30 0 [-3.9,3.9]                   ! vary Y0 coordinate in OBJET (OBJET is element 32, follos MARKER)
2 31 0 [-9,9]                                                                                   ! T0,
2 32 0 [-1.9,1.9]                                                                               ! Z0,
2 33 0 [-9,9]                                                                                   ! P0.
4    1e-10                                                                       ! 4 constraints:
3.1 1 2 #End 0. 1. 0                                                            ! Yfinal = Y0,
3.1 1 3 #End 0. 1. 0                                                            ! Tfinal = T0,
3.1 1 4 #End 0. 1. 0                                                            ! Zfinal = Z0,
3.1 1 5 #End 0. 1. 0                                                            ! Pfinal = P0.
'FAISCEAU'  atSVDOC
'DRIFT'                                          ! just for greater accuracy on coordinates.
0.
'SVDOC'
1  0  zgoubi.SVD.out
PUH{PUH} PUV{PUV}       ! Names of the 12 PU-H, 12 PU-V families; no PU-HV here (max each =MPULAB/3=1).
CRH{HKIC} CRV{VKIC}                               ! Name of 12 H- and 12 V-corrector families;
1e-4     1e-4                                     ! kick values for SVD matrix coputation (rad).
1  2  5      ! "1" triggers ERRORS; "2" data are changed prior to that, following 2 lines; 5 trials.
ERRORS  12                        ! parameter 12 in ERRORS: value of HKIC dipole field: will set H orbit,
ERRORS  22                        ! parameter 22 in ERRORS: value of VKIC dipole field: will set V orbit.
!                                                  step 1: A^-1 is applied to correctors.
!               Step 2: when back down to FIT (from here), FIT will get the corrected periodic orbit.
'SYSTEM'
4
gnuplot <./gnuplot_SVDOrbits.gnu
!cp gnuplot_SVDOrbits.eps gnuplot_SVDOrbits_24PUx24COR.eps
sed -i 's@zgoubi.fai none@ zgoubi.fai all @g' zgoubi.FIT.out.dat
sed -i "s@'SVDOC'@'END'" 'SVDOC'@g" zgoubi.FIT.out.dat
!~/zgoubi/current/zgoubi/zgoubi -in zgoubi.FIT.out.dat
!gnuplot <./gnuplot_zgoubi.fai_YZ.vs.s.gnu
!cp ./gnuplot_zgoubi.fai_YZ.vs.s.eps ./gnuplot_zgoubi.fai_24PUx24COR.eps

'MARKER' SatII_RING_SVDOC_E ! Just for edition purposes.
'END'
```

The reason for this iteration is that, changing the field scaling to match the tunes, also changes the orbit across the combined function dipoles.

Results are found under TWISS in the execution listing zgoubi.res. They are also logged in zgoubi.TWISS.out header:

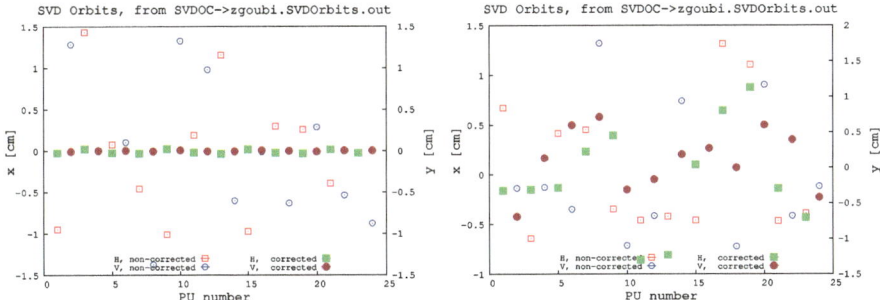

Fig. 9.27 Defect closed orbit observed at PUs, and corrected orbit obtained using SVDOC, both H and V cases. These data are read from the SVDOC procedure output file zgoubi.SVDOrbits.out. Left: case of 24 defects (at QF and QD quadrupoles) and 24 correctors at the same locations. Right: case where every other corrector is removed, leaving only 12 for orbit correction

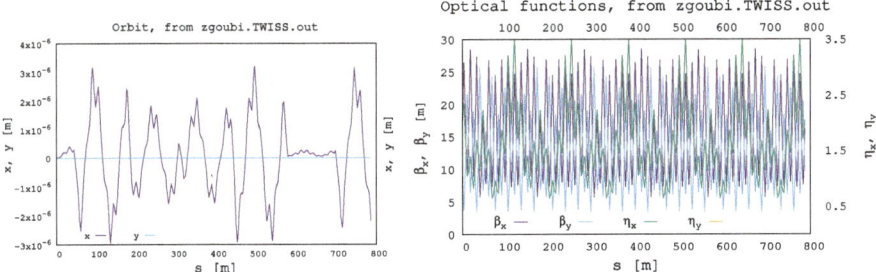

Fig. 9.28 Quasi-zero closed orbit, and the optical functions in Cornell RCS, from a TWISS command (Table 9.8)

```
@ LENGTH       %le   786.9469435
@ ALFA         %le   0.1378691509E-01
@ ORBIT5       %le              -0
@ GAMMATR      %le   8.516603910
@ Q1           %le   0.61992974E+00   9  [frac., int.]
@ Q2           %le   0.81998850E+00  13  [frac., int.]
@ DQ1          %le  -12.89613552
@ DQ2          %le  -15.83690059
@ DXMAX        %le   3.48440265E+00    6.38648731E-01   @ DXMIN
@ DYMAX        %le   0.00000000E+00    0.00000000E+00   @ DYMIN
@ XCOMAX       %le   3.18809892E-04   -2.91707847E-06   @ XCOMIN
@ YCOMAX       %le   0.00000000E+00    0.00000000E+00   @ YCOMIN
@ BETXMAX      %le   2.84026479E+01    5.66721825E+00   @ BETXMIN
@ BETYMAX      %le   2.56839723E+01    3.38044706E+00   @ BETYMIN
@ XCORMS       %le   1.27054100E-06
@ YCORMS       %le   0.    not computed
@ DXRMS        %le   7.04853516E-01
@ DYRMS        %le   0.00000000E+00
```

TWISS logs the optical functions in zgoubi.TWISS.out. A gnuplot script taken from the sourceforge repository [49] is used to plot them, Fig. 9.28.

(b) An acceleration cycle, without radiative energy loss.

The input data file is given in Table 9.21, it is derived from that in Table 9.8, *mutatis mutandis*.

Note in particular "CornellSynch" argument under CAVITE. This causes cavite.f program to resort to the hard-coded Cornell RCS voltage function [46]

Table 9.21 Simulation input data file for an acceleration cycle in Cornell RCS, from 320 MeV to ~8 GeV

```
Cornell RCS. Track w/o SR, 320 MeV to ˜7 GeV.
 'MCOBJET'
1.0674037436922512 * 1E3          | 320 MeV.
3
1                                 | Just 1 particle.
2 2 2 2  1 1
0. 0. 0. 0. 0.  1.
0.089591   23.410413   25.e-6  3.
-0.145912   6.924692   25.e-6  3.
0.          1e-6        0.     1
123456 234567 345678
'PARTICUL'
POSITRON

| Uncomment to get spins tracked.
| 'SPNTRK'
|3


'FAISTORE'
zgoubi.fai
1


| Uncomment to account for SR loss.
|'SRLOSS'
|1
|MULTIPOL
|0 123456

'SCALING'
1  3
MULTIPOL
-1
1.0674037436922512
1
MULTIPOL F_BEND
-1
0.99292280 * 1.0674037436922512
1
MULTIPOL D_BEND
-1
1.1294084 * 1.0674037436922512
1
```

```
'INCLUDE'
1
6 * superCell.inc[superCell_S:superCell_E]

 'FAISCEAU'
 'SRPRNT'

! Comment when accounting for SR loss.
 'CAVITE'
2
786.9469435  1800
6e6  2.61799387799          | DE = 3MV/turn.

! Uncomment to account for SR loss.
! 'CAVITE'
!11    CornellSynch
!786.9469435  1800
!0. 1.0471975512 9.48145321E-6

 'FAISCEAU'
 'SRPRNT'
 'REBELOTE'
3000 0.2 99

 'SYSTEM'
1
gnuplot < ./gnuplot_Zfai_phaseSpaces_multiplot.gnu
!                   A gnuplot script found in
!                   zgoubi-code/toolbox/gnuplotFiles/.
 'END'
```

$$V(t) = 4.4 \sin(2\pi f t) + 8.8 \sin^8(2\pi f t/2), \quad f = 60 \, \text{Hz}$$

SCALING[NT = −1] ensures that magnetic fields follow the turn by turn rigidity increase imparted by CAVITE.

The expected particle motion is summarized in Fig. 9.29.

(c, d) Simulation of an acceleration cycle with radiation loss, and spins.
Indications are provided here: detailed simulations are left to the reader. Further guidance, including methods and data as well as their outcomes, can be found in [46].

The input data file in Table 9.21 is used: comment and uncomment as indicated, in order to change from ignoring SR to accounting for it. Typical particle dynamics outcomes expected are displayed in Fig. 9.30.

Non-systematic spin resonances can be excited by introducing a random gradient defect in the combined function dipoles. This is achieved by adding, before INCLUDE, an ERRORS command, as follows:

```
'ERRORS'
1 1 123466      !      Relative  Uniform         dB(kG)
MULTIPOL{B}     2  BP  R         U        0.d0    1.e-2    0
```

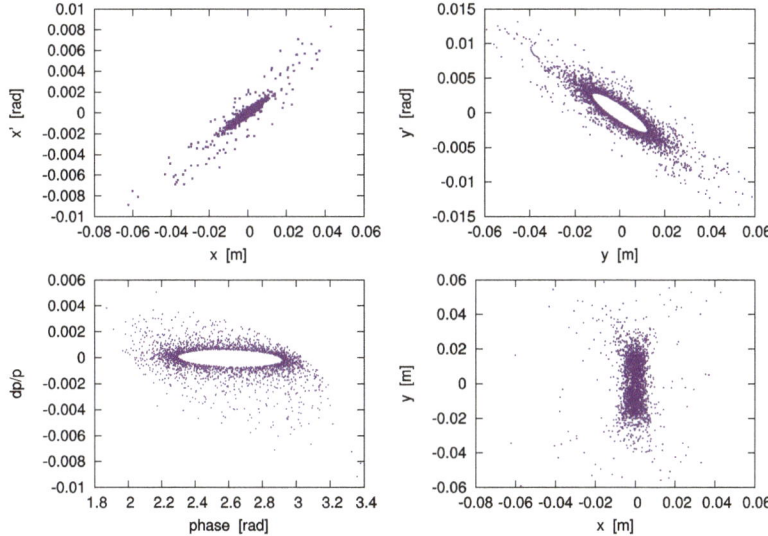

Fig. 9.29 Transverse and longitudinal phase spaces, 2,300 turns from 0.32 to 7.2 GeV, SR ignored. Motion damping is apparent in the three phase spaces

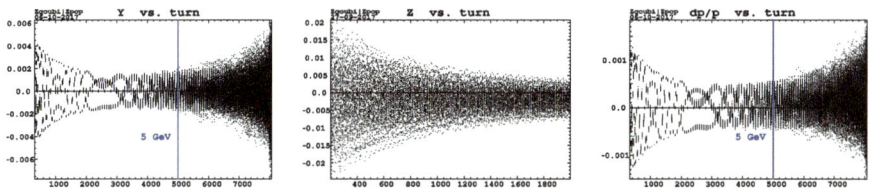

Fig. 9.30 Particle motion during acceleration in Cornell RCS, from injection energy, 0.32 GeV, to up to 8 GeV [46]. The radial motion features anti-damping

The amplitude of the random defect, 10^{-2} relative, here, can be changed to control the strength of the resonances.

Tracking is expected to show drops of average polarization, at $a\gamma$ locations which coincide with the resonance line spectrum. The latter can be computed using Eq. 9.53 (with $a \equiv G = 1.159652 \times 10^{-3}$). The optical functions and quadrupole strengths needed for that are read from the zgoubi.TWISS.out file produced in question (a). In the present simplified lattice, the first strong depolarizing resonance is at $a\gamma = \nu_Z = 13.82$. Figure 9.31 displays a simulation of resonance crossing during acceleration, typical of expected outcomes.

9.5 Coupling in a Light Source Storage Ring

A JINST article, "*Simulation of radiation damping in rings, using stepwise ray-tracing methods*" [48], provides all necessary details regarding the working hypotheses, and guidance regarding damping simulations, based on the early ESRF lattice.

Fig. 9.31 Depolarizing resonance crossing in Cornell RCS, case of a few electrons, launched on invariant $\beta\gamma\varepsilon_y/\pi =$ 100 μm [46]. The lattice includes errors: a 1.2 mm rms vertical closed orbit, a gradient defect $dK1/K1$ in $\pm 1\%$ and a K_0 roll in $\pm 0.02°$, both random uniform, in all main bends

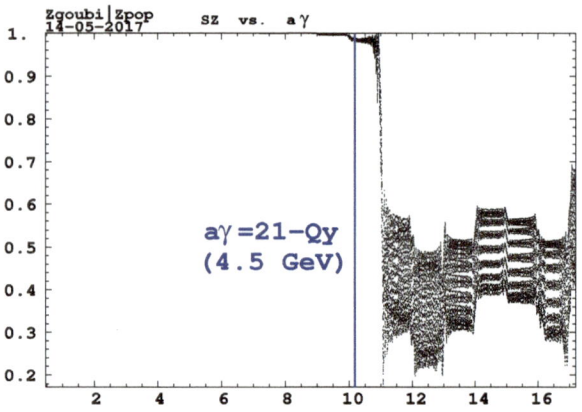

Regarding the present case of a coupled lattice, Sect. 5 in that article is concerned. Under graphical form, Figs. 7 and 8 should be reproduced. If so, then raytracing outcomes are in accord with Eqs. 24–26.

In addition to referring to Zgoubi Users' Guide [50], general guidance regarding the present simulations, the way to handle zgoubi i/o files for instance, the various options, including SRLOSS, can be found in the previous exercises, and in simulation examples found in zgoubi sourceforge repository, in the folder https://sourceforge.net/p/zgoubi/code/HEAD/tree/branches/exemples/.

Working the details of the present simulations is left to the reader.

9.6 SR Electric Impulse and Interference in a Miniwiggler

A program can be written for the calculation of the electric field impulse and its spectrum, based on the equations given in Sect. 8.2.3. Particle data can be read from zgoubi.plt. On the other hand, zpop does perform SR computations (its menu 8/16), it is used here. Zpop's file sref.f manages the computation of the electric field impulse, using Eq 8.42, accounting for Eq. 8.37 for the retarded time; the file srdw.f manages the computation of the spectral brightness, using Eqs. 8.38 and 8.39.

(a) Simulation of a pair of dipoles, configured as in the LEP miniwiggler.

The input data file is given in Table 9.22. The trajectory so obtained is displayed in Fig. 9.32.

(a) Single dipole. Electric field impulse, spectral brightness.

Using Table 9.22 data file, a way to account for radiation from a single dipole is to set MULTIPOL[IL=2] in one or the other. Or insert END (or FIN, in French), following the first dipole.

The electric field impulse $E_\sigma(\phi = \psi = 0, t)$ from the first dipole, in observer time, is displayed in Fig. 9.33. It would have opposite sign from the second dipole. $E_\pi(t) = 0$ in the bend plane.

Table 9.22 Simulation input data file for a pair of dipoles. Initial electron incidence is $\alpha/2$, so that the observation direction $\phi = \psi = 0$ is tangent to the trajectory arc in the center of a dipole

```
LEP MINI WIGGLER
'OBJET'
150103.8428                                    Rigidity(kG.cm) - 45 GeV electrons.
2
1  1                                                          ! Defines a single electron.
0.  0.4 0. 0. 0. 1. '0'   ! Initial coordinates of electron. Incidence is half the dipole bend angle.
1
'MULTIPOL'                                                    ! First dipole.
2                         ! IL=2 logs stepwise particle data in zgoubi.plt, for further plotting.
52.602  10.  2.282861  0. 0. 0. 0. 0. 0. 0. 0.               ! Length(cm), field(kG).
10. 10. 0. 0. 0. .0 .0  0. 0. 0. 0.                          ! Entrance fringe field,
6  .1122 6.2671 -1.4982 3.5882 -2.1209 1.723                    ! Enge coefficients.
10. 10. 0. 0. 0. .0 .0  0. 0. 0. 0.                          ! Entrance fringe field,
6  .1122 6.2671 -1.4982 3.5882 -2.1209 1.723
0. 0. 0. 0. 0. 0. 0. 0. 0.  0.
#200|1000|200                                 ! 1000 steps in body, 200 in end fields if any.
1 0. 0. 0.
'FAISCEAU'
! 'END'
'ESL'
23.098  split 10 2       ! Distance d. DRIFT[split,IL=2] causes log of particle data to zgoubi.plt.
'MULTIPOL'                                                    ! Second dipole.
2                         ! IL=2 logs stepwise particle data in zgoubi.plt, for further plotting.
52.602  10.  -2.282861  0.  0.  0.  0.  0. 0. 0. 0. 0.
10. 10. 0. 0. 0. .0 .0  0. 0. 0. 0.
6  .1122 6.2671 -1.4982 3.5882 -2.1209 1.723
10. 10. 0. 0. 0. .0 .0  0. 0. 0. 0.
6  .1122 6.2671 -1.4982 3.5882 -2.1209 1.723
0. 0. 0. 0. 0. 0. 0. 0. 0.  0. 0. 0. 0. 0. 0.
#200|1000|200                                 ! 1000 steps in body, 200 in end fields if any.
1 0. 0. 0.
'FAISCEAU'
'END'
```

Fig. 9.32 Electron trajectory through a pair of dipoles of opposite signs. A graph obtained using zpop: menu 7; 1/1 to open zgoubi.plt; 2/[6,2] for Y versus s; 7 to plot

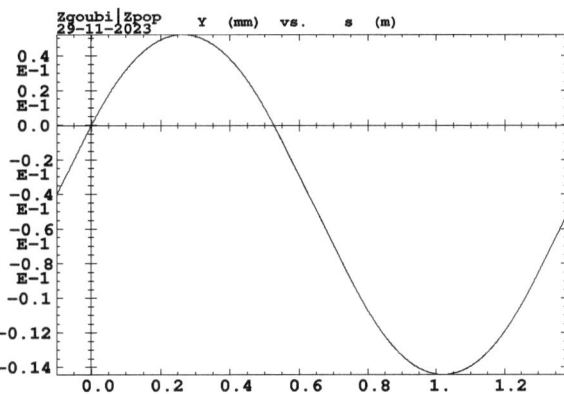

Expected duration of the impulse: the typical width of the impulse in observer time, in the bend plane ($\psi = 0$) is (Eq. 8.44) $\Delta\tau_c = 2\rho/(3\gamma^3 c) \approx (2 \times 657)/(3(2000\,E_{[GeV]})^3 c) \approx 3 \times 10^{-21}$ s. This is consistent with $E_\sigma(\tau)$ duration in Fig. 9.33. The spectrum in the bend plane is expected to peak at (Eq. 8.44) $\hbar\omega_c = \hbar/(2\pi\Delta\tau_c) \approx 0.3_{[s]}\hbar/e \approx 0.2$ MeV. This is consistent with the spectrum result in Fig. 9.33.

(b) Interference between two dipoles. Electric field impulse, spectral brightness. Case of bend plane radiation: the results are displayed in Fig. 9.34.

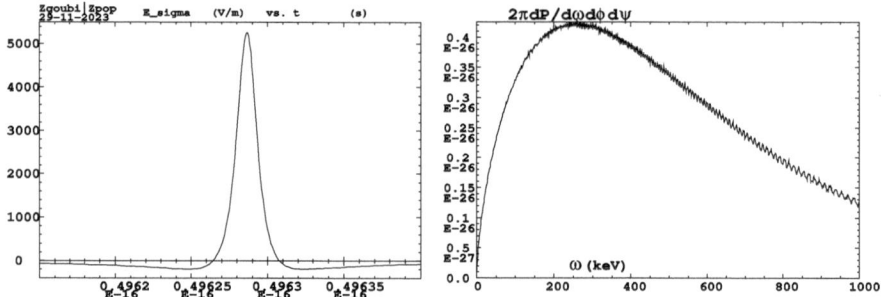

Fig. 9.33 Single dipole electric impulse $E_\sigma(\tau)$ observed in the direction $\phi = 0$, $\psi = 0$, and its spectrum. $E_\pi(\tau) \equiv 0$ in the bend plane. Graphs obtained using zpop: menu 8; 16 will open zgoubi.plt; 2/[1,E] for particle 1, electron; 5 for electric impulse or 6 for spectrum

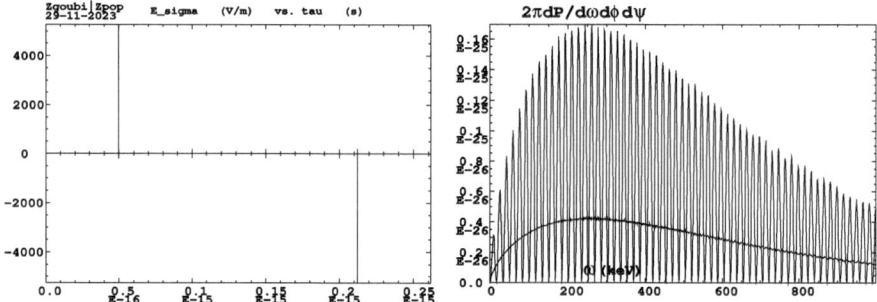

Fig. 9.34 Dipole doublet electric impulse observed in the direction $\phi = 0$, $\psi = 0$ (left), and its interferencial spectrum - the single dipole spectrum is superimposed for comparison (right). Graphs obtained using zpop: menu 8; 16 will open zgoubi.plt; 2/[1,E] for particle 1, electron; 5 for electric impulse or 6 for spectrum

Case of radiation observed in the direction $\phi = 0$, $\psi = 0.2$ mrad: the results are displayed in Fig. 9.35.

(b) Interference. Same sign dipoles.

The input data file of Table 9.22 is used, changing the magnetic field sign in the first dipole, to negative, and the launch angle of the particle under OBJET, to -0.8 mrad. The resulting trajectory is displayed in Fig. 9.36.

The resulting electric impulses and spectra in the direction $\phi = 0$, $\psi = 0.2$ mrad are displayed in Fig. 9.37.

9.7 Depolarizing Resonances in SATURNE 2

Crossing $G\gamma = 7 - \nu_Z$ non-systematic spin resonance

The input data file is copied from Table 9.15, with the following modifications:

– use OBJET[KOBJ=2] to define individual particles. It does not matter if initial coordinates are not taken on respective chromatic closed orbits, as the resonance strength does not depend on the horizontal invariant value;

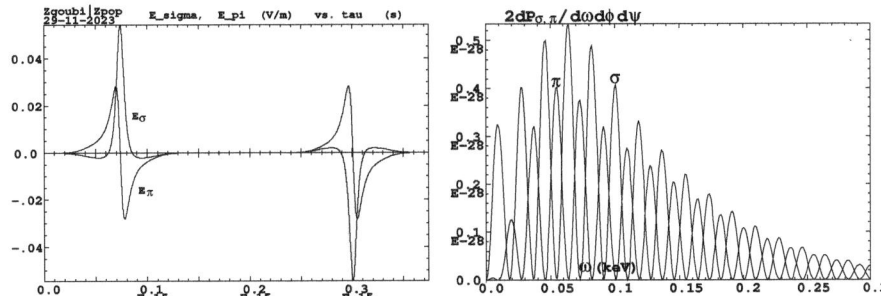

Fig. 9.35 Left: electric impulse components $E_\sigma(\tau)$ and $E_\pi(\tau)$ from a dipole doublet in observer time, in the direction $\phi = 0$, $\psi = 0.2$ mrad. Right: their interferencial spectra

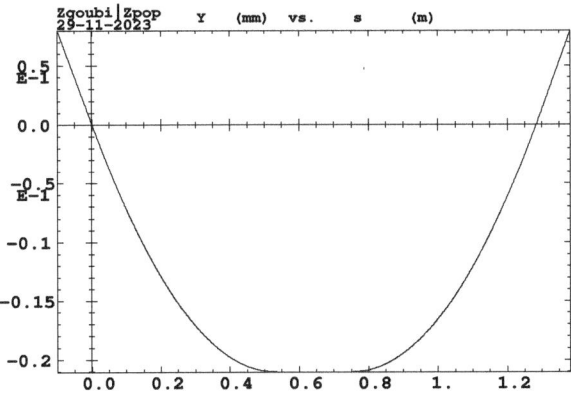

Fig. 9.36 Electron trajectory through a pair of same sign dipoles. A graph obtained using zpop: menu 7; 1/1 to open zgoubi.plt; 2/[6, 2] for Y versus s; 7 to plot

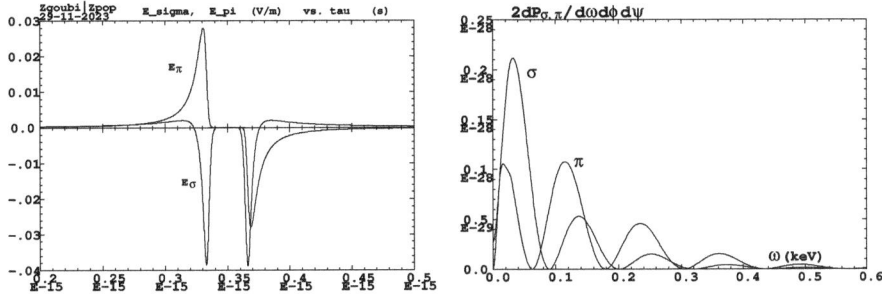

Fig. 9.37 Left: electric impulse components $E_\sigma(\tau)$ and $E_\pi(\tau)$ from a doublet of same sign dipoles, in observer time, in the direction $\phi = 0$, $\psi = 0.2$ mrad. Right: their interferencial spectra. This is an instance of edge radiation, interference is between impulse from downstream end of upstream dipole and from upstream end of downstream dipole

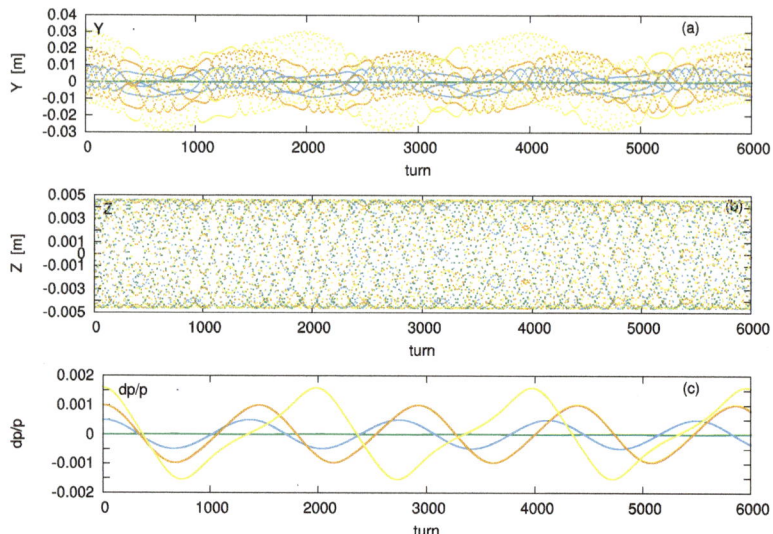

Fig. 9.38 Acceleration through $G\gamma = 7 - \nu_Z$ spin resonance: Y, Z particle motion, and $D - 1 = \delta p/p$, as a function of turn. Turn-by-turn data read from zgoubi.fai

- change the rigidity to 5.01 T m, and accordingly the SCALING factors to the same value; this sets the starting point of the tracking a few 1,000 turns upstream of the resonance, where its effect is not felt;
- add SPNTRK[KSO=3] to set initial spin of all particles to vertical;
- add a thin lens to break the 4-periodicity—necessary condition to obtain any depolarization at traversal of $G\gamma = 7 - \nu_Z$, as it is a non-systematic spin resonance.

The resulting resonance crossing input data file is given in Table 9.23.

Tracking results are summarized in Figs. 9.38 and 9.39. Spin motion features multiple resonance crossing in the three cases $dp/p \neq 0$ (Fig. 9.40), an effect of the synchrotron motion.

Table 9.23 Simulation input data file: acceleration through $G\gamma = 7 - \nu_Z$ spin resonance. To avoid any time-consuming logging to zgoubi.plt, OPTIONS[.plt=0] imposes $IL = 0$ in all magnets

```
File name: Saturne_Gg7-NuZ.INC.dat
! Acceleration through G.gamma=7-nu_Z in SATURNE 2 Ring.
'MARKER'    SatII_Gg7-NuZ_S                              ! Just for edition purposes.
'OBJET'
5.01 * 1e3              ! Reference Brho: about 1,000 tuns upstream fo Ggamma=7-nu_Z.
2
4  1
0.     0.     .458  0.  0.  1.        'o'
0.     0.     .458  0.  0.  1.0005    '1'
0.     0.     .458  0.  0.  1.001     '2'
0.     0.     .458  0.  0.  1.00159   '3'             ! Near longitudinal separatrix.
1 1 1 1
'PARTICUL'
PROTON       ! Necessary data in order to allow acceleration (CAVITE needs to know mass and charge).
'SPNTRK'
3                                                    ! Set all spins vertical.
'OPTIONS'
1 1
.plt 0                         ! Inhibits possible logging to zgoubi.plt by any optical element.
'FAISTORE'
zgoubi.fai
1
'SCALING'                      ! Scale field in the following 3 optical elements, during acceleration.
1  4                 ! The field in these magnets must be set for a rigidity of 1 Tm, which is the
QUADRUPO                                   ! case in their respective INCLUDEd files.
-1      ! This option '-1' causes the field to be ramped in proportion of rigidity increase by CAVITE.
5.01                                    ! Initial value of the field scaling factor.
1
MULTIPOL
-1
5.01
1
MULTIPOL  DFCT                                       ! Break 4-periodicity (0 or 5.01).
-1
5.01  ! * 0.  to zero it!
1
BEND
-1
5.01
1

'MULTIPOL' DFCT ! Break 4-periodicity: A thin QD quadrupole, with 1/10th of a QD integrated strength.
0
48.6273e-4  10.  0.  -.765533e3 0. 0. 0. 0.  0. 0. 0. 0.
0. 0. 0. 0. 0.  0. 0. 0. 0.  .0 .0
6  .1122 6.2671 -1.4982 3.5882 -2.1209 1.723
0. 0. 0. 0. 0.  0. 0. 0. 0.  0. 0.
6  .1122 6.2671 -1.4982 3.5882 -2.1209 1.723
0. 0. 0. 0. 0. 0. 0. 0. 0. 0.
#40|40|40
1 0. 0. 0.
'INCLUDE'
1
4 * saturneCell.inc[satCell_S:satCell_E]

'CAVITE'     ! IOPT=2 accounts for RF pahse at particle arrival (use IOPT=3 for constant energy-kick).
2            ! PRINT added here would log RF data to zgoubi.cavite.Out (for post-treatment, plot, debug ...).
105.555614  3.                ! Length of closed orbit (taken from former TWISS run); harmonic number.
6e3     0.2363176                                    ! Peak voltage; synchronous phase.

'REBELOTE'
6000 0.3 99
'FAISCEAU'

'SYSTEM'
1
gnuplot < ./gnuplot_Zfai_YZD.vs.turn_multiplot.gnu          ! Produce Y, Z, D vs. turn graphs.
'MARKER'    SatII_Gg7-NuZ_E                                 ! Just for edition purposes.
'END'
```

Fig. 9.39 Spin motion through $G\gamma = 7 - \nu_Z$ spin resonance. A graph obtained using zpop: menu 7; 1/5 to open zgoubi.fai; 2/[39,23] for S_Z versus turn; 7 to plot

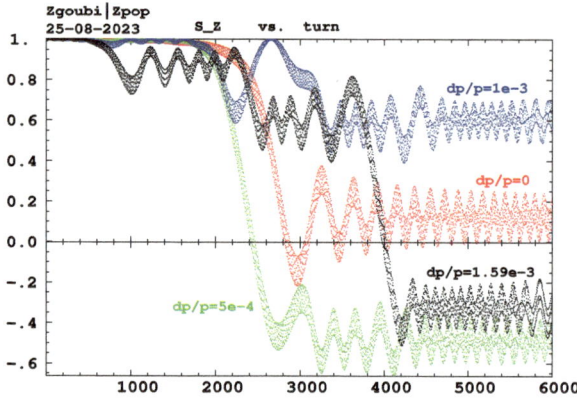

Fig. 9.40 Multiple resonance crossing due to energy oscillation. The spin resonance is at $G\gamma = 7 - \nu_Z \approx 3.39$, i.e. $E \approx 836\,\text{MeV}$ (horizontal line)

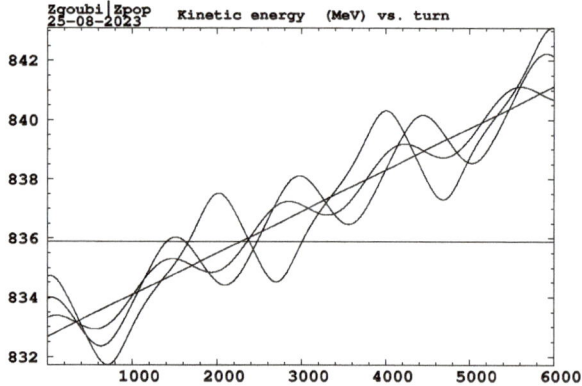

References

1. N. Christofilos, Focussing system for ions and electrons. US Patent Office Application filed March 10, 1950, Serial No. 148,920. https://patentimages.storage.googleapis.com/fa/bb/52/0ce28e28b492a6/US2736799.pdf
2. Credit: Brookhaven National Laboratory. https://www.flickr.com/photos/brookhavenlab/8495311598/in/album-72157611796003039/
3. E.D. Courant, M.S. Livingston, H.S. Snyder, The strong-focusing synchrotron - a new high energy accelerator. Phys. Rev. **88**, 1190 (1952)
4. E.D. Courant, H.S. Snyder, Theory of the alternating-gradient synchrotron. Ann. Phys. **3**, 1–48 (1958)
5. SATURNE 2 photo: credit CEA Saclay. Archives historiques du CEA. Copyright CEA/Service de documentation
6. Credit: Lawrence Berkeley National Laboratory. The Regents of the University of California, Lawrence Berkeley National Laboratory
7. Radial focusing in the linear accelerator. Phys. Rev. **88**(5) (1952)
8. M. Benedikt, F. Zimmermann, Status of the future circular collider study, in *TUYMH01 Proceedings of RuPAC2016*, St. Petersburg, Russia. https://accelconf.web.cern.ch/rupac2016/papers/tuymh01.pdf

9. F. Méot, et al., Progress on the optics modeling of BMI's ion rapid-cycling medical synchrotron at BNL, in *THPMP050, 10th International Particle Accelerator Conference IPAC2019*, Melbourne, Australia. https://accelconf.web.cern.ch/ipac2019/papers/thpmp050.pdf Copyrights under license CC-BY-3.0, https://creativecommons.org/licenses/by/3.0; no change to the material

10. N. Nishimori, A new compact 3 GeV light source in Japan, in *13th International Particle Accelerator Conference IPAC2022*, Bangkok, Thailand. https://accelconf.web.cern.ch/ipac2022/papers/thixsp1.pdf

11. F. Méot, eRHIC ERL modeling in Zgoubi. BNL-111832-2016-TECH; EIC/49;BNL-111832-2016-IR. https://technotes.bnl.gov/PDF?publicationId=38865. https://www.osti.gov/biblio/1335396

12. C. Benabderrahmane, Status of the ESRF-EBS magnets, in *WEPMK009, 9th International Particle Accelerator Conference, IPAC2018*, Vancouver, BC, Canada. https://accelconf.web.cern.ch/ipac2018/papers/wepmk009.pdf

13. F. Méot, et al.: Plans for polarized bunch R&D at Cornell rapid-cycling synchrotron, in *MOPMF013, 9th International Particle Accelerator Conference IPAC2018*, Vancouver, BC, Canada. https://accelconf.web.cern.ch/ipac2018/papers/mopmf013.pdf Figure 9.7: Copyrights under license CC-BY-3.0, https://creativecommons.org/licenses/by/3.0; no change to the material

14. R.R. Wilson, The 10 to 20 GeV Cornell electron synchrotron. Report CS-33, Laboratory of Nuclear Studies, Cornell University, Ithaca, New York (May 1, 1967)

15. D.L. Rubin, et al., Upgrade of the Cornell electron storage ring as a synchrotron light source, in *WEPOB36, Proceedings of NAPAC2016*, Chicago, IL, USA. https://accelconf.web.cern.ch/napac2016/papers/wepob36.pdf

16. CERN Courier: Partnership yields big wins for the EIC. 27 September 2021 issue. https://cerncourier.com/a/partnership-yields-big-wins-for-the-eic/

17. https://www.bnl.gov/eic/

18. https://www.slac.stanford.edu/gen/grad/GradHandbook/slac.html

19. https://www.desy.de/~teslatdr/tdr_web/pages/latest_version.html

20. https://www-project.slac.stanford.edu/lc/

21. https://muoncollider.web.cern.ch/node/25

22. https://linearcollider.org/

23. N.V. Litvinenko, N. Bachhawat, M. Chamizo-Llatas, Y. Jing, F. Méot, I. Petrushina, T. Roser, The ReLiC: recycling linear e+e- Collider. arXiv:2203.06476 [hep-ex]

24. G. Jackson (Ed.), Fermilab recycler ring technical design report. Rev. 1.1. FERMILAB-TM-1981 (July 1996). http://inspirehep.net/record/424541/files/fermilab-tm-1981.PDF

25. F. Méot, *On the Effects of Fringe Fields in the Recycler Ring*. FERMILAB-TM-2016 (Aug. 1997). http://inspirehep.net/record/448603/files/fermilab-tm-2016.PDF

26. G. Leleux, Compléments sur la Physique des Accélérateurs. DEA "Physique et Technologie des Grands Instruments", Université Paris VI. Rapport interne LNS//86-101, CEA Saclay (1986)

27. F. Méot, et al., Beam dynamics validation of the Halbach technology FFAG cell for Cornell-BNL energy recovery Linac. Nuclear Inst. Methods Phys. Res. A **896**, 60–67 (2018)

28. H. Bruck, Accélérateurs circulaires de particules. Presses Universitaires de France (1966)

29. G. Leleux, Accélérateurs Circulaires. INSTN lectures, internal report CEA Saclay (1978), unpublished

30. B.J. King, Further studies on the prospects for many-TeV muon colliders, in *Proceedings of PAC 2001 Particle Accelerator Conference*, 18–22 June 2001, Chicago, IL, USA. https://accelconf.web.cern.ch/p01/PAPERS/RPPH314.PDF

31. A. Hofmann, *The Physics of Synchrotron Radiation*. Cambridge Monographs on Particle Physics, Nuclear Physics and Cosmology (20) (Cambridge University Press, 2004)

32. G. Leleux, Rayonnement synchrotron (Aspect machine). Note technique, Laboratoire National SATURNE, CEA Saclay (1993) (unpublished)

33. A. Hofmann, F. Méot, Optical resolution of beam cross-section measurements by means of synchrotron radiation. Nucl. Inst. Methods **203**, 483–493 (1982)

34. R. Bossart, et al., Proton beam profile measurements with synchrotron light. CERN-SPS-80-8-ABM, 18 June 1980
35. F. Méot, Mesure de profil par rayonnement ondulateur des faisceaux de protons et antiprotons. Ph.D. Thesis. Report CERN/SPS 81-21 (ABM) 30 October 1981
36. L. Ponce, R. Jung, F. Méot, LHC proton beam diagnostics using synchrotron radiation. Yellow Report CERN-2004-007
37. F. Méot, L. Ponce, N. Ponthieu, Low frequency interference between short SR sources. PRST-AB **4**, 062801 (2001)
38. F. Méot, A theory of low frequency far-field synchrotron radiation. Part Accel **62**, 215–239 (1999)
39. F. Méot, Synchrotron radiation interferences at the LEP miniwiggler. CERN SL/94-22 (AP) (1994)
40. F. Méot, *Polarized Beam Dynamics and Instrumentation in Particle Accelerators*. USPAS Summer 2021 Spin Class Lectures, Open Access (Springer Nature, 2023). https://link.springer.com/book/10.1007/978-3-031-16715-7
41. S.Y. Lee, *Spin Dynamics and Snakes in Synchrotrons* (World Scientific, 1997)
42. The 20 years of the synchrotron SATURNE-2, in *Proceedings of the Colloquium*, Paris, France, 04–05 May 1998, ed. by A. Boudard, P.-A. Chamouard. Edited By CEA - Laboratoire National SATURNE & CEN Saclay, France. https://doi.org/10.1142/3965
43. Plus d'anneaux autour de SATURNE (pp. 33–34) Published in: Courrier CERN Volume 39, Num. 2, Mars 1999. https://cds.cern.ch/record/1740121
44. E. Grorud, J.L. Laclare, G. Leleux, Crossing of depolarization resonances in strongly modulated structures. IEEE Trans. Nucl. Sci. **NS-26**(3) (1979). https://accelconf.web.cern.ch/p79/PDF/PAC1979_3209.PDF
45. J.P. Aknin, et al., Status report on rejuvenating SATURNE and future aspects. PAC 1979 Conference. IEEE Tans. Nucl. Sci. **NS-26**(3) (1979). https://accelconf.web.cern.ch/p79/PDF/PAC1979_3138.PDF
46. F. Méot, Polarized e-bunch acceleration at Cornell RCS: tentative tracking simulations. Tech. Note BNL-114452-2017-TECHEIC/57;BNL-114452-2017-IR (2017). https://technotes.bnl.gov/PDF?publicationId=42654https://www.osti.gov/biblio/1408712
47. USPAS Summer 2021 Spin Class Lectures, *Polarized Beam Dynamics and Instrumentation in Particle Accelerators*, ed. by F. Méot et al. (Springer, Particle Acceleration and Detection, 2022). https://doi.org/10.1007/978-3-031-16715-7
48. F. Méot, Simulation of radiation damping in rings, using stepwise ray-tracing methods. 2015 JINST 10 T06006. http://iopscience.iop.org/1748-0221/10/06/T06006
49. Gnuplot scripts to plot optical functions, reading from zgoubi.TWISS.out, can be found at https://sourceforge.net/p/zgoubi/code/HEAD/tree/trunk/toolbox/gnuplotFiles/gnuplot_TWISS/
50. F. Méot, Zgoubi users' guide. https://www.osti.gov/biblio/1062013-zgoubi-users-guide. Sourceforge latest version: https://sourceforge.net/p/zgoubi/code/HEAD/tree/trunk/guide/Zgoubi.pdf. The betaFromPlt.f program is available here: https://sourceforge.net/p/zgoubi/code/HEAD/tree/trunk/toolbox/betaFromPlt/

Chapter 10
FFAG, Scaling

Abstract This chapter is an introduction to Fixed-Field Alternating Gradient (FFAG) cyclic accelerators. It begins with a brief reminder of the historical and technological context, and continues with the theoretical material needed for the simulation exercises. It relies on charged particle optics and acceleration concepts introduced in the previous cyclotron and synchrotron chapters. Furthermore it addresses

– design aspects of scaling FFAGs,
– beam dynamics in radial- and spiral-sector rings,
– synchrotron acceleration and various other acceleration techniques.

Simulations introduce dedicated keywords providing an analytical modeling of the field: FFAG (radial sector dipole) and FFAG-SPI (spiral sector). They otherwise use optical elements met in the previous chapters: DIPOLE[S], TOSCA, CAVITE, data input/output keywords such as FAISCEAU, FAISTORE, the SYSTEM keyword, etc. Beam dynamics simulations include

– particle trajectories through multiple-dipole FFAG cells,
– closed-orbit finding, from multi-turn raytracing or using FIT,
– deriving ancillary outcomes from rays, such as

  • transport matrices using MATRIX,
  • periodic optical functions and their transport using TWISS,

– finding dynamical aperture.

Notations Used in the Text

A	Sector angle of a dipole
$\vec{B}$; $B_{x,y,s}$; B_0	Field vector; components in moving frame; field at reference radius R_0
$B\rho$; $B\rho_0$	Particle rigidity: $B\rho = p/q$; for reference momentum p_0
C; C_0	Closed orbit length: $C = \oint ds = 2\pi R$; for reference momentum p_0

© The Author(s) 2024

F. Méot, *Understanding the Physics of Particle Accelerators*, Particle Acceleration and Detection, https://doi.org/10.1007/978-3-031-59979-8_10

ds	Path length increment: $ds = R d\theta$
E; E_{xtr}; E_{inj}; E_s	Particle energy, $E = \gamma m_0 c^2$; at extraction; injection; synchronous
EFB	Effective field boundary
f, $\mathcal{F}(\theta)$, $\mathcal{F}(r,\theta)$	Flutter factor
f_{rev}, f_{rf}	Revolution and RF voltage frequencies
g; g_0	Gap of a dipole magnet, a function of R; $g(R = R_0)$
h	Harmonic number, an integer, $h = f_{rf}/f_{rev}$
I_1	Fringe field integral
k	Geometric field index: $k = \frac{R}{B}\frac{\partial B}{\partial R} \approx -n\frac{R}{\rho}$
$\mathcal{L}$	Magnetic length
m; m_0; M	Particle mass; rest mass; mass in eV/c^2 units
N	Number of cells in a ring
n	Local focusing index: $n = -\frac{\rho}{B}\frac{\partial B}{\partial x} \approx = -\frac{\rho}{R}k$
pf	Packing factor: $pf = \mathcal{L}/C$
p; p_0; δp, Δp	Particle momentum; reference momentum; offset
q	Particle charge
R	Average closed orbit radius: $R = C/2\pi$
R; R_0	Radial coordinate, from center of ring; reference
RF	Radio-Frequency
s	Path variable
v	Particle velocity
V_{rf}; $\hat{V}_{rf}$	Accelerating voltage; peak value
x, x', y, y'	Particle coordinates in the moving frame $\left[(*)' = d(*)/ds\right]$
α	Momentum compaction, or trajectory deviation
$\beta = v/c$; β_0; β_s	Normalized velocity; reference; synchronous
β_u, α_u, γ_u; η_u	Optical functions ($u : x$, y, l); dispersion function
γ	Lorentz relativistic factor: $\gamma = E/m_0 c^2 = E[eV]/M$
δ	Relative momentum offset: $\delta = \delta p/p$
ε	Wedge angle
ϵ_R	Strength of a depolarizing resonance
ε_u	Courant-Snyder invariant; beam emittance ($u : x, y, l$)
μ_u	Betatron phase advance per period, $\mu_u = \int_{period} \frac{ds}{\beta_u(s)}$ ($u : x, y$)
ν_u	Wave numbers, horizontal, vertical, synchrotron ($u : x, y, l$)
ζ	Spiral angle of a spiral sector dipole EFB
η	Phase-slip factor
θ	Azimuthal angle
κ	Gap shape index
λ	Fringe field extent
ω_{rf}	Accelerating voltage angular frequency: $\omega_{rf} = 2\pi f_{rf}$
ρ	Local curvature radius
φ	Scalloping angle
ϕ_s	Synchronous RF phase

10.1   Introduction

The Fixed field alternating gradient (FFAG) concept was devised in the early 1950s [1–6]. Electrostatic accelerators, cyclotrons, betatrons, synchrotrons were part of the landscape at the time, as instruments for nuclear physics research, medical and industrial applications, X-ray generators, etc. Higher energies were driving accelerator technology R&D, and strong focusing, pulsed synchrotron cascades and collider rings were on their way to take over. The FFAG concept was explored as an alternate implementation of strong focusing, liable to allow high intensity beams due to their—synchrocyclotron-like—capability of very fast cycling resulting from the fixed magnetic field, and to the large momentum and geometrical acceptance of strong focusing zero-chromaticity optics. Three electron models were built and operated in the 1953–1967 period (Fig. 10.1), by the Midwestern Universities Research Association [1]. These early FFAG studies produced a wealth of theoretical and computational contributions to beam theory and beam manipulation in cyclic accelerator magnets and RF systems.

The interest in FFAG technologies resurrected in the late 1990s in the context of high energy physics R&D programs and the acceleration of short-lived beams, with potential spin-offs such as medical and other high power proton and electron beam accelerators [7–9] (Fig. 10.2). Several proton and electron machines were built in Japan from the 1990s on [7]. These developments included an ADS-Reactor

Fig. 10.1 The third FFAG electron model at MURA, a two-way 50 MeV electron ring [5]. The ring was operated in a collider mode with two counter-rotating beams, in the early 1960s

Fig. 10.2 PoP, the first proton FFAG, a 500 keV Proof-of-Principle ring operated at KEK from 1999 on [10]. Beam from the 50 kV H^+ source (in the background) is steered across the ring vacuum chamber onto the inner radius injection orbit. The high gradient RF cavity can be seen between dipoles to the right, with its power supply to its right

installation where first experiments in the world for basic ADS research have been undertaken in 2009, an internal target experiment, and more [11, 12]. A prototype FFAG spiral sector dipole was built in 2005, as part of a multiple-beam protontherapy ring design study [13] (Fig. 10.3).

During acceleration orbit spirals out in an FFAG, as in cyclotrons and synchro-cyclotrons. FFAGs optics is non-isochronous: the radial field index k is constant to ensure constant tunes as the beam spirals out (isochronism requires $k \propto \gamma^2$, Eq. 4.1). FFAGs are normally operated as synchrocyclotrons: the RF is cycled, voltage frequency is modulated during the ramp; repetition rates of tens of kHz are potentially achievable with today's RF system technologies. High power and fast acceleration R&D have produced alternate acceleration and RF manipulation techniques, including quasi-isochronous optics and CW acceleration [14–18].

Fig. 10.3 Full scale prototype (left), and pole (right), of the spiral sector dipole of a proton FFAG designed for proton-therapy use [13]

10.2 Basic Concepts and Formulæ

Consider Hill's equations

$$
\begin{cases}
\dfrac{d^2 x}{d\theta^2} + \dfrac{R^2}{\rho^2}(1 - n)x = 0 \\[2ex]
\dfrac{d^2 y}{d\theta^2} + \dfrac{R^2}{\rho^2}ny = 0
\end{cases}
\tag{10.1}
$$

These equations assume that orbit scalloping around the ring is marginal, i.e. a quasi-circular closed orbit [19]. At a given angle θ in an FFAG sector, constant radial and axial focusing is equivalent to

$$
\frac{d}{dp}\left((1 - n)\frac{R^2}{\rho^2}\right)\bigg|_\theta = 0 \quad \text{and} \quad \frac{d}{dp}\left(n\frac{R^2}{\rho^2}\right)\bigg|_\theta = 0
\tag{10.2}
$$

A sufficient condition for Eq. 10.2 is

$$
\frac{\partial}{\partial p}\left(\frac{R}{\rho}\right)\bigg|_\theta = 0 \quad \text{and} \quad \frac{\partial n}{\partial p}\bigg|_\theta = 0
\tag{10.3}
$$

The first condition yields constant ratio *particle radial position/local curvature radius*: the geometrical scaling property, orbits of different momenta scale with energy, the center of similitude is the center of the ring. The second condition yields momentum-independent local focusing index: the zero-chromaticity property.

Geometrical similarity results in a constant geometrical field index

$$k = \frac{R}{B(R)} \frac{\partial B(R)}{\partial R}\bigg|_R \approx -\frac{R}{\rho}n = constant \qquad (10.4)$$

Note that the approximation $k \approx -\dfrac{R}{\rho}n$ results from the hypothesis of a circular orbit, neglecting the orbit scalloping [19, Eq. 8]. Integration yields the R-dependence of the field in an FFAG dipole,

$$B(R) = B_0 \left(\frac{R}{R_0}\right)^k, \quad k > 0 \qquad (10.5)$$

From $k = constant$ (Eq. 10.4) it results that reversing the sign of the curvature radius ρ reverses the sign of the field index n. Radial FFAG lattices combine such alternating index dipoles, Fig. 10.4 and Sect. 10.2.1. A way to obtain such radial field distribution is by shaping the dipole gap, according to

$$g(R) \approx g_0 \left(\frac{R_0}{R}\right)^\kappa \quad \text{with } \kappa \approx k \qquad (10.6)$$

The gap is greater (lower) at lower (greater) energy and radius (Fig. 10.4). Another way is by distributed current windings along the poles of a parallel gap dipole [1, 20, 21]. More generally, in a lattice comprised of bends and field-free sections, the magnetic field along an orbit in the median plane ($y = 0$) satisfies

$$B(R, \theta) = B_0 \left(\frac{R}{R_0}\right)^k \mathcal{F}(R, \theta) \qquad (10.7)$$

where the $2\pi/N$-periodic flutter factor $\mathcal{F}(R, \theta)$ describes the azimuthal modulation of the field along an orbit (in a similar way to the modeling of the AVF cyclotron, *cf.* Chap. 4, Eqs. 4.4, 4.5 and 4.11).

Orbits

It results from the scaling field (Eq. 10.5) that the average orbit radius and the orbit length satisfy the momentum dependence, respectively,

$$\frac{R(p)}{R_0} = \left(\frac{B\rho}{B\rho_0}\right)^{1/(k+1)} = \left(\frac{p}{p_0}\right)^{1/(k+1)} \quad \text{and} \quad C(p) = C_0 \left(\frac{p}{p_0}\right)^{\frac{1}{k+1}} \qquad (10.8)$$

The R- and ρ-radius arcs share a common cord (Fig. 10.5), which writes

$$R \sin(A/2) = \rho \sin(\pi/N) \qquad (10.9)$$

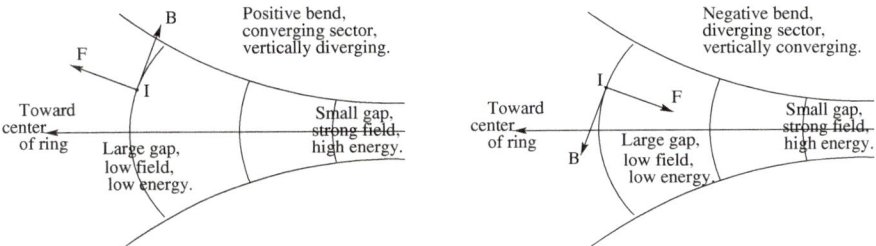

Fig. 10.4 Focusing and defocusing FFAG sectors

Fig. 10.5 Geometrical parameters in a radial (straight EFBs) or spiral (dashed lines; spiral angle ζ) FFAG sector dipole. O is the center of the ring and the EFBs form a sector angle A. The R-radius arc is a line of constant field (Eq. 10.5). The closed orbit is approximately along the ρ-radius arc. Both arcs share the same cord

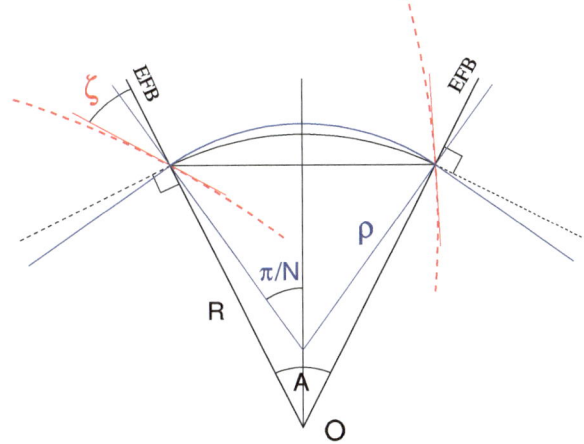

A packing factor can be defined,

$$pf = \frac{\text{magnetic length}}{\text{orbit length}} = \frac{\mathcal{L}}{C} = \frac{N \times A}{2\pi} \qquad (10.10)$$

In a general manner, closed orbits have to be computed numerically, searching for the momentum-dependent closed solution over a period. They feature small amplitude scalloping in the vicinity of an average circular path with radius $R(p)$ (Eq. 10.8), thus the initial radius value for a numerical search can be taken as $R \approx R(p)$, whereas the incidence dR/ds is null with proper choice of the origin.

The orbit excursion, from injection to extraction momentum, satisfies (Eq. 10.8)

$$R_{xtr} - R_{inj} = R_{xtr} \left(1 - \left(\frac{p_{inj}}{p_{xtr}} \right)^{\frac{1}{1+k}} \right) \qquad (10.11)$$

Focusing

There are two ways that the FFAG technique implements strong focusing,

- one consists in alternating strong transverse gradients (large $|n|$, Eq. 10.4), which is achieved by alternating positive- and negative-bend magnets (Sect. 10.2.1), with the detrimental effect of increased circumference of the ring and decreased packing factor (Eq. 10.10);
- a second method consists in using positive bend only, and relying on spiral EFBs and Thomas (AVF) focusing: a large spiral angle (strong axial focusing, radially defocusing) compensates the large field index (strong radial focusing, axially defocusing). A logarithmic spiral edge [3] has the virtue of ensuring constant wedge angle (*cf.* Sect. 10.2.2).

Fringe fields may have a noticeable effect on the effective axial focusing (*cf.* Sect. 14.4.1): Eq. 14.20 indicates that, in the case of constant wedge angle (spiral scaling dipoles), the axial focusing correction for the fringe field extent, ψ, is constant *iff* $\lambda \propto R$, which requires λ (a measure of the the gap height) to increase linearly with radius. In the gap shaping method (Eq. 10.6) the gap decreases with radius instead, thus leading to an increase in axial wave number with energy: overcoming that effect requires proper counter-measures such as for instance a specific design of the chamfers, and field clamps [22].

Wave Numbers

The wave numbers of a radial sector lattice can be derived from the method of averages, and are given to reasonable accuracy, accounting for orbit scalloping, by [19]

$$
\begin{aligned}
\nu_R &= \left[1 + k + \left(k + \tfrac{3}{2}\right)^2 \langle \phi^2 \rangle + \tfrac{9}{8} N^2 \langle \phi^2 \rangle^2 \right]^{1/2} \\
\nu_y &= \left[-k + \langle \phi'^2 \rangle + \left(k + \tfrac{1}{2}\right)^2 \langle \phi^2 \rangle \right]^{1/2}
\end{aligned}
\tag{10.12}
$$

where N is the number of periods, $\phi(\theta)$ is the scalloping angle (the local angle, at azimuth $s = R\theta$, between the radial axis of the moving frame and the radius to the center of the ring [19, Fig. 2]), $\phi' = d\phi/d\theta$, and $< * >$ denotes the average value over a closed orbit.

Keeping the first oder terms only, and accounting for the spiral focusing, leads to the approximations for the cyclotron, namely (Sects. 4.2.1 and 4.2.2)

$$
\nu_R \approx \sqrt{1 + k}, \quad \nu_y \approx \sqrt{-k + F^2(1 + 2\tan^2 \zeta)}
\tag{10.13}
$$

(with spiral angle $\zeta = 0$ in the case of a radial sector). The later underestimate the value of ν_R and overestimate the value of ν_y. However they may be found helpful in evaluating the relative effects of a small change in the flutter F, in the geometrical field index k, or in the spiral angle ζ in the case of a spiral sector.

10.2.1 Radial Sector

A radial sector scaling FFAG facility is displayed in Fig. 10.6 [23]: a 150 MeV ring
built and operated at KEK in the early 2000s. The ring is comprised of 12 defocusing-
focusing-defocusing (DFD) dipole triplets. The radial dependence of the magnetic
field in the D and F sectors satisfies the scaling law, Eq. 10.5, as a result of the gap
shape, Eq. 10.6. The main parameters of the ring are summarized in Table 10.1.

Hall-probe measurements of an isolated dipole triplet are displayed in Fig. 10.7.
Mutual influence in the ring actually produces a 200 Gauss field across the drift
between two triplets. Figure 10.7 shows the field from OPERA computation in the
periodic hypothesis [24].

Transverse Acceptance

Scaling FFAG optics features large dynamical transverse acceptance [25], see the
case of KEK FFAG ring in Fig. 10.8.

10.2.2 Spiral Sector

The spiral sector FFAG was devised by the MURA group, and an electron model
was operated in 1957 (Fig. 10.9). Compared to the radial sector it had an advantage
of compactness, as focusing relies on the sector edges rather than on alternating
gradient and curvature.

A typical design of a proton spiral sector scaling FFAG is shown in Fig. 10.10: a
variable energy and multiple extraction ring, aimed at cancer tumor treatment [26,
27]. Table 10.2 summarizes the parameters of the dipole magnet and of the ring. The
ring is comprised of 10 spiral sector cells. The radial dependence of the magnetic field
in the spiral dipole satisfies Eq. 10.5 and results from the gap shape which follows
Eq. 10.6. The dipole can be operated up to 2 T on the extraction radius, corresponding

Fig. 10.6 Left: KEK 150 MeV 12-cell scaling FFAG ring, and its cyclotron injector [10]. Right:
Its lattice cell magnet, a DFD dipole triplet. The gap shape follows Eq. 10.6 so ensuring the scaling
field law (Eq. 10.5) [23]

Table 10.1 Parameters of KEK 150 MeV radial sector FFAG [23]

Injection—extraction energy (proton)	MeV	12–150
Injection—extraction radius	m	4.7–5.2
Lattice		DFD
Number of cells N		12
Maximum β_R; β_z max.	m	3.8; 1.3
Wave numbers, ν_R; ν_z		3.7; 1.2
Magnet		
Type		Radial sector DFD triplet
Sector angle A_D; A_F	deg	3.43; 10.24
Injection—extraction gap height	cm	20–4
Scaling index $k_D = k_F$		7.6
B_D; B_F, on 150 MeV orbit	T	−1.21745; 1.69056
Acceleration		
Frequency swing	MHz	1.5–4.6
Harmonic		1
Voltage, peak-to-peak	kV	19
Cycle time	ms	4
Maximum repetition rate	Hz	250
Equivalent dB/dt	T/s	280
Synchrotron tune ν_s		0.039–0.012

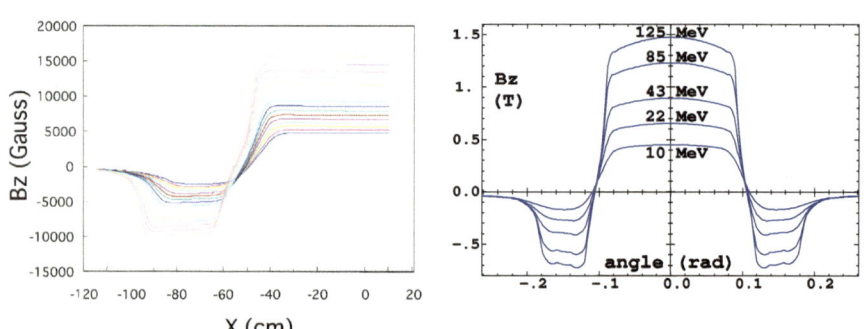

Fig. 10.7 Left: measured vertical magnetic field $B_Z(X)$, along R = constant arcs across one half of the radial sector dipole triplet [23]. The X coordinate is along an axis normal to the vertical symmetry plane of the triplet (the X = 0 plane). The field on the plateau accurately follows the R^k scaling law (Eq. 10.5), lower (greater) field at lower (greater) energy, greater (lower) dipole gap. Right: field along the periodic orbits across the cell at various energies (proton), from an OPERA field map of the KEK FFAG dipole triplet cell [24]

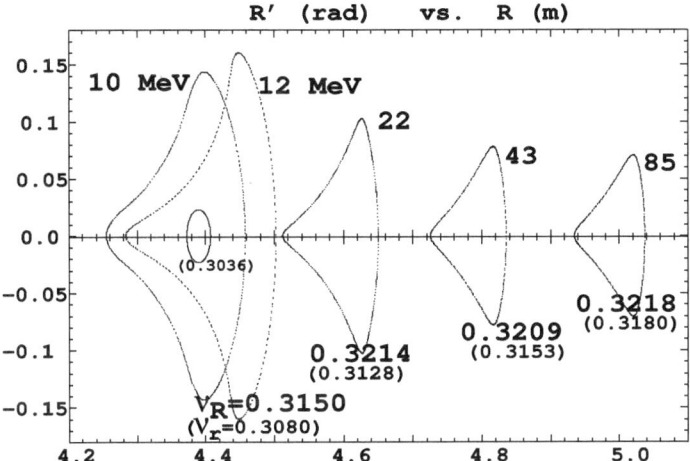

Fig. 10.8 Radial motion 1,000-turn stability limit at various energies. The small ellipse within the 10 MeV stability invariant on the left is for a nominal $\varepsilon_R = 0.04$ πcm beam at injection energy. The single-cell radial wave number at stability limit (ν_R) and paraxial (ν_r, between parentheses) give an idea of the amplitude detuning [24]

Fig. 10.9 The second (of three) electron model of an FFAG, a spiral sector design by the MURA group, operated in 1957 [5]

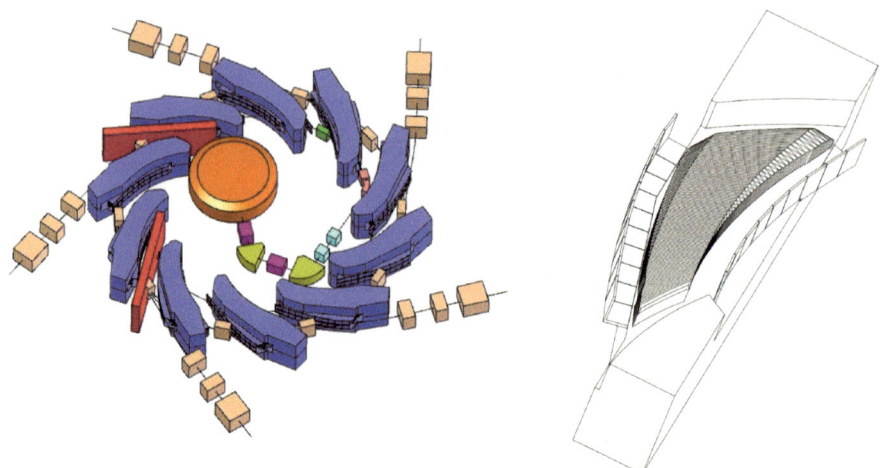

Fig. 10.10 Left: RACCAM proton therapy scaling FFAG ring design, including a variable energy H$^-$ cyclotron injector. Right: a scheme of its spiral dipole half-yoke, showing the gap shaping pole piece with its variable width chamfers and the EFB field clamps, two features that result in quasi-constant axial wave number [26]. The EFB has a constant spiral angle $\zeta = 53.7°$

to 230 MeV extraction energy [22, 28]. Hall-probe field measurements are displayed in Fig. 10.11, together with fields from OPERA computation which are in accord within a few percent.

Constant Wedge Angle

A spiral EFB ensures constant wedge angle and thus R-independent focusing, while compensating the axially defocusing effect of a strong field index (Eq. 10.13). A spiral sector field boundary is defined by

$$\theta - \tan \zeta \, \ln \frac{R}{R_0} = \text{constant}, \quad \text{i.e.,} \quad R = R_0 \exp\left(\frac{\theta}{\tan \zeta}\right) \tag{10.14}$$

Note that an $\frac{R}{R_0}$-homothety, $\frac{2\pi}{N}$-rotation orbit similarity results (which reduces to a simple homothety in a radial lattice as $\zeta = 0$). The median plane field in a spiral sector can be written under the form

$$B(R, \theta) = B_0 \left(\frac{R}{R_0}\right)^k \mathcal{F}\left(\tan \zeta \, \ln \frac{R}{R_0} - \theta\right) \tag{10.15}$$

wherein $\mathcal{F}\left(\tan \zeta \, \ln \frac{R}{R_0} - \theta\right)$ is a $\frac{2\pi}{N}$-periodic function of θ. A convenient model for the azimuthal modulation assumes a sinusoidal dependence of the field,

$$\mathcal{F}(R, \theta) = 1 + f \, \sin\left(N\left(\tan \zeta \, \ln \frac{R}{R_0} - \theta\right)\right) \tag{10.16}$$

Table 10.2 Design parameters of the RACCAM proton therapy spiral sector scaling FFAG ring. Some of the parameter values vary with variable operation energy: values given here concern the extraction energy range 70 to 180 MeV

		Injection		Extraction
Energy, variable	MeV	$5.55 \rightarrow 15$		$70 \rightarrow 180$
$B\rho$	T.m	$0.341 \rightarrow 0.562$		$1.231 \rightarrow 2.030$
$B\rho_{\text{extr.}}/B\rho_{\text{inj.}}$			3.612	
$\beta\gamma$		$0.109 \rightarrow 0.180$		$0.393 \rightarrow 0.648$
Lattice type			Spiral, scaling	
Number of cells N			10	
Packing factor pf			0.34	
Drift length	m	1.15		1.42
Orbit radius R	m	2.794		3.460
Orbit excursion (Eq. 10.11)	m		0.667	
Wave numbers ν_R; ν_y			2.76; $1.55 \rightarrow 1.60$	
Transition gamma γ_{tr}			2.45	
Magnet				
Type			Spiral sector	
Sector angle A	deg.		12.24	
Spiral angle ζ	deg.		53.7	
Scaling index k			5	

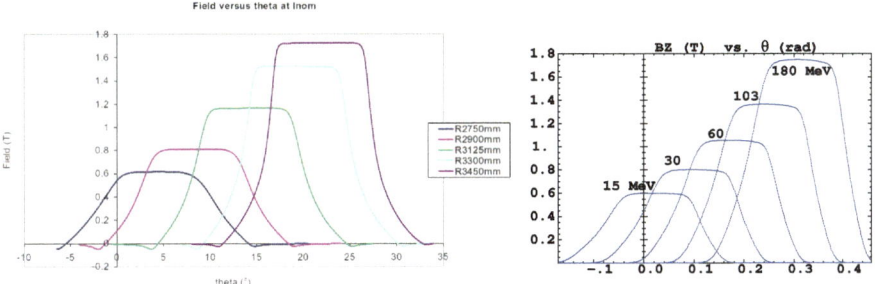

Fig. 10.11 Left: measured vertical magnetic field $B_Z(\theta)$, along R = constant arcs across RACCAM spiral sector dipole [28]. The field on the plateau satisfies the R^k scaling law (Eq. 10.5), lower (greater) field at lower (greater) energy, greater (lower) dipole gap. Right: field along closed orbits across the dipole, from OPERA field map computations

Transverse Acceptance

Spiral sector scaling FFAG optics features large dynamical transverse acceptance. As an illustration of that property, the radial dynamical acceptance of RACCAM spiral sector FFAG ring (Fig. 10.10) is displayed in Fig. 10.12. The latter has been obtained from raytracing in a theoretical field model built from the EFB geometry and the R^k dependence of the field, whereas the azimuthal dependence is modeled using Eq. 10.7 and Enge's style fall-off (Eq. 14.11) [27].

10.2.3 Longitudinal Motion. Acceleration

Given the orbit length (Eq. 10.8), the revolution period can be written

$$T_{\text{rev}} = \frac{C}{\beta c} = T_{\text{rev},0} \left(\frac{p}{p_0}\right)^{\frac{1}{k+1}} \frac{\beta_0}{\beta} = T_{\text{rev},0} \left(\frac{p}{p_0}\right)^{\frac{-k}{k+1}} \frac{E}{E_0} \tag{10.17}$$

The momentum compaction and transition γ write, respectively,

$$\alpha = \frac{\Delta C/C}{\Delta p/p} = \frac{1}{1+k} \quad \text{and} \quad \gamma_{tr} = \sqrt{1/\alpha} = \sqrt{1+k} \tag{10.18}$$

Synchrotron Acceleration

Longitudinal focusing and synchronous acceleration in a scaling FFAG proceed as in synchro-cyclotrons and synchrotrons, as addressed in Sect. 9.2.4.

A practical injection to extraction cycle includes single-bunch or multiturn injection, RF capture, synchronous acceleration, and single-turn kicker-septum extraction.

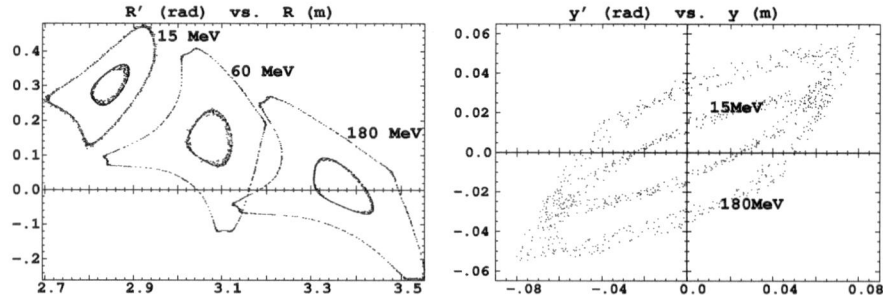

Fig. 10.12 Large stability limit (1,000-turn), at various energies, in a spiral sector FFAG. Left: radial motion; the outer invariants are for pure radial motion, they are several 10^3 π mm mrad, inner invariants are the stability limit in the presence of small amplitude axial motion, dynamical acceptance decreases to $\approx 10^3$ π mm mrad, an effect of non-linear coupling. Right: 15 MeV case (inner, elliptical shaped distribution) and 180 MeV case (outer distribution); the dynamical acceptance is about 600 π mm mrad and 2000 π mm mrad, respectively

Fixed-field allows fast cycling, with repetition rate up to hundreds of Hz provided an appropriate amount of accelerating voltage.

Betatron Damping

In the presence of acceleration (or deceleration) the equations of transverse motion write [31]

$$
\begin{cases}
x'' + \dfrac{(\gamma\beta)'}{(\gamma\beta)}x' + \dfrac{1-n}{\rho^2}x = 0 \\[3mm]
y'' + \dfrac{(\gamma\beta)'}{(\gamma\beta)}y' + \dfrac{n}{\rho^2}y = 0
\end{cases}
\tag{10.19}
$$

In the adiabatic approximation (damping is slow, compared to betatron motion oscillation) the solutions can be written

$$
\begin{matrix} x(s) \\ y(s) \end{matrix} = \frac{1}{\sqrt{|h_{\substack{x\\y}}|}}\frac{1}{\sqrt{\beta\gamma}}\left[A_{\substack{x\\y}} \exp\left(\int h_{\substack{x\\y}}\,ds\right) + B_{\substack{x\\y}} \exp\left(-\int h_{\substack{x\\y}}\,ds\right)\right]
\tag{10.20}
$$

with

$$
\left.\begin{matrix} h_x^2(s) = -\dfrac{1-n}{\rho^2} \\[3mm] h_y^2(s) = -\dfrac{n}{\rho^2} \end{matrix}\right\} + \frac{1}{2}\frac{d}{ds}\left[\frac{(\gamma\beta)'}{(\gamma\beta)}\right] + \frac{1}{4}\left[\frac{(\gamma\beta)'}{(\gamma\beta)}\right]^2
\tag{10.21}
$$

and $A_{\substack{x\\y}}$ and $B_{\substack{x\\y}}$ constants depending upon the initial conditions. Considering that $\rho \propto R$ (Eq. 10.4), assuming stable periodic motion, and dropping the $(\beta\gamma)'$ terms in Eq. 10.21 (i.e., h(s) varying slowly), it results from Eq. 10.20 that the transverse particle oscillations satisfy

$$
x, y \propto \frac{\sqrt{R}}{\sqrt{\beta\gamma}}, \qquad x', y' \propto \frac{1}{\sqrt{R}\sqrt{\beta\gamma}}
\tag{10.22}
$$

thus the damping of betatron oscillations is R-dependent. An invariant ensemble average results,

$$
\beta\gamma\,\varepsilon_{rms} = \beta\gamma\left[<x^2><x'^2> - <xx'>^2\right]^{1/2} = \text{constant}
\tag{10.23}
$$

i.e. transverse emittances of the accelerated motion are damped (or grow if deceleration) according to $\varepsilon_{rms} \propto 1/\beta\gamma$.

Fast Acceleration

Beyond synchrotron acceleration, alternate methods have been proposed for fast acceleration in scaling FFAG rings. They are aimed at short-lived particles, high

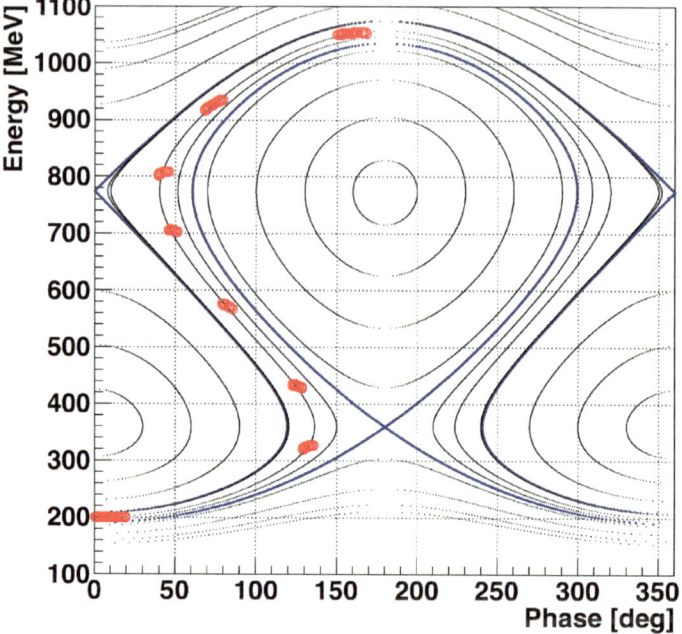

Fig. 10.13 A simulation of serpentine acceleration of a proton bunch from 1.4 to 2.2 GeV (and deceleration) in a scaling FFAG [32]

Fig. 10.14 Fast stationary
bucket acceleration of a
muon bunch, from 10 to
20 GeV

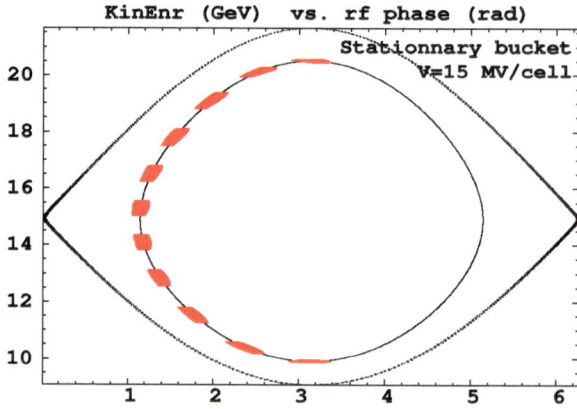

average current, longitudinal phase rotation. Several of these techniques have been subject to a proof-of-principle, including quasi-synchronous serpentine accelera-tion [15] (Fig. 10.13); bucket acceleration [34] (Fig. 10.14); multiple-bunch accelera-tion [17]; fast longitudinal phase rotation [18]; induction acceleration using a betatron core [20]; hybrid betatron-synchrotron acceleration [29, 30]; harmonic-jump [16].

10.3   Exercises

The following exercises address the two types of scaling FFAG lattices discussed above: radial sector and spiral sector. Because scaling optics dipoles have a wide gap, fringe field extent and overlapping may be a concern: the technique described in Sect. 14.3.3 is used to handle this aspect of the optics.

Note: some of the input data files for these simulations are available in `zgoubi` sourceforge repository at

[pathTo]/branches/exemples/book/zgoubiMaterial/FFAG_scaling/.

10.3.1   A 150 MeV, Proton, Radial Sector FFAG

The 150 MeV radial sector FFAG operated at KEK in the early 2000 (Fig. 10.6) is the subject of the simulations in this series of exercises. Its parameters are given in Table 10.1, the cell geometry is sketched in Fig. 10.15, the ring geometry and a few orbits (an outcome of the present exercises) are displayed in Fig. 10.16.

10.1   Field in a Radial Sector Dipole Triplet
Solution 10.1.

The FFAG keyword is based on Eqs. 10.7 and 14.15 to provide an analytical model of the combined field from N neighboring dipoles, at particle location during the stepwise integration of motion. The field flutter factor $\mathcal{F}_i(R, \theta)$ (Eq. 14.15) is based on the fringe field model described in Sect. 14.3.3 (Eq. 14.11).

(a) Using FFAG keyword, and accounting for the magnet and cell parameters of Table 10.1, produce an input data file for the simulation of a cell.
(b) Produce a graph of the median plane field $B_Z(R, \theta)$ in the dipole triplet. The keyword OBJET[KOBJ=1] can be used to generate a dense set of parallel trajectories in the median plane, and OPTIONS[CONSTY=ON] to force them on constant radii as they are pushed through the dipoles. Use option FFAG[IL=2]

Fig. 10.15 Geometry of a 30° DFD cell. The center of the ring is at O. F and D are the focusing and defocusing sectors of the dipole triplet, respectively 10.24° and 3.43°. The shadowed 4.75° "E" regions represent a half of the interval between two dipole triplets, a region of ≈200 G stray field [24, 35]

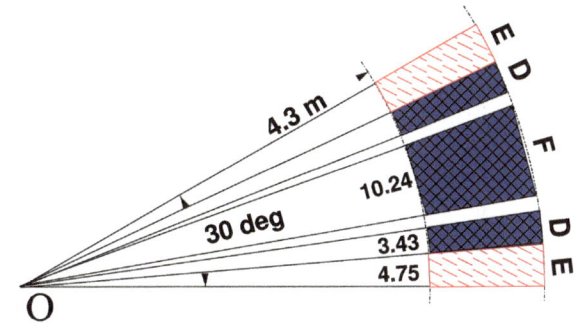

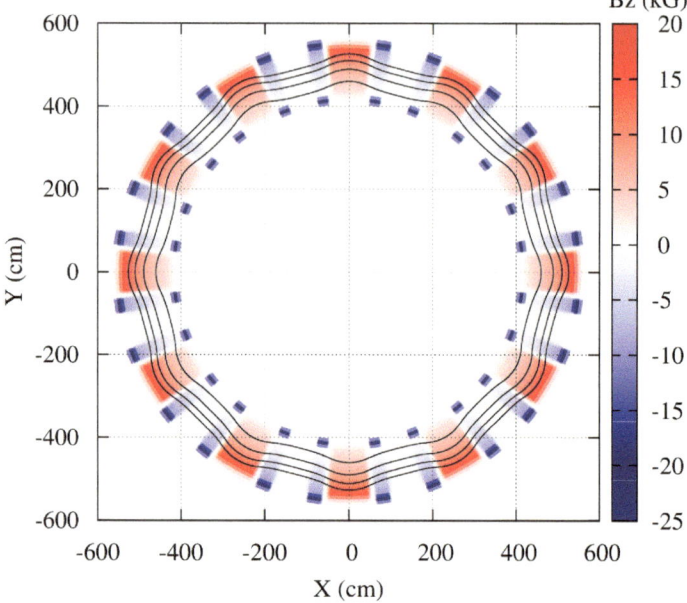

Fig. 10.16 A simulation of KEK 150 MeV FFAG ring, and a few closed orbits obtained using the keyword FFAG. A graph obtained using gnuplot: geometrical data taken from zgoubi.dat, orbit coordinates read from zgoubi.plt [31]

for a record of step by step trajectory data in zgoubi.plt. Field data can be read from the latter, to produce a 2D graph of the field $B_Z(R, \theta)$.

(c) While we are here… Using the process in (b) it is possible to generate a mid-plane field map, on an even 2D meshing, which TOSCA could possibly use and track through. This requires (i) proper particle sampling for constant ΔR between trajectory arcs so ray-traced, (ii) proper integration step size FFAG[XPAS] to cover in an evenly fashion the 30° angular sector.

Field data can subsequently be read from zgoubi.plt and re-written in a field map ascii file with proper formatting for TOSCA to handle.

Work this out, and re-do question (b) to check the identity of the raytracing outcomes.

10.2 Orbits, Scalloping
Solution 10.2.

The input data file of Exercise 10.1 can be used as a starting point in this exercise.

(a) Compute a scan of the periodic orbits $R(\theta)$ across the cell, for a few proton energies ranging in $12 \leq E \leq 200$ MeV. REBELOTE[IOP=1, N=1] can be used to loop on the energy (by changing the relative particle rigidity D under OBJET), preceded by FIT to find the periodic orbit at the energy of concern.

Give a graph of these orbits $R(\theta)$, and on a separate graph the field $B(\theta)$ along the orbits. These data can be read from zgoubi.plt, filled using FFAG[IL=2].

(b) Give a graph of the previous orbits around the ring. Show graphically that these orbits are similar, check the similarity ratio.

(c) By tracking, show that orbit excursion over an energy range $12 \leq E \leq 200\,\mathrm{MeV}$ (average radius spans from R_{inj} to R_{xtr}), satisfies Eqs. 10.8, 10.11. Particle coordinates at some azimuth along the ring can be logged in that aim in zgoubi.res using FAISCEAU (a linux "grep" can then grab them for plotting), or in an ancillary zgoubi.fai file using FAISTORE.

(d) Evaluate the orbit scalloping, i.e., the maximum value of $|R(\theta) - R|/R$. Give a graph of the latter as a function of energy.

10.3 Zero-Chromaticity
Solution 10.3.

This exercise investigates the momentum dependence of the wave numbers.

(a) Compute and give a graph of the momentum dependence of the radial and axial wave numbers in the 12-cell ring (Fig. 10.6). Use for that either one of the following two methods to obtain the wave number values:

  (i) From the cell transport matrix, using MATRIX[IORD=1, IFOC=11]. REBELOTE[IOP=1, N=1] can be used in that case to repeat on momentum values.
  (ii) from Fourier analysis of small amplitude motion.

  Compare the results with theory (Eq. 10.13).

(b) It can be observed that the radial wave number is constant with momentum/orbit radius R, this is expected from the scaling law (Eq. 10.5); however the axial wave number is R-dependent.

  In the field model, using FFAG[κ] EFB parameter, introduce an R-dependence of the gap height (Eq. 10.6): this is equivalent to introducing an R-dependence of the fringe field extent, or equivalently of the field form factor $\mathcal{F}(R, \theta)$ (Eq. 10.7), proper to change the R-dependence of the axial focusing. Find the value of κ which minimizes the change of ν_y over the energy interval $12 \leq E \leq 150\,\mathrm{MeV}$, provide a simulation to show the efficiency of the method.

(c) Compute the value of the momentum compaction and transition γ_{tr} at two sample energies, 12 and 150 MeV. TWISS can be used for that, with OBJET[KOBJ=5]. Check their relationship to the radial wave number.

10.4 Beam Envelopes; Phase Space
Solution 10.4.

(a) Produce a graph of the trajectories of a beam bundle across the cell, at 12 and 150 MeV. Take initial coordinates evenly distributed on initial paraxial invariants. OBJET[KOBJ=8] can be used to define that set of particles.

(b) Perform single particle tracking, over many turns, using REBELOTE. Consider two cases, separately: paraxial motion, and large excursion motion. Show that large excursion phase space motion features non-linear coupling.

10.5 Acceleration. Transverse Betatron Damping
Solution 10.5.

(a) Produce a simulation of the transverse and longitudinal motions of a particle taken on a small initial invariant, over a $10 \rightarrow 150\,\text{MeV}$ acceleration cycle in the 12-cell ring. Assume the following RF parameters: peak voltage $\hat{V} = 40\,\text{kVolts}$, synchronous phase $\phi_s = 20°$, harmonic $h = 1$. Acceleration uses CAVITE[IOPT=6], which imposes defining the particle type, with PARTICUL; multiturn is obtained using REBELOTE. SCALING[IOPT=−2] is used to read RF program data for CAVITE[IOPT=6], from an ancillary file (with default file name zgoubi.freqLaw.In).

(b) Show graphically that the transverse betatron oscillation damping satisfies the R-dependence of Eq. 10.22.

(c) Accelerate a bunch of a few tens of particles. Check the beam emittance damping of Eq. 10.23.

10.3.2 RACCAM Proton Therapy Spiral Sector FFAG

This series of exercises is based on the 180 MeV spiral sector FFAG design of Fig. 10.10. The parameters of concern are given in Table 10.2 [22, 26–28]. The cell geometry is sketched in Fig. 10.17.

Fig. 10.17 A sketch of RACCAM spiral sector dipole and $2\pi/10$ cell. O is the center of the ring and the EFBs form a sector angle A. The reference orbit is not circular, the bending radius is not constant along the trajectory over the $2\pi/N$ arc—a line of constant field is an arc of a circle, centered at O

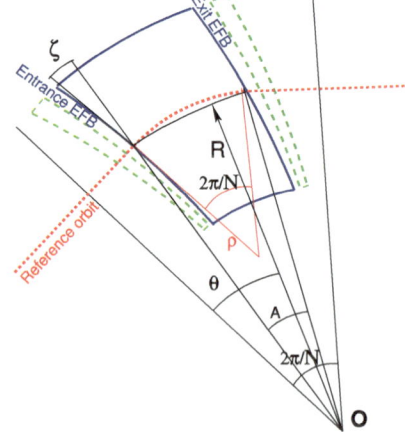

10.6 Field in a Spiral Sector Dipole
Solution 10.6.

The FFAG-SPI keyword is based on Eq. 10.15 to generate the field from a spiral sector dipole (or several sectors side-by-side within an AT angular extent [33, Sect. FFAG-SPI]) at particle location, while motion proceeds across the magnet. FFAG-SPI has provision for the modeling of the azimuthal form factor $\mathcal{F}_i(R, \theta)$, and the overlapping of neighboring fringe fall-offs, based on the method described in Sect. 14.3.3 (Fig. 10.18).

(a) Using FFAG-SPI keyword, produce a graph of the median plane field $B_Z(R, \theta)$ in RACCAM spiral sector dipole. OBJET[KOBJ=1] can be used to generate a trajectory sample, and OPTIONS[CONSTY=ON] to force these trajectories on constant radii. Option FFAG-SPI[IL=2] will log in zgoubi.plt the step by step trajectory field data; the latter can subsequently be plotted.

(b) While we are here... Using the process in (a) it is possible to generate a midplane field map, on an even 2D meshing, which TOSCA[$IZ > 1$, $MOD = 25$] could possibly use and track through. This requires (i) proper particle sampling in OBJET for constant ΔR between trajectory arcs so ray-traced, (ii) proper integration step size FFAG[XPAS] to cover in an evenly fashion the 36° angular sector.

Field data can subsequently be read from zgoubi.plt and logged in a field map file with proper formatting for TOSCA[$IZ > 1$, $MOD = 25$] to handle [33, cf. TOSCA].

Work this out, and re-do question (b) using TOSCA, to check the accord of the raytracing outcomes with the FFAG-SPI case.

Fig. 10.18 A simulation of RACCAM FFAG ring, including a few orbits, using the keyword FFAG-SPI. A graph obtained using gnuplot, geometrical data taken from zgoubi.dat, and particle data read from zgoubi.plt

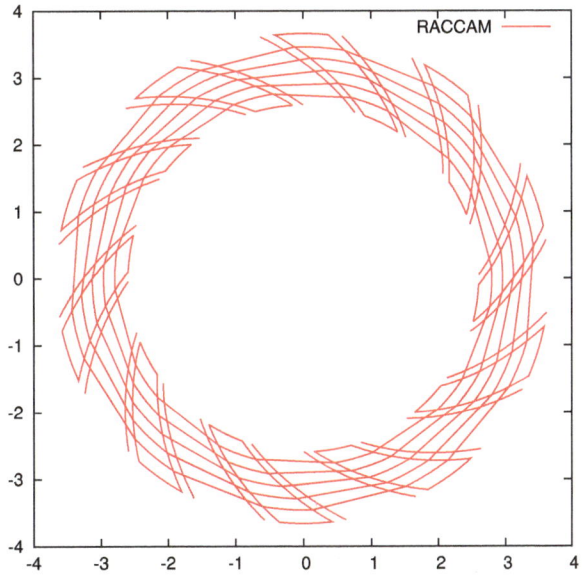

(c) CYCLOTRON keyword could be used as well: it allows some sophistication in field modeling compared to using FFAG-SPI, such as accounting for an R-dependence of the geometrical field index k (a capability which is used in Exercise 4.6—Relativistic Cyclotron Chapter, to adjust the isochronism), and of the fringe field extent, via an R-dependent gap shape index $\kappa(R)$ (Eq. 10.6).

Move the FFAG modeling of (a) to CYCLOTRON keyword. Try some radial dependence of both k and fringe extent, with the constraint of maintaining constant radial and axial tunes over the spiral orbit radial excursion. FIT can be used for this optimization of tune constancy.

10.7 Orbits, Scalloping
Solution 10.7.

Characterizing the focusing properties of the lattice (say, over the radial span of the accelerated orbit) first requires finding the periodic orbits. The radius—or momentum—dependence of optical functions may then be found (Exercise 10.9), as well as the radius dependence of time of flight for further acceleration (Exercise 10.13), etc.

(a) Compute a scan of the periodic orbits $R(\theta)$ across the cell, for a few proton energies ranging in $15 \leq E \leq 180\,\mathrm{MeV}$. REBELOTE[IOPT=1] can be used to loop on the energy (by changing the relative particle rigidity D under OBJET), preceded by FIT to find the periodic orbit at the energy of concern.
    Give a graph of these orbits $R(\theta)$, and on a separate graph the field $B(\theta)$ along the orbits. These data can be logged in zgoubi.plt during ray-tracing, using FFAG-SPI[IL=2].
(b) Show graphically the homothety-rotation of the orbits.
(c) By tracking, show that orbit excursion over an energy range $15 \leq E \leq 180\,\mathrm{MeV}$ (average radius spans from R_{inj} to R_{xtr}), satisfies Eqs. 10.8 and 10.11.

10.8 Zero-Chromaticity
Solution 10.8.

(a) Compute and give a graph of the momentum dependence of the radial and axial wave numbers in the 10-cell ring (Fig. 10.10). Use for that either one of the following two methods to obtain the wave number values:

    (i) from the cell transport matrix,
    (ii) from Fourier analysis of paraxial motion.

    Compare with expectations (Eq. 10.13).
(b) It can be observed that the radial wave number is constant with momentum, or equivalently with the orbit radius R, this is expected from the scaling law (Eq. 10.5). However the axial wave number is R-dependent. Explain why.
(c) In the field model, introduce an R-dependence of the gap height following Eq. 10.6: this is equivalent to introducing an R-dependence of the fringe field extent, or equivalently of the field form factor $\mathcal{F}(R, \theta)$ (Eq. 10.7), proper to

change the R-dependence of the axial focusing. Using the FIT procedure, compute the value of κ which minimizes the change of ν_y over the energy interval $15 < E < 180$ MeV.

(d) Compute the value of the momentum compaction and transition γ_{tr}, at 15 and 180 MeV. Check their relationship to the radial wave number.

10.9 Beam Envelopes, Optical Functions
Solution 10.9.

From single-particle multiturn tracking, produce graphs of radial and axial excursion across the cell, at 15 and 180 MeV. From these multiturn data, derive the betatron function amplitudes.

10.10 Periodic Stability Domain
Solution 10.10.

Vary the scaling index k and spiral angle ζ of the spiral dipole, in FFAG-SPI:

– produce a two-dimensional (ν_R, ν_y) wave number scan diagram, covering the motion stability area resulting from varying k and ζ.
– produce the corresponding (k, ζ) stability limit diagram.

10.11 Motion Stability Limit
Solution 10.11.

Tracking single particle radial motion in the OPERA field map yields at stability limit the phase space portrait of Fig. 10.12 [26]. Re-produce a similar phase space graph at stability limit, using the analytical field model FFAG-SPI.

10.12 Dynamic Aperture
Solution 10.12.

Extend the previous stability limit search (Exercise 10.11) to producing the dynamic aperture in (Y, Z) space, at 15, 57 and 180 MeV.

10.13 Acceleration. Transverse Betatron Damping
Solution 10.13.

Produce a simulation of a $15 \rightarrow 180$ MeV acceleration cycle in RACCAM ring, for a single particle with paraxial radial and axial motions. Take an acceleration rate of 10 kVolts per turn. Acceleration uses CAVITE[IOPT=6], which imposes defining the particle type, with PARTICUL; multiturn is obtained using REBELOTE[K=99]. SCALING[NFAM=CAVITE[NT=−2]] takes care of reading the RF program from zgoubi.freqLaw.In.

Show the betatron damping, graphically, check Eqs. 10.22, 10.23.

10.3.3 FFAG Acceleration Methods

Regarding the lattice, the following three exercises are based on a similar radial sector triplet FFAG to that studied in detail in the Sect. 10.3.1 exercise series. Thus earlier simulation input data files can be resorted here, and will only require minor adaptations.

Regarding beam acceleration, the input data files and methods developed in Exercises 10.13, 10.5 can be used to set up the present acceleration simulation input data.

10.14 Hybrid Acceleration
Solution 10.14.

Produce a simulation of hybrid acceleration in the $35\,\text{keV} \rightarrow 7\,\text{MeV}$, C^{4+}, FFAG injector addressed in Ref. [29, Slides 17–18]. It is suggested to proceed with staged simulations in the following order:

(a) Build the input data file for a $k = 0.7$, radial sector DFD cell, and subsequently for an 8-cell ring. Check the paraxial parameters, using OBJET[KOBJ=5] and MATRIX[IORD=1, IFOC=11] or TWISS, or OBJET[KOBJ=6] and MATRIX [IORD=2, IFOC=11]. The methods of the exercises in Sect. 10.3.1 can be used to construct the cell.
(b) Add acceleration, using a single, 5 kV RF cavity, simulated using CAVITE;
(c) add betatron-style acceleration, using CAVITE[IOPT]—proper choice of IOPT option is to be determined.

10.15 Bucket Acceleration
Solution 10.15.

Produce a simulation of bucket acceleration of a short-lived muon bunch (*cf.* Fig. 10.14), from 3.6 to 12.6 GeV, following Ref. [16, pp. 4507–4508]. It is suggested to proceed with staged simulations in the following order:

(a) Set up a 225-cell, $k = 1390$, DFD ring. Check its parameters. The methods of the radial lattice exercises (Sect. 10.3.1) can be used to construct the cell. Check the paraxial parameters, using OBJET[KOBJ=5] and MATRIX[IORD=1, IFOC=11] or TWISS, or OBJET[KOBJ=6] and MATRIX[IORD=2, IFOC=11].
(b) Add acceleration, 1.8 GV per turn, using 225 (one per cell), 8 MV, 200 MHz RF cavities, harmonic $h = 675$. Re-produce Figs. 5, 6 of Ref. [16, p. 4508].
(c) Add Monte Carlo in-flight decay (MCDESINT keyword): find the muon survival rate over the acceleration cycle. Check against theoretical expectations (*cf.* Sect. 12.2.3).

Table 10.3 Parameters of K. Yamakawa's 1.1 GeV quasi-isochronous scaling FFAG ring [32]

Lattice	FDF triplet
Number of cells	To be determined[a]
k-value	1.45
Transition energy [MeV]	530
Equivalent mean radius at 200 MeV [m]	3
Equivalent mean radius at 1 GeV [m]	5.9
Stationary kinetic energy below transition [MeV]	360
rf voltage [MV/turn]	15 (h = 1)
rf frequency [MHz]	9.6 (h = 1)

[a] As part of the exercise

10.16 Serpentine Acceleration
Solution 10.16.

Produce a simulation of 0.38–1.1 GeV proton beam serpentine acceleration, in 10 turns, following the hypotheses found in Ref. [32] for an MW power Accelerator Driven System. Parameters are recalled in Table 10.3.

It is suggested to proceed with staged simulations in the following sequence:

(a) Set up a data file for a ring, according to parameters in Table 10.3. The methods of the radial lattice exercises (Sect. 10.3.1) can be used to construct the cell. Check consistency with Table 10.3 data. Produce the paraxial parameters using OBJET[KOBJ=5] and MATRIX[IORD=1, IFOC=11] or TWISS, or OBJET[KOBJ=6] and MATRIX[IORD=2, IFOC=11].

(b) Add acceleration, using a single, 60 MV RF cavity, harmonic h = 10. CAVITE[IOPT=7] can be used. Produce the longitudinal phase space of an acceleration cycle (similar to Fig. 10.13).

10.4 Solutions of Exercises of This Chapter: FFAG, Scaling

10.4.1 A 150 MeV, Proton, Radial Sector FFAG

10.1 Field in a Radial Sector Dipole Triplet

(a) An input data file to simulate a 150 MeV scaling FFAG cell.

The input data file, including comments for guidance, is given in Table 10.4. Geometry and field data under FFAG are from Table 10.1, possibly slightly varied according to the purposes of subsequent questions.

(b) A graph of $B_Z(R, \theta)|_{Z=0}$ in a radial sector scaling FFAG dipole.

In view of generating the mid-plane field from raytracing outcomes, a set of 25 trajectories is launched, with initial coordinates Y_0, T_0, Z_0, P_0 and relative momentum $D_0 = p/p_{\text{ref}}$, defined using OBJET[KOBJ=1] (Table 10.4). Initial radii Y_0 are evenly spaced over the useful field region, namely

Table 10.4 Simulation input data file SFFAGCell.inc: 150 MeV KEK FFAG dipole triplet cell, a 30° sector. The FFAG keyword allows defining up to 5 independent dipoles in that $AT = 30°$ angular sector, only three are needed for the present DFD triplet. OPTIONS[CONSTY ON] allows to raytrace a set of trajectories on *constant radii*. This file also defines the segment #S_SFFAG150Cell to #E_SFFAG150Cell, for use in INCLUDEs in subsequent exercises. The step size value $XPAS = 3.0079078598$ cm here is for field map fabrication, change to ~1 cm for multiturn tracking Note: this file is available in zgoubi sourceforge repository at https://sourceforge.net/p/zgoubi/code/HEAD/tree/branches/exemples/book/zgoubiMaterial/FFAG_scaling/radialSectorTriplet_KEK/accelerationCycle

```
SFFAGCell.inc file. FFAG triplet cell of a 12-cell 150MeV ring.
'MARKER' SFFAGProbFieldMap_S
 'OBJET'
1839.090113                                      ! Rigidity, case of 150 MeV proton.
1
25 1 1 1 1 1                      ! This creates 25 trajectories, corresponding to as many radial steps,
3. 0. 0. 0. 0. 0.                   ! whereas theta-steps result from the integration step size, below.
482. 0. 0. 0. 0. 1.              ! Yo is taken half-way between a minimum 442 cm (injection region) and
!                                                  a maximum 516 cm (extraction region).
'OPTIONS'
1 1
CONSTY ON                          ! Forces particles on Y=Yo and Z=Zo, in subsequent raytracing.

'MARKER' #S_SFFAG150Cell
!!!!!!!!!!!!!!!!!!!                                Note the pattern, repeating 3 times under FFAG.
'FFAG'
0    ! Set IL=2 to store particle and field data along trajectories (in zgoubi.plt) for further plotting.
3  30.  540.              ! Number of dipoles over AT; AT=tetaF+2tetaD+2Atan(XFF/R0); reference radius R0.
6.465  0.  -12.1744691  7.6   !!!!!!!!!!!!!! Dipole 1 (D type): ACNT; unused; B0; geometrical index k.
6.3   3.         ! EFB 1 : lambda~gap size; kappa=gap shape index: detemines radius-dependent fringe extent.
4  .1455   2.2670  -.6395  1.1558  0. 0.  0.
1.715 0.    1.E6  -1.E6  1.E6  1.E6
6.3   3.         ! EFB 2 : lambda~gap size; kappa=gap shape index: detemines radius-dependent fringe extent.
4  .1455   2.2670  -.6395  1.1558  0. 0.  0.
-1.715 0.   1.E6  -1.E6  1.E6  1.E6
0. -1                                                        ! EFB 3 : inhibited by iop=0.
0  0. 0. 0. 0. 0. 0. 0.
0  0. 0. 0. 0. 0. 0. 0.
15.  0.  16.9055873  7.6   !!!!!!!!!!!!!!!!!!!!!!!!! Dipole 2 (F type): the pattern repeats a first time.
6.3   3.         ! EFB 1 : lambda~gap size; kappa=gap shape index: detemines radius-dependent fringe extent.
4  .1455   2.2670  -.6395  1.1558  0. 0.  0.
5.12  0.    1.E6  -1.E6  1.E6  1.E6
6.3   3.         ! EFB 2 : lambda~gap size; kappa=gap shape index: detemines radius-dependent fringe extent.
4  .1455   2.2670  -.6395  1.1558  0. 0.  0.
-5.12  0.   1.E6  -1.E6  1.E6  1.E6
0. -1
0  0. 0. 0. 0. 0. 0. 0.
0  0. 0. 0. 0. 0. 0. 0.
23.535 0.  -12.1744691  7.6   !!!!!!!!!!!!!!!!!!!!! Dipole 3 (D type): the pattern repeats a second time.
6.3   3.         ! EFB 1 : lambda~gap size; kappa=gap shape index: detemines radius-dependent fringe extent.
4  .1455   2.2670  -.6395  1.1558  0. 0.  0.
1.715 0.    1.E6  -1.E6  1.E6  1.E6
6.3   3.         ! EFB 2 : lambda~gap size; kappa=gap shape index: detemines radius-dependent fringe extent.
4  .1455   2.2670  -.6395  1.1558  0. 0.  0.
-1.715 0.   1.E6  -1.E6  1.E6  1.E6
0. -1                                                        ! EFB 3 : lateral face, unused.
0  0. 0. 0. 0. 0. 0. 0.
0  0. 0. 0. 0. 0. 0. 0.
0 2 ! Field computation analytic, 2nd order (take KIRD=2, 25 or 4 for flying grid interpolation instead).
.1       ! 3.0079078598=Integration step size determines angular step=3.007907/R0*30[deg], for proper number
2 0. 0. 0. 0.    ! of integration steps along the 30 deg sector. USE 1~2cm RATHER, FOR MULTITURN TRACKING.
'MARKER' #E_SFFAG150Cell

'SYSTEM'
2                                                       ! A 'call system' for next 2 commands:
gnuplot < ./gnuplot_Zplt_fieldMap.gnu        ! plot field data logged in / read from zgoubi.plt;
! okular gnuplot_Zplt_fieldMap.eps &          ! view graph - requires a .eps viewer.

'MARKER' SFFAGProbFieldMap_E
 'END'
```

gnuplot script gnuplot_Zplt_fieldMap.gnu, for Fig. 10.19:

```
#gnuplot_Zplt_fieldMap.gnu
set xlabel 'X [cm]' ; set ylabel 'Y [cm]' ; set zlabel 'B [kG]' ; set hidden3d
# Plot field:
splot "zgoubi.plt" u ($10*cos($22)):($10*sin($22)<520? $10*sin($22):1/0):($25) w p pt 5 ps .8 lc palette notit; pause 1
# Plot field data:
set size ratio -1; plot "zgoubi.plt" u ($10*cos($22)):($10*sin($22)) with p pt 4 ps .2 notit
```

$$Y_0 : r_1 \to r_{25}, \text{ step } \Delta r \quad \text{with } r_1 = 446\,\text{cm}, \ r_{25} = 518\,\text{cm}, \ \Delta r = 3\,\text{cm}$$

This Y_0 range results from Eq. 10.8, given $k = 7.6$ (Table 10.1) and $R_0 = 540$ cm (the choice of R_0 is arbitrary). As a mid-plane field is desired, axial motion is taken null, $Z_0 = 0$ and $P_0 = 0$. By means of OPTIONS[CONSTY ON] each particle is forced to maintain constant radius throughout the dipole. Note that a 3D map of the vector field instead, $\vec{B}(R, \theta, Z)$, can be generated if desired, by adding a Z_0 sampling, as CONSTY also forces Z to maintain its initial value Z_0.

The integration step size along the reference arc R_0 is $\Delta s = 3$ cm, resulting in 94 steps over

$$\theta : \theta_1 \to \theta_{95}, \text{ step } \Delta\theta = \Delta s / R_0, \quad \text{with } \theta_1 = 0, \ \theta_{95} = 30°, \ \Delta\theta = 0.31915°$$

with the 30° triplet sector opening including half-drifts on both sides to make up a cell (Fig. 10.15). In that manner, during the stepwise raytracing process, the mid-plane field $B_Z(R, \theta)$ is computed at particle locations which are made to coincide with the $N_r \times N_\theta = 25 \times 95$ nodes of a 2D meshing.

Note: although not necessary as far as the present question (b) is concerned, this meshing has been tailored to be uniform, and to exactly cover the 30° sector, in view of the next question.

The magnetic field vector experienced along the trajectory across the dipoles is part of the particle data logged in zgoubi.plt during the stepwise integration, as an effect of the flag FFAG[IL=2] (Table 10.4). A graph of $B_Z(R, \theta)$ data so obtained, read from zgoubi.plt, is given in Fig. 10.19.

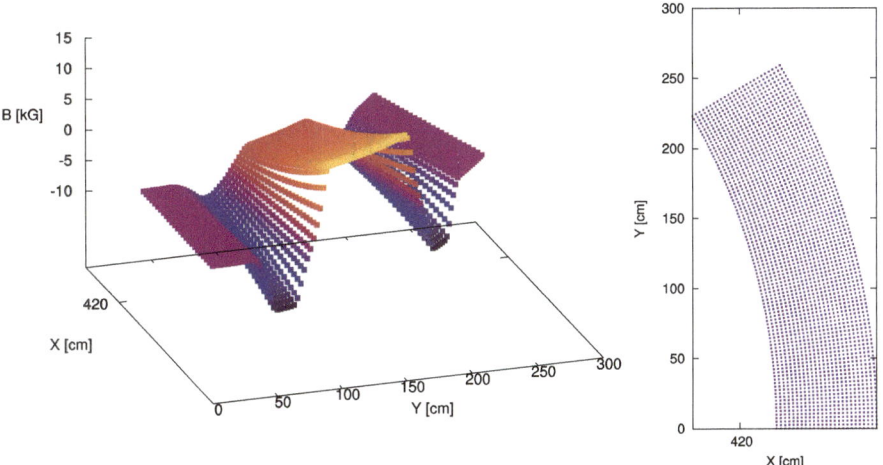

Fig. 10.19 Using the analytical field model provided by the FFAG keyword (Eq. 10.7), the 2D mid-plane field of KEK 150 MeV DFD dipole triplet is produced (left), over a uniform 2D polar meshing (right) defined by OBJET[δY] particle sampling and by the step size FFAG[XPAS] ($\Delta\theta$ sampling)

(c) Generating a 2D mid-plane field map (which TOSCA could possibly handle).

A straightforward way to generate a field map for possible use by TOSCA is to get field data on a uniform 2D mesh from the raytracing, using appropriate integration step size and particle sampling.

This has been accounted for in the input data file for the previous question, Table 10.4: the radial increment ΔR was defined to be constant using OBJET [KOBJ=1]. The angular increment $\Delta \theta = (\pi/6)/94$[steps] is constant by definition; what matters, and accounted for in (b), is ensuring that the last step (downstream boundary of the mesh) is on the exit border of the 30° sector: this is ensured taking an integration step size $XPAS = R_0$[540 cm] $\times (\pi/6)/94$[steps] $=$ 3.0079078598 cm.

From this, it results evenly distributed $(R_{i,j}, \theta_{i,j})$ particle locations during the raytracing; step-by-step particle data logged in zgoubi.plt include coordinates and the field values $B_Z(R_{i,j}, \theta_{i,j}, Z = 0)$, they are read to be re-written in an ASCII file with proper formatting for TOSCA to handle and track through. The appropriate formatting can be found in [33, Table 1].

Questions 3.1, 4.1 and their solutions may be resorted to in working out the details of the present question. This is left to the reader.

10.2 Orbits, Scalloping

(a) Periodic orbits.

Based on REBELOTE do-loop, the input data file in Table 10.5 produces 10 closed orbits consecutively (FIT finds them, one after the other) for as many different momenta (REBELOTE repeats the sequence for 10 different momenta) ranging in (relative to $p_{\text{ref}} = 551.345$ MeV/c, 150 MeV proton) p/p_{ref} : $0.2730426 \rightarrow$ 1.168858 (12–200 MeV).

The input data file ends with a SYSTEM command which results in the two graphs displayed in Fig. 10.20.

(b) Homothetic orbits.

Homothetic orbits around the ring, obtained from the previous question, read from zgoubi.plt, are plotted in Fig. 10.21.

It is found from Fig. 10.20 that the scalloping is about $0.2/5.4 \approx 3.7\%$ for the high energy closed orbit, about $0.2/4.5 \approx 4.4\%$ for the low energy closed orbit.

The similarity ratio $\frac{R}{\rho}(B\rho)$ is expected to be close to constant, within a few %, so justifying the oft-met assumption that, at all azimuth, $R \approx C/2\pi$. It can be computed for these 10 different rigidities: orbit radius R and field B_Z in the region $\theta \approx 15°$ are read from the step by step $R(\theta)$, $B_Z(\theta)$ data logged in zgoubi.plt, the latter yielding $\rho(\theta = 15°) = B\rho/B(\theta = 15°)$. Results are displayed in Fig. 10.22.

(c) Orbit excursion.

Figure 10.23 displays the numerical and theoretical (Eq. 10.8) values of the average orbit radius, they appear in good accord. It results from this that Eq. 10.11 is satisfied, with similar accuracy, $\approx 1\%$.

Table 10.5 Simulation data file to find the closed orbit for a series of different momenta. The INCLUDE grabs the FFAG dipole triplet segment defined in Table 10.4

```
Orbit scan.
'MARKER'  SFFAGProbOrbits_S                          ! Just for edition purposes.
'OBJET'
1839.090113                                ! Reference rigidity (150MeV proton).
2
1 1
445.234  0.  0.  0. 0.  0.273042677097  'o'   ! 12MeV, Brho=502.1500877. To=0 due to sector geometry.
1
'PARTICUL'
938.27231  1.60217733D-19 0. 0. 0.                                ! PROTON would do as well.

'INCLUDE'                                   ! Include the 30 degree sector dipole triplet,
1                                           ! within dedicated LABEL1s.
./SFFAGCell.inc[#S_SFFAG150Cell:#E_SFFAG150Cell]

 'FIT2'
1
2 30 0  2.
2  1e-9 99                                  ! A penalty value controls the accuracy
3.1 1 2 #End 0. 1. 0                        ! of the convergence to periodic coordinates.
3.1 1 3 #End 0. 1. 0                        ! 3.1 is the constraint for periodicity.
 'MARKER' afterFIT                          ! FAISTORE above applies here.

 'REBELOTE'                                 ! Repeat the previous sequence,
10 0.1 0  1                                 ! 10 times; prior to repeating, change
1                                           ! 1 parameter, namely parameter 35 under OBJET: the
OBJET 35 0.27304263798:1.1688582876         ! relative momentum D=p/p_ref, in the range 0.27304...:1.16...

 'SYSTEM'
3
gnuplot < ./gnuplot_Zplt_XY.gnu
gnuplot < ./gnuplot_Zplt_XB.gnu             ! Call gnuplot scripts once that tracking is completed.
gnuplot < ./gnuplot_Zplt_orbits.gnu

 'MARKER'  SFFAGProbOrbits_E                              ! Just for edition purposes.
 'END'
```

gnuplot scripts for the graphs of Fig. 10.20 (excerpt):

```
# gnuplot <./gnuplot_Zplt_XY.gnu
set xlabel "{/Symbol q} [deg]" ; set ylabel "R [m]"; r2d = 180./ (4.*atan(1.)) ; cm2m = 0.01 ; npass=11 ; FITlast=1
plot for [FITnb=2:npass] 'zgoubi.plt' u \
($49==FITnb && $51==FITlast ? $22*r2d :1/0):($10*cm2m):($49) w p pt 4 ps .2 lc palette notit ; pause 1

# gnuplot <./gnuplot_Zplt_XB.gnu
set xlabel "{/Symbol q} [deg]" ; set ylabel "R [m]"; r2d = 180./ (4.*atan(1.)) ; kG2T= 0.1 ; npass=11 ; FITlast=1
plot for [FITnb=2:npass] 'zgoubi.plt' u \
($49==FITnb && $51==FITlast ? $22*r2d :1/0):($25 *kG2T):($49) w l lc palette notit ; pause 1
```

gnuplot script for Fig. 10.21 (excerpt):

```
# ./gnuplot_Zplt_orbits.gnu
set size ratio 1 ; set polar; pi = 4.*atan(1.) ; set grid polar 2*pi/12. ; cm2m = 0.01 ; Ncell = 12 ; FITLast=1
plot  for [i=1:Ncell] "zgoubi.plt" u ($51==FITLast ? $22 +(i-1)*2*pi/Ncell :1/0):(cm2m* $10)  w p pt 6 ps .09 notit
```

gnuplot script for Fig. 10.22:

```
# gnuplot_Rovrho.gnu
set xlabel "{/Symbol q} [deg]" ; set ylabel "R/{/Symbol r}" ; r2d = 180./ (4.*atan(1.)) ; npass=11 ; FITlast=1
plot for [FITnb=2:npass] 'zgoubi.plt' u ($49==FITnb && $51==FITlast && ($22*r2d>12 && $22*r2d<18) ? \
$22*r2d :1/0):($10 / ($40*(1.+$2)/$25)) w lp ps .6 notit; pause 1
# Replace  '($10 / ($40*(1.+$2)/$25))' by '($40*(1.+$2)/$25' for \rho(\theta) graph.
```

(d) Orbit scalloping

A gnuplot script can plot the scalloping $|R(\theta) - R|/R$. $R(\theta)$ is read from zgoubi.plt (column 10 therein) as in (b). The value of the average orbit radius $R = C/2\pi$ can be formulated using Eq. 10.11 with $p = qB\rho = q \times D \times BORO$, with D and BORO both read from zgoubi.plt (columns 2 and 40, respectively [33, Sect. 8.3]).

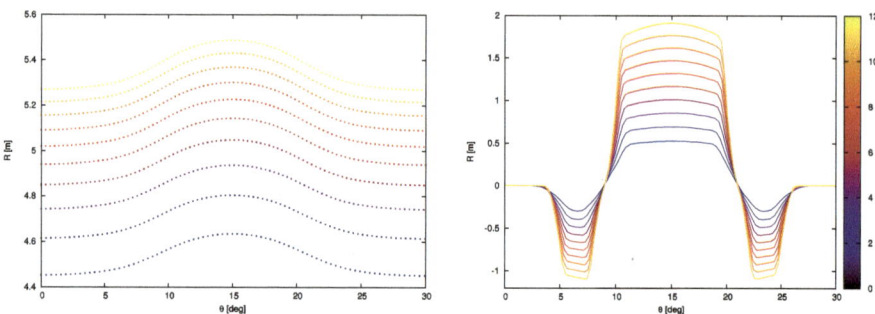

Fig. 10.20 Left: orbit scalloping across the AT=30° angular extent encompassed in FFAG keyword simulation, for 10 different proton energies ranging in 12–200 MeV (from bottom, smaller radius, to top, greater radius). Right: field experienced along these orbits, increasing with radius

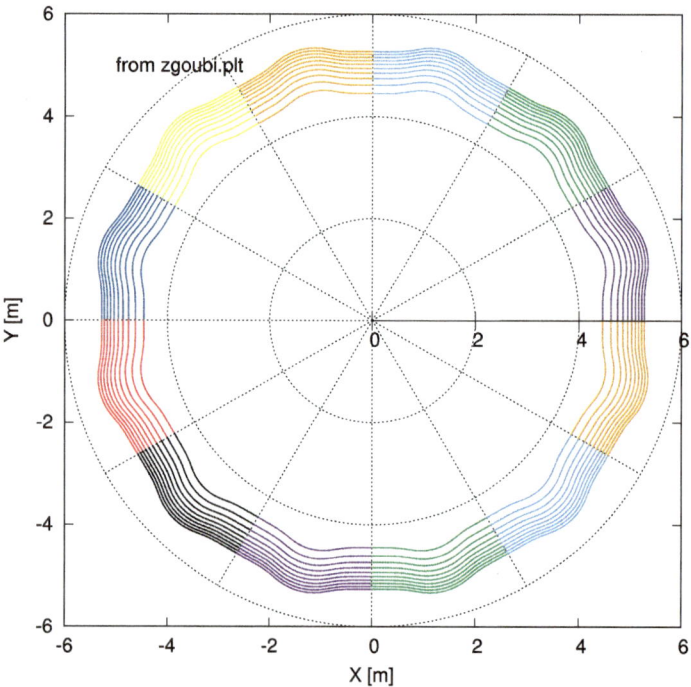

Fig. 10.21 Ten closed orbits, from 12 to 200 MeV, around the 12-cell radial sector FFAG ring. A graph obtained using gnuplot, all necessary data read from zgoubi.plt

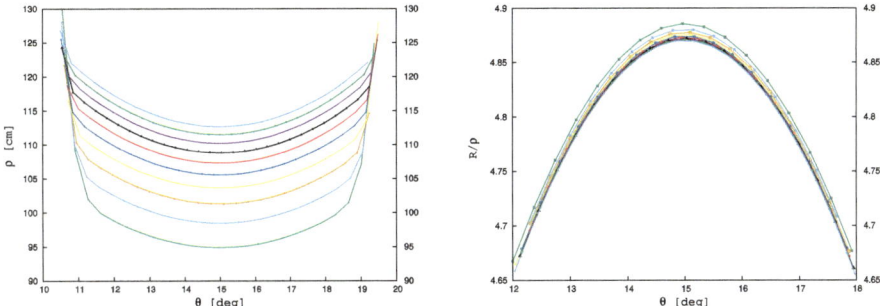

Fig. 10.22 Left: curvature radius $\rho(\theta)$ across the focusing dipole of the DFD triplet, for ten closed orbits in the energy range $12 < E < 200$ MeV. Right: $R/\rho(\theta)$ in the central region of the focusing dipole; the variation is $\approx \pm 0.1/4.75 \approx \pm 2\%$

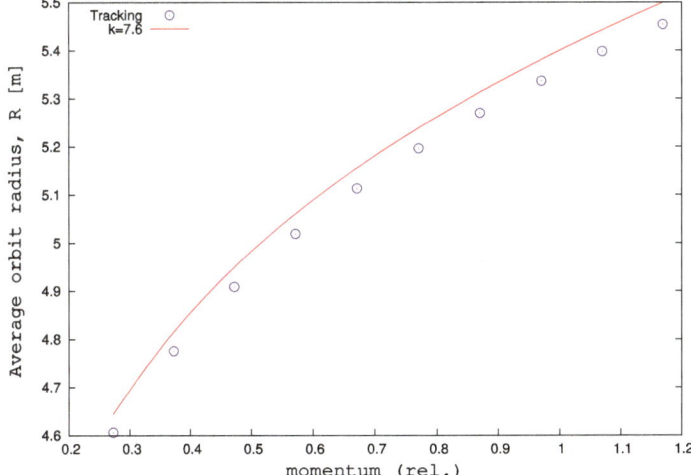

Fig. 10.23 Dependence of the average closed orbit radius $R = C/2\pi$ on the relative momentum, from 12 to 200 MeV. Markers are from one-turn raytracing. The solid line is from theory, for comparison, after Eq. 10.8 taken for $k = 7.6$ (Table 10.1) and reference momentum $p_{ref} = 551.3$ MeV/c (150 MeV, radius $R_0 = 5.4$ m). They only differ by $\approx 1\%$

10.3 Zero-Chromaticity

(a) Momentum dependence of tunes.

MATRIX is used to compute the wave numbers, REBELOTE[K=0;IOPT=1] is used to repeat with a different momentum, the input data file is given in Table 10.6.

OBJET[KOBJ=5] defines a set of 13 particles with proper initial coordinate sampling for matrix computation, by MATRIX. Prior to matrix computation, the momentum-dependent closed orbit is found by FIT. REBELOTE then changes the relative momentum D in OBJET, and repeats the sequence. The command

Table 10.6 Simulation input data file: compute the first order transport matrix of the DFD cell, for a series of momenta. Prior to matrix computation, the closed orbit is found by FIT. Next to that, REBELOTE repeats for a new momentum value. The INCLUDE grabs the dipole triplet segment of Table 10.4

```
Scan transport matrix of 150 MeV FFAG dipole triplet
'MARKER'  SFFAGZroChro_S                               ! Just for edition purposes.
'OBJET'
1839.090113                                            ! Reference rigidity (150MeV proton).
5                                                      ! Option for MATRIX computation.
.01 .001 .01 .001 0. 0.0001                            ! 13-trajectory sampling for MATRIX computation.
445.234  0.  0.  0. 0.   0.273042677097  'o'           ! Reference trajectory (number 1 in the 11-set).
1

'INCLUDE'                                              ! Include the 30 degree sector dipole triplet,
1                                                      !   within dedicated LABEL1s.
./SFFAGCell.inc[#S_SFFAG150Cell:#E_SFFAG150Cell]

 'FIT'
1  noFinal
2 30 0  2.
2  1e-8  79                                            ! A penalty value controls the accuracy
3.1 1 2 #End 0. 1. 0                                   ! of the convergence of the constraints to the target values.
3.1 1 3 #End 0. 1. 0                                   ! 3.1 is the constraint for periodicity.

'MATRIX'
1 11  PRINT                                            ! PRINT causes log of transport coefficients to zgoubi.MATRIX.out.

'REBELOTE'                                             ! Repeat the previous sequence,
20 0 0  1                                              ! 20 times; prior to repeating, change
1                                                      ! 1 parameter: parameter 35 under OBJET, namely,
OBJET 35 0.27304263798:1.                              ! the relative momentum, D=p/p_ref, in the range 0.27304263798:1.

'SYSTEM'
1
gnuplot <./gnuplot_tunes.gnu                           ! Call  gnuplot script once that tracking is completed.

'MARKER'  SFFAGZroChro_E                               ! Just for edition purposes.
'END'
```

---

gnuplot script gnuplot_tunes.gnu, for Fig. 10.24:

```
# gnuplot_tunes.gnu
set xlabel "relative momentum p/p_{ref}"; set ylabel "{/Symbol n}_R,  {/Symbol n}_Z"; set y2label "penalty"
set xtics; set ytics nomirror; set y2tics nomirror; set key t 1; set key maxrow 2; set logscale y2
# A system command to extract penalty values from zgoubi.res execution listing
system "grep 'Fit reached penalty value ' zgoubi.res | cat > grep.out"
k=7.6 ; Ncell = 12; set yrange [:4.5]; set y2range [1e-12:1e-4]
plot "zgoubi.MATRIX.out" u (0+$47):(Ncell* $56) w  p  pt 6 ps .9 tit "{/Symbol}_R", \
"zgoubi.MATRIX.out" u (0+$47):(Ncell* $57) w  p  pt 7 ps .9 tit "{/Symbol}_Z", \
(x<1.03 ? sqrt(1+k) :1/0) w l lw 2 tit "\sqrt{(1+k)}" ,  "grep.out" u :5 axes x2y2
```

MATRIX[PRINT] logs the transport coefficients into zgoubi.MATRIX.out, together with the beam matrix and tunes which are obtained from the hypothesis of periodicity (MATRIX[IFOC=11]), namely by identifying (Sect. 14.5.2)

$$[T_{ij}] = I \cos \mu + J \sin \mu$$

Outcomes are plotted in Fig. 10.24. The radial tune ν_R is constant as expected from the zero-chromaticity resulting from the scaling law (momentum-independent index, Eq. 10.3). Such is not the case for the axial tune, ν_Z, this is due to the first order effect of fringe field extent on the axial focusing (see Sect. 14.4.1): fringe extent varies with orbit radius in correlation with gap height in the "gap shaping" hypothesis (Eq. 10.6; smaller gap at greater radius/greater energy).

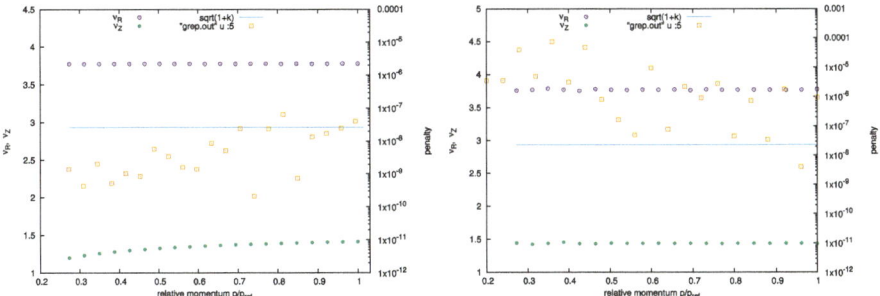

Fig. 10.24 Radial (v_R) and axial (v_Z) tunes (left vertical scale) of the 12-cell ring, as a function of relative momentum ($p/p_{ref} = 1$ for 150 MeV). The penalty values (scattered squares, right vertical scale) monitor the FIT runs: a small penalty indicates convergence of FIT. Left: case of decreasing gap height with radius (equivalent to a decreasing fringe extent): $g(R) = g_0 \left(\frac{R_0}{R}\right)^3$. Right: case of linear increase of gap height with radius (equivalent to a linear increase of fringe extent): $g(R) = g_0 \left(\frac{R}{R_0}\right)$. v_Z is now about constant over the momentum span

(b) Momentum-dependent axial tune.

The variation of the axial tune with momentum, as observed in (a), is due to the fringe field extent decreasing with radius, following in that the gap height which induces the scaling field law $B(R) \propto R^k$ (Eq. 10.5): as a matter of fact the gap shape index value in the present radial sector model is $\kappa = 3$ (Table 10.4), resulting in a gap height (Eq. 10.4)

$$g(R) = g_0 \left(\frac{R_0}{R}\right)^3$$

The axial tune can be made constant using instead a gap law (Eq. 10.6)

$$g(R) = g_0 \left(\frac{R}{R_0}\right) \qquad (\kappa = -1)$$

i.e., gap height proportional to momentum/R [22, 26]. The simulation is obtained by changing the lines of concern under FFAG keyword (Table 10.4), namely

change "6.3 3. ! EFB [etc.]"   to   "6.3 − 1. ! EFB [etc.]"

at the 6 EFBs. The resulting tunes are displayed in Fig. 10.24: v_R is marginally affected, whereas v_Z is now about constant.

A FIT procedure can be further attempted, in order to try and improve things, proceeding in the following way:

– vary κ
– constrain v_Z = constant at a few energies in the 12–150 MeV range.

Table 10.7 Simulation input data file: TWISS command, to obtain beam matrix, momentum compaction, chromaticities, etc. The initial reference coordinates, under OBJET, are for 12 MeV; reference coordinates for 150 MeV (to substitute to the previous ones) have been added as a comment. TWISS is an implicit do-loop, it proceeds in 3 stages: it first computes tunes of an on-momentum particle, then for $\pm \delta p/p$ off-momentum particles; at each stage, FIT ensures that the reference particle (1st particle of the 11-set) is on the closed orbit

```
Compute momentum compaction.
'MARKER' SFFAG_TWISS_S
 'OBJET'
1839.090113                                           ! Reference rigidity (150MeV proton).
5                                                  ! Option for MATRIX computation.
.01 .001 .01 .001 0. 0.0001                        ! 13-trajectory sampling for TWISS computation.
445.234 0. 0. 0. 0. 0.273042677097  'o'    ! 12 MeV reference trajectory (traj. number 1 in the 11-set).
1

! 150 MeV reference trajectory, should replace the 12 MeV data line above:
! 517.4981 0.   0.  0. 0.   1.  'o'

'INCLUDE'                                          ! Include the 30 degree sector dipole triplet,
1                                                  ! within dedicated  LABEL1s.
./SFFAGCell.inc[#S_SFFAG150Cell:#E_SFFAG150Cell]
'FIT'
1  noFinal
1 30 0  2.
2  1e-8 79   ! A penalty value controls accuracy of convergence of the constraints to the target values.
3.1 1 2 #End 0. 1. 0     ! 3.1 is the constraint for periodicity; coordinate 2 (Y) of particle 1, here;,
3.1 1 3 #End 0. 1. 0                        ! coordinate 3 (T) of particle 1, here;.
'TWISS'
2 1. 1.
'MARKER' SFFAG_TWISS_E
'END'
```

However, as expected from theory, it is found that the constraint is already fairly satisfied with $\kappa = -1$.

(d) Momentum compaction and transition γ_{tr}

TWISS keyword is used concurrently with OBJET[KOBJ=5], to compute various first and second order optical parameters including chromaticities. MATRIX[IORD=2, IFOC=11] concurrently with OBJET[KOBJ=6] can be used as well. A typical input file is given in Table 10.7. Results are given in Table 10.8.

It is expected for the momentum compaction to satisfy $\alpha = \frac{\Delta C}{C} / \frac{\Delta p}{p} = 1/(1+k)$ (Eq. 10.18). In the present design $k = 7.6$ in all three dipoles of the triplet (Table 10.4), yielding $\alpha = 0.11628$. This approximation appears valid at high energy where ray-tracing yields $\alpha = 0.11620$ (Table 10.7), however it is not the case at low energy where the numerical integration yields $\alpha = 0.42528$. This discrepancy might result from the greater flutter at smaller orbit radius (shorter magnet body compared to fringe field extent)—further investigation is left to the reader.

Table 10.8 Outcomes of TWISS computation out of zgoubi.res execution listing, including beam matrix, tunes, momentum compaction, chromaticities

Case of 12 *MeV optics*:

```
Reference, before change of frame (particle #  1  - D-1,Y,T,Z,s,time)  :
-7.26957323E-01   4.45233802E+02   6.40323244E-05   0.00000000E+00   0.00000000E+00   2.41165083E+02   5.07793867E-02

Reference, after change of frame (particle #  1  - D-1,Y,T,Z,s,time)  :
-7.26957323E-01   0.00000000E+00   0.00000000E+00   0.00000000E+00   0.00000000E+00   2.41165083E+02   5.07793867E-02

        Beam matrix (beta/-alpha/-alpha/gamma) and periodic dispersion (MKSA units)
         0.734624    0.000003    0.000000    0.000000    0.000000    0.514568
         0.000003    1.361240    0.000000    0.000000    0.000000   -0.000000
         0.000000    0.000000    4.398783   -0.000040    0.000000   -0.000000
         0.000000    0.000000   -0.000040    0.227336    0.000000    0.000000

                    Betatron  tunes  (Q1 Q2 modes)
             NU_Y =  0.31422833      NU_Z =  0.99804552E-01

                    Momentum compaction :
                    dL/L / dp/p =  0.42528937

                    Transition gamma  =  1.53340804E+00

                    Chromaticities :
       dNu_y / dp/p =  1.92734237E-04          dNu_z / dp/p =  2.19919557E-02
```

Case of 150 *MeV optics*:

```
Reference, before change of frame (particle #  1  - D-1,Y,T,Z,s,time)  :
 0.00000000E+00   5.17498491E+02  -2.99045520E-05   0.00000000E+00   0.00000000E+00   2.80426451E+02   1.84634162E-02

Reference, after change of frame (particle #  1  - D-1,Y,T,Z,s,time)  :
 0.00000000E+00   0.00000000E+00   0.00000000E+00   0.00000000E+00   0.00000000E+00   2.80426451E+02   1.84634162E-02

Beam matrix (beta/-alpha/-alpha/gamma) and  periodic  dispersion  (MKSA units)
         0.853519   -0.000046    0.000000    0.000000    0.000000    0.600422
        -0.000046    1.171620    0.000000    0.000000    0.000000   -0.000009
         0.000000    0.000000    4.322948    0.000012    0.000000   -0.000000
         0.000000    0.000000    0.000012    0.231324    0.000000    0.000000

                    Betatron  tunes  (Q1 Q2 modes)
             NU_Y =  0.31448874      NU_Z =  0.11757537

                    Momentum compaction :
                    dL/L / dp/p =  0.11619673

                    Transition gamma  =  2.93361449E+00

                    Chromaticities :
       dNu_y / dp/p = -5.78695533E-04          dNu_z / dp/p =  4.78678315E-03
```

10.4 Beam Envelopes; Phase Space

(a) Motion envelopes.

There is various possibilities to get the beam envelopes along the cell. One consists in raytracing a few particles with initial coordinates taken on an ellipse. Another method tracks a single particle, for a few tens of turns. A third possibility consists in pushing the initial beam matrix $\sigma(s_0)$ through the cell, using $\sigma(s) = T(s \leftarrow s_0)\sigma(s_0)$ $\tilde{T}(s \leftarrow s_0)$ (Sect. 14.5.2), by computing $T(s \leftarrow s_0)$ from the stepwise particle coordinates in the option OBJET[KOBJ=5]—a tool to push $\sigma(s)$ can be found in zgoubi toolbox [36].

In any case, linear envelopes, with maximal excursion $\sqrt{(\frac{\varepsilon_Y}{\pi}\beta_Y(s))}$ and $\sqrt{(\frac{\varepsilon_Z}{\pi}\beta_Z(s))}$, require paraxial motion.

The first method is retained, here. Set IL=2 under FFAG to have particle data logged, step by step, in zgoubi.plt. Graphs of the trajectories of the beam bundle across the cell, at 12 and 150 MeV are given in Fig. 10.25.

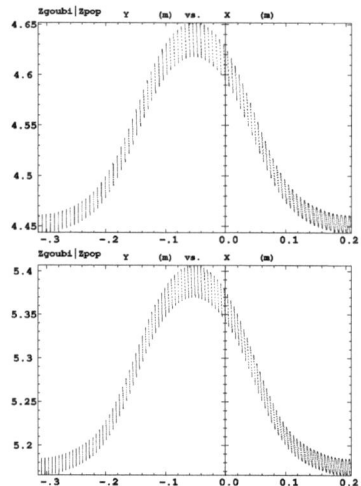

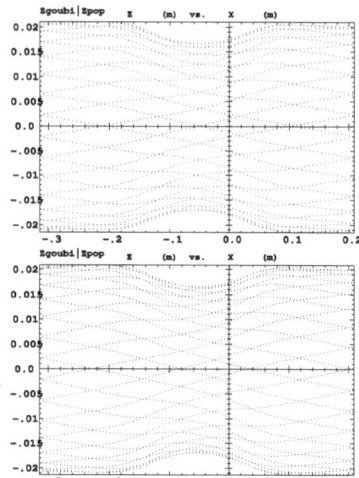

Fig. 10.25 Radial (left) and axial (right) beam bundle trajectories across the FFAG triplet cell; top row: 12 MeV, bottom row: 150 MeV. Graphs obtained using `zpop`: menu 7; 1/1 to open zgoubi.plt; 2/[8, 2] to select Y (radius) versus X (azimuthal angle) [or 2/[8, 4] to select Z (axial coordinate)]; 7 to plot

Table 10.9 Simulation input data file: proper OBJET, using KOBJ=8, to define a beam bundle by initial reference orbit coordinates, and Courant invariant values

Case of 12 MeV optics:

```
'OBJET'
1839.090113                                        !  Reference rigidity (150MeV proton).
8                                                  !  Option for particles on an ellipse.
30 30 1                          ! 30 particle evenly spaced on Y-T ellipse, same on Z-P ellipse.
4.45234 0. 0. 0. 0. 0.273042677097 'o'             ! Reference trajectory (note the units: units: m, rad).
0. 0.734624 1e-4                                   ! alpha_Y, beta_Y, invariant value.
0. 4.399057 1e-4                                   ! alpha_Z, beta_Z, invariant value.
0. 1. 0.                                           ! alpha_X, beta_X, invariant value.
```

Case of 150 MeV optics:

```
'OBJET'
1839.090113                                        !  Reference rigidity (150MeV proton).
8                                                  !  Option for particles on an ellipse.
30 30 1                          ! 30 particle evenly spaced on Y-T ellipse, same on Z-P ellipse.
517.4981 0. 0. 0. 0. 1. 'o'                        ! Reference trajectory (note the units: units: m, rad).
0. 0.853818 1e-4                                   ! alpha_Y, beta_Y, invariant value.
0. 4.323902 1e-4                                   ! alpha_Z, beta_Z, invariant value.
0. 1. 0.                                           ! alpha_X, beta_X, invariant value.
```

Particle trajectories result from initial coordinates taken on a small invariant value (an ellipse), in both radial and axial planes, namely, $\varepsilon_Y/\pi = \varepsilon_Z/\pi = 0.1$ mm (Table 10.9). The definition of the initial coordinates on an ellipse uses the keyword OBJET[KOBJ=8]. The ellipse parameters, $\alpha_{Y,Z}$, $\beta_{Y,Z}$, are taken from TWISS outcomes of the previous exercise (Table 10.8).

Note that the rather large initial invariant values (0.1π mm) result in the two motions to be slightly coupled (in the presence of non-zero axial motion), the initial elliptical invariant is not strictly preserved during the propagation, some smear shows, see next question.

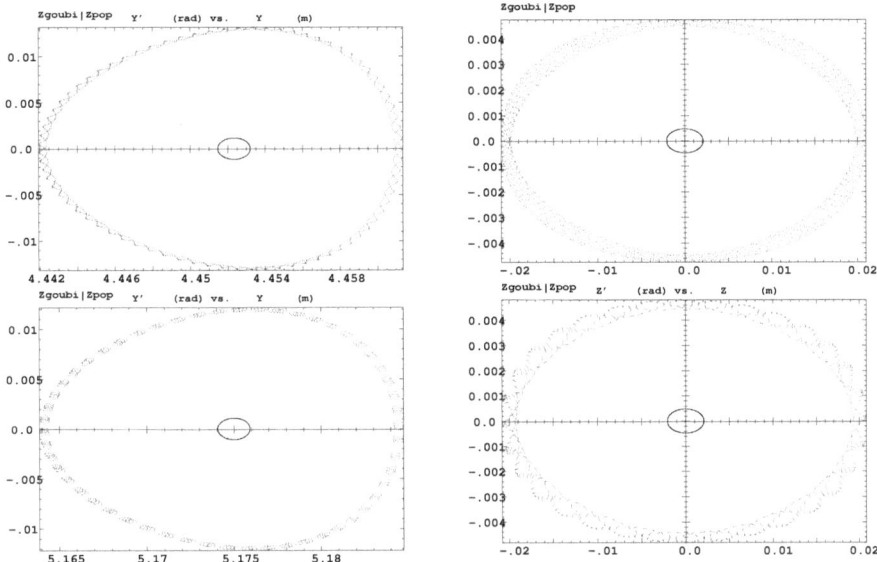

Fig. 10.26 Multiturn tracking: radial (left) and axial (right) phase space motion, observed at the end (middle of the drift) of an FFAG triplet cell; top row: 12 MeV, bottom row: 150 MeV. The central quasi-elliptical motion is for $\varepsilon_Y/\pi = \varepsilon_Z/\pi = 1\,\mu$m, the outer motion, distorted and coupled, is for $\varepsilon_Y/\pi = \varepsilon_Z/\pi = 0.1$ mm. Graphs obtained using zpop: menu 7; 1/5 to open zgoubi.fai; 2/[2, 3] to select T versus Y (or, 2/[4, 5] to select P versus Z); 7 to plot

(b) Multiturn tracking, phase space.

The definition of the initial coordinates of the particle to be tracked uses the keyword OBJET[KOBJ=8] as in the previous method. However, considering the input data in Table 10.9,

$$\text{change} \quad \text{"30 30 1"} \quad \text{to} \quad \text{"1 1 1"}$$

Multiturn tracking reveals that 0.1 mm motion invariants are large enough that (i) they are distorted by field non-linearities (compared to ellipses in the case of parax-ial motion), and (ii) Y and Z motions feature non-linear coupling. This shows in Fig. 10.26 which displays phase space motion in the two cases of initial coordi-nates taken on an $\varepsilon_Y/\pi = \varepsilon_Z/\pi = 0.1$ mm ellipse and on an $\varepsilon_Y/\pi = \varepsilon_Z/\pi = 1\,\mu$m ellipse. In the first case the motion is not elliptical, whereas in the second case, parax-ial conditions, the phase space portrait is an ellipse, with no visible coupling (a thin ellipse, in both planes).

10.5 Acceleration. Transverse Betatron Damping

(a) Acceleration cycle in an FFAG ring, using an RF program.

A typical prior check: it is a good idea to first validate the input data file to be used for acceleration, by producing and checking a few closed orbits, or one-turn

Table 10.10 Simulation input data file: checking the file set up for acceleration. This data file is derived from the acceleration input file in Table 10.12, and provides periodic matrix computation at 10 MeV and 200 MeV, combining OBJET[KOBJ=5] (to allow computation by MATRIX), FIT (find the orbit) and REBELOTE (repeat for an additional momentum)

```
Check inut data prior to acceleration cycle
'OBJET'
1839.090113 150MeV
5
.001 .01 .001 .01 .001 .0001
440.54197   0.   0.000  0. 0.   0.2491207073   'o'                      ! 10MeV proton, Brho=458.155.
'PARTICUL'
PROTON
'INCLUDE'                                          ! Include the 30 degree sector dipole triplet,
1                                                  ! within dedicated LABEL1s.
./SFFAGCell.inc[#S_SFFAG150Cell:#E_SFFAG150Cell]
'FIT2'
1 nofinal
1 30 0 2.
1 1e-9 99          ! A penalty value controls the accuracy of the convergence to periodic coordinates.
3.1 1 2 #End 0. 1. 0                                 ! 3.1 is the constraint for periodicity.
'FAISCEAU'                                           ! Log particle coordintes in zgoubi.res, here.
'MATRIX'                                             ! Periodic matrix.
1 11
'REBELOTE'
1 0 0  1                                                               ! Repeat just once;
1                               ! prior to repeat, change parameter 35 under OBJET, which is D=p/p_ref,
OBJET 35 1.1688582876:1.1688582876                                     ! to the value D=1.1688582876.
'END'
```

beam matrices. An input data file in that aim is given in Table 10.10. Excerpts from zgoubi.res execution listing so obtained are given in Table 10.11, they show periodic orbits at expected radii, respectively $Y_0 = 440.5$ cm (10 MeV) and $Y_0 = 526.9$ cm (200 MeV), and expected beam matrices (compare to Table 10.8).

Following these preliminary checks, an input data file set for acceleration from 10 to 200 MeV is derived, Table 10.12. Note the "12 *" under INCLUDE, for 12 cells. The cell, a dipole triplet with half-drifts on both sides, is as in Table 10.4, yet making sure for the values of the following three parameters:

(i) IL is set to 0, to inhibit output to zgoubi.plt as this saves on computing time— another possibility is to use OPTIONS[.plt 0]

(ii) the integration step size is set to $\Delta s = 1$ cm, for accuracy over the 13,000 turn acceleration cycle

(iii) FFAG allows a few methods for the numerical integration, KIRD=0 was used in Table 10.4, whereas the method retained here instead, for a change (an interesting exercise would consist in comparing the outcomes from the two methods), is KIRD=2; as a consequence

"0 2 ! *Field computation analytic 2nd order (take KIRD = 2, 25 or 4 etc.*"

in Table 10.4, is changed to (comments, beyond the exclamation mark, have no effect)

"2 10. ! *Field computation method: 3 * 3 flying grid, 2nd deg interpolation.*"

Table 10.11 Output file: checking the file set up for acceleration. This table shows excerpts from zgoubi.res execution listing, following from the input data file in Table 10.10, namely, the closed orbit, matrix and tunes for 10 MeV (relative momentum D=0.2491, top part) and for 200 MeV (D=1.1689, bottom part). Particle 1 is the reference particle for the computation of the transport matrix from which the beam matrix is deduced

```
************************************************************************************
                              TRACE DU FAISCEAU
                           (follows element #      6)
                              13 TRAJECTOIRES
                   OBJET                                  FAISCEAU
      D      Y(cm)    T(mr)   Z(cm)   P(mr)    S(cm)   D-1     Y(cm)   T(mr)   Z(cm)   P(mr)    S(cm)
 0  1  0.2491 440.542  0.000   0.000   0.0000   0.0000 -0.7509 440.542  0.000   0.000   0.000  2.386113E+02
       Time of flight (mus) :  5.49502499E-02 mass (MeV/c2) :   938.272

       Beam matrix  (beta/-alpha/-alpha/gamma) and  periodic  dispersion  (MKSA units)
             0.726851     0.000000     0.000000     0.000000     0.000000     0.508871
             0.000000     1.375798     0.000000     0.000000     0.000000    -0.000000
             0.000000     0.000000     4.498812    -0.000001     0.000000     0.000000
             0.000000     0.000000    -0.000001     0.222281     0.000000     0.000000

                           Betatron  tunes
                 NU_Y =  0.31421037      NU_Z =  0.96625558E-01
************************************************************************************
                              TRACE DU FAISCEAU
                           (follows element #      6)
                              13 TRAJECTOIRES
                   OBJET                                  FAISCEAU
      D      Y(cm)    T(mr)   Z(cm)   P(mr)    S(cm)   D-1     Y(cm)   T(mr)   Z(cm)   P(mr)    S(cm)
 0  1  1.1689 526.956  0.000   0.000   0.000    0.0000  0.1689 526.956 -0.000   0.000   0.000  2.855586E+02
       Time of flight (mus) :  1.68242271E-02 mass (MeV/c2) :   938.272

       Beam matrix  (beta/-alpha/-alpha/gamma) and  periodic  dispersion  (MKSA units)
             0.869721    -0.000000     0.000000     0.000000     0.000000     0.611749
            -0.000000     1.149794     0.000000     0.000000     0.000000    -0.000000
             0.000000     0.000000     4.384709    -0.000001     0.000000     0.000000
             0.000000     0.000000    -0.000001     0.228065     0.000000     0.000000

                           Betatron  tunes
                 NU_Y =  0.31439931      NU_Z =  0.11802584
************************************************************************************
```

The consequence is that the field and derivatives [33, Sects. 1.2 and 1.2.1] are computed using a 3 × 3 node flying grid technique (that is what '2' stands for, '10' stands for the grid mesh size, taken equal to $\Delta s/10$), whereas in the previous exercises, given KIRD=0, field and derivatives are computed from (hard-coded) analytical expressions [33, Part B, FFAG] [35].

Setting up the acceleration, now:

- PARTICUL[PROTON] is necessary as CAVITE is used: it allows converting energy change in rigidity change (zgoubi pushes particles using rigidity),
- SCALING[IOPT=-2], akin to a "power supply rack", provides the RF program to CAVITE, by reading it from the ancillary file zgoubi.freqLaw.In, see below,
- CAVITE[IOPT=6] boosts the particle(s) at each pass, following that pre-defined RF program,
- REBELOTE sends the execution pointer back to the top of the input data file, for multiturn tracking. The number of turns results from a peak voltage $\hat{V} = 40$ kV and synchronous phase 20° (Table 10.12), the 10 to 200 MeV acceleration range is covered in $(200 - 10) \times 10^3/[40 \times \sin(20°)] \approx 13900$ turns.

For simplicity the RF program is limited in the present case to the turn dependence of RF frequency (peak voltage and synchronous phase maintained constant).

Table 10.12 Simulation input data file: proton acceleration from 10 MeV to about 200 MeV using zgoubi.freqLaw.In RF program file. Note: the step size must be in the centimeter range for multiturn accuracy, in such non-linear field (make sure to substitute "1." (cm) to "3.0079078598", in the INCLUDE file SFFAGCell.inc, Table 10.4). Note: this file, and all INCLUDE files it resorts to, are available in zgoubi sourceforge repository at https://sourceforge.net/p/zgoubi/code/HEAD/tree/branches/exemples/book/zgoubiMaterial/FFAG_scaling/radialSectorTriplet_KEK/accelerationCycle

```
Acceleration from 10 to 150 MeV.
'MARKER' SFFAGParoAccel_S                         ! Just for edition purposes.
'OBJET'
1839.08991465                                     ! 150MeV proton.
2
1 1
440.54197 3. 0. 1. 0. 0.24912070392  'o'          ! 10MeV proton, Brho=1839.090113 * 0.2491207073.
1                                                 ! Y_0 is +2 cm wrt. closed orbit R=445.234; Z_0=1 cm.
'PARTICUL'                                         ! Particle data will allow compute energy, time of flight.
PROTON                                            ! Acceleration requires particle data.
'SCALING'                                         ! SCALING is used to control CAVITE.
1 1
CAVITE
-2   ! Option -2 causes read of RF law for CAVITE from external file (default is zgoubi.freqLaw.In).
1                                                 ! unused with option -2.
1                                                 ! unused with option -2.
'PICKUPS'
1
#E
'FAISTORE'
zgoubi.fai AftCAV       ! (use "b_" for binary storage: b_zgoubi.fai). Storage is at LABEL1=AftCAV.
'INCLUDE'                                          ! Include the 30 degree sector dipole triplet,
1                                                 ! as comprised within dedicated LABEL1s.
12 * SFFAGCell.inc[#S_SFFAG150Cell:#E_SFFAG150Cell] ! Build the ring from 12 cell INCLUDEs.
'CAVITE'
6 PRINT ! IOPT=6 -> read RF program from external file (dflt is zgoubi.freqLaw.In). PRINT for checks.
0. 10.                                            ! unused; kinetic energy at start (MeV).
40000.  0.349066                                  ! Peak voltage; synchronous phase (rad), 20 degree.
'FAISCEAU'                                         ! Log particle coordintes in zgoubi.res, here.
'MARKER' AftCAV                                    ! Storage by FAISTORE is effective here.
'REBELOTE'                                         ! Repeat sequence 12499 times. 0.2: code to video every
12499 0.2 99                                       ! other 10 turns. '99': ignore OBJET at subsequent passes.

'SYSTEM'
1
gnuplot <./gnuplot_Zfai_phaseSpaces_multiplot.gnu  ! Plot 3 phase spaces and (x,y) space.
'MARKER' SFFAGParoAccel_E                          ! Just for edition purposes.
'END'
```

Elaborating zgoubi.freqLaw.In (essentially two columns: RF frequency versus turn number) requires the following steps:

(i) run a search for closed orbits for a few tens of energies in the range 10–200 MeV. The corresponding data file is given in Table 10.13. Time of flight, derived from path length and particle velocity, is part of the outcome of orbit computation as PARTICUL provides necessary particle data in that aim. That yields the turn dependence of RF frequency, namely, $f_{RF} = h \times f_{rev} = v/C$ ($h = 1$, here). Details below. Note that in the absence of orbit defects or other tailored bumps, the frequency law may be obtained, more simply, from the theoretical orbit length $C(p) = C_0(p/p_0)^{\frac{1}{k+1}}$ (Eq. 10.8);

(ii) store these turn-frequency data, properly formatted, in zgoubi.freqLaw.In (Table 10.14). zgoubi.freqLaw.In does not need to contain every turn, zgoubi will interpolate from the set of values provided.

(b) Transverse motion.

The acceleration simulation file is that of Table 10.12. Longitudinal and transverse motion samples are displayed in Fig. 10.27. The integration step size is $\Delta s = 1$ cm in these simulations. Taking KIRD=0 instead (see remarks above), all the rest unchanged, would result in a marginal difference.

Table 10.13 Simulation input data file: search closed orbits for a few tens of energies in the range 10–200 MeV, for fabrication of zgoubi.freqLaw.In RF program file

```
FFAG triplet. 150MeV machine. Find orbits from 10 to 200 MeV.
'OBJET'
1839.08991465                                                    ! 150MeV proton.
2
1 1
440.54197   0.   0.000  0. 0.  0.2491207073   'o'                ! 10MeV proton, Brho=458.155.
1
'PARTICUL'
PROTON
'INCLUDE'                                  ! Include the 30 degree sector dipole triplet,
1                                          ! within dedicated LABEL1s.
SFFAGCell.inc[#S_SFFAG150Cell:#E_SFFAG150Cell]        ! INCLUDE the FFAG triplet sector.
'FIT'                                      ! Find initial coordinate uch that final = initial.
1
1 30 0 1.
1
3.1 1 2 #End 0. 1. 0
'FAISTORE'
RFprogram_orbits.In
1
'REBELOTE'                                 ! Repeat 50 times, from 10 to 200 MeV.
50 0.2 0 1
1
OBJET  35  0.24912070392:1.16885844252     ! Change relative momentum value, from 10 to 200MeV.
'END'
```

Table 10.14 Top and bottom parts of the RF program file zgoubi.freqLaw.In, as read by zgoubi when using CAVITE[IOPT=6]. Zgoubi actually only requires turn number, column 1, and revolution time which is computed from the cumulated time-of-flight across the cells, column 4

	cell Time of flight,	cumulated phase	cumulated time of flight,	kinetic energy,	Test. Expected one
turn#	tauCell	phi	oclock	Ekin	
1.0000E+00	5.49502566E-02	0.0000E+00	6.59403079E-01	1.0000E+01	1.0000E+00
2.0000E+00	5.49185204E-02	6.28318531E+00	1.31844010E+00	1.00136808E+01	9.99962422E-01
3.0000E+00	5.48868211E-02	1.25663706E+01	1.97711108E+00	1.00273616E+01	9.99925969E-01
4.0000E+00	5.48551589E-02	1.88495559E+01	2.63541601E+00	1.00410424E+01	9.99890628E-01
5.0000E+00	5.48235338E-02	2.51327412E+01	3.29335493E+00	1.00547232E+01	9.99856388E-01
6.0000E+00	5.47919462E-02	3.14159265E+01	3.95092786E+00	1.00684040E+01	9.99823237E-01
7.0000E+00	5.47603962E-02	3.76991118E+01	4.60813483E+00	1.00820848E+01	9.99791162E-01
8.0000E+00	5.47288840E-02	4.39822972E+01	5.26497589E+00	1.00957656E+01	9.99760153E-01
....					
1.3881E+04	1.67887669E-02	8.72106121E+04	3.95518038E+03	1.99889584E+02	1.00389444E+00
1.3882E+04	1.67931495E-02	8.72168952E+04	3.95538164E+03	1.99903264E+02	1.00342779E+00
1.3883E+04	1.67975355E-02	8.72231784E+04	3.95558299E+03	1.99916945E+02	1.00296144E+00
1.3884E+04	1.68019247E-02	8.72294616E+04	3.95578442E+03	1.99930626E+02	1.00249542E+00
1.3885E+04	1.68063170E-02	8.72357448E+04	3.95598595E+03	1.99944307E+02	1.00202973E+00
1.3886E+04	1.68107123E-02	8.72420280E+04	3.95618757E+03	1.99957988E+02	1.00156440E+00
1.3887E+04	1.68151104E-02	8.72483112E+04	3.95638927E+03	1.99971668E+02	1.00109943E+00
1.3888E+04	1.68195111E-02	8.72545944E+04	3.95659107E+03	1.99985349E+02	1.00063485E+00
1.3889E+04	1.68239142E-02	8.72608775E+04	3.95679295E+03	1.99999030E+02	1.00017066E+00

(c) Track a particle bunch.

Use MCOBJET[KOBJ=3] here, to populate an initial Gaussian distribution with random transverse coordinates. The periodic optical functions needed in MCOBJET are taken from the 10 MeV beam matrix in Table 10.11, namely, $\alpha_x = \alpha_y = 0$, $\beta_x = 0.7268$ m, $\beta_x = 4.4988$ m. This yields

```
'MCOBJET'
1839.08991465                                          ! 150MeV proton.
3
200                                                    ! 200 particles.
2 2 2 1 1                                              ! Gaussian radial and axial distributions.
4.4054197   0.   0.000  0. 0.  0.24912070392            ! 10MeV closed orbit.
0.  0.7268  1e-8 3                                     ! alpha_Y, beta_Y, epsilon_Y/pi, cut-off.
0.  4.4988  1e-8 3                                     ! alpha_Z, beta_Z, epsilon_Z/pi, cut-off.
0.  1.      0.   3                                     ! alpha_X, beta_X, epsilon_X/pi, cut-off.
```

to substitute to OBJET in Table 10.12.

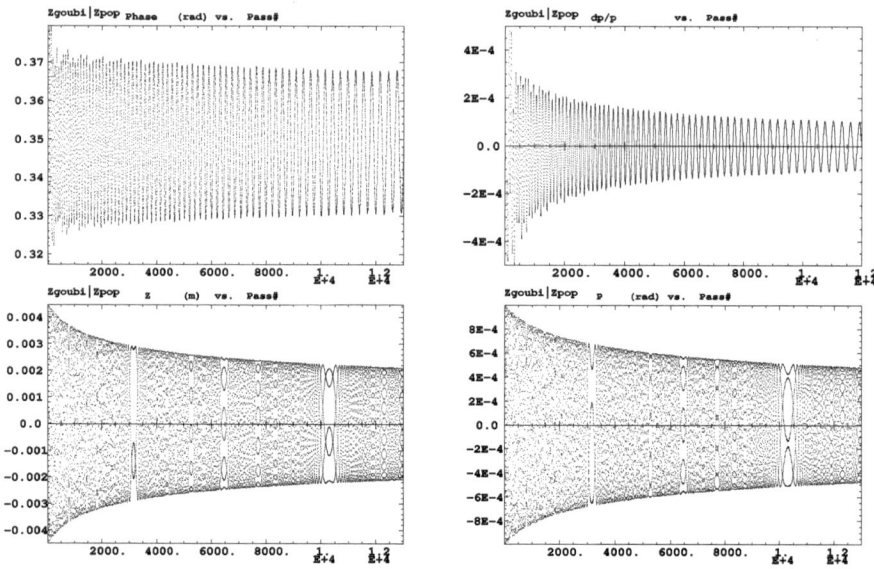

Fig. 10.27 An acceleration from 10 to 200 MeV. Top row: RF phase (left) and relative momentum (right) as a function of turn number, over an acceleration cycle. Bottom row: vertical excursion (left) and vertical angle (right). Motion damping is given by Eq. 10.22, namely, Z damping $\propto \sqrt{R/p}$, P damping $\propto 1/\sqrt{Rp}$, and normalized invariant $p\varepsilon_Z = $ constant

10.4.2 RACCAM Proton Therapy Spiral Sector FFAG

This series of exercises concerns the 180 MeV spiral sector proton therapy FFAG design displayed in Fig. 10.10, and its simulation using FFAG-SPI. The design parameters of the ring and of its cell dipole are given in Table 10.2 [22, 26–28]. The cell geometry is sketched in Fig. 10.28, orbits through a pair of cells are sketched in Fig. 10.29 as an illustration. Note the presence of field clamps on both sides of the dipole, these can be simulated in FFAG-SPI, by adding narrow, negative field spiral sectors adjacent to the main dipole [27].

10.6 Field in a Spiral Sector Dipole

(a, b) Generating a 2D field map. Using TOSCA.

The mid-plane magnetic field can be generated from Eq. 10.15, with geometrical and field data taken from Table 10.2. This is the FFAG-SPI field model, used here to produce the field along trajectories. It is based on the field modeling technique described in Sect. 10.2.2. The input data file is given and commented in Table 10.15, which can be referred to for details.

The mid-plane field data $B_Z(R, \theta)|_{Z=0}$ are arranged under the form of a 2D even meshing, this is in order to allow possible handling by TOSCA or POLARMES

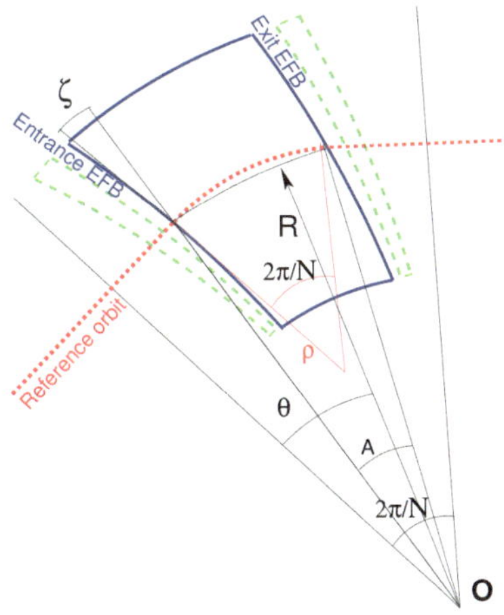

Fig. 10.28 A sketch of RACCAM spiral sector dipole and $2\pi/10$ cell. O is the center of the ring and the EFBs form a sector angle A. Note that the reference orbit is not strictly circular, the bending radius is not constant along the trajectory over the $2\pi/N$ arc (a line of constant field is an R-radius arc, centered on O). Field clamps can be seen represented (dashed lines), on both sides of the dipole

Fig. 10.29 A simulation of a pair of cells of RACCAM FFAG ring in zgoubi, including a few orbits, using the keyword FFAG-SPI. The simulation includes field clamps on both sides of the dipoles

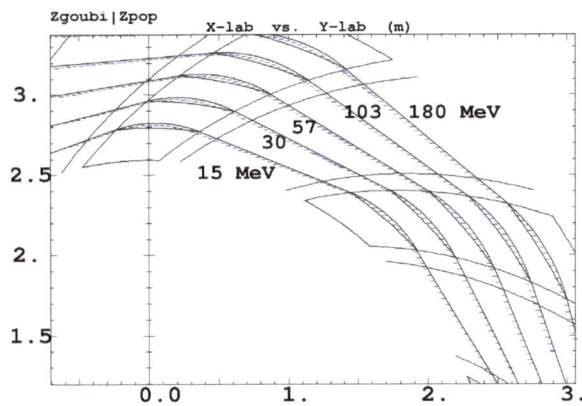

keywords. In order to generate a field map, from particle raytracing, a set of 29 trajectories is launched, with initial coordinates Y_0, T_0, Z_0, P_0 and relative momentum $D = p/p_{\text{ref}}$ defined using OBJET[KOBJ=1]: they all have initial incidence $T_0 = 0$, normal to the $AT = 45.83662°$ angular sector which contains the magnet, whereas initial radii Y_0 are evenly spaced over the useful field region, namely

$$Y_0 : r_1 \rightarrow r_{29}, \text{ step } \Delta r \quad \text{with } r_1 = 282\,\text{cm}, \ r_{29} = 340\,\text{cm}, \ \Delta r = 1\,\text{cm}$$

Axial motion is taken null, $Z_0 = 0$ and $P_0 = 0$. For each particle, the motion is forced to maintain constant radius, $r \in \{r_1, r_{29}\}$, throughout the dipole, using OPTIONS[CONSTY ON]. The integration step size is $\Delta s = 3.46\,\text{cm}$, resulting in 81 steps over

$$\theta : \theta_1 \rightarrow \theta_{81}, \text{ step } \Delta\theta = \Delta s/R_0, \quad \text{with } \theta_1 = 0, \ \theta_{81} = AT, \ \Delta\theta = 0.572906°$$

Table 10.15 Simulation input data file RACCAMCell.inc: RACCAM 36° cell, comprised of a spiral sector dipole. The FFAG-SPI keyword allows defining up to 5 independent dipoles in an AT angular sector; only one is defined in the present case, in order to generate field in a single dipole (note: as FFAG-SPI can house 5 dipoles, field clamps could be simulated by adding a reversed-field narrow sector on each side of the main dipole). OPTIONS[CONSTY ON] allows to generate a field map by raytracing a set of trajectories with *constant radius*. The present file also defines the sequence segment #S_RACCAMCell to #E_RACCAMCell, for use in INCLUDEs in subsequent exercises

```
RACCAMCell.inc file. 36 deg Spiral sector.
'MARKER'  RACCAMProbFieldMap_S
 'OBJET'
2029.47926                                                    ! Rigidity of 180 MeV proton.
1
29 1 1 1 1 1                         ! This creates 29 trajectories, corresponding to as many radial steps,
1. 0. 0. 0. 0. 0.                    ! whereas theta-steps result from the integration step size, below.
311. 0. 0. 0. 0. 1.                  ! Yo is half-way between a minimum  282 cm  and     a maximum 340 cm.
                                     ! (injection region, 15MeV)  (extraction region, 180MeV)
'PARTICUL'                                                    ! Allows computation of particle energy.
PROTON
'OPTIONS'
1 1
CONSTY ON                            ! Forces particles on Y=Yo and Z=Zo, in subsequent raytracing.

'MARKER'  #S_RACCAMCell
'FFAG-SPI'
0    ! Set IL=2 to store particle and field data along trajectories (in zgoubi.plt) for further plotting.
1  45.83662  346.031                 ! Number of dipoles within AT; AT; R0. Bend angle is AT+TE-TS = 36 deg.
12 0.  17.   5.                      ! ACN, dR0, B0, geometrical index k.
8.95 -0.52   ! EFB 1 : lambda~gap size, kappa=gap shape index: detemines radius-dependent fringe extent.
6  .1455   2.2670  -.6395  1.1558  0. 0.  0.
6.120000E+00  5.370000E+01 1.E6  -1.E6  1.E6  1.E6            ! Entrance face azimuth; spiral angle.
8.95 -0.52   ! EFB 2 : lambda~gap size, kappa=gap shape index: detemines radius-dependent fringe extent.
6  .1455   2.2670  -.6395  1.1558  0. 0.  0.
-6.120000E+00  5.370000E+01 1.E6  -1.E6  1.E6 1.E6            ! Exit face azimuth; spiral angle.
0. -1
0 0.  0.  0.  0.  0. 0. 0.
0.  0.  0.  0.   0. 0.
2  10.                               ! KIRD = 2: 3*3 flying grid interpolation, grid size is step size/Resol.
3.46                                             ! Step size (cm). Use ~1cm for multiturn tracking.
2
0. -0.295280245 0. -0.123598776                  ! RE, TE, RS, TS. Total deviation is AT+TE-TS = 36 deg.
'MARKER'  #E_RACCAMCell

'SYSTEM'
2                                                            ! A 'call system' for next 2 commands:
gnuplot < ./gnuplot_Zplt_fieldMap.gnu             ! plot field data logged in / read from zgoubi.plt;
! okular gnuplot_Zplt_fieldMap.eps &                        ! view graph - requires a .eps viewer.

'MARKER'  RACCAMProbFieldMap_E
'END'
```

---

gnuplot script gnuplot_Zplt_fieldMap.gnu, to obtain Fig. 10.30:

```
#gnuplot_Zplt_fieldMap.gnu
set xlabel 'X [cm]' ; set ylabel 'Y [cm]' ; set zlabel 'B [kG]'
set xtics 340, 40, 420 ; set hidden3d ; set view 63, 94 ; set zrange[0:22]
splot "zgoubi.plt" u ($10*cos($22)):($10*sin($22)>-50? $10*sin($22) :1/0):($25) w p pt 5 ps .8 lc palette notit; pause 1
```

with radius $R_0 = 346.031$ cm, and $AT = 45.83662°$ the sector opening. AT includes extra extent, beyond the 36° angular extent of a period, in order to avoid cutting off field tails. In doing so, the angles TE and TS of FFAG-SPI's KPOS parameters, are used to re-establish the 36° periodicity. This generates the mid-plane field $B_Z(R, \theta)$ over a $N_r \times N_\theta = 29 \times 81$ node 2D meshing, as displayed in Fig. 10.30. Note that if a 3D map is desired instead, a Z_0 sampling can be added in OBJET, as CONSTY also forces Z to its initial value Z_0.

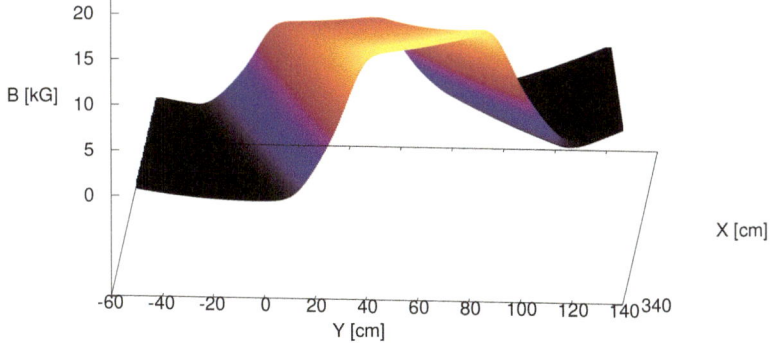

Fig. 10.30 Using FFAG-SPI theoretical modeling: mid-plane magnetic field in RACCAM spiral sector dipole, in the laboratory frame ($X = R\cos\theta$ and $Y = R\sin\theta$). The meshing geometry is obtained by ray-tracing 29 particles forced on circular trajectories evenly spaced in radius with constant angular integration step size. FFAG-SPI uses the spiral sector analytical field model of Eq. 10.15

Using TOSCA.

Formatting the field map for TOSCA[$IZ > 1$, $MOD = 25$], and raytracing using the latter, is left to the reader.[1] A dedicated table in [33, *TOSCA*] explains the choice [$IZ > 1$, $MOD = 25$], and other options available in the case of a polar field map mesh. See also Exercises 3.1, 4.1 for guidance.

A similar problem is treated, its input data file and field map are provided, in zgoubi sourceforge repository, at

https://sourceforge.net/p/zgoubi/code/HEAD/tree/branches/exemples/FFAG/ KEK150MeV/OPERAMapModel

(c) Moving the model to CYCLOTRON.

This exercise is left to the reader. Refer to Exercise 4.6 and to Zgoubi User's Guide [33, *cf.* CYCLOTRON] to work out this simulation.

A CYCLOTRON simulation can be found, input data file is available, in zgoubi sourceforge repository, at

https://sourceforge.net/p/zgoubi/code/HEAD/tree/branches/exemples/cyclotron/ PSI/usingCYCLOTRON.

10.7 Orbits, Scalloping

(a) Periodic orbits.

The input data file in Table 10.16 will produce 10 closed orbits (found one by one by FIT) for as many different momenta (REBELOTE repeats the sequence) ranging in

---

[1] Note that, as a guidance, TOSCA simulations of all sorts, with various types of data formatting— IZ, MOD and MOD2 options—can be found in zgoubi sourceforge repository, https://sourceforge. net/p/zgoubi/code/HEAD/tree/branches/exemples/KEYWORDS/TOSCA/ folder.

Table 10.16 Simulation input data file: find the closed orbit for a set of different momenta. The INCLUDE is the FFAG-SPI spiral dipole segment from Table 10.15, within LABEL1s #*S_RACCAMCell* and #*E_RACCAMCell*. Once REBELOTE do-loop is completed, SYSTEM launches a subsequent zgoubi job, plotOrbits.dat, in an *ad hoc* temporary folder, ./tempo

```
Orbit scan.
'MARKER'  RACCAMProbOrbits_S                                    ! Just for edition purposes.
'OBJET'
2029.47926                                          ! Reference rigidity (180MeV proton).
2
1  1
2.848940E+02  3.005184E+02  0.0E+00  0. 0.  0.276852600 'i'              ! 15 MeV proton.
1
'PARTICUL'
PROTON
'FAISTORE'
orbits.fai afterFIT  ! Coordinates are saved in orbits.fai. This is effective at 'afterFIT' label, below.
1

'INCLUDE'                                         ! Include the 36 degree spiral sector dipole,
1                                                         ! within dedicated LABEL1s.
./RACCAMCell.inc[#S_RACCAMCell:#E_RACCAMCell]
'FIT2'
2  nofinal
2 30 0  2.
2 31 0  2.
2  1e-9 99                                        ! A penalty value controls the accuracy
3.1 1 2 #End 0. 1. 0                              ! of the convergence to periodic coordinates.
3.1 1 3 #End 0. 1. 0                              ! 3.1 is the constraint for periodicity.
'MARKER' afterFIT                                 ! FAISTORE above applies here.

'REBELOTE'                                        ! Repeat the previous sequence,
10 0 0  1                                         ! 10 times; prior to repeating, change
1                                                 ! 1 parameter: parameter 35 under OBJET, namely,
OBJET 35 0.276852600:1.          ! the relative momentum, D=p/p_ref, in the range  0.276852600:1.

'SYSTEM'
6
(mkdir -p tempo; cp zgoubi.res tempo/temp.res; cp orbits.fai tempo)       ! In a separate folder,
ln -s /home/meot/zgoubi/current/zgoubi zgoubi              ! establish a link to \zgoubi\ executable.
(cp plotOrbits.dat tempo ; cd tempo; ./zgoubi -in plotOrbits.dat)      ! track orbits using initial
mv tempo/zgoubi.plt .                                ! coordinates as read from orbits.fai.
gnuplot <./gnuplot_Zplt_XY.gnu                 ! Call gnuplot scripts once that tracking is completed.
gnuplot <./gnuplot_Zplt_XB.gnu

'MARKER' RACCAMProbOrbits_E                                     ! Just for edition purposes.
'END'
```

plotOrbits.dat file, to track orbits using initial coordinates as read from orbits.fai (note: in the sub-folder 'tempo', a link to zgoubi *executable is required to run this file):*

```
plotOrbits.dat
'OBJET'
2029.47926                                          ! Reference rigidity (180MeV proton).
3
1 999 1
1 999 1
1. 1. 1. 1. 1. 1. 1. '*'
0. 0. 0. 0. 0. 0. 0.
0
orbits.fai

'INCLUDE'
1
./temp.res[#S_RACCAMCell:#E_RACCAMCell]

'FAISCEAU'
'END'
```

gnuplot scripts gnuplot_Zplt_XY.gnu and gnuplot_Zplt_XB.gnu, for Fig. 10.31:

```
#gnuplot_Zplt_XY.gnu
set xtics mirror ; set ytics mirror ; set xlabel "X_{Lab} [m]" ;set ylabel "Y_{Lab} [m]"
set polar; unset colorbox; cm2m = 0.01; set xrange [0:3.5]; set yrange [-1.75:1.75]; set size square
plot for [i=2:11] "zgoubi.plt" u ($22):(cm2m* $10):($19) w p ps .4  lc palette notit; pause 1

#gnuplot_Zplt_XB.gnu
set xtics; set ytics; set xlabel "Angle  [deg]"  ; set ylabel "Field  [T]"  ; r2d = 180./ (4.*atan(1.)) ; kG2T=0.1
plot for [i=1:11] "zgoubi.plt" u ($19== i ? $22 *r2d : 1/0):($25 *kG2T) w lp lw 2 ps .2 notit; pause 1
```

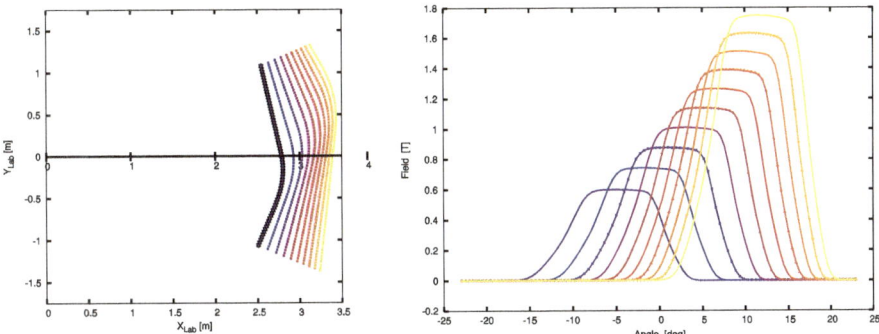

Fig. 10.31 Periodic orbits and field across RACCAM FFAG spiral sector dipole. Left: orbit scalloping across AT=45.83° arc extent in FFAG-SPI simulation (a cell is 36°), for different proton energies ranging in 15–180 MeV. Right: field experienced along these orbits, increasing with energy

(relative to $p_{ref} = 608.422$ MeV/c, 180 MeV proton) $p/p_{ref} = 0.2768526 : 1$ (15–180 MeV). For each closed orbit, coordinates are stored in orbits.fai file (at MARKER with LABEL1=afterFIT right after the FIT, prior to REBELOTE repeat).

The input data file ends with a SYSTEM command which, once zgoubi is done with finding/storing the periodic orbits, launches a subsequent computation ("cd tempo; ./zgoubi -in plotOrbits.dat" command) which performs the following:

– first, the periodic orbit coordinates are read from orbits.fai,
– they are then pushed through the magnet, and the trajectories are logged in zgoubi.plt (the effect of IL=2 under FFAG) for further post-treatment or plotting.

A plot is launched by the next two gnuplot commands under SYSTEM, outcomes are displayed in Fig. 10.31.

A plot of orbits around the ring can be obtained from the previous raytracing, for instance using a loop in gnuplot to increment the polar angle by steps of 36°, reading particle data across the cell from zgoubi.plt; it is displayed in Fig. 10.32.

(b) Homothety-rotation of the orbits.

The orbit scalloping is apparent in Figs. 10.31 and 10.32. Step By Step values can be drawn from zgoubi.plt and show that the scalloping $\delta R/R$ is in the % range. Rotation of the closed orbit pattern is also apparent in Figs. 10.31 and 10.32; the radius dependence of the rotation angle satisfies Eq. 10.14. The expected value from the latter can be checked against closed orbit data from zgoubi.plt.

(c) Figure 10.33 compares the numerical and theoretical (Eq. 10.8) values of the average orbit radius $R = C/2\pi$, both in good accord.

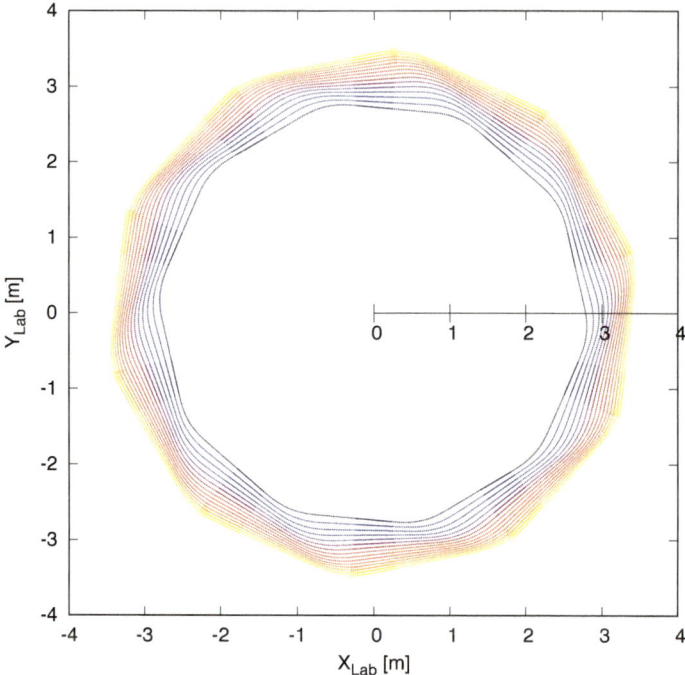

Fig. 10.32 Fifteen closed orbits around the 10-cell spiral sector RACCAM ring, in the range 15 to 180 MeV

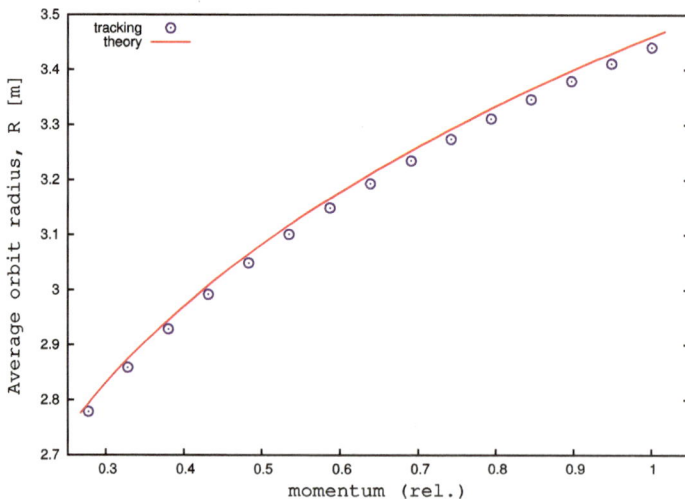

Fig. 10.33 Dependence of the average closed orbit radius $R = C/2\pi$ on the relative momentum. Markers are from one-turn raytracing. The solid line is from theory, for comparison, after Eq. 10.8 taken for $k = 5$ (Table 10.2) and reference momentum $p_{\text{ref}} = 608.422 \, \text{MeV/c}$ (180 MeV, radius $R_0 = 3.46031 \, \text{m}$). They differ by $\approx 1\%$

10.8 Zero-Chromaticity

(a) Momentum dependence of tunes.

The input data file in Table 10.17 computes momentum-dependent transport matrices of the cell, $[T_{ij}](p)$, for a series of different momenta.

OBJET[KOBJ=5] defines a set of 13 particles with proper initial coordinates for matrix computation, by MATRIX. Prior to matrix computation, the momentum-dependent closed orbit is found by FIT. REBELOTE changes the relative momentum D in OBJET, and repeats the procedure. MATRIX[PRINT] logs the transport coefficients to zgoubi.MATRIX.out, together with the beam matrix and tunes which are obtained from the hypothesis of periodicity (specified via MATRIX[IFOC=10+1 period]), namely from the identification (Sect. 14.5.2)

$$[T_{ij}] = I \cos \mu + J \sin \mu$$

Table 10.17 Simulation input data file: compute the first order transport matrix of the cell for a series of momenta. Prior to matrix computation, the closed orbit is found by FIT, or FIT2. The INCLUDE is the FFAG-SPI spiral dipole segment from Table 10.15, within LABEL1s #$S\_RACCAMCell$ and #$E\_RACCAMCell$

```
Scan transport matrix of 180 MeV RACCAM cell
'OBJET'
2029.47926                                             | Reference rigidity, 180 MeV proton.
5                                                      | Option for MATRIX computation.
.01 .001 .01 .001 0. 0.0001                            | 13-trajectory sampling for MATRIX computation.
284.894   300.516 0. 0. 0. 0.276852600  'o'            | Reference trajectory (number 1 in the 11-set).
1

'INCLUDE'                                              | Include the 30 degree sector dipole triplet,
1                                                      | within dedicated  LABEL1s.
./RACCAMCell.inc[#S_RACCAMCell:#E_RACCAMCell]

'FIT2'
2 noFinal
1 30 0  2.
1 31 0  2.
2  1e-6 79                                             | A penalty value, 1e-9 here, controls the accuracy
3.1 1 2 #End 0. 1. 0                                   | of the convergence of the constraints to the target values.
3.1 1 3 #End 0. 1. 0                                   | 3.1 is the constraint for periodicity.

'MATRIX'
1 11  PRINT                                            | PRINT causes log of transport coefficients to zgoubi.MATRIX.out.

'REBELOTE'                                             | Repeat the previous sequence,
20 0 0  1                                              | 20 times; prior to repeating, change
1                                                      | 1 parameter: parameter 35 under OBJET, namely,
OBJET 35 0.276852600:1.                                | the relative momentum, D=p/p_ref, in the range  0.276852600:1.

'SYSTEM'
1
gnuplot <./gnuplot_tunes.gnu                           | Call gnuplot script once that tracking is completed.
 'END'
```

gnuplot script for Fig. 10.34:

```
# gnuplot_tunes.gnu
system "grep 'Fit reached penalty value ' zgoubi.res | cat > grep.out"   # extract penalty values from zgoubi.res
plot "zgoubi.MATRIX.out" u (0+$47):(Ncell* $56) w  p  pt 6 ps .9 tit "{/Symbol}_R", \
"zgoubi.MATRIX.out" u (0+$47):(Ncell* $57) w  p  pt 7 ps .9 tit "{/Symbol}_Z", \
(x<1.03 ? sqrt(1+k) :1/0) w l lw 2 tit "\sqrt{(1+k)}" , "grep.out" u :5 axes x2y2  ; pause 1
```

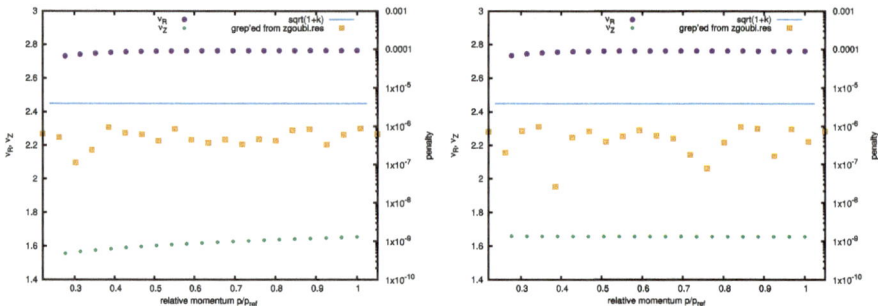

Fig. 10.34 Radial (ν_R) and axial (ν_Z) tunes of the 10-cell ring (left vertical scale), as a function of relative momentum ($p/p_{\text{ref}} = 1$ for 180 MeV). The penalty values (scattered squares, right vertical scale) monitor the FIT runs, they have to be small to confirm the convergence of FIT. Left: case of slowly increasing gap height (thus increasing fringe extent) with radius: $g(R) = g_0 \left(\dfrac{R_0}{R}\right)^{-0.52}$. Right: case of linear increase of gap height (thus linear increasing of fringe extent) with radius: $g(R) = g_0 \left(\dfrac{R}{R_0}\right)$, ν_Z is now constant

These data are then read and plotted (Fig. 10.34). The radial tune ν_R is constant (apart from a slight depression towards lower momenta where the dipole width reduces due to the spiral shape) as expected from the zero-chromaticity resulting from the scaling law (momentum-independent index, Eq. 10.3). Such is not the case for the axial tune, ν_Z.

(b) R-dependence of axial tune.

The variation of the axial tune with momentum, as observed in (a), is due to the fact that the fringe field extent does not increase fast enough with radius. Indeed the gap shape index in the present spiral sector model is $\kappa = -0.52$ (*cf.* simulation input data file in Table 10.15), resulting, in the FFAG-SPI modeling, in a gap height

$$g(R) = g_0 \left(\frac{R_0}{R}\right)^{-0.52}$$

whereas constant wedge angle focusing requires the fringe field extent to be proportional to R, i.e. $\kappa = -1$, yielding

$$g(R) = g_0 \left(\frac{R}{R_0}\right)$$

(c) Constant axial tune.

The axial tune is constant if the gap height is proportional to R [22, 26]. The simulation is obtained by changing the lines concerned in FFAG-SPI keyword input data list (Table 10.15), namely

change "8.95 − 0.52 ! EFB [etc.]"   to "8.95 − 1. ! EFB [etc.]"

at the 2 EFBs. The resulting tunes, for this $g(R) \propto R$ gap shape, are displayed in Fig. 10.34.

(d) Momentum compaction and transition γ_{tr}

TWISS keyword is used concurrently with OBJET[KOBJ=5], to compute various first and second order optical parameters including chromaticities. MATRIX [IORD=2, IFOC=11] concurrently with OBJET[KOBJ=6] can be used as well. A typical input file is given in Table 10.18. Results are given in Table 10.19.

The momentum compaction is expected to satisfy $\alpha = \frac{\Delta C}{C} / \frac{\Delta p}{p} = 1/(1 + k)$. In the present design $k = 5$, yielding $\alpha = 0.1666$. This approximation is acceptable at high energy where computed $\alpha = 0.1666$ (Table 10.19), fairly close to $1/(1 + k)$, however it is not at low energy where the numerical integration yields $\alpha = 0.60$. This discrepancy might result from the greater flutter at smaller orbit radius (shorter magnet body compared to fringe field extent)—further investigation is left to the reader.

Table 10.18 Simulation input data file: TWISS command, to obtain the periodic beam matrix, momentum compaction, chromaticities, etc. The initial reference coordinates, under OBJET, are for 15 MeV. TWISS proceeds in 3 stages: it first computes tunes of an on-momentum particle, then for $\pm\delta p/p$ off-momentum particles; at each stage, FIT ensures that the reference particle (1st particle of the 11-set) is on the closed orbit. The INCLUDE uses the segment #S_RACCAMCell to #E_RACCAMCell defined in Table 10.15

```
RACCAM cell, TWISS computation
'MARKER' RACCAM_TWISS_S
 'OBJET'
2029.47926                                      ! Reference rigidity, 180 MeV proton.
5                                               ! Option for MATRIX computation.
.01 .001 .01 .001 0. 0.0001                     ! 13-trajectory sampling for TWISS computation.
284.894    300.516 0. 0. 0. 0.276852600  'o'    ! Reference trajectory (number 1 in the 11-set).
1

'INCLUDE'                                       ! Include the 30 degree sector dipole triplet,
1                                               ! within dedicated LABEL1s.
./RACCAMCell.inc[#S_RACCAMCell:#E_RACCAMCell]

 'FIT2'                                          ! Find closed orbits, on- and off-momentum.
2  noFinal
1 30 0  2.
1 31 0  2.
2 1e-6 79                                       ! A penalty value, 1e-6 here, controls the accuracy
3.1 1 2 #End 0. 1. 0                             ! of the convergence of the constraints to the target values.
3.1 1 3 #End 0. 1. 0                             ! 3.1 is the constraint for periodicity.

'TWISS'
2 1. 1.
'MARKER' RACCAM_TWISS_E
'END'
```

Table 10.19 Outcomes of TWISS computation out of zgoubi.res execution listing, including beam matrix, tunes, momentum compaction factor, and near-zero chromaticities

Case of 15 *MeV optics*:

```
Reference, before change of frame (particle #  1  - D-1,Y,T,Z,s,time)  :
 -7.23147400E-01   2.84894063E+02   3.00515234E+02   0.00000000E+00   0.00000000E+00   1.74585982E+02   3.29572993E-02

Reference, after change of frame (particle #  1  - D-1,Y,T,Z,s,time)  :
 -7.23147400E-01   0.00000000E+00   0.00000000E+00   0.00000000E+00   0.00000000E+00   1.74585982E+02   3.29572993E-02

               TRANSFER  MATRIX  ORDRE  1  (MKSA units)
     -0.881744     0.852462      0.00000      0.00000      0.00000     0.685431
     -1.75188      0.559596      0.00000      0.00000      0.00000     0.358061
      0.00000      0.00000      -1.16322      3.85003      0.00000     0.00000
      0.00000      0.00000      -0.950162     2.28518      0.00000     0.00000
      0.575188     -7.833153E-02  0.00000      0.00000      1.00000     4.164093E-02
      0.00000      0.00000       0.00000      0.00000      0.00000     1.00000

     DetY-1 =    -0.0000068160,    DetZ-1 =    -0.0000160216

 Beam matrix  (beta/-alpha/-alpha/gamma) and  periodic  dispersion  (MKSA units)
      0.863743      0.730208     0.000000     0.000000     0.000000     0.261440
      0.730208      1.775068     0.000000     0.000000     0.000000    -0.226953
      0.000000      0.000000     4.650813     2.082825     0.000000    -0.000000
      0.000000      0.000000     2.082825     1.147791     0.000000     0.000000

                    Betatron  tunes  (Q1 Q2 modes)
            NU_Y =  0.27574797      NU_Z =  0.15521091

                  Momentum compaction :
                  dL/L / dp/p =  0.59900604

                  Chromaticities :
            dNu_y / dp/p = -1.70372346E-03        dNu_z / dp/p =  7.26717831E-03
```

Case of 180 *MeV optics*:

```
Reference, before change of frame (particle #  1  - D-1,Y,T,Z,s,time)  :
  0.00000000E+00   3.36987609E+02   9.05358647E+00   0.00000000E+00   0.00000000E+00   2.16232505E+02   1.32569144E-02

Reference, after change of frame (particle #  1  - D-1,Y,T,Z,s,time)  :
  0.00000000E+00   0.00000000E+00   0.00000000E+00   0.00000000E+00   0.00000000E+00   2.16232505E+02   1.32569144E-02

               TRANSFER  MATRIX  ORDRE  1  (MKSA units)
      0.551741      1.04666      0.00000      0.00000      0.00000     0.485854
     -1.42011      -0.881528     0.00000      0.00000      0.00000     0.359691
      0.00000      0.00000      -0.446750     2.01463      0.00000     0.00000
      0.00000      0.00000      -0.820807     1.46308      0.00000     0.00000
      0.879375      0.804775     0.00000      0.00000      1.00000     5.070098E-02
      0.00000      0.00000       0.00000      0.00000      0.00000     1.00000

     DetY-1 =    -0.0000041477,    DetZ-1 =    -0.0000094508

 Beam matrix  (beta/-alpha/-alpha/gamma) and  periodic  dispersion  (MKSA units)
      1.061187     -0.726582     0.000000     0.000000     0.000000     0.553967
     -0.726582      1.439822     0.000000     0.000000     0.000000    -0.226945
      0.000000      0.000000     2.339177     1.108747     0.000000    -0.000000
      0.000000      0.000000     1.108747     0.953036     0.000000     0.000000

                    Betatron  tunes  (Q1 Q2 modes)
            NU_Y =  0.27636408      NU_Z =  0.16516182

                  Momentum compaction :
                  dL/L / dp/p =  0.16657894

                  Chromaticities :
            dNu_y / dp/p =  4.34427556E-04        dNu_z / dp/p =  7.34661292E-03
```

10.9 Beam Envelopes, Optical Functions

Beam envelopes can be obtained by single particle raytracing over a few tens of passes through a cell. A proper data file for that is given in Table 10.20, it uses OBJET[KOBJ=2] to define the particle.

Multiturn raytracing results are given in Fig. 10.35, which also displays the square of the transverse particle excursion, normalized to the motion invariant, to yield betatron function amplitudes, namely, $\beta_Y = Y^2/\varepsilon_Y/\pi$ and $\beta_Z = Z^2/\varepsilon_Z/\pi$.

Table 10.20 Simulation input data file: raytrace two particles with different momenta through a cell, over 110 passes. Initial radial and axial coordinates are taken on $\varepsilon_{Y,Z}/\pi = 10^{-8}$ m invariants. The INCLUDE uses the segment [#S_RACCAMCell:#E_RACCAMCell] defined in Table 10.15

```
Track beam envelops
'OBJET'
2029.47926                                    ! Rigidity of 180 MeV proton.
2
2 1
2.84903389E+02 300.593197 2.1566E-02 9.658E-02 0. 0.2768526 'o'  ! 15 MeV, on 1e-8m Y nd Z invariants.
3.37019202E+02 8.84910511 1.5294E-02 7.249E-02 0. 1.      'i'    ! 180 mEv, on 1e-8m Y nd Z invariants.
1 1

'INCLUDE'                                      ! Include the 30 degree sector dipole triplet,
1                                              ! within dedicated LABEL1s.
./RACCAMCell.inc[#S_RACCAMCell:#E_RACCAMCell]

'REBELOTE'             ! Repeat sequence 109 times; '0.1': log pass-by-passes count to video;
109 0.1 99             ! '99': ignore OBJET at subsequent passes.
'END'
```

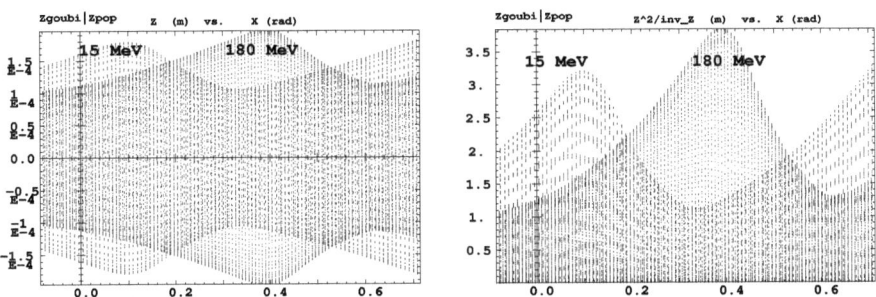

Fig. 10.35 Left: axial excursion $Z(s)$ of two particles, one 15 MeV and the other 180 MeV, over 110 passes across a 45.83° arc (the extent of the field region, AT, in FFAG-SPI, whereas a period is 36°). Both particles are taken on an axial invariant $\varepsilon_Z/\pi = 10^{-8}$ m. This multiple pass plot generates the beam envelopes: the extrema of particle excursion. Right: a graph of $Z(s)^2/\varepsilon_Z$, whose extreme value represents the betatron function amplitude $\beta_Z(s)$. Graphs obtained using zpop, left: menu 7; 1/1 to open zgoubi.plt; 2/[8,4] to select Z versus X (azimuthal angle); 7 to plot; right: menu 7: 3/14 to change the axial coordinate to Z^2/constant

An alternate technique to get the optical functions at all s across the cell is by transporting the beam matrix from the origin (s_0)

$$\sigma(s) = T(s \leftarrow s_0)\,\sigma(s_0)\,\tilde{T}(s \leftarrow s_0)$$

The transport matrix $T(s \leftarrow s_0)$ can be computed step by step from the particle coordinates stored in zgoubi.plt during the raytracing. A tool in zgoubi toolbox just does that, betaFromPlt [36], it requires using the 13-particle OBJET[KOBJ=5], and logging stepwise particle data in zgoubi.plt, using FFAG-SPI[IL=2] or equivalently OPTIONS[.plt 2]. This method is used in various other exercises, which can be referred to.

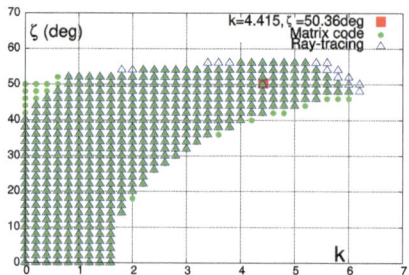 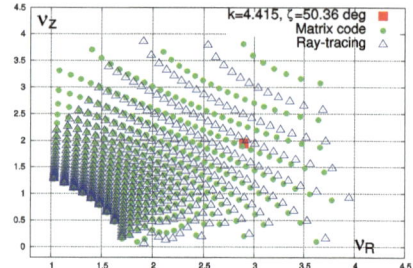

Fig. 10.36 A scan of k and ζ. Left: (k, ζ) stability domain, right: corresponding (ν_R, ν_Z) stability domain. In both diagrams a particular working point, $(k, \zeta) = (4.415, 50.36)$, is shown for illustration (different from the working point in these exercises, which is $(k, \zeta) = (5, 53.7)$) A matrix code was also used for this scan, results differ noticeably, they are displayed here for comparison

10.10 Periodic Stability Domain

The stability domain in the tune diagram, and the corresponding (k, ζ) domain, are obtained by a (k, ζ) scan, performed for some arbitrary momentum, for instance half-way between injection and top energy. A similar simulation input data file to that in Table 10.17 is used. The process is the following:

(1) a FIT and MATRIX sequence finds closed orbit and related tunes, for a given (k, ζ) value,
(2) REBELOTE then varies k and repeats the FIT & MATRIX sequence (1),
(3) from that scan results a series of (ν_R, ν_Z) couples, corresponding to a series of (k, ζ) couples at fixed spiral angle ζ.

This (1)–(3) sequence is repeated for a series of ζ values—an external program can be used to perform that iteration on ζ.

This results in (k, ζ) and (ν_R, ν_Z) stability diagrams displayed in Fig. 10.36. The correlation comes out to be, mostly, an increase of ν_R with k and increase of ν_Z with ζ, however the two quantities are not fully decoupled, increasing k (respectively, ζ) has a slight effect on ν_Z (resp. ν_R).

10.11 Motion Stability Limit

The input data file in Table 10.21 can be used:

– raytracing is performed for one particle at a time (namely, for a particular energy taken in [15 MeV, 180 MeV]),
– REBELOTE performs a multiturn raytracing.

Then push the initial coordinate (Y_0 for radial stability limit, Z_0 for axial), up to stability limit. An external program available in `zgoubi` toolbox, searchStabLim, can be used for that [37].

This will result in the horizontal and vertical phase space portraits displayed in Fig. 10.12.

Table 10.21 Simulation input data file: track one particle, for 1000 turns. Push Y_0 up to find the stability limit. Using the complete ring (10 cells) allows introducing non-systematic errors if desired (random field or positioning errors for instance, using ERRORS keyword). If only systematic errors or non-linearities are of interest, then a single cell is used

```
Track one particle
'OBJET'
2029.47926                                              | Rigidity of 180 MeV proton.
2
2 1
284.894   300.516 0. 0. 0. 0.276852600  'o'           | Reference trajectory (number 1 in the 11-set).
1 1
'INCLUDE'                                              ! Include the 30 degree sector dipole triplet,
1                                                          ! within dedicated LABEL1s.
10* ./RACCAMCell.inc[#S_RACCAMCell:#E_RACCAMCell]              ! 10 cells, a complete ring.
'REBELOTE'                    | Repeat sequence 999 times; '0.3': log count every other 100 pass to video;
999 0.3 99                                         | '99': ignore OBJET at subsequent passes.
'END'
```

10.12 Dynamic Aperture Scan

A DA scan is obtained by repeating the previous (Exercise 10.11) stability limit search:

– first, look for the maximal radial extent $[x_{min}, x_{max}]$ of stable horizontal motion, at quasi-zero axial invariant,
– second, look for the maximum stable axial amplitude, at various values $x \in [x_{min}, x_{max}]$.

An external program available in zgoubi toolbox does that, searchDA [38]. The exercise results in the DA graph of Fig. 10.37.

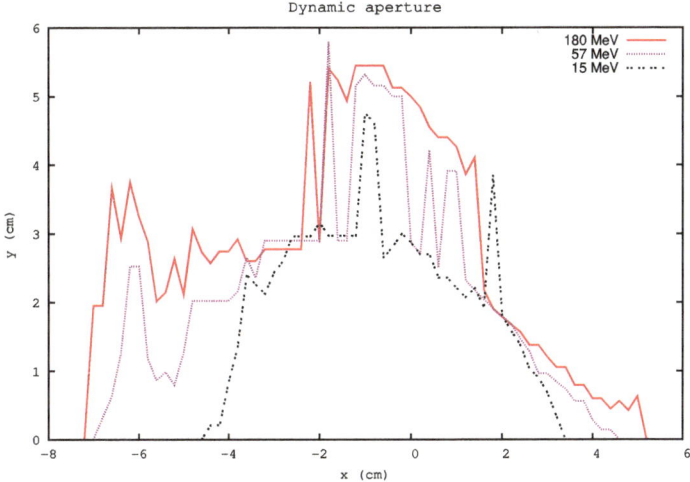

Fig. 10.37 Dynamic aperture in the (Y,Z) space, at 15, 57 and 180 MeV. The origin here, x = 0, is on the closed orbit at the momentum of concern

10.13 Acceleration, Transverse Betatron Damping

A similar problem is solved for the radial FFAG, it can be referred to (Exercise 10.5), in addition to the guidance below.

Setting up the acceleration requires the following:

- PARTICUL[PROTON] is necessary as CAVITE is used: it allows converting energy change in rigidity change (zgoubi pushes particles using rigidity),
- SCALING[IOPT=-2] provides the RF program to CAVITE (by reading it from an external file, zgoubi.freqLaw.In),
- CAVITE[IOPT=6] boosts the particle(s) at each pass, following that pre-defined RF program,
- REBELOTE do-loop sends the execution pointer back to the top of the input data file, for multiturn tracking. A 10 kV acceleration rate per turn may be obtained from constant peak voltage $\hat{V} = 20$ kV and synchronous phase $\phi_s = 30°$, this determines the number of passes for a $15 \rightarrow 180$ MeV cycle, namely, $(180 - 15)/0.01 = 16500$, hence a number of repeat REBELOTE[NPASS=16499],
- FAISTORE[FNAME=zgoubi.fai, IP=1] stores turn-by-turn particle data in zgoubi.fai, every turn.

For simplicity the RF program may be limited to the turn dependence of RF frequency (peak voltage and synchronous phase maintained constant). Elaborating zgoubi.freqLaw.In (essentially, here, determining revolution period versus turn number) requires the following steps (Exercise 10.5 may also be referred to, see zgoubi.freqLaw.In file formatting and sample content in Table 10.14):

(i) run a search for closed orbits for a few tens of energies in the range 15 to 180 MeV. Time of flight, derived from path length and particle velocity, is part of the outcome of orbit computation as PARTICUL provides necessary particle data in that aim. That yields the turn dependence of revolution period, namely, $f_{RF} = h \times f_{rev} = v/C$ (assuming RF harmonic h = 1). Note that in the absence of orbit defects or other tailored bumps, the time of flight may be obtained with good accuracy from the theoretical orbit length $C(p) = C_0 (p/p_0)^{\frac{1}{k+1}}$ (Eq. 10.8);

(ii) re-write these turn-time of flight data, properly formatted (*cf.* Table 10.14), in zgoubi.freqLaw.In. Note that zgoubi.freqLaw.In does not need to contain all turns, zgoubi will interpolate.

The resulting radial and axial phase spaces are displayed in Fig. 10.38. The betatron damping satisfies Eqs. 10.22, 10.23. The homothety-rotation of the geometrical scaling is apparent in the axial phase space portrait.

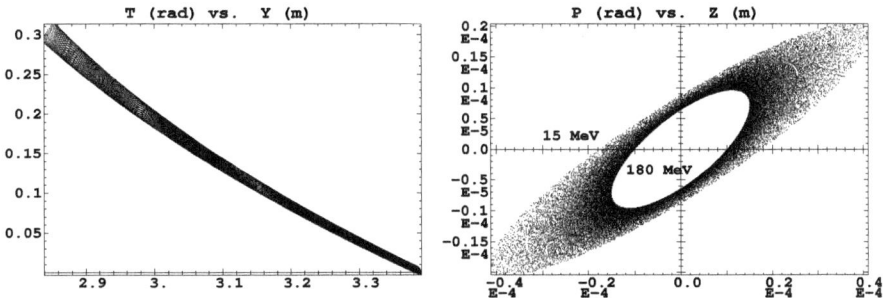

Fig. 10.38 Acceleration from 15 to 180 MeV. Radial and axial phase spaces. Graphs obtained using zpop: menu 7; 1/5 to open zgoubi.fai; 2/[2, 3] to select T (angle) versus Y (radius) [or 2/[4, 5], for P versus Z]; 7 to plot

10.4.3 FFAG Acceleration Methods

In the following three exercises, solutions are based on input data files worked out in Sect. 10.3.1 exercise series, with minor adaptations. Regarding beam acceleration, input data files developed as part of Exercises 10.5, 10.13 are used.

10.14 Hybrid Acceleration

Refer to [29, 30] where all necessary details regarding working hypotheses, and partial results, including numerical simulations, can be found.

Exercise 11.1-c may also be used as a guidance, it simulates serpentine acceleration in the linear FFAG ring EMMA.

10.15 Bucket Acceleration

Refer to [39] where all necessary details regarding working hypotheses, and partial results, including numerical simulations, can be found.

Regarding in-flight decay method and outcomes, refer to the exercises in Sects. 12.3.3, 12.3.4.

10.16 Serpentine Acceleration

Refer to [40, 41] where all necessary details regarding working hypotheses, and partial results, including numerical simulations, can be found.

References

1. F.T. Cole, O Camelot, a Memoir of the MURA Years. Cyclotron Conference, East Lansing, USA, May 13–17, 2001. https://accelconf.web.cern.ch/accelconf/c01/cyc2001/extra/Cole.pdf
2. A.A. Kolomensky, et al., Some questions of the theory of cyclic accelerators. Ed. AN SSR, 1955, p. 7, PTE, No 2, 26 (1956)

3. K.R. Symon et al., Fixed-field alternating-gradient particle accelerators. Phys. Rev. **103**, 1837 (1956)

4. T. Ohkawa, Two-beam fixed field alternating gradient accelerator. Rev. Sci. Instrum. **29**, 108 (1958); A concept presented at a meeting of the Physical Society of Japan in 1953 (1967)

5. K.R. Symon, MURA days, in *Proceedings of the 2003 Particle Accelerator Conference*. https:// accelconf.web.cern.ch/p03/PAPERS/WOPA003.PDF Figs. 10.1 and 10.9: Copyrights under license CC-BY-3.0, https://creativecommons.org/licenses/by/3.0; no change to the material

6. L. Jones, F. Mills, A. Sessler, K. Symon, D. Young, *Innovation Was Not Enough; A History of the Midwestern University Research Association (MURA)* (World Scientific, 2010)

7. Y. Mori, Developments of FFAG accelerator, in *17th International Conference on Cyclotrons and Their Applications 2004*, Tokyo (Japan), 18–22 October 2004. https://www.osti.gov/ etdeweb/biblio/20676358

8. M. Craddock, The rebirth of the FFAG. CERN Courier (27 July 2004). https://cerncourier. com/a/the-rebirth-of-the-ffag/

9. J. Collot, The rise of the FFAG. CERN Courier (19 August 2008). https://cerncourier.com/a/ the-rise-of-the-ffag/

10. S. Machida, Muon (FFAG) accelerators. THYAB01 talk; *PAC 2007 Accelerator Conference*, June 25–29, 2007, Albuquerque, NM, USA. https://accelconf.web.cern.ch/p07/TALKS/ THYAB01_TALK.PDF. Figures 10.2, 10.6: Copyrights under license CC-BY-3.0, https:// creativecommons.org/licenses/by/3.0; no change to the material

11. Y. Ishi, Status of KURRI facility, in *Proceedings of the FFAG 2016 Workshop*, Imperial College, London (2016). https://indico.cern.ch/event/543264/contributions/2295846/ attachments/1333675/2005286/FFAG16_LONDON_ishi.pdf

12. C.H. Pyeon et al., First injection of spallation neutrons generated by high-energy protons into the Kyoto University critical assembly. J. Nucl. Sci. Technol. **46**, 1091 (2009)

13. F. Méot, A multiple-room, continuous beam delivery, hadron-therapy installation. Phys. Procedia **66**, 361–369 (2015). https://www.sciencedirect.com/science/article/pii/ S1875389215001984

14. E. Yamakawa, et al., High intensity proton FFAG ring with serpentine acceleration for ADS, in *MOP209 Proceedings of HB2012*, Beijing, China. https://accelconf.web.cern.ch/HB2012/ papers/mop209.pdf

15. E. Yamakawa et al., Serpentine acceleration in zero-chromatic FFAG accelerators. Nucl. Instrum. Methods Phys. Res., Sect. A **716**(11), 46–53 (2013). (July)

16. T. Planche, et al., New approaches to Muon acceleration with zero-chromatic FFAGs, in *THPD093, Proceedings of IPAC'10*, Kyoto, Japan (2010). http://accelconf.web.cern.ch/ AccelConf/IPAC10/papers/thpd093.pdf

17. Y. Mori, et al., Multi-beam acceleration in FFAG synchrotron, in *Proceedings of the PAC 2001 Accelerator Conference*, pp. 588–590 (2001). http://accelconf.web.cern.ch/AccelConf/ p01/PAPERS/ROPA010.PDF

18. A. Sato, et al., FFAG as phase rotator for the PRISM project, in *Proceedings of the EPAC 2004 Accelerator Conference*, pp. 713–715 (2004). http://accelconf.web.cern.ch/AccelConf/ e04/PAPERS/MOPLT070.PDF

19. M. Haj Tahar, F. Méot, Tune compensation in nearly scaling fixed field alternating gradient accelerators. Phys. Rev. Accel. Beams **23**, 054003 (2020). https://journals.aps.org/prab/ abstract/10.1103/PhysRevAccelBeams.23.054003

20. K. Okabe, et al., Development of H-injection of proton-FFAG at Kurri, in *THPEB009 Proceedings of IPAC'10*, Kyoto, Japan. https://accelconf.web.cern.ch/IPAC10/papers/thpeb009. pdf Fig. 5.1: Copyrights under license CC-BY-3.0, https://creativecommons.org/licenses/by/ 3.0; the photo has been trimmed to mostly leave the 2.5 MeV injector

21. D. Neuvéglise, F. Méot, An alternative design for the RACCAM magnet with distributed conductor, in *FR5REP095, Proceedings of the PAC09 Conference*, pp. 5002–5004, Vancouver, BC, Canada (2009). https://accelconf.web.cern.ch/PAC2009/papers/fr5rep095.pdf

22. T. Planche et al., Design of a prototype gap shaping spiral dipole for a variable energy proton therapy FFAG. NIMA **604**, 435–442 (2009)

23. M. Aiba, et al., Development of 150-MeV FFAG at KEK, in *FFAG03 Workshop*, KEK, Japan, July 7–12, 2003. https://ffag.pp.rl.ac.uk/FFAG/FFAG03_HP/index.html

24. M. Aiba, F. Méot, Determination of KEK 150 MeV FFAG parameters from ray-tracing in TOSCA field maps, CERN-NUFACT-NOTE-140; CARE-Note-2004-030-BENE; CEA-DAPNIA-2004-188-2004. http://cds.cern.ch/record/806545/files/note-2004-030-BENE.pdf

25. M. Aiba, Tracking study for FFAG, in *FFAG Accelerator Workshop*; FFAG02, KEK, Tsukuba February 13–15, 2002. https://ffag.pp.rl.ac.uk/FFAG/FFAG02_HP/2002_02_13/20020213_M.Aiba.pdf

26. S. Antoine et al., Principle design of a proton therapy, rapid-cycling, variable energy spiral FFAG. NIM A **602**, 293–305 (2009)

27. J. Fourrier, F. Martinache, F. Méot, J. Pasternak, Spiral FFAG lattice design tools, application to 6-D tracking in a proton-therapy class lattice. NIM A **589**, 133–142 (2008)

28. F. Méot, RACCAM: a status including magnet prototyping and magnetic measurements, in *International conference on FFAGs*, Fermilab, 21–25 September 2009. https://indico.fnal.gov/event/2672/contributions/77834/attachments/48652/58457/FMeot1-090921.pdf

29. H. Tanaka, Feasibility study of hybrid accelerator and superconducting FFAG, in *FFAG04 Accelerator Workshop*, KEK, Tsukuba (October 13–16, 2004). http://130.246.92.181/FFAG/FFAG04_HP/index.html

30. H. Tanaka, et al., Hybrid accelerator using an FFAG injection scheme, in *Cyclotrons 2004 Conference*, Tokyo, Japan (October 18–22, 2004). http://accelconf.web.cern.ch/AccelConf/c04/data/CYC2004_papers/19C6.pdf

31. M. Haj Tahar, *High Power Ring Methods and Accelerator Driven Subcritical Reactor Application*. Ph.D. thesis dissertation, BNL and University Grenoble-Alpes (January 2017). https://www.bnl.gov/isd/documents/94721.pdf. https://www.osti.gov/biblio/1351800

32. E. Yamakawa, et al., Serpentine acceleration in scaling FFAG, in *Proceedings of FFAG12 workshop*, Osaka, 2012. https://indico.cern.ch/event/194713/contributions/1473080/attachments/282693/395230/FFAG12Slides.pdf Additional details in: https://repository.kulib.kyoto-u.ac.jp/dspace/bitstream/2433/174929/2/D_Yamakawa_Emi.pdf

33. F. Méot, Zgoubi Users' Guide. https://www.osti.gov/biblio/1062013-zgoubi-users-guide. An up-to-date version of the guide can be found at: https://sourceforge.net/p/zgoubi/code/HEAD/tree/trunk/guide/Zgoubi.pdf

34. F. Lemuet, *Collection and Muon Acceleration in the Neutrino Factory Project*. Ph.D. Thesis dissertation, CEA and CERN, Paris-Saclay University, April 2007. https://inis.iaea.org/collection/NCLCollectionStore/_Public/42/013/42013892.pdf

35. F. Méot, F. Lemuet, Developments in the ray-tracing code Zgoubi for 6-D multiturn tracking in FFAG rings. NIM A **547**, 638–651 (2005)

36. From Zgoubi toolbox, part of the sourceforge package: a Fortran tool to compute optical functions from a zgoubi.plt output file, and some related gnuplot scripts: https://sourceforge.net/p/zgoubi/code/HEAD/tree/trunk/toolbox/betaFromPlt/

37. From Zgoubi toolbox, part of the sourceforge package: a Fortran tool to perform a dynamic aperture scan, and some related gnuplot scripts: https://sourceforge.net/p/zgoubi/code/HEAD/tree/trunk/toolbox/searchStabLim/

38. From Zgoubi toolbox, part of the sourceforge package: a Fortran tool to perform a dynamic aperture scan, and some related gnuplot scripts: https://sourceforge.net/p/zgoubi/code/HEAD/tree/trunk/toolbox/searchDA/

39. T. Planche, et al., New approaches to muon acceleration with zero-chromatic FFAGs, in *THPD093, Proceedings of IPAC'10*, Kyoto, Japan (2010). http://accelconf.web.cern.ch/AccelConf/IPAC10/papers/thpd093.pdf

40. E. Yamakawa, Serpentine acceleration in zero-chromatic FFAG with long straight section, in *International Workshop on FFAG Accelerator (FFAG'10)*, Kyoto University Research Reactor Institute, Osaka, Japan (28–31 October 2010). http://130.246.92.181/FFAG/FFAG10_HP/slides/Sat/Sat14Yamakawa.pdf

41. E. Yamakawa et al., Serpentine acceleration in zero-chromatic FFAG accelerators. Nucl. Instrum. Methods Phys. Res., Sect. A **716**(11), 46–53 (2013). (July)

Chapter 11
FFAG, Linear

Abstract This chapter is an introduction to linear Fixed-Field Alternating Gradient (FFAG) cyclic accelerators. It begins with a brief reminder of the historical and technological context, and continues with the theoretical material needed for the simulation exercises. It relies on charged particle optics and acceleration concepts introduced in the previous synchrotron and scaling FFAG chapters and further addresses

– design aspects of linear FFAGs,
– diverse specificities of linear FFAG optics,
– beam dynamics in rings,
– serpentine acceleration.

Simulations do not require specific keywords, they use optical elements met in the previous chapters, including MULTIPOL DIPOLE[S], CAVITE, data input/output keywords such as FAISCEAU, FAISTORE, the SYSTEM keyword, etc. Beam dynamics simulations include

– computation of optical parameters,
– particle trajectories through a cell or a ring,
– closed-orbit finding, from multi-turn raytracing or by coordinate matching,
– deriving ancillary outcomes from rays, such as

  • transport matrices using MATRIX,
  • periodic optical functions and their transport using TWISS,

– finding dynamical aperture,
– serpentine fast acceleration, and more.

© The Author(s) 2024   445
F. Méot, *Understanding the Physics of Particle Accelerators*, Particle Acceleration
and Detection, https://doi.org/10.1007/978-3-031-59979-8_11

Notations Used in the Text

a	gyromagnetic anomaly of electron, $a = 1.15965 \times 10^{-3}$
$\mathbf{B}$; $B_{x,y,s}$; B_0	field vector; components in moving frame; reference R_0
$B\rho$; $B\rho_0$	particle rigidity: $B\rho = p/q$; for reference momentum p_0
C; C_0	orbit length: $C = \oint ds = 2\pi R$; for reference momentum p_0
E	particle energy
f_{rev}, f_{rf}	revolution and accelerating voltage frequencies
$\mathcal{H}$	longitudinal Hamiltonian
h	harmonic number, an integer, $h = f_{\text{rf}}/f_{\text{rev}}$
m; m_0; M	particle mass; rest mass; mass in eV/c^2 units
N	number of cells in a ring
p; p_0; δp; Δp	particle momentum; reference momentum; offset
q	particle charge
R	average orbit radius: $R = C/2\pi$
RF	Radio-Frequency
s	path variable
v	particle velocity
V; $\hat{V}$	accelerating voltage; peak value
x', y'	particle coordinates in the moving frame $\left[(*)' = d(*)/ds \right]$
α	momentum compaction, or trajectory deviation
$\beta = v/c$; β_0; β_s	normalized velocity; reference; synchronous
β_u, α_u, γ_u, D_x	optical functions ($u : x, y, l$), dispersion
γ	Lorentz relativistic factor: $\gamma = E/m_0 c^2 = E[eV]/M$
δ; $\delta p/p$	momentum offset; relative
ε_u	Courant-Snyder invariant; beam emittance ($u : x, y, l$)
η; η_i	phase-slip factor; coefficients of polynomial in δ ($i = 0, 1, 2, ...$)
θ	azimuthal angle
ν_u	wave numbers, horizontal, vertical, synchrotron ($u : x, y, l$)
ω_{rf}	angular RF, $\omega_{\text{rf}} = 2\pi f_{\text{rf}}$
ρ	curvature radius
ϕ	RF phase

11.1   Introduction

The concept of linear FFAG is one of the latest innovations in cyclic accelerators. It was devised in the late 1990s, in the context of lattice R/D and rapid acceleration regarding short-lived muon beams for Higgs and other Neutrino Factory.

A linear FFAG ring for rapid acceleration leans on

- the use of linear magnets: quadrupoles and/or combined function dipoles [1, 2] (Fig. 11.1)—by contrast with scaling FFAG optics which uses highly non-linear AG dipoles, Chap. 10. As in the latter, the magnetic field is fixed during acceleration;
- quasi-isochronous acceleration using high gradient and fixed RF. A technique known as serpentine acceleration, or gutter acceleration [3], allowing cyclotron-style CW operation.

Light-source style chromatic invariant minimization allows minimizing the dispersion function and thus dipole size [4].

As a consequence of the use of fixed field linear magnets, the betatron tunes decrease during acceleration. Furthermore, a large cell tune variation may be sought, within a half integer range, in order to allow largest momentum sweep over the acceleration range. A consequence of tune variation during acceleration is that many horizontal and vertical integer tune values have to be crossed, which essentially limits the use of this optics to rapid acceleration systems.

Two outcomes of the linear AG method are, a large transverse dynamical acceptance intrinsic to a linear lattice, and, as a result of the small dispersion function, a small transverse extent of magnets in comparison to scaling FFAG technology.

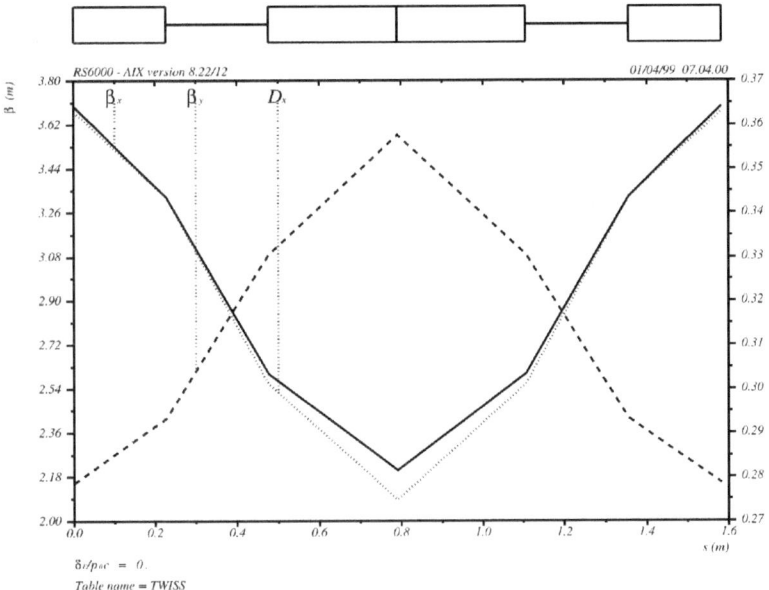

Fig. 11.1 An early $\frac{1}{2}F D\frac{1}{2}F$ fixed-field AG FODO cell design for a 4–16 GeV rapid muon accelerator, for injection into a $\mu^+ - \mu^-$ Higgs Factory collider [2]. Case of 4 GeV optics here—optical functions change with energy as the field is fixed

Fig. 11.2 A schematic of a neutrino factory [5]. Proton bunches in the GeV range hit a pion production target at a ~50 Hz repetition rate. Pions decay into muons in a few tens of meters. They are distributed in a huge 6D phase space (see Beam Lines Chap., Figs. 12.17 and 12.18). Bunching and cooling follow, prior to 200 MeV to 5 GeV acceleration in linacs, and up to 10–30 GeV in FFAG1 and FFAG2. The latter are filled with high gradient 200 MHz SCRF cavities

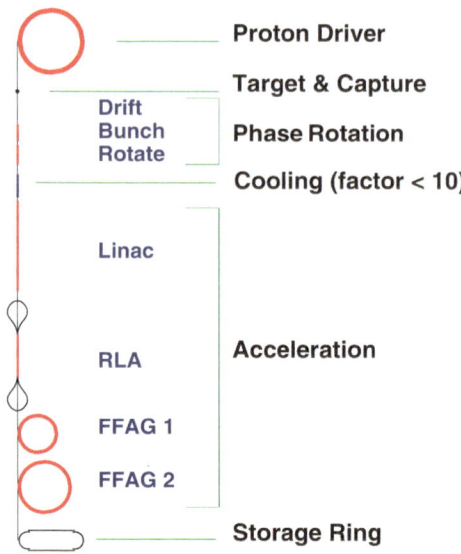

A design of a Neutrino Factory based on the linear FFAG technology for the rapid acceleration of muon beams was completed in the early 2000s (Fig. 11.2) [5]. An experimental electron model of a linear FFAG ring was proposed in that context [7], it was built and commissioned at Daresbury Laboratory in the 2007–2012 period [8, 9], Fig. 11.3.

Further outcomes of the linear FFAG method include the design of the circular return arcs of a multiple-pass energy recovery linac, eRHIC, to be located in RHIC heavy ion collider tunnel, providing 21 GeV electron bunches for an electron-ion linac-ring collider at the Brookhaven National Laboratory [10–12] (Fig. 11.4). A prototyping of eRHIC FFAG ERL return loop was proposed and led to the CBETA

Fig. 11.3 Left: EMMA FD cell, and 1.3 GHz RF cavity insert at the center. Right: 42-cell, 16.57 m circumference EMMA ring [6]. Nineteen 1.3 GHz RF cavities allow 2 MV of acceleration voltage. They are distributed every other cell (left picture), except for two intervals dedicated to injection and extraction systems. EMMA was designed for 10–20 MeV electron acceleration in a few turns

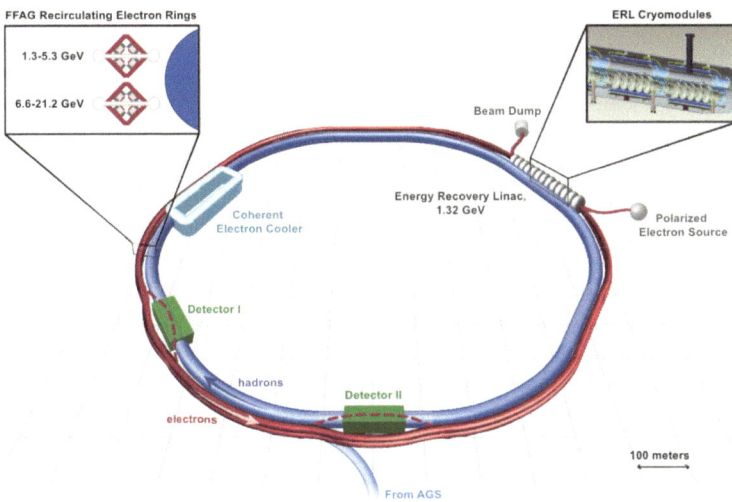

Fig. 11.4 eRHIC 21.2 GeV ERL project of a linac-ring collider, with its two linear FFAG recirculation loops, respectively 1.3–5.3 GeV (FFAG1) and 6.6–21.2 GeV (FFAG2), alongside RHIC ion rings. The 1.322 GeV SCRF linac is located in RHIC interaction region 2, it is connected to the FFAG loops by a merger (respectively spreader) section at its upstream (resp. downstream) end

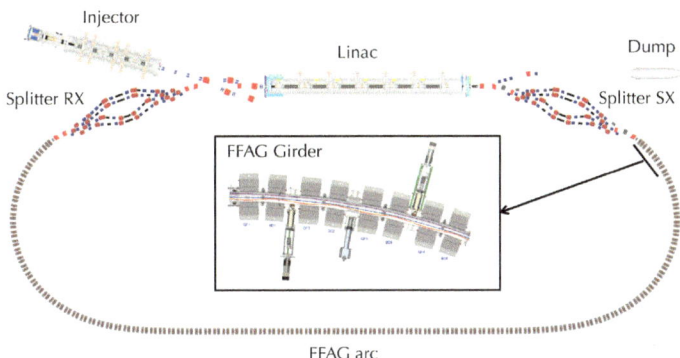

Fig. 11.5 CBETA ERL [14]. The permanent magnet 8-pass linear FFAG loop recirculates the 42, 78, 114 and 150 MeV beams in a single 2-in diameter vacuum pipe, up and down for energy recovery

ERL experiment [13, 14] (Fig. 11.5). CBETA, comprised of a proof-of-principle 8-pass (4 up, 4 down) permanent magnet return loop [15], was built and commissioned in the 2015–2020 period, it was aimed at demonstrating a factor of 4 in energy, and full energy recovery.

Studies in the same line are carried on today, regarding application of the linear FFAG return arc method to a 24 GeV energy upgrade at CEBAF 12 GeV RLA [16].

11.2 Basic Concepts and Formulæ

A linear FFAG cell is comprised of quadrupoles and/or combined function dipoles, with alternating field gradients in the range of T/m [1, 2]. This makes it akin to alternating gradient (AG) strong focusing optical systems. The theoretical material of transverse periodic stability in beam lines and rings is that of AG focusing (Chap. 9).

Acceleration in rings is based on the quasi-isochronous serpentine method [3], akin to fixed-frequency, high gradient acceleration in cyclotrons (Chaps. 3 and 4), yet using higher RF.

In the following sections some specificities of beam optics and acceleration in linear FFAGs are addressed.

11.2.1 Linear FFAG Cell

A linear FFAG cell is generally comprised of a doublet (FODO, FD) or triplet (FDF, DFD) of quadrupoles with some radial offset, and/or combined function dipoles. Drift spaces allow room for instrumentation, RF systems, etc. FFAG cells are designed to feature a large momentum acceptance, p_{max}/p_{min} up to a factor 4 or more. Any periodic orbit within the momentum acceptance of the cell experiences a curvature $2\pi/N$, so that N cells make up a ring. Depending on momentum, the orbit curvature may have the same sign, or opposite signs in the cell magnets. Periodic orbits with increasing momentum move almost everywhere from an inner to an outer radial excursion. These basic aspects are illustrated in Fig. 11.6 with the case of the FD cell of the Neutrino Factory 5→ 10 GeV FFAG ring [18]. More details regarding the design of linear FFAG cell magnets, including raytracing in OPERA field maps as part of optical parameter optimization, can be found in [17].

Two main constraints in designing a cell are, the tune excursion, and quasi-isochronism of the orbits:

– the natural chromaticity causes orbits of increasing momentum to feature decreasing paraxial tune values (Eqs. 9.18 and 9.19); the rule here is to minimize tune excursions within a [0, 0.5] interval (Fig. 11.7);
– in the quasi-isochronous hypothesis the momentum dependence of the time of flight is tailored to be an about symmetric parabola, centered on the center of the momentum range (Fig. 11.8), with minimized time excursion so to minimize RF phase excursion.

Additional design constraints include orbit excursion, field value and gradients, magnet apertures; fringe fields may require special attention as, often in these structures, the length/aperture ratio of the magnets may be on the small side. Theoretical material regarding linear FFAG cell design methods can be found in the Neutrino Factory technical reports [19] and FFAG workshops [20].

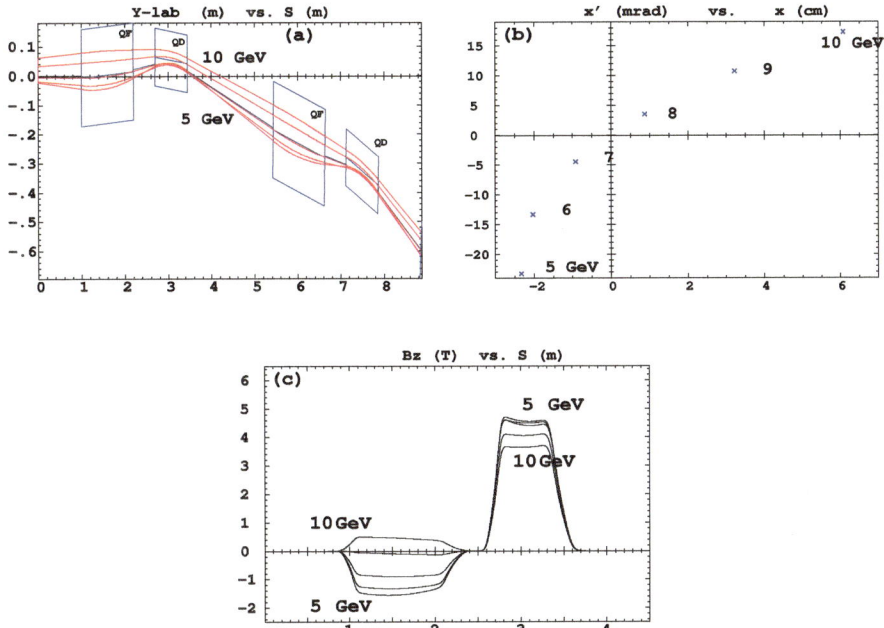

Fig. 11.6 Optical properties of the Neutrino Factory 10 GeV FFAG ring [18]. **a** periodic orbits across a pair of cells; **b** their coordinates in phase space at $s = 0$; **c** magnetic field along the periodic orbits across a cell

Fig. 11.7 Cell tunes (ν_x, ν_y), decreasing with increasing beam energy

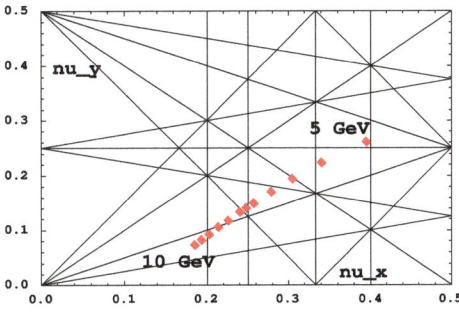

Fig. 11.8 Quadratic dependence of time of flight on energy, in the neutrino factory $5 \to 10$ GeV muon ring

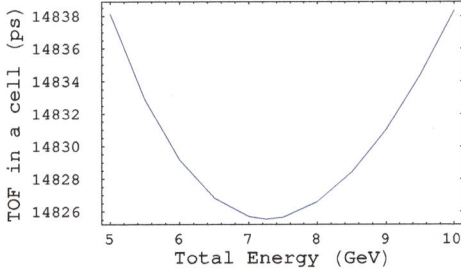

11.2.2 Quasi-Isochronous Serpentine Acceleration

In cyclotrons, isochronism has to be achieved to a level of typically 0.01%, meaning a revolution time shifting by 0.01% over injection to extraction energy range. Quasi-isochronous fixed RF acceleration in an FFAG is based on lattice isochronism at typically 0.1% level in large rings (Fig. 11.8) or higher in smaller rings (as EMMA, see Exercise 11.1).

In the quasi-isochronism hypothesis the expression of the phase slip factor (Eq. 8.33) has to include higher order terms in the momentum deviation [21], namely,

$$\eta = \eta_0 + \eta_1 \delta + \eta_2 \delta^2 + \cdots \tag{11.1}$$

Under these conditions, Eqs. 9.27 take the form

$$\frac{d\phi}{dt} = \frac{2\pi h}{T_s}(\eta_0 \delta + \eta_1 \delta^2 + \eta_2 \delta^3 + \cdots), \qquad \frac{d\delta}{dt} = \frac{e\hat{V}}{T_s \beta_s^2 E_s}[\sin\phi - \sin\phi_s] \tag{11.2}$$

From this the Hamiltonian to the second order in δ can be inferred, namely,

$$\mathcal{H}(\phi, \delta, t) = 2\frac{\delta^2}{\Delta^2} + \frac{4}{3}\frac{\eta_1}{\eta_0}\frac{\delta^3}{\Delta^2} + [\cos\phi - \cos\phi_s + \phi\sin\phi_s] \tag{11.3}$$

where

$$\Delta = \left(\frac{2e\hat{V}}{\pi h |\eta| \beta_s^2 E_s}\right)^{1/2} \tag{11.4}$$

Due to the quadratic dependence of the time of flight (Fig. 11.8), two new longitudinal phase space fixed points appear, namely (Fig. 11.9),

$$\phi_{\text{fps}_1} = \pi - \phi_s, \quad \delta_{\text{fps}_1} = 0 \qquad \text{stable fixed points}$$

$$\phi_{\text{fps}_2} = \phi_s, \quad \delta_{\text{fps}_2} = -\frac{\eta_0}{\eta_1}$$

$$\phi_{\text{fpi}_1} = \pi - \phi_s, \quad \delta_{\text{fpi}_2} = -\frac{\eta_0}{\eta_1} \qquad \text{unstable fixed points}$$

$$\phi_{\text{fpi}_2} = \phi_s, \quad \delta_{\text{fpi}_1} = 0$$

This results in two possibilities to accelerate a bunch, namely,

(i) carrying out a half synchrotron oscillation inside a bucket (in the manner of Fig. 10.14, Sect. 10.2.3);

Fig. 11.9 Contour of the
longitudinal gutter motion,
$\mathcal{H}(\phi, \delta) = $ constant [18],
and the stable (pfS$_{1,2}$) and
unstable (pfI$_{1,2}$) fixed points.
Case here of $\phi_s = 0$, $\Delta = 1$,
and $\eta_1/\eta_0 = 1$

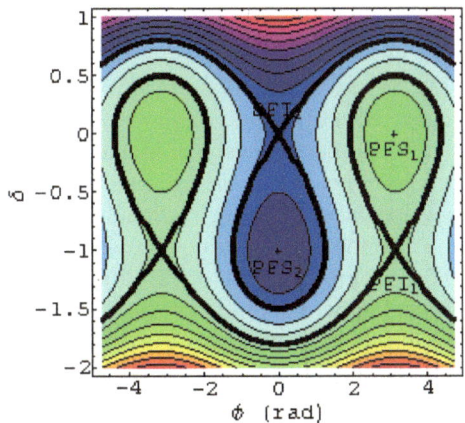

(ii) following the channel between buckets, a method called gutter acceleration, or serpentine acceleration.

The channel width defined by the separatrices, and the energy reach, depend on Δ (Eq. 11.4). Figure 11.10 shows longitudinal phase space portraits for various Δ values.

Experiments conducted in the EMMA ring (Fig. 11.3) have demonstrated serpentine acceleration in a linear FFAG, and absence of emittance growth upon fast crossing of integer resonances (nine are crossed on the way from 10 to 20 MeV/c) at a high acceleration rate (>1 MV/turn). Difficulties lie in the large chromaticity, which causes transverse decoherence and thus emittance growth [8], as well as significant time-of-flight dependence on transverse amplitude which may result in longitudinal emittance growth [22].

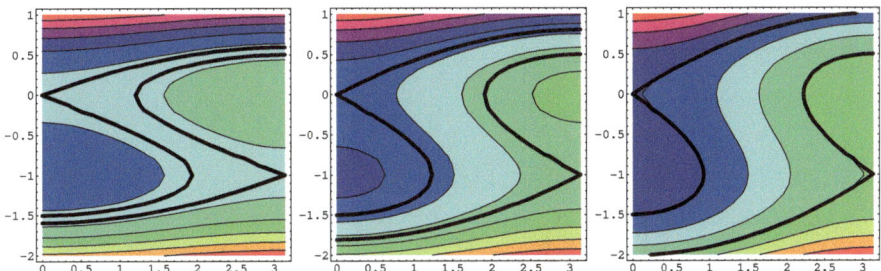

Fig. 11.10 Contour of the longitudinal gutter motion ($\phi_s = 0$, $\eta_1/\eta_0 = 1$) for, from left to right, $\Delta = 0.7$, 1 and 1.3

11.2.3 Synchrotron Radiation

The SR energy loss along an arc has been introduced in Chap. 5, Betatron, Sect. 5.2.3, as the effect was first observed in that cyclic accelerator. This theoretical material applies to beam lines, that includes FFAG lines. Beam emittance perturbations are proportional to the energy spread (Eq. 5.13) $\sigma_E/E \propto \gamma^{5/2}/\rho$, which may result in substantial emittance growth at high energy.

Assume a bunch traveling along a linear FFAG lattice, for instance an eRHIC ERL return loop [10], or FFA@CEBAF RLA recirculation arc [16]. Radiation will occur in the dipole field regions along the trajectory. For simplicity a DF doublet is considered below, this can easily be extended to a triplet. In the present formal approach, these magnets can be, indifferently, either offset quadrupoles, or combined function dipoles.

Over an arc $\Delta\theta$ with $1/\rho$ the curvature, assumed constant, the energy loss (Eq. 5.12) and energy spread (Eq. 5.13) can be written, respectively [11]

$$\frac{\overline{\Delta E}}{E_{\text{ref}}} = 1.9 \times 10^{-15} \frac{\gamma^3 \Delta\theta}{\rho}, \qquad \frac{\sigma_E}{E_{\text{ref}}} = 3.8 \times 10^{-14} \frac{\gamma^{\frac{5}{2}}\sqrt{\Delta\theta}}{\rho} \qquad (11.5)$$

Take for average radius, in the focusing magnet (QF) and in the defocusing magnet (BD) respectively,

$$\rho_{\text{BD}} \approx \frac{s_{\text{BD}}}{\Delta\theta_{\text{BD}}}, \qquad \rho_{\text{QF}} \approx \frac{s_{\text{QF}}}{\Delta\theta_{\text{QF}}}$$

with s_{BD} and s_{QF} the arc lengths. Consider in addition, with l_{BD}, l_{QF} the magnet lengths,

$$s_{\text{BD}} \approx l_{\text{BD}}, \qquad s_{\text{QF}} \approx l_{\text{QF}}$$

This yields, over a cell,

$$\overline{\Delta E}[MeV] \approx 0.96 \times 10^{-15}\gamma^4 \left(\frac{l_{\text{BD}}}{\rho_{\text{BD}}^2} + \frac{l_{\text{QF}}}{\rho_{\text{QF}}^2}\right) \qquad (11.6)$$

Taking in addition $< (1/\rho)^2 > \approx 1/ < \rho^2 >$, an estimate of the energy spread is

$$\sigma_E \approx 1.94 \times 10^{-14}\gamma^{7/2}\sqrt{\frac{l_{\text{BD}}}{|\rho_{\text{BD}}^3|} + \frac{l_{\text{QF}}}{|\rho_{\text{QF}}^3|}} \qquad (11.7)$$

The bunch lengthening over a $[s, s_f]$ distance, resulting from the stochastic energy loss, can be written [23]

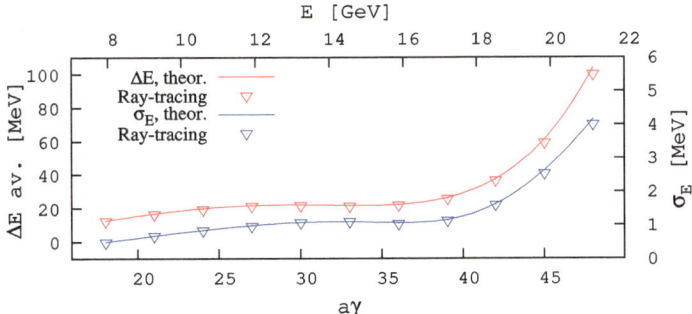

Fig. 11.11 Average energy loss (left axis) and energy spread (right axis). Solid lines: theory, respectively Eqs. 11.6 and 11.7. Markers: E : $6.322 \rightarrow 21.164$, step 1.322 GeV recirculations, from tracking with Monte Carlo SR. Lower horizontal scale: $a = 1.16 \times 10^{-3}$ is the electron gyromagnetic anomaly

$$\sigma_l = \left(\frac{\sigma_E}{E} \right) \left[\frac{1}{L_{\text{bend}}} \int_s^{s_f} \left(D_x(s) T_{51}(s_f \leftarrow s) + D'_x(s) T_{52}(s_f \leftarrow s) - T_{56} \right)^2 ds \right]^{1/2}$$

with the integral taken over the bends, D_x and D'_x the dispersion function and its derivative, T_{5i} the trajectory lengthening coefficient of the first order mapping ($i = 1, 2, 5, 6$ stand for respectively $x, x', \delta l, \delta p/p$ coordinates).

As an illustration, Fig. 11.11 shows the case of the 3.8 km long eRHIC ERL (energy recovery linac) return loop, comprised of 6 arcs, 120 cells per arc, based on the cell studied in Exercise 11.2. Eleven beams are circulated in the ring, with energies E : $6.322 \rightarrow 21.164$, step 1.322 GeV. The energy dependence of energy loss shows a local minimum in the $a\gamma = 30 - 35$ region, a different behavior from the classical γ^4 dependence in an isomagnetic lattice (Eq. 5.14), due to the large variation of the curvature radius over the $7.9 \rightarrow 21.2$ GeV energy range (Fig. 11.11).

11.2.4 Polarization

Spin dynamics in magnetic fields has been introduced in Sect. 3.2.5. Over long beam lines and in particular conditions, spin motion may be subject to resonance with the betatron motion [24] (resonant spin motion is introduced in Sects. 4.2.5, 8.2.4 and 9.2.7). However this should generally not be the case in a beam line and is not considered here.

Spins of electrons traveling in a vertical guiding field precess by an angle $a\gamma\alpha$ ($a = 1.16 \times 10^{-3}$ is the electron gyromagnetic anomaly) as the velocity vector precesses by an angle α (Fig. 3.16). The resulting precession rate in a ring is $a\gamma$ per turn (Eq. 3.33).

Fig. 11.12 Spin diffusion out of the bend plane, at each pass (not cumulated). Case of a 5,000 electron bunch tracking (empty circles), and from Eq. 11.8 (dashed curve)

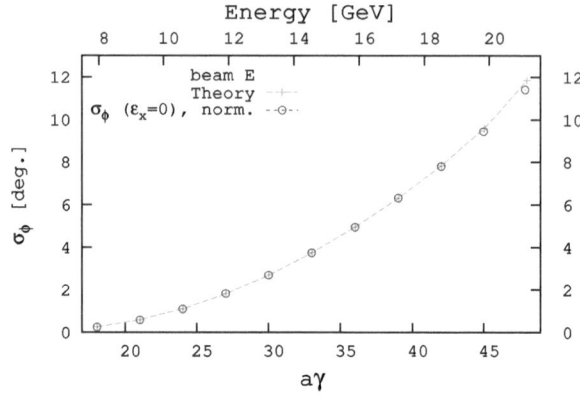

Considering the aforementioned eRHIC linear FFAG [10], where polarization orientation is in the bend plane, momentum spread in a bunch is one cause of depolarization. Under the effect of radial and longitudinal field components, vertical motion (orbital or betatron) causes spins to leave the median plane, also causing depolarization. The stochastic emission of photons induces spin diffusion which also contributes to depolarizing the beam. The effect is introduced in Sect. 9.2.7. As far as numerical integration is concerned, spin diffusion is a sub-product of the step-wise integration of spin motion, accounting for Monte Carlo simulation of photon emission. The theoretical evolution of the spin diffusion in eRHIC satisfies [25]

$$
\begin{pmatrix} \overline{\Delta E^2} \\ \overline{\Delta E \Delta \phi} \\ \overline{\Delta \phi^2} \end{pmatrix} = \begin{pmatrix} 1 & 0 & 0 \\ \alpha s & 1 & 0 \\ \alpha^2 s^2 & 2\alpha s & 1 \end{pmatrix} \begin{pmatrix} \overline{\Delta E^2} \\ \overline{\Delta E \Delta \phi} \\ \overline{\Delta \phi^2} \end{pmatrix}_{s=0} + \omega \times \begin{pmatrix} s \\ \alpha s^2/2 \\ \alpha^2 s^3/3 \end{pmatrix} \quad (11.8)
$$

where s is the distance in the field, $\omega = C\bar{\lambda}_c r_e \gamma^5 E^2/\rho^3 \approx 1.44 \times 10^{-27} \gamma^5 E^2/\rho^3$, $\alpha = a/\rho E_0 \approx 2.27/\rho$ (with $\bar{\lambda}_c = \hbar/m_e c$ the electron Compton wavelength, $C = 110\sqrt{3}/144$, $E_0 = m_e c^2/e$ the electron rest mass).

Assuming a starting state $\begin{pmatrix} \overline{\Delta E^2} \\ \overline{\Delta E \Delta \phi} \\ \overline{\Delta \phi^2} \end{pmatrix}_{s=0} = 0$ (this is the case for each energy

for instance in Fig. 11.12) yields $\sigma_E = \overline{\Delta E^2}^{1/2} = \sqrt{\omega s}$ (which in passing identifies with the familiar $\sigma_E/E = 3.8 \times 10^{-14} \frac{\gamma^{5/2}}{\rho^{3/2}} \sqrt{s}$), so that

$$
\sigma_\phi = \overline{\Delta \phi^2}^{1/2} = \sqrt{\frac{\omega \alpha^2 s^3}{3}} = \frac{\alpha s}{\sqrt{3}} \sigma_E \quad (11.9)
$$

or, given $s = 2\pi\rho$, $\quad \dfrac{\sigma_\phi}{\sigma_E} = 8.23$ [rad/GeV/turn]

As an illustration, Fig. 11.12 displays the evolution of the *rms* value σ_ϕ of the angle of the spin with respect to the bend plane, in the aforementioned 3.8 km long eRHIC return loop. σ_ϕ is calculated at each pass, for the 12 different energies $E : 6.322 \to$ 21.164, step 1.322 GeV. This is an outcome of Exercise 11.2.

11.3 Exercises

The first exercise deals with EMMA lattice and ring. It is concluded with a 6D bunch acceleration simulation.

The second exercise deals with the DF cell of the second, high energy, eRHIC 21 GeV ERL recirculating loop (FFAG2). Simulations include synchrotron radiation and spin diffusion.

11.1 EMMA Ring

Solution 11.1
In this exercise EMMA cell and 42-cell ring (Fig. 11.3) input data files are built (after [26, 27]), their optical parameters are produced. Accelerating gaps are installed and a rapid acceleration cycle is simulated.

Figure 11.13 displays a synoptic of EMMA ring, an outcome of the present exercise.

(a) Construct EMMA cell. Parameters are given in Table 11.1.
Consider three different simulations of the quadrupoles:

– use MULTIPOL and hard-edge model;
– use MULTIPOL accounting for fringe fields, take MULTIPOLE[XE = XS = 5 cm, $\lambda_E = \lambda_S = 2$ cm] ($X_{E,S} \gg \lambda_{E,S}$ to ensure that the magnetic field at the extrem-

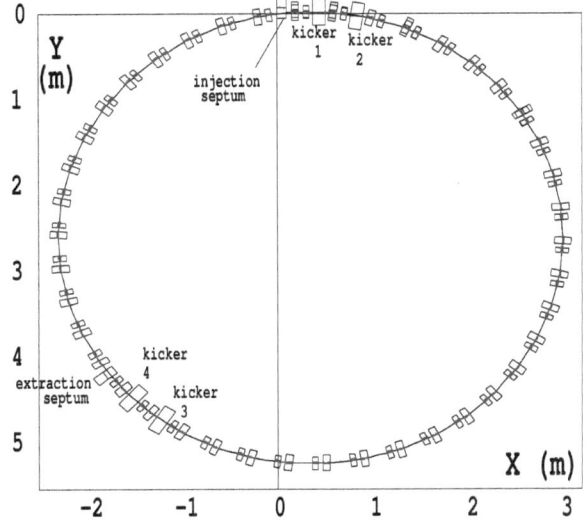

Fig. 11.13 EMMA ring and its injection and extraction kicker-septum systems, and the 15 MeV orbit (thick line in the quadrupoles, blue). A graph obtained using zpop, menu 7: 1/1 to open zgoubi.plt; 2/[42, 48] for X, Y laboratory coordinates; 7 to plot

Table 11.1 Parameters of EMMA ring, a 42 edge polygon. The cell is straight, the 360/42 deg polygon corners are at the interface between long drift and QD quadrupole

Energy range	MeV	10–20
Number of turns		<16
Circumference	m	16.568
Lattice		D/F doublet
No of cells		42
RF frequency and range	GHz; MHz	1.3; 5.6
No of RF cavities		19
RF voltage	kV/cavity	20–120
EMMA cell:		
– length	cm	39.448
– drifts, short/long	cm	5/21
– length, QD/QF	cm	7.57/5.878
– gradient, QD/QF	T/m	4.704/6.695
– QD/QF offsets	cm	3.4048/0.7514

ity of the fall-offs (where it is truncated) is negligible compared to the field in quadrupole body);
– use DIPOLES[$RM \approx 10^5$], with both quadrupoles comprised within DIPOLES[AT] aperture so to allow linear superimposition (addition) of fringe fields in the overlapping region.
For these three different models, repeat the following: plot the evolution with energy $E : 10 \rightarrow 20$ MeV, step 1 MeV, of
– closed orbits in laboratory frame; field along the latter;
– betatron functions and dispersion along the cell;
– tunes;
– time-of-flight.
Use FIT embedded in REBELOTE[IOPT = 1] do-loop for the closed orbit energy scan. Add MATRIX[IORD = 1, IFOC = 11] within the loop (just before REBELOTE) to get tunes and periodic beam matrix at each loop.

(b) Give phase space diagrams of the maximum stable amplitudes, horizontal and vertical.

(c) Serpentine acceleration.

Construct EMMA ring—at this stage Fig. 11.13 can be produced. Add one accelerating cavity per cell, using CAVITE[IOPT = 7] (note that EMMA only had 19 cavities, as two series of three cells were taken for the injection system: two kickers and a septum, and the extraction system: a septum and two kickers [26, 27]). From Sect. 11.2.2 figure a proper RF voltage for acceleration from 10 to 20 MeV in $\sim 15-20$ turns.

Table 11.2 Parameters of a DF quadrupole doublet cell of eRHIC ERL FFAG2 return loop [12, Appendix A]. Both BD and QF are offset quadrupoles, treated here as combined function dipoles

Top recirculation energy	GeV	21.164
Energy of first recirculation	GeV	6.622
Cell length	m	3.3624
Number of cells per sextant		120
Cell structure, in that order:		
Upstream drift length	m	0.09652479
BD combined function dipole:		
– length	m	1.129301
– dipole field	T	0.00293364
– gradient	T/m	−0.5225857
Middle drift length	m	0.19304957
QF quadrupole:		
– length	m	1.847002
– dipole field	T	0.00293364
– gradient	T/m	0.3728876
Downstream drift length	m	0.09652479

Simulate an acceleration cycle of a hollow bunch (i.e., all electrons on a common invariant; OBJET[KOBJ = 8] can be used for that) to 20 MeV, followed by deceleration back to 10 MeV. Produce the turn-by-turn phase space trajectory. Storage of turn-by-turn data can use FAISTORE.

11.2 eRHIC ERL FFAG2 loop. Synchrotron Radiation. Polarization
Solution 11.2

The eRHIC project of a linac-ring electron-ion collider (Fig. 11.4) is based on the existing RHIC, and on an electron ERL comprised of two linear FFAG recirculation loops (FFAG1 and FFAG2) alongside RHIC ring. The ERL takes an electron bunch through a 1.322 GeV linac, up to 21.164 GeV and back down to 1.3 GeV for energy recovery [11, 12].

In this simulation exercise the 6.322–21.164 GeV FFAG2 recirculating loop is considered. Optical properties of the quadrupole doublet cell are produced, as well as effects of synchrotron radiation on electron dynamics and on polarization. More simulations and their outcomes can be found in [11, 12].

In a similar periodicity to RHIC, the electron ERL return loop is comprised of 6 sextants, interleaved with 6 long straights. Each sextant is comprised of 120 cells. One of the straights includes a 1.322 GeV linac.

(a) Build the DF quadrupole doublet cell input data file, following its parameters given in Table 11.2. Use MULTIPOL and a hard-edge model to simulate both quadrupoles.

(b) Consider the cell. Produce graphs of the evolution with recirculation energy $E : 6.322 \rightarrow 21.164$ GeV, step 1.322 GeV, of:

- periodic orbits in BD and QF quadrupoles;
- field along the latter;
- time-of-flight;
- tunes and chromaticities;
- bend-plane components of spins across BD and QF (assumed entering the cell with $\mathbf{S} = \mathbf{S}_l$).

Use FIT embedded in REBELOTE[K = 0, IOPT = 1] do-loop for that closed orbit energy scan. Add MATRIX within the loop to get tunes and periodic beam matrix. MATRIX[PRINT] can be used to log MATRIX computation outcomes in zgoubiMATRIX.out.

Produce graphs of the betatron functions and dispersion along the cell, at 6.622 and 21.164 GeV.

(c) Consider FFAG2 recirculation loop, assumed here only comprised of six arcs (no long straights), 120 cells each. Introduce synchrotron radiation, the switch for that is SRLOSS[KWRD = MULTIPOL].

Note: field scaling as discussed in Sect. 9.2.5, which adapts beam line strength to upstream SR loss, is not considered here. The reason is that, (i) the present FFAG beam line encompasses 12 different energies, and (ii) the effect, orbit spiraling mostly affecting the highest energy, 21.164 GeV, is within transverse acceptance.

Produce a graph of the evolution of energy loss per particle, and of the *rms* energy of the radiated photons, as a function of energy, from 6.322 to 21.164 GeV, over the recirculation loop.

Check against theoretical expectations.

(d) Add spin, all 5,000 longitudinal at start. Produce a graph of the evolution of the average and *rms* values of vertical spin angle with energy, over the 12 recirculations from 6.322 to 21.164 GeV.

Check against theoretical expectations.

Give the evolution with energy of the horizontal and longitudinal emittances.

11.3 eRHIC 22 GeV ERL, From Start to End
Solution 11.3

This simulation recirculates a beam 23 times in the single linear FFAG2 loop, via spreaders, combiners and linac. Beam energy increases first, from 6.622 to 21.164 GeV, and decreases next, back down to 6.622 GeV. It is a good idea here, to refer to [12, 31].

(a) Simulation of eRHIC from start to end requires building its 120 m, 1.322 GeV linac. The latter is comprised of 42 modules. A module is approximated as DRIFT [L = 53.249249 cm]-CAVITE[K = 0, IOPT = 10]-DRIFT[L = 53.249249 cm]. CAV-ITE[K = 0, IOPT = 10] is a point transform, yet its data include length so to allow updating the distance and time of flight of particles [28, *cf.* CAVITE]. Parameters are as follows: length 1.7749 m; RF 422.26 MHz; peak voltage 31476190 V; RF phase $\pi/2$; matrix model option +1 [[28] cf. CAVITE[IOPT = 10, IOP = +1]]. With a peak voltage of 31.47619 MVolts and synchronous phase of $\pi/2$, 42 modules total 1.322 GeV.

Distribute a few tens of electrons on same value horizontal and vertical invariants $\epsilon_Y = \epsilon_Z = 25\,\pi\,\mu$m, normalized. Orient the invariant for a waist in the middle of the linac, both planes. MCOBJET[KOBJ = 3] can be used. Transport these particles through the linac, produce graphs of the initial and final invariants, horizontal and vertical.

(b) This start to end eRHIC ERL simulation uses the results of the previous exercise. When set up, it takes the beam from 6.622 upto 21.164 GeV, and then back down to 6.622 GeV.

The simulation requires the use of the keywords

- REBELOTE[IOPT = 1] in combination with GOTO[PASS#] and GOTO [GOBACK] to switch the beam to proper subsystems (return loops, spreader, combiner and linac),
- REBELOTE[IOPT = 1, LMNT = DRIFT[DltaPhase]] to flip the RF phase by pi at entrance to the linac, for energy recovery from pass number 14 on,
- INCLUDE in abundance, so to allow separate data files for the recirculating channel(s), spreaders, combiners and linac—which has the additional merit of simplifying the main input data file,
- FINISH, which ends the job, in lieu of END which cannot be used in this context.
- AUTOREF possibly, for fine-centering of the beam at entrance of the sub-systems.

Moreover, spreaders and recombiners input data files have to be set up. They can be found as modules in https://sourceforge.net/p/zgoubi/code/HEAD/tree/branches/exemples/didacticExercises/LR-eRHIC/upDown.dat file.

11.4 Solutions of Exercises of This Chapter: FFAG, Linear

11.1 EMMA Ring
(a) EMMA Cell and optical parameters.
- *For a hard-edge model of the quadrupoles:* use MULTIPOL[$XE = XS = 0$, $\lambda_E = \lambda_S = 0$]. The input data file is given in Table 11.3.

- *For a model of the quadrupoles accounting for fringe fields:* use MULTIPOL[$XE = XS = 5$ cm, $\lambda_E = \lambda_S = 2$ cm] (the fringe extents $\lambda_{E,S}$ are taken commensurate with quadrupole bore diameter). Thus, in Table 11.3 replace

```
0.   0.   1.00 1.00 1.00 1.00 1.00 1. 1. 1. 1.
```

by

```
5.   2.   1.00 1.00 1.00 1.00 1.00 1. 1. 1. 1.
```

- *A model of the cell using DIPOLES[$RM \rightarrow \infty$]:* The input data file can be found in [26, Table 3], and the magnetic field it yields in [26, Fig. 5].

Table 11.3 Input data file: EMMA cell EMMACell.inc. This file defines the double-cell segment #S_cell to #E_cell for use in further questions. Run as it is, it computes the first order mapping at 11 different energies, from 10 to 20 MeV, step 1 MeV

```
'MARKER'  EMMACellParameters_S                              ! Just for edition purposes.
! EMMA cell. Corner with angle 2pi/42 at long drift to QD interface.
! Shifted QD and QF use MULTIPOL[B1.ne.0] (dY = B1/(B2/R0)).
'OBJET'
51.71103865922                                             ! 15 MeV kinetic energy.
5
.001 .01 .001 .01 0. .0001
-0.55038013 41.86602 0.0 0.0 0.0 0.677214420               ! 10. MeV kinetic.
'PARTICUL'
POSITRON                    ! ELECTON (e<0) would do as well: change sign of BORO and fields then.
'MARKER'   #S_cell
'DRIFT' H-LD
10.5
'CHANGREF'
ZR  -8.571428571429
'MULTIPOL' QD
0
7.5699 5.3 1.60161792 -2.49312 0. 0. 0. 0. 0. 0. 0. 0.   ! B1=- offset(3.4048)*gradient(-2.493120/R0).
0.  0.  1.00 1.00 1.00 1.00 1.00 1. 1. 1. 1.
4  .1455   2.2670  -.6395  1.1558 0. 0. 0.
0.  0.  1.00 1.00 1.00 1.00 1.00 1. 1. 1. 1.
4  .1455   2.2670  -.6395  1.1558 0. 0. 0.
0. 0. 0. 0. 0. 0.00 0.00 0.00 0.00
0.7                                                      ! Step size ~ length/10 for accuracy.
1 0.00 0.00 0.00
'DRIFT' SD
5.00
'MULTIPOL' QF
0
5.8782 3.70 -0.5030623 2.47715 0. 0. 0. 0. 0. 0. 0. 0.  ! B1=- offset( 0.7514)*gradient(2.477150/R0).
0.  0.  1.00 1.00 1.00 1.00 1.00 1. 1. 1. 1.
4  .1455   2.2670  -.6395  1.1558 0. 0. 0.
0.  0.  1.00 1.00 1.00 1.00 1.00 1. 1. 1. 1.
4  .1455   2.2670  -.6395  1.1558 0. 0. 0.
0. 0. 0. 0. 0. 0.00 0.00 0.00 0.00
0.6                                                      ! Step size ~ length/10 for accuracy.
1 0.00 0.00 0.00
'DRIFT' H-LD
10.5
'MARKER'   #E_cell

'FIT2'
2
2 30 0 [-10,10]                                          ! Two variables: initial Y0 and T0.
2 31 0 [-100,100]
2    1e-15                          ! Two constraints: Y0=Y_end,  and T0=T_end; penalty = 1e-15.
3.1 1 2 #End 0. 1. 0
3.1 1 3 #End 0. 1. 0

'MATRIX'              ! PRINT => at each REBELOTE loop, will log MATRIX[IFOC=11] outcomes (beta
1 11  PRINT                        ! function values, tunes, etc.) in zgoubi.MATRIX.out.
'FAISCEAU'

'REBELOTE'
10 0.1 0 1                         ! Re-run the above sequence 9 times, over rigidity range below.
1                                  ! One parameter to be changed prior to REBELOTE.
OBJET 35 0.7417895338:1.32265941425  ! Relative regidity relative to BORO, at 11 and 20 MeV kinetic.

'MARKER' EMMACellParameters_E                              ! Just for edition purposes.
'END'
```

With either one of these three different models: use FIT embedded in REBELOTE do-loop [28, cf. REBELOTE], to find the periodic Y, T horizontal conditions at cell end, for 11 different energies from 10 MeV to 20 MeV. MATRIX[PRINT] stacks matrix computation outcomes in zgoubi.MATRIX.out.

Raytracing outcomes are displayed in Fig. 11.14, 11.15, 11.16 and 11.17. Optical functions, Fig. 11.16, are computed using the [pathTo]/zgoubi-code/toolbox/beta FromPlt tool found in zgoubi package [29].

Completing these results accounting for all three models is left to the reader. Similar cases may be found in various examples in zgoubi sourceforge repository, https://sourceforge.net/p/zgoubi/code/HEAD/tree/branches/exemples/ folder.

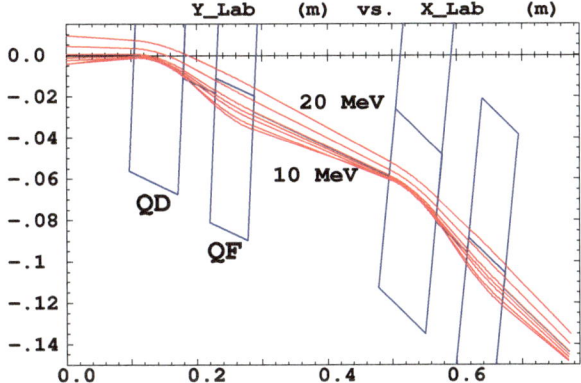

Fig. 11.14 Closed orbits along a pair of cells. A graph obtained using zpop: menu 7; 1/1 to open zgoubi.plt; 2/[42,48] for X_{lab} versus Y_{lab}; 7 to plot

Fig. 11.15 Field along closed orbits, in the fringe field model. A graph obtained using zpop: menu 7; 1/1 to open zgoubi.plt; 2/[6,32] for BZ versus s; 7 to plot

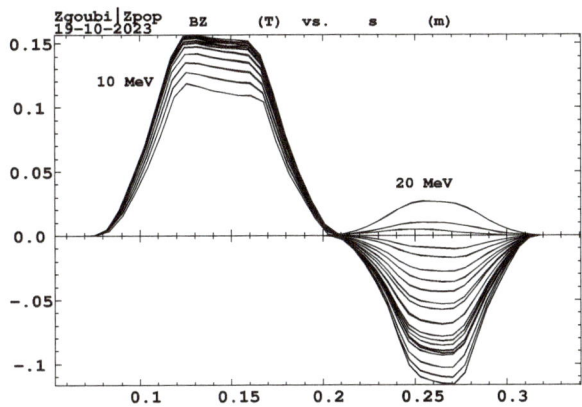

The periodic beam matrix at injection is needed in question (c). Its value out of the first step in the REBELOTE scan is:

```
Beam matrix (beta/-alpha/-alpha/gamma) and periodic dispersion (MKSA units)
 0.235199  -1.627054   0.000000   0.000000   0.000000  -0.001659
-1.627054  15.507340   0.000000   0.000000   0.000000  -0.074719
 0.000000   0.000000   0.233834   1.318588   0.000000  -0.000000
 0.000000   0.000000   1.318588  11.712066   0.000000   0.000000

              Betatron tunes (Q1 Q2 modes)
       NU_Y = 0.35434753     NU_Z = 0.30712658
```

(b) Maximum stable amplitudes.

They are computed using the [pathTo]/zgoubi-code/toolbox/searchStabLim tool found in zgoubi package [30]. Results are given in Figs. 11.18 and 11.19.

(c) A serpentine acceleration-deceleration cycle.

An input data file for this simulation is given in Table 11.4: a 300 electron bunch is tracked over 17 turns at an acceleration rate of 68 kV per cavity. Particle data are logged turn by turn in zgoubi.fai. Tracking outcomes are displayed in Fig. 11.20.

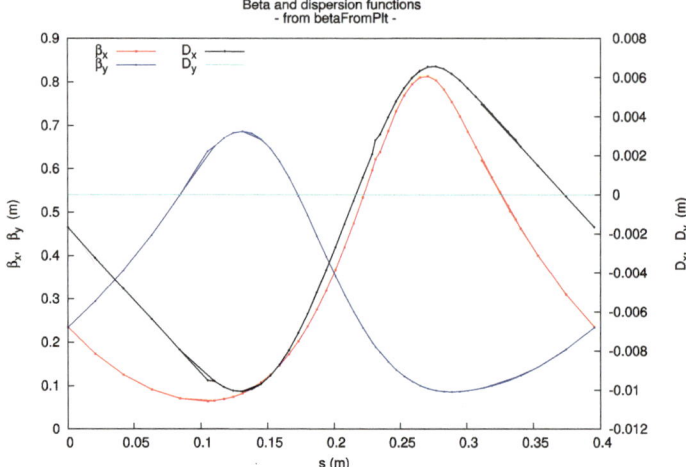

Fig. 11.16 Betatron functions and dispersion along a cell (with some artifacts in overlapping fringe field regions)

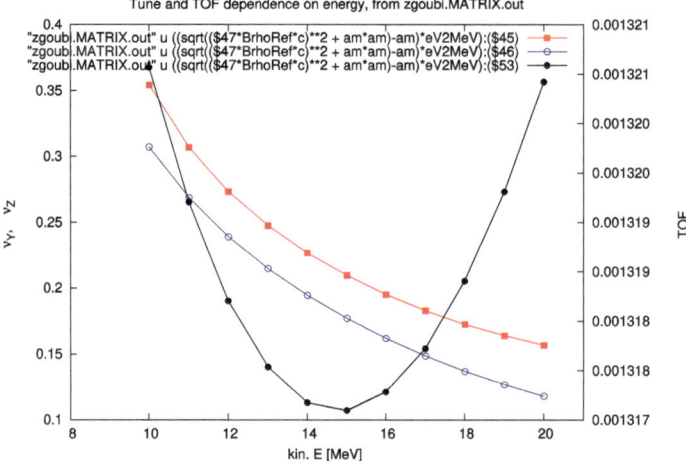

Fig. 11.17 Tunes and time-of-flight parabola as a function of particle energy

11.2 eRHIC ERL FFAG2 loop. Synchrotron Radiation. Polarization

(a) Input data file for eRHIC FFAG2 DF cell [12].

The input data file, constructed from the parameters of Table 11.2, is given in Table 11.5. The orbital angle through a cell is explicit, as MULTIPOL[ALE] parameter, thus the total deviation for 6 arcs, i.e., times 120 cells, is

$$720 \times 2 \times (\underbrace{1.6564835 \times 10^{-3}}_{\text{-ALE in BD}} + \underbrace{2.7090305 \times 10^{-3}}_{\text{-ALE in QF}}) \approx 2\pi \text{ rad}$$

Fig. 11.18 Maximum stable horizontal amplitudes, from inner to outer invariant at 10, 13, 15, 17 and 20 MeV

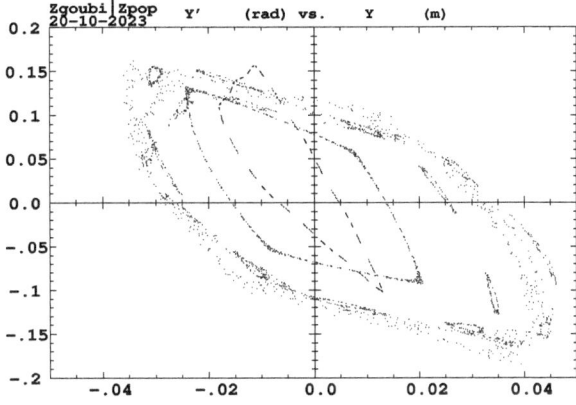

Fig. 11.19 Maximum stable vertical amplitudes, from inner to outer invariant at 10, 13, 15, 17 and 20 MeV

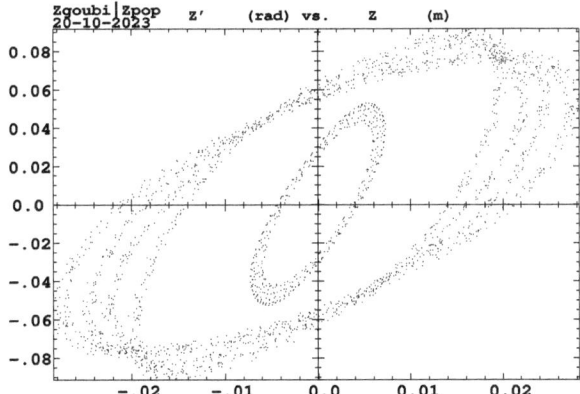

in agreement with the present working hypotheses (things are a little different in the actual ring design, which is comprised of six 102 cell arcs, dispersion suppressor FFAG sections between arcs [11], and merger and combiner sections at linac ends). It is a good idea to first check in zgoubi.res the length of the cell, as obtained when running this file.

(b) Optical parameters of the DF cell.

Scans take care of producing cell parameters for each one of the 12 passes. A typical input data file is given in Table 11.6, it loops on MATRIX computation, so producing a scan of the wave numbers and other parameters, logged in zgoubi.MATRIX.out. The latter is read and plotted using a gnuplot script (Table 11.6). Results are given in Fig. 11.21.

Use OBJET[KOBJ = 6] and MATRIX[IORD = 2, IFOC = 11] to add computation of chromaticities. Or as well, use a TWISS command, however it takes longer, as it does more than MATRIX.

Periodic orbits: OPTIONS[.plt, 2] (Table 11.6) causes log of stepwise particle data across the quadrupoles, in zgoubi.plt. These data include periodic orbit

Table 11.4 Left input data file, EMMA2Cells+CAV.inc: EMMA double cell and accelerating cavity. This file defines the cell segment #S_2cells-withCAV to #E_2cells-withCAV INCLUDEd in the complete ring, right file. Right input data file: a 300 electron bunch accelerated-decelerated in EMMA ring

```
                                        'MARKER' EMMARing21CAVITE_S                      ! Just for edition purposes.
                                        ! EMMA ring with 21 CAVITEs. serpentine acceleration/deceleration.
                                        'MCOBJET'
                                        51.71103865921708                               ! 15MeV.
                                        3
                                        300
                                        3 3 3 2 2
                                        -0.55038013e-2 41.86602e-3  0.0  0.0  0.0  0.677214420    ! 10. MeV kinetic.
                                         1.627 0.235199  0e-6  2        ! Correlation to momentum locally via Dx and
                                        -1.318 0.233834  0e-6  2        ! Dx' is weak, ignored in this MCOBJET.
  ! EMMA2Cells+CAV.inc.                 -5.    3.       2E-4  -.4  .40001      ! Generate a thin s-dp disk.
  'MARKER'  #S_2cells-withCAV           123456 234567 345678
  'INCLUDE'                             'PARTICUL'
  1                                     POSITRON
  2 *./EMMACell.inc[#S_cell:#E_cell]
  'CAVITE'                              'CAVITE'       ! This CAVITE with zero voltage is to intilize particles' phase.
  7                                     7
  0.00    1.51750e9                     0.0  1.51730e9
  68e3    0.                            0.    0.
  'MARKER'  #CAV
  'MARKER'  #E_2cells-withCAV           'FAISTORE'
  'END'                                 zgoubi.fai                ! Particle coordinates here, are logged in zgoubi.fai.
                                        1

                                        'INCLUDE'
                                        1
                                        21 *./EMMA2Cells+CAV.inc[#S_2cells-withCAV:#E_2cells-withCAV]

                                        'REBELOTE'
                                        16 0.2 99                                      ! 16 additional turns.
                                        'MARKER' EMMARing21CAVITE_E                     ! Just for edition purposes.
                                        'END'
```

Fig. 11.20 Longitudinal phase space of an acceleration-deceleration cycle over $10 \leq E \leq 20\,\mathrm{MeV}$ range in EMMA ring. Turn-by-turn bunch records over 17 turns are shown here. A graph obtained using zpop: menu 7; 1/5 to open zgoubi.fai; 2/[18, 19] for E versus RF phase; 7 to plot

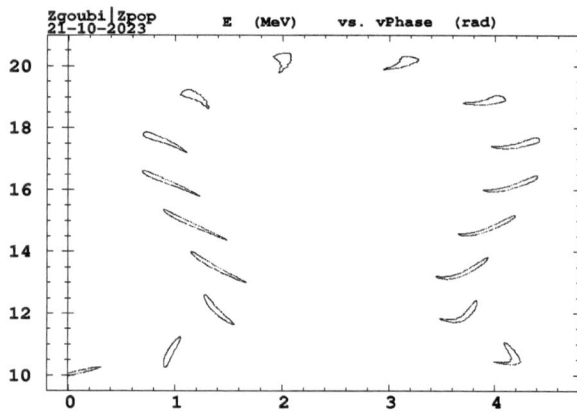

coordinates, as well as magnetic field along the latter. To get graphs, a possibility is to use gnuplot scripts from zgoubi toolbox in sourceforge, in the folder [pathTo]/zgoubi-code/toolbox/gnuplot_Zplt. $B_Z(s)$ for instance uses gnuplot_Zplt_sBZ.gnu. Trajectories use gnuplot_Zplt_sYZ.gnu. The gnuplot commands are reproduced in Table 11.6. Expected outcomes are given in Fig. 11.22.

Betatron functions: a similar question is treated in the previous exercise (cf. Fig. 11.16), apply the same method here. The expected result is displayed in Fig. 11.23.

Table 11.5 Input data file: eRHIC DF quadrupole doublet cell. As it is, it produces a transport matrix and the periodic beam matrix for the DF cell, at 6.622 GeV. This file defines the segment S_eRHIC_BDCell:E_eRHIC_BDCell for INCLUDEs in subsequent input data files

```
'OBJET'
55.038075681301081 *1e3                                        ! Reference BRho, 16.5 GeV.
5
.001 .01 .001 .01 0. .0001
-1.154400E+00  -9.149457E+00    0.0E+00   0.0E+00   0.0E+00   4.01364290E-01 'o'    ! 6621.999980 MeV.
'MARKER'  S_eRHIC_BDCell
'DRIFT'
9.652479
'MULTIPOL' BD
0  !  .Dip        b_0 (kG)      b_1 (kG)
112.9301   10.00    0.0293364    -0.5225857    0. 0.0 0.0 0.0 0.0 0.0 0.0 0.0 0.0
0. 0. 10.00  4.0  0.8 0. 0.00 0.00 0.00 0. 0. 0. 0       .       ! Entrance fringe field (hard edge).
4  .1455   2.2670  -.6395  1.1558  0. 0.  0.
0. 0. 10.00  4.0  0.8 0. 0.00 0.00 0.00 0. 0. 0. 0       .       ! Entrance fringe field (hard edge).
4  .1455   2.2670  -.6395  1.1558  0. 0.  0.
0. 0. 0. 0. 0. 0. 0. 0. 0.
1.  | Dip BD2_A01_2                                              ! Integration step size (cm).
3   0.   0.  -1.6564835000E-03             ! KPOS=3 -> automatic positioning at an angle theta_D/2.
'DRIFT'
19.304957
'MULTIPOL' QF
0  !  .Dip        b_0 (kG)      b_1 (kG)
184.7002   10.00    0.0293343    0.3728876    0. 0.0 0.0 0.0 0.0 0.0 0.0 0.0 0.0
0. 0. 10.00  4.0  0.8 0. 0.00 0.00 0.00 0. 0. 0. 0       .       ! Entrance fringe field (hard edge).
4  .1455   2.2670  -.6395  1.1558  0. 0.  0.
0. 0. 10.00  4.0  0.8 0. 0.00 0.00 0.00 0. 0. 0. 0.            ! Exit fringe field (hard-edge).
4  .1455   2.2670  -.6395  1.1558  0. 0.  0.
0. 0. 0. 0. 0. 0. 0. 0. 0.
1.  | Dip QF2_A01_2                                              ! Integration step size (cm).
3   0.   0.  -2.7090305000E-03             ! KPOS=3 -> automatic positioning at an angle theta_D/2.
'DRIFT'     DLHH2_A01_ DRIF
9.652479
'MARKER'  E_eRHIC_BDCell
'MATRIX'
1 11  PRINT
'END'
```

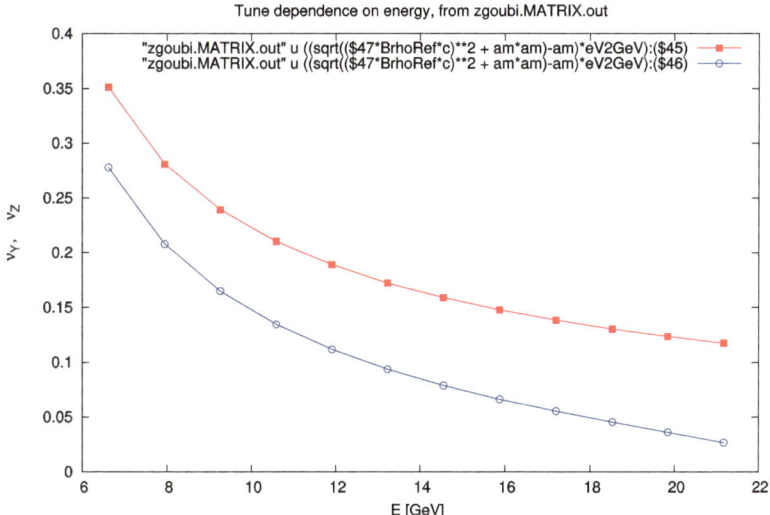

Fig. 11.21 A scan of cell tunes ν_Y (square markers) and ν_Z (circles) for the 12 recirculated energies in $6.622 \rightarrow 21.164$ GeV eRHIC FFAG2 loop. The solid line is to guide the eye

Table 11.6 Input data file: a do-loop scan on MATRIX computation. Outcomes are logged in zgoubi.MATRIX.out (a consequence of MATRIX[PRINT]). This file INCLUDEs the S_eRHIC_BDCell:E_eRHIC_BDCell segment grabbed from the input data file of Table 11.5

```
A loop on MATRIX computation for a series of OBJET[D values]
 'OBJET'
55.038075681301081 *1e3                                                    ! Reference BRho, 16.5 GeV.
5
.001 .01 .001 .01 0. .0001
-1.276 9.648   0.0E+00   0.0E+00   0.0E+00   4.01364290E-01 'o'             | 6621.999980 MeV.
 'SCALING'
1 1
MULTIPOL
-1
55.038075681301081
1

 'OPTIONS'
1 1
.plt  2                     | Sets IL value under MULTIPOL, for log of stepwise particle data in zgoub.plt.

 'INCLUDE'
1
eRHIC_BDCell.inc[S_eRHIC_BDCell:E_eRHIC_BDCell]

 'FIT2'
2                                                                          ! Vary Y0 and T0 in OBJET.
1 30 0 [-10,10]
1 31 0 [-50,50]
2  1e-15                                    | Request Y=Y0 and T=T0 at end of optical sequence.
3.1 1 2 #End 0. 1. 0
3.1 1 3 #End 0. 1. 0

 'FAISCEAU'
 'MATRIX'
1 11  PRINT

 'REBELOTE'
12 0.1 0 1
1
OBJET 35 4.01364290E-01:1.28269760E+00   | Change parameter 35 in OBJET (i.e., D), prior to looping.

 'SYSTEM'
3
gnuplot <./gnuplot_Zplt_sYZ.gnu
gnuplot <./gnuplot_Zplt_sB.gnu
gnuplot <./gnuplot_MATRIX_Qxy.gnu

 'END'
```

gnuplot scripts to obtain Figs. 11.21 *and* 11.22:

```
#gnuplot_MATRIX_Qxy.gnu
BORO = 55.038075681301081;am = 0.51099892;  c = 2.99792458e8; BrhoRef = BORO *1e-3    # T.m; eV2GeV = 1e-6
plot \
"zgoubi.MATRIX.out" u ((sqrt(($47*BrhoRef*c)**2 + am*am)-am)*eV2GeV):($45) w lp pt 5 lt 1 lw .5 lc rgb "red"   ,\
"zgoubi.MATRIX.out" u ((sqrt(($47*BrhoRef*c)**2 + am*am)-am)*eV2GeV):($46) w lp pt 6 lt 3 lw .5 lc rgb "blue"

# gnuplot <./gnuplot_Zplt_sYZ.gnu
cm2m = 0.01 ; TRAJ1 = 1 ; TRAJ2 = 1 ; FITLST = 1 # col. 51 in zgoubi.plt, 1 if FIT completed.
plot for [i=TRAJ1:TRAJ2] 'zgoubi.plt' u ($51==FITLST && $19== i? $14*cm2m :1/0):($10 *cm2m) w l lw 2.5 lc rgb "red" ,\
        for [i=TRAJ1:TRAJ2] 'zgoubi.plt' u ($51==FITLST && $19== i? $14*cm2m :1/0):($12 *cm2m) axes x1y2 w l lw 2 lc rgb "blue"

# gnuplot <./gnuplot_Zplt_sB.gnu
cm2m = 0.01 ; kG2T= 0.1 ; MeV2eV = 1e6 ; TRAJ1 =1 ;  TRAJ2 =1 ; FITLST =1  # col. 51 in zgoubi.plt, 1 if FIT completed.
plot for [i=TRAJ1:TRAJ2]  'zgoubi.plt' u ($51==FITLST && $19== i ? $14 *cm2m : 1/0):($25 *kG2T) w l lw 2
```

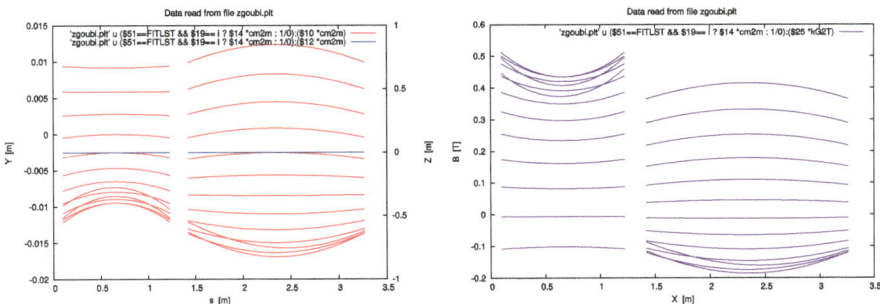

Fig. 11.22 Left, red curves: transverse excursion of the 12 periodic orbits across the FFAG cell quadrupoles (vertical is null). Right: magnetic field experienced along these 12 orbits, in a hard-edged model

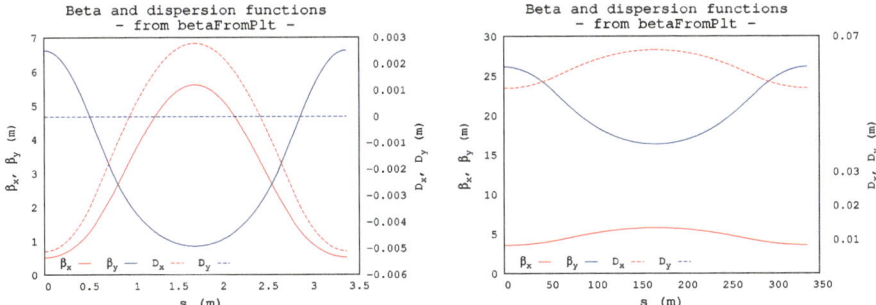

Fig. 11.23 Optical functions at 6.622 GeV (left) and 21.164 GeV (right), from stepwise ray-tracing across FFAG2 cell

Spin components: add PARTICUL[POSITRON] and SPNTRK[KSO = 1] after OBJET, to define initial spin coordinates. They will be tracked and logged in zgoubi.plt. S_X, S_Y, and S_Z are respectively columns 33, 34 and 35 [28, Sect. 8.3].

(c) SR energy loss.

The input data file for this simulation is given in Table 11.7. The way it works:

- MCOBJET[KOBJ = 3] defines a 6D bunch, comprised of 5,000 electrons, with null 6D emittance here. Due to REBELOTE[K = 0], MCOBJET generates a new set of initial coordinates (namely, the periodic orbit coordinates Y and T for a new D value) at the start of each new pass
- defining PARTICUL is necessary as mass and charge are needed for computation of SR, and for integration of spin motion
- the cell is INCLUDEd 720 times, that makes up the ring
- REBELOTE loops 12 times, that takes care of the 12 passes (the lowest energy happens to be treated twice, not an issue—REBELOTE could as well start from the second energy for 11 passes, instead)
- prior to launching the next pass, REBELOTE[NPRM = 45, 40, 41] changes the relative reference momentum of the particles (D parameter, number 45 under

Table 11.7 Input data file: transport a 5000 electron bunch, accounting for SR energy loss. Initial 6D emittance is null. The ring is comprised of 720 cells (no long straights)

```
SR loss
'MCOBJET'
55.038075681301d3
3
5000
2 2 2 2 1 1           ! Next line defines periodic orbit coordinates Y ad T, for a particular D value.
-1.276E-02 9.648E-03 0. 0. 0.   4.01364290E-01           ! Watchout! Y, T are in m, rad in MCOBJET.
0. 1. 0.      9.
0. 1. 0.      9.
0. 1. 0.      9.
123456 234567 345678

'PARTICUL'                                            ! Necessary so to compute SR.
5.10998902E-01 1.60217653e-19 1.15965213627E-03 0. 0.
'SPNTRK'
1                                                         ! Used in question (d).
'SRLOSS'                                                  ! SR loss set, in all MULTIPOL.
1  srLoss
MULTIPOL
1 123456
'FAISTORE'                                                ! Used in question (d),
zgoubi.fai          ! stores particle and spin adta at end of each pass, at LABEL1=#EndRing.
1
'SCALING'
1  1
MULTIPOL
-1
55.038075681301
1

'OPTIONS'
1  1                     ! This is required in order to allow loging in zgoubi.res at all passes
WRITE ON                                     (otherwise  just 1st and last pass are logged).

'INCLUDE'
1
720 * eRHIC_BDCell.inc[S_eRHIC_BDCell:E_eRHIC_BDCell]                        ! A 720 cell ring.

'MARKER'    #EndRing
'FAISCEAU'
'SRPRNT'

'REBELOTE'            ! Run this problem 12 times. Prior to next loop, change D, Y and T in OBJET.
12 0.1  0 1
3                    ! Watchout! 40 and 41 lines should acutally be just 2 lines, not 4! Both arriage
MCOBJET 45   4.01364290E-01:1.28269760E+00               ! returned just for edition purposes.
MCOBJET 40  -1.276E-02 -1.288E-02 -1.241E-02 -1.142E-02 -9.968E-03 -8.104E-03 -5.873E-03
-3.314E-03 -4.577E-04  2.667E-03  6.035E-03  9.625E-03
MCOBJET 41   9.648E-03  8.102E-03  6.670E-03  5.344E-03  4.113E-03  2.969E-03  1.903E-03
 9.081E-04 -2.318E-05 -8.963E-04 -1.716E-03 -2.488E-03

'SYSTEM'
1
gnuplot < ./gnuplot_Zres_SRLoss.gnu

'END'
```

A gnuplot script to obtain Fig. 11.24:

```
# gnuplot_Zres_SRLoss.gnu
system("grep ' Average energy loss per particle per pass' zgoubi.res | cat > gnuplot_temp")
set xlabel "E [GeV]" ; set ylabel "SR loss per pass  [MeV]"; dE = 1.322; E1=6.622 ; keV2MeV=1e-3
plot 'gnuplot_temp'   u  ($0>0 ? E1+dE*($0 -1) :1/0):($9 *keV2MeV) w lp pt 5
```

MCOBJET) and the periodic coordinates (Y and T, number 40 and 41 under MCOBJET) accordingly. The latter are taken from the outcomes of question (b).

The informations of interest happen to be in zgoubi.res—no need of any other log file. Namely, "Average energy loss per particle per pass", and "Average energy of radiated photon", found under SRPRNT. A graph is displayed in Fig. 11.24, obtained using a gnuplot script given in Table 11.7. Checking against theoretical expectations is left to the reader.

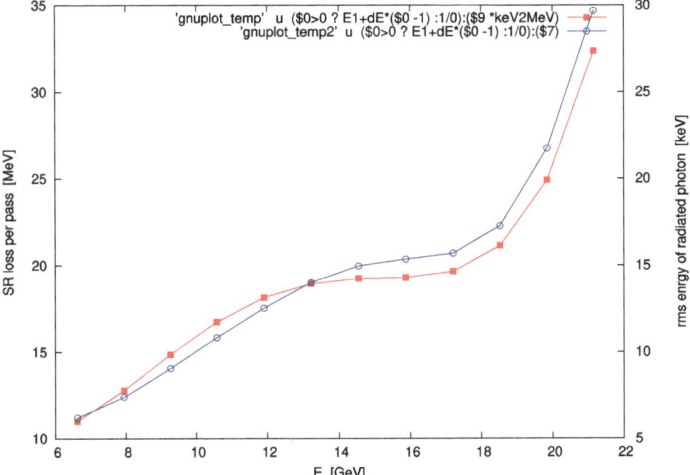

Fig. 11.24 Energy dependence of the average energy loss (left axis, solid squares) and *rms* energy of radiated photon (right axis, empty circles), over the FFAG2 recirculation loop, at the 12 recirculated energies. The solid lines are to guide the eye

FAISCEAU computes the concentration ellipses, for 5,000 particles here. Thus graphs of the emittance growth due to synchrotron radiation can be obtained in a similar way to the SR loss graph (cf. gnuplot script, Table 11.7), by a grep of '(Y, T)', '(Z, P)' and '(t, K)' as found under FAISCEAU in zgoubi.res, for instance (case of last, 21.164 GeV recirculation):

```
--------------- Concentration ellipses :
surface/pi         alpha      beta        <X>           <XP>        numb. of prtcls ratio   space pass#
                                                                   n ellips, out
2.8901E-11 [m.rad]   1.6524E+00  7.4334E+00  9.285078E-03 -2.419357E-03  5000   3993  0.7986   (Y,T)  13
0.0000E+00 [m.rad]   0.0000E+00  1.0000E+00  0.000000E+00  0.000000E+00  5000   5000  1.000    (Z,P)  13
3.3939E-07 [mu_s.MeV] -1.7004E+00  7.4845E-08  8.075575E+00  2.104242E+04  5000  3942  0.7884   (t,K)  13
```

(d) Adding spin.

SPNTRK[KSO = 1] (initial spin along the X axis, longitudinal) is included in the input data file used in (c), Table 11.7. Running it also tracked the spins. Particle and spin data are logged in zgoubi.fai, as a result of FAISTORE command.

Thus, it just remains to produce a graph from the previous question outcomes. There are various ways to compute the *rms* spin angle from the 5000 electron sample at the end of the loop. An ancillary program can be used, awk commands in a gnuplot script is another possibility. The outcome is displayed in Fig. 11.25, which also displays histograms of the longitudinal spin component at 3 different passes, 6, 8 and 11, accounting for an initial vertical jitter.

11.3 eRHIC 22 GeV ERL, From Start to End

It is a good idea here, to refer to [11, 12].

(a) eRHIC linac.

eRHIC 120 m, 1.322 GeV linac is essentially based on CAVITE. CAVITE [IOPT = 10] transports particles using a Chambers matrix [28, cf. CAVITE]. Transverse damping can be accounted for, or not, using the $IOP = \pm 1$, ± 2 flag.

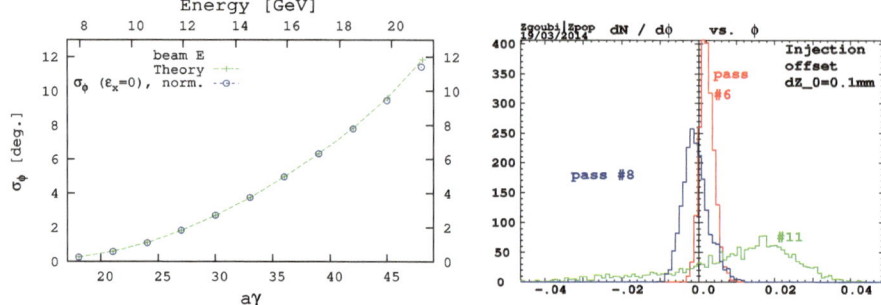

Fig. 11.25 Left: polarization loss due to energy spread, from tracking (markers) and from Eq. 11.9 (dashed line). Right: an histogram of the vertical spin angle at pass 6, 8 and 11, in the presence of an initial $\delta Z = 0.1$ mm vertical beam jitter

Table 11.8 Input data files. Left: linacModule.inc, defines a module of eRHIC linac. Right: linac.INC.dat, defines the 42 module, 120 m long 1.322 GeV linac

```
                                              linac.INC.dat file.
                                              'MCOBJET'
                                              55.03807568130ld3
                                              3
                                              299
                                              2 2 2 2 2
                                              0. 0.  0. 0. 0. 1.20257640E+00  1.28269760E+00
                                              1. 120.  6.7337413106e-10 9.           ! 25mu_m norm. at 19.82GeV
'MARKER'    S_linacModule                     1. 120.  6.7337413106e-10 9.
'DRIFT'     CAV_UP                            0.   8.  .5e-6 1.
1.41994249041e2                               123456 234567 345678
'DRIFT'
-88.745                                       'PARTICUL'
'CAVITE'                                      0.51099892 1.60217653e-19 1.15965218076e-3 0.0 0.0
10          PRINT
1.7749      422260000.0                       'FAISTORE'
31476190.4762       1.57079632679 +1          zgoubi.fai  #StartL CAV_DO  #EndL
'DRIFT'                                       1
-88.745
'DRIFT'     CAV_DO                            'MARKER'    #StartL
1.41994249041e2                               'FAISCEAU'
'MARKER'    E_linacModule
'END'                                         'DRIFT'     DltaPhase
                                              0.

                                              'INCLUDE'
                                              1
                                              42 * linacModule.inc[S_linacModule:E_linacModule]

                                              'FAISCEAU'   #EndL
                                              'END'
```

eRHIC linac simulation input data file is given in Table 11.8. Tracking results are displayed in Fig. 11.26.

(b) eRHIC, from start to end.

Guidance in setting up the input data file can be found in an existing complete recirculating linac simulation, an energy recovery linac (ERL), 12 passes up 11 passes down, in https://sourceforge.net/p/zgoubi/code/HEAD/tree/branches/exemples/didacticExercises/LR-eRHIC/ folder. This ERL simulation is based on the keywords:

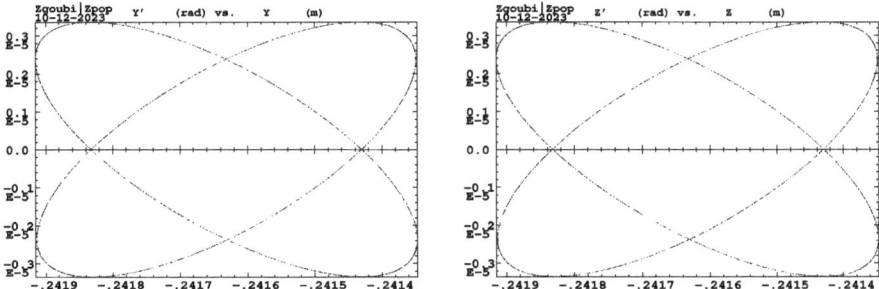

Fig. 11.26 An initial 1000-particle bunch with all transverse particle coordinates taken on a given invariant $\gamma u^2 + 2\alpha u u' + \beta u'^2 = \epsilon_u/\pi$ is tracked from entrance to exit of the linac (u stands for Y or Z; $\epsilon_Y = \epsilon_Z = 25\,\pi\mu$m, normalized, 6.622 GeV here). The figure shows horizontal (left plot) and vertical (right plot) phase spaces at linac entrance (converging ellipse) and exit (diverging ellipse)

- REBELOTE[IOPT = 1] in combination with GOTO[PASS#] and GOTO [GOBACK] to switch the beam to proper subsystems (return loops, spreader, combiner and linac),
- REBELOTE[IOPT = 1, LMNT = DRIFT[DltaPhase]] to flip the RF phase by pi at entrance to the linac, for energy recovery from pass number 14 on,
- abundant use of INCLUDE so to allow separate data files for the recirculating channel(s), spreaders, combiners and linac—which has the merit additional of simplifying the main input data file,
- FINISH, which ends the job, in lieu of END which cannot be used in this context.

References

1. F.E. Mills, C. Johnstone, *Proceedings of the 4th International Conference on Physics Potential & Development of $\mu^+ - \mu^-$ Colliders, San Francisco, CA* (UCLA, Los Angeles, CA, 1999), pp. 693–698
2. C. Johnstone, et al., Fixed field circular accelerator designs, in *Proceedings of the 1999 Particle Accelerator Conference*, New York (1999). https://accelconf.web.cern.ch/p99/PAPERS/ THP50.PDF. Fig. 11.1: Copyrights under license CC-BY-3.0, https://creativecommons.org/ licenses/by/3.0; no change to the material
3. S. Koscielniak, C. Johnstone, Longitudinal dynamics in an FFAG accelerator under conditions of rapid acceleration and fixed, high RF, in *Proceedings of the 2003 Particle Accelerator Conference*. https://accelconf.web.cern.ch/p03/PAPERS/TPPG009.PDF
4. D. Trbojevic, et al., Fixed field alternating gradient lattice design without opposite bend, in *Proceedings of EPAC 2002, Paris, France*. https://accelconf.web.cern.ch/e02/PAPERS/ WEPLE051.pdf
5. J.S. Berg, et al., Cost-effective design for a neutrino factory. Phys. Rev. Spec. Topics - Accel. Beams **9**, 011001 (2006). https://journals.aps.org/prab/pdf/10.1103/PhysRevSTAB.9.011001
6. S. Machida, et al., What we learned from EMMA, in *MO2PB01 Proceedings of Cyclotrons 2013, Vancouver, BC, Canada*
7. R. Edgecock, EMMA - the world's first non-scaling FFAG, in *THOBAB01 Proceedings of PAC07, Albuquerque, New Mexico, USA*

8. S. Machida, et al., Acceleration in the linear non-scaling fixed-field alternating-gradient accelerator EMMA. Nat. Phys. **8**, 243–247 (2012). https://www.nature.com/articles/nphys2179

9. J.S. Berg, The EMMA lattice NIM A **596**, 276–284 (2008)

10. D. Trbojevic, et al., ERL with non-scaling fixed field alternating gradient lattice for eRHIC, in *IPAC2015, Richmond, VA, USA*. https://accelconf.web.cern.ch/IPAC2015/papers/tupty047.pdf

11. F. Méot, et al., Tracking studies in eRHIC energy-recovery recirculator. Tech Note C-A/eRHIC/45, BNL C-AD, Upton, LI, NY 11973 (2015). https://www.osti.gov/biblio/1210189

12. F. Méot, et al., eRHIC ERL modeling in zgoubi. Tech Note C-A/eRHIC/49, BNL C-AD, Upton, LI, NY 11973 (2016). https://technotes.bnl.gov/PDF?publicationId=38865. https://www.osti.gov/biblio/1335396

13. D. Trbojevic, CBETA Technical Report. Tech Note BNL-211697-2019-TECHCBETA/031 (2018). https://www.osti.gov/servlets/purl/1515414

14. C. Mayes, CBETA Multipass lattice design, in *Proceedings of the ERL Workshop, CERN*, June 20 (2017). https://accelconf.web.cern.ch/erl2017/talks/tuidcc003_talk.pdf. Fig. 11.5: Copyrights under license CC-BY-3.0, https://creativecommons.org/licenses/by/3.0; no change to the material

15. F. Méot, et al., Simulation of the CBETA 4-pass FFAG ERL using field-maps - exclusively. Tech Note BNL-211811-2019-JAAM, BNL (2019). https://www.osti.gov/servlets/purl/1529889

16. S.A. Bogacz, et al., 20-24 GeV FFA CEBAF energy upgrade, in *12th International Particle Accelerator Conference IPAC2021, Campinas, SP, Brazil*. https://accelconf.web.cern.ch/ipac2021/papers/mopab216.pdf

17. F. Méot, N. Tsoupas, S. Brooks, D. Trbojevic, Beam dynamics validation of the halbach technology FFAG cell for cornell-BNL energy recovery Linac. Nuclear Inst. Methods Phys. Res. A **896**, 60–67 (2018)

18. F. Lemuet, Collection and muon acceleration in the neutrino factory project. Ph.D. Thesis dissertation, CEA and CERN, Paris-Saclay University (2007). https://inis.iaea.org/collection/NCLCollectionStore/_Public/42/013/42013892.pdf

19. https://cds.cern.ch/collection/Neutrino%20Factory%20Notes?as=1&ln=ru

20. https://www.bnl.gov/ffaworkshop/events/

21. E. Keil, A.M. Sessler, Muon acceleration in FFAG rings. CERN-NUFACT-note-139, CERN-AB-2004-033 (ABP) (2004). https://cds.cern.ch/record/784564/files/nufact-note-139.pdf

22. J.S. Berg, Amplitude dependence of time of flight and its connection to chromaticity. NIMA **570**(1), 15–21 (2007)

23. G. Leleux, et al., Synchrotron radiation perturbations in long transport lines, in *Proceeding of the PAC 1991 Accelerator Conference*. https://accelconf.web.cern.ch/p91/PDF/PAC1991_0517.PDF

24. F. Méot, V. Litvinenko, T. Roser, Polarization transport in the ERL-ERL FCC e+ e- Collider. Tech Note C-A/AP/684 BNL C-AD (2022)

25. V. Ptitsyn, Electron polarization dynamics in eRHIC, in *EIC'14 Workshop*. http://appora.fnal.gov/pls/eic14/agenda.full

26. F. Méot, Y. Giboudot, et al.: Beam dynamics simulations regarding the experimental FFAG EMMA, using the on-line code. THPD023, Proceedings of IPAC'10, Kyoto, Japan. https://accelconf.web.cern.ch/IPAC10/papers/thpd023.pdf

27. F. Méot, The ray-tracing code Zgoubi - status. Nuclear Instrum. Methods Phys. Res. A **767**, 112–125 (2014)

28. F. Méot, Zgoubi users' guide. https://www.osti.gov/biblio/1062013-zgoubi-users-guide An up-to-date version of the guide: https://sourceforge.net/p/zgoubi/code/HEAD/tree/trunk/guide/Zgoubi.pdf

29. From zgoubi toolbox, part of the sourceforge package: a Fortran tool to compute optical functions from a zgoubi.plt output file, and some related gnuplot scripts, see the README and example, there: https://sourceforge.net/p/zgoubi/code/HEAD/tree/trunk/toolbox/betaFromPlt/

30. From `zgoubi` toolbox, part of the sourceforge package: a Fortran tool to perform a dynamic aperture scan, and some related gnuplot scripts, see the README and example, there: https://sourceforge.net/p/zgoubi/code/HEAD/tree/trunk/toolbox/searchStabLim/

31. F. Méot, et al., Tracking studies in eRHIC energy-recovery recirculator. Tech Note C-A/eRHIC/45, BNL C-AD, Upton, LI, NY 11973 (2015). https://www.osti.gov/biblio/1210189

Chapter 12
Beam Lines

Abstract This chapter introduces beam transport and manipulations in beam lines. It provides the theoretical material resorted to in the simulation exercises, leaning on charged particle optics and beam manipulation concepts introduced in earlier Chapters. The simulation of beam lines and specific functionalities they ensure require new optical elements, such as WIENFILTER, EBMULT, high order multipoles, etc. Particle monitoring resorts to keywords introduced in the previous Chapters, including FAISCEAU, FAISTORE, possibly PICKUPS, and some others. Spin motion computation and monitoring resort to SPNTRK, SPNPRT, FAISTORE. Optics matching and optimization use FIT[2]. INCLUDE is sometimes resorted to, mostly in order to modularize and/or simplify the input data files. SYSTEM is used to, mostly, call gnuplot scripts so as to end simulations with some specific graphs, or animations. These scripts read their data from output files filled up during the execution of the code, such as zgoubi.fai (resulting from the use of FAISTORE), zgoubi.plt (resulting from IL $= 2$ flag), or other zgoubi.*.out files resulting from a PRINT command.

12.1 Introduction

Beam lines ensure diverse functions such as

- beam transfer between accelerators,
- injection into, and extraction from accelerators,
- beam transport to experimental areas, purification systems, mass separators,
- beam delivery to collision points,
- beam expansion and uniformization,
- microprobes, microscope columns,
- medical rotating gantries,

and more.

Moreover, in designing large accelerators, specific sections with specific functionalities, "optical modules", may be considered separately, and in doing so handled as beam lines. This includes arcs, dispersion suppressors, spin rotators in polarized electron storage rings, recirculating loops in energy recovery linacs, etc.

© The Author(s) 2024   477
F. Méot, *Understanding the Physics of Particle Accelerators*, Particle Acceleration
and Detection, https://doi.org/10.1007/978-3-031-59979-8_12

Particle transport in beam lines in the Gauss approximation can be described using elementary laws: parabolic, sinusoidal or hyperbolic. The complexity of a beam line, in a first order approach, results from the variety of possible combinations of these laws in the optical sequence: a particle will for instance describe an arc of a circle, followed by a parabola and finally a sine or hyperbolic arc. This makes it possible to design complex optical systems with a reduced number of basic optical elements.

The simulation exercises proposed in Sect. 12.3 use stepwise raytracing as this provides detailed insight in fields and particle motion. Stepwise integration allows dealing with optical elements featuring complex geometry and/or **E** and/or **B** field structure, it allows for large transverse acceptance, large momentum offset, strong field non-linearities, in an accurate manner.

12.2 Basic Concepts and Formulæ

In the following sections, a few different types of beam lines, with specific functionalities, are addressed. The underlying theory is introduced. Corresponding simulation exercises complete the landscape.

12.2.1 Beam Expander

Beam uniformization is used for biological irradiation [1], at high power neutron targets [2], it has been foreseen for hadrontherapy [3], radioactive waste treatment and other material irradiation [4, 5]. Depending on the application the transverse beam size needed may be in the centimeter (e.g., radio-biology) to meter range (e.g., nuclear waste irradiation). Transverse uniformization is achieved by means of octupoles ($B_y|_{y=0} \propto x^3$), and possibly enhanced by adding higher odd-order components ($B_y|_{y=0} \propto x^5$: dodecapole, ...) [6–8]. The technique is generally implemented in so-called beam expanders.

A two-dimensional beam expander, designed for nuclear waste irradiation, is schemed in Fig. 12.1 [5]. It uniformizes the transverse density of a 700 MeV proton beam with 0.6 μm emittances, over a typically $1 \times 1\, m^2$ area. Quadrupoles Q1–Q7 ensure beam focusing at the irradiation target located further down the line at $s = 37\, m$. Beam uniformization over a rectangular beam cross-section at the target requires [7, 8]

   (i) an non-linear lens OH (respectively, OV), for uniformization of the transverse horizontal (respectively, vertical) particle density distribution;
  (ii) flat horizontal and vertical phase-space beam ellipses at the location of, respectively, the horizontal and the vertical lens;
 (iii) the derivative dy/dy'_l of the polynomial in Eq. 12.2 (see Fig. 12.2 and notations below) to change sign at least once.

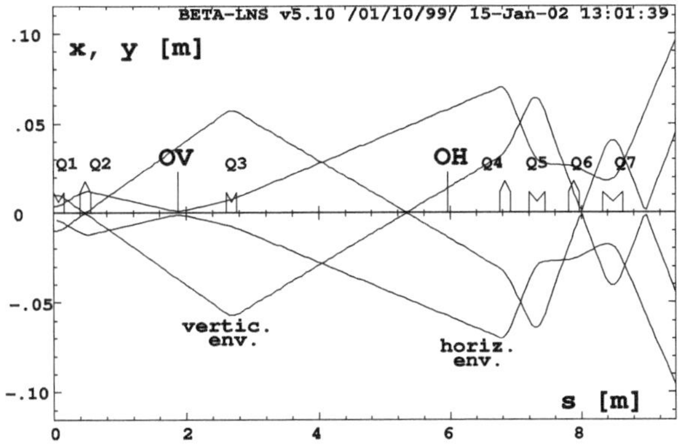

Fig. 12.1 A beam expander, including two non-linear lenses OH and OV for transverse beam uniformization. Quadrupoles Q4–Q7 control the size of that rectangle. Quadrupoles Q1–Q3 control the envelopes at the non-linear lenses. The graph displays the linear beam envelopes $\sqrt{\beta_x \varepsilon_x/\pi}$ and $\sqrt{\beta_y \varepsilon_y/\pi}$

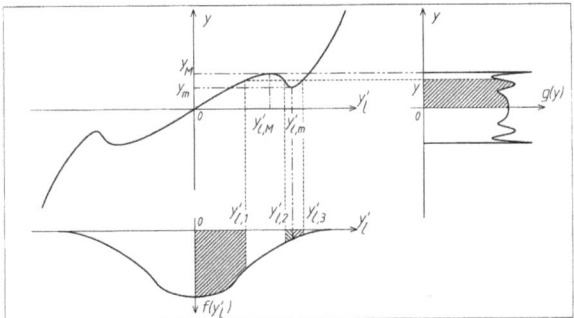

Fig. 12.2 A geometrical representation of transverse density mapping, from a Gaussian density upstream of the lens (bottom), to uniformized density at some distance s down the line (top right; Eq. 12.3), via Eq. 12.2 mapping (top left; $y(s) = \lambda_1(s)\, y'_l + \lambda_3(s)\, y'^3_l + \lambda_5(s)\, y'^5_l$ here) [8]

Given hypothesis (i) above, uniformization can be tuned independently in each plane. Hypothesis (ii) above requires $\beta_l \gamma_l$ at the lens to be large (with β_l, $\alpha_l = -\beta'_l/2$, $\gamma_l = (1 + \alpha_l^2)/\beta_l$ the local optical functions at the lens). This results in the beam ellipse at the non-linear lens to degenerate into a line, which can be written

$$\gamma_l(s)y^2 + 2\alpha_l(s)yy' + \beta_l(s)y'^2 = \varepsilon_y/\pi \overset{\beta_l \gamma_l \gg 1}{\to} y_l = r_l y'_l \quad (r_l = -\beta_l/\alpha_l) \quad (12.1)$$

In such conditions, the transverse position y(s) of a particle at arbitrary $s > s_l$ downstream of the lens is related to its incidence angle y'_l at entrance of the lens by the polynomial relationship

$$y(s) = \sum_{p=0}^{n} \lambda_{2p+1}(s)\, y'^{2p+1}_l \quad \text{with} \quad \begin{cases} n = 1 : \text{octupole alone} \\ n = 2 : \text{octupole} + \text{dodecapole} \\ \text{etc.} \end{cases} \quad (12.2)$$

wherein $\lambda_1(s) = r_l \sqrt{\beta(s)/\beta_l}\cos(\Delta\phi)$, $\lambda_{2p+1}(s) = -K_{2p+1}Lr_l^{2p+1}\sqrt{\beta(s)\beta_l}$ $\sin(\Delta\phi)$, $\Delta\phi = \phi(s) - \phi_l$ is the phase advance from the lens, and $K_{2p+1}L$ ($p = 1$ to $n \geq 1$) are the integrated strengths of the n odd-order non-linear components present in the lens.

Uniformized density

Let $f(y_l')$ be the probability density of the angle variable y_l' at the lens. Thus the density of the position variable y at arbitrary $s > s_l$ can be written

$$g(y(s)) = \sum_{i=1}^{N} \frac{f\left(y_{l_i}'(y(s))\right)}{\left| \sum_{p=0}^{n} (2p+1)\,\lambda_{2p+1}(s)\,y_{l_i}'^{\,2p}(y(s)) \right|} \tag{12.3}$$

wherein y_{l_i}' $(i = 1, N)$ are the $N \leq 2n + 1$ real roots of the polynomial Eq. 12.2 (to be found analytically for an octupole lens, $n = 1$, numerically for octupole + dodecapole and beyond, $n \geq 2$).

The third hypothesis (iii) above ensures that the image $g(y)$ of y by the Eq. 12.2 mapping presents a discontinuity of the first kind (a sufficient condition is $\lambda_1 \lambda_3 < 0$). In the case of an octupole lens it occurs at

$$\pm y_M = \pm \frac{2}{3}\lambda_1 \left(-\frac{\lambda_1}{3\lambda_3}\right)^{1/2} \tag{12.4}$$

and in immediate neighboring if higher orders are added for improved uniformization.

Assuming that the incoming beam has a Gaussian density

$$f(y_l') = \frac{1}{\sigma_{y_l'}\sqrt{2\pi}} \exp\left(-\frac{y_l'^2}{2\sigma_{y_l'}^2}\right) \tag{12.5}$$

an additional condition for near-uniform $g(y)$ is for the beam divergence $\sigma_{y_l'} = \sqrt{\gamma_{y,l}\,\varepsilon_y/\pi}$ to satisfy

$$\sigma_{y_l'} \lesssim y_{lM}' \tag{12.6}$$

This ensures that the folding of the Gaussian projected beam density occurs in the region $y' \approx \sigma_{y_l'}$ (Fig. 12.2).

Proper multipole lens strengths for that are

$$\text{octupole:} \quad K_3 L = \frac{4}{27\,y_M^2} \frac{\beta(s)}{\beta_l^2} \frac{\cos^3(\phi(s) - \phi_l)}{\sin(\phi(s) - \phi_l)} \tag{12.7}$$

$$\text{added dodecapole:} \quad K_5 L = -\frac{(K_3 L)^2}{4}\,\beta_l \tan(\phi(s) - \phi_l) \tag{12.8}$$

Figure 12.3 displays the typical 2D-uniform beam cross-section so obtained at the downstream end of the line, $s = 37\,\text{m}$ (Fig. 12.1).

Non-linear beam envelopes

First order beam optics deals with envelopes in terms of the transport of the optical functions, and the *rms* beam size in dispersion free regions is $\sigma(s) = \sqrt{\beta(s)\varepsilon/\pi}$.

Fig. 12.3 Two-dimensional
uniform beam and projected
densities

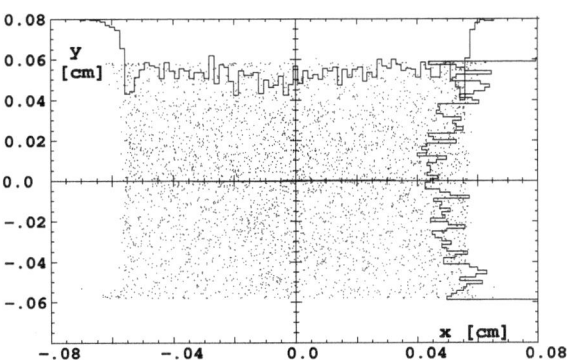

Transverse apertures (chamber, collimators, etc.) are usually defined in terms of the
$k\sigma$ envelope, with k a few units depending on the loss tolerances. The parameter of
concern in this respect is the loss rate, or relative population beyond the $k\sigma$ transverse excursion boundary, $\tau = 1 - \mathrm{erf}(k\sigma)$. In a beam expander the non-linearities
introduced by the uniformization lens are so strong that it is no longer relevant to
address τ in terms of the $k\sigma$ -envelope as the transverse density $g(y(s))$ is way too
far from Gaussian. On the other hand, non-linear lenses may induce a substantial
change of the beam loss boundary compared to the linear envelope (Fig. 12.4).

  Given some loss rate τ (e.g., of the order of $10^{-7} - 10^{-9}$ in high power installations) [5], the $(2n + 1)$-th order non-linear envelope $Y_{2n+1}(s)$ in a beam expander is
obtained by solving for $Y_{2n+1}(s)$ the integral equation

$$\int_0^{Y_{2n+1}(s)} g(y)\,dy \equiv \sum_{i=1}^{N} \int_{y'_{l_i}(y=0)}^{y'_{l_i}(y=Y_{2n+1}(s))} f(y'_{l_i}) \left| dy'_{l_i} \right| = 1 - \tau \qquad (12.9)$$

Assuming a Gaussian incoming beam density (Eq. 12.5), this expression takes the
simpler form

$$\mathrm{sign}(\lambda_1) \sum_{i=1}^{N} (-1)^i \, \mathrm{erf}\left\{ y'_{l_i} \left[Y_{2n+1}(s) \right] \right\} = 1 - \tau \qquad (12.10)$$

which can be solved numerically. Typical non-linear envelopes so obtained are displayed in Fig. 12.4.

  A possible numerical procedure to transport non-linear envelops in that manner
is the following:

- first, the roots y'_{l_i} $(y = Y_{2n+1}(s))$ of Eq. 12.2 are calculated (analytically in the
  case $2n + 1 = 3$, numerically if $2n + 1 \geq 5$),
- next, given τ, Eq. 12.10 is solved numerically for $Y_{2n+1}(s)$.

  That procedure also provides the local transverse density $g(y(s))$ at arbitrary s
along the beam line as illustrated in Fig. 12.5.

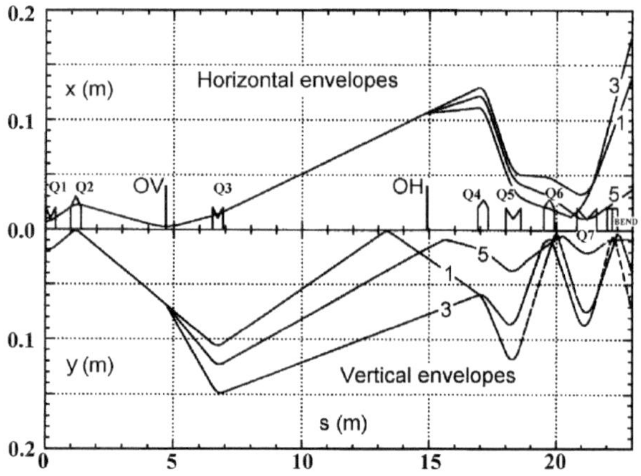

Fig. 12.4 This graph shows beam envelopes defined as a $\tau = 10^{-7}$ loss boundary (Eq. 12.10), horizontal [$x(s)$, upper curves] and vertical [$y(s)$, lower curves]. Three cases are plotted: linear envelopes (marker "1"); case of octupole at OH and at OV ("3"); case of octupole+dodecapole at OH and at OV ("5") [8]

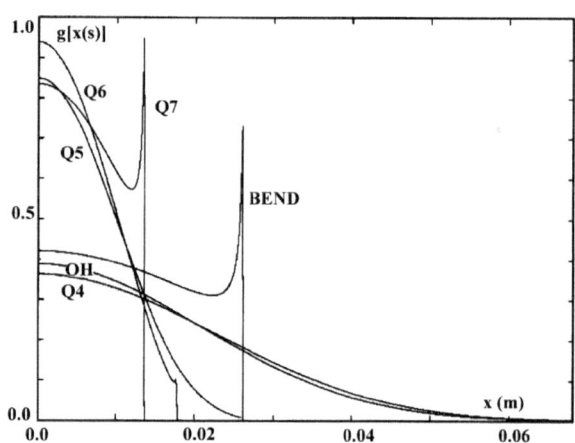

Fig. 12.5 Transverse horizontal beam density profiles $g(x)$ (a sub-product of the transport of non-linear envelopes using Eq. 12.10) at various optical elements along the beam line of Fig. 12.4, given a sole octupole component ($n = 1$) at OH [8]

12.2.2   Nano-Probe Final Focus Achromat

A magneto-electrostatic quadrupole combines electric and magnetic quadrupole components, superimposed (Fig. 12.6) [9].

It presents the property that, in the case of non-relativistic beams, it can be made achromatic to the second order so long as the electric and magnetic strengths

$$K_e = \frac{\phi}{a^2\,W} \quad \text{and} \quad K_m = \frac{B}{a\,B\rho} \tag{12.11}$$

satisfy the relationship

$$K_m = -2K_e \tag{12.12}$$

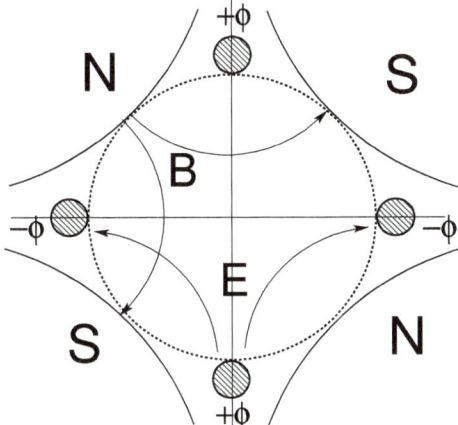

Fig. 12.6 A principle sketch of a combined magnetic-electrostatic quadrupole lens. The conducting pole tips (filled circles) are at 45° to the magnetic pole tips—and at the same pole tip radius, here

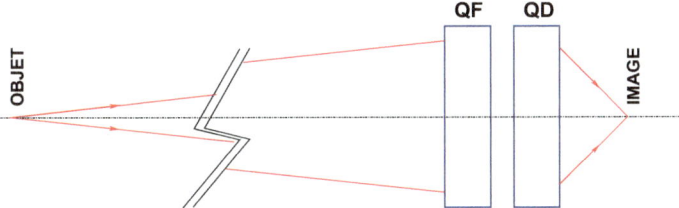

Fig. 12.7 Principle scheme of a final focus second-order achromat for a nano-probe ion beam. It is comprised of a pair of **E, B** quadrupoles (Fig. 12.6) tuned to ensure same focal distance in both transverse planes

There, ϕ is the potential at the electrodes, a is the inner radius at the electrodes and at the magnetic poles, $W = p^2 / 2m$ is the kinetic energy of the particle of mass m and momentum p, assumed non-relativistic, B is the magnetic field at pole tip. This property of second order achromatism makes the doublet of **E, B** quadrupoles an option for a final focus doublet in a nano-probe [10] (Fig. 12.7).

12.2.3 A Muon Collect Channel

A Neutrino Factory accelerator complex [11] is aimed at the production of high flux neutrino beams based on muon decay

$$\mu^{\pm} \rightarrow e^{\pm} + \nu_e \, (\overline{\nu_e}) + \overline{\nu_\mu} \, (\nu_\mu)$$

in a racetrack storage ring. A muon collect channel is part of the Neutrino Factory design [12, 13]. Muon bunches are the product of pion decay

$$\pi \to \mu + \nu$$

Pions bunches are obtained from multi-MW, GeV range proton bunches hitting a production target. A funneling optical system downstream of a set of targets steers the parent pion beams onto a common axis, into a long (a few tens of meters) muon collect channel, which can be a solenoid, or a FODO lattice (see section "Neutrino Factory Muon Collect Channel"). Following the muon collect channel, muon bunches are cooled and then accelerated to multi-GeV energy for storage in a racetrack decay ring. One of the long straight sections in the latter directs the neutrino beam towards a far distant detector [11].

In a first part in the following, the momentum and time distributions of the parent pion bunch as it decays, and of the muon bunch as it builds up, are examined. In a second part a Funnel + FODO lattice design of the Neutrino Factory muon collect channel is described.

Momentum and Energy Spectra

In-flight decay is sketched in Fig. 12.8 which also defines the scattering angles θ and ϕ. Indices π and μ in the following designate respectively the pion and muon particles. A superscript "*" denotes quantities taken in the center of mass of the two-body decay. Useful quantities in the following include,

Regarding pions:
- Momentum; energy; normalized velocity $\quad p_\pi; \; E_\pi; \; \beta_\pi = p_\pi/E_\pi$
- Life time at rest; corresponding path length $\quad \tau_\pi, \; s_\pi = \beta_\pi \gamma_\pi c \tau_\pi = p_\pi/\eta$
- Decay law $\quad N(s) = N_0 e^{-\eta s/p_\pi} \quad (\eta = m_\pi/c\tau_\pi)$

Regarding muons:
- Momentum; energy; normalized velocity $\quad p_\mu; \; E_\mu; \; \beta_\mu = p_\mu/E_\mu$
- Center-of-mass energy $\quad E_\mu^* = (m_\pi^2 + m_\mu^2)/2m_\pi$
- Energy in laboratory $\quad E_\mu = \gamma_\pi \left(E_\mu^* + \beta_\pi p_\mu^* \cos\theta_\mu^*\right)$
- Laboratory decay angles $\quad \phi_\mu = \phi_\mu^*, \; \tan\theta_\mu = \dfrac{1}{\gamma_\pi} \dfrac{\sin\theta_\mu^*}{\frac{\beta_\pi}{\beta_\mu^*} + \cos\theta_\mu^*}$

Parent Pion Bunch

The density of survived pions with initial momentum p_π, as a function of flight distance s, writes

$$g_{s|p_\pi}(s, p_\pi) = (\eta/p_\pi)\exp(-\eta s/p_\pi) \tag{12.13}$$

Given $g_{p_\pi}(p_\pi)$ the initial momentum density (at $s = 0$), one gets the 2D density at arbitrary $s > 0$

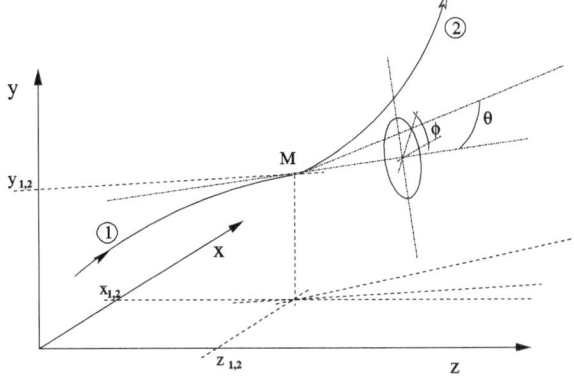

Fig. 12.8 In-flight decay of particle 1 at $M(x_1, y_1, z_1)$. Particle 2 is emitted at angles θ, ϕ

$$g_{s,p_\pi}(s, p_\pi) = g_{s|p_\pi} \times g_{p_\pi} \qquad \left(\text{and } \int_{s=0}^{\infty} \int g_{s,p_\pi}(s, p_\pi)\, ds\, dp_\pi = 1\right) \quad (12.14)$$

For the sake of simplicity a uniform initial pion momentum density may be considered,

$$g_{p_\pi}(p_\pi) = \mathbf{1}_{\Delta p_\pi}(p_\pi) = \begin{cases} 1/(p_{\pi_2} - p_{\pi_1}) & \text{if } p_\pi \in [p_{\pi_1}, p_{\pi_2}] \\ 0 & \text{otherwise} \end{cases} \quad (12.15)$$

The resulting form of $g_{s,p_\pi}(s, p_\pi)$ is shown in Fig. 12.9, given a pion bunch launched at $s = 0$ with zero size and $p_\pi \in [100, 500]$ MeV/c. Integrating Eq. 12.14 with respect to s yields the p_π-density of the decayed parent pions at distance s,

$$g_{p_\pi}(p_\pi)\big|_s = \int_0^s g_{s,p_\pi}(s, p_\pi)\, ds = \mathbf{1}_{\Delta p_\pi}(p_\pi)\left[1 - \exp\left(-\frac{\eta s}{p_\pi}\right)\right] \quad (12.16)$$

The p_π-density of the *non-decayed* pion population ensues,

$$\tilde{g}_{p_\pi}(p_\pi)\big|_s = (g_{p_\pi}(p_\pi) - g_{p_\pi}(p_\pi)\big|_s) = \mathbf{1}_{\Delta p_\pi}(p_\pi) \exp\left(-\eta s/p_\pi\right) \quad (12.17)$$

The s-dependence of the pion bunch average momentum follows, namely (Fig. 12.10)

$$\bar{p}_\pi(s) = \frac{\int p\, \tilde{g}_{p_\pi}(p)\big|_s\, dp}{\int \tilde{g}_{p_\pi}(p)\big|_s\, dp} = \frac{\sum_{i=1,2}(-1)^i \cdot \dfrac{p_{\pi_i}^2 - \eta s\, p_{\pi_i} - \eta^2 s^2\, e^{\frac{\eta s}{p_{\pi_i}}} \operatorname{Ei}(-\frac{\eta s}{p_{\pi_i}})}{2\, e^{\eta s/p_{\pi_i}}}}{\sum_{i=1,2}(-1)^i \dfrac{p_{\pi_i} + \eta s\, e^{\frac{\eta s}{p_{\pi_i}}} \operatorname{Ei}(-\frac{\eta s}{p_{\pi_i}})}{e^{\eta s/p_{\pi_i}}}}$$

$$(12.18)$$

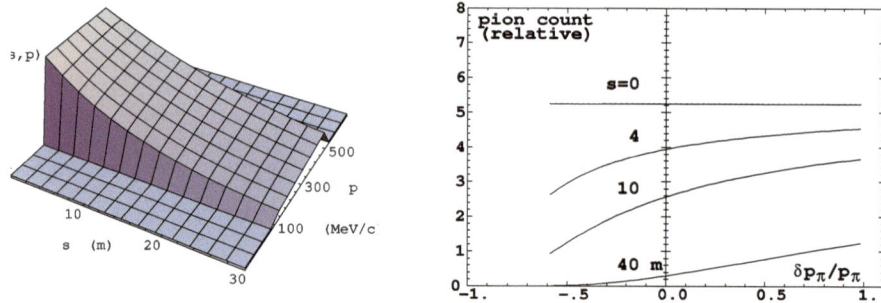

Fig. 12.9 Left: pion bunch momentum density along the 30 m decay channel (Eq. 12.14), assuming uniform initial momentum density over [100, 500] MeV/c (Eq. 12.15). Right: *s*-sections of the former, at various distances along the collect channel

Fig. 12.10 Average momentum of pion and muon bunches over 60 m flight distance (Eq. 12.18)

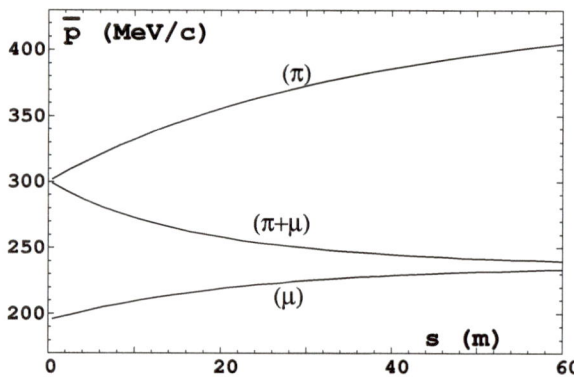

Muon bunch

Given a fixed muon momentum $p_\mu \in [\gamma_\pi (\beta_\pi E_\mu^* - p_\mu^*), \gamma_\pi (\beta_\pi E_\mu^* + p_\mu^*)]$, the decay muon momentum (Fig. 12.11) satisfies a p_π-conditional density which writes

$$g_{p_\mu | p_\pi}(p_\mu) = g_{\theta^*}(\theta_\mu^*) \left| d\theta_\mu^* / dp_\mu \right| = \frac{m_\pi}{2 p_\pi p_\mu^*} \frac{p_\mu}{\sqrt{p_\mu^2 + m_\mu^2}} = \frac{m_\pi}{2 p_\pi p_\mu^*} \beta_\mu \quad (12.19)$$

wherein (Fig. 12.11-right) $g_{\theta^*}(\theta_\mu^*) = \sin \theta_\mu^* / 2 \ \ (\theta_\mu^* \in [0, \pi])$ is the decay angle density. Outside the p_μ interval $g_{p_\mu | p_\pi}(p_\mu) \equiv 0$. The muon energy density at fixed p_π writes

$$g_{E_\mu | p_\pi}(E_\mu) = \frac{m_\pi}{2 p_\pi p_\mu^*} \quad \text{with } E_\mu \in [\gamma_\pi (E_\mu^* - \beta_\pi p_\mu^*), \gamma_\pi (E_\mu^* + \beta_\pi p_\mu^*)] \quad (12.20)$$

A non-zero pion bunch momentum extent is accounted for by multiplying the p_π-conditional density $g_{p_\mu | p_\pi}(p_\mu)$ (Eq. 12.19) by the muon density at s at given p_π,

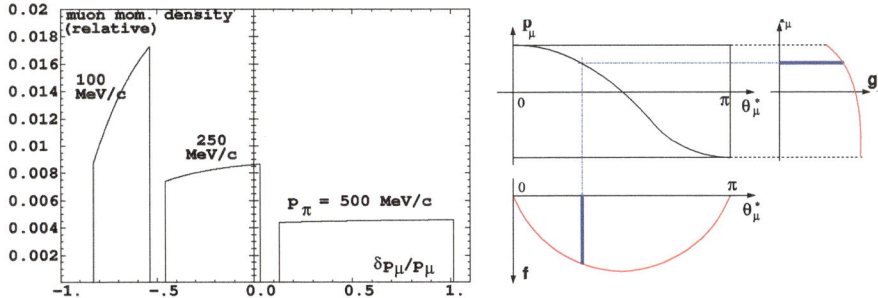

Fig. 12.11 Left: density $g_{p_\mu | p_\pi}(p_\mu)$ (Eq. 12.19) for initial pion momenta $p_\pi = 100$, 300 and 500 MeV/c. Right: geometrical representation of its build-up from $g_{\theta^*}(\theta_\mu^*)$ in the change of variable $\theta_\mu^* \to p_\mu$

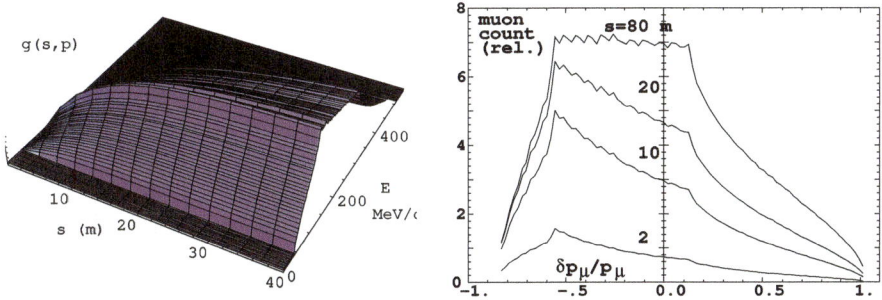

Fig. 12.12 Left: muon momentum density along the decay channel (from an analytical primitive of the integral in Eq. 12.21 [13]). Right: s-sections of the former, at various distances along the collect channel, from rough numerical computation of the integral

$g_{s,p_\pi}(s, p_\pi)$ (Eq. 12.14). (The muon decay is not taken into account in the following for simplicity, doing so would mean accounting for an s-dependent muon survival additional factor.) This yields the muon momentum spectrum at s under the integral form

$$g_{p_\mu}(p_\mu)|_s = \int_{\Delta p_\pi} g_{p_\mu | p_\pi} \, dp_\pi \int_0^s g_{s,p_\pi}(s, p_\pi) \, ds \quad (\lim_{s \to \infty} g_{p_\mu}(p_\mu)|_s = 1) \quad (12.21)$$

The Δp_π integration interval is a function of p_μ following the dependence given in Eq. 12.19 (not all pions can produce a muon of momentum p_μ). A similar integral holds for the energy spectrum $g_{E_\mu}(E_\mu)|_s$, given $g_{E_\mu | p_\pi}$ (Eq. 12.20).

The summation in Eq. 12.21 can be viewed as a superposition of the fixed-p_π muon spectra of Fig. 12.11, this is the way the muon spectra shown in Fig. 12.12 have been numerically calculated.

The determination of the mean momentum $\bar{p}_\mu(s)$ and energy $\bar{E}_\mu(s)$ of the muons, and as well the momentum of the center of gravity of the hybrid bunch follow similar

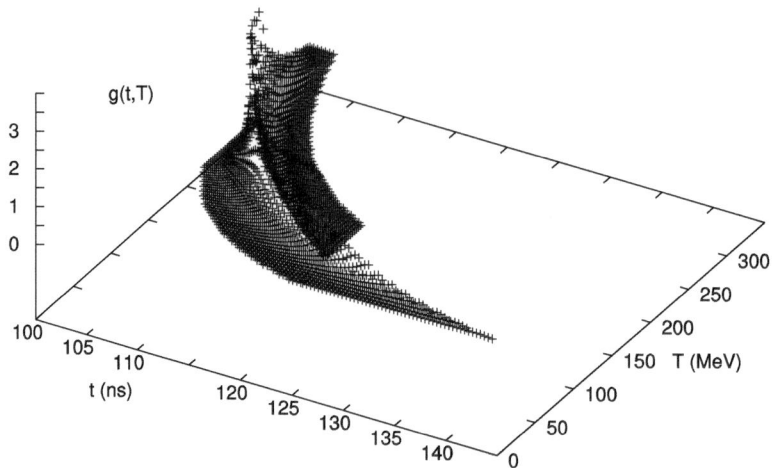

Fig. 12.13 Muon bunch time-kinetic energy density (Eq. 12.22, arbitrary units) at $s = 30\,\mathrm{m}$, in the case $p_\pi \in [200, 400]\,\mathrm{MeV/c}$

calculations as for the pion bunch, above, they are represented in Fig. 12.10. Average momenta of both the pion and the muon bunches are increasing functions of the distance, because the lower energy parent pions decay faster, whereas the average momentum of the $\pi + \mu$ bunch decreases monotonically (Fig. 12.10), a behavior that can be accounted for to maintain constant focusing strength in tuning the decay channel.

Muon bunch longitudinal phase space

The p_π-conditional time density $g_{t_\mu | p_\pi}(p_\mu)$ of the muons at arbitrary s is derived from $g_{\theta^*}(\theta^*_\mu)$ through a change of variable

$$\theta^*_\mu \quad \rightarrow \quad t_\mu = s_\pi/c\beta_\pi + (s - s_\pi)/c\beta_\mu$$

This yields the time-energy two-dimensional density under the integral form (Fig. 12.13)

$$g_{t_\mu, E_\mu}(t_\mu, E_\mu)|_s = \int_{s_\pi=0}^{s} \int_{\Delta p_\pi} g_{t_\mu | p_\pi}(t_\mu)\, g_{s, p_\pi}(s_\pi, p_\pi)\, dp_\pi\, ds_\pi \qquad (12.22)$$

The time boundaries of the distribution, namely,

$$ct_\mu^{\pm} = (s - s_\pi)\frac{(E_2^* \pm \beta_\pi p_\mu^*)}{(\beta_\pi E_2^* \pm p_\mu^*)} \qquad (12.23)$$

correspond to the fastest and slowest muon emitted by respectively the fastest and slowest pion. The first two moments of $g_{t_\mu}(t_\mu)|_s$ can be calculated so as to derive capture efficiencies as was done for the energy spectra.

The muon time density $g_{t_\mu}(t_\mu)|_s$ (Eq. 12.22) has an explicit dependence on s. Note that, given the pion energies in concern here, the flight distance s can be considered in good approximation as the position along the channel length.

The muon population at arbitrary s can also be reconstructed from Eq. 12.22: the (t_μ, E_μ) space can be meshed, $N_0 g_{t_\mu, E_\mu}(t_\mu, E_\mu)|_s \Delta p_\mu \Delta E_\mu$ gives the local number of points on the mesh. Monte Carlo simulations of longitudinal phase-space at distance s along a drift confirm these results (see Exercise 12.10).

Neutrino Factory Muon Collect Channel

A target system of a neutrino factory requires the incident proton driver MW bunches to be distributed over several targets [12]. The pion bunches produced at the targets are focused by a horn system into a few meter long funnel, which merges them along a common axis (Fig. 12.14). A straight pion decay/muon collect channel follows which may be either a FODO line or a long solenoid. The main goal in designing a muon collect channel is to maximize the pion and muon transmission. The overall length of that funnel+FODO collect channel depends on the muon momentum, ≈ 30 m typically for parent pion beam momentum $\approx 200 \sim 400$ MeV/c (Fig. 12.15).

As an illustration, Fig. 12.15 shows sample parent pion and decay muon trajectories, in the case of a funnel+FODO channel tuned on 300 MeV/c and with initial pion bunch densities parabolic in x and y coordinates, total emittances $\varepsilon_x/\pi = \varepsilon_y/\pi = 0.01$ m, initial pion momentum distribution $p \in [200, 400]$ MeV/c, uniform. Tracking trials in these hypotheses show that pion losses are marginal along a 0.3 m aperture radius funnel channel, muon losses are marginal as well along a 0.5 m aperture radius FODO line. The effect of beam line aperture is illustrated in Fig. 12.16.

The latter result assumes that all pions and muons that make it to the end of the line ($s = 30$ m) are counted. In actuality, downstream lines (muon bunching, accelerators, etc.) only allow so much admittance, accounted for this is illustrated in Fig. 12.17: an admittance is defined, namely $\varepsilon_x/\pi = 0.01$, 0.02, or 0.04 m, and a best matching ellipse is computed, i.e., the surface of the ellipse ε_x is imposed, beam parameters $\alpha_{x,y}$ and $\beta_{x,y}$ are free, they are obtained by matching with the sole criterion of best transmission, i.e., greatest count within the ellipse boundary. As part of that matching procedure, the FODO channel strength may be fine-tuned so that the matching ellipse find itself on-axis at the end of the line. The vertical phase-space is subjected to a similar treatment.

The muon beam longitudinal emittance (Fig. 12.13) may be subjected to a boundary as well, to account for the longitudinal admittance of the downstream lines: the surface ε_l of the ellipse is imposed, its center and beam parameters α_l and β_l are free, the constraint is maximum count through the ellipse boundary (Fig. 12.18).

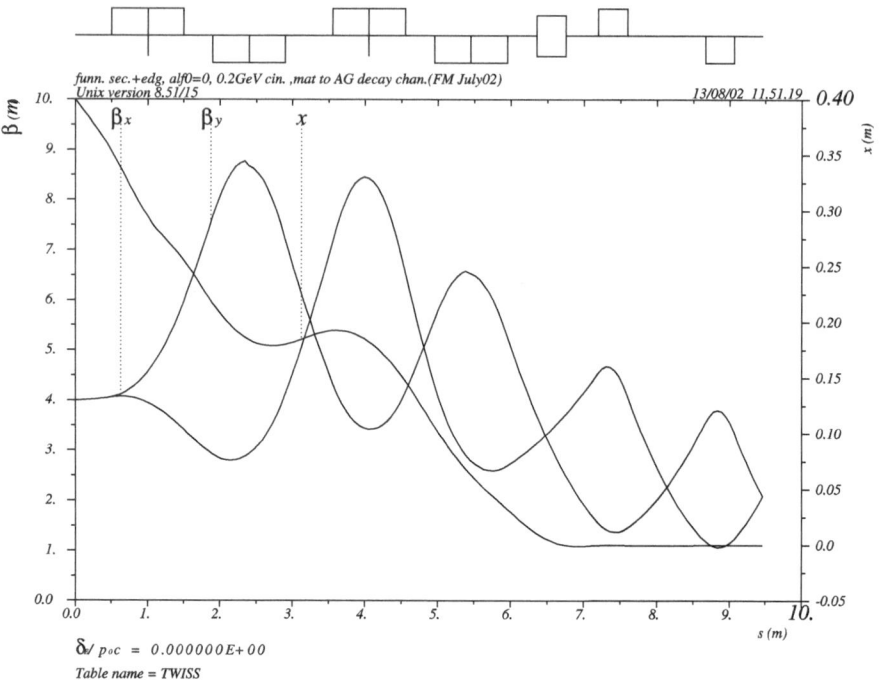

Fig. 12.14 Funnel section of the upstream part of a multiple-target pion and muon collect channel [12]. On the top synoptic the boxes on both sides of the axis represent the quadrupoles (focusing, up, or defocusing, down), boxes straddling the axis represent bending dipoles. The pion production target is at the left end, a long muon collect channel follows to the right (Fig. 12.15). This graph shows the off axis trajectory of the beam centroid and the optical functions. A graph obtained using the MAD program [14]

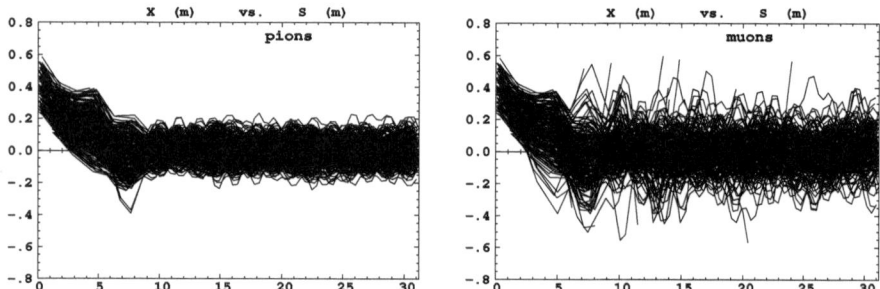

Fig. 12.15 Sample rays in a muon collect channel, including an upstream 9 m long funnel and a downstream 32 m long FODO line. Left: parent pions. Right: decay muons. Trajectories end where particles hit the vacuum pipe. The latter is 0.3 m in radius along the funnel, 0.5 m in radius along the FODO line

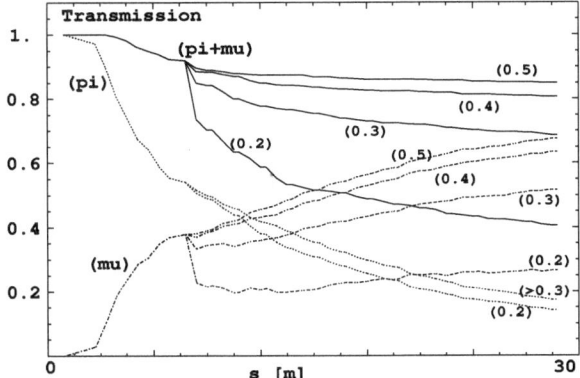

Fig. 12.16 Transmission efficiency of the the pion, muon or pion+muon beams, versus distance along a $\pi + \mu$ collect channel (relative to the total number of parent pions, at s=0), for either 0.2, 0.3, 0.4 or 0.5 m radius collimation along the FODO line, and given an aperture radius > 0.3 m along the funnel. A 0.5 m aperture radius FODO line yields 85% transmission at s $= 30$ m

Fig. 12.17 Horizontal phase space at end of the 0.5 m radius collimation FODO channel (s $= 30$ m). The ellipses are the best match to the $\pi + \mu$ beam, with emittances imposed, namely, $\varepsilon_x/\pi = 0.01,\ 0.02$ or 0.04 m. The FODO channel strength has been fine-tuned to center the beam ellipses

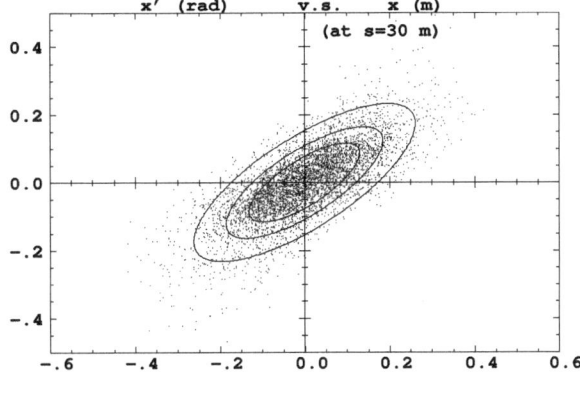

Fig. 12.18 Time-energy phase space at the end of the FODO channel (s $= 30$ m), and best matching ellipse, for the four different admittance values $\varepsilon_l = 0.1,\ 0.5,\ 1$ and $2\,\pi$ eV.s

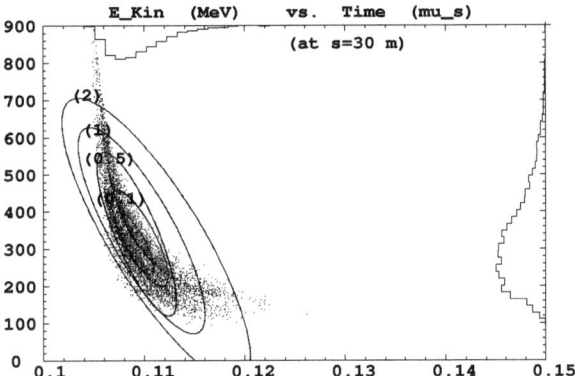

Fig. 12.19 Transmission
efficiency (relative to the
total number of parent pions,
at s = 0) for various
transverse admittance at
s = 30 m, as a function of the
longitudinal admittance ε_l/π
at s = 30 m

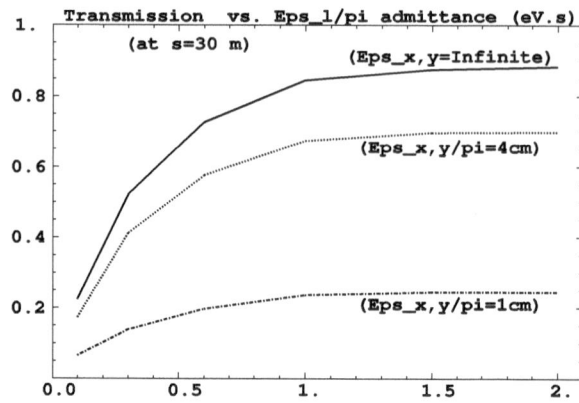

Results may be summarized as in Fig. 12.19 which shows the transmission efficiency as a function of the longitudinal acceptance ε_l/π at line end (s = 30 m), given $\varepsilon_x/\pi = \varepsilon_y/\pi = 0.01$ m, or 0.04 m, or infinite transverse acceptance. Parent pions have been launched with $p \in [200, 400]$ MeV/c, uniform, and with $\varepsilon_x/\pi = \varepsilon_y/\pi = 0.01$ m. The FODO section of the AG channel has a 0.5 m radius collimation, the beam line is tuned to 240 MeV/c for optimum transmission.

12.2.4 Low Energy Spin Rotator

In a polarized beam installation, prior to injection into downstream stages, a linac for instance, polarization generally needs be oriented with respect to the beam propagation axis, from longitudinal at the polarized electron source (Fig. 12.20) [15, 16].

In electron installations, Wien filters may be used as part of the polarization rotation equipment.

An expression for the spin rotation over a distance L may be obtained by analogy between the Lorentz equation and the spin motion equation. Namely, from

$$\frac{d\mathbf{v}}{ds} = \mathbf{v} \times \frac{\mathbf{B}}{B\rho} \quad \Rightarrow \quad \text{trajectory deviation} = \frac{BL}{B\rho} \tag{12.24}$$

wherein **B** is the velocity precession vector, one infers

$$\frac{d\mathbf{S}}{ds} = \mathbf{S} \times \frac{\omega}{B\rho} \quad \Rightarrow \quad \theta_{sp} = \frac{\omega L}{B\rho} \tag{12.25}$$

with ω the spin precession vector. The spin angular momentum results from two torques: from the magnetic field and from the electric field. The respective precession vectors can be expressed under the form

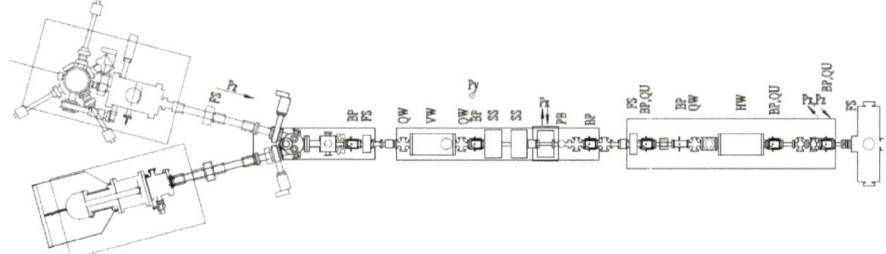

Fig. 12.20 Low energy (~150 keV) beam line at CEBAF. The first Wien filter (VW, vertical) downstream of the photo-guns rotates the polarization from longitudinal to vertical. The second Wien filter (HW, horizontal) rotates the polarization in-plane to compensate precession of CEBAF transport magnets. Solenoids (SS) ensure additional polarization rotation requirements [15]

Fig. 12.21 A sketch of the reference frame used and electron motion in pure **E** field (dots, red) or pure **B** field (dashes, blue). Under the conjugate effect of both, with $E_Y = vB_Z$, the electron trajectory is along the X axis (green)

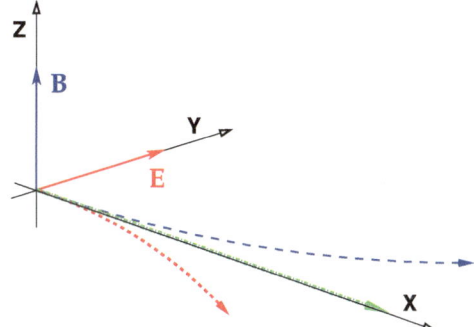

$$\omega_B = (1+a)\,\mathbf{B}_\parallel + (1+a\gamma)\,\mathbf{B}_\perp \tag{12.26}$$

$$\omega_E = \gamma\left(\frac{1}{\gamma+1} + a\right)\frac{\mathbf{E}\times\mathbf{v}}{c^2}$$

with $\mathbf{B}_\parallel$ the projection of $\mathbf{B}$ on the velocity direction and $\mathbf{B}_\perp = \mathbf{B} - \mathbf{B}_\parallel$ normal to the latter. The anomalous magnetic moment for the electron is $a = 1.15965 \times 10^{-3}$.

Considering the hypotheses defined in Fig. 12.21, take $\mathbf{E} \parallel \mathbf{Y}$ and $\mathbf{B} \parallel \mathbf{Z}$. The condition for a straight electron trajectory is

$$E_Y = v\,B_Z \tag{12.27}$$

In the present $\mathbf{E}$, $\mathbf{B}$ configuration, in particular $\mathbf{B} \perp \mathbf{v}$ always, one has

$$\omega = \omega_B + \omega_E = (1+a\gamma)\,\mathbf{B}_\perp + \gamma\left(\frac{1}{\gamma+1} + a\right)\frac{\mathbf{E}\times\mathbf{v}}{c^2} \tag{12.28}$$

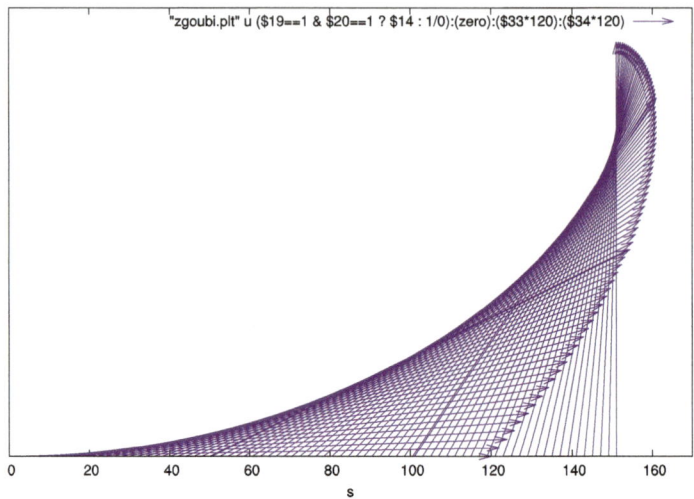

"zgoubi.plt" u ($19==1 & $20==1 ? $14 : 1/0):(zero):($33*120):($34*120) ⟶

Fig. 12.22 Spin motion in the $\mathbf{B}^\perp$ plane (the (X,Y) plane in Fig. 12.21) over the straight electron trajectory, from longitudinal at s=0 to transverse at s = 1.5 m

Taking $\mathbf{z}$ unitary vector along Z, accounting in addition for $\mathbf{E} \perp \mathbf{v}$, one has

$$\mathbf{B} = B_Z\, \mathbf{z}, \quad \mathbf{E} \times \mathbf{v} = -v E_Y\, \mathbf{z} = -v^2 B_Z\, \mathbf{z}, \quad \omega = \omega\, \mathbf{z} \qquad (12.29)$$

With these ingredients ω can be substituted in the θ_{sp} expression above (Eq. 12.25) to yield

$$\theta_{sp,th} = \underbrace{(1+a\gamma)\frac{B_Z L}{B\rho}}_{\text{from }\mathbf{B}} - \underbrace{\gamma\left(\frac{1}{\gamma+1}+a\right)\beta^2\frac{B_Z L}{B\rho}}_{\text{from }\mathbf{E}} = 30° \qquad (12.30)$$

This is illustrated in Fig. 12.22 which shows a 90° Y-rotation of the spin, from $\mathbf{S} \parallel \mathbf{v}$ to $\mathbf{S} \perp \mathbf{v}$, along the straight trajectory of an electron through the Wien filter $\mathbf{E},\ \mathbf{B}$ field.

12.2.5 Synchrotron Radiation

Electron beams radiate in beam lines, which include the long beam lines which recirculating linear accelerators are. This entails energy loss with effects on beam dynamics, such as spiraling in bends and emittance growth. On the other hand, visible radiation is used for beam diagnostics in circular accelerators, at high energy in the case of hadron rings. The diagnostics system in these rings may be based on a main magnet edge, on one or more dedicated short dipoles, or a short wiggler, or else, and can be dealt with as a stand alone short beam line.

SR and its simulation in these beam line style of systems has been addressed in earlier chapters. Topics addressed there are summarized below, together with further additional material relevant to numerical raytracing simulations, possibly documented in zgoubi sourceforge repository examples.

SR Energy Loss

Energy loss by synchrotron radiation has been introduced in Chap. 5, Sect. 5.2.3, as its effects were first observed, and overcome, in a betatron. SR loss in a linear FFAG beam line, and its effects on spin motion, are addressed in Chap. 11, Sect. 11.2.3. First order perturbation methods [17] can be resorted to in order to validate raytracing outcomes in long beam lines, such as longitudinal and transverse beam matrix perturbation, and transverse and longitudinal emittance growth.

The 1.2 km long final focus delivery system of the 250 GeV TESLA linear collider test facility (TTF) for instance, has been the object of such cross-checks. An ad hoc Tech. Note details the method and outcomes [18]. Zgoubi simulation files are also available [19].

SR Poynting

Visible SR and its use for beam diagnostics have been introduced in Chap. 8, Sect. 8.2.3, as the first observation happened at a weak focusing synchrotron. Several publications are available which the reader can refer to, reporting zgoubi simulations and/or analytical modeling of SR in short sections of rings. From the point of view of the SR electric field impulse and it spectrum, these can be handled and treated as short beam line segments. Published material includes:

- 45 GeV electron beam diagnostics and SR interference, at LEP in the 1990s, from a series of four short dipoles, Sect. 9.2.6 and Refs. [20–22]
- diagnostics at the LHC, using visible SR from a dipole edge+wiggler string [23]
- 270 GeV proton beam diagnostics at the SPS, using edge radiation, Refs. [24, 25], a case of edge observation similar to the two-dipole configuration of Exercise 9.6
- visible proton SR from an undulator at the SPS [26], a case which can be simulated using the UNDULATOR element in zgoubi.

12.3   Exercises

Note: some of the input data files for these simulations (the long files!) are available in zgoubisourceforge repository at https://sourceforge.net/p/zgoubi/code/HEAD/tree/branches/exemples/book/zgoubiMaterial/beamLines/

12.3.1   Beam Expander

This exercise concerns optical tests planned at the G4 line of GANIL experimental areas (Fig. 12.23) [3], regarding beam uniformization for hadrontherapy application [3].

The linear optics of the line is first established, including proper locations for horizontal and vertical non-linear lenses. The latter are then powered to uniformize the transverse beam density, in one or both planes. The exercise is concluded with the exploration of a uniform ribbon sweeping technique.

Fig. 12.23 G4 beam line enclosure in GANIL experimental areas [28]. Beam goes from bottom left to top right. The first four quadrupoles in the present simulation sequence (Q21–Q24, Table 12.1) are located in the upstream part of the line, Q21 is next to the shielding at the lower left corner of the room

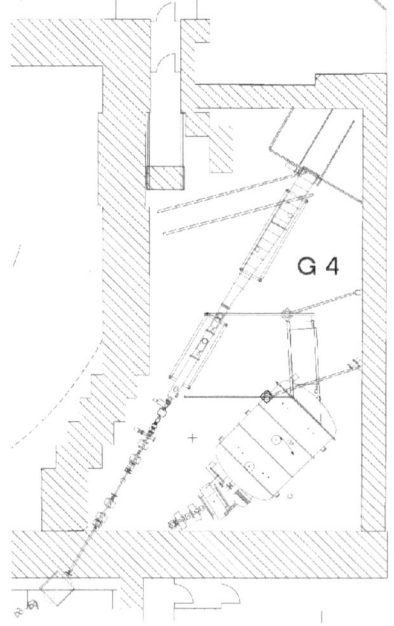

12.1 Construct a Beam Expander Line
Solution 12.1.

Consider Fig. 12.24. The beam expander section of the line follows G4's Q21–Q24 quadrupole section. The optics is set to create a vertical waist and large β_x at the horizontal uniformization octupole, OH, and a horizontal waist and large β_y at the vertical uniformization octupole, OV. The four quadrupoles downstream of OV are tuned to provide the desired size of the square image at the target, including $\alpha_x = \alpha_y = 0$.

The optical parameters of the optical sequence are detailed in Table 12.1.

(a) Explain how the beam uniformization line operates.

(b) Produce a simulation input data file of this expander, from the data in Table 12.1. QUADRUPO can be used to simulate the quadrupoles, or MULTIPOL as well; the latter is used to simulate the octupoles; BEND can simulate the final horizontal and vertical scanning dipoles. The octupole lenses and final dipoles are assumed off, for the moment.

(c) Produce the transport coefficients of the line from quadrupole Q21 to TARGET, to third order. OBJET[KOBJ = 6.1] can be used to generate a proper object for 3rd order matrix computation by MATRIX.

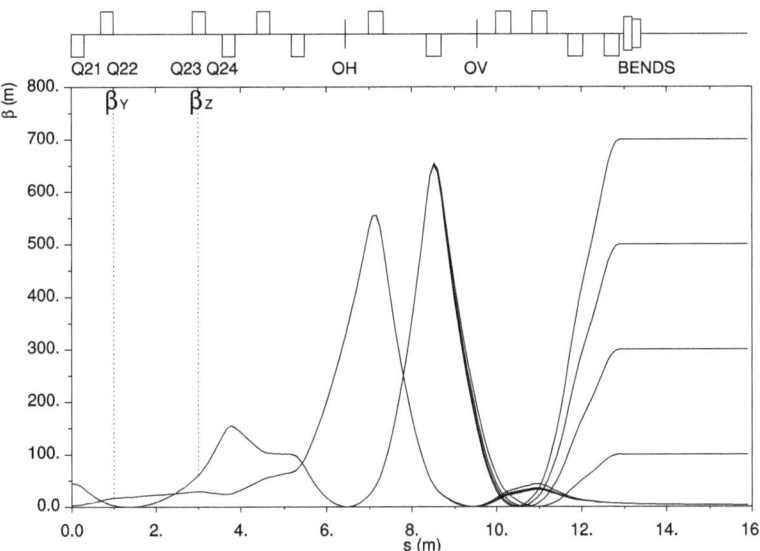

Fig. 12.24 A beam expander/beam uniformization line. Optical functions β_Y and β_Z along the line, including a scan of β_Z at target, from 10 to 700 m. The first four quadrupoles (boxes above or under the axis) are part of GANIL G4 beam line. The following eight ones are part of the expander line extension. The rightmost two scanning dipoles (boxes straddling the axis) are aimed at allowing a sweeping of the uniformized beam spot across the target, located at $s = 15.9$ m

Table 12.1 G4 beam line and added expander section. Fields in quadrupoles are at 10 cm radius

Element	Name	Length (cm)	Field (kG)	
Final section of GANIL G4 beam line:				
QUADRUPO	Q21	30.4	−3.674189	
DRIFT	DR12	39.6		
QUADRUPO	Q22	30.4	2.067729	
DRIFT	DR23	187.5		
QUADRUPO	Q23	30.4	1	
DRIFT	DR34	39.6		
QUADRUPO	Q24	30.4	−2.791886	
DRIFT	SDP5	50		
Additional tuning knobs:				
QUADRUPO	Q25	30.4	1	
DRIFT	SDP5	50		
QUADRUPO	Q26	30.4	−3	
DRIFT	SDP5-4	90		
Expander/uniformization section:				
MULTIPOL	OH	10.000	TBD[(*)]	Horizontal octupole
DRIFT	SDP5	50		
QUADRUPO	QV1	35.000	2.805048	
DRIFT	SD1	100		
QUADRUPO	QV	35.000	−3.034423	
DRIFT	SDP8	80		
MULTIPOL	OV	10.000	TBD[(*)]	Vertical octupole
DRIFT	SDP4	40		
QUADRUPO	QPL3	35.000	1.896253	
DRIFT	SDP5	50		
QUADRUPO	QPL2	35.000	2.464301	
DRIFT	SDP5	50		
QUADRUPOQ	QPL1	35.000	−0.1375607	
DRIFT	SDP5	50		
QUADRUPO	QPL0	35.000	−3.357716	
DRIFT	SD1LB	10		
H. BEND	BH	20	0	Scanning dipoles
V. BEND	BV	20	0	
DRIFT	SD1LA	250		
MARKER	TARGET			

(*) Field value to be determined, as part of the exercise

12.2 Expander Line Optics
Solution 12.2.

The horizontal and vertical optical functions at entrance of Q21 in GANIL G4 line are set to, respectively,

$$\alpha_Y = -1.435, \ \beta_Y = 3.00933\,\text{m}, \ \varepsilon_Y/\pi = .05 \times 10^{-6}\,\text{m}$$
$$\alpha_Z = -6.129, \ \beta_Z = 44.3186\,\text{m}, \ \varepsilon_Z/\pi = 10^{-6}\,\text{m} \tag{12.31}$$

(a) Give a graph of a few tens of trajectories through the line, in the horizontal and vertical planes, taking initial coordinates at random in a Gaussian distribution with beam parameters given in Eq. 12.31. MCOBJET[KOBJ = 3] can be used for that.

(b) Compute and plot the optical functions along the beam line, using the $\beta_Z(\text{TARGET}) = 10\,\text{m}$ expander section settings of Table 12.1. Produce the beam matrix at target.

Compute and plot the optical functions for the additional four settings, $\beta_Z = 100, 300, 500$, and $700\,\text{m}$, Table 12.2. Check that β_Z values at TARGET are as expected.

12.3 Transverse Uniformization; Beam Scan

Solution 12.3.

Consider the expander/uniformization section optical settings of Table 12.2: they sample the quadrupole strengths (the vertical octupole strengths are to be determined as part of the exercise), necessary to achieve the rectangle beam spot sweep at target displayed in Fig. 12.25. Note that the vertical octupole only is used, here, uniformization is in the vertical direction only, the horizontal density remains that defined by MCOBJET, i.e., Gaussian.

(a) Determine the OV octupole field values for a uniform vertical density at target, in the five different β_Z cases.

Table 12.2 Strength in expander/uniformization section quadrupoles for five different values of β_Z at target. A continuous sweep of the rectangular spot across the target is obtained by interpolation, from these discrete values, of the field in the final lenses and in the BH and BV bends of the line

		β_Z at target (m)				
		10	100	300	500	700
QV	m^{-2}	−3.0342	−3.2068	−3.2996	−3.3529	−3.3933
OV octupole	m^{-4}	Strengths to be determined				
QPL3	m^{-2}	1.8962	2.884847	3.207578	3.230446	3.211739
QPL2	m^{-2}	2.4643	2.3276	2.0705	2.1048	2.1589
QPL1	m^{-2}	−0.13756	−1.4562	−1.1073	−1.1073	−1.1073
QPL0	m^{-2}	−3.3577	−1.1889	−1.0442	−0.95639	−0.91447

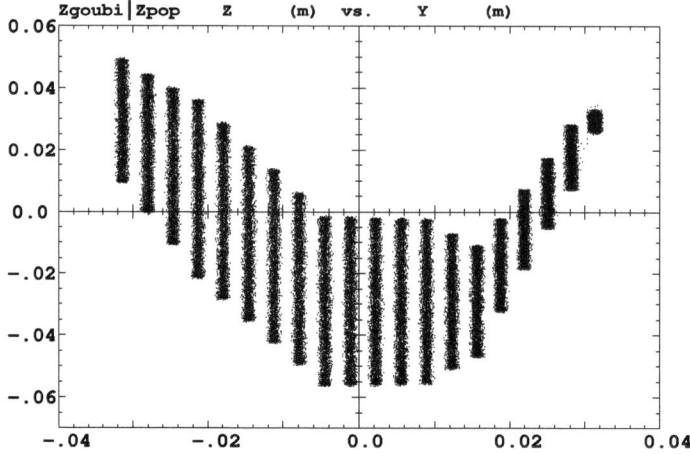

Fig. 12.25 Simulation of a continuous rectangle sweep. Twenty steps are shown here. The rectangle beam density is uniform in the vertical direction, Gaussian in the horizontal direction

Determine the dipole field values in the scanning dipoles:

– in the case of BH, at spots number 1 (rightmost), 6 and 20 (leftmost),
– in the case of BV, at spots number 1 (rightmost), 6, 12 and 20 (leftmost),

proper to position the rectangular spot as displayed in Fig. 12.25, at the five β_Z (TARGET) values.

Reproduce Fig. 12.25, using REBELOTE[NRBLT = 19] to iterate, and SCALING to vary the expander/uniformization quadrupole settings, the BH and BV fields, and the vertical octupole field, over the 20 iterations.

(b) Using the results of question 12.3(a), produce a 2D-uniformized $6 \times 6 \, \text{cm}^2$ beam spot at target. Give a graph of the transverse cross-section.

(c) Reproduce some of the transverse beam density patterns downstream of the octupoles, in a similar way to Fig. 12.5. HISTO[PRINT] can be used, it prints out its histograms to zgoubi.res, and to zgoubi.HISTO.out whose content can then be plotted (using gnuplot, for instance).

12.4 Animation: Transverse Scan by a Uniform Beam
Solution 12.4
Produce an animation of a transverse scan by a 1,000-particle rectangle with uniform vertical density.

12.3.2 Nano-Probe E, B Final Doublet Achromat

12.5 Construct a Final Focus Doublet

Solution 12.5.

(a) Construct a beam optics model of the optical sequence sketched in Fig. 12.26 leaving first the electrostatic component zero, with B set to provide the image distance indicated in the figure. The field-free sections can use DRIFT, the two quadrupoles can be simulated using EBMULT. Note that, as long as the E (respectively B) component is zero, QUADRUPO (respectively ELMULT) can be used instead.

Produce the first and second order transport matrices of this system from paraxial trajectory tracking, using MATRIX, together with OBJET[KOBJ = 6] to define a proper set of paraxial rays. Check the second order geometric coefficients: they are expected to be null [29]. Check the second order chromatic coefficients: the non-coupled ones are expected to be non-zero [29].

Plot the horizontal Y(s) and vertical Z(s) projections of the paraxial trajectories used to compute the transport coefficients.

(b) Assume a point object with 0.2 mrad rms divergence, Gaussian, in both planes, and take 10^{-3} rms momentum spread, uniform. MCOBJET[KOBJ = 1] can be used to define this initial coordinate distribution.

Check the location of the vertical and horizontal waists, in the image space. IMAGE and IMAGEZ can be used for that.

Provide the histograms of these initial random distributions, using HISTO, possibly HISTO[PRINT].

Produce a plot of the beam cross-section, of the horizontal and vertical phase spaces, and of the projected coordinate histograms, downstream of the final 25 cm drift plane.

(c) Repeat (a) and (b) using an electrostatic doublet instead. This requires using ELMULT or EBMULT.

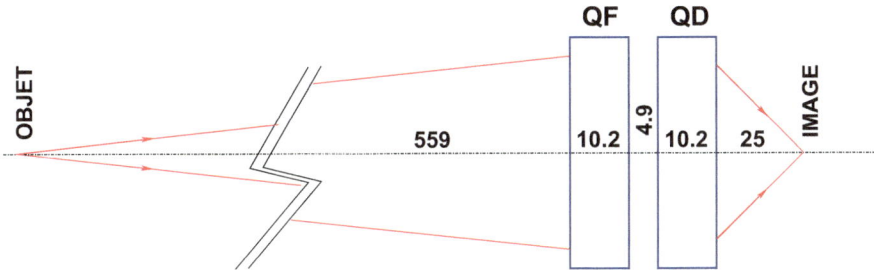

Fig. 12.26 Principle scheme of a QF-QD (focusing-defocusing) final focus second-order achromat for a nano-probe ion beam. The scheme includes design lengths and distances used in the exercises: 559 cm from object to entrance of first quadrupole, 4.9 cm between quadrupoles, 25 cm from exit of second quadrupole to image, and 10.2 cm quadrupole length

12.6 E, B Lens, Hard-Edge Model

Solution 12.6.

(a) Given Exercise 12.5 conditions, which assume hard-edge modeling of the two lenses, now switch on the electric component (this imposes using EBMULT), setting ϕ and B to fulfill the following two constraints:

    (i) preserve the focal distance,

    (ii) ensure cancellation of second order chromatic aberrations.

    Produce the second order transport matrix, compare with the previous E=0 case.

    (b) Track 5,000 particles through the achromat. Assume a point object with 0.2 mrad rms divergence, Gaussian, in both planes, and take 10^{-3} rms momentum spread, uniform.

    Produce the beam cross-section (point-spread function), the horizontal and vertical phase spaces, and the projected coordinate histograms, downstream of the final 25 cm drift. Compare with the conditions of Exercise 12.5. FAISTORE or FAISCNL can be used to store particle data at that location.

12.7 E, B Lens With Fringe Fields

Solution 12.7.

Add fringe fields, considering an unrealistically large radius at pole tip of 5 cm (for the sake of enhanced possible adverse effects). The same geometry is maintained (Fig. 12.26).

    (a) Find the new values of the E and B components of the lenses, for a paraxial image at the same location together with cancelled chromatic coefficients. FIT can be used for that matching.

    Produce the second order transport matrix, compare with the hard-edge case.

    (b) Plot some of the $E(s)$ and $B(s)$ field components experienced along a straight line across the doublet, e.g., $Y = 1$ cm or $Z = 1$ cm. EBMULT[IL = 2] can be used to store particle data during stepwise raytracing across the lens. Plot some of the first and second order field derivatives as well.

12.8 Animation: Dynamical Squeeze of Point-Spread Function

Solution 12.8.

Build the following graphic animation: a dynamic squeeze of the image by slowly switching on the E, B parameters, while preserving the focal distance of the final quadrupole doublet. REBELOTE[K = 99, IOPT = 1] can be used to repeat transport through the optical sequence with EBMULT parameters varied, pass after pass.

12.3.3 $\pi \rightarrow \mu + \nu$ In-Flight Decay, Bunch Densities

In a first part in the following simulation exercises, the evolution of pion and muon beam distributions resulting from in-flight pion decay

$$\pi \rightarrow \mu + \nu$$

is investigated. A Monte Carlo method is used to simulate the process. Unless otherwise specified, a parent bunch comprised of $N_0 = 10^4$ to 10^6 pions is considered. The bunch is launched at $s = 0$ with zero transverse emittances, zero length, and $p_\pi \in [100, 500]$ MeV/c (either a set of initial discrete p_π values, or some random distribution). MCDESINT defines the daughter particle and switches on the decay process. PARTICUL is needed prior, in order to define the parent particle.

In a second part, a muon collect beam line is simulated [11–13]. Its optical properties and collect efficiency are investigated.

12.9 Parent Pion Bunch

Solution 12.9.

(a) Figure 12.9 shows the theoretical pion momentum density at various distances along the muon collect channel (Eq. 12.14), assuming initial random uniform $p_\pi \in [100, 500]$ MeV/c. Produce these distributions from Monte Carlo simulations.

MCOBJET[KOBJ = 3] can be used to generate the initial bunch. A series of successive DRIFTs allows to transport the beam over distance increments; HISTO[PRINT] can be used, histogram plots can be obtained reading from zgoubi.HISTO.out which the PRINT command produces. An other possibility is to use DRIFT[split, IL = 2] and plot the histograms from particle data so logged in zgoubi.plt.

(b) Produce the distance dependence of the pion bunch average momentum (Fig. 12.10). This can use DRIFT[split, IL = 2]

12.10 Muon Bunch

Solution 12.10.

(a) Figure 12.11 shows muon momentum density plots (Eq. 12.19), for three different initial discrete pion momenta values $p_\pi = 100, 250, 500$ MeV/c. Produce these distributions from Monte Carlo simulations.

(b) Figure 12.12 shows the theoretical muon momentum density at various distances along the muon collect channel (Eq. 12.21). Reproduce these distributions from Monte Carlo simulations.

(c) Produce the distance dependent average momentum of the muon bunch (Fig. 12.10).

(d) Reproduce the final longitudinal muon bunch phase space of Fig. 12.13. FAI-STORE can be used to store particle data at the end of the line.

12.3.4   Pion Funnel and Muon FODO Collect Channel

The Neutrino Factory muon collect channel of section "Neutrino Factory Muon Collect Channel" is considered here. The funnel section starts at s = 0 (downstream of the pion production target and focusing horn system [12], not concerned, here). The orbit excursion and betatron functions along are displayed in Fig. 12.14. In this exercise, the funnel is followed by a 14-cell, $\approx$ 20 m long FODO channel.

The beam line sequence and nominal parameters are detailed in Table 12.3. Initial values of the optical functions are given in Table 12.4. Using ray-tracing methods allows accurate modeling of magnetic fields in the large aperture optical elements traversed by the beam (up to 1 m radius for some of the upstream ones in the funnel), and on the other hand allows detailed simulation of on-flight $\pi \rightarrow \mu$ decay based on stepwise Monte Carlo.

Table 12.3 Design parameters of the funnel+FODO collect channel. Field values assume a reference pion momentum p = 300 MeV/c. B_0 is the dipole component in combined function dipoles. The quadrupole field B_1 is at 10 cm radius

Element	Name	Length (cm)	B_0 (kG)	B_1 (kG)
Funnel				
DRIFT	DRF1	50		
MULTIPOL	DQ1	100	−0.838529167	0.291349611
DRIFT	DRF2	40		
MULTIPOL	Q2	100	0	−0.617467029
DRIFT	DRF3	65		
MULTIPOL	Q3	100	−0.250226990	0.81443677
DRIFT	DRF2	40		
MULTIPOL	Q4	100	0	−0.73629954
DRIFT	DRF4	40		
MULTIPOL	DIP	40	0	−1.35257476
DRIFT	DRFM1	44		
MULTIPOL	QM1	40	0	1.79773547
DRIFT	DRFM2	107		
MULTIPOL	QM2	40	0	−2.40617729
DRIFT	DRFM3	40		
FODO cell				
DRIFT	DRF	18.66435		
MULTIPOL	QF	40.	0	3.00591
DRIFT	DRF	37.3287		
MULTIPOL	QD	40.	0	−3.00591
DRIFT	DRF	18.66435		

Table 12.4 Reference trajectory coordinates and optical functions at funnel entrance (s = 0)

x_{co}/x'_{co}	m/rad	0.3943/0.09036
D_x/D'_x	m/rad	0/0
β_x/β_z	m	4/4
α_x/α_z		0/0

12.11 Beam Line Simulation

Solution 12.11.

Refer to the design parameters given in Table 12.3.

(a) Simulate this beam line, produce a graph of the reference trajectory. Produce the transport matrix.

A proper 13-particle object for MATRIX computation can be defined using OBJET[KOBJ = 5]. The simulation of straight sections uses DRIFT, dipoles and combined function dipoles in the funnel can use MULTIPOL. Quadrupoles can use QUADRUPO, or MULTIPOL (the latter would allow introducing multipole defects, if desired).

(b) Optical functions at s = 0 are given in Table 12.4, produce a graph of the optical functions along the line.

Give their values at end of funnel and at end of FODO channel.

OPTICS can be used to transport the optical functions, with initial values defined using OBJET[KOBJ = 5.1].

(c) Produce a graph of paraxial horizontal and vertical phase space invariants at start and end of the channel, at 250, 300 and 400 MeV/c, given initial elliptical invariant values $\varepsilon_x = \varepsilon_y = 2.5\,\pi$ mm at the entrance of the funnel.

12.12 Bunch Densities

Solution 12.12.

A 10^4 pion bunch is considered, launched at entrance of the funnel with beam and emittance parameters as defined in Table 12.4. The bunch can be defined using the random coordinate generator MCOBJET[KOBJ = 3].

In-flight decay requires additional keywords: it requires specifying the parent particle, pions, this is done using PARTICUL, as well as the daughter particle, muon, this is part of the data of the Monte Carlo in-flight decay switch MCDESINT. Collimation is introduced to stop decay muons which reach the channel aperture, long collimation can use CHAMBR, local collimators can use COLLIMA.

(a) Plot the transverse coordinates of a few particles, along the line, in a similar way to Fig. 12.15.

(b) Extend the channel to 100 m distance, take a 0.5 m radius collimation all the way.   Take $\varepsilon_x/\pi = \varepsilon_y/\pi = 0.01$ cm   initial   pion   beam   emittances,   and $p \in [200, 400]$ MeV/c uniform.

Produce a graph of the average momentum of respectively π, μ and $\pi + \mu$ beams.

Plot transverse pion bunch densities at start, 20 and 50 m in a similar way to Fig. 12.9, and transverse muon bunch densities at 100 m in a similar way to Fig. 12.12. Check *qualitatively* by comparison with these theoretical densities.

(c) For four different longitudinal ellipse surfaces: $\varepsilon_l = 0.1$, 0.5, 1 and 2π eV s, determine the phase space ellipse parameters (centering and shape) which correspond to a maximum acceptance at the end of the collect channel, in a similar way to Fig. 12.18. Phase space collimation with COLLIMA[IFORM = 11–16] can be used for that.

(d) Indicate how the FIT procedure can be used to optimize the collect channel design parameters for maximized transmission.

12.3.5 Low Energy Spin Rotator

This exercise series addresses electron spin dynamics in a spin rotator based on a Wien filter. Stepwise ray-tracing techniques allow detailed insight in the dynamics of spin motion through that **E, B** device.

The analytical modeling WIENFILTER is used in these exercises (TOSCA could be used instead, would a **E, B** field map be available). **E** and **B** fields are first taken hard-edge so to allow tight comparison with theory, fringe field are added, next, their effect is assessed and compensated.

The reference frame in the exercises is that of Fig. 12.21, i.e., longitudinal axis X, horizontal and vertical transverse axes Y and Z.

The WIENFILTER considered is comprised of three 0.5 m long segments, rotating the spin by 30° each. A hard edge field model is considered first, then fringe fields, which have a noticeable effect, are added.

12.13 Spin Rotation, Spin Matrix

Solution 12.13.

A hard-edge field model WIENFILTER segment is considered here.

(a) Calculate from theory, and check from raytracing, the value of the electric and magnetic field components in a 0.5 m long Wien filter, to obtain a straight electron trajectory and 30° spin rotation. FIT can be used to obtain the required E and B values.

(b) Produce the spin rotation matrix. OBJET[KOBJ = 2, IMAX = 3] and SPN-PRT[MATRIX] can be used for that.

12.14 Integration Step Size

Solution 12.14.

The effect of integration step size on the convergence of the numerical integration is addressed in this exercise. In both simulations below, the do-loop REBELOTE[IOPT = 1] can be used to repeat spin transport through the optical sequence, concurrently with changing the step size.

(a) Produce a graph of the variation of particle position and angle at the exit of the Wien filter, as well as spin rotation angle, as a function of integration step size, with E_Y and B_Z maintained constant (at their theoretical values).

(b) Produce a graph of the required variation of E_Y and B_Z (found using FIT) to recover straight trajectory and $30°$ spin rotation, as a function of integration step size.

12.15 Add Fringe Fields

Solution 12.15.

To simulate fringe fields, use Enge fringe field coefficient values

$$C_0 = 0.2401, \ C_1 = 1.8639, \ C_2 = -0.5572, \ C_3 = 0.3904, \ C_4 = C_5 = 0$$

(a) First, take $\lambda_E = \lambda_B = 5\,\text{cm}$, at entrance and exit of a $50\,\text{cm}$ segment.
Determine a proper value of the numerical integration extents X_E and X_S, at both ends.
Determine an integration step size.

(b) λ_E and λ_B are in general different, owing in particular to different gap heights. A consequence is that the theoretical field values, and in particular the condition $E_Y = v\,B_Z$, no longer ensure a straight trajectory.

Compute the dependence of the $\delta E_Y/E_Y$ and $\delta B_Z/B_Z$ field adjustments, proper to recover particle trajectory and $30°$ spin rotation, on the field fall-off extent ratio λ_B/λ_E. FIT can be used to find E_Y and B_Z, and REBELOTE can be used to repeat the sequence with prior change of λ_B.

Produce a graph of sample reference trajectories, $Y(X)$ along the Wien filter, for various values of λ_B/λ_E. Check the spin rotation angle, as per the FIT procedure.

12.16 Transport of a 6D Polarized Bunch

Solution 12.16.

Launch through the Wien filter a 10^4 electron bunch with normalized *rms* emittances $\beta\gamma\varepsilon_Y = \beta\gamma\varepsilon_Z = 10\pi\,\mu\text{m}$, Gaussian transverse densities truncated at $4\,\text{sigma}$, momentum spread uniform over $[-10^{-4}, +10^{-4}]$. MCOBJET[KOBJ=3] can be used to generate this bunch.

Produce sample trajectories and $E_Y(s)$ and $B_Z(s)$ fields, across the Wien filter.

Produce the phase spaces and projected densities, spin rotation and densities, at the downstream end of the complete, 3-segment, $1.5\,\text{m}$ Wien filter.

12.3.6 Synchrotron Radiation

12.17 Synchrotron Radiation in a Long FFAG Beam Line

See "eRHIC 22 GeV ERL. Synchrotron Radiation. Polarization", a linear FFAG beam line, Exercise 11.2.

12.4 Solutions of Exercises of This Chapter: Beam Lines

12.4.1 Beam Expander

12.1 Construct a Beam Expander Line
(a) Expander line operation.

The way the expander/uniformizer line described in Table 12.1 operates is as follows (see Sect. 12.2.1).

- At the downstream end of the line (TARGET):

  - The beam has zero divergence, its transverse dimensions are a few millimeters to centimeters (Fig. 12.25). Four quadrupoles, QPL3-0, are tuned to adjust four variables: $\alpha_{Y,Z}, \beta_{Y,Z}$.
  - A final 3 m long section, SD1LB-SD1LA, includes two scanning dipoles, horizontal (BH) and vertical (BV).

- Beam uniformization at the target:

  - one octupole is needed per transverse dimension, they are tuned individually;
  - the horizontal (respectively vertical) uniformization octupole is located at a vertical beam waist in a region of large β_Y (respectively, horizontal waist and large β_Z).

- Upstream end of the line: it includes the downstream end of GANIL G4 line, its role is to match GANIL beam to the downstream expander/uniformizer section. Six quadrupoles are available for that, the G4 quadrupoles Q21–Q24 and two additional ones, Q25 and Q26 for optics setting flexibility.

(b) A simulation of the optical sequence

The optical sequence resulting from Table 12.1 data is given in Table 12.5. It starts with OBJET[KOBJ = 6.1] which creates a 102 particle sample, for computation of the transport coefficients of the line up to 3rd order, by MATRIX. The latter can be found at the end of the sequence. The reference rigidity is set to $B\rho_{\text{ref}} \equiv BORO = 1000\,\text{kG cm}$, as the actual rigidity of the beam does not matter here. One advantage in taking $BORO = 1000\,\text{kG cm}$ is that, with $R_0 = 10\,\text{cm}$ the reference radius of the quadrupoles, the field at the pole identifies with the strength of the element:

$$\text{Strength K}\,[\text{m}^{-2}] = \frac{B_{\text{pole}}[\text{kG}]\,/\,R_0[\text{cm}]}{B\rho[\text{kG cm}]}$$

(c) Transport coefficients.

The transport coefficients can be found in zgoubi.res execution listing, under MATRIX at the bottom of the file. An excerpt is given in Table 12.6. Note that transport coefficients up to second order can be obtained with OBJET[KOBJ = 6], which generates an ad hoc 61 particle sample for computation by MATRIX. OBJET[KOBJ = 6.1] generates a 102 particle sample for the computation of transport coefficients up to third order.

12.2 Expander Line Optics

(a) A few trajectories, from start to end.

The simulation input data file of Table 12.5 is used to raytrace and plot particle coordinates along the line. The only change needed is to replace OBJET by the following MCOBJET:

```
'MCOBJET'
1000.                                                       ! Reference rigidity.
3
80                                                ! 80 particles defined here.
2 2 2 2 2 2                       ! All six coordinates are sorted at random in a Gaussian density.
0. 0. 0. 0. 0. 1.                 ! Reference coordinates, at start of the structure, including D=1.
-1.435 3.00933   .05E-6  3                   ! alpha_Y; beta_Y; epsilon_Y/pi; cut-off.
-6.129 44.3186   1.E-6   3                   ! alpha_Z; beta_Z; epsilon_Z/pi; cut-off.
0.     1.        1.E-96  3                   ! alpha_l; beta_l; epsilon_l/pi; cut-off.
323456 134567 545679                  ! Three seeds used by the random number generator.
```

In order to have particle coordinates logged in zgoubi.plt during raytracing, IL is set to 2 in all lenses using the global command OPTIONS[IOPT = 1, NBOP=1, .plt IL= 2]. For that very reason, a split command has been added in some of the drifts using DRIFT[split, IL = 2]. Graphs of 80 trajectories through the line, in the horizontal and vertical planes, are given in Fig. 12.27.

Table 12.5 Simulation input data file: expander/uniformization beam line G4_beamLine.inc. Uniformization octupoles and BENDs are off at this stage.   This input data file defines two segments (using LABEL1's): S_G4 to OH_dwn on the one hand, S_G4 to Target on the other hand,   for use in INCLUDEs in subsequent exercises. *Note* this file is available in zgoubi sourceforge repository at https://sourceforge.net/p/zgoubi/code/HEAD/tree/branches/exemples/book/zgoubiMaterial/beamLines/expander/MATRIX_O3/

```
G4 beam line with expander. File name: G4_beamLine.inc.
'MARKER'   G4_MATRIX_S      ! Just for edition purposes.
'OBJET'
1000.                        ! Reference rigidity (kG.cm)
6.1            ! A set of 102 rays to allow computation of 3rd
0.01 .001 0.01 .001 0.01 1.E-4  ! order transport matrices.
0. 0. 0. 0. 0. 1.
'SCALING'
1 1
MULTIPOL  OH
1
0.               ! B3 component of MULTIPOL[LABEL1=OH] set to 0.
1
'MARKER'  S_G4
'QUADRUPO'     Q21
2   ! IL, for log in zgoubi.plt
30.400 10.000  -3.674189
0. 0.
4    0.2490   5.3630  -2.4100   0.9870  0. 0.
0. 0.
4    0.2490   5.3630  -2.4100   0.9870  0. 0.
#10|15|10
1  0. 0. 0.
'DRIFT'     DR12
39.60000
'QUADRUPO'     Q22
2   ! IL, for log in zgoubi.plt
30.400 10.000  2.067729
[same as in Q21]
'DRIFT'     DR23
187.5 split 10 2   ! Split drift in 10; log to zgoubi.plt.
'QUADRUPO'     Q23
2   ! IL, for log in zgoubi.plt
30.400 10.000  1.0000
[same as in Q21]
'DRIFT'     DR34
39.60000
'QUADRUPO'     Q24
2   ! IL, for log in zgoubi.plt
30.400 10.000  -2.791886
[same as in Q21]
'DRIFT'     SDP5
50.00000
'QUADRUPO'     Q25
2   ! IL, for log in zgoubi.plt
30.400 10.000  1.0000
[same as in Q21]
'DRIFT'     SDP5
50.00000
'QUADRUPO'     Q26
2   ! IL, for log in zgoubi.plt
30.400 10.000  -3.0000
[same as in Q21]
'DRIFT'     SDP5-4
90.00000   split 10 2
'MULTIPOL'   OH       Octupole dodecapole
0   ! r @ pole         B3 @ pole    B5 @ pole
10.   10.    0. 0. 0.   1.    0.      0.   0. .0 .0 .0
.0 .0 .0 .0 .0 .0 .00 .0 .0 .0
4   .2490  5.3630  -2.4100   .9870   .0   .0
.0 .0 .0 .0 .0 .0 .00 .0 .0 .0
4   .2490  5.3630  -2.4100   .9870   .0   .0
0. 0. 0. 0. 0. 0. 0 .0 .0 .0
1.
1  0. 0. 0.
'MARKER'    OH_dwn
'DRIFT'     SDP5
50.00000   split 10 2
'QUADRUPO'    QV1
2   ! IL, for log in zgoubi.plt
35.000 10.000  2.805048
[same as in Q21]
```

```
'DRIFT'     SD1
100.00000
'QUADRUPO'    QV
2   ! IL, for log in zgoubi.plt
35.000 10.000  -3.034423
'DRIFT'     SDP8
80.00000 split 10 2
'MULTIPOL'   OV       Octupole dodecapole
0   ! r @ pole         B3 @ pole    B5 @ pole
10.   10.    0. 0. 0.   9.    0.      0.   0. .0 .0 .0
.0 .0 .0 .0 .0 .0 .0 0 .0 .0 .0
4   .2490  5.3630  -2.4100   .9870   .0   .0
.0 .0 .0 .0 .0 .0 .0 .0 .0 .0
4   .2490  5.3630  -2.4100   .9870   .0   .0
0. 0. 0. 0. 0. 0. 0 .0 .0 .0
1.
1  0. 0. 0.
'DRIFT'     SDP4
40.00000 split 10 2
'QUADRUPO'    QPL3
2   ! IL, for log in zgoubi.plt
35.000 10.000   1.896253
[same as in Q21]
'DRIFT'     SDP5
50.00000
'QUADRUPO'    QPL2
2   ! IL, for log in zgoubi.plt
35.000 10.000   2.464301
[same as in Q21]
'DRIFT'     SDP5
50.00000
'QUADRUPO'    QPL1
2   ! IL, for log in zgoubi.plt
35.000 10.000  -0.1375607
[same as in Q21]
'DRIFT'     SDP5
50.00000
'QUADRUPO'    QPL0
2   ! IL, for log in zgoubi.plt
35.000 10.000  -3.357716
[same as in Q21]
'DRIFT'     SD1LB
10.00000
'BEND'     BH
2   ! IL, for log in zgoubi.plt
20. 0. 0.
8. 4. .0
4  .2401  1.8639  -.5572  .3904 0. 0. 0.
8. 4. .0
4  .2401  1.8639  -.5572  .3904 0. 0. 0.
#30|10|30
1 0 0 0
'BEND'     BV
2   ! IL, for log in zgoubi.plt
20.  1.5707963268  0.
8. 4. .0
4  .2401  1.8639  -.5572  .3904 0. 0. 0.
8. 4. .0
4  .2401  1.8639  -.5572  .3904 0. 0. 0.
#30|10|30
1 0 0 0
'DRIFT'     SD1LA
250.00000    split 10 2
'MARKER'  Target
'MATRIX'
3 0  ! Computes transport coefficients up to 3rd order.
'MARKER'   G4_MATRIX_E      ! Just for edition purposes.
'END'
```

Table 12.6 Outcomes of the simulation file of Table 12.5. These are excerpts of zgoubi.res output file, truncated for shortness

```
Reference particle (#    1), path length :   1589.1000     cm  relative momentum :    1.00000

                              *1st order TRANSPORT coeffs.* (Units m, rad)
            0.94547  -2.77904   0.00000   0.00000   0.00000   0.00000
            0.48939  -0.38083   0.00000   0.00000   0.00000   0.00000
            0.00000   0.00000  -0.17501  -2.15098   0.00000   0.00000
            0.00000   0.00000   0.29449  -2.09426   0.00000   0.00000
            0.00000   0.00000   0.00000   0.00000   1.00000   0.00000
            0.00000   0.00000   0.00000   0.00000   0.00000   1.00000

                              *2nd order TRANSPORT coeffs.* (units m, rad) (excerpt)
      1 11  0.000E+00
      1 12  0.000E+00    1 22  0.000E+00
      1 13  0.000E+00    1 23  0.000E+00    1 33  0.000E+00
      1 14  0.000E+00    1 24  0.000E+00    1 34  0.000E+00    1 44  0.000E+00
      1 15  0.000E+00    1 25  0.000E+00    1 35  0.000E+00    1 45  0.000E+00    1 55  0.000E+00
      1 16 -2.002E+02    1 26 -3.478E+02    1 36  0.000E+00    1 46  0.000E+00    1 56  0.000E+00    1 66  0.000E+00

      2 11  0.000E+00
      2 12  0.000E+00    2 22  0.000E+00
      2 13  0.000E+00    2 23  0.000E+00    2 33  0.000E+00
      2 14  0.000E+00    2 24  0.000E+00    2 34  0.000E+00    2 44  0.000E+00
      2 15  0.000E+00    2 25  0.000E+00    2 35  0.000E+00    2 45  0.000E+00    2 55  0.000E+00
      2 16 -5.574E+01    2 26 -9.683E+01    2 36  0.000E+00    2 46  0.000E+00    2 56  0.000E+00    2 66  0.000E+00

                              *3rd order TRANSPORT coeffs.* (units m, rad) (excerpt)
    1 111  1.012E+03
    1 112  5.366E+03
    1 122  9.503E+03    1 222  5.618E+03
    1 113  0.000E+00
    1 123  1.540E-02    1 223  0.000E+00
    1 133  6.367E+02    1 233  1.118E+03    1 333  0.000E+00

    2 111  2.833E+02
    2 112  1.503E+03
    2 122  2.663E+03    2 222  1.575E+03
    2 113  0.000E+00
    2 123  4.364E-03    2 223  0.000E+00
    2 133  1.778E+02    2 233  3.127E+02    2 333  0.000E+00
    2 114  0.000E+00
    2 124  2.010E-01    2 224  0.000E+00
    2 134 -7.512E+02    2 234 -1.319E+03    2 334  0.000E+00
    2 144  7.969E+02    2 244  1.395E+03    2 344  0.000E+00    2 444  0.000E+00

    3 116  0.000E+00
    3 126  0.000E+00    3 226  0.000E+00
    3 136  4.454E-04    3 236 -1.521E-04    3 336  0.000E+00
    3 146 -7.988E-04    3 246 -2.279E-03    3 346 -8.105E-03    3 446  0.000E+00
    3 156  0.000E+00    3 256  0.000E+00    3 356  0.000E+00    3 456  0.000E+00    3 556  0.000E+00
    3 166  0.000E+00    3 266  0.000E+00    3 366  7.075E+02    3 466 -1.689E+03    3 566  0.000E+00    3 666  0.0E+00

                              * 4th order TRANSPORT coefficients *    (excerpt)
  1 1111  0.000E+00
  1 1122  0.000E+00    1 2222  0.000E+00
  1 1133  0.000E+00    1 2233  0.000E+00    1 3333  0.000E+00
  1 1144  0.000E+00    1 2244  0.000E+00    1 3344  0.000E+00    1 4444  0.000E+00
  1 1155  0.000E+00    1 2255  0.000E+00    1 3355  0.000E+00    1 4455  0.000E+00    1 5555  0.000E+00
  1 1166  0.000E+00    1 2266  0.000E+00    1 3366  0.000E+00    1 4466  0.000E+00    1 5566  0.000E+00    1 6666  0.0E+00

  5 1111 -2.638E+03
  5 1122 -6.750E+06    5 2222 -1.417E+12
  5 1133 -1.066E+02    5 2233  3.553E+05    5 3333 -4.896E+03
  5 1144 -1.084E+07    5 2244 -1.084E+11    5 3344 -1.137E+07    5 4444 -1.445E+12
  5 1155  0.000E+00    5 2255  0.000E+00    5 3355  0.000E+00    5 4455  0.000E+00    5 5555  0.000E+00
  5 1166  5.276E+03    5 2266  6.217E+06    5 3366  6.786E+03    5 4466  1.332E+06    5 5566  0.000E+00    5 6666  0.0E+00
```

(b) Optical functions.

Different methods can be resorted to, to compute the optical functions. Two possibilities are detailed in the following.

- first method

OPTICS[all, PRINT] allows transporting initial values of the optical functions. OPTICS works in a similar way to TWISS, except for the fact that TWISS assumes the sequence to be a periodic structure and as a consequence computes the periodic functions at the start of the sequence (using Eq. 14.49) whereas OPTICS does

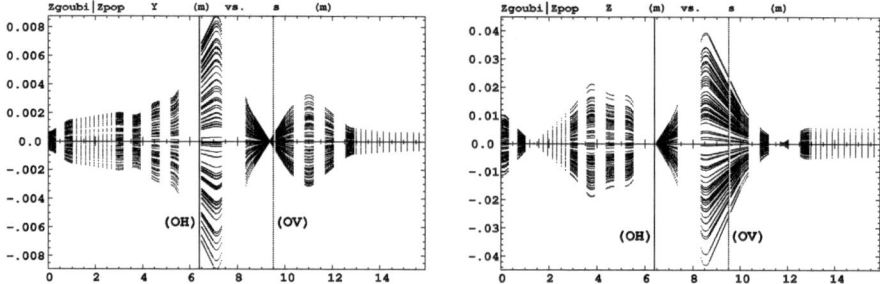

Fig. 12.27 Trajectories through the expander beam line. Left: horizontal plane; right: vertical plane. Note the different vertical scales. The horizontal uniformization octupole (OH) is at a location of large horizontal excursion, small vertical excursion, the reverse holds for the vertical octupole (OV). A graph obtained using zpop: menu 7; 1/1 to open zgoubi.plt; 2/[6, 2] to select Y versus distance (or 2/[6, 4] to select Z); 7 to plot. Option 3/10/10 would plot trajectories as continuous lines, whereas the present default 3/10/9 plots dots, does not connect the stepwise data

not make that assumption and thus requires these to be provided. OBJET[KOBJ = 5.1] is used to generate a 13 particle sample, proper for MATRIX computation, as OPTICS uses the same Fortran routines as MATRIX to compute transport matrices. OBJET[KOBJ = 5.1] also allows providing the initial values of the optical functions. OPTICS[all, PRINT] causes computation of optical functions after (almost) all keywords in the optical sequence; the PRINT argument causes these to be logged in zgoubi.OPTICS.out (in a similar way that TWISS causes the optical functions to be logged in zgoubi.TWISS.out). The input data file for this simulation is given in Table 12.7.

A gnuplot script file found in zgoubi toolbox, gnuplot_OPTICS.gnu [31], can be used to plot zgoubi.OPTICS.out content; this script is called using a SYSTEM command (Table 12.7). OPTICS computation is also an opportunity to check

– the value of the reference orbit, with respect to which transport coefficients and the resulting optical functions are computed,
– the degree of non-symplecticity of the numerical integration of the motion, via the first order mapping determinants, for instance.

The graphs resulting from the execution of Table 12.7 simulation file, via the SYSTEM command, are displayed in Fig. 12.28.

- second method

The input simulation file is the same as in the first method, Table 12.7. Set OPTIONS[.plt, IL = 2], uncomment (in case it is commented) the split command DRIFT[split, IL = 2] in various drifts where it is present, this will cause stepwise data to be logged in zgoubi.plt. This allows using the post-treatment tool betaFromPlt, found in zgoubi toolbox [32], to compute optical functions from the stepwise data read in zgoubi.plt. The second gnuplot script under SYSTEM, Table 12.7, found as well in zgoubi toolbox [32], performs the two tasks, first a system call to

Table 12.7 Simulation input data file: computation of the optical functions along the expander/uniformization beam line, using OPTICS. This input data file INCLUDEs the segment from the G4_beamLine.inc file of Table 12.5 The first gnuplot script under SYSTEM plots the optical functions as read from zgoubi.OPTICS.out (Fig. 12.28); the second gnuplot script (found in zgoubi toolbox) resorts to betaFromPlt to transport the optical functions, and plots the latter (Fig. 12.29)

```
Expander. Compute optica functions.
'MARKER'    G4_OPTICS_S                                    ! Just for edition purposes.
'OBJET'
1000.
5.1              ! Generate 13 particles. '.1' suffix allows providing initial optical function values.
0.01 .001 0.01 .001 0.01 .0001
0. 0. 0. 0. 0. 1.                         ! Initial coordinates of reference trajectory.
-1.435 3.00933 -6.129 4.43186E+01 0. 1. 0. 0. 0. 0.         ! Initial optical function values.
'OPTICS'                                           ! Causes computation of optical functions,
1.1 all PRINT           ! at all elements in the sequence. PRINT causes log in zgoubi.OPTICS.out.
'OPTIONS'
1 1
.plt  IL=0
'SCALING'
1  1
MULTIPOL  OH
1
0.
1

'INCLUDE'
1
./G4_beamLine.inc[S_G4:Target]

'SYSTEM'
2
gnuplot <./gnuplot_OPTICS.gnu                    ! Plot optical functions as read from zgoubiOPTICS.out.
gnuplot <./gnuplot_betaFromPlt.gnu    ! Plot optical functions, computed from zgoubi.plt data; this
                                      ! script prior resorts to a system call to betaFromPlt.
'MARKER'    G4_OPTICS_E                                     ! Just for edition purposes.
'END'
```

A gnuplot script (excerpt) to obtain Figs. 12.29, 12.30:

```
# gnuplot_betaFromPlt.gnu. Excerpt.
  system './betaFromPlt'
  plot \
       "betaFromPlt.out" u ($13 * cm2m):($2)  axes x1y1 w lp ps .2 pt 4 lc rgb "red"   tit "{/Symbol b}_Y"  ,\
       "betaFromPlt.out" u ($13 * cm2m):($4)  axes x1y1 w lp ps .2 pt 5 lc rgb "blue"  tit "{/Symbol b}_y"  ,\
       "betaFromPlt.out" u ($13 * cm2m):($7)  axes x1y2 w lp ps .2 pt 6 lc rgb "black" tit "  D_Y"  ,\
       "betaFromPlt.out" u ($13 * cm2m):($9)  axes x1y2 w lp ps .2 pt 7 lc rgb "cyan"  tit "  D_y"
```

Fig. 12.28 Left: optical functions along the expander line, case of $\beta_Z = 10$ m at target. Right: the first order mapping determinants only weakly differ from one—note that a contribution to the difference to 1 is the limited accuracy of the interpolation method [30, cf. MATRIX] that computes the transport coefficients from the coordinates of the 13 particle sample

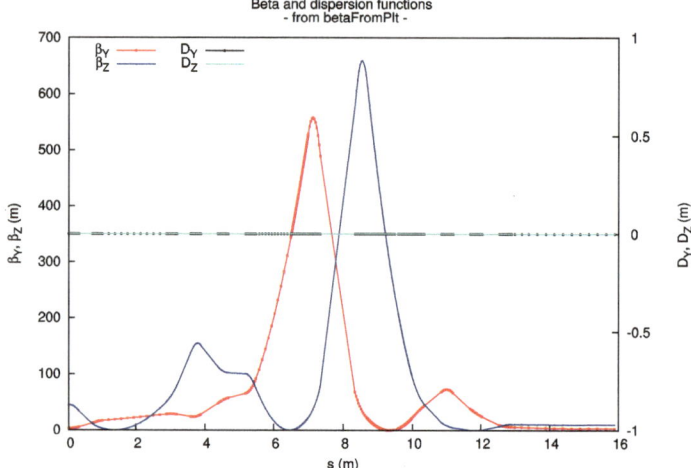

Fig. 12.29 Optical functions along the expander line. Stepwise particle coordinates across optical elements are read in zgoubi.plt, they allow computing the transport matrix $[T_{ij}](s \leftarrow s_0)$ from the origin s_0 to s, which in turn allows transporting the beam matrix $\sigma(s) = [T_{ij}](s \leftarrow s_0)\sigma(s_0)[\tilde{T}_{ij}](s \leftarrow s_0)$

betaFromPlt, and then plotting the content of the resulting betaFromPlt.out. The resulting graph is displayed in Fig. 12.29.

Beam matrix at target

This is an outcome of OPTICS[all]: this releases beam matrices at all optical elements along the beam line. The beam matrix at target is found at the bottom of zgoubi.res execution listing:

```
 37  Keyword, label(s) :  MARKER      Target

 Reference particle (#    1), path length :   1589.1000 cm    relative momentum : 1.

     BEAM  MATRIX (beta/alpha/alpha/gamma),                 FINAL
         3.00026        -2.153829E-05     0.00000            0.00000
        -2.153829E-05    0.333284         0.00000            0.00000
         0.00000         0.00000          9.99788            9.805930E-05
         0.00000         0.00000          9.805930E-05       0.100014
```

The vertical betatron function takes the value $\beta_Z = 9.997\,\text{m}$ at the target, this determines the vertical size of the rectangle in the presence of the vertical octupole (Eq. 12.2). It can be observed that $\alpha_Y = \alpha_Z = 0$, beam ellipses for the linear setting (octupoles off) are upright.

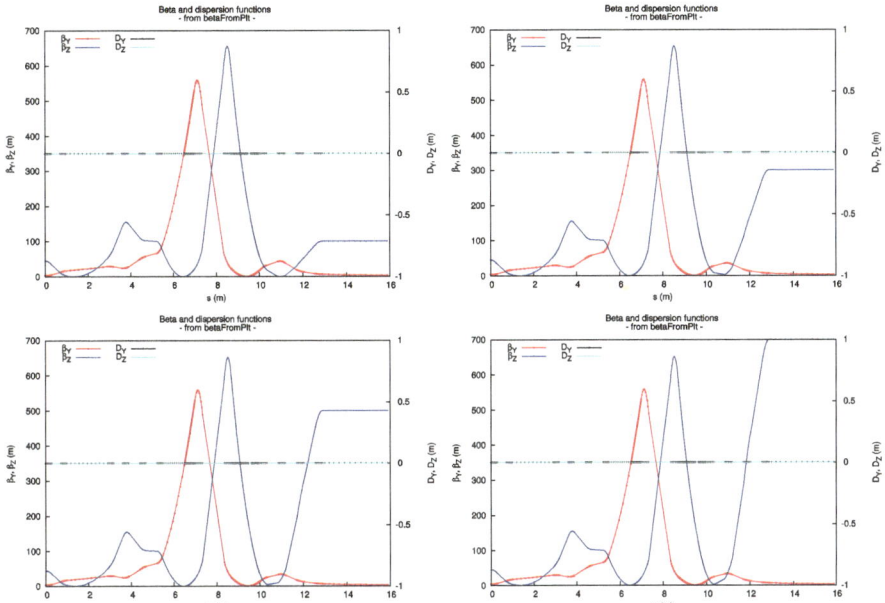

Fig. 12.30 Optical functions along the expander line, for $\beta_Z = 100, \ 300, \ 500$ and $700\,$m at target, using zgoubi.plt stepwise raytracing data to compute the beam matrix transport $\sigma(s) = T(s \leftarrow s_0)\sigma(s_0)\tilde{T}(s \leftarrow s_0)$

Optical functions for $\beta_Z = 100, \ 300, \ 500$ *and* 700 m at target

They are produced as above, using the simulation input file of Table 12.7. The field values in the quadrupoles of the expander/uniformization section (Table 12.5) are given in Table 12.2. The simulation needs be repeated for each value of β_Z at target. Results are given in Fig. 12.30.

Beam matrices so obtained from OPTICS[all, PRINT] are logged in zgoubi.OPTICS.out, their values at TARGET can be found at the bottom of the file, they can be found as well at the bottom of zgoubi.res execution listing, and checked against the expected values, namely,

$$\alpha_Y = \alpha_Z = 0; \ \ \beta_Y = 3\,\text{m in all cases}; \ \beta_Z = 100, \ 300, \ 500 \text{ and } 700\,\text{m}$$

12.3 Transverse Uniformization; Beam Scan

The beam at target retains its Gaussian horizontal density as the horizontal octupole is off, with *rms* width $\sigma_x = (\beta_x \varepsilon_x/\pi)^{1/2} \approx (3 \times 0.05 \times 10^{-6})^{1/2} = 0.4 \times 10^{-3}\,$m ; the vertical beam profile is uniformized due to the vertical octupole, and with total extent ΔZ in millimeter to several centimeter range depending on β_z (Eq. 12.4).

Table 12.8 In complement to Table 12.2: octupole strengths for beam uniformization

		β_z at target (m)				
		10	100	300	500	700
Octupole	m^{-4}	0.2	-3.8	-2.49539	-2.8696	-4.1106

(a) Vertical octupole field value.

The OV octupole integrated strength required to obtain a uniform vertical density at target is determined using

$$K_3 L|_{\mathrm{OV}} = \frac{1}{12\varepsilon_z \beta_l^2} \frac{\cos^3 \Delta\phi}{\sin \Delta\phi}$$

a simplified form of Eq. 12.7. The values of the betatron amplitude β_l at the octupole lens, and of the vertical phase advance $\Delta\phi$ from the octupole to the target, in the five different cases of β_Z value at target (i.e., five different cases of spot size), are taken from the optical function values, as found in zgoubi.OPTICS.out or zgoubi.res files produced in the question (d). The field values so determined can be found under MULTIPOL[LABEL1 = OV], they are recapped in Table 12.8; they appear under SCALING[MULTIPOL OV] in the simulation input data file, Table 12.9.

Field values in the scanning dipoles (Table 12.5) are determined from the geometrical data, Fig. 12.25. These values are found under BEND/BH and BEND/BV, under SCALING keyword.

The input data file for this sweep simulation is given in Table 12.9. The resulting graph of the sweep is displayed in Fig. 12.25.

(b) Two-dimensional uniform expansion.

A proper setting of QV and QPL3 to QPL0 lenses has to be re-computed, to achieve a $6 \times 6 \,\mathrm{cm}^2$ square beam cross-section. In particular it has to satisfy Eq. 12.4 for both horizontal and vertical planes. The octupole strengths are derived from Eq. 12.7. The field values so obtained can be found under SCALING in the simulation data file (Table 12.10).

(c) Histograms along the line

Graphs of transverse beam densities along the line are obtained using HISTO[PRINT], which causes a log of particle density histograms to zgoubi.HISTO.out. Three HISTO have been inserted in the simulation data file (Table 12.10), their outcomes are plotted using gnuplot, as part of the simulation file input (SYSTEM keyword), Fig. 12.32. These histograms are logged in zgoubi.res execution listing as well, Table 12.11.

Table 12.9 Simulation input data file: a rectangle beam spot sweep at target (Fig. 12.25). SCAL-ING is introduced to allow interpolating the field in magnets from pass 1 to pass 20, from field values at five different timings. The number of passes is specified as REBELOTE[NRBLT = 19]. The INCLUDed segment [S_G4:OH_dwn] is grabbed from G4_beamLine.inc file (Table 12.5). Fields in all quadrupoles from QV on, as well as the B3 (octupole) component in MULTI-POL[LABEL1 = OV], are set to 1, and scaled to proper value using SCALING: this allows varying these values at each one of the 19 passes, for proper image size and uniformized verti-cal density. *Note* this file, and the gnuplot script for the animation in Exercise 12.4, are available in zgoubi sourceforge repository at https://sourceforge.net/p/zgoubi/code/HEAD/tree/branches/ exemples/book/zgoubiMaterial/beamLines/expander/rectangleSweep/

```
G4 beam line with expander. Rectangle beam sweep at target.
'MARKER'    G4_SWEEP_S      ! Just for edition purposes.
'MCOBJET'
1000.
3
1000
2 2 2 2 2 2
0. 0. 0. 0. 0. 1.
-1.435 3.00933      .05E-6  3
-6.129 4.43186E+01  1.E-6   3
0.    1.            0.      3
323456 134567 545679
'SCALING'
1 9
BEND   BH
3
-.2 -.1 .2
1   6   20
BEND   BV
4
-.4  .4  .4   -.4
1   6  12   20
QUADRUPO  QV
5           ! beta_Z = 10, 100,500,700,300
-3.034423 -3.206831  -3.352989 -3.393327     -3.299631
1  2  8   17  20
QUADRUPO  QPL3
5
1.896253   2.884847   3.230446   3.211739     3.207578
1  2  8   17  20
QUADRUPO  QPL2
5
2.464301   2.327635   2.104845   2.158918     2.070500
1  2  8   17  20
QUADRUPO  QPL1
5
-1.375607E-01  -1.456258   -1.107302  -1.107302    -1.107302
1  2  8   17  20
QUADRUPO  QPL0
5
-3.357716  -1.188920 -9.563942E-01  -9.144798E-01 -1.044225
1  2  8   17  20
MULTIPOL  OV
5
2.  -38. -28.6963391 -41.1060953  -24.9539018
1  2  8   17   20
MULTIPOL  OH
1
0.
1
'OPTIONS'
1 1       ! Force IL=0 in all optical elements: no print
.plt 0                        ! out to zgoubi.plt.
'INCLUDE'
1
./G4_beamLine.inc[S_G4:OH_dwn]

'DRIFT'      SDP5
50.00000
'QUADRUPO'     QV1
0
35.000  10.000   2.805048
0.  0.
4   0.2490   5.3630  -2.4100   0.9870  0.  0.
0.  0.
4   0.2490   5.3630  -2.4100   0.9870  0.  0.
#10|10|10
1  0. 0. 0.
'DRIFT'    SD1
100.00000   !
```

```
'QUADRUPO'     QV
0
35.000  10.000  1.
[same as in QV1]
'DRIFT'      SDP2
80.00000
'MULTIPOL' OV        Ocupole Z
0   | r @ pole        B3 @ pole    B5 @ pole
10.  10.   0. 0. 0.    1.   0.   0.  0. .0 .0
.0 .0 .0 .0 .0 .0 .0 0 .0 .0
4   .2490  5.3630 -2.4100   .9870   .0   .0
.0 .0 .0 .0 .0 .0 .0 0 .0 .0
4   .2490  5.3630 -2.4100   .9870   .0   .0
0. 0. 0. 0. 0. 0. 0 .0 .0 .0
1.
1  0. 0. 0.
'DRIFT'       SD4
40.00000
'QUADRUPO'     QPL3
0
35.000  10.000  1.
[same as in QV1]
'DRIFT'      SD5
50.00000
'QUADRUPO'     QPL2
0
35.000  10.000  1.
[same as in QV1]
'DRIFT'      SD5
50.00000
'QUADRUPO'     QPL1
0
35.000  10.000  1.
[same as in QV1]
'DRIFT'      SD5
50.00000
'QUADRUPO'     QPL0
0
35.000  10.000  1.
[same as in QV1]
'DRIFT'       SD1LB
10.00000
'BEND'   BH
0
20.  0.  2.8
8. 4. .0
4 .2401  1.8639  -.5572  .3904 0. 0. 0.
8. 4. .0
4 .2401  1.8639  -.5572  .3904 0. 0. 0.
#30|10|30
1 0 0 0
'BEND'   BV
00  00
20.  1.5707963268  1.4
[same as in BH]
'DRIFT'      SD1LA
250.00000
'FAISCNL'
zgoubi.fai            ! Sweep step logged in zgoubi.fai.
'REBELOTE'
19 0.1 0

'SYSTEM'                 ! This produces an animation of
1                        ! the sweep, see next exercise.
gnuplot < ./gnuplot_sweep.gnu

'MARKER'   G4_SWEEP_E    ! Just for edition purposes.
'END'
```

Table 12.10 Simulation input data file: uniform 2D beam cross-section at target (Fig. 12.31). SCALING is used to set the proper field values in the various magnets. The INCLUDed segment [S_G4:OH_dwn] is grabbed for G4_beamLine.inc file (Table 12.5). The HISTO[PRINT] commands cause a log of particle density histograms to zgoubi.HISTO.out

```
G4 beam line with expander. 2D expander. G4_2DUniform.inc
'MARKER'   G4_2DUniform_S  ! Just for edition purposes.
'MCOBJET'
1000.
3
10000
2 2 2 2 2 2
0. 0. 0. 0. 0. 1.
-1.435 3.00933      .8E-6  3
-6.129 4.43186E+01  .8E-6  3
0.       1.        0.  3
323456 134567 545679
'SCALING'
1 7
QUADRUPO QV
1
-2.8
1
QUADRUPO  QPL3
1
-7.402098E+00        beta= 2010
1
QUADRUPO  QPL2
1
3.717142E+00
1
QUADRUPO  QPL1
1
-3.929684E+00
1
QUADRUPO  QPL0
1
1.696915E+00
1
MULTIPOL  OH
1
-4.
1
MULTIPOL  OV
1
60.
1

'INCLUDE'
1
./G4_beamLine.inc[S_G4:OH_dwn]

'DRIFT'      SDP5
50.00000
'QUADRUPO'       QV1
2
35.000 10.000   2.805048
0.  0.
4   0.2490   5.3630  -2.4100   0.9870  0.  0.
0.  0.
4   0.2490   5.3630  -2.4100   0.9870  0.  0.
#10|10|10
1 0. 0. 0.
'DRIFT'      SD1
100.00000   !
```

```
'QUADRUPO'       QV
2
35.000 10.000  1.
[same as in QV1]
'DRIFT'      SDP2
80.00000
'MULTIPOL'  OV        Ocupole Z
0   ! r @ pole         B3 @ pole    B5 @ pole
10.  10.   0. 0. 0.   1.    0.   0.  0. .0 .0 .0
.0 .0 .0 .0 .0 .0 .0 .0 .0 .0 .0
4   .2490  5.3630 -2.4100   .9870   .0   .0
.0 .0 .0 .0 .0 .0 .0 .0 .0 .0 .0
4   .2490  5.3630 -2.4100   .9870   .0   .0
0. 0. 0. 0. 0. 0. 0 .0 .0 .0
1.
1 0. 0. 0.
'DRIFT'      SD4
40.00000
'QUADRUPO'       QPL3
2
35.000 10.000   1.
[same as in QV1]
'DRIFT'      SDP5
50.00000
'QUADRUPO'       QPL2
2
35.000 10.000  1.
[same as in QV1]
'DRIFT'      SDP5
50.00000
'HISTO'
2 -10. 10. 40 1 PRINT
20 Y 0 P
'QUADRUPO'       QPL1
2
35.000 10.000  1.
[same as in QV1]
'DRIFT'      SDP5
50.00000
'HISTO'
2 -10. 10. 40 2 PRINT
20 Y 0 P
'QUADRUPO'       QPL0
2
35.000 10.000  1.
[same as in QV1]
'HISTO'
2 -10. 10. 40 3 PRINT
20 Y 0 P
'DRIFT'      SD1LB-to-Target
300.00000
'FAISCNL'
b_zgoubi.fai
'SYSTEM'
1
gnuplot <./gnuplot_HISTO.gnu
'MARKER'   G4_2DUniform_E  ! Just for edition purposes.
'END'
```

A gnuplot script (excerpt) to obtain Fig. 12.32:

```
# gnuplot_HISTO.gnu
set xlabel "Y [cm]"; set ylabel "count"; set xtics mirror; set ytics mirror; set key t r; set xrange [-7:7]
plot for [i=1:3] zgoubi.HISTO.out  u ($9==i? $1 :1/0):($2) axes x1y1 w lp ps .8 tit 'Histo '.i; pause 1
```

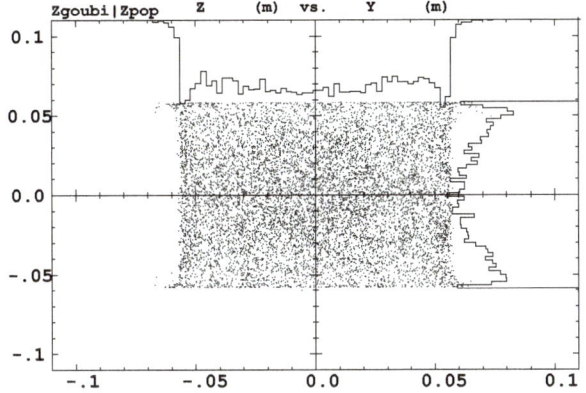

Fig. 12.31 Two-dimensional uniformized beam cross-section at target, and projected histograms

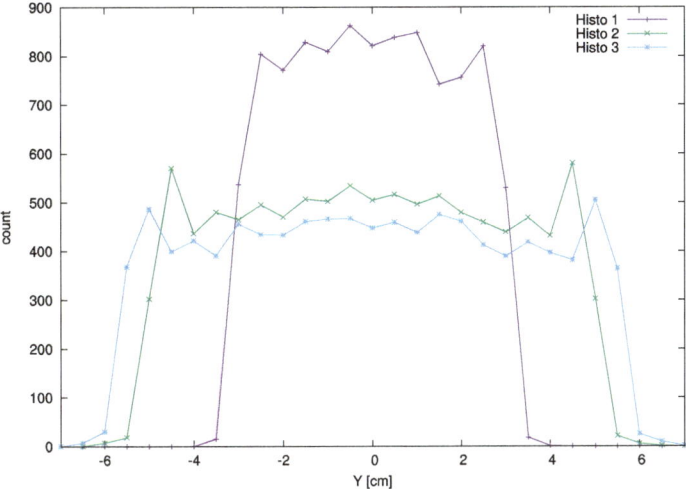

Fig. 12.32 Horizontal beam density histograms at entrance of QPL1 (1), entrance (2) and exit (3) of QPL0

12.4 Animation: Transverse Scan by a Uniform Beam

An animation of the frozen rectangle sweep in Fig. 12.25 is obtained by plotting the content of the zgoubi.fai file produced by running the sweep simulation of Table 12.9. A possible gnuplot script for that is:

```
ip1=1; ip2=20 ; set xrange [-4:4] ; set yrange [-7:7]
do for [ip=ip1:ip2] {
plot "zgoubi.fai" u ($38<=ip ? $10 :1/0):($12)  w p
     } ; pause 4
```

Table 12.11 Outcome of the HISTO keyword, in zgoubi.res execution listing. The histogram at exit of QPL0, here

```
 35 Keyword, label(s) : HISTO                                                           IPASS= 1

                        HISTOGRAMME  DE  LA  COORDONNEE  Y
                        DANS  LA  FENETRE :   -10.00    /    10.00     (CM)
                        Request to print out histogram to zgoubi.HISTO.out.

20                              0000000000000000000000000
19                              YYYYYYYYYYYYYYYYYYYYYYYYYY
18                              YYYYYYYYYYYYYYYYYYYYYYYYYY
17                              YYYYYYYYYYYYYYYYYYYYYYYYYY
16                              YYYYYYYYYYYYYYYYYYYYYYYYYY
15                              YYYYYYYYYYYYYYYYYYYYYYYYYY
14                              YYYYYYYYYYYYYYYYYYYYYYYYYY
13                              YYYYYYYYYYYYYYYYYYYYYYYYYY
12                              YYYYYYYYYYYYYYYYYYYYYYYYYY
11                              YYYYYYYYYYYYYYYYYYYYYYYYYY
10                              0000000000000000000000000
 9                              YYYYYYYYYYYYYYYYYYYYYYYYYYY
 8                              YYYYYYYYYYYYYYYYYYYYYYYYYYY
 7                             YYYYYYYYYYYYYYYYYYYYYYYYYYYY
 6                             YYYYYYYYYYYYYYYYYYYYYYYYYYYY
 5                             YYYYYYYYYYYYYYYYYYYYYYYYYYYY
 4                             YYYYYYYYYYYYYYYYYYYYYYYYYYYY
 3                            YYYYYYYYYYYYYYYYYYYYYYYYYYYYY
 2                            YYYYYYYYYYYYYYYYYYYYYYYYYYYYY
 1                           YYYYYYYYYYYYYYYYYYYYYYYYYYYYYY

                        12345678901234567890123456789012345678 90
                              5         6         7

   TOTAL  COMPTAGE  TOUS  BINS    :   10000  PARTICULES  SUR  UN  TOTAL  DE    10000  LANCEES
   NUMERO   DU  BIN  MOYEN        :      61
   COMPTAGE  AU   "      "        :     447
   VAL. PHYS. AU   "      "       :     0.00    (CM)

   PARAMETRES  PHYSIQUES  DE  LA  DISTRIBUTION :
      Found min. =   -6.619   , found max. =  7.293   , MAX-MIN =   13.91  (CM) , DX PAR BIN = 0.5
      MOYENNE = -2.0382E-02 (CM)           SIGMA =    3.282    (CM)
```

12.4.2 Nano-Probe E, B Final Doublet Achromat

12.5 Construct a Final Focus Doublet

(a) Optical sequence. Trajectories and transport matrix.

The simulation input data file is given in Table 12.12, and follows the scheme given in Fig. 12.26. Note the use of KOBJ = 6 in OBJET, which creates 61 rays, so allowing computation of transport coefficients to second order in the coordinates, by MATRIX (using KOBJ = 6.1 instead would create 102 rays and allow additional computation of the 3rd order transport coefficients).

Out of zgoubi.res execution listing, the resulting first order transport matrix $[R_{ij}]$ at the image is the following:

```
-0.142195      4.125875E-07    0.00000        0.00000        0.00000    0.00000
-1.63380      -7.03261         0.00000        0.00000        0.00000    0.00000
 0.00000       0.00000        -2.986608E-02   6.239409E-08   0.00000    0.00000
 0.00000       0.00000        -5.94705      -33.4828         0.00000    0.00000
 0.00000       0.00000         0.00000        0.00000        1.00000    0.00000
 0.00000       0.00000         0.00000        0.00000        0.00000    1.00000
```

Table 12.12 Simulation input data file: final doublet achromat. The FIT procedure is used to adjust the field values in both quadrupoles in order to ensure 25 cm image distance. Once FIT is done, the execution pointer proceeds in sequence, to MATRIX, DRIFT, and whatever may follow. *Note* this file is available in zgoubi sourceforge repository at https://sourceforge.net/p/zgoubi/code/HEAD/ tree/branches/exemples/book/zgoubiMaterial/beamLines/nanoProbeAchromat/animation/

```
E, B final doublet achromat. File Q12_hardEdge.inc
'MARKER' EBdoublet_B_S
'OBJET'
20.435                                                       ! H+ ions, 20keV.
6       ! KOBJ=6 creates 61 rays, to allow 2nd order in/out parametrization by MATRIX.
.01 .01 .007 .01 .01 .0001
0. 0. 0. 0. 0. 1.
'PARTICUL'           ! Necessary keyword, due to EBMULT (possibly) including E components.
PROTON

'DRIFT'    drift_1
559. split 10 2      ! 'split' and IL=2, for print out of trajectories to zgoubi.plt.

'MARKER'  Q1_S
'EBMULT'  Q1                                                 ! First quadrupole.
2                                   ! IL=2 for step-by-step print out to zgoubi.plt.
10.2  10. 0. 0. 0. 0. 0. 0. 0. 0. 0. 0.    ! Electric q-pole component is 0.
0.  0.  0.  0.  0.  0.  0.  0. 0. 0. 0.             ! Entrance EFB, sharp edge.
6  .1122 6.2671 -1.4982 3.5882 -2.1209 1.723
0.  0.  0.  0.  0.  0.  0.  0. 0. 0. 0.                 ! Exit EFB, sharp edge.
6  .1122 6.2671 -1.4982 3.5882 -2.1209 1.723
0. 0. 0. 0. 0. 0. 0. 0. 0. 0.
10.2  10. 0. 0.94747616 0. 0. 0. 0. 0. 0. 0. 0.    ! Magnetic q-pole component set
0.  0.  0.  0.  0.  0.  0.  0. 0. 0. 0.      ! for image located 25 cm downstream
6  .1122 6.2671 -1.4982 3.5882 -2.1209 1.723          ! of quadrupole exit.
0.  0.  0.  0.  0.  0.  0.  9. 0. 0. 0.             ! Entrance EFB, sharp edge.
6  .1122 6.2671 -1.4982 3.5882 -2.1209 1.723
0. 0. 0. 0. 0. 0. 0. 0. 0. 0.
1.                                        ! Integration step size is 1 cm.
1 0. 0. 0.

'MARKER'  Q1_E

'DRIFT'    drift_2
4.9 split 10 2       ! 'split' and IL=2, for print out of trajectories to zgoubi.plt.
'MARKER'  Q2_S

'EBMULT'  Q2                                                 ! Second quadrupole.
2                                   ! IL=2 for step-by-step print out to zgoubi.plt.
10.2  10. 0. 0. 0. 0. 0. 0. 0. 0. 0. 0.    ! Electric q-pole component is 0.
0.  0.  0.  0.  0.  0.  0.  0. 0. 0. 0.
6  .1122 6.2671 -1.4982 3.5882 -2.1209 1.723
0.  0.  0.  0.  0.  0.  0.  0. 0. 0. 0.
6  .1122 6.2671 -1.4982 3.5882 -2.1209 1.723
0. 0. 0. 0. 0. 0. 0. 0. 0. 0.
10.2  10. 0. -1.4079640 0. 0. 0. 0. 0. 0. 0. 0.    ! Magnetic q-pole component set
0.  0.  0.  0.  0.  0.  0.  0. 0. 0. 0.      ! for image located 25 cm downstream
6  .1122 6.2671 -1.4982 3.5882 -2.1209 1.723          ! of quadrupole exit.
0.  0.  0.  0.  0.  0.  0.  0. 0. 0. 0.
6  .1122 6.2671 -1.4982 3.5882 -2.1209 1.723
0. 0. 0. 0. 0. 0. 0. 0. 0. 0.
1.                                        ! Integration step size is 1 cm.
1 0. 0. 0.
'MARKER'  Q2_E
'DRIFT'    drift_3
25. split 20 2       ! 'split' and IL=2, for print out of trajectories to zgoubi.plt.
     ! In this data list, B values in Q1 & Q2 are already matched for image location here,
           ! otherwise, uncomment the FIT keyword and data, if further matching is desired.
! 'FIT'
! 2                                                         ! 2 variables:
! 6 63  0 .1                                            ! B field in first quad,
! 8 63  0 .1                                           ! and B field in second quad.
! 2 1E-12 50 ! 2 constraints. Optional: penalty and #iter. values (default otherwise).
! 1 1 2 #End 0. 1. 0     ! R11=0 at end of sequence (i.e., after the final25 cm drift),
! 1 3 4 #End 0. 1. 0                          ! R34=0 at end of sequence.

'MATRIX'         ! Local value of the 1st and 2nd order tranport matrices, from OBJET.
2  0

'DRIFT'    ! Add short drift just for graphic purposes: extend rays beyond focal plane.
20. split 20 2       ! 'split' and IL=2, for print out of trajectories to zgoubi.plt.
'MARKER' EBdoublet_B_E
'END'
```

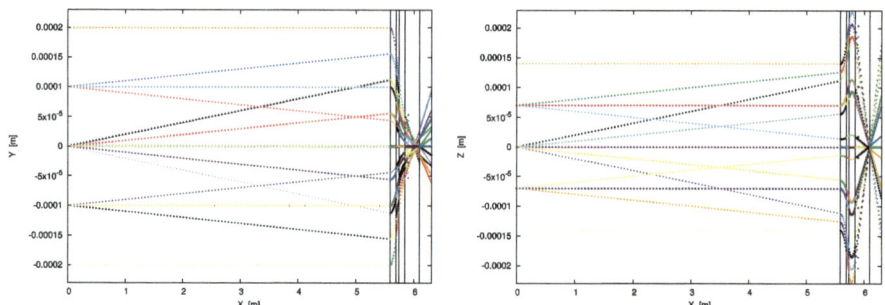

Fig. 12.33 Projected coordinates Y(X) and Z(X), across the optical system from object to 20 cm downstream of the image plane, of the 61 paraxial rays used to compute the transport coefficients to 2nd order (X is the distance along the optical axis). The vertical bars materialize the quadrupoles and (rightmost bar) the location of the paraxial image plane

It can be verified that $R_{12} \approx 0$ and $R_{34} \approx 0$, conditions for an image located 25 cm downstream of the quadrupole doublet. These two constraints are realized by a FIT procedure (Table 12.12), by varying the magnetic field in the quadrupoles.

The second order transverse geometric coefficients, under MATRIX in zgoubi.res execution listing, appear to all be null, as expected. The non-coupled chromatic coefficients $T_{1-2,1-2,6}$, $T_{3-4,3-4,6}$, and $T_{1-4,6,6}$ are non-zero, namely, out of zgoubi.res:

1 16	1.15	1 26	6.27	1 36	0.00	1 46	0.00	1 56	0.00	1 66	0.00
2 16	2.85	2 26	15.4	2 36	0.00	2 46	0.00	2 56	0.00	2 66	0.00
3 16	0.00	3 26	0.00	3 36	1.02	3 46	5.89	3 56	0.00	3 66	0.00
4 16	0.00	4 26	0.00	4 36	4.72	4 46	27.0	4 56	0.00	4 66	0.00

Figure 12.33 displays the projected coordinates of the 61 paraxial rays used to compute the transport coefficients to 2nd order, across the optical system. The gnuplot script used for that is given in Table 12.13.

(b) Optical imaging. Locations of waist.

Table 12.14 gives the input data file to generate the required random object and perform the tracking.

The gnuplot script given in Table 12.15 is used to produce Fig. 12.34, which shows the initial horizontal and vertical Gaussian angle distributions, and the momentum uniform distribution, resulting from MCOBJET keyword.

Table 12.13 A gnuplot script to obtain Fig. 12.33: projected coordinates of paraxial rays along the optical system

```
set xlabel 'X [m]' ; set ylabel 'Y [m]' ; set xtics mirror ; set ytics mirror
set key maxrow 3 ; set key t c ; cm2m = 0.01

set arrow from 5.59, graph 0 to 5.59, graph 1 nohead
set arrow from 5.692, graph 0 to 5.692, graph 1 nohead
set arrow from 5.741, graph 0 to 5.741, graph 1 nohead
set arrow from 5.843, graph 0 to 5.843, graph 1 nohead

set arrow from 6.093, graph 0 to 6.093, graph 1 nohead
x1 = 559 ; x2 = x1 + 10.2 ; x3 = x2 + 4.9 ; x4 = x3 + 10.2 ; x5 = x4 + 25.
set xrange [:6.3]; set yrange [-2.3e-4:2.3e-4]

plot \
for [i=1:61] 'zgoubi.plt' u ($19== i && $42== 5 ? $22    *cm2m : 1/0):($10 *cm2m) w p ps .2 notit ,\
for [i=1:61] 'zgoubi.plt' u ($19== i && $42== 6 ? ($22+x1) *cm2m : 1/0):($10 *cm2m) w p ps .2 notit ,\
for [i=1:61] 'zgoubi.plt' u ($19== i && $42== 7 ? ($22+x2) *cm2m : 1/0):($10 *cm2m) w p ps .2 notit ,\
for [i=1:61] 'zgoubi.plt' u ($19== i && $42== 8 ? ($22+x3) *cm2m : 1/0):($10 *cm2m) w p ps .2 notit ,\
for [i=1:61] 'zgoubi.plt' u ($19== i && $42== 9 ? ($22+x4) *cm2m : 1/0):($10 *cm2m) w p ps .2 notit ,\
for [i=1:61] 'zgoubi.plt' u ($19== i && $42==11 ? ($22+x5) *cm2m : 1/0):($10 *cm2m) w p ps .2 notit

pause 1

set ylabel 'Z [m]'

set arrow from 5.59, graph 0 to 5.59, graph 1 nohead
set arrow from 5.692, graph 0 to 5.692, graph 1 nohead
set arrow from 5.741, graph 0 to 5.741, graph 1 nohead
set arrow from 5.843, graph 0 to 5.843, graph 1 nohead

set arrow from 6.093, graph 0 to 6.093, graph 1 nohead

plot \
for [i=1:61] 'zgoubi.plt' u ($19== i && $42== 5 ? $22    *cm2m : 1/0):($12 *cm2m) w p ps .2 notit ,\
for [i=1:61] 'zgoubi.plt' u ($19== i && $42== 6 ? ($22+x1) *cm2m : 1/0):($12 *cm2m) w p ps .2 notit ,\
for [i=1:61] 'zgoubi.plt' u ($19== i && $42== 7 ? ($22+x2) *cm2m : 1/0):($12 *cm2m) w p ps .2 notit ,\
for [i=1:61] 'zgoubi.plt' u ($19== i && $42== 8 ? ($22+x3) *cm2m : 1/0):($12 *cm2m) w p ps .2 notit ,\
for [i=1:61] 'zgoubi.plt' u ($19== i && $42== 9 ? ($22+x4) *cm2m : 1/0):($12 *cm2m) w p ps .2 notit ,\
for [i=1:61] 'zgoubi.plt' u ($19== i && $42==11 ? ($22+x5) *cm2m : 1/0):($12 *cm2m) w p ps .2 notit
```

The IMAGE and IMAGEZ keywords allow confirming the location of the image, which appears to be on-axis ($Y = Z = 0$), and, for respectively the horizontal and vertical image, $\Delta X = -47\,\mu\mathrm{m}$ and $\Delta X = -13\,\mu\mathrm{m}$ upstream of the design position, a marginal difference compared to the expected 25 cm distance from the second quadrupole. Additional information regarding the positioning of the horizontal and vertical images can be found under respectively IMAGE and IMAGEZ in zgoubi.res execution listing, for instance:

```
Keyword, label(s) : IMAGE
   RECHERCHE DU POINT DE FOCALISATION HORIZONTAL DE  10000 TRAJECTOIRES (SUR  10000)
   POINT DE FOCALISATION HORIZONTAL SUR L ORBITE MOYENNE    X = -4.669011E-03 CM    Y = 2.017584E-05 CM
   LARGEUR IMAGE, A MI-HAUTEUR =  3.363129E-04 CM,  TOTALE =  1.606970E-03 CM

Keyword, label(s) : IMAGEZ
   RECHERCHE DU POINT DE FOCALISATION VERTICAL   DE  10000 TRAJECTOIRES (SUR  10000)
   POINT DE FOCALISATION VERTICAL   SUR L ORBITE MOYENNE    X = -1.316689E-03 CM    Z = -1.060501E-04 CM
   HAUTEUR IMAGE, A MI-HAUTEUR =  3.103773E-04 CM,  TOTALE =  1.311241E-03 CM
```

Figure 12.35 shows the horizontal and vertical phase spaces at the image plane. They feature a wide Y-extent, due to the beam momentum spread.

(c) Using an electrostatic doublet.

The exercise can use the simulation input data file of Table 12.12, with proper values for the E component of the field in the **E, B** quadrupoles, whereas B is set to zero. The same procedures can be applied to reproduce (a) and (b) exercises and their outcomes.

Table 12.14 Simulation input data file: case of regular magnetic quadrupole doublet, ray-trace from object to image, 5×10^3 H+ ions. Coordinates at the image are logged in zgoubi.fai. The INCLUDEd file Q12_hardEdge.inc is given in Table 12.12

```
E, B final doublet achromat
'MCOBJET'
20.435                                                             ! H+, 20keV.
1                                                   ! Particle distribution in a window.
5000                                                      ! Number of particles.
2   2   2   2   1   1          ! Densities: Y, T, Z, P Gaussian; s, D uniform.
0.  0.  0.  0.  0.  1.                             ! Central value of coordinates.
0.  .2E-3 .0  .2E-3 0.  0.001                      ! Gaussian 1/2width; delta_D=1e-3.
2   2   2   2   1   1   ! Cut-offs of Gaussian distributions - unused if uniform.
9  9. 9. 9. 9.                          ! Data for exponential D density - unused.
186387 548728 472874                            ! Random generator seeds.
'PARTICUL'
938.2723 1.60217733E-19 0. 0. 0.                                   ! H+ data.
'FAISCEAU'                              ! Allows local check of particle coordinates.
'ESL'
559.
'INCLUDE'
1
Q12_hardEdge.inc[Q1_S:Q1_E]      ! Grab segment [Q1_S:Q1_E] from file Q12_hardEdge.inc.
'ESL'
4.9
'INCLUDE'
1
Q12_hardEdge.inc[Q2_S:Q2_E]      ! Grab segment [Q2_S:Q2_E] from file Q12_hardEdge.inc.
'ESL'
25.
'IMAGE'                                 ! Compute the location of the horizontal image.
'IMAGEZ'                                ! Compute the location of the vertical image.

'HISTO'
13 -.62  .62  120  2  PRINT                          ! Histogram of To coordinate,
20  Y  1  Q                                          ! and log to zgoubi.HISTO.out.
'HISTO'
15 -.62  .62  120  2  PRINT                          ! Histogram of Po coordinate,
20  Z  1  Q                                          ! and log to zgoubi.HISTO.out.
'HISTO'
1  .9989 1.0011 120  2  PRINT                        ! Histogram of D coordinate,
20  D  1  Q                                          ! and log to zgoubi.HISTO.out.

'FAISCNL'                               ! Log final coordinates (and more) to zgoubi.fai.
zgoubi.fai
'MARKER' beamLinesAchro_OneB_E
'END'
```

Table 12.15 A gnuplot script to obtain Fig. 12.34: initial horizontal and vertical Gaussian distributions, and the uniform momentum distribution. The data are read from zgoubi.HISTO.out, produced by the HISTO commands (Table 12.14)

```
set xlabel "dp/p_{ref}" font "roman,18"; set ylabel "count" font "roman,18"
set xtics mirror font "roman,11"; set ytics mirror font "roman,11"
set key  t r font "roman,10"  ; set key maxcol 1

# MONTE CARLO TRACKING DATA:
npass = 1 ; npart = 1e4 ; nbin = 120 ; dpBin = (1.0011-0.9989) / nbin

plot  [0.9989:1.0011] \
  'zgoubi.HISTO.out'  u ($12==1 ? $1 : 1/0):($2/npart/dpBin) w lp ps .4 pt 10 lw 2 lc rgb "black" notit

set xlabel "T_0 [mrad]" font "roman,18"
plot  \
    'zgoubi.HISTO.out'  u ($12==13 ? $1 : 1/0):($2/npart/dpBin) w lp ps .4 pt 10 lw 2 lc rgb "red" notit

set xlabel "P_0 [mrad]" font "roman,18"
plot  \
    'zgoubi.HISTO.out'  u ($12==15 ? $1 : 1/0):($2/npart/dpBin) w lp ps .4 pt 10 lw 2 lc rgb "blue" notit
```

Fig. 12.34 From top to bottom: Gaussian distributions of the initial horizontal and vertical angles; uniform momentum distribution as per MCOBJET[KOBJ = 1] definition in Table 12.14

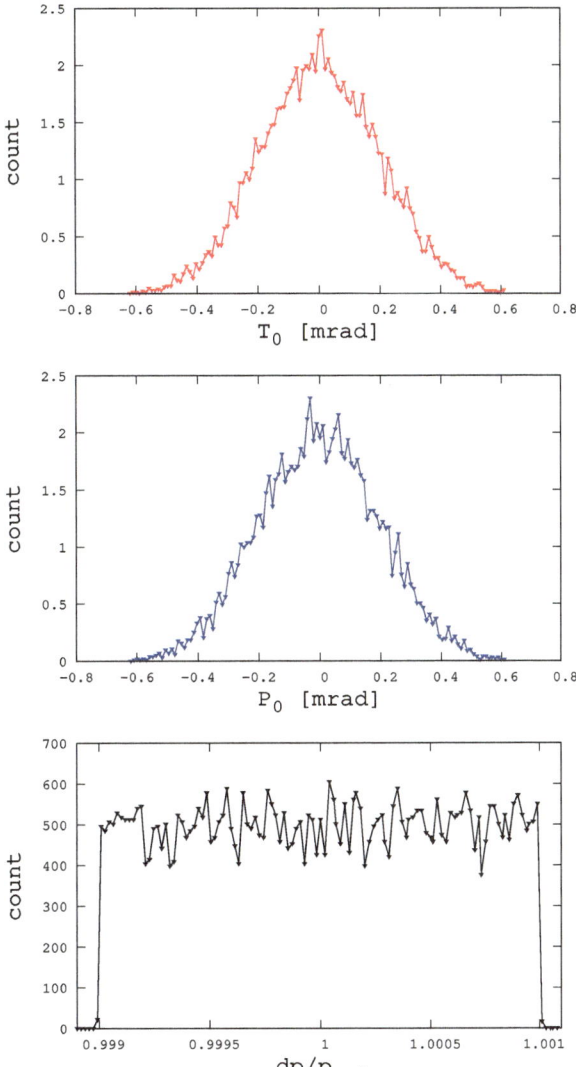

12.6 E, B Lens, Hard-Edge Model

(a) E and B values. Transport matrix.

The beam optics model with both E and B fields set is given in Table 12.16, and follows the geometry given in Fig. 12.26.

Finding initial (ϕ_F, B_F), (ϕ_D, B_D) electric potential and magnetic field values for respectively QF and QD quadrupoles is a matter of solving 2 equations with 2 unknowns, for each one of the two quadrupoles, whereas exercise 12.5 tells the

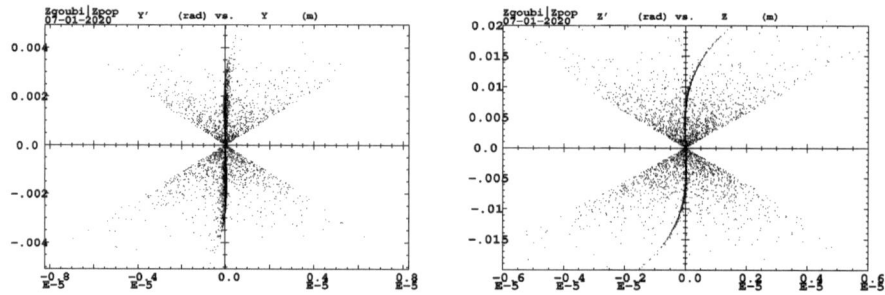

Fig. 12.35 Horizontal (left) and vertical (right) phase spaces at the image plane. V-shaped, wide extent portraits: case of regular magnetic quadrupole doublet. S-shaped portraits: case of the **E**, **B** doublet, the second order chromatic aberration has been canceled, only remains a third order geometric aberration typical of quadrupoles. Graphs obtained using zpop: menu 7; 1/5 to open zgoubi.fai; 2/[2,3] to select Y, Y' coordinates, 2[4/5] to select Z, Z' coordinates; 7 to plot

required combined E, B focusing strength to get proper positioning of the image plane (25 cm downstream of QD). *In fine*, one needs to essentially double B compared to the $\phi = 0$ case (Exercise 12.5), by virtue of Eq. 12.12.

Following this preliminary setting, a FIT then takes care of a fine adjustment of (ϕ_F, B_F) and (ϕ_D, B_D), With focusing strengths calculated according to the hypotheses of Table 12.16, namely, protons, $B\rho = 20.435\,\mathrm{T\,m}$ and thus $W = 20\,\mathrm{keV}$, radius at pole tip taken to be $a = 0.1\,\mathrm{m}$, then the FIT procedure yields potentials $\phi_{F,D}$ and potential and field $B_{F,D}$ such that

$$
\begin{cases} \phi_F = -927.33\,\mathrm{V} \\ B_F = 0.18950\,\mathrm{T} \end{cases} \Rightarrow \begin{cases} K_{F,e} = \dfrac{\phi_F}{aW} = -46.367\,m^{-2} \\ K_{F,m} = \dfrac{B_F}{aB\rho} = 92.732\,m^{-2} \end{cases} \Rightarrow K_{F,m} = -2K_{F,e}
$$

$$
\begin{cases} \phi_D = 1378\,\mathrm{V} \\ B_D = -0.28159\,\mathrm{T} \end{cases} \Rightarrow \begin{cases} K_{D,e} = \dfrac{\phi_D}{aW} = 68.9017\,m^{-2} \\ K_{D,m} = \dfrac{B_D}{aB\rho} = -137.80\,m^{-2} \end{cases} \Rightarrow K_{D,m} = -2K_{D,e}
$$

Out of zgoubi.res execution listing, the resulting first order transport matrix $[R_{ij}]$ at the image is the following:

```
-0.142191      7.125658E-06   -1.084862E-16   -4.836514E-16    0.00000    0.00000
-1.63377       -7.03252       -1.678508E-16   -5.217183E-16    0.00000    0.00000
4.992819E-18   -4.132432E-16   -2.986418E-02    5.779779E-06    0.00000    0.00000
1.706437E-16   -2.899205E-16   -5.94705        -33.4828         0.00000    0.00000
0.00000         0.00000         0.00000          0.00000         1.00000    0.00000
0.00000         0.00000         0.00000          0.00000         0.00000    1.00000
```

It can be verified that $R_{12} = 0$ and $R_{34} = 0$, conditions for an image located 25 cm downstream of the quadrupole doublet.

Table 12.16 Simulation input data file: final **E, B** doublet achromat, in the hard-edge quadrupole field model. The FIT procedure adjusts the E and B field values in both quadrupoles in order to ensure exact 25 cm image distance and zero second order chromatic coefficients

```
E, B final doublet achromat
'MARKER' beamLinesAchro_EBDoublet_S
'OBJET'
20.435                                                  ! H+ ions, 20keV.
6         ! KOBJ=6 creates 61 rays, to allow 2nd order in/out parametrization by MATRIX.
.01 .01 .007 .01 .01 .0001
0. 0. 0. 0. 0. 1.
'PARTICUL'      ! Necessary keyword, due to EBMULT (possibly) including E components.
PROTON

'MARKER'        ! OBJET is located 559 cm upstream of entrance of first quadrupole.
'DRIFT'   drift_1
559.
'INCLUDE'
1
Q12_hardEdge.inc[Q1_S:Q1_E]      ! Q1 data  are grabbed from   Q12_hardEdge.inc file.
'ESL'
4.9
'INCLUDE'
1
Q12_hardEdge.inc[Q2_S:Q2_E]      ! Q2 data  are grabbed from   Q12_hardEdge.inc file.
'DRIFT'   drift_3
25.

'FIT'                ! Constrints on T_ijk transport coefficients impose OBJET/KOBJ=6.
4                                               ! 4 variables:
  7  5  0 .1                                    ! E field in first quad,
  7 63  0 .1                                    ! B field in first quad,
 11  5  0 .1                                    ! and E field in second quad.
 11 63  0 .1                                    ! and B field in second quad.
 6 1E-12 50  ! 6 constraints. Optional: penalty and #iter. values (default otherwise).
 1 1  2 #End 0. 1. 0      ! R11=0 at end of sequence (i.e., after the final25 cm drift),
 1 3  4 #End 0. 1. 0                            ! R34=0 at end of sequence.
 2 2 16 #End 0. 1. 0                            ! T_216=0 at end of sequence.
 2 2 26 #End 0. 1. 0                            ! T_226=0 at end of sequence.
 2 4 36 #End 0. 1. 0                            ! T_436=0 at end of sequence.
 2 3 46 #End 0. 1. 0                            ! T_446=0 at end of sequence.

'MATRIX'                     ! Once FIT done: local value of the 1st and 2nd order
2  0                         ! tranport matrices, from OBJET.
'MARKER' beamLinesAchro_EBDoublet_E
'END'
```

The second order matrix out of zgoubi.res execution listing shows that all chromatic coefficients are now zero,

```
1 16  1.678E-07    1 26  7.882E-07    1 36  1.517E-14    1 46 -1.116E-12    1 56  0.00    1 66  0.00
2 16  2.293E-07    2 26  7.515E-07    2 36  3.840E-14    2 46 -3.400E-12    2 56  0.00    2 66  0.00
3 16  1.631E-13    3 26 -1.349E-13    3 36  1.079E-06    3 46  6.189E-06    3 56  0.00    3 66  0.00
4 16  4.462E-13    4 26 -8.443E-13    4 36  4.015E-06    4 46  2.299E-05    4 56  0.00    4 66  0.00
```

(b) Point-spread function.

Now 5,000 particles are tracked through the achromat, with the goal of highlighting the effect of canceled second order chromatic aberrations in the presence of momentum spread in the beam.

The input data file is similar to that of Exercise 12.5, yet with updated ϕ and B components (Table 12.17).

Figure 12.35 shows the horizontal and vertical phase spaces at the image plane, where they have been superposed to those resulting from the magnetic quadrupole doublet case for comparison. The former feature a much narrower Y-extent as a consequence of the cancellation of the second order chromatic aberrations by the combined **E, B** field. The IMAGE and IMAGEZ keywords detail the smaller image sizes, respectively horizontal and vertical:

Table 12.17 Simulation input data file: case of an achromat quadrupole doublet, ray-tracing from object to image, 5×10^3 H+ ions. Coordinates at the image are logged in zgoubi.fai. Details of the potential and B field settings in Q1 and Q2 quadrupoles are given at the bottom of the Table

```
E, B final doublet achromat
'MCOBJET'
20.435                                               ! H+, 20keV.
1                                    ! Particle distribution in a  window.
5000                                          ! Number  of  particles.
2    2    2    2    1    1         ! Densities: H, T, Z, P Gaussian; s, D uniform.
0.   0.   0.   0.   0.   1.             ! Central  value of coordinates.
0.   .2E-3 .0  .2E-3 0.   0.001        ! Gaussian 1/2width; delta_D=1e-3.
2    2    2    2    1    1   ! Cut-offs of Gaussian distributions - unused if uniform.
9   9. 9. 9.                       ! Data for exponential D density - unused.
186387 548728 472874                      ! Random generator seeds.
'PARTICUL'
938.2723 1.60217733E-19 0. 0. 0.                           ! H+ data.
'FAISCEAU'                        ! Allows local check of particle coordinates.
'ESL'
559.
'INCLUDE'
1
Q12_hardEdge.inc[Q1_S:Q1_E]       ! Q1 data  are grabbed from   Q12_hardEdge.inc file.
'ESL'
4.9
'INCLUDE'
1
Q2_hardEdge.inc[Q2_S:Q2_E]        ! Q2 data  are grabbed from   Q2_hardEdge.inc file.
'ESL'
25.
'IMAGE'                                  ! Compute the location of the horizontal image.
'IMAGEZ'                                 ! Compute the location of the vertical image.

'FAISCNL'                              ! Log final coordinates (and more) to zgoubi.fai.
zgoubi.fai
'END'
```

```
Potential and B field settings in Q1 quadrupole (''Q12_hardEdge.inc'' file in the INCLUDE above):

10.2  10. 0. -9272.986  0. 0. 0. 0. 0. 0. 0. 0.        ! Electric q-pole component.
10.2  10. 0. 1.89493 0. 0. 0. 0. 0. 0. 0. 0.           ! Magnetic q-pole component.


Potential and B field settings in Q2 quadrupole (''Q12_hardEdge.inc'' file in the INCLUDE above):

10.2  10. 0. 13779.90 0. 0. 0. 0. 0. 0. 0. 0.          ! Electric q-pole component.
10.2  10. 0. -2.81592 0. 0. 0. 0. 0. 0. 0. 0.          ! Magnetic q-pole component.
```

```
Keyword, label(s) :  IMAGE
    RECHERCHE DU POINT DE FOCALISATION HORIZONTAL DE   5000 TRAJECTOIRES (SUR   5000)
    POINT DE FOCALISATION HORIZONTAL SUR L ORBITE MOYENNE   X = -1.005774E-03 CM    Y = -7.940402E-07 CM
    DECALAGE DU CENTRE DE GRAVITE EN Y =    7.439999E-07 CM
    LARGEUR IMAGE, A MI-HAUTEUR = 1.698952E-05 CM,  TOTALE = 1.735729E-04 CM

Keyword, label(s) :  IMAGEZ
    RECHERCHE DU POINT DE FOCALISATION VERTICAL   DE   5000 TRAJECTOIRES (SUR   5000)
    POINT DE FOCALISATION VERTICAL   SUR L ORBITE MOYENNE   X = -1.803303E-03 CM    Z =  6.816746E-06 CM
    DECALAGE DU CENTRE DE GRAVITE EN Z =   -6.865026E-06 CM
    HAUTEUR IMAGE, A MI-HAUTEUR = 2.867521E-05 CM,  TOTALE = 2.715059E-04 CM
```

12.7 E, B Lens, With Fringe Fields

Fringe fields are added, considering an unrealistically large, 5 cm bore for the sake of enhanced possible adverse effects. Otherwise, the same geometry is maintained (Fig. 12.26).

(a) E and B values; transport matrix

Finding the new values of the E and B components of the lenses, for a paraxial image at the same location together with cancelled chromatic coefficients, requires re-running the FIT procedure.

The corresponding input file is given in Table 12.18. Here again the complete FIT (location of the paraxial image plane concurrently with cancelled second order chromatic aberrations) is performed in one go, taking starting field values from the earlier matched hard-edge model.

The resulting first order transport matrix $[R_{ij}]$ at the image, found under MATRIX in zgoubi.res execution listing, is the following:

```
-0.143799       -4.091498E-08   -5.931279E-15   -1.426745E-14   0.00000    0.00000
-1.62606        -6.95399        -2.192285E-14   -5.254189E-14   0.00000    0.00000
-2.913262E-17   -1.290835E-16   -2.958046E-02   -4.487652E-08   0.00000    0.00000
1.282855E-16    6.591733E-16    -6.00354        -33.8052        0.00000    0.00000
0.00000         0.00000         0.00000         0.00000         1.00000    0.00000
0.00000         0.00000         0.00000         0.00000         0.00000    1.00000
```

It can be verified that $R_{12} = 0$ and $R_{34} = 0$, conditions for an image located 25 cm downstream of the quadrupole doublet.

The second order matrix out of zgoubi.res execution listing shows that all chromatic coefficients are zero:

```
1 16  4.076E-07    1 26 -7.286E-07    1 36 -4.989E-10    1 46 -1.960E-09    1 56  0.00    1 66  0.00
2 16  1.408E-06    2 26 -3.008E-07    2 36 -1.796E-09    2 46 -6.978E-09    2 56  0.00    2 66  0.00
3 16  4.352E-14    3 26  9.639E-13    3 36  4.999E-07    3 46 -3.647E-07    3 56  0.00    3 66  0.00
4 16  1.149E-13    4 26  2.278E-12    4 36  4.268E-07    4 46 -7.152E-06    4 56  0.00    4 66  0.00
```

(b) Field and derivatives across EBMULT

$\mathbf{E}_Z(s)$ and $\mathbf{B}_Z(s)$ field components along, respectively, $(Y = 0, Z = 1\,\text{cm})$ and $(Y = 1\,\text{cm}, Z = 0)$ lines across the doublet are displayed in Fig. 12.36. For the purpose of these graphs, fields are logged in zgoubi.plt using EBMULT[IL = 2]. (Table 12.18). Two particle trajectories are tracked for that, forced to straight lines using OPTIONS[CONSTY ON] keyword:

```
'OBJET'
20.435
2
2 1
1. 0. 0. 0. 0. 1. 'Y'                    ! Y0, T0, Z0, P0, X0 (unused) and D0 coordinates;
0. 0. 1. 0. 0. 1. 'Z'              ! D >> 1 would also force particles on straight trajectories,
1 1                                                     ! in lieu of Options[CONSTY].
'PARTICUL'
PROTON
'OPTIONS'
1 1
CONSTY ON
 'DRIFT'
559.
etc.
```

Table 12.18 Simulation input data file: final doublet achromat, accounting for E and B fringe fields ($X_E = X_S = 9$ cm, $\lambda_E = \lambda_S = 5$ cm, $E_2 = E_3 = 1$, for both E and B fields, both quadrupoles). The FIT procedure adjusts the field values in the quadrupoles with the constraints of ensuring, concurrently, 25 cm image distance and cancelled chromatic aberrations—MATRIX computation follows once FIT is done, to allow checking the second order chromatic coefficients, expected null

```
E, B final doublet achromat
'MARKER' EBwithFF_S
'OBJET'
20.435                                                          ! H+ ions, 20keV.
6          ! KOBJ=6 creates 61 rays, to allow 2nd order in/out parametrization by MATRIX.
.01  .01 .007 .01 .01 .0001
0. 0. 0. 0. 0. 1.
'PARTICUL'        ! Necessary keyword, due to EBMULT (possibly) including E components.
PROTON

'MARKER'         ! OBJET is located 559 cm upstream of entrance of first quadrupole.
'DRIFT'    drift_1
559.

'EBMULT'                                                        ! First quadrupole.
0                                     ! IL=2 for step-by-step print out to zgoubi.plt.
10.2  10. 0. -9446.2968  0. 0. 0. 0. 0. 0. 0. 0.     ! Electric q-pole component.
9.  5.  1.  1.  0.  0.  0. 0. 0. 0. 0.              ! Entrance EFB, field fall-off.
6  .1122 6.2671 -1.4982 3.5882 -2.1209 1.723
9.  5.  1.  1.  0.  0.  0. 0. 0. 0. 0.                  ! Exit EFB, field fall-off.
6  .1122 6.2671 -1.4982 3.5882 -2.1209 1.723
0. 0. 0. 0. 0. 0. 0. 0. 0. 0.
10.2  10. 0. 1.9303418  0. 0. 0. 0. 0. 0. 0. 0.      ! Magnetic q-pole component.
9.  5.  1.  1.  0.  0.  0. 0. 0. 0. 0.            ! for image located 25 cm downstream
6  .1122 6.2671 -1.4982 3.5882 -2.1209 1.723              ! of quadrupole exit.
9.  5.  1.  1.  0.  0.  0. 0. 0. 0. 0.
6  .1122 6.2671 -1.4982 3.5882 -2.1209 1.723
0. 0. 0. 0. 0. 0. 0. 0. 0. 0.
1.                                                 ! Integration step size is 1 cm.
1 0. 0. 0.
'DRIFT'    drift_2
4.9
'EBMULT'                                                       ! Second quadrupole.
0                                     ! IL=2 for step-by-step print out to zgoubi.plt.
10.2  10. 0. 14048.719  0. 0. 0. 0. 0. 0. 0. 0.      ! Electric q-pole component.
9.  5.  1.  1.  0.  0.  0. 0. 0. 0. 0
6  .1122 6.2671 -1.4982 3.5882 -2.1209 1.723
9.  5.  1.  1.  0.  0.  0. 0. 0. 0. 0
6  .1122 6.2671 -1.4982 3.5882 -2.1209 1.723
0. 0. 0. 0. 0. 0. 0. 0. 0. 0.
10.2  10. 0. -2.8708418  0. 0. 0. 0. 0. 0. 0. 0.     ! Magnetic q-pole component.
9.  5.  1.  1.  0.  0.  0. 0. 0. 0. 0.            ! for image located 25 cm downstream
6  .1122 6.2671 -1.4982 3.5882 -2.1209 1.723              ! of quadrupole exit.
9.  5.  1.  1.  0.  0.  0. 0. 0. 0. 0.
6  .1122 6.2671 -1.4982 3.5882 -2.1209 1.723
0. 0. 0. 0. 0. 0. 0. 0. 0. 0.
1.                                                 ! Integration step size is 1 cm.
1 0. 0. 0.
'DRIFT'    drift_3
25.

'FIT'               ! Constrints on T_ijk transport coefficients impose OBJET/KOBJ=6.
4                                                  ! 4 variables:
  6  5  0 .1                                       ! E field in first quad,
  6 63  0 .1                                       ! B field in first quad,
  8  5  0 .1                                       ! and E field in second quad.
  8 63  0 .1                                       ! and B field in second quad.
6 1E-15 50  ! 6 constraints. Optional: penalty and #iter. values (default otherwise).
1 1  2 #End 0. 1. 0   ! R11=0 at end of sequence (i.e., after the final25 cm drift),
1 3  4 #End 0. 1. 0                                ! R34=0 at end of sequence.
2 2 16 #End 0. 1. 0                                ! T_216=0 at end of sequence.
2 2 26 #End 0. 1. 0                                ! T_226=0 at end of sequence.
2 4 36 #End 0. 1. 0                                ! T_436=0 at end of sequence.
2 3 46 #End 0. 1. 0                                ! T_446=0 at end of sequence.

'MATRIX'         ! Local value of the 1st and 2nd order tranport matrices, from OBJET.
2  0
'MARKER' EBwithFF_E
'END'
```

Fig. 12.36 Field
components along
($Y = 1\,\mathrm{cm}$, $Z = 0$) and
($Y = 0$, $Z = 1\,\mathrm{cm}$) straight
lines across the **E**, **B**
quadrupole doublet,
simulated with EBMULT
and including fringe fields

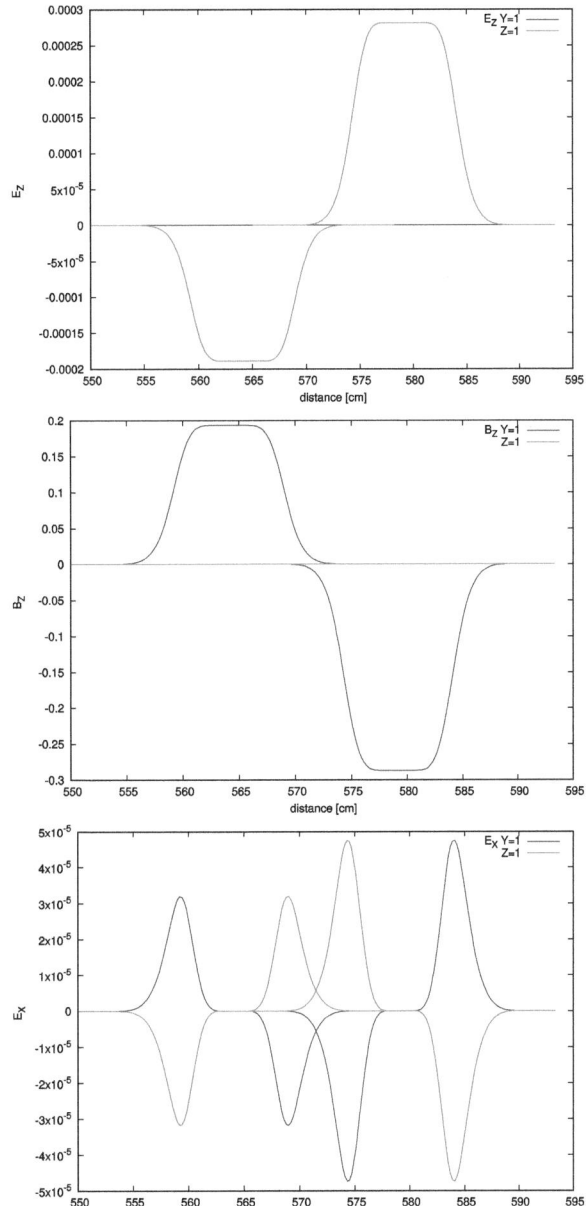

Note that an alternate way to force straight trajectories through the optical element is to give the particles a high rigidity, e.g., $D \approx 10^{10}$ under OBJET.

With the option EBMULT[IL = 7], field derivatives $d^{i+j+k} \, \mathbf{E}/dX^i dY^j dZ^k$ and $d^{i+j+k} \, \mathbf{B}//dX^i dY^j dZ^k$ across the magnetic quadrupole doublet are logged in zgoubi.impdev.out during the step-by-step integration of motion. Some of these derivatives are displayed in Fig. 12.37, produced using the gnuplot script given in Table 12.19.

12.8 Animation: Dynamical Squeeze of Point-Spread Function

The input file in Table 12.20 performs a squeeze of the image of a point object with momentum spread (uniform in $dp/p \in [-10^{-3}, +10^{-3}]$). The squeeze is achieved by slowly switching on the electric component in the quadrupoles, while slowly decreasing the magnetic component, toward their optimum values, in 30 steps (as per REBELOTE[NRBLT = 30, IOPT = 1]). IOPT = 1 option allows the NPRM = 4 parameters concerned to be varied, as specified by the four "KEYWORD{LABEL1} parameter number, range" subsequent lines under REBELOTE.

The gnuplot script for this animation can be found at the bottom of Table 12.20. It reads its data from zgoubi.fai. The animation cycles on a series of 31 images. As an illustration, the first image (maximal size, E = 0) and final image (minimal size, E and B set to cancel chromatic aberrations) together with two intermediate steps, are displayed in Fig. 12.38.

12.4.3 $\pi \rightarrow \mu + \nu$ In-Flight Decay, Bunch Densities

12.9 Parent Pion Bunch

(a) Pion momentum density

The input data file to simulate transverse beam densities along the line is given in Table 12.21. This is the parent pion bunch case of Fig. 12.9 with initial random uniform $p_\pi \in [100, 500]$ MeV/c. The data list includes a series of field free drifts aimed at generating the s-dependence of pion momentum densities.

The non-decayed pion count at successive distances s is obtained from zgoubi.HISTO.out file generated by HISTO[PRINT], and represented in Fig. 12.39. The latter is produced using the gnuplot script given in Table 12.22. The theoretical pion count in a bin $[p_\pi, p_\pi + \delta p_\pi[$ writes (Eq. 12.17)

$$\delta N_{p_\pi}\big|_s = \mathbf{1}_{\Delta p_\pi}(p_\pi) \, \exp(-\eta s/p_\pi) \, N_0 \, \delta p_\pi \tag{12.32}$$

and yields the theoretical curve $\delta N_{p_\pi}(p_\pi) / N_0 \delta p_\pi$, superposed to the tracking outcomes in Fig. 12.39.

Fig. 12.37 Field derivatives along $(Y = 1\,\text{cm},\ Z = 0)$ and $(Y = 0,\ Z = 1\,\text{cm})$ lines across the **E**, **B** quadrupole doublet, simulated with EBMULT and including fringe fields. The two derivatives $\partial B_Y/\partial Z$ and $\partial B_Z/\partial Y$ are computed independently in zgoubi, they do superimpose, as expected; same observation for $\partial E_Z/\partial X$ and $\partial E_X/\partial Z$, as well as for the three derivatives $\partial^2 B_Z/\partial X^2$, $\partial^2 B_X/\partial X dZ$ and $\partial^2 B_X/\partial Z \partial X$. The interval between markers is an integration step

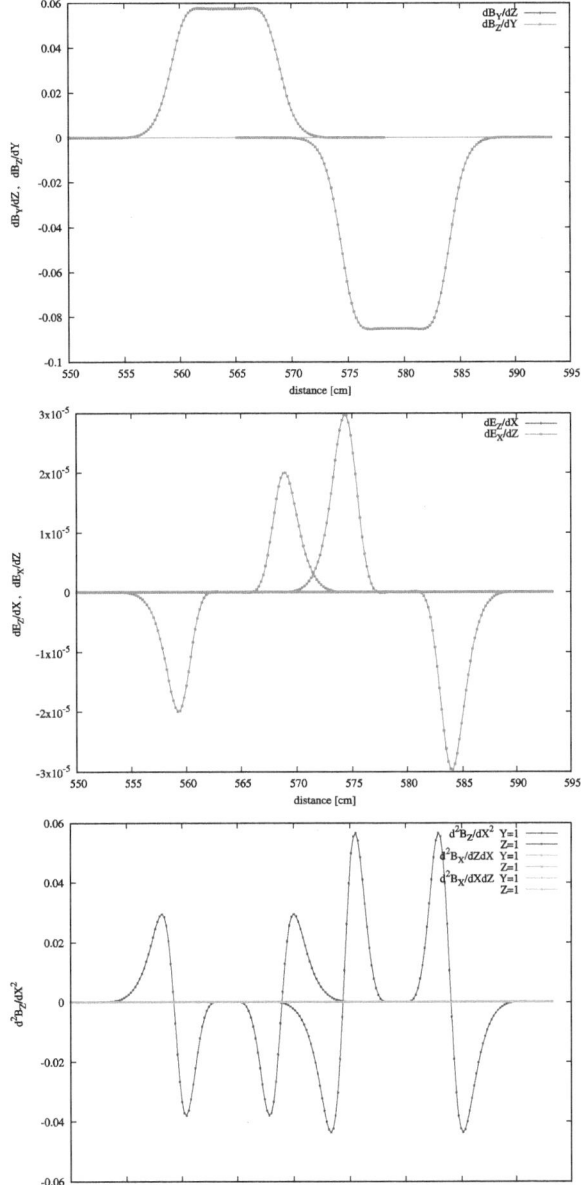

Table 12.19   A gnuplot script to produced Fig. 12.37, field derivatives across the **E**, **B** quadrupole doublet

```
# dBY/dZ and dBZ/dY:
plot 'zgoubi.impdev.out' u ($7):($16 /Brho) w lp ps .4 tit 'dB_Y/dZ' ,\
     'zgoubi.impdev.out' u ($7):($18 /Brho) w lp ps .4 tit 'dB_Z/dY'

# dEZ/dX and dEX/dZ:
plot 'zgoubi.impdev.out' u ($7):($380 /Brho) w lp ps .4 tit 'dE_Z/d_X' ,\
     'zgoubi.impdev.out' u ($7):($376 /Brho) w lp ps .4 tit 'dE_X/dZ'

# d2BZ/dX2 and d2BX/dXdZ:
plot 'zgoubi.impdev.out' u ($410 ==1 ? $7 : 2/0):22 w lp ps .3 lc 1 tit 'd^2B_Z/dX^2  Y=1' ,\
     'zgoubi.impdev.out' u ($410 ==1 ? $7 : 2/0):22 w lp ps .3 lc 1 tit '             Z=1' ,\
     'zgoubi.impdev.out' u ($410 ==2 ? $7 : 2/0):26 w lp ps .3 lc 3 tit 'd^2B_X/dZdX  Y=1' ,\
     'zgoubi.impdev.out' u ($410 ==2 ? $7 : 2/0):26 w lp ps .3 lc 3 tit '             Z=1' ,\
     'zgoubi.impdev.out' u ($410 ==2 ? $7 : 2/0):38 w lp ps .3 lc 3 tit 'd^2B_X/dXdZ  Y=1' ,\
     'zgoubi.impdev.out' u ($410 ==2 ? $7 : 2/0):38 w lp ps .3 lc 3 tit '             Z=1' ,\
```

Note in passing, the decay distance from this 10^6 pion Monte Carlo simulation with random uniform $p_\pi \in [100, 500]\,\mathrm{MeV/c}$ comes out to be $< s_\pi > = 16.725\,\mathrm{m}$, so satisfying the theoretical expectation

$$< s_\pi > = \frac{\int_0^\infty s\,ds \int_{p_{\pi 1}}^{p_{\pi 2}} \tilde{g}_{p_\pi}(p)|_s\,dp}{\int_0^\infty ds \int_{p_{\pi 1}}^{p_{\pi 2}} |_s\,dp}$$

(with $\tilde{g}_{p_\pi}(p)|_s\,dp$ from Eq. 12.17).

(b) Pion bunch momentum

The evolution of the average momentum of a pion bunch over 60 m flight distance from Monte Carlo simulations is obtained using REBELOTE[NRBLT = 11, IOPT = 1] to repeat the tracking with increasing s. The input data file for that simulation is given in Table 12.23. Results are displayed in Fig. 12.40.

12.10   Muon Bunch

(a) Muon momentum density at fixed p_π. Reproducing Fig. 12.11

The input data file to simulate muon momentum densities is given in Table 12.24. While REBELOTE[NRBLT, IOPT = 1] repeats (NRBLT = 2 additional passes, for a total of three different passes, i.e. three different initial pion bunch momenta), HISTO cumulates particle counts, pass after pass.

The muon count for various initial p_π values is obtained from the zgoubi. HISTO.out   file generated by the HISTO keyword, and represented in Fig. 12.41 (produced using the gnuplot script given in Table 12.25), consistent with theoretical expectations, Fig. 12.11.

(b) Muon momentum density, s-dependence. Reproducing Fig. 12.12

The input data file to simulate the muon momentum densities along the line, as muons decay, is derived from that of Exercise 12.9, Table 12.21, changing "40 D 0 P" to "40

Table 12.20 Simulation input data file: case of achromat quadrupole doublet, ray-trace from object to image, 5×10^3 H+ ions. Coordinates at the image are logged in zgoubi.fai. Details of the potential and B field settings in these **E, B** Q1 and Q2 quadrupoles are given at the bottom of the Table. *Note* this file, with its INCLUDE expanded, and the gnuplot script for the animation, are available in zgoubi sourceforge repository at https://sourceforge.net/p/zgoubi/code/HEAD/tree/branches/exemples/book/zgoubiMaterial/beamLines/nanoProbeAchromat/animation/

```
E, B final doublet achromat. Animation: image squeeze.
'MARKER'   achromatAnim_S                                  ! Just for edition purposes.
'MCOBJET'
20.435                                                              ! H+, 20keV.
1                                                  ! Particle distribution in a  window.
1000                                                       ! Number of particles.
2   2   2   2   1   1                 ! Densities: H, T, Z, P Gaussian; s, D uniform.
0.  0.  0.  0.  0.  1.                              ! Central  value of coordinates.
0.  .2E-3 .0  .2E-3 0.  0.005                       ! Gaussian 1/2width; delta_D=5e-3.
2   2   2   2   1   1          ! Cut-offs of Gaussian distributions - unused if uniform.
9   9. 9. 9.                                        ! Data for exponential D density - unused.
186387 548728 472874                                      ! Random generator seeds.
'PARTICUL'
938.2723 1.60217733E-19 0. 0. 0.                                   ! H+ data.
'FAISCEAU'                                          ! Allows local check of particle coordinates.
'OPTIONS'
1 1     ! Force IL=0 in all optical elements: no print
.plt 0                          ! out to zgoubi.plt.
'ESL'
559.
'INCLUDE'
1
Q12_hardEdge.inc[Q1_S:Q1_E]                        ! Q1 data  are grabbed from   Q12_hardEdge.inc file.
'ESL'
4.9
'INCLUDE'
1
Q12_hardEdge.inc[Q2_S:Q2_E]                        ! Q2 data  are grabbed from   Q12_hardEdge.inc file.
'ESL'
25.
'IMAGE'                                             ! Compute the location of the horizontal image.
'IMAGEZ'                                            ! Compute the location of the vertical image.

'FAISTORE'                                          ! Log final coordinates (and more) to zgoubi.fai.
zgoubi.fai
1
'REBELOTE'
30 0  0  1                                          ! Iterate 30 times from a to b (see below).
4                                  ! Change data in the following 4 elements in the sequence;
EBMULT{Q1}  5      0:-9272.986     ! {Q1} restricts to EBMULT with LABEL1=Q1; parameter numbering is
EBMULT{Q1} 63 0.947465:1.89493             ! as for FIT, with parameters 5 and 63 are respectively
EBMULT{Q2}  5      0:13779.90                         ! electric potential and magnetic field;
EBMULT{Q2} 63 -1.40796:-2.81592            ! a:b stands for the variation interval [a,b].

'SYSTEM'
1
gnuplot <./gnuplot_image_squeeze.gnu
'MARKER'   achromatAnim_E                                  ! Just for edition purposes.
'END'
```

gnuplot script "./gnuplot_image_squeeze.gnu" called by the SYSTEM command in the input data file above:

```
set xtics mirror ; set ytics mirror; set xlabel "Y [cm]" ; set ylabel "Z [cm]"
set key t c ;  set k b l; set size ratio -1

while (1) {
   do for [i=2:31]{ plot [-4e-3:4e-3] [-4e-3:4e-3] \
   "zgoubi.fai" u ($38 ==i ? $10 : 1/0):($12)  w p ps .4 pt 4 lc i tit "pass=".i
      pause .1
   }
pause 1
}
```

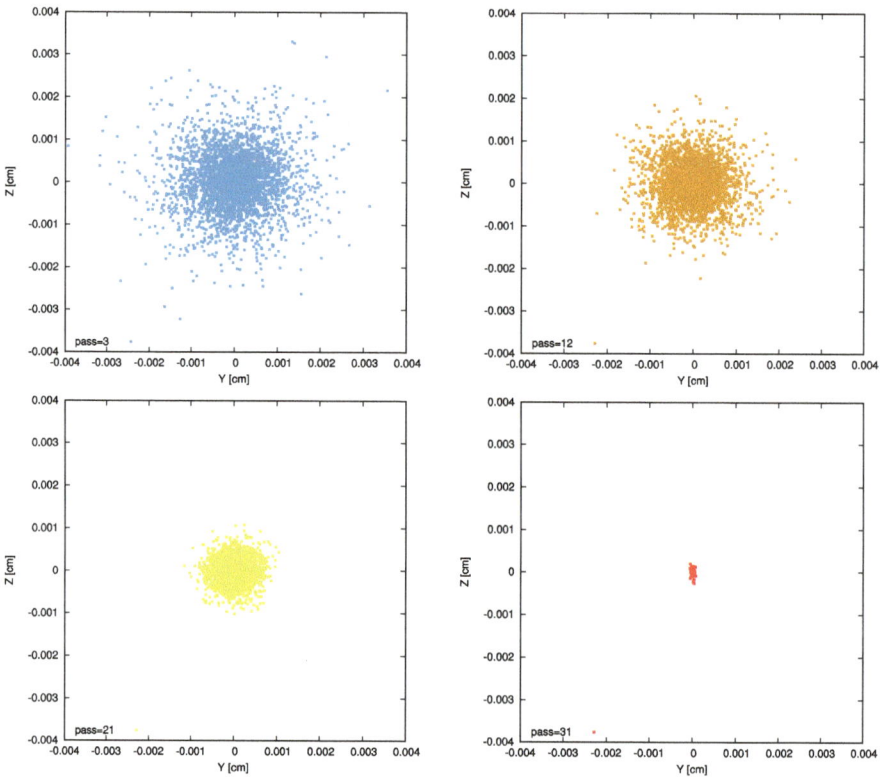

Fig. 12.38 A series of four successive views of the image of a point object, squeezed from a maximal size with E = 0, to an image of minimal size, free of chromatic aberration with optimized E and B values. These snap shots are taken from the 30-series of the animation (Table 12.20)

D 0 S" under HISTO, for an histogram of S-econdary particles, instead of P-rimary, and changing the drift lengths. The resulting simulation file is given in Table 12.26.

The muon count at various distances s is obtained from the zgoubi.HISTO.out file generated by HISTO[PRINT]. The result is represented in Fig. 12.42, produced using the gnuplot script of Table 12.26, consistent with Fig. 12.12.

(c) Muon bunch momentum, average

The evolution of the average momentum of a muon bunch over 60 m flight distance from Monte Carlo simulations is shown in Fig. 12.40.

It is obtained by repeating a s-distance tracking, with increasing s up to 60 m, using the pion case data file in Table 12.23 where "40 D 0 P" under HISTO (for Parent particles) is changed to "40 D 0 S" (for Secondary particles).

Table 12.21 Simulation input data file for the computation of $\tilde{g}_{p_\pi}(p_\pi)\big|_s$ density (Eq. 12.18). Track 10^4 pions over 40 meters. Initial pion momentum distribution is random, uniform in $[100, 500]$ MeV/c

```
Pion beam in-flight decay to muon.
'MARKER' piMuDec_pbgpi_s_S
'MCOBJET'
833.910238                         ! Reference rigidity BORO -> 250 MeV/c pions,
3                                  ! defines the reference momentum p_pi = q*BORO.
10000
2 2 2 2 1 1         ! distribution densities: Gaussian in Y, T, Z, P. Uniform in s, D.
.0   .0   .0    .0   0.  1.2       ! Central momentum: 1.2*q*BORO = 300 MeV/c.
0.   1.   0.0   1
.0   1.   0.0   1
0.   1.   0.64  1   ! Momentum spread= +/-sqrt(0.64)= +/-0.8 => p \in [100,500]MeV/c,
123456 234567 345678               ! => p_pi1/p_pi=0.4, p_pi2/p_pi=2.
'PARTICUL'
PION+

'MCDESINT'
105.66  0.  1.e6          ! PION -> MUON + NEUTRINO. Muon life time set to infnty.
136928 768370 548375

'HISTO'      S_0m                          ! Histogram at start (s=0).
1 .3 2.1 120 1 PRINT         ! D coordinate; in [0.3,2.1]; 100 bins; histo. #1; print
40 D 0 P      ! to zgoubi.HISTO.out. 40 lines; tag; non-norm.; Primary particles only.

  'DRIFT'    S_4m
400.
  'HISTO'                                  ! Histogram at s= 4m.
1 .3 2.1 120 2 PRINT
40 D 0 P
  'DRIFT'    S_10m
600.
  'HISTO'                                  ! Histogram at s= 10m.
1 .3 2.1 120 3 PRINT
40 D 0 P
  'DRIFT'    S_40m
3000.
  'HISTO'                                  ! Histogram at s= 40m.
1 .3 2.1 120 4 PRINT
40 D 0 P

  'REBELOTE'                     ! Repeat 99 times to total 10^6 particles for
99  0.1 0                        ! statistics purposes.

  'SYSTEM'
1
gnuplot <./gnuplot_HISTO.cmd

  'MARKER' piMuDec_pbgpi_s_E
  'END'
```

Table 12.22 gnuplot script for $\tilde{g}_{p_\pi}(p_\pi)\big|_s$ pion density plot, reading data from zgoubi.HISTO.out file

```
set xlabel "dp_{/Symbol p}/p_{ref}" font "roman,18"; set ylabel "{/Symbol d}N/{/Symbol d}p /N_0" font "roman,18"
set xtics mirror font "roman,11"; set ytics mirror font "roman,11"
set key  t r font "roman,10"  ; set key maxrow 1

# HYPOTHESES AND THEORETICAL PARAMETERS
c = 2.99792458e8 ; m_pi=139.571e6 ; tau=2.6033e-8; p_pi1 = 100e6 ; p_pi2 = 500e6 ; pRef = 250.e6   # Units: MeV/c
One = 1./(p_pi2/pRef - p_pi1/pRef) ; eta = m_pi / (c*tau) ; s_pi = pRef / eta; nBin = 1000.; dp = 1./One/nBin
g0(x) = One ; g4(x) = One* exp(-eta * 4. / ((1.+x)*pRef))
g10(x) = One* exp(-eta * 10. / ((1.+x)*pRef))  ; g40(x) = One* exp(-eta * 40. / ((1.+x)*pRef))

# MONTE CARLO TRACKING DATA:
npass = 100 ; npart = 1e6 ; nbin = 120 ; dpBin = (2.1 - .3) / nbin

plot  [-.8:1.2] [0:.75] \
(x>p_pi1/pRef -1. && x< p_pi2/pRef -1. ? g0(x) : 1/0) w l lw 2 notit ,\
(x>p_pi1/pRef -1. && x< p_pi2/pRef -1. ? g4(x) : 1/0) w l lw 2 notit ,\
(x>p_pi1/pRef -1. && x< p_pi2/pRef -1. ? g10(x) : 1/0) w l lw 2 notit ,\
(x>p_pi1/pRef -1. && x< p_pi2/pRef -1. ? g40(x) : 1/0) w l lw 2 notit ,\
'zgoubi.HISTO.out' u ($10== 5 && $11==npass && $6>8 && $6< 114? $1-1.:1/0):($2/npart/dpBin) w p ps .0 pt 6 notit  ,\
'zgoubi.HISTO.out' u ($10== 5 && $11==npass && $6>8 && $6< 114? $1-1.:1/0):($2/npart/dpBin) w p ps .8 pt 6 tit " s= 0" ,\
'zgoubi.HISTO.out' u ($10== 7 && $11==npass && $6>8 && $6< 114? $1-1.:1/0):($2/npart/dpBin) w p ps .8 pt 6 tit " s= 4m" ,\
'zgoubi.HISTO.out' u ($10== 9 && $11==npass && $6>8 && $6< 114? $1-1.:1/0):($2/npart/dpBin) w p ps .8 pt 8 tit " s=10m" ,\
'zgoubi.HISTO.out' u ($10==11 && $11==npass && $6>8 && $6< 114? $1-1.:1/0):($2/npart/dpBin) w p ps .8 pt 10 tit " s=40m"
```

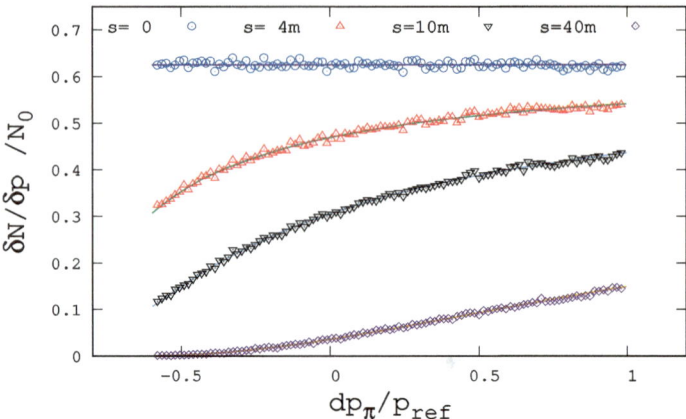

Fig. 12.39 Pion momentum density at successive s along the decay channel, from Monte Carlo simulations (markers) superposed on the theoretical curves, Eq. 12.32 (solid lines). $dp_\pi = p_\pi - p_{\text{ref}}$, $p_{\text{ref}} = 250\text{MeV/c}$

Table 12.23 Simulation input data file: for $\bar{p}_\pi(s)$ computation. Track 10^3 pions over increasing length drift. Initial pion momentum distribution is random, uniform in $[100, 500]\,\text{MeV/c}$

```
Pion beam in-flight decay to muon.
'MARKER' piMuDec_pbgmu_s_S
'MCOBJET'
833.910238                         ! Reference rigidity BORO -> 250 MeV/c pions,
3                                  ! defines the reference momentum p_pi = q*BORO.
1000
2 2 2 2 1 1        ! distribution densities: Gaussian in Y, T, Z, P. Uniform in s, D.
.0   .0   .0    .0   0.  1.2       ! Central momentum: 1.2*q*BORO = 300 MeV/c.
0.   1.   0.0   1
.0   1.   0.0   1
0.   1.   0.64  1   ! Momentum spread= +/-sqrt(0.64)= +/-0.8 => p \in [100,500]MeV/c,
123456 234567 345678               ! => p_pi1/p_pi=0.4, p_pi2/p_pi=2.
'PARTICUL'
PION+

'MCDESINT'
105.66  0.  1.e6          ! PION -> MUON + NEUTRINO. Muon life time set to infnty.
136928 768370 548375

'HISTO'     S_0m                              ! Histogram at start (s=0).
1 .3 2.1 120 1 PRINT       ! D coordinate; in [0.3,2.1]; 100 bins; histo. #1; print
40 D 0 P     ! to zgoubi.HISTO.out. 40 lines; tag; non-norm.; Primary particles only.

 'DRIFT'     S_4m
 500.
'HISTO'                                       ! Histogram at s= 4m.
1 .3 2.1 120 2 PRINT
40 D 0 P

 'REBELOTE'
11  0.1 0 1                                   ! Repeat 11 times.
1                          ! This changes parameter #1 in element #5, i.e., drift length,
5 1 10:60                  ! from 10 to 60m, step 5m.

'SYSTEM'
1
gnuplot <./gnuplot_piAverageMomentum.cmd

'MARKER' piMuDec_pbgmu_s_E
'END'
```

Fig. 12.40 Average
momentum of pion and
muon bunches over 60 m
flight distance from
Monte Carlo simulations
(markers). For comparison,
superposed: solid lines from
Eq. 12.18

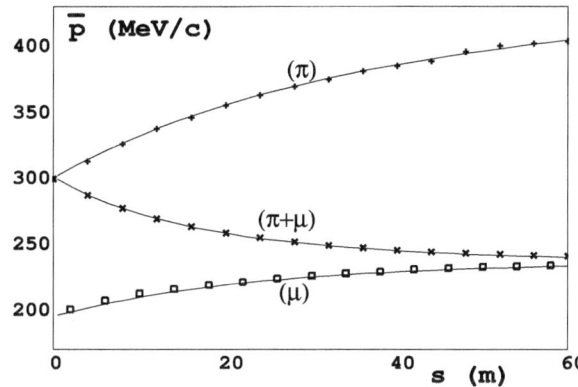

Table 12.24 Simulation input data file: for $g_{p_\mu | p_\pi} (p_\mu)$ computation. Track 10^4 pions with zero momentum spread, over a distance $s \gg s_\pi$, in three different cases of initial momentum $p_\pi =$ 100, 250 and 500 MeV/c

```
Pion beam in-flight decay to muon.
'MARKER' piMuDec_PbOne_S
'MCOBJET'
833.910238                          ! Reference rigidity BORO -> 250 MeV/c pions,
3                                   ! defines the reference momentum p_pi = q*BORO.
10000
2 2 2 2 1 1       ! distribution densities: Gaussian in Y, T, Z, P. Uniform in s, D.
.0  .0 .0   .0  0. 0.4    ! First pass with central momentum: 0.4*q*BORO = 100 MeV/c.
0.  1.  0.0  1
.0  1.  0.0  1
0.  1.  0.   1          ! Momentum spread = sqrt(0.64) = +/-0.8 => p \in [100,500]MeV/c,
123456 234567 345678               ! => p_pi1/p_pi=0.4, p_pi2/p_pi=2.
'PARTICUL'
PION+

'MCDESINT'
105.66  0.  1.e6            ! PION -> MUON + NEUTRINO. Muon life time set to infnty.
136928 768370 548375

'HISTO'
11 .1 2.1 120 1 PRINT
40 D 0 P

! Let all pions decay => take drift >> s_pi = 16.7m
'DRIFT'
1.e5
'HISTO'    s>>s_pi                                  ! Histogram at s>>17m.
1 .1 2.1 120 1 PRINT         ! D coordinate; in [0.3,2.1]; 100 bins; histo. #1; print
40 D 0 S   ! to zgoubi.HISTO.out. 40 lines; tag; non-norm.; Secondary particles only.

'REBELOTE'                    ! Repeat twice, with central momentum successively
2  0.1 0 1                    ! 1.*q*BORO = 250 MeV/c  and  2.*q*BORO = 500 MeV/c.
1
MCOBJET 45 1. 2.                        ! This changes parameter # 45 = D, undr MCOBJET.
'SYSTEM'
1
gnuplot <./gnuplot_HISTO_mu.cmd
'MARKER' piMuDec_PbOne_E
'END'
```

Table 12.25 Gnuplot script for the plot of muon density, $g_{p_\mu | p_\pi}(p_\mu)$, case of parent beam momentum 100, 250 or 500 MeV/c, reading data from zgoubi.HISTO.out file

```
set xlabel "dp/p_{ref}" font "roman,18"; set ylabel "{/Symbol d}N/{/Symbol d}p /N_0" font "roman,18"
set xtics mirror font "roman,11"; set ytics mirror font "roman,11"
set key  t r font "roman,10"  ; set key maxcol 1

# MONTE CARLO TRACKING DATA:
npass = 1 ; npart = 1e4 ; nbin = 120 ; dpBin = (2.1 - .1) / nbin

plot [-1.:1.1] 'zgoubi.HISTO.out' \
u ($10==7 && $11==3 ? $1-1. : 1/0):($2/npart/dpBin) w lp ps .4 pt 10 lw 2 lc rgb "black" notit; pause 1
```

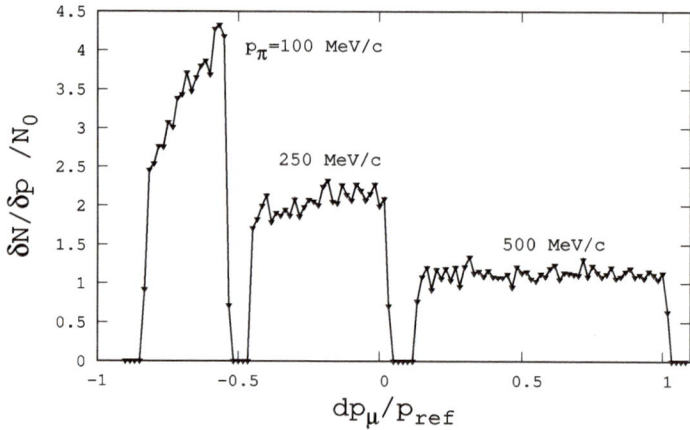

Fig. 12.41 Muon momentum density $g_{p_\mu | p_\pi}(p_\mu)$, for $p_\pi = 100,\ 300$ and 500 MeV/c, from Monte Carlo simulations

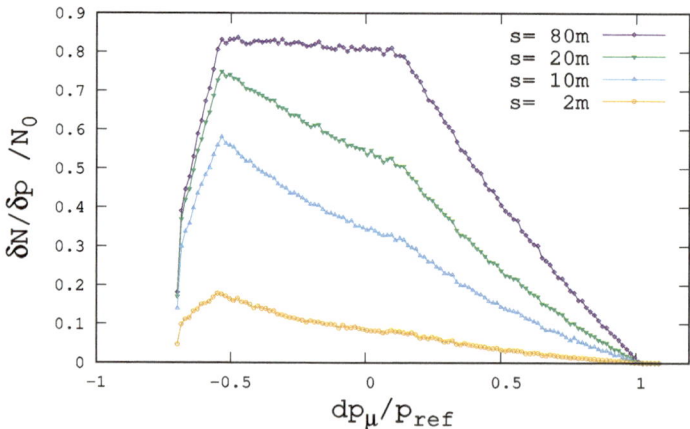

Fig. 12.42 Muon momentum density at various s along the decay channel from Monte Carlo simulations. $dp_\mu = p_\mu - p_{\text{ref}}$, $p_{\text{ref}} = 250$ MeV/c

Table 12.26 Simulation input data file for the computation of $g_{p_\mu}(p_\mu)\big|_s$. Decay of 10^4 pions is tracked over 80 meters. Initial pion momentum distribution is random, uniform in [100, 500] MeV/c

```
300 MeV tuning. NuFactory. Pion capture. In-flight decay into muon
'MARKER' piMuDec_Pbgpi_s_S
'MCOBJET'
833.910238                               ! Reference rigidity -> 250 MeV/c pions.
3                                        ! defines the reference momentum p_pi = q*BORO.
10000
2 2 2 2 1 1        ! distribution densities: Gaussian in Y, T, Z, P. Uniform in s, D.
.0   .0  .0    .0   0.  1.2             ! Central momentum: 1.2*Q*BORO = 300 MeV/c.
0.   1.   0.0   1
.0   1.   0.0   1
0.   1.   0.64  1   ! Momentum spread = sqrt(0.64) = +/-0.8 => p \in [100,500]MeV/c,
123456 234567 345678                    ! => p_pi1/p_pi=0.4, p_pi2/p_pi=2.
'PARTICUL'
PION+
'MCDESINT'
105.66  0.  99999.         ! PION -> MUON + NEUTRINO. Muon life time set to infnty.
136928 768370 548375
'DRIFT'   S_2m
200.
'HISTO'                                       ! Histogram at s=2m.
1 .3 2.1 120 1 PRINT         ! D coordinate; in [0.3,2.1]; 100 bins; histo. #1; print
40 D 0 S     ! to zgoubi.HISTO.out. 100 lines; tag; non-norm.; Primary particles only.
'DRIFT'   S_10m
800.
'HISTO'                                       ! Histogram at s= 10m.
1 .3 2.1 120 2 PRINT
40 D 0 S
'DRIFT'   S_20m
1000.
'HISTO'                                       ! Histogram at s= 20m.
1 .3 2.1 120 3 PRINT
40 D 0 S
'DRIFT'   S_80m
6000.
'HISTO'                                       ! Histogram at s= 80m.
1 .3 2.1 120 4 PRINT
40 D 0 S
'REBELOTE'                       ! Repeat 99 times to total 10^6 particles for
99  0.1 0                        ! statistics purposes.
'SYSTEM'
1
gnuplot <./gnuplot_HISTO_S.cmd
'MARKER' piMuDec_Pbgpi_s_S
'END'
```

A gnuplot script to obtain Fig. 12.42:

```
set xlabel "dp_{/Symbol m}/p_{ref}" font "roman,18"; set ylabel "{/Symbol d}N/{/Symbol d}p /N_0" font "roman,18"
set xtics mirror font "roman,11"; set ytics mirror font "roman,11"
set key  t r font "roman,14"; set key maxcol 1
# MONTE CARLO TRACKING DATA:
npass = 100 ; npart = 1e6 ; nbin = 120 ; dpBin = (2.1 - .3) / nbin
plot  [-1.:1.2]  \
    'zgoubi.HISTO.out'  u ($10==12 && $11==npass? $1-1. : 1/0):($2/npart/dpBin) w lp ps .4 pt 12  tit " s= 80m"  ,\
    'zgoubi.HISTO.out'  u ($10==10 && $11==npass? $1-1. : 1/0):($2/npart/dpBin) w lp ps .4 pt 10  tit " s= 20m" ,\
    'zgoubi.HISTO.out'  u ($10== 8 && $11==npass? $1-1. : 1/0):($2/npart/dpBin) w lp ps .4 pt 8  tit " s= 10m" ,\
    'zgoubi.HISTO.out'  u ($10== 6 && $11==npass? $1-1. : 1/0):($2/npart/dpBin) w lp ps .4 pt 6  tit " s=  2m"
pause 1
```

(d) Longitudinal muon phase space. Reproducing Fig. 12.13

This simulation is performed using the input data file given in Table 12.26, *mutatis mutandis*, as follows:

- change the parent pion initial momentum interval to $p_\pi \in [200, 400]$ MeV/c,
- introduce and additional 20 m DRIFT right after DRIFT[LABEL1 = s_20m],
- followed by FAISCNL or FAISTORE[FNAME = zgoubi.fai] in order to get particle data at s = 40 m in zgoubi.fai.

A graph of the longitudinal bunch population $g_{t_\mu, E_\mu}(t_\mu, E_\mu)\big|_s$ at $s = 40$ m so obtained is given in Fig. 12.43, consistent with Fig. 12.13.

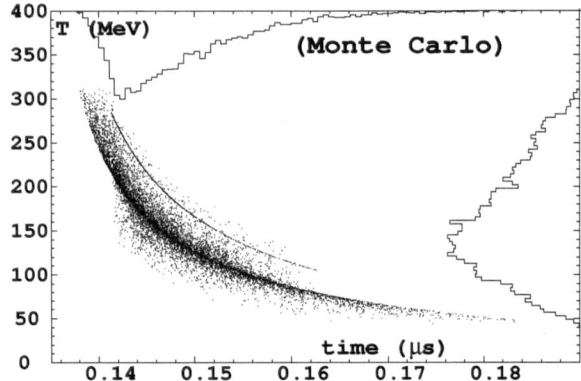

Fig. 12.43 Muon longitudinal phase-space $g_{t_\mu, E_\mu}(t_\mu, E_\mu)|_s$ at $s = 40$ m, from Monte Carlo simulation, and projected densities along the axes. The thin arc on the right of the muon bunch is the survived pion bunch

12.4.4 Pion Funnel and Muon FODO Collect Channel

12.11 Beam Line Simulation

(a) Reference trajectory, transport matrix, phase space invariants.

The simulation input data file for the complete collect channel, funnel+FODO line, described in Table 12.3, is given in Table 12.27. The data include OBJET[BORO = 833.91, KOBJ = 5] for transport coefficient computation by MATRIX. The reference rigidity BORO = 833.91 kG.cm corresponds to 250 MeV pions; however the optical axis (i.e., the off-centered optical axis in the funnel, and the reference axis along the center of the quadrupoles in the FODO channel) corresponds to 1.2×833.91 kG.cm, or 300 MeV/c pions (hence D=1.2 relative momentum of the 13-particle set, under OBJET[KOBJ = 5], Table 12.27).

Figure 12.44 shows the reference trajectory along the funnel section, it is zero beyond, along the FODO channel. Out of zgoubi.res execution listing, the resulting first order transport matrix $[R_{ij}]$ at the end of the line ($s = 31.1291475$ m) is the following:

```
Reference, before change of frame (particle #  1  - D-1,Y,T,Z,s,time)  :
   2.00000000E-01   3.58088321E-04  -1.40109121E-02   0.00000000E+00   0.00000000E+00   3.11291475E+03   5.47393263E-02

               TRANSFER  MATRIX  ORDRE  1   (MKSA units)
      0.157313        2.64910         0.00000         0.00000         0.00000       4.179761E-02
     -0.264285        1.88869         0.00000         0.00000         0.00000       2.649820E-02
      0.00000         0.00000         0.490598        2.23057         0.00000         0.00000
      0.00000         0.00000        -0.560230       -0.508769        0.00000         0.00000
      0.105474       -8.683260E-03    0.00000         0.00000         1.00000      -1.724655E-02
      0.00000         0.00000         0.00000         0.00000         0.00000        1.00000
```

Table 12.27 Simulation input data file for a beam line including an upstream pion funnel section, and a downstream muon FODO collect channel. This sequence INCLUDEs 14 copies of the basic cell of the FODO channel, taken from the file FODOCell_piCollect.inc (Table 12.28), in which the segment [#S_muonFODOCell:#E_muonFODOCell] is defined using LABEL1s. This file is resorted to in subsequent INCLUDEs under the name opticalSequence.inc. For possible further data analysis or graphs, use OPTIONS[.plt, IL = 2] and/or DRFIT[split, IL = 2] to log particle data in zgoubi.plt, step-by-step as numerical integration proceeds. *Note* this file is available in zgoubi sourceforge repository at https://sourceforge.net/p/zgoubi/code/HEAD/tree/branches/exemples/book/zgoubiMaterial/beamLines/muonCollectChannel/

```
Optical sequence of collect channel. opticalSequence.inc.
'MARKER' NFmuonCollect_S    ! Just for edition purposes.
'OBJET'
833.910238          ! BORO corresponds to 250 MeV/c pion.
5   ! Create a 13 particle sample for MATRIX computation.
0.001 0.01 0.001 0.01 0.  .0001            ! Sampling.
39.43 -90.36 .0 .0 .0 1.2 ! Reference trajectory offset.
'OPTIONS'
1 1
.plt 0
'MARKER' S_NF_Funnel
'DRIFT'    DRIF    DRF1
50.
'MULTIPOL'   FUN  HKIC     DQ1
20  .kicker
100. 10.  -0.838529167 0.291349611 0. 0. 0. 0. 0. 0. 0.
29. 18.  1.00 0.00 0.00 0.00 0.00 0. 0. 0. 0.
6 -.010967 5.464823 .996848 1.568787 -5.671630 18.505734
29. 18.  1.00 0.00 0.00 0.00 0.00 0. 0. 0. 0.
6 -.010967 5.464823 .996848 1.568787 -5.671630 18.505734
0. 0. 0. 0. 0. 0. 0. 0. 0. 0.
2.
1 0. 0. 0.
'DRIFT'    DRIF    DRF2
40.  split 10 2       ! Drift is split for graph purposes.
'MULTIPOL'   FUN  QUAD     Q2
20  .Quad
100. 10.  0.0 -0.617467029 0. 0. 0. 0. 0. 0. 0. 0.
29. 18.  1.00 0.00 0.00 0.00 0.00 0. 0. 0. 0.
6 -.010967 5.464823 .996848 1.568787 -5.671630 18.505734
29. 18.  1.00 0.00 0.00 0.00 0.00 0. 0. 0. 0.
6 -.010967 5.464823 .996848 1.568787 -5.671630 18.505734
0. 0. 0. 0. 0. 0. 0. 0. 0. 0.
2.   Quad  Q2
1 0. 0. 0.
'DRIFT'    DRIF    DRF3
65.
'MULTIPOL'   FUN  QUAD        Q3
20  .Quad
100. 10.  -0.250226990  0.81443677 0. 0. 0. 0. 0. 0. 0. 0.
29. 18.  1.00 0.00 0.00 0.00 0.00 0. 0. 0. 0.
6 -.010967 5.464823 .996848 1.568787 -5.671630 18.505734
29. 18.  1.00 0.00 0.00 0.00 0.00 0. 0. 0. 0.
6 -.010967 5.464823 .996848 1.568787 -5.671630 18.505734
0. 0. 0. 0. 0. 0. 0. 0. 0. 0.
2.   Quad  Q3
1 0. 0. 0.
'DRIFT'    DRIF    DRF2
40.  split 10 2
'MULTIPOL'   FUN  QUAD     Q4
20  .Quad
100. 10.  0.0 -0.73629954 0. 0. 0. 0. 0. 0. 0. 0.
29. 18.  1.00 0.00 0.00 0.00 0.00 0. 0. 0. 0.
6 -.010967 5.464823 .996848 1.568787 -5.671630 18.505734
29. 18.  1.00 0.00 0.00 0.00 0.00 0. 0. 0. 0.
6 -.010967 5.464823 .996848 1.568787 -5.671630 18.505734
0. 0. 0. 0. 0. 0. 0. 0. 0. 0.
2.   Quad  Q4
1 0. 0. 0.   !


'DRIFT'    DRIF    DRF4
40.  split 10 2
'MULTIPOL'   FUN  HKIC     DIP4
20  .kicker
40. 10.  -1.35257476 0. 0. 0. 0. 0. 0. 0. 0. 0.
12. 8.  1.00 0.00 0.00 0.00 0.00 0. 0. 0. 0.
6 .015527 3.874961 -2.362230 2.978209 12.604429 15.025689
12. 8.  1.00 0.00 0.00 0.00 0.00 0. 0. 0. 0.
6 .015527 3.874961 -2.362230 2.978209 12.604429 15.025689
0. 0. 0. 0. 0. 0. 0. 0. 0. 0.
1.
1 0. 0. 0.
'DRIFT'    DRIF    DRFM1
44.
'MULTIPOL'   FUN     QM1
20  .Quad
40. 10.   0.  1.79773547 0. 0. 0. 0. 0. 0. 0. 0.
12. 8.  1.00 0.00 0.00 0.00 0.00 0. 0. 0. 0.
6 -.010967 5.464823 .996848 1.568787 -5.671630 18.505734
12. 8.  1.00 0.00 0.00 0.00 0.00 0. 0. 0. 0.
6 -.010967 5.464823 .996848 1.568787 -5.671630 18.505734
0. 0. 0. 0. 0. 0. 0. 0. 0. 0.
2.   Quad  QM1
1 0. 0. 0.
'DRIFT'    DRIF    DRFM2
107.  split 10 2
'MULTIPOL'   FUN     QM2
20  .Quad
40. 10.   0. -2.40617729 0. 0. 0. 0. 0. 0. 0. 0.
12. 8.  1.00 0.00 0.00 0.00 0.00 0. 0. 0. 0.
6 -.010967 5.464823 .996848 1.568787 -5.671630 18.505734
12. 8.  1.00 0.00 0.00 0.00 0.00 0. 0. 0. 0.
6 -.010967 5.464823 .996848 1.568787 -5.671630 18.505734
0. 0. 0. 0. 0. 0. 0. 0. 0. 0.
2.   Quad  QM2
1 0. 0. 0.
'DRIFT'    DRIF    DRFM3
40.  split 10 2
'MARKER' E_NF_Funnel

'MARKER' S_NF_FODO
'INCLUDE'                      ! Include the 14-cell FODO line.
1
14*FODOCell_piCollect.inc[#S_muonFODOCell:#E_muonFODOCell]

'MARKER' E_NF_FODO
'MATRIX'
1 0
'MARKER' NFmuonCollect_E   ! Just for edition purposes.
'END'
```

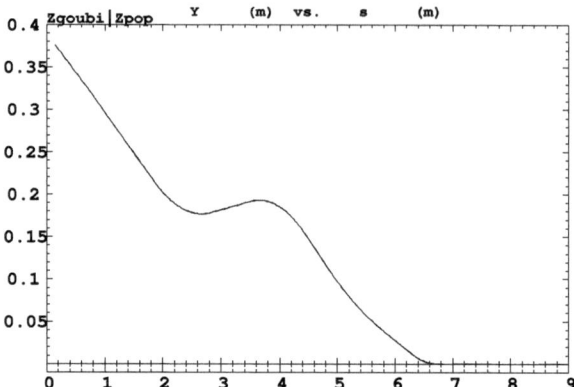

Fig. 12.44 Reference 300 MeV/c (pions) trajectory along the funnel section of the muon collect channel. A graph obtained using zpop: menu 7; 1/1 to open zgoubi.plt; 2/[6, 2] to select Y versus distance

Table 12.28 Simulation input data file for one cell of the 14-cell muon collect channel. This is the FODO cell segment [#S_muonFODOCell:#E_muonFODOCell] subject to INCLUDE in Table 12.27

```
! FODOCell_piCollect.inc
'MARKER'    #S_muonFODOCell
'DRIFT'     DRF
18.66435 split 3 2
'MULTIPOL' QF
20 .Quad
40. 10. 0.  3.00591 0. 0. 0. 0. 0. 0. 0.
0.0 0.0 1.00 0.00 0.00 0.00 0.00 0. 0. 0. 0.
6  .1122 6.2671 -1.4982 3.5882 -2.1209 1.723
0.0 0.0 1.00 0.00 0.00 0.00 0.00 0. 0. 0. 0.
6  .1122 6.2671 -1.4982 3.5882 -2.1209 1.723
 0. 0. 0. 0. 0. 0. 0. 0. 0. 0.
2.  Quad  Q3
1 0. 0. 0.
'DRIFT'    DRF37
37.3287 split 3 2
'MULTIPOL' QD
20 .Quad
40. 10. 0. -3.00591 0. 0. 0. 0. 0. 0. 0.
0.0 0.0 1.00 0.00 0.00 0.00 0.00 0. 0. 0. 0.
6  .1122 6.2671 -1.4982 3.5882 -2.1209 1.723
0.0 0.0 1.00 0.00 0.00 0.00 0.00 0. 0. 0. 0.
6  .1122 6.2671 -1.4982 3.5882 -2.1209 1.723
 0. 0. 0. 0. 0. 0. 0. 0. 0. 0.
2.  Quad  Q3
1 0. 0. 0.
'DRIFT'    DRF
18.66435 split 3 2
'MARKER'    #E_muonFODOCell
 'END'
```

(b) Optical functions,

They are computed and logged in zgoubi.OPTICS.out by the simulation data file of Table 12.29. A graph is given in Fig. 12.45, it resorts to gnuplot_OPTICS.gnu, a script found in zgoubi toolbox [31]. It is a good idea in this derivation of transport coefficients, and optical functions, from particle trajectories, to keep an eye on the determinants of the first order transport matrix $T (s \leftarrow 0)$ from s=0 to current s, along the line; they are displayed in Fig. 12.46 and appear to marginally differ from 1, as

Table 12.29 Simulation input data file for the transport of optical functions along the funnel+FODO channel. This sequence INCLUDEs the segment from occurrence of LABEL1 = S_NF_Funnel to occurrence of LABEL1 = E_NF_FODO, as defined in Table 12.27. The SYSTEM command produces a graph of the optical functions along the line (Fig. 12.45)

```
NF muon collect channel.
'MARKER'  NFmuonColl_optics_S                        ! Just for edition purposes.
'OBJET'
833.910238                                           ! 250 MeV/c pion.
5.1                              ! Create a 13 particle sample for MATRIX computation.
0.001 0.01 0.001 0.01 0. .0001                                          ! Sampling.
39.43 -90.36 .0 .0 .0 1.2    ! Reference trajectory offset. D=1.2 for 300 MeV/c momentum particles.
0. 4. 0. 4. 0. 1. 0. 0. 0. 0.         ! Optical functions: (alpha, beta)_{Y,Z,1}, (eta, eta')_{Y,Z}.
'FAISCEAU'
'OPTICS'      ! This will cause transport of optical functions, using tarnsport matrix computed from
1 all PRINT                           ! particle coordinates, at 'all' optical elements along the line.

'OPTIONS'
1 1                                          ! This inhibits any possible logging to zgoubi.plt,
.plt 0                                       ! as it is not needed here (saves on CPU time, file volume.

'INCLUDE'
1
opticalSequence.inc[S_NF_Funnel:E_NF_FODO]
'FAISCEAU'
'SYSTEM'
1
gnuplot <./gnuplot_OPTICS.gnu

'MARKER'  NFmuonColl_optics_E                        ! Just for edition purposes.
'END'
```

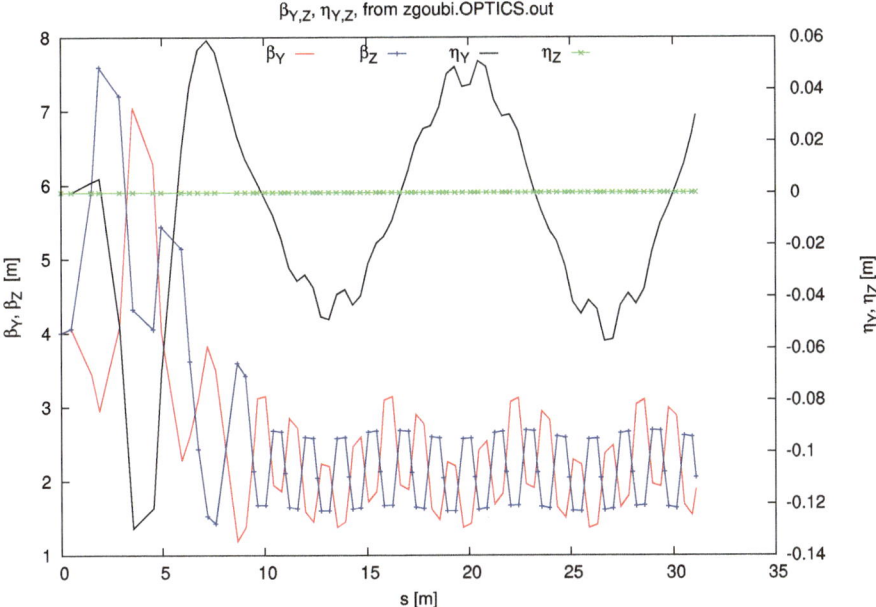

Fig. 12.45 Optical functions along the muon collect channel, for the reference rigidity 833.91 kG.cm (300 MeV/c pions). A graph obtained using gnuplot_OPTICS.gnu [31]

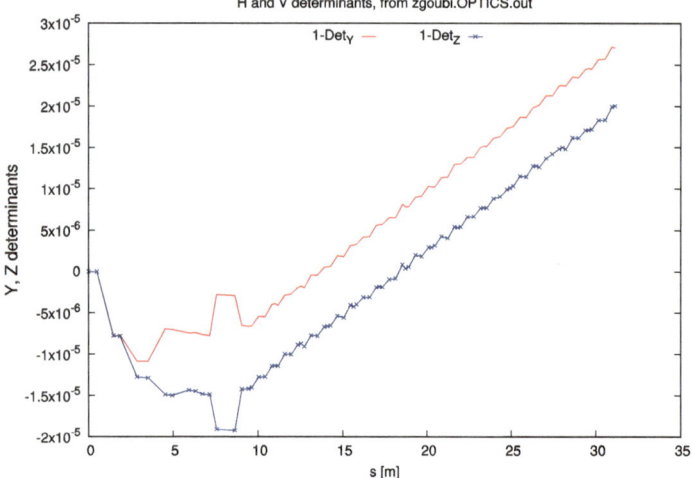

Fig. 12.46 Evolution of the horizontal and vertical determinants of the first order transport matrix $T(s \leftarrow 0)$, from s=0 to current abscissa s along the line. They are computed by OPTICS, from the 13 trajectories generated by OBJET[KOBJ = 5.1]

expected. Note that first order coefficients are computed by polynomial interpolation, which contributes from that departure from a value of 1.

The values the optical functions at end of funnel and end of FODO channel are given in Table 12.30.

(c) Phase space ellipses.

Motion invariants remain ellipses if particle motion is paraxial. However large initial invariant values are considered here: $\varepsilon_Y/\pi = \varepsilon_Z/\pi = 2.5$ mm, as well as large momentum offsets (250 and 400 MeV/c are tracked, for a reference 300 MeV/c). Distortion of the ellipse invariant along the line is expected.

OBJET[KOBJ = 8] is used to generate the appropriate object, namely, 300 particles on $\varepsilon_Y = 2.5\,\pi$ mm invariant. In addition to a first pass at 250 MeV/c, REBELOTE is used to repeat the tracking twice, for D = 1.2 and 1.6 (300 and 400 MeV/c). The simulation input data file is given in Table 12.31, resulting invariants at the end of the muon collect channel are displayed in Fig. 12.47.

12.12 Bunch Densities

The simulation input data file for the transport of a 10^4 pion bunch and its daughter decay muons, along the funnel+FODO channel, is taken from Table 12.27 (using INCLUDE), it is just required to substitute the Monte Carlo object definition MCOBJET[KOBJ=3] to the former OBJET[KOBJ = 5], whereas MATRIX at the end of the file may be removed. The resulting simulation data file is given in Table 12.32.

Table 12.30 Optical functions at start of the line, end of funnel, and end of FODO channel, out of zgoubi.res execution listing

At s=0:

```
Reference, before change of frame (particle #  1  - D-1,Y,T,Z,s,time)  :
 2.00000000E-01  3.94300000E+01 -9.03600000E+01  0.00000000E+00  0.00000000E+00  0.00000000E+00  0.00000000E+00

Reference, after change of frame (particle #  1  - D-1,Y,T,Z,s,time)  :
 2.00000000E-01  0.00000000E+00  0.00000000E+00  0.00000000E+00  0.00000000E+00  0.00000000E+00  0.00000000E+00

                BEAM  MATRIX (beta/alpha/alpha/gamma, D,D'),          FINAL
      4.00000        -0.00000        0.00000        0.00000        0.00000        0.00000
     -0.00000         0.250000       0.00000        0.00000        0.00000        0.00000
      0.00000         0.00000        4.00000       -0.00000        0.00000        0.00000
      0.00000         0.00000       -0.00000        0.250000       0.00000        0.00000

                Betatron phase advances (fractional), phi_y/2pi, phi_z/2pi :
                    0.000000E+00      0.000000E+00
```

At s=9.4771 m:

```
Reference, before change of frame (particle #  1  - D-1,Y,T,Z,s,time)  :
 2.00000000E-01  3.67371273E-02 -1.13085032E+00  0.00000000E+00  0.00000000E+00  9.47711792E+02  1.73800023E-02

Reference, after change of frame (particle #  1  - D-1,Y,T,Z,s,time)  :
 2.00000000E-01  0.00000000E+00  0.00000000E+00  0.00000000E+00  0.00000000E+00  9.47711792E+02  1.73800023E-02

                BEAM  MATRIX (beta/alpha/alpha/gamma, D,D'),          FINAL
      2.45082        -1.64770        0.00000        0.00000        0.00000        6.643128E-03
     -1.64770         1.51580        0.00000        0.00000        0.00000       -1.540450E-02
      0.00000         0.00000        2.13220        1.34596        0.00000        0.00000
      0.00000         0.00000        1.34596        1.31866        0.00000        0.00000

                Betatron phase advances (fractional), phi_y/2pi, phi_z/2pi :
                    5.096201E-01      4.245413E-01
```

At s=31.1291 m:

```
Reference, before change of frame (particle #  1  - D-1,Y,T,Z,s,time)  :
 2.00000000E-01  2.70732032E-01  2.17977764E+00  0.00000000E+00  0.00000000E+00  3.11291761E+03  5.47392102E-02

Reference, after change of frame (particle #  1  - D-1,Y,T,Z,s,time)  :
 2.00000000E-01  0.00000000E+00  0.00000000E+00  0.00000000E+00  0.00000000E+00  3.11291761E+03  5.47392102E-02

                BEAM  MATRIX (beta/alpha/alpha/gamma, D,D'),          FINAL
      1.89336        -1.05958        0.00000        0.00000        0.00000        3.000664E-02
     -1.05958         1.12110        0.00000        0.00000        0.00000        3.886735E-02
      0.00000         0.00000        2.05439        1.32603        0.00000        0.00000
      0.00000         0.00000        1.32603        1.34265        0.00000        0.00000

                Betatron phase advances (fractional), phi_y/2pi, phi_z/2pi :
                    2.076168E-01      1.387201E-01
```

(a) Sample pion and muon trajectories.

Sample pion and muon trajectories along the line are displayed in Fig. 12.48.

(b) Bunch densities.

Tracking is performed using the simulation input data file of Table 12.32. The number of pions under MCOBJET is changed to 10^4, transverse emittances are Gaussian, cut-off at one-σ, momentum density is uniform, in [200, 400] MeV/c. In this file, following the funnel,

INCLUDE[5 * opticalSequence.inc[S_NF_FODO:E_NF_FODO]]

defines a 108 m FODO channel.

The evolution of the average momentum of π, μ and $\pi + \mu$ beams so obtained are displayed in Fig. 12.49; the theoretical curves, (after Eq. 12.13 for the s-dependence of the pion density, Eq. 12.21 for the muon density) are superimposed for comparison. The higher momentum on average from the raytracing, is due to the loss of low energy muons in the walls defined by CHAMBR.

Table 12.31 Simulation input data file for the transport, along the funnel+FODO channel, of a set of 300 particles on horizontal or vertical invariants defined (under OBJET) by $\varepsilon_Y/\pi = \varepsilon_Z/\pi = 2.5$ mm and (Tables 12.4, 12.30) $\alpha_Y = \alpha_Z = 0$, $\beta_Y = \beta_Z = 4$ m. This sequence INCLUDEs the segment from occurrence of LABEL1 = S_NF_Funnel to occurrence of LABEL1 = E_NF_FODO, as defined in Table 12.27

```
NF muon collect channel.
'MARKER' NFmuonColl_invrnt_S                              ! Just for edition purposes.
'OBJET'
833.910238                                               ! 250 MeV/c pion.
8                                                        ! Create particles on vertical invariant.
1 300 1                        ! '60 1 1" for particles on horizontal invariant, '1 60 1' for vertical.
0.3943 -0.09036 .0 .0 .0 1.    ! Reference optical axis offset; D=1 for 1st pass at 250 MeV/c.
0. 4. 0.
0. 4. 0.0025                                              ! Initial vertical invariant parameters and Value.
0. 0. 0.
! 300 1 1                       ! Substitute the following lines to what precedes, for horizontal.
! 0.3943 -0.09036 .0 .0 .0 1.
! 0. 4. 0.0025
! 0. 4. 0.
! 0. 0. 0.

'FAISTORE'
zgoubi.fai E_NF_FODO                         ! Locations where particle data are to be logged, namely,
1                                            ! the optical element featruing LABEL1=E_NF_FODO.

'INCLUDE'
1
opticalSequence.inc[S_NF_Funnel:E_NF_FODO]

'REBELOTE'
2 0 0 1                                      ! 2 more passes; include prior changes as specified below.
1
OBJET 35 1.2:1.6                             ! Repeat 300 particle tracking, for D=1.2 and D=1.6 under OBJET.

'MARKER' NFmuonColl_invrnt_E                 ! Just for edition purposes.
'END'
```

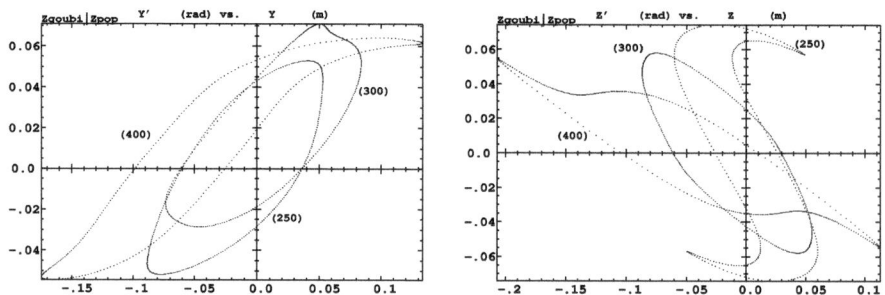

Fig. 12.47 A hundred particles on their distorted invariants at the end of the FODO channel, at 250, 300 and 400 MeV/c. A graph obtained using zpop: menu 7; 1/5 to open zgoubi.fai; 2/[2, 3] to select (Y, T) phase space or 2/[4, 5] to select (Z, P) phase space

Transverse pion bunch densities at start, 20 and 50 m, together with transverse muon bunch densities at 100 m, are displayed in Fig. 12.50. The theoretical densities are superimposed, for comparison.

(c) Phase space collimation.

Phase space ellipse parameters (centering and shape) which correspond to maximized acceptance at the end of the collect channel can be determined using COLLIMA[IFORM = 14–16]. In the case of time-energy phase space and an ellipse surface $\varepsilon_l = 0.5\,\pi$ eV s (cut-off at 1σ), COLLIMA takes the form

Table 12.32 Simulation input data file for the transport of a bunch of 100 pions, and decay muons as it proceeds, along the funnel followed by a 108 m FODO channel. Parent and daughter particles are specified under PARTICUL and MCDESINT, respectively. CHAMBR defines the vacuum pipe opening. The SYSTEM command at the bottom of the file produces a graph of Y and Z coordinates of the particles along the line, using the gnuplot script gnuplot_Zplt_sYZ.gnu, found in zgoubi toolbox [33]

```
NF muon collect channel.
'MARKER'  NFDensities_S                          ! Just for edition purposes.
'MCOBJET'
833.910238
3
100
2 2 2 2 1 1                    ! distribution densities: Gaussian in Y, T, Z, P. Uniform in s, D.
0.3943 -0.09036 .0  .0  0.   1.2           ! Reference coordinates at funnel entrance.
0.   4.   0.01  1.00                       ! Initial Y,T emittance. Cut-off at 1-sigma.
0.   4.   0.01  1.00                       ! Initial Z,P emittance. Cut-off at 1-sigma.
0.   1.   0.16  1.00                       ! Initial s, 1-D emittance. Cut-off at 1-sigma.
123456 234567 345678
'PARTICUL'                                        ! Define MCOBJET particles as pions.
139.567 1.60217733D-19 0. 26.03E-9 0.
! PION+                                          ! Can be substituted to the previous line.

'MCDESINT'
105.66  0.                                        ! Pions decay to muon.
136928 768370 548375

'FAISTORE'                        ! Log particle data at LABEL1s as specified, in zgoubi.fai.
zgoubi.fai S_NF_Funnel E_NF_Funnel E_NF_FODO
1

'COLLIMA' #START                              ! Scrape particles beyond
1 collim.                                     ! 40cm +/- 20cm horizontal,
2 20.00 20.00 40. 0.                          ! +/- 20cm vertical.

'INCLUDE'
1
opticalSequence.inc[S_NF_Funnel:E_NF_Funnel]
'CHAMBR'                           ! Defines elliptical aperture limit +/-50cm both planes.
1
2 50.00 50.00 0. 0.
'INCLUDE'
1
5 * opticalSequence.inc[S_NF_FODO:E_NF_FODO]              ! A total length of 5*21.7 m.
'SYSTEM'
1
gnuplot <./gnuplot_Zplt_sYZ.gnu
'MARKER'  NFDensities_E                          ! Just for edition purposes.
'END'
```

Fig. 12.48 Horizontal and vertical trajectories of π and μ particles, along the collect channel. A pion track stops when the particle either decays, or hits the vacuum pipe, whose transverse aperture is defined by CHAMBR. A muon track stops when the particle hits the pipe

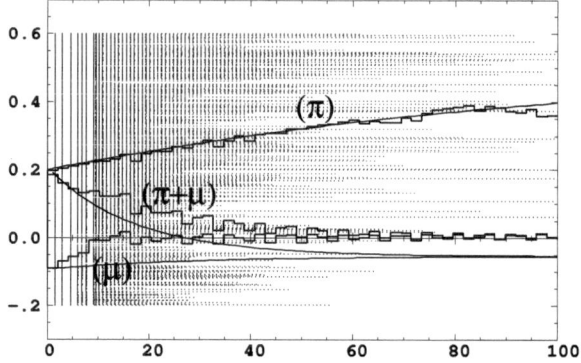

Fig. 12.49 Average beam momentum of π, μ and $\pi + \mu$ beams (relative to 250 MeV/c). Dots represent the survived pions, at the exit of optical elements. Smooth curves are from theory, assuming loss-free propagation. Histograms (broken lines) are from raytracing, obtained using zpop, which reads particle data at exit of optical elements, from zgoubi.fai

Fig. 12.50 Momentum histograms of pions at entrance of the funnel and at 20 and 50 m, and of (triangular shape distribution) muon beam at 100 m. Smooth curves are from theory. Histograms (broken lines) are from raytracing, obtained using zpop: menu 7; 1/5 to open zgoubi.fai; 2/[1,28] for an histogram of p/p_{ref}; 7 to plot

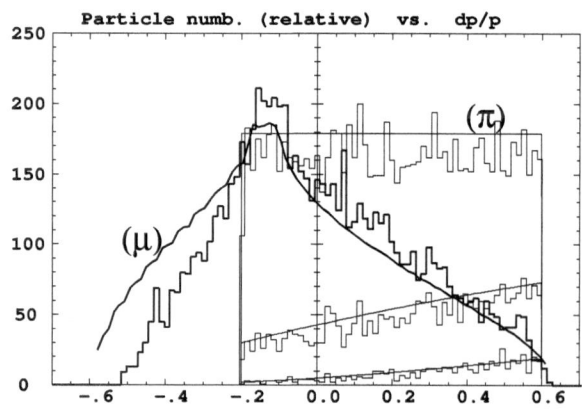

```
'COLLIMA'
1
16 0. 0. .5 1 0. 0.
```

It is placed at the end of the optical sequence of Table 12.32. Figure 12.51 shows four different ellipses so determined, with surface $\varepsilon_l = 0.1$, 0.5, 1, and 2π eV s.

(d) Maximization of the transmission.

It can use FIT[IC = 5, I = 4–6], which computes the *rms* ellipse that matches the outgoing beam, and the transmission through it, while the ellipse surface is left free (its value is an outcome of the rms match). Maximization of the transmission may use also FIT[IC = 5, I = 1–3]; in that case the FIT still delivers the ellipse parameters $\alpha_{Y,Z,s}$ and $\beta_{Y,Z,s}$, but the surface of the ellipse is part of the constraints.

Fig. 12.51 Optimum ellipse shape and positioning in time-energy phase space, for maximized acceptance, at the end of the muon collect channel, for the four cases of emittances $\varepsilon_l = 0.1,\ 0.5,\ 1,$ and $2\,\pi\,\mathrm{eV\,s}$

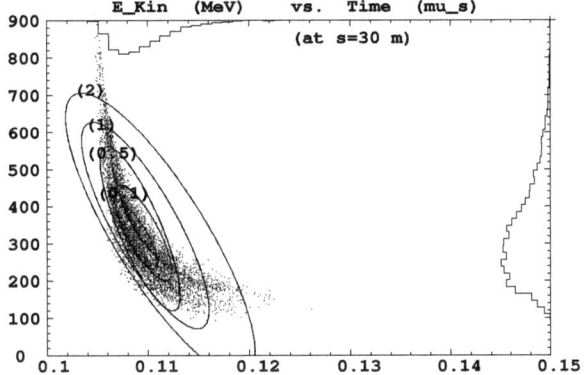

12.4.5   Low Energy Spin Rotator

12.13   Spin Rotation, Spin Matrix

(a) E and B settings.

A 0.5 m long `WIENFILT` is set with magnetic and electric E and B field values proper to ensure straight electron trajectory and $\theta_{\mathrm{sp}} = 30°$ spin Y-rotation. E and B are determined from Eq. 12.27 (straight trajectory) and from Eq. 12.30 yielding the following contributions

$$\theta_{\mathrm{sp}} = \underbrace{(1+a\gamma)\frac{B_Z L}{B\rho}}_{50.588°\ \text{from}\ B_Z\ \text{torque}} \underbrace{-\gamma\left(\frac{1}{\gamma+1}+a\right)\beta^2\frac{B_Z L}{B\rho}}_{20.588°\ \text{from}\ E_Y\ \text{torque}} = 30° \qquad (12.33)$$

given the field values

$$B_Z = \frac{B\rho\,\theta_{\mathrm{sp}}}{L}\Big/\left[1+a\gamma-\gamma\left(\frac{1}{\gamma+1}+a\right)\beta^2\right] = 0.00407378\,\mathrm{T}$$
$$E_Y = vB_Z = 982939\,\mathrm{V/m} \qquad (12.34)$$

The simulation input file is given in Table 12.33. It includes a FIT procedure, aimed in this first exercise, where a hard-edge field model is used, at confirming the theoretical E and B values.

Table 12.33 Simulation input data file WFSegment_FIT.inc: 350 keV electron straight trajectory, together with 30° spin rotation, across one 0.5 m segment of a 3-segment Wien filter. The initial E and B values may be taken from Eq. 12.34, they are about doubled here for the sole purpose of highlighting the effect of the FIT procedure. This file includes spin matrix computation, following FIT, so only occurring when the FIT is completed. The file defines the optical sequence segment #S_WFSegment to #E_WFSegment for INCLUDEs in subsequent exercises. Note the $C_0 - C_5$ Enge coefficient values in the Enge fringe field model [30, WIENFILT], set to values used in subsequent exercises. The field model is hard-edged, here, due to the fall-off parameter values $X_E = \lambda_{E_E} = \lambda_{B_E} = 0$ and $X_S = \lambda_{E_S} = \lambda_{B_S} = 0$

```
Wien filter used as spin rotator. WFSegment_FIT.inc
'MARKER' WFSegment_FIT_S                                      | Just for edition purposes.
'OBJET'
2.31147953865                                                 | Rigidity of a 350 keV electron.
2
3  1                        | 3 electrons, same initial coordinates, for spin matrix computation by SPNPRT.
0.  0.  0.  0.  0.  1. 'o'
0.  0.  0.  0.  0.  1. 'o'
0.  0.  0.  0.  0.  1. 'o'
1 1 1
'PARTICUL'
POSITRON

'SPNTRK'                                    | Allows getting the rotation of all 3 spin components
4                                           | (they are computed independently), for matrix computation by SPNPRT.
1. 0. 0.
0. 1. 0.
0. 0. 1.

'MARKER'  #S_WFSegment
'WIENFILT'
20
0.5  -19E5 8E-03 1                          | Doubled E (V/m) and B (T) field values: will be tuned by the FIT.
0. 0. 0.    | Hard-edge entrance face, XE=lambdaE_e=lambdaB_e=0. Substitue 20. 5. 5. for fringe fields.
0.2401  1.8639  -0.5572  0.3904 0. 0.          | Fringe field coefficients in Enge model.
0.2401  1.8639  -0.5572  0.3904 0. 0.
0. 0. 0.       | Hard-edge exit face, XS=lambdaE_s=lambdaB_s=0. Substitue 20. 5. 5. for fringe fields.
0.2401  1.8639  -0.5572  0.3904 0. 0.
0.2401  1.8639  -0.5572  0.3904 0. 0.
.2                          | Integration step size. It may be given different values, depending on the exercise.
1. 0. 0. 0.
'MARKER'  #E_WFSegment

'FIT2'
2
6 11 0 .8                                                    | Vary E in WIENFILT.
6 12 0 .8                                                    | Vary B in WIENFILT.
6 1E-15                                              | Six constraints; penalty 1e-15:
3      1 2 #End 0. 1. 0                                         | straight trajectory,
3      1 3 #End 0. 1. 0                                              | ibidem,
10.2 1 1 #End 0.52359877559 1. 0                          | 30 deg spin rotation,
10     1 4 #End 1. 1. 0                                | |S|=1, for all 3 electrons.
10     2 4 #End 1. 1. 0
10     3 4 #End 1. 1. 0

'FAISCEAU'                                            | Log trajectory coordinates in zgoubi.res.
'SPNPRT' MATRIX                           | Log spin outcomes, including spin rotation matrix.
'SYSTEM'
1
gnuplot < ./gnuplot_spin.gnu                    | Produce a graph of spin motion along the Wien filter.
'MARKER' WFSegment_FIT_E                                 | Just for edition purposes.
'END'
```

A gnuplot script to obtain Fig. 12.52:

```
# gnuplot_spin.gnu
set xlabel "s [m]"; set xtics mirror; unset ytics; set k t 1; zero = 0.
set arrow from 50,0 to 50,-130 nohead; set arrow from 100,0 to 100,-130 nohead; set arrow from 150,0 to 150,-130 nohead
# in zgoubi.plt, test $19: particle number; test $51: FIT stage number; $14: s; $33: SX
plot [:170] [0:-130]  \
       "zgoubi.plt" u ($19==1 & $51==0 ? $14 : 1/0):(zero):($33*120):($34*120)  w vector lw .5 lc rgb "red" ,\
       "zgoubi.plt" u ($19==1 & $51==1 ? $14 : 1/0):(zero):($33*120):($34*120)  w vector lw 3 lc rgb "blue"
```

The outcome of the FIT procedure, in zgoubi.res execution listing, is as follows:

```
**********************************************************************************************
 STATUS OF VARIABLES  (Iteration #     0 /    999 max.)
 LMNT VAR PARAM  MINIMUM     INITIAL          FINAL         MAXIMUM        STEP        NAME    LBL1
    6   1    11 -3.420E+06 -1.900E+06  -982938.94      -3.800E+05  0.163        WIENFILT   -
    6   2    12  1.600E-03  8.000E-03 4.07378127E-03  1.440E-02  6.744E-10     WIENFILT   -
 STATUS OF CONSTRAINTS (Target penalty =   1.0000E-15)
 TYPE  I   J LMNT#    DESIRED        WEIGHT         REACHED        KI2      NAME    LBL1
    3   1   2     7  0.000000E+00  1.000E+00   2.891169E-10  Infinity MARKER    #E_WFSegment
    3   1   3     7  0.000000E+00  1.000E+00   1.129879E-08  Infinity MARKER    #E_WFSegment
   10   1   1     7  5.235988E-01  1.000E+00   5.235988E-01  Infinity MARKER    #E_WFSegment
   10   1   4     7  1.000000E+00  1.000E+00   1.000000E+00      NaN MARKER    #E_WFSegment
   10   2   4     7  1.000000E+00  1.000E+00   1.000000E+00      NaN MARKER    #E_WFSegment
   10   3   4     7  1.000000E+00  1.000E+00   1.000000E+00      NaN MARKER    #E_WFSegment
 Fit reached penalty value   4.8086E-16
**********************************************************************************************
```

The resulting "STATUS OF VARIABLES: FINAL" field values
$$E = -982938 \text{ V/m} \quad \text{and} \quad B = 0.00407378 \text{ T}$$
are in accord to high accuracy with the theoretical ones (Eq. 12.34) as expected in this hard-dege model. A graph of the 30° Z-rotation of the spin in the $\mathbf{B}^{\perp}$ plane along a 50 cm Wien filter segment, is given in Fig. 12.52

(b) Spin matrix.

The simulation data file of Table 12.33 includes spin matrix computation. Three electrons are needed for SPNPRT[MATRIX] to operate. All three have identical

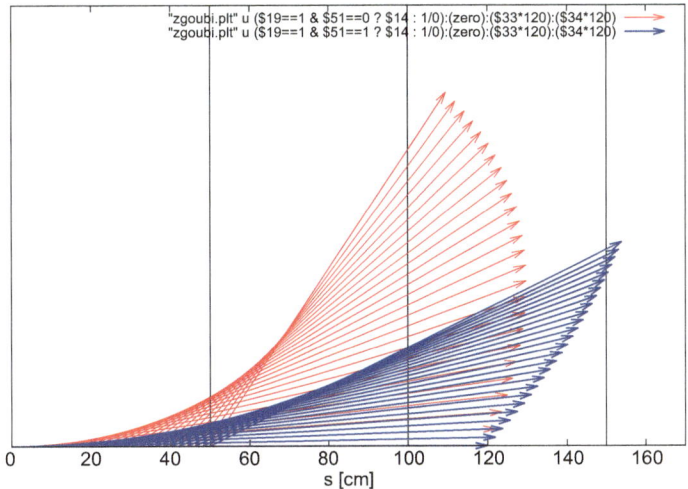

Fig. 12.52 Spin motion in the $\mathbf{B}^{\perp}$ plane, from longitudinal at s = 0 to 30° to the X axis at s = 50 cm. The thin vector series (red) is before any FIT, when field values are about twice the optimum ones. The thick vector series (blue) is at the last pass of the FIT procedure, when fields have their matched values E = −982939 V/m and B = 0.00407378 T. The vertical bars materialize the three Wien filter segments, the first one only is tracked, here

initial coordinates and rigidity, and their spins (defined under SPNTRK) have to form a direct triedra, namely, for electrons 1, 2, 3 respectively, $\mathbf{S}_0(1) \equiv \mathbf{S}_X$, $\mathbf{S}_0(2) \equiv \mathbf{S}_Y$, $\mathbf{S}_0(3) \equiv \mathbf{S}_Z$.

So obtained spins, spin matrix (a Z-rotation, as expected) and rotation precession angle (30°, as expected) at the end of the Wien filter segment, in zgoubi.res execution listing, are as follows:

```
****************************************************************************************************
   10  Keyword, label(s) :  SPNPRT       MATRIX

                        -- 1  GROUPS  OF  MOMENTA  FOLLOW   --
         ----------------------------------------------------------------------------
                    Momentum  group  #1 (D=   1.000000E+00;  particles  1  to 3 ;

              Spin  components  of  the      3  particles,  spin  angles :
                      INITIAL                                         FINAL

        SX     SY     SZ     |S|       SX        SY        SZ       |S|    GAMMA   |Si,Sf|   (Z,Sf_yz) (Z,Sf)
                                                                                  (deg.)    (deg.)    (deg.)
                                                                         (Sf_yz : projection of Sf on YZ plane)
   o  1  1.000 0.000  0.000  1.000   0.866025 -0.500000 -0.000000  1.000000  1.6849   30.000   90.000   90.000
   o  1  0.000 1.000  0.000  1.000   0.500000  0.866025  0.000000  1.000000  1.6849   30.000   90.000   90.000
   o  1  0.000 0.000  1.000  1.000   0.000000 -0.000000  1.000000  1.000000  1.6849    0.000   00.000    0.000

              Spin transfer matrix, momentum group # 1 :
                 0.866025        0.500000       1.933926E-17
                -0.500000        0.866025      -1.508203E-17
                -2.429799E-17    3.436238E-18   1.00000

   Determinant =    1.0000000000
   Trace =    2.7320507888;  spin precession acos((trace-1)/2) =    30.0000010762 deg
   Precession axis :  ( 0.00, 0.00,-1.00) -> angle to (X,Y) plane, X axis, Z axis :   -90.00, 90.00, 180.00 deg
   Spin precession/2pi (or Qs, fractional) :    8.3333E-02
****************************************************************************************************
```

12.14 Integration Step Size

Numerical convergence of both particle and spin motion may be affected in a non-negligible manner if the integration step size is taken too large. On the other hand the step size should be taken as large as tolerable to save on computation time, if repeated thousands of electrons trials are to be performed, for statistics purposes for instance.

(a) Perturbation of exit coordinates.

The input data file to explore the effect of step size on electron and spin coordinates at the exit of the Wien filter, leaving field values unchanged, is given in Table 12.34. Figure 12.53 shows that, for E_Y and B_Z maintained constant (at their theoretical values), varying the step size up to 10 cm only has a small effect on electron position and angle at the exit of the Wien filter segment, whereas the relative error on the spin rotation θ_s remains below 10^{-3} if the step size remains below 8 cm.

As a consequence, electrons can be pushed through the Wien filter in a quick 20 steps.

(b) Recovering proper coordinates, step size

The input data file to explore the effect of step size on field values, to recover null electron coordinates and 30° spin rotation, is given in Table 12.35.

A scan of the step size shown in Fig. 12.54 indicates that it can be several centimeters with minor impact on nominal E or B field amplitudes: these would need to

Table 12.34 Simulation input data file: integration step size scan using REBELOTE[IOPT=1]. The focus here is the effect of the step size on the value of final position and angle of the electron on the one hand, and of the spin rotation angle on the other hand. Fields are set to their theoretical values, $E_Y = -982938.94$ V/m and $B_Z = 0.00407378127$ T

```
Wien filter used as spin rotator. Step size scan.
'MARKER'  WFSegment_ScanTraj_S                            ! Just for edition purposes.
'OBJET'
2.31147953865                                  ! Rigidity of a 350 keV electron.
2
3 1                      ! 3 electrons, same initial coordinates, for spin matrix computation by SPNPRT.
0.  0. 0. 0. 0. 1. 'o'
0.  0. 0. 0. 0. 1. 'o'
0.  0. 0. 0. 0. 1. 'o'
1 1 1
'PARTICUL'
POSITRON

'SPNTRK'                              ! Allows getting the rotation of all 3 spin components
4                            ! (they are computed independently), for matrix computation by SPNPRT.
1. 0. 0.
0. 1. 0.
0. 0. 1.

'MARKER'  #S_WFSegment_EBth
'WIENFILT'
20
0.5 -982939 0.00407378  1                           ! Theoretical E (V/m) and B (T) field values.
0. 0. 0.   ! Hard-edge entrance face, XE=lambdaE_e=lambdaB_e=0. Substitue 20. 5. 5. for fringe fields.
0.2401  1.8639  -0.5572  0.3904 0. 0.             ! Fringe field coefficients in Enge model.
0.2401  1.8639  -0.5572  0.3904 0. 0.
0. 0. 0.   ! Hard-edge exit face, XS=lambdaE_s=lambdaB_s=0. Substitue 20. 5. 5. for fringe fields.
0.2401  1.8639  -0.5572  0.3904 0. 0.
0.2401  1.8639  -0.5572  0.3904 0. 0.
.2                   ! Integration step size. It may be given different values, depending on the exercise.
1. 0. 0. 0.
'MARKER'  #E_WFSegment_EBth

'FAISCEAU'                                   ! Log local particle coordinates in zgoubi.res.
'FAISTORE'
zgoubi.fai                                   ! Log local particle coordinates in zgoubi.res.
1
 'REBELOTE'
400 0.1 0 1                    ! Repeat the previous sequence, 400 times, and prior to each repeat,
1                              ! change value of one parameter,
WIENFILT 80 0.01:10.           ! namely, number 80 (integration step size) in WIENFILT.

 'SYSTEM'
2
gnuplot <./gnuplot_scanTraj.gnu
!okular ./gnuplot_scanTraj.eps &
'MARKER'  WFSegment_ScanTraj_E                           ! Just for edition purposes.
 'END'
```

A gnuplot script to obtain Fig. 12.53:

```
# gnuplot_scanTraj.gnu
set xlabel "step size [cm]"; set ylabel " x [cm], x' [mrad]"; set y2label "{/Symbol d q}_s/{/Symbol q}_s"
set xtics mirror; set ytics nomirror; set y2tics; set k c l  font "roman, 10"; set logscale x; set logscale x2
# data to compute step size, for abscissa axis, and relative change of spin rotation, vertical axis:
stpi = 0.01; stpf = 10.; nrblt = 400; dstp = (stpf-stpi)/(nrblt-1); nbtraj = 3; ttaInf = -30.

plot  \
    "zgoubi.fai" u ($38>1 && $26==1? stpi+dstp*$0/nbtraj :1/0):($10) axes x1y1 w p pt 5 ps .7 lc rgb "red"  ,\
    "zgoubi.fai" u ($38>1 && $26==1? stpi+dstp*$0/nbtraj :1/0):($11) axes x1y1 w lp pt 6 ps .7 lc rgb "blue"  ,\
    "zgoubi.fai" u ($38>1 && $26==1? stpi+dstp*$0/nbtraj :1/0):(-(atan($21/$20)/(4.*atan(1.))*180. -ttaInf)/ttaInf) '\
    axes x1y2 w lp pt 7 ps .7 lc rgb "green" tit "spin-R (right axis)" ; pause 1
```

be tweaked by less than 10^{-3} (relative) compared to their theoretical values, in order to recover $30°$ spin rotation and straight trajectory.

12.15 Add Fringe Fields

(a) Fringe field extent; field profile; step size Δs.

The integration region extents at both ends of WIENFILT can be fixed at $XE = XS = 20$ cm. At such large distance from the effective field boundary, given the fringe field extents considered, λ_E and λ_B of a few centimeters, the fields from the

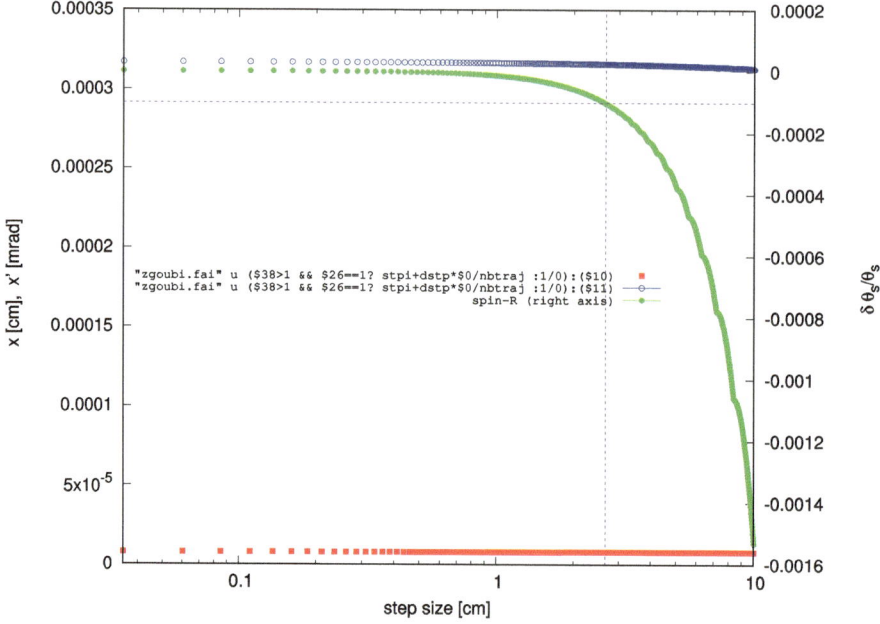

Fig. 12.53 Final electron position (square markers) and angle (empty circles), left vertical axis, and relative spin rotation angle error, right vertical axis, as a function of step size. E_Y and B_Z are maintained constant, at their theoretical values

exponential Enge fall-off model are down to negligible amplitude: the loss of field integral matters, and it is negligible in these conditions. A sample field profile along the Wien filter, obtained using WIENFILT[IL = 2] which logs step by step particle data in zgoubi.plt, is displayed in Fig. 12.55.

Convergence of the numerical integration can be checked using the simulation data file of Table 12.35, with proper fringe field parameters under WIENFILT. The latter is INCLUDEd from Table 12.33, thus that is where $X_E = 20$ cm, $\lambda_{E_E} = 5$ cm, $\lambda_{B_E} = 5$ cm has to be substituted to $X_E = \lambda_{E_E} = \lambda_{B_E} = 0$, and $X_S = 20$ cm, $\lambda_{E_S} = 5$ cm, $\lambda_{B_S} = 5$ cm to $X_S = \lambda_{E_S} = \lambda_{B_S} = 0$. The integration step size Δs has to be small enough to ensure accuracy of the numerical integration of the equation of motion in the fringe field regions, namely $\Delta s < \min[\lambda_E, \lambda_B]$ (a greater Δs value can be used in the Wien filter body where fields are uniform, see [30, Sect. 7.10]). For this reason, the variation range under the do-loop REBELOTE[IOPT=1] is limited to $0.1 \leq \Delta s \leq 2.5$ cm.

Results are displayed in Fig. 12.56: if the integration step size Δs is taken a centimeter and below, recovering null trajectory coordinates at the ends of the Wien filter, and exact 30° spin rotation, requires adjusting E and B by less than $\approx 10^{-4}$, relative.

Δs in centimeter range allows fast computation, convenient for statistics trials: with $\Delta s = 1$ cm, pushing 10,000 electrons through the 3-segment, 1.5 m long Wien filter takes about 20 seconds on a 3 GHz CPU.

Table 12.35 Simulation input data file for a step size scan of $\delta E/E$ and $\delta B/B$. This file uses the optical sequence segment [WFSegment_FIT_S:#E_WFSegment] defined in Table 12.33. REBE-LOTE[IOPT=1] is added and changes the step size value in WIENFILT, in NRBLT=20 steps in the range $0.01 \leq \Delta s \leq 10$ cm. A SYSTEM call to an external gnuplot script produces a graph of the results of this scan

```
Wien filter used as spin rotator. Step size scan.
'MARKER'  WFSegment_ScanS_S                                           ! Just for edition purposes.
'INCLUDE'
1
WFSegment_FIT.inc[WFSegment_FIT_S:#E_WFSegment]

'FIT2'
2
7 11 0 .8                                                    ! Vary E in WIENFILT.
7 12 0 .8                                                    ! Vary B in WIENFILT.
6 1E-15                                           ! Six constraints; penalty 1e-15:
3    1 2 #End 0. 1. 0                                        ! straight trajectory,
3    1 3 #End 0. 1. 0                                        ! ibidem,
10.2 1 1 #End 0.52359877559 1. 0                             ! 30 deg spin rotation,
10   1 4 #End 1. 1. 0                                        ! |S|=1, for all 3 electrons.
10   2 4 #End 1. 1. 0
10   3 4 #End 1. 1. 0

'REBELOTE'
20 0.1 0 1              ! Repeat sequnce above, NPASS=20 times; prior, change parameter as specified below.
1
WIENFILT 80 0.01:10.                         ! Step size range; step increment is (10. - 0.01)/(NPASS-1)

'SYSTEM'
2
gnuplot <./gnuplot_scanEB.gnu
okular ./gnuplot_scanEB.eps &
'MARKER'  WFSegment_ScanS_E                                           ! Just for edition purposes.
'END'
```

A gnuplot script to obtain Fig. 12.54:

```
# gnuplot\_scanEB.gnu
set xlabel "step size [cm]"; set ylabel "{/Symbol d}E/E"; set y2label "{/Symbol d}B/B"
set xtics mirror; set ytics nomirror; set y2tics; set k t l; set log x; set k font "roman, 11"
#grab FIT variables data in zgoubi.res execution listing:
system "rm -f grep.res4E.out"; system "grep '7    1    11 ' zgoubi.res |  cat > grep.res4E.out"
system "rm -f grep.res4B.out"; system "grep '7    2    12 ' zgoubi.res |  cat > grep.res4B.out"
# data to compute step size, for abscissa axis, and relative change of field values, vertical axes
stpi = 0.01; stpf = 10.; nrblt = 20; dstp = (stpf-stpi)/(nrblt-1); Etheor=-982939.4; Btheor=4.073783E-03

plot \
 "grep.res4E.out" u (stpi + dstp*($0-1)):(abs(($6-Etheor)/Etheor)) axes x1y1 w p pt 5 ps 1 lc rgb "red" ,\
 "grep.res4B.out" u (stpi + dstp*($0-1)):(abs($6-Btheor)/Btheor) axes x1y2 w l lw 2 lt 1 lc rgb "blue" ; pause 1
```

(b) Unequal fall-off extents.

Different values of the electrostatic condenser and magnetic dipole gaps result in different fringe field extents λ_E and λ_B, which in turn jeopardizes the Wien filter rule $E_Y = vB_Z$. However it is possible in that case, using the two knobs δE_Y and δB_Z, to recover simultaneously zero trajectory coordinates at both ends of the Wien filter and exact 30° spin rotation. An input data file to simulate this is given in Table 12.36: it performs a scan of the λ_B/λ_E ratio and, using FIT, finds proper re-tuned field values. In this simulation $\lambda_E = 5$ cm is fixed, whereas λ_B is varied between 3 and 7 cm.

The result of that scan is displayed in Fig. 12.57, where the required adjustments of E_Y and B_Z are plotted as a function of the ratio λ_B/λ_E. The "penalty" value monitors the FIT convergence, it comes out to be $< 10^{-15}$, $\forall \lambda_B/\lambda_E$, as requested.

Unequal E-field and B-field fall-off extents, $\lambda_E \neq \lambda_B$, cause a shift of the reference trajectories inside the Wien filter. The reference trajectory coordinates are anyway zeroed at the exit of the device, together with maintaining $\theta_s = 30°$, by adjusting E and B amplitudes as discussed above. Such sample reference trajectories, as displayed in Fig. 12.58, come as a sub-product of the simulation file of Table 12.36, as step by step particle data are logged in zgoubi.plt due to the option WIENFILT[IL=2].

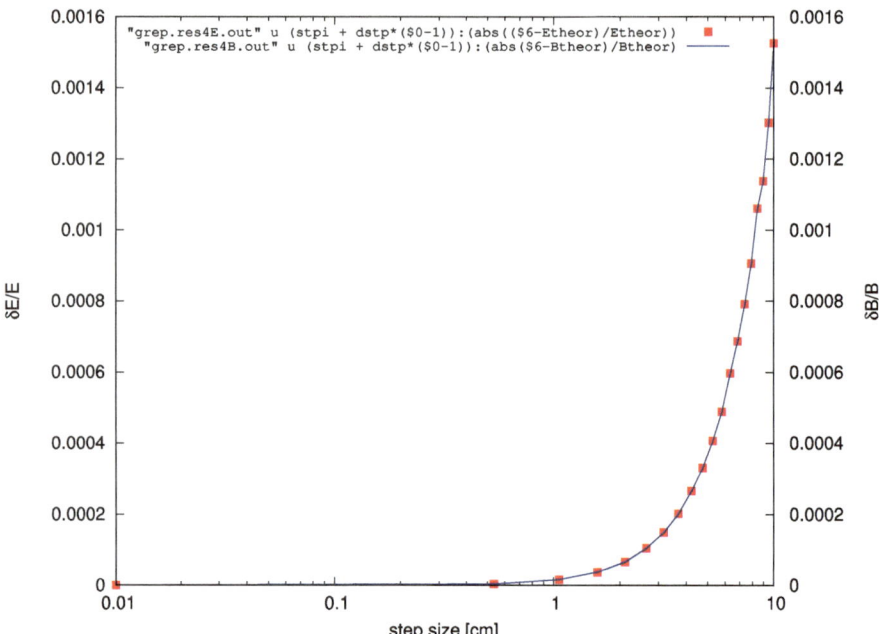

Fig. 12.54 An integration step size scan. This graph shows, as a function of step size, the required variation of the E_Y and B_Z fields, relative to their theoretical value, to get exact 30° spin rotation and straight trajectory

Fig. 12.55 Field shape across a 50 cm Wien filter segment, on-axis. The same for both E_Y and B_Z components here, obtained with $\lambda_E = \lambda_B = 5$ cm. A graph obtained using zpop: menu 7; 1/1 to open zgoubi.plt; 2/[6,35] to select E_Y versus distance and 2/[6, 32] to select B_Z versus distance; 7 to plot

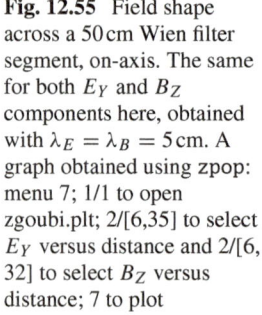

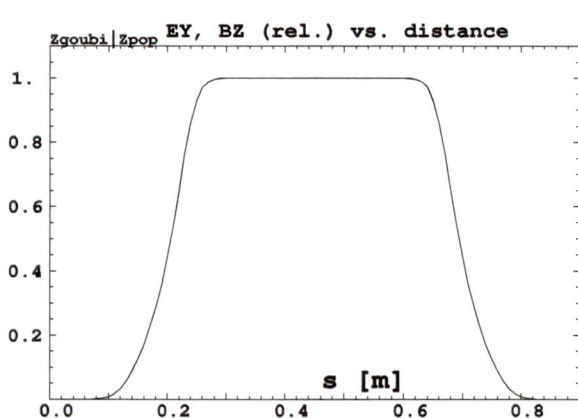

During the execution, spin data are logged to zgoubi.res listing on the one hand, and to zgoubi.SPNPRT.Out on the other hand as an effect of the command (Table 12.36)

```
'SPNPRT' PRINT MATRIX
```

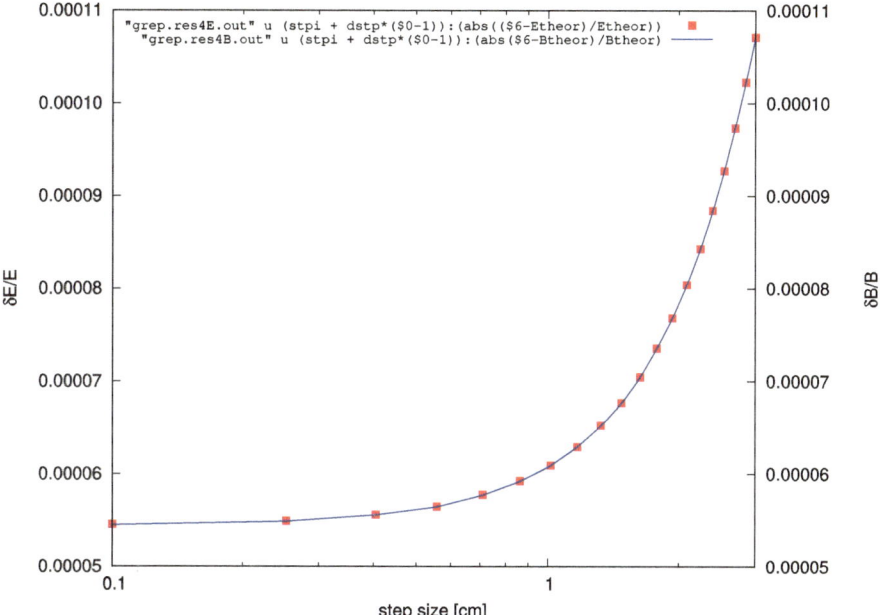

Fig. 12.56 An integration step size scan, in the presence of fringe fields. This graph shows, as a function of step size, the required variation of the E_Y and B_Z fields, relative to their theoretical value, to get exact 30° spin rotation and straight trajectory. The Enge model normalization coefficients are taken equal, $\lambda_E = \lambda_B = 5$ cm, in this case

There is various ways to check the spin rotation angle. One consists in a linux '*grep*' from zgoubi.res execution listing, namely the very line under SPNPRT which displays the spin rotation angle of the S_X spin coordinate (which, at s = 0, identifies with the spin vector: $\mathbf{S}(s=0) \equiv \mathbf{S}_X(s=0)$), like so:

```
grep -n -c ' o  1  1.000000  0.000000 ' zgoubi.res
```

The result is the following sequence of 38 lines (as per REBELOTE[NRBLT = 37], Table 12.36):

		INITIAL COORDINATES				CURRENT COORDINATES				ROTATION ANGLE			
254: o	1	1.000	0.000	0.000	1.000	0.866025	-0.5000	-0.000	1.000	1.6849	30.000	90.000	90.000	1
342: o	1	1.000	0.000	0.000	1.000	0.866025	-0.5000	-0.000	1.000	1.6849	30.000	90.000	90.000	1
424: o	1	1.000	0.000	0.000	1.000	0.866025	-0.5000	-0.000	1.000	1.6849	30.000	90.000	90.000	1
506: o	1	1.000	0.000	0.000	1.000	0.866025	-0.5000	-0.000	1.000	1.6849	30.000	90.000	90.000	1
....														
2966: o	1	1.000	0.000	0.000	1.000	0.866025	-0.5000	-0.000	1.000	1.6849	30.000	90.000	90.000	1
3048: o	1	1.000	0.000	0.000	1.000	0.866025	-0.5000	-0.000	1.000	1.6849	30.000	90.000	90.000	1
3130: o	1	1.000	0.000	0.000	1.000	0.866025	-0.5000	-0.000	1.000	1.6849	30.000	90.000	90.000	1
3212: o	1	1.000	0.000	0.000	1.000	0.866025	-0.5000	-0.000	1.000	1.6849	30.000	90.000	90.000	1
3530: o	1	1.000	0.000	0.000	1.000	0.866025	-0.5000	-0.000	1.000	1.6849	30.000	90.000	90.000	1

Table 12.36 Simulation input data file: a scan of λ_B/λ_E fringe field extent ratio. FIT finds the E_Y and B_Z values which recover null trajectory coordinates at both ends of the Wien filter, together with 30° spin rotation. The accuracy of the solution is quantified by the penalty, requested to be $< 10^{-15}$. This input data file defines (using LABEL1's) the optical segment #S_WFSegment_FF to #S_WFSegment_FF for use in INCLUDE commands in a subsequent exercise

```
Wien filter with fringe fields. File WFSegment_FF.inc
'MARKER'  WFSegment_FF_ScanS_S                                  ! Just for edition purposes.
'OBJET'
2.31147953865                                                   ! Rigidity of a 350 keV electron.
2
3  1                             ! 3 electrons, same initial coordinates, for spin matrix computation by SPNPRT.
0.  0. 0. 0. 0. 1. 'o'
0.  0. 0. 0. 0. 1. 'o'
0.  0. 0. 0. 0. 1. 'o'
1 1 1
'PARTICUL'             ! Electron mass, charge and gyromagneitc factor are ecessary due to spin tracking.
POSITRON
'SPNTRK'                                  ! Allows getting the rotation of all 3 spin components
4                                         ! (they are computed independently), for matrix computation by SPNPRT.
1. 0. 0.
0. 1. 0.
0. 0. 1.

'MARKER'  #S_WFSegment_FF
'WIENFILT'
2
0.5  -979009.12 0.00406037568  1
20. 5. 7.                                  ! Entrance fringe model: XE=20cm; lambdaE_e=5cm; lambdaB_e=7cm.
0.2401  1.8639  -0.5572  0.3904 0. 0.      ! Enge coefficients, eletrostatic gap.
0.2401  1.8639  -0.5572  0.3904 0. 0.      ! Enge coefficients, magnetic gap.
20. 5. 7.                                  ! Exit fringe model: XE=20cm; lambdaE_e=5cm; lambdaB_e=7cm.
0.2401  1.8639  -0.5572  0.3904 0. 0.
0.2401  1.8639  -0.5572  0.3904 0. 0.
.5                                                               ! Integration step size.
1. 0. 0. 0.                                                      ! Positioning.
'MARKER'  #E_WFSegment_FF

'FIT2'
2   nofinal
6 11 0 1.05                                     ! vary EY     in attempt to recover straight;
6 12 0 1.05                                     ! vary BZ     trajectory and 30deg spin rotation.
6 1E-15                                         ! 6 constraints; penalty 1e-15.
3 1 2 #End 0. 1. 0                              ! Trajctory position Y=0 at end of sequence,
3 1 3 #End 0. 1. 0                              ! Trajctory angle T=0 at end of sequence,
10.2 1 1 #End 0.5235987755 1. 0    ! 30deg spin rotation
10 1 4 #End 1. 1. 0
10 2 4 #End 1. 1. 0
10 3 4 #End 1. 1. 0

!                  For possible post-treatment or dta analysis: PRINT saves outcomes to zgoubi.SPNPRT.Out;
'SPNPRT' PRINT MATRIX                          ! MATRIX computes spin matrix & logs in zgoubi.res.

'REBELOTE'
37 0.1 0 1                      ! NPASS is of the form int*(7[cm]-3[cm])+1 to allow for lambdaB/lambdaE=1.
2
WIENFILT 22 3.:7.                               ! vary lambda_B at entrance EFB from 3 to 7 cm.
WIENFILT 52 3.:7.                               ! vary lambda_B at exit EFB from 3 to 7 cm.

'SYSTEM'
2
gnuplot <./gnuplot_scanEB_FFratio.gnu
okular ./gnuplot_scanEB_FFratio.eps &
'MARKER'  WFSegment_FF_ScanS_E                  ! Just for edition purposes.
'END'
```

A gnuplot script to obtain Fig. 12.57:

```
set xlab "Fringe ratio {/Symbol l}_B/{/Symbol l}_E"; set ylab "|{/Symbol d}E/E|, |{/Symbol d}B/B|"; set y2lab "penalty"
set k c t spacin 1; set xtics mirror; set ytics nomirror; set y2tics; set logscale y2; set format y2 "%.0s*10^{%T}"
# grep status of FIT variables, in zgoubi.res execution listing:
system "grep '   1    11 ' zgoubi.res | cat > grep.res4E.out"
system "grep '   2    12 ' zgoubi.res | cat > grep.res4B.out"
system "grep 'penalty value ' zgoubi.res | cat > grep.penalty.out"
# Parameters of the lambda_B/lambda_E scan, to determine the horizontal scale:
NPASS = 37; lmdaBi = 3.; lmdaBf = 7.; lmdaE = 5.; dlmda = (lmdaBf-lmdaBi)/(NPASS-1); Eth= -982939.444; Bth= 4.0737834E-03

plot [.58:1.42] [:.0042] \
"grep.res4E.out" u ($0>0? (lmdaBi+dlmda*($0-1))/lmdaE:1/0):(abs(($6-Eth)/Eth)) axes x1y1 w lp pt 4 lc rgb "red" tit "E" ,\
"grep.res4B.out" u ($0>0? (lmdaBi+dlmda*($0-1))/lmdaE:1/0):(abs(($6-Bth)/Bth)) axes x1y1 w lp pt 6 lc rgb "blue" tit "B",\
"grep.penalty.out" u ($0>0? (lmdaBi+dlmda*($0-1))/lmdaE:1/0):($5) axes x1y2 w l dt 2 lc rgb "black" tit "penalty"; pause 1
```

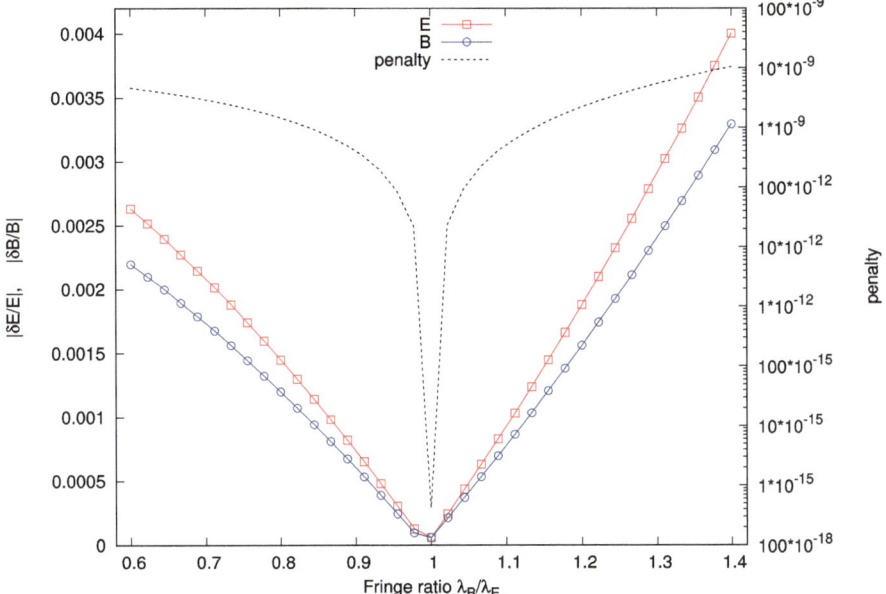

Fig. 12.57 Variation of the E_Y and B_Z fields (relative to their hard-edge model values), required to recover (i) exact 30° spin rotation, (ii) straight trajectory, when the λ_B/λ_E ratio is varied (B fringe field extent is varied, while E fringe field extent is kept constant, here). The fit "penalty" quantifies the distance to these two constraints, steadily small here

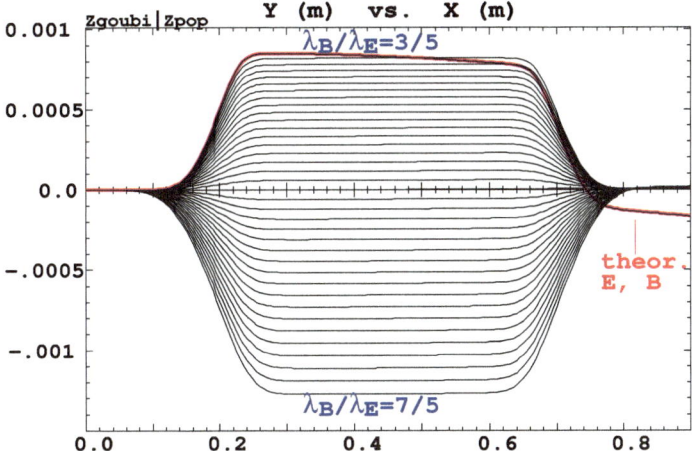

Fig. 12.58 A scan of the on-momentum reference trajectories across the Wien filter, with varying ratio λ_B/λ_E. The former have zero coordinates at entrance by hypothesis, and at exit (together with $\theta_s \equiv 30°$) as a result of the FIT on E and B amplitudes. A graph obtained using zpop: menu 7; 1/1 opens zgoubi.plt; 2/[8, 2] selects Y versus X; 7 to plot

The first number in these lines is the line number in zgoubi.res execution listing. The rotation angle of the spin, from $\mathbf{S}(s = 0)$ to $\mathbf{S}(s = 50\,\text{cm})$ is the 13th data in each line, 30° as expected from penalty $< 10^{-15}$ at each iteration of FIT (Fig. 12.57). A second possibility is to use zgoubi.SPNPRT.Out data, they include the spin coordinates, from which the rotation angle can be computed. There exists actually a third possibility, which is to add FAISTORE[FNAME=zgoubi.fai] in the data file, in lieu of SPNPRT just before REBELOTE: this logs particle data in zgoubi.fai, these include spin coordinates.

12.16 Transport of a 6D Polarized Bunch

A 10^4 electron bunch is considered, *rms* emittances $\beta\gamma\varepsilon_Y = \beta\gamma\varepsilon_Z = 10\pi\,\mu\text{m}$, Gaussian transverse densities truncated at 4 sigma, momentum spread uniform over $[-10^{-4}, +10^{-4}]$.

The Wien filter model includes fringe fields and a $\lambda_E/\lambda_B = 7/5$ ratio between E and B fall-off extents. Adjusting the fields to $E_Y = -979009.12\,\text{V/m}$ and $B_Z = 0.00406037568$ T (this is actually the first case of the simulation file of Table 12.36, where these updated values are taken from) ensures zero reference trajectory coordinates at exit, straight trajectory parallel to the X axis in the Wien filter body, and 90° spin rotation over the 1.9 m distance (which accounts the Wien filter effective length, 1.5 m, and the entrance and exit field integration extents $XE = XS = 0.2\,\text{m}$) (Table 12.37).

Sample trajectories and fields along, obtained using zpop, are displayed in Fig. 12.59. Phase spaces and projected densities, including spin, at the downstream end of the 3-segment Wien filter are displayed in Fig. 12.60.

Table 12.37 Simulation input data file: launching a 10^4 electron bunch through the Wien filter. Fringe fields are included

```
Track a bunch
'MARKER'  TrackBunch_S                                    ! Just for edition purposes.
'MCOBJET'
2.3114795386518345                                        ! Rigidity of a 350 keV electron.
3
50
2 2 2 2 1 1                    ! Distribution densities: Gaussian in Y, T, Z, P. Uniform in s, D.
0.  0. 0. 0. 0. 1.
0. 3.4 7.37e-6  4
0. 3.4 7.37e-6  4
0. 1. 1e-8  2
123456 234567 345678
'PARTICUL'
0.510998946  1.602176487D-19 1.159652181D-3 0. 0.
! POSITRON                                                ! Can be substituted to the previous line.

'SPNTRK'
4.1
1. 0. 0.                                     ! All electrons have initial spin along X axis.
'INCLUDE'
1
3* ./WFSegment_FF.inc[#S_WFSegment_FF:#E_WFSegment_FF]
'FAISTORE'
zgoubi.fai
1
'FAISCEAU'                                                ! Get trajectory coordinates, here.
'SPNPRT'                                                  ! Get spin coordinates, here.
'MARKER'  TrackBunch_E                                    ! Just for edition purposes.
'END'
```

Fig. 12.59 Sample vertical trajectories and E_Y and B_Z fields along, across the Wien filter. Graphs obtained using zpop: menu 7; 1/1 opens zgoubi.plt; 2/[6,4] selects Z versus distance (or 2/[6,35] for E_Y versus distance, or 2/[6,32] selects B_Z versus distance); 7 to plot

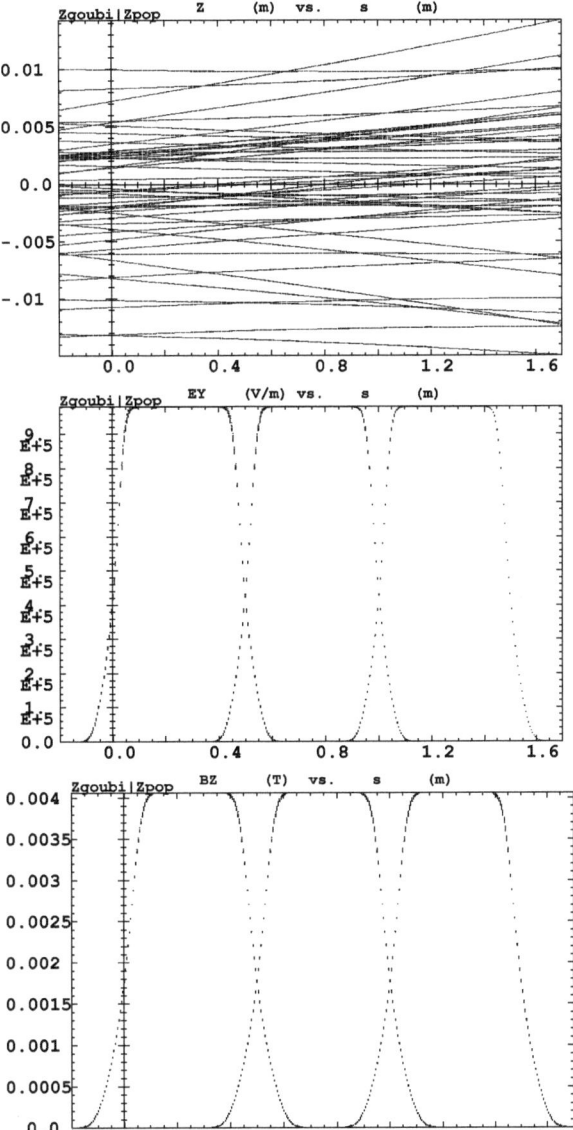

Fig. 12.60 Horizontal phase space (top) and spin angle histogram (center) at the exit of the 3-segment Wien filter, case of a 6D bunch with *rms* emittances $\beta\gamma\varepsilon_Y = \beta\gamma\varepsilon_Z = 10\pi\,\mu\text{m}$, Gaussian transverse densities truncated at 4 sigma, momentum spread uniform in $\pm10^{-4}$. The spin angle spreading (bottom) stems from the horizontal emittance: spin angle spreading under the effect of 10^{-4} momentum spread, and with $\varepsilon_Y = \varepsilon_Z = 0$, amounts to $\approx \pm0.01\,\text{deg}$

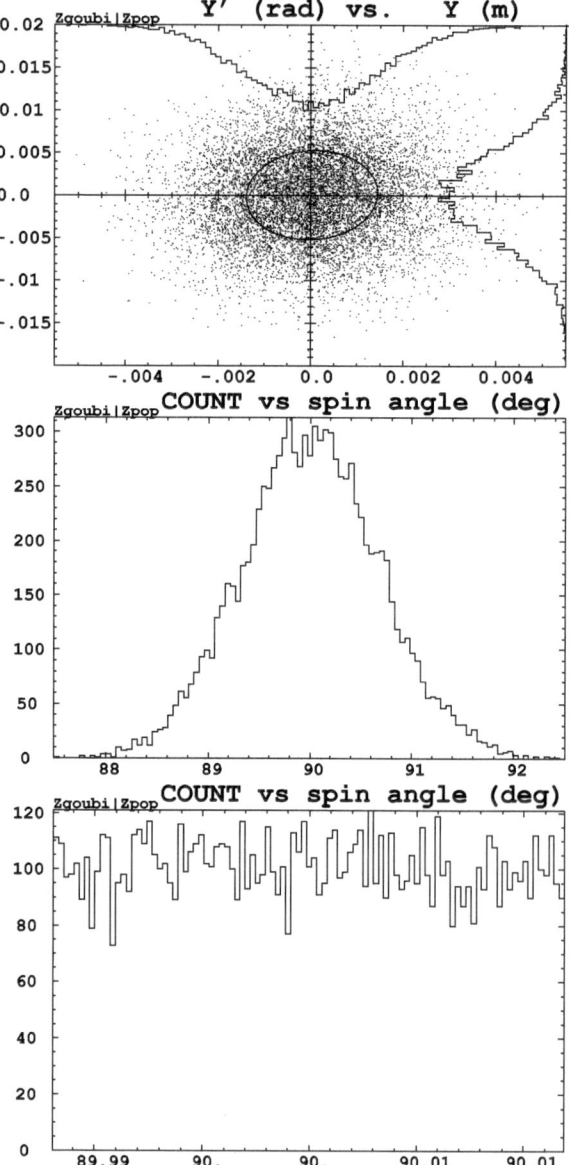

References

1. N. Tsoupas, et al., Results from the commissioning of the NSRL beam line at BNL, in *Proceedings of the EPAC 2004 Conference, Lucerne, Switzerland*
2. S. Meigo, et al., High power target instrumentation at J-PARC for neutron and muon source, in *Proceedings of HB2016 (WEPM2X01)* (Malm_o, Sweden), pp. 391–396
3. F. Méot, Uniform, variable size rectangle beam scanning. Application to hadrontherapy. Nucl. Instrum. Methods Phys. Res. A **564**, 108–114 (2006)
4. B. Blind, Generation of a rectangular beam distribution. Report MS H811, LANL, Los Alamos, NM 87545
5. F. Méot, T. Aniel, Non-linear tuning and halo transport in beam expanders, in *Proceedings of EPAC 1996, Sitges, Spain*. http://accelconf.web.cern.ch/e96/PAPERS/MOPL/MOP074L.PDF
6. C.H. Johnson, A ring lens for focusing ion beams to uniform densities. NIM **127**, 163–171 (1975); P.F. Meads, IEEE Trans. Nucl. Sci. NS-30 (1983); B. Kashy, B. Sherrill, A method for the uniform charged particle irradiation of large targets. Nucl. Instr. Meth. B **26**(4), 610–613 (1987)
7. F. Méot, T. Aniel, Principles of the non-linear tuning of beam expanders. Nucl. Instrum. Methods Phys. Res. A **379**, 196–205 (1996)
8. F. Méot, T. Aniel, Calculation of non-linear envelopes in beam expanders. Phys. Rev. Spec. Top. - Accel. Beams **3**, 103501 (2000)
9. S.Y. Yavor, et al., Achromatic quadrupole lenses. Nucl. Instrum. Methods **26**, 13–17 (1964)
10. F. Méot, Generalization of the Zgoubi method for ray-tracing to include electric fields. NIM A **340**, 594–604 (1994)
11. Study of a European Neutrino Factory Complex, CERN NUFACT Note 122 (2002). https://cds.cern.ch/record/610249/files/ps-2002-080.pdf
12. B. Autin, F. Méot, A. Verdier, efficiency of an alternating gradient muon collection channel CERN NUFACT Note 128 (2003). https://cds.cern.ch/collection/Neutrino20Factory20Notes
13. B. Autin, F. Méot, Time-energy densities in $\pi \to \mu$ decay. CERN NUFACT Note 136 (2004). https://cds.cern.ch/record/703869/files/nufact-note-136.pdf
14. C. Iselin, H. Grote, The MAD Program. http://mad8.web.cern.ch/mad8/
15. J. Grames, et al., Two Wien filter spin flipper, in *TUP025, Proceedings of 2011 Particle Accelerator Conference, New York, NY, USA*. http://accelconf.web.cern.ch/PAC2011/papers/tup025.pdf Figure 12.20: Copyrights under license CC-BY-3.0, https://creativecommons.org/licenses/by/3.0; No change to the material
16. F. Méot, Spin simulations in eRHIC Wien filter. Tech. Note BNL-212123-2019-TECH, EIC/68 (2019). https://www.osti.gov/servlets/purl/1566292
17. G. Leleux, et al., Synchrotron radiation perturbations in long transport lines, in *Proceedings of the PAC 1991 Accelerator Conference*. https://accelconf.web.cern.ch/p91/PDF/PAC1991_0517.PDF
18. F. Méot, J. Payet, Numerical tools for the simulation of synchrotron radiation loss and induced dynamical effects in high energy transport lines. Internal report CEA DSM DAPNIA/SEA-00-01 (CEA Saclay, 2000). https://sourceforge.net/p/zgoubi/code/HEAD/tree/branches/publications/SR/TESLA_BDS/
19. The input data file for the simulation of the 1.2 km long TESLA BDS is available at https://sourceforge.net/p/zgoubi/code/HEAD/tree/branches/exemples/TESLA_BDS/SRalongBDS
20. Méot, F., Synchrotron radiation interferences at the LEP miniwiggler, CERN SL/94-22 (AP) (1994)
21. F. Méot, L. Ponce, N. Ponthieu, Low frequency interference between short SR sources. PRST-AB **4**, 062801 (2001)
22. F. Méot, A theory of low frequency far-field synchrotron radiation. Part. Accel. **62**, 215–239 (1999)
23. L. Ponce, R. Jung, F. Méot, LHC proton beam diagnostics using synchrotron radiation. Yellow Report CERN-2004-007

24. R. Bossart et al., Observation of visible synchrotron radiation emitted by a high-energy proton beam at the edge of a magnetic field. NIM **164**(2), 375–380 (1979)
25. R. Coïsson, Angular-spectral distribution and polarization of synchrotron radiation from a "short" magnet. Phys. Rev. A **20**, 524 (Published 1 Aug 1979); R. Coïsson, On synchrotron radiation in non-uniform magnetic fields. Opt. Commun. **22**(2), 135–137 (1977)
26. F. Méot, Mesure de profil par rayonnement ondulateur des faisceaux de protons et antiprotons. Ph.D. Thesis. Report CERN/SPS 81-21 (ABM) 30 Oct 1981
27. Lookup   https://sourceforge.net/p/zgoubi/code/HEAD/tree/branches/exemples/ SRDiagnostics/SPS_undulator
28. G4 beam line enclosure in GANIL experimental areas. *Private communication* (Bernard Bru, GANIL, 2002)
29. K. Brown, A first- and second-order matrix theory for the design of beam transport systems and charged particle spectrometers. SLAC Report-75 (1982). https://cds.cern.ch/record/283218/ files/SLAC-75.pdf
30. F. Méot, Zgoubi Users' Guide. https://www.osti.gov/biblio/1062013-zgoubi-users-guide. Sourceforge latest version: https://sourceforge.net/p/zgoubi/code/HEAD/tree/trunk/guide/ Zgoubi.pdf
31. gnuplot_OPTICS.gnu: a gnuplot script which plots optical functions, as read from zgoubi.OPTICS.out.   https://sourceforge.net/p/zgoubi/code/HEAD/tree/trunk/toolbox/ gnuplotFiles/ gnuplot_OPTICS/gnuplot_OPTICS.gnu
32. betaFromPlt: a post-processing Fortran tool to transport betatron functions step-by-step, using stepwise raytracing data logged in zgoubi.plt; outputs are logged in betaFromPlt.out. A gnu-plot_betaFromPlt.gnu script, found therein as well, plots the content of betaFromPlt.out. https:// sourceforge.net/p/zgoubi/code/HEAD/tree/trunk/toolbox/betaFromPlt/
33. gnuplot_Zplt_sYZ.gnu a gnuplot script which plots particle coordinates, etc., as read from zgoubi.plt.   https://sourceforge.net/p/zgoubi/code/HEAD/tree/trunk/toolbox/gnuplotFiles/ gnuplot_Zplt/ gnuplot_Zplt_sYZ.gnu

Chapter 13
Spectrometer; Mass Separator

Abstract This chapter is an introduction to nuclear physics spectrometer and mass separator optical systems. Optical specificities are discussed, leaning on concepts and on theoretical material introduced in earlier chapters, or found in Optical Elements and Keywords (Chap. 14). The chapter begins with a brief overview of mass separation techniques and their use, and continues with a reminder of underlying beam optics principles. The simulation of mass separation requires essentially three optical elements from `zgoubi` library. DIPOLE[S] which allows for curved EFBs and field index; it also allows several, extended shimming pole plates; detector planes, in case they are located in the fringe field region of DIPOLE[S], can be introduced using the option IDRT (all this, inherited from the *ab initio* spectrometer design capabilities of `zgoubi`). QUADRUPOLE (or MULTIPOL) to simulate focusing elements, and DRIFT to simulate field-free sections. Beam monitoring relies on FAISCEAU, FAISTORE, IL = 2 flag, etc. Graphic treatment is performed using zpop or gnuplot, possibly resorted to via a SYSTEM call.

Notations Used in the Text

A	atomic mass
B; B_0	magnetic field; reference
$\mathbf{B}$; B_x, B_y, B_s	field vector; radial, axial, longitudinal components
$B\rho = p/q$	magnetic rigidity
c	velocity of light
D_x, D_x'	dispersion, its derivative $D_x' = dD_x/ds$
EFB	Effective Field Boundary
$\mathbf{F}$	Lorentz force, $\mathbf{F} = dm\mathbf{v}/dt = q(\mathbf{E} + \mathbf{v} \times \mathbf{B})$
f; f_x, f_y	focal distance; radial, axial
M; $M_{x,y}$	magnification; radial, axial
n	radial field index
q	particle charge
x', y'	radial and axial coordinates in the moving frame $\left[(*)' = d(*)/ds \right]$
R	orbit radius in dipole field

© The Author(s) 2024

F. Méot, *Understanding the Physics of Particle Accelerators*, Particle Acceleration
and Detection, https://doi.org/10.1007/978-3-031-59979-8_13

R_{ij}	first order transport coefficients
s	path variable
$\mathbf{v}; v_x, v_y, v_s$	velocity vector; its components in the moving frame
α	trajectory deviation
β	normalized velocity
$\Delta x, \Delta y$	radial, axial excursion off a reference axis
$\delta p; \delta p/p$	momentum offset; relative
$\varepsilon_x, \varepsilon_y$	beam emittances
ε	wedge angle
ρ	local curvature radius

13.1 Introduction

Magnetic spectrometers populate the experimental areas of nuclear physics facilities since the 1960s. They used to be large instruments of nuclear physics, as electron and proton beams used for nuclear reactions in a target were being boosted to GeV range by synchrotrons, linacs, microtrons. They are today the technological cathedrals of the particle physics TeV collider facility, the LHC.

Spectrometers are aimed at momentum analyzes of reaction products from a target. Space and time separation of the reaction products is achieved in an optical assembly. This chapter only addresses the optical part of these systems, other components as target and detectors are out of the scope. Nuclear physics magnetic spectrometry assemblies sometimes include an analyzer section: a magnetic structure similar to, possibly symmetric of the spectrometer optical system proper, whose role is to match the parent beam to the latter, so allowing to present a cleaned, low momentum spread parent beam at the reaction target. It took some time for multipoles to appear in nuclear physics spectrometers, nothing beyond dodecapole whatsoever, and wide aperture bulky magnets anyway for the purpose of geometrical acceptance. Mutipoles were complex components to design in the early times, 1950–1960s, as little was available in the way of computer aided magnet design—several years would still be needed until the inverse problem, from field shape to magnet, would benefit from routinely computing tools.

High resolution mass separators are based on similar optical methods to spectrometers—after all, adding a detector at the focal plane of a mass separator makes it a spectrometer. Their role is to purify beams sent to experiments, which in some cases happen to be spectrometry. High order multipoles, electrostatic or magnetic, are of common usage in mass separators, for the compensation of optical aberrations, with some designs resorting to up to 20-pole lenses.

Mass separators and spectrometers today populate radioactive ion beam (RIB) factories at accelerator facilities worldwide: CERN, FRIB, GANIL, GSI, RIKEN, TRIUMF, ... (Figs. 13.1, 13.2, 13.3, 13.4 and 13.5).

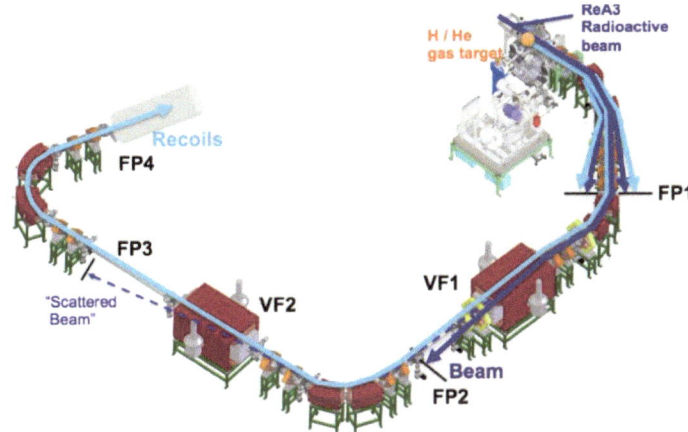

Fig. 13.1 The recoil spectrometer SECAR at MSU NSCL, for astrophysics studies [2]. An upstream mass separator purifies the beam

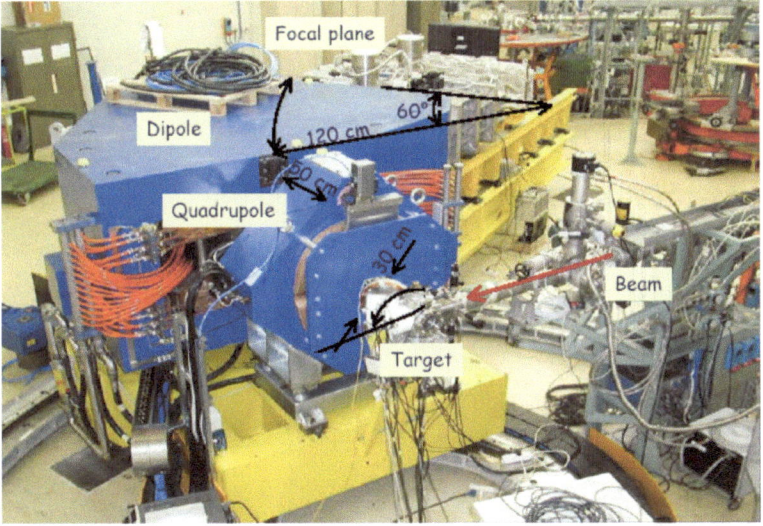

Fig. 13.2 PRISMA heavy ion spectrometer at INFN/LN Legnaro [3]

13.2 Basic Concepts and Formulæ

Mass separation in transverse space uses magnetic fields to discriminate particles by their momentum. Determination of mass and charge of target reaction products is performed concurrently. Separation and discrimination methods include time of flight measurements from hodoscopes, dE/dx energy loss. Beam size control along these optical assemblies resorts to quadrupoles, multipole correctors, filtering,

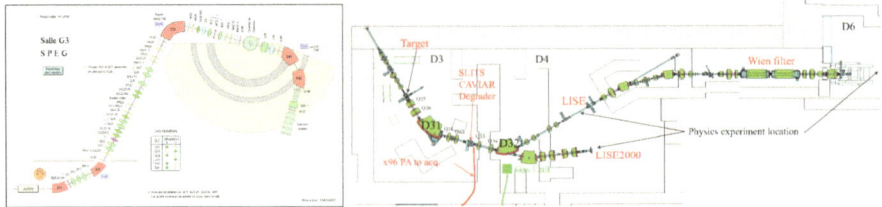

Fig. 13.3 Magnetic spectrometers at GANIL. Left: SPEG, in operation since 1985, shown here with its upstream analysis section. SPEG was designed using zgoubi, by Birien and Saby Valero [4], early developers of the code. Right: LISE [5], used for atomic physics and isotope studies

Fig. 13.4 A 3-spectrometer vertical assembly for cross-section measurements of hadron knock-out and meson production at MAMI microtron facility [6] including the KAOS spectrometer [7]

collimation. Due to their selectivity, incident background in spectrometers, such as beam scattering on collimation systems, neutrons, fission fragments, hard radiation, the nuclear reaction parent beam proper, can in a large measure be removed from the desired beam or detected signal.

Magnetic Spectrometers [1]

Magnetic spectrometers are dispersive optical systems, they separate particles according to their q/A ratio. Separation results from a difference in particle trajectory deflection

Fig. 13.5 HRS mass separator for the DESIR experimental hall at GANIL [8], under commissioning at LP2IB. The red arrow materializes the beam path [9]

$$\alpha = \frac{q}{p} \int_{\mathcal{L}} B_y(s) ds \qquad (13.1)$$

over paths $\mathcal{L}$ through the magnetic assembly.

Spectrometers are often imaging systems, comprised of guiding and focusing fields, designed to provide the image on a focal surface of an object which is the beam-target interaction volume. The 'rays' in this imaging process are the trajectories of the particles of interest.

The dispersion $D_x = \Delta x / \delta p / p$ caused by dipole fields determines the distance Δx between images in the focal plane resulting from a relative difference δp in momentum. A spectrometer is characterized by its resolution

$$R = \Delta x / D_x \qquad (13.2)$$

where Δx is the minimum necessary separation which permits to discriminate images at the focal plane.

Spectrometers are large geometrical and momentum acceptance devices, comprised of large gap, wide radial extent dipoles, featuring extended field fall-offs. Dipole EFBs may be designed with large wedge angles, as well as second and higher order curvatures (Fig. 13.6) in order to mitigate the effect of optical aberrations on the resolution. As a consequence the magnetic field across the dipoles is essentially non uniform. The ability to discriminate particles resides in the accuracy on the field integral along a trajectory, rather than on an accurate local knowledge of the field. As a matter of fact, for a given field integral, a local field inhomogeneity δB_y over a small extent Δs along a trajectory, resulting in a perturbation $\delta B_y \Delta s / B\rho$ of the

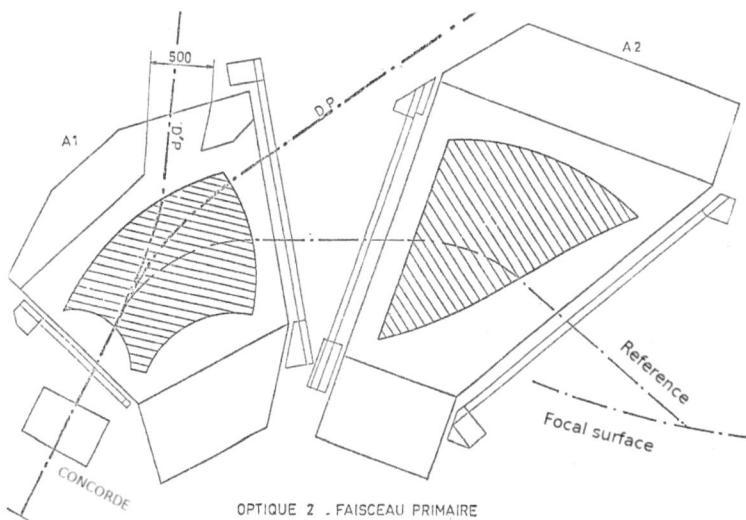

OPTIQUE 2 . FAISCEAU PRIMAIRE

Fig. 13.6 SPES II QDD spectrometer at SATURNE 3 GeV synchrotron [10]. CONCORDE quadrupole, A1 and A2 dipoles are large acceptance magnets. DP, and D'P for reversed field, are escape axes for the primary beam. A central trajectory defines a reference optical axis, from target to focal surface

bend angle, finds its compensation $-\delta B_y \Delta s$ elsewhere along the path. The global effect of local field inhomogeneities is an offset of the image at the focal plane, and does not affect the resolution.

Charged particles traveling in the magnetic fields are subjected to the Lorentz force

$$\mathbf{F} = q\mathbf{v} \times \mathbf{B} \tag{13.3}$$

which determines their trajectories. Solving the problem in terms of paraxial optics in the moving frame (Fig. 1.2), with $\mathbf{v} = (v_x, v_y, v_s)$ the velocity vector,

$$v_x/v \ll v, \quad v_y/v \ll v, \quad v_s \approx v$$

requires defining a reference optical axis, or a set of reference optical axes, in general trajectories at reference momenta properly distributed over the momentum acceptance of the optical system (Fig. 13.6). Focusing and optical aberration properties are determined with respect to the latter, following the transport coefficient technique reminded in Sect. 14.5.2.

The paraxial coordinate transport method, using theoretical modeling of the magnet fields and fringe fields, holds for a resolution of the order of 10^{-3}. For momentum resolution to reach a few 10^{-4} and better, compensation of optical aberrations to high order is critical, this requires raytracing: numerical resolution of the Lorentz force equation, including the use of computed or measured 3D field maps of the magnets.

The fact that the accuracy on field measurements for the goal resolution bears on the field integral, not on the local field, relaxes on fabrication and shimming tolerances, and on the complexity of the magnetic measurement apparatus.

Spectrometer properties

Various parameters characterize the device, including,

- its focal surface: the surface defined by sweeping the image over the momentum acceptance of the spectrometer (Fig. 13.6);
- the optical magnification: $M = $ (image width)/(object width); this applies to both planes, horizontal and vertical;
- dispersion, $D_x = \Delta x / \delta p / p$;
- resolution (Eq. 13.2): the smallest $\delta p / p$ which can be measured; the limit is given by the typical criterion (image width) < (image offset $\Delta x(\delta p)$), so that the ultimate resolution is

$$R \approx \frac{M_x}{D_x} \times \text{(object width)}$$

- solid angle, of the order of $10^{-3} - 10^{-2}$ srd; a larger solid angle increases optical aberrations and makes their compensation more difficult;
- momentum acceptance $\Delta p/p$: it can be up to a few tens of a percent; this determines the radial extent of the focal surface;
- transverse extent of the focal surface, both planes; it determines the volume of the spectrometer detection system;
- transport coefficients (a paraxial approach, Eq. 14.41, Sect. 14.5.2); they characterize various optical properties, for instance:
- R_{11}, R_{33}: M_x and M_y magnifications, respectively;
- R_{12} and R_{34} coefficients: they characterize the focusing; point-to-point focusing if null;
- R_{21}, R_{43}: inverse of the focal distance: $R_{21} = -1/f_x$, $R_{43} = -1/f_y$:
- R_{16}, R_{26}: horizontal dispersion and its derivative; if $R_{26} = 0$, beams with different momenta exit the optical system parallel.

Tailoring $\int B_y(s)\, dl$

The field integral $\int B_y(s)\, dl$ determines the resolution of a spectrometer. Whereas reaching field uniformity beyond $10^{-3} \sim$ a few 10^{-4} in a large dipole is a delicate process, accuracy on the field integral instead can more easily reach 10^{-4} and beyond. Local field fluctuations along the path do not change the field integral, nor the focusing properties.

Diverse methods may be used to tailor $\int B_y(s)\, dl$, and by this means adjust focusing properties and compensate optical aberrations, as follows.

At order zero:

- shimming at dipole ends, or, at a greater scale,
- introducing pole pieces [4, Fig. 6]; they reduce the gap height, so changing $\int B_y(s)\, dl$ at percent level (Fig. 13.7).

Fig. 13.7 Pole pieces reduce
the gap height, so modifying
B_y as a function of x

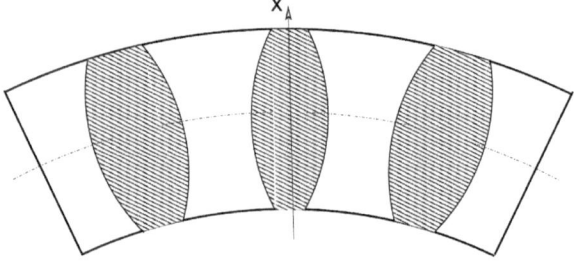

Fig. 13.8 A wedge angle
introduces a quadrupole
effect, $B_y \Delta l =$
$B_y x \tan \varepsilon \propto x$

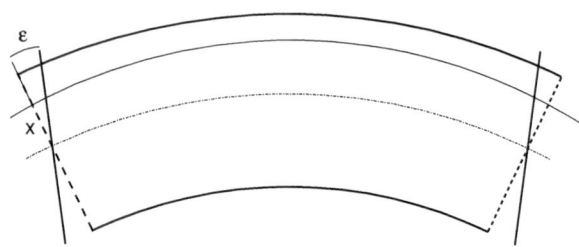

At order 1:

– if the spectrometer includes quadrupoles: for a trajectory at about constant excursion x in a quadrupole, $\Delta \left(\int B_y(s)\,dl \right) \approx \Delta B_y \int dl \approx G x\, L$;
– shaping the gap in dipole entrance and/or exit regions (an hyperbolic gap) for a local field index n (as in Fig. 9.11, Sect. 9.2.1). This introduces a quadrupole effect at magnet ends of the form $B_y = B_0 \left(1 - n\frac{x}{R} \right)$;
– a wedge angle ε at dipole entrance and/or exit EFB (Fig. 13.8): $\Delta \left(\int B_y(s)\,dl \right) = B_0 x \tan \varepsilon$ (more in Sect. 14.4.1).

At order 2 and beyond, compensation of aberrations:

– shaping the gap to introduce sextupole and higher order terms in the field correction, yielding $B_y = B_0 \left(1 - n\frac{x}{R} + \frac{n(n-1)}{2} \frac{x^2}{R^2} + \cdots \right)$;
– as mentioned above, pole pieces modify B_y by modifying the gap (Fig. 13.7); x^k-dependence (k = 1, 2, ...) of the effect also results from shaping their boundaries;
– curving entrance and/or exit dipole EFBs, Fig. 13.9; this allows for sextupole effect (parabolic curvature): $\Delta \left(\int B_y(s)\,dl \right) = bx + cx^2$; octupole effect (cubic curvature): $\Delta \left(\int B_y(s)\,dl \right) = bx + dx^3$; and beyond, cf. FRIB's SECAR spectrometer magnets, they feature 4th degree polynomial EFB profiling (Fig. 13.10).

Trajectory reconstruction

Diverse approaches are possible to determine the angle at which a particle with given momentum has left the target. In any case the transport coefficient method does not allow large enough momentum and angle acceptance. The method used at SPES II for instance, was the following [10, 11].

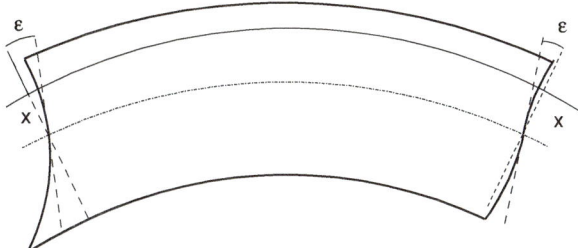

Fig. 13.9 Curved EFBs. Left end of the dipole: wedge and parabolic curvature, $B_y \Delta l = Ax + Cx^2$.
Right end: wedge and cubic curvature, $B_y \Delta l = Dx + Ex^3$

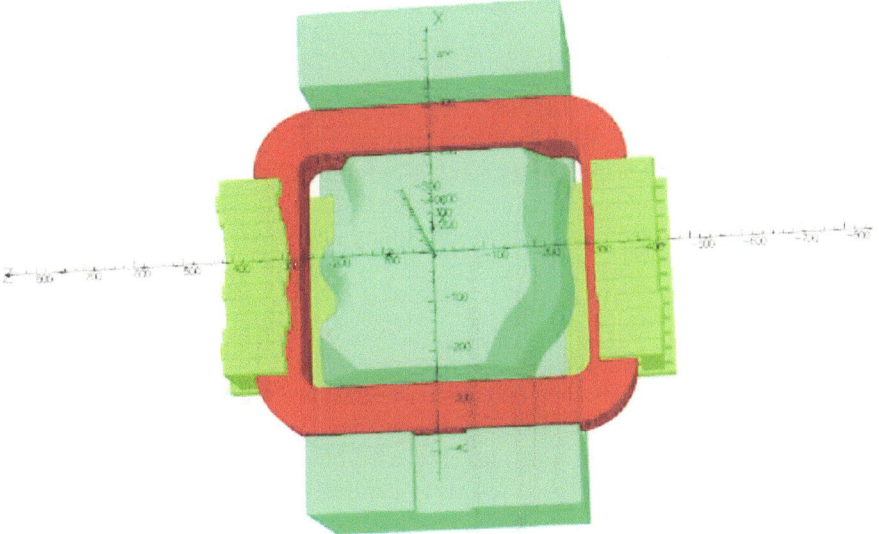

Fig. 13.10 A dipole magnet whose entrance and exit EFBs have been designed using a 4th degree
polynomial modeling [2]

SPES II measured field maps were used. For a trajectory defined by its target
coordinates p, y_t, θ_t, ϕ_t, assuming $x_t \approx 0$ (a "narrow" target), coordinates at the
focal, x_f, y_f, θ_f, ϕ_f can be computed. A polynomial development in the final coordinates recovers the former, to some accuracy, from the detector measurement data.
The coefficients of that interpolation polynomial were determined from trajectory
computation, using some fitting method. On the other hand, particles detected at the
focal have a large dispersion in time of flight and momentum, a consequence of the
large momentum and horizontal acceptance of the spectrometer. With the momentum
p known from the previous parameterization (p is given by the intersection of the
trajectory and the focal surface), and with the trajectory length dependence $L(x_f, \theta_f)$
determined in a similar approach, the time of flight along the trajectory is derived,
from time measurement using hodoscopes, with 1 ns resolution. Determination of

the vertical coordinate y_t from the reconstruction allowed in addition background rejection to some extent.

Mass Separators

Mass separators purify the beam of interest from polluting products. This requires separation of the constituents with a high mass resolution, of 1/20,000 and beyond. The mass resolution can be defined as [8]

$$R = \frac{D_m}{2\sigma_0 M_x + \sigma_A} \tag{13.4}$$

where

$$D_m = \frac{\delta x}{\delta m/m} = \frac{1}{2}\frac{\delta x}{\delta p/p} = \frac{1}{2}D_x \tag{13.5}$$

is the mass dispersion, σ_0 the object width, M_x the horizontal magnification, and σ_A the increase of the image width due to aberrations. For this function, mass separators comprise electrostatic elements, such as lenses for mass independent optics, or Wien filters. Electrostatic elements are introduced in Chap. 2, the Wien filter is introduced in Sect. 12.2.4. Charged particles traveling along these systems are subjected to the force

$$\mathbf{F} = q(\mathbf{E} + \mathbf{v} \times \mathbf{B}) \tag{13.6}$$

which determines their trajectories.

Much of the aforementioned optical constraints, properties, and design methods regarding spectrometer optics hold concerning mass separators. After all, the goal is similar in both causes: separating particles based on their mass, energy or charge.

The design in a first approach resorts to paraxial transport techniques in a similar way to spectrometers, with limited achievement regarding mass resolution. In order for a design to achieve high mass resolution, accurate knowledge of optical aberrations, with both kinematic and field origins, is needed. All field defects matter: inhomogeneities, fringe fields, overlapping fields, ..., and their knowledge is critical. This requires resorting to computed 3D magnetic and electrostatic field maps, and to numerical resolution of the Lorentz force Eq. 13.6.

13.3 Exercises

Note: some of the input data files for these simulations are available in zgoubi sourceforge repository at https://sourceforge.net/p/zgoubi/code/HEAD/tree/branches/exemples/book/zgoubiMaterial/spectrometers/.

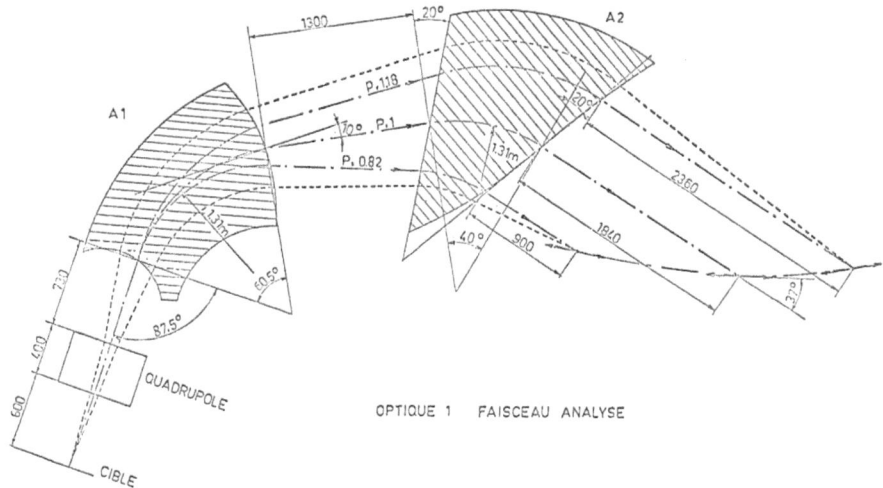

Fig. 13.11 SPES II QDD spectrometer [10]. Optical axes at three different reference momenta are represented, they run from the target on the left ('CIBLE'), along the CONCORDE quadrupole axis, down to the focal surface on the right. A1 and A2 are respectively 45 tons and 60 tons, including coils

13.1 A Nuclear Physics Spectrometer, SPES II
Solution 13.1

SPES II QDD spectrometer (Figs. 13.6 and 13.11) used to be operated at SATURNE 2 (see experimental areas in Fig. 9.3) and CERN PS. Its double-D structure is aimed at allowing the correction of optical aberrations using EFB curvatures. SPES II was well adapted to three-body reactions, as two particles emitted in the forward direction with close to zero degree incidence, with comparable momenta, can be detected simultaneously at the focal surface. Three planes of wire chambers allowed localizing particles exiting the spectrometer.

In this exercise, SPES II is installed in zgoubi. Subsequent numerical simulations reproduce its optical properties, resolution, etc. Analytical field modeling is used here to simulate SPES II quadrupole and dipoles, however similar simulations using field maps (field measurements dating from the early 1970s) are available in zgoubi examples sourceforge repository [12] (see also [13, PART C, Sect. 1]).

Entrance and exit EFBs of SPES II dipoles A1 and A2 (Fig. 13.11) are oriented to contribute focusing, and shaped for compensation of aberrations at the focal surface over the angular acceptance. The optics is such that trajectories at different momenta with null incidence (i.e. traveling along the quadrupole axis) come out of A2 essentially parallel over an extended momentum range.

(a) Install SPES II in zgoubi, using Fig. 13.11 and Table 13.1 data. For the simulation of A1 and A2, use DIPOLE or DIPOLE-M, indifferently (the latter generates a field map—this is transparent to the user—which can be printed out using DIPOLE-

M[IC = 1 or 2]). Set DIPOLE[IL = 2] to get magnetic field along the proton paths logged in zgoubi.plt [13, *cf.* DIPOLE].

Regarding the simulation of A1 and A2 fringe fields: take a fall-off extent commensurate with their 20 cm gap. Enge coefficients given in Sect. 14.3.3 (Eq. 14.12) can be used.

Consider 670 MeV/c as the central momentum (so, the 670 MeV/c trajectory coincides with the reference optical axis, "P.1" in Fig. 13.11).

Check the central trajectory length and exit angle from A2. Fine-tune as needed, including actual dipole field (power A1 and A2 dipoles separately if necessary, in this exercise). IMAGES for instance can be used to localize the waist. Or AUTOREF, for an automatic move of the moving frame to the waist.

Using IL = 2 flag, produce a graph of the magnetic field across A1 and A2, in the momentum regions 670 MeV/c and $\pm 18\%$. Make sure there is no truncation of the field at A1 or A2 boundaries.

Using OBJET[KOBJ = 1] and IMAGES, find the angle of the focal plane with respect to the central reference axis, over $p_{ref} \pm 18\%$. Give a graph of the footprint of the focal surface in the horizontal plane. Repeat, using AUTOREF[IL = 2].

Check $(\partial x'/\partial p/p)$ value at exit of A2.

(b) Use the multi-dimensional grid Monte Carlo object MCOBJET[KOBJ = 2] to create Gaussian beamlets, say 200 particles each, at three momenta, p_0 and $p_0 \pm \delta p$ with $\delta p/p_0 \ll 1$, with some divergence $\Delta \theta \times \Delta \phi$. Produce a graph of the three images at the focal plane, for $p_o = p_{ref} = 670$ MeV/c. Produce the horizontal phase space portrait.

Using this object, find the resolution at the center of the focal surface for an initial beam solid angle $\Delta \theta \times \Delta \phi = \pm 50$ mrad $\times \pm 100$ mrad (take distance between images equal to their half-width, as an image resolution criterion).

Determine the maximum horizontal divergence $\Delta \theta$ and vertical divergence value $\Delta \phi$, proper to ensure 5×10^{-4} resolution at the center of the focal. Repeat at $\pm 18\%$ momentum offsets.

13.2 SPES III Spectrometer
Solution 13.2

SATURNE's SPES III is an Elbeck spectrometer, Fig. 13.12. It was used for the study of kaon and pion rare decay channels. Although a normal conducting dipole, it used to be powered up to saturating field, 3 Tesla, for operation up to, typically, 3.4 T m beam rigidity.

Most SPES III parameters are given in Table 13.2. Question marks have been introduced, they are to be answered as part of the exercise, based on appropriate simulations and their outcomes. Refer to SPES II data Table 13.1 to clarify Table 13.2 questions, as needed.

Regarding the geometrical acceptance, to be determined: horizontally, take an estimate of the radial size of the dipole, from Fig. 13.12; vertically, take a 20 cm dipole gap. Use CHAMBR to reject particles that hit the chamber walls.

Produce isomagnetic field lines of SPES III in a laboratory frame, as in Fig. 13.12. DIPOLE-M[IC = 2] can be used to get a field map from its $B(R, \theta)$ analytical

Table 13.1 SPES II parameters, in complement to Fig. 13.11. The downstream edge of A1 60.5 deg sector is parallel to the upstream edge of A2 40 deg sector, the central optical axis is normal to both; it results that the total deviation by A1 is 2.5 + 60.5 = 63 deg. In A2 the exit EFB combined radii R_1, R_2 and straight sections U_1, U_2 simulate a third degree curvature

Momentum resolution		$\approx 5 \times 10^{-4}$
Momentum range	%	±18
Angle acceptance $\Delta\theta$ (horiz.), $\Delta\phi$ (vertic.)	mrad, mrad	±50, ±100
Solid angle $\Delta\theta \times \Delta\phi$	msrd	20
Horizontal magnification		0.5
Nominal field, dependences		
Nominal field[a]	T	1.7
Central momentum	MeV/c	670
Focal extent	MeV/c	550–790
Angle of focal surface[b]	deg	≈ 37
Focal surface curvature		Small
Non-nominal field, dependences		
Maximum field	T	1.9
Case of 1.8 T field		
– Central momentum	MeV/c	707
– Focal point distance[c] at 800 MeV/c	mm	40
Quadrupole, vertically focusing		
Nominal gradient	T/m	−6.155
Dipole A1[d]		
Central trajectory deviation	deg	60 + 2.5
Entrance EFB wedge angle[e] θ	deg	17.5
Entrance EFB curvature radii $R_1 = R_2$	cm	−57.24
Exit EFB wedge angle θ	deg	−10
Exit EFB curvature radii R_1, R_2	cm	−158, −126
Dipole A2[d]		
Central trajectory deviation	deg	40
Entrance EFB wedge angle θ	deg	20
Exit EFB curvature radii R_1, R_2	cm	−350, +800
Exit EFB straight sections U_1, U_2	cm	−26.5, +20.5

[a] The two dipoles were powered in series. They are individually in this exercise
[b] With respect to central momentum axis
[c] To center of focal surface
[d] Zgoubi notations and sign conventions are used here [13, Fig. 9]
[e] Accounts for the 2.5 deg tilt of A1 *wrt.* incoming reference axis

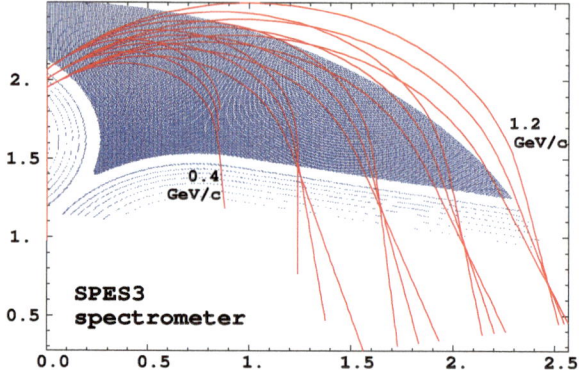

Fig. 13.12 A schematic of SPES III Elbeck spectrometer, and a few groups of momenta converging on the focal surface

Table 13.2 SPES III parameters, in complement to Fig. 13.12

Momentum resolution		?
Momentum range at maximum field		?
Angle acceptance $\Delta\theta$ (horiz.), $\Delta\phi$ (vertic.)	mrad, mrad	?, ?
Horizontal magnification		?
Upstream drift, from target to dipole	cm	77.3627
Dipole data, at 3 T		
Central momentum	MeV/c	?
Focal extent	MeV/c	?
Angle of focal surface[a]	deg	?
Focal surface curvature		?
Dipole data[b]		
Central trajectory radius RM	cm	200
Central trajectory deviation	deg	80
Entrance EFB wedge angle θ	deg	0
Entrance EFB straight segments U_1, U_2	cm	0
Entrance EFB curvature radii $R_1 = R_2$	cm	−65
Exit EFB wedge angle θ	deg	+69
Exit EFB straight segments U_1, U_2	cm	0, ∞
Exit EFB curvature radii R_1, R_2	cm	85, ∞

[a]With respect to central momentum axis
[b]Zgoubi notations and sign conventions used here [13, Fig. 9]

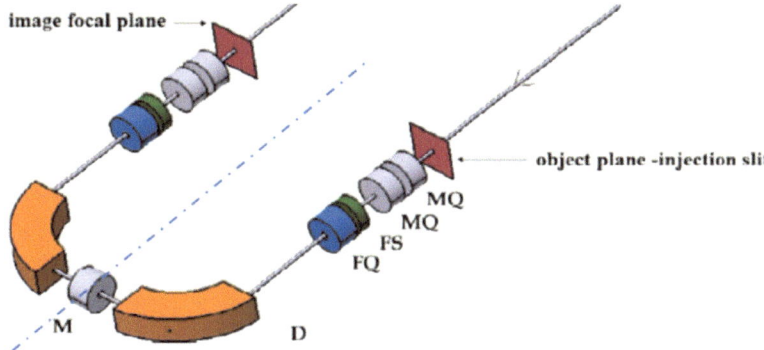

Fig. 13.13 A schematic of LP2IB HRS for DESIR experimental hall at GANIL. The beam line is symmetric with respect to its mid-distance transverse plane, at M

model. An alternate method consists in using OPTIONS[CONSTY] and raytracing an appropriate set of rays in the bend plane. Superimpose pion trajectories on this iso-B schematic.

13.3 A High-Resolution Mass Separator
Solution 13.3

Bordeaux LP2IB[1] high resolution mass separator (Figs. 13.5 and 13.13), a beam delivery system intended for the DESIR experimental hall at GANIL, has been designed for a resolving power $\Delta m/m = 31,000$ for a $1\pi\,\mu$rad m beam emittance. The beam line is comprised of two 90 degree magnetic dipoles, and electrostatic focusing and correction lenses, in a symmetric arrangement to minimize aberrations [8]. In this exercise, the HRS is installed in zgoubi. Ion transport and separation at the focal plane is simulated. These numerical simulations are based on the parameters given in Table 13.3.

The beam line from object to image plane has the configuration

OBJECT – d1 – MQ1 – d2 – MQ2 – d3 – FS1 – d4 – FQ1 – d5 – D – d6 –

M – d6 – D – d5 – FQ2 – d4 – FS2 – d3 – MQ3 – d2 – MQ4 – d1 – IMAGE

In this sequence, d1-6 are drifts, MQ denotes a matching electrostatic quadrupole, FS a sextupole, FQ a focusing quadrupole, both electrostatic, D a 90 deg magnetic dipole with 36 deg wedge angles. M is an electrostatic 20-pole lens aimed at the correction of aberrations, located mid-way between the two dipoles. The line is symmetric with respect to the latter location, for the minimization of aberrations.

(a) Install the HRS 90 deg bending magnet in zgoubi. BEND or DIPOLE can be used, indifferently. Take fringe fields into account (coefficients for the Enge

---

[1] Laboratoire de Physique des 2 Infinis de Bordeaux.

Table 13.3 HRS-DESIR parameters. The Enge coefficients for the bend have been obtained by matching from an OPERA field map of the magnet

Projectile		
Species		$^{132}Sn^+$
Mass	keV	122.95721
Kinetic energy	keV	60
$\beta = v/c$		0.9879×10^{-3}
Beam line length	cm	1004
Drift lengths, d1–d6	cm	42, 10, 55, 6, 95.5, 75
Dipoles[a]		
Bending radius	cm	85
Angle	deg	90
Wedge angles	deg	36
Quadrupoles[b], *length, voltage*[c]		
MQ1, MQ4	cm, V	18.5, –680
MQ2, MQ3	cm, V	18.5, 770
FS	cm, V	11, tbd[d]
FQ	cm, V	22, –860
M	cm, V	30, 0

Enge coefficients for the fringe fields
[a]Dipoles: $C_{0-5} = 0.498959, 1.911289, -1.185953, 1.630554, -1.082657, 0.318111$
fringe field extent $\lambda = 6$ cm $\approx$ gap size
[b]Quadrupoles: $C_{0-5} = 0.296471, 4.533219, -2.270982, 1.068627, -0.036391, 0.022261$
fringe field extent $\lambda = 4$ cm $\approx$ bore diameter
[c]Approximate value. Setting to be determined as part of the exercise
[d]Voltage to be determined as part of the exercise

model (*cf.* Sect. 14.3.3) are given in Table 13.3). Produce its transport matrix. Give a graph of the magnetic field along the reference orbit.

(b) Install the d1-MQ1-d2-MQ2-½d3 doublet section. ELMULT is used to simulate these quadrupoles. Take fringe fields into account (Enge coefficients are given in Table 13.3). Using FIT[2], find the voltage setting for a double focus at 1.165 m downstream of the object; give the transport matrix.

Produce the electrostatic field along $Y = Z = 1$ cm lines across the quadrupoles, and a few $^{132}Sn^+$ trajectories over the 1.165 m distance.

(c) Assemble the upstream half of the line, from objet to middle of multipole M. The FQ1 lens causes the beam to diverge horizontally and converge vertically. The horizontal divergence causes the beam to occupy the entire dipole magnet acceptance and so maximizes mass dispersion. The combined effects of the entrance and exit wedge angles of the dipoles produce a parallel beam in the horizontal direction.

The focusing condition at the mid-plane is point-to-parallel in Y, and, in Z, point-to-point and parallel-to-parallel. Find the proper MQ1, MQ2 and FQ1 voltages for that.

(d) Assemble the complete line, from object to image focal plane. The second (symmetric) half of the separator refocuses the beam at the image focal plane.

Constrain horizontal and vertical magnifications to ± 1 (sign to be determined) using MQ1, MQ2, FQ1 coupled (for equal voltage) with respectively MQ4, MQ3, FQ2.

Find the new MQs and FQs settings. Produce the transport matrices, at the intermediate double-focus, at the middle of the line, and at the image plane.

Check the value of the mass dispersion D_m, compare to expectations.

(e) Raytrace a 6000 particle object comprised of three Gaussian beamlets, centered respectively at $\Delta p/p = 0$ and $\Delta p/p = \pm 5 \times 10^{-4}$. For each beamlet, take *rms* $\delta p/p = 0$, *rms* horizontal width 0.5 mm, and *rms* divergence 2 mrad. Take zero vertical size.

Minimize the aberrations at the image plane, by means of the central multipole, using/combining any of its up to 20-pole components.

Use these simulations to determine the spectrometer resolution (Eq. 13.4) depending on M settings.

(f) Build an animation of the squeeze of the image at the final focus while the M multipole field components are slowly incremented. Use REBELOTE[IOPT = 1] to loop on that slow increment, together with FAISTORE[zgoubi.fai] to log particle data at each increment (gnuplot can be used for the animated graph, reading from zgoubi.fai).

13.4 Solutions of Exercises of This Chapter: Spectrometer; Mass Separator

13.1 A Nuclear Physics Spectrometer, SPES II

(a) Zgoubi input data file for SPES II. Central momentum/optical axis.

CONCORDE quadrupole is simulated using QUADRUPO. Fringe fields are not accounted for here, an arbitrary choice, they could be, yet without major effect expected.

DIPOLE-M and DIPOLE are both used, for respectively A1 and A2. Input data only differ by the former requiring some field map meshing data (RMIN, RMAX, etc.).

A1 and A2 dipole data necessary for DIPOLE[-M] and deduced from Table 13.1 are detailed in Table 13.4, using Zgoubi Users' Guide notation, *cf.* Fig. 14.13, Chap. 14. These data include up- and down-stream negative drift compensation for the AT extra extent beyond 60.5 deg for A1, and beyond 40 deg for A2.

In particular, with $u = 87.5°$ the angle positioning of A1 (2.5 deg from the incoming optical axis, Fig. 13.11), and accounting for the respective AT and ACENT values for A1 and A2, one has

– A1 and A2 entrance angle $T_E = -ACENT + \omega^+$,

Table 13.4 Parameters of SPES II A1 and A2 in zgoubi formalism

		A1	A2
Field	kG	17	17 ($\times 1.0159$)
$\omega^+ - \omega^-$	deg	60.5	40
ω^+	deg	31	20
ω^-	deg	−29.5	−20
AT	deg	107	83
ACENT	deg	50	44
RM	cm	131	131
T_E	deg (rad)	−19 (−0.3316)	−24 (−0.41888)
T_S	deg (rad)	27.5 (0.47997)	19 (0.3316)
Entrance drift compensation	cm	−44.48117	−58.3249
Exit drift compensation	cm	−68.19428	−45.1069
R_E	cm	136.496	138.5483
R_S	cm	147.687	138.5483

- A1 exit angle $T_S = AT - ACENT + \omega^- - 2.5$,
- A2 exit angle $T_S = AT - ACENT + \omega^-$,
- A1 entrance drift compensation amounts to $-RM \sin u \tan(T_E - u)$,
- A2 entrance drift compensation: $-RM \tan T_E$,
- A1 and A2 exit drift compensation: $-RM \tan T_S$,
- A1 entrance radius $R_E = RM \sin u / \sin(ACENT - \omega^+ + u)$,
- A2 entrance radius $R_E = RM / \cos T_E$,
- A1 and A2 exit radius R_S: expected close to $RM / \cos T_S$, eventually fine-tuned so that the radial coordinates (position Y and angle T) of the reference orbit come out null at dipole exit.

The input data file resulting from these settings is given in Table 13.5.

With these geometrical settings, and with a 17 kG field in A1 and A2, together with the initial conditions defined in OBJET[KOBJ = 1], the central momentum (set to $p_{\text{ref}} = 670$ MeV/c, here) exits A2 in the vicinity of, and almost parallel to the reference axis.

Fine-tuning

Ultimately, a fine tuning of the field in A2, by a factor 1.0159 (SCALING factor for DIPOLE family, Table 13.5; DIPOLE-M left unchanged with a SCALING factor of 1.) ensures that the 670 MeV/c trajectory is close to the expected central axis downstream of A2, and parallel to the latter (A1 and A2 can be varied, possibly in series, to do so uncomment FIT, down the data file in Table 13.5). The following excerpt from zgoubi.res shows that this fine tuning yields, at the image location, $Y = -1.993$ cm (distance to the central axis) and $T = 0.001$ mrad (angle with respect to the central axis):

Table 13.5 Simulation input data file SPES2_IMAGES.inc: SPES II spectrometer, compute images at the focal surface . This file also defines the segment SPES2_S to SPES2_E for INCLUDE purposes in subsequent exercises

Note: this file is available in `zgoubi` sourceforge repository at
https://sourceforge.net/p/zgoubi/code/HEAD/tree/branches/exemples/book/zgoubiMaterial/spectrometers/SPES2

```
'MARKER'   SPES2_IMAGES_S
! SPES2_IMAGES.inc file.  670M MeV/c setting.
'OBJET'
2.23487943783 *1e3                    ! 670 MeV/c for proton.
1
1  3  1  1  1  3   ! 3 momentum groups: 670MeV/c & +/-18%;
0. 20. 0. 0. 0.18  ! groups have 3 different horizontal
0. 0. 0. 0. 1.     ! take-off angles: 0 & +/-20mrad.

'MARKER' SPES2_S

'SCALING'          ! Used to change field in magnets, when
1  3               ! reference rigidity OBJET[BORO] is changed.
QUADRUPO
-1
1.
1
DIPOLE-M           ! Scale A1 field.
-1
1.                 ! left unchanged.
1
DIPOLE             ! Scale A2 field.
-1
1.0159  ! Fine-tuning of field brings 670\,MeV/c trajectory
1       ! close to the expected central axis downstream of A2.

'ESL'
60.
'QUADRUPO'  CONCORDE
0
40.  10. -6.1549699
0.  0.
6  .1122 6.2671 -1.4982 3.5882 -2.1209 1.723
0.  0.
6  .1122 6.2671 -1.4982 3.5882 -2.1209 1.723
3.
1 0. 0. 0.
'ESL'
73.
'CHANGREF'
ZR -2.5            ! A1 has a 2.5 deg tilt wrt optical axis.
!----start A1 sector
'ESL'
-44.48117 ! Correction for A1 entrance fringe field extent.
'DIPOLE-M'    A1
2 0 2                           ! NFACE, IC, IL.
700  300
17.  0. 0. 0.
107. 50.  131. 70. 180.    ! AT, ACENT, RM, Rmin, Rmax.
36. -1.
4  .14552 5.21405 -3.38307 14.0629 0. 0. 0.
31.      17.5   -57.24 0.    0. -57.24
36. -1.
4  .14552 5.21405 -3.38307 14.0629 0. 0. 0.
-32 -10.    -158.   0.   0. -126.
0
2                         ! Order of numerical integration.
1.                        ! Integration step size (cm).
2                         ! KPOS.
136.496269599   -0.331612557879  147.687  0.479965544298
'ESL'
```

```
-68.1942836223                    ! =RM*tan(27.5deg).
!----end A1 sector
FAISCEAU             ! Particle coordinates at exit of A1.
'ESL'     ! 24.21
130.    ! Y=0 and T=0 expected here, on central momentum.
'FAISCEAU'
!----start A2 sector
'ESL'
-58.3249577754                    ! -RM*tan(24deg).
'DIPOLE'     A2
2
83. 131.                              ! AT, RM.
44. 17.  0. 0. 0.   ! 15.273   ! ACENT, B0, N, B, G.
36. -1.                            ! Entrance EFB.
4  .14552 5.21405 -3.38307 14.0629 0. 0. 0.
20. 20.    1.E6   -1.E6    1.E6   1.E6
36. -1.                            ! Exit EFB.
4  .14552 5.21405   -3.38307 14.0629 0. 0. 0.
-20. 20.    -350.   -26.5 20.5 800.
0. 0.                  ! Lateral EFB (unused).
0. 0.    0. 0.    0.    0. 0. 0.
0. 0.  0.   0.   0. 0. 0.
2 10  ! Order of numerical integration; flying grid sizing.
1.
2
138.54831  -0.41887902  138.54831 0.331612557879
'ESL'
-45.1069173409                    ! -RM*tan(19deg).
!----end A2 sector
FAISCEAU             ! Particle coordinates at exit of A2.
'ESL'   ! Expected distance from A2 EFB to focal surface,
184.         ! along central momentum reference axis.
'MARKER' SPES2_E
'FAISCEAU'  ! Coordinates, in plane normal to central mom.

!'FIT2'   ! Uncomment for adjustment of A1 and/or A2 field.
! 2
! 4 8  0.2
! 4 12  0.2
! 2 1e-7
! 3 1 2 #End 0. 1. 0
! 3 1 3 #End 0. 1. 0

'IMAGES'          ! Find location (X,Y,S) of images. There are
!     as many as momentum groups defined by OBJET[KOBJ=1].
'MARKER'   SPES2_IMAGES_E
'END'
```

```
22 Keyword, label(s) : IMAGES
RECHERCHE DU POINT DE FOCALISATION HORIZONTAL DE       3 TRAJECTOIRES (SUR     3)
POINT DE FOCALISATION HORIZONTAL SUR L ORBITE MOYENNE X= -3.71452E+00 CM   Y= -1.9929E+00 CM   ATAN(Y/X)= 2.821E+01 DEG
DECALAGE DU CENTRE DE GRAVITE EN Y =   6.274219E-02 CM
LARGEUR IMAGE, A MI-HAUTEUR =  0.104271        CM,  TOTALE =  9.493492E-02 CM
                                      ........         ........
                                    TRACE DU FAISCEAU
                                    3 TRAJECTOIRES
                                  OBJET                                          FAISCEAU
             D      Y(cm)   T(mr)   Z(cm)   P(mr)   S(cm)    D-1    Y(cm)   T(mr)   Z(cm)   P(mr)    S(cm)
O  1  1.0000  0.000   0.000   0.000   0.000   0.0000  0.0000  -1.993   0.001  0.000  0.000  7.139608E+02   1
A  1  1.0000  0.000  20.000   0.000   0.000   0.0000  0.0000  -2.078 -48.459  0.000  0.000  7.292358E+02   2
B  1  1.0000  0.000 -20.000   0.000   0.000   0.0000  0.0000  -1.710  51.063  0.000  0.000  6.991264E+02   3
```

This fine-tuning yields the following coordinates in the interval between A1 and A2 (an excerpt from zgoubi.res):

– under FAISCEAU at exit of A1:

D-1	Y(cm)	T(mr)	Z(cm)	P(mr)	S(cm)
0.0000	3.649	-4.023	0.000	0.000	3.118677E+02

– under FAISCEAU at entrance of A2:

D-1	Y(cm)	T(mr)	Z(cm)	P(mr)	S(cm)
0.0000	3.3604	-4.023	0.000	0.000	3.8354334E+02

These data show a residual radial offset: $Y \approx 3.5$ cm, a residual incidence: $T \approx -4$ mrad, with respect to the theoretical reference optical axis between A1 and A2. A small enough difference to be ignored (the objective of the exercise is not a fine analysis of a three-body rare decay reaction!).

Finally, from IMAGES one also gets a distance of 7.139608 m from the target (at OBJET) to the focal plane, whereas Fig. 13.11 for the 670 MeV/c reference shows a path length of

$$600 + 400 + 730 + 1310 \, (60.5 + 40) \frac{\pi}{180} + 1300 + 1840 = 7167.8 \, \text{mm}$$

The agreement between both distances is at $< 0.5\%$ level.

Distances from target to focal image are summarized in Table 13.6, based on further raytracing results detailed below. A greater relative difference is observed for the lower momentum, this may be an effect of fringe fields: lower momenta travel a relatively longer distance in fringe field regions compared to higher momenta, Fig. 13.14.

Table 13.6 Distance from target to image (in mm), from raytracing, first row, and from Fig. 13.11 for comparison, second row

Momentum	−18%	670 MeV/c	+18%
Raytracing	6787	7140	7515
Figure 13.11	6228	7168	7678

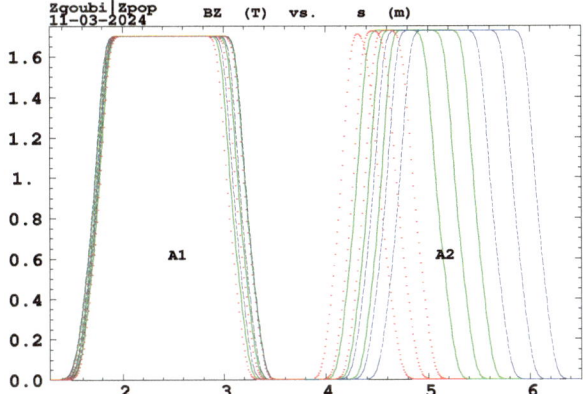

Fig. 13.14 Magnetic field across A1 and A2 dipoles, along trajectories at momenta $p_{ref} = 670$ MeV/c and $p_{ref} \pm 18\%$ (respectively green/solid curve, red/dotted, blue/dashed), three horizontal take-off angles from target in each case: T = 0 and ± 20 mrad. Lower momenta experience a strong field inhomogeneity through A2. A graph obtained using zpop, menu 7: 1/1 to open zgoubi.plt; 2/[6, 32] for B_Z versus s; 7 to plot

Field along trajectories

In OBJET[KOBJ = 1], Table 13.5, add a few groups of momenta at $\pm 18\%$, like so:

```
'OBJET'
2.23487943783 *1e3                            ! 670 MeV/c for proton.
1
1   3   1   1   1   3      ! 3 groups of momenta, at 0 and +/-18% from 670MeV/c central momentum.
0.  20.  .0  0.  0.  0.18           ! Each group has 3 different take-off horizontal angles, 0 and
0.  0.  0.  0.  0.  1.                ! +/-20mrad so to allow finding the focus region by IMAGES.
```

Set IL = 2 under DIPOLE[–M]. A graph of the stepwise field data so logged in zgoubi.plt shows reasonable $B_Z(s)$ across the dipole doublet, Fig. 13.14. No truncation of the field at A1 or A2 boundaries is observed within an horizontal beam opening of ± 20 mrad, in the momentum regions $p_{ref} = 670$ MeV/c and $p_{ref} \pm 18\%$.

Tilt angle of the focal surface, using IMAGES

Using OBJET[KOBJ = 1], IMAGES will recognize the momentum groups and will produce the location of their respective images.

Information concerning each momentum group includes the radial distance of its image, Y coordinate, and distance of the latter to the plane normal to the reference, X coordinate. The horizontal angle under FAISCEAU, T, of trajectory number 1 is the angle of the momentum of concern, to the central optical axis. For instance:

– case D = 1.18:

```
RECHERCHE DU POINT DE FOCALISATION HORIZONTAL DE      3 TRAJECTOIRES (SUR      3)
POINT DE FOCALISATION HORIZONTAL SUR L ORBITE MOYENNE X=  6.2595E+01 CM  Y= 5.09007E+01 CM ATAN(Y/X) =  3.912E+01 DEG
DECALAGE DU CENTRE DE GRAVITE EN Y =   3.670310E-02 CM
LARGEUR IMAGE, A MI-HAUTEUR = 6.098960E-02 CM,  TOTALE =  5.508049E-02 CM
                                   ..........                 ..........
                                          TRACE DU FAISCEAU
                                          3 TRAJECTOIRES
                               OBJET                                      FAISCEAU
         D       Y(cm)    T(mr)    Z(cm)    P(mr)    S(cm)    D-1     Y(cm)    T(mr)    Z(cm)    P(mr)    S(cm)
O 1  1.1800    0.000    0.000    0.000    0.000    0.0000   0.1800   51.058   -2.518    0.000    0.000  7.514704E+02
A 1  1.1800    0.000   20.000    0.000    0.000    0.0000   0.1800   54.410  -55.135    0.000    0.000  7.689737E+02
B 1  1.1800    0.000  -20.000    0.000    0.000    0.0000   0.1800   47.826   49.965    0.000    0.000  7.346724E+02
```

– case D = 0.82:

```
22  Keyword, label(s) :  IMAGES                                                                                          IPASS= 1
RECHERCHE DU POINT DE FOCALISATION HORIZONTAL DE        3 TRAJECTOIRES (SUR      3)
POINT DE FOCALISATION HORIZONTAL SUR L ORBITE MOYENNE X= -1.1143E+02 CM  Y= -5.55146E+01 CM ATAN(Y/X)=  2.648E+01 DEG
DECALAGE DU CENTRE DE GRAVITE EN Y =  -7.185363E-03 CM
LARGEUR IMAGE,  A MI-HAUTEUR =  1.194944E-02 CM,  TOTALE =  1.102654E-02 CM
                                     .........            .........
                                            TRACE DU FAISCEAU
                                            3 TRAJECTOIRES
                               OBJET                                              FAISCEAU
          D      Y(cm)     T(mr)     Z(cm)     P(mr)     S(cm)     D-1     Y(cm)     T(mr)     Z(cm)     P(mr)     S(cm)
O 1  0.8200    0.000     0.000     0.000     0.000   0.0000   -0.1800  -52.421    27.757    0.000    0.000   6.787251E+02
A 1  0.8200    0.000    20.000     0.000     0.000   0.0000   -0.1800  -58.842   -29.753    0.000    0.000   6.923017E+02
B 1  0.8200    0.000   -20.000     0.000     0.000   0.0000   -0.1800  -45.069    93.564    0.000    0.000   6.657433E+02
```

In the following OBJET, the number of groups is increased to 61 in order to get a dense set of (X, Y) image location coordinates:

```
'OBJET'
2.23487943783 *1e3                                                          ! 670 MeV/c for proton.
1
1    3    1   1   1    61      ! 61 groups of momenta, from 0 to +/-18% of 670MeV/c central momentum.
0.  20.  .0   0.  0.  0.6E-2   ! Each group has 3 different take-off horizontal angles, 0 and
0.   0.   0.  0.  0.  1.       ! +/-20mrad so to allow finding the focus region by IMAGES.
```

Running the input data file of Table 13.5 with these 61 momentum groups produces the plot of Fig. 13.15 (an effect of the SYSTEM call to gnuplot, bottom of the graph).

Using AUTOREF

AUTOREF[I = 2] may be used for similar results, as it causes a positioning of the moving frame at the location of the waist, with its longitudinal axis aligned on the velocity vector of particle 1 [13, lookup INDEX, AUTOREF]. The simulation data file of Table 13.5 is used, changing to OBJET[KOBJ = 5] and adding AUTOREF at the end. MATRIX is also added following the latter, to get the R_{26} first order transport coefficients. Three different momenta are considered: $p_{ref} = 670 \, \text{MeV/c}$ and $p_{ref} \pm 18\%$, The file is run three times (REBELOTE[IOPT = 1] could be used instead, to repeat over a larger momentum set) with reference momentum in OBJET[KOBJ = 5] consecutively taken as 1, 1.18 and 0.82 respectively. AUTOREF and MATRIX outcomes are as follows (excerpts):

● Momentum $p_{ref} + 18\%$.

```
19  Keyword, label(s) :  AUTOREF
Change  of  reference,  horizontal,    XC =    62.82642826 cm , YC =    50.90018683 cm ,    A =   -0.144265986 deg
TRAJ 1 IEX,D,Y,T,Z,P,S,time :  1   1.180      1.0270E-15   0.000        0.000      0.000       814.30

                TRANSFER  MATRIX  ORDRE   1  (MKSA units)
         -0.355969      -1.573645E-06     0.00000          0.00000          0.00000          3.26950
         -1.68097       -2.62716          0.00000          0.00000          0.00000         -8.186927E-02
```

● Momentum $p_{ref} = 670 \, \text{MeV/c}$

```
20  Keyword, label(s) :  AUTOREF
Change  of  reference,  horizontal,    XC =    -3.63959164 cm , YC =    -1.99291260 cm ,    A =    0.000050933 deg
TRAJ 1 IEX,D,Y,T,Z,P,S,time :  1   1.000      7.6575E-17   0.000        0.000      0.000       710.32

   21  Keyword, label(s) :  MATRIX
Reference particle (#       1), path length :    710.32118      cm  relative momentum :    1.00000

                TRANSFER  MATRIX  ORDRE   1  (MKSA units)
         -0.403897       9.457240E-08     0.00000          0.00000          0.00000          3.02912
         -1.72732       -2.48611          0.00000          0.00000          0.00000          1.529908E-03
```

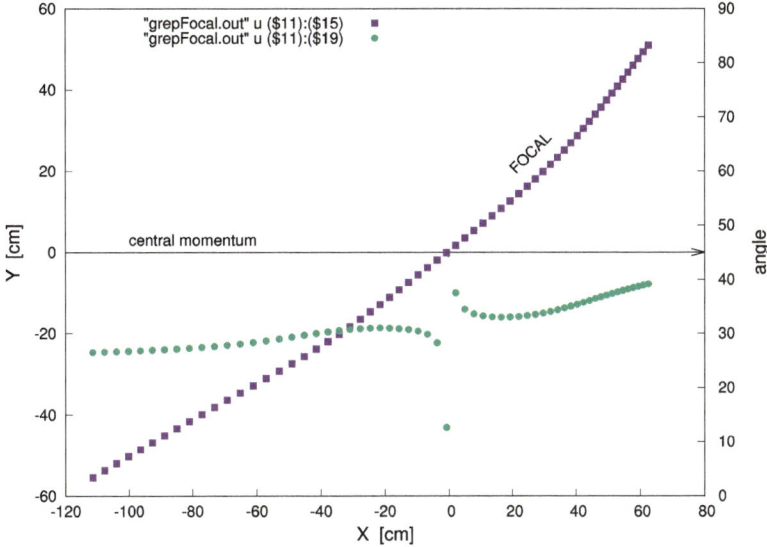

A gnuplot script to produce Fig. 13.15, reading data from zgoubi.res execution listing:

```
r2d = 180/(4.*atan(1.))
system("grep 'POINT DE FOCALISATION HORIZONTAL SUR' zgoubi.res | cat > grepFocal.out")
set y2range [0:90]; set arrow 1 from -120, 0 to 80, 0; set label "central momentum" at -100, 3
set label "FOCAL" at 20, 20 rotate by 45
plot "grepFocal.out" u ($11):($15) w p pt 5 , "grepFocal.out" u ($11):($19) axes x1y2 w p pt 7
```

Fig. 13.15 Footprint of the focal surface in the (X, Y) bend plane (squares), and its angle to the central momentum axis (circles, right vertical axis)—the discontinuity in the region $X \approx 0$ is due to both X and $Y \rightarrow 0$. The gnuplot file below is used to produce this graph, taking data from zgoubi.res

• Momentum $p_{ref} - 18\%$.

```
   19 Keyword, label(s) :  AUTOREF
Change  of  reference,  horizontal,   XC =  -110.69095638 cm ,  YC =  -55.49409770 cm ,   A =    1.590332936 deg
TRAJ 1 IEX,D,Y,T,Z,P,S,time :  1  0.8200    2.2204E-15  0.000     0.000     0.000        567.99


   20 Keyword, label(s) :  MATRIX
Reference particle (#    1), path length :    567.99147    cm  relative momentum :    0.820000

             TRANSFER  MATRIX  ORDRE  1  (MKSA units)
     -0.334343      1.246995E-07  0.00000      0.00000      0.00000      2.20578
     -2.27095      -3.06684       0.00000      0.00000      0.00000     -0.442606
```

This tells that
– the reference trajectories at the two momenta $p_{ref} = 670 \pm 18\%$ MeV respectively, are at an angle of -0.144265986 deg and 1.590332936 deg to the $p_{ref} = 670$ MeV/c central trajectory. This is a small angle, and $R_{26} = \partial x'/(\partial p/p) \approx 0$, small at all three momenta correlatively, as expected. This confirms the expected property of parallel exit trajectories over $\pm 18\%$ momentum span at the focal plane;

Table 13.7 Simulation input data file: SPES II spectrometer. Images at the focal surface

```
SPES2  670M MeV/c setting.  Compute matrices to get D_x' after A2.
'OBJET'
2.23487943783 *1e3                                          | 670 MeV/c for proton.
5.20                          | Generate 20 groups of 13 particles, ready for use by MATRIX.
.001 .01 .001 .01 .001 .0001
0.  0.  0.  0.  0.  1.18    ! Obviously, a little Fortran coding in objet.f would easily allow to
0.  0.  0.  0.  0.  1.16               ! reduce these 20 lines to a single one ...
........ step 0.02 decrement                  | the exercise is left to the reader.
0.  0.  0.  0.  0.  0.82
0.  0.  0.  0.  0.  0.80

'OPTIONS'
1 1                          ! It is not a bad idea to log stepwise data in zgoubi.plt, to allow
.plt 2                                        | checking fields along paths.

'INCLUDE'
1
SPES2_IMAGES.inc[SPES2_S:SPES2_E]     ! Grab appropriate segment from earlier SPES2_IMAGES.inc file.

'FAISCEAU'                   | Final coordinates, in a plane normal to the central momentum.
'MATRIX'                     ! As a result of OBJET[KOBJ=5.N], N=20, MATRIX computes 20 matrices.
1 1
'SYSTEM'
1
gnuplot <./gnuplot_Zres_R26.gnu

'END'
```

A gnuplot script which 'greps' MATRIX data from zgoubi.res, to obtain Fig. 13.16:

```
# gnuplot_Zres_R26.gnu
system("grep -A 3 'TRANSFER MATRIX ORDRE' zgoubi.res | cat > gnuplot_temp")
set xlabel "p/p_{ref} " ; set ylabel "R_{16} [m]"
plot 'gnuplot_temp' every 5::1 u ((1.18-$0*0.02)):6 w lp pt 5
```

– ignoring the curvature and other deformation of the focal surface, its angle to the central ray is near

$$\text{atan}\frac{YC}{XC} = \text{atan}\frac{54.49 - 2}{110.6 - 3.6} \approx 26° \quad \text{and atan}\frac{YC}{XC} = \text{atan}\frac{50.9 + 2}{62.8 + 3.6} \approx 38°$$

on respectively the low momentum and high momentum side of the central axis, about as expected from the first method, Fig. 13.15. A momentum scan may be performed in a similar way, using REBELOTE[IOPT = 1] to iterate through the spectrometer, changing the D coordinate (relative rigidity) in OBJET at each iteration.

$(\partial x'/\partial p/p)$ *value at exit of A2.*

$R_{26} = (\partial x'/\partial p/p)$ is expected to be zero, or close, over a large momentum range. This is the condition for different momenta to come out about parallel. The previous simulation provides the answer.

Out of curiosity, a different approach for a similar outcome is given in Table 13.7. That simulation computes a series of transport matrices over a 670 MeV/c ± 18% range, using OBJET[5.N] (N = 20, here) concurrently with MATRIX. Each one of the N = 20 matrices is computed with reference the trajectory of the first particle in each 13 particle set. That particle leaves OBJET with null transverse coordinates, which is what is required here.

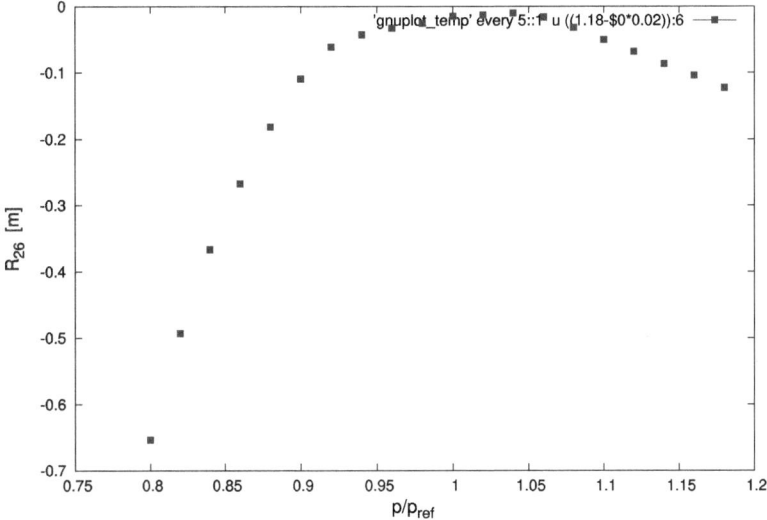

Fig. 13.16 Evolution of SPES II R_{26} (from target to focal surface) over $p = 670$ MeV/c $\pm 18\%$

The transport matrix can be found in zgoubi.res, for each one of the 5 momenta considered, $p_{ref} = 670$ MeV/c, $p_{ref} \pm 9\%$, $p_{ref} \pm 18\%$. (another possibility is to have the transport coefficients R_{ij} logged to zgoubi.MATRIX.out, using MATRIX [PRINT]). The rightmost term in the second line of a matrix is the dispersion derivative $R_{26} = D'_x$. A gnuplot script given in Table 13.7 plots $R_{26}(p/p_{ref})$, Fig. 13.16.

(b) Spectrometer resolution.

The simulation data file of Table 13.5 can be used and modified/completed as follows:

– to OBJET[KOBJ = 1] substitute MCOBJET[KOBJ = 2], formatted like so:

```
'MCOBJET' 2335.
! Reference rigidity. 2
!   Distribution on a grid. 10000
!   Numger of particles. 1   1   1   1   1   1
!   Uniform distributions, 0.  0.  0.  0.  0.  1.
!   Central values of bins. 1   1   1   1   1   5
! Number of bins in momentum. 0.  0.  0.  0.  0.  .001
!   Relative spacing (Brho/BORO) between momentum bins. 0.  50.e-3
0.  50.e-3 0.  0.
!   Width of bins. 1.  1.  1.  1.  1.  1.
!   Sorting cut-offs (unused). 9   9. 9. 9. 9.
!   For p(D) (unused). 186387 548728 472874
!   Seeds.
```

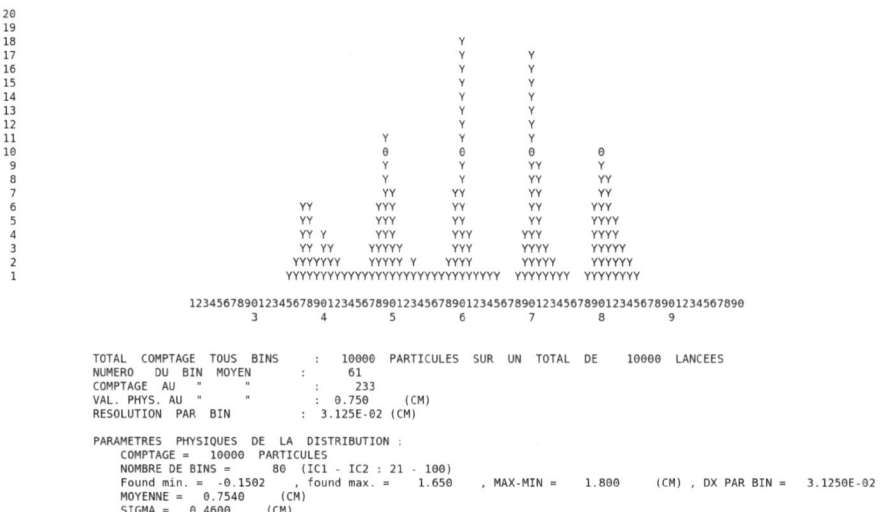

Fig. 13.17 HISTO outcome, in zgoubi.res execution listing. A histogram of particle momentum at the focal surface [13, PART C, Sect.1] [12]

– move the observation location at the image, by introducing FOCALE[XL = distance to image] prior to HISTO;
– add HISTO keywords to get histograms of the coordinates, at that location, printed out in zgoubi.res (Fig. 13.17).

Histograms may also be obtained using zpop, or some other graphic tool, reading final coordinates (at the focal surface) from zgoubi.fai. The latter is generated by adding FAISTORE[FNAME = zgoubi.fai] at the bottom of the input data file.

From these simulations, with MCOBJET[KOBJ = 2] above completed considering SPES II geometrical acceptance data (from Table 13.1), and assuming some image separation criterion such as distance between two images ≥ image width, the spectrometer resolution at the focal surface can be deduced. The expected result of this simulation is $\Delta p/p \approx 10^{-3} \sim 5 \times 10^{-4}$ over the ±18% momentum range.

For additional details, please refer to Zgoubi Users' Guide [13, PART C, Sect. 1].

13.2   SPES III Spectrometer

Zgoubi input data file for SPES III is given in Table 13.8. DIPOLE-M is used, it first fabricates a field map through which Zgoubi then pushes the pions. DIPOLE-M[IC = 2] logs that field map in zgoubi.map, which zpop can read to generate various magnetic field graphs, including isofield lines.

Zpop both computes and plots the isomagnetic field lines. It allows superimposing trajectories read from zgoubi.plt, outcomes are displayed in Fig. 13.18.

Filling up Table 13.2 with the missing data is left to the reader—the previous SPES II exercise can be used as a guidance.

Table 13.8 SPES III simulation input data file. In-flight pion decay can be triggered if desired, by uncommenting PARTICUL and adding MCDESINT[M@=muon mass] command. If so, muons from the decay will be tracked as well

```
SIMULATION OF PION TRACKS
'OBJET'
3360.                                           ! Reference rigidity BORO (kG.cm, pion).
1
1    3    1    1    1   5                                    ! 3*5 sample particles,
0.   50.   0.   0.    0.   0.2     ! at 3 different angles: 200, 200+/-50 mrad and 5 different
0.   200.  0.   0.    0.   .8      ! rigidities: 0.8*BORO, (0.8+/-0.20)*BORO, (0.8+/-0.40)*BORO.
!'PARTICUL'                        ! Pion mass and life time. Necessary if MCDESINT is used (in-flight decay),
!139.6000 0. 0. 26.03E-9 0.                                 ! useless/commented here.
'ESL'
77.3627
'CHAMBR'        ! Stops particles out of transverse limits (usefull if in-flight decay is simulated).
1
1   100. 10. 245.  0.
'DIPOLE-M'
2 2 2 ! NFACE=2 EFBs defined. IC=2 logs fieldmap in zgoubi.map; IL=2 logs prticle data in zgoubi.plt.
400  200                                  ! Azimuthal and radial number of nodes of the mesh.
30    0. 0. 0.                                            ! Field and field indices.
80. 18. 200.    140. 320.             ! AT, ACENT, RM, and field map limits RMIN, RMAX.
46. -1.                                          ! Entrance fringe field extent; gap shape.
4 .14552 5.21405 -3.38307 14.0629 0. 0. 0.                       ! Enge coefficients.
 15.   0.   -65. 0.    0.   -65.        ! omega+, wedge angle, , R1, U1 , U2 , R2.
46. -1.                                             ! Exit fringe field extent; gap shape.
4 .14552 5.21405 -3.38307 14.0629 0. 0. 0.                       ! Enge coefficients.
-15. 69.   85. 0.    1.E6 1.E6        ! omega-, wedge angle, , R1, U1 , U2 , R2.
0
2                                       ! Order of polynomial interpolation.
4.                                          ! Integration step size.
2                                       ! Positioning of downstream frame.
180.755 .479966 233.554 -.057963
'FAISCEAU'                                                 ! Pion coordinates, here.
'END'
```

Fig. 13.18 Isomagnetic fields lines, and pion tracks at five different momenta, using DIPOLE-M. A graph obtained using zpop, menu 7: 1/1 to open zgoubi.plt; 2/[48, 42] for Y(X), lab frame coordinates; 7 to plot. Superimpose field lines using menu 8, opening zgoubi.map as produced by DIPOLE-M[IC = 2]

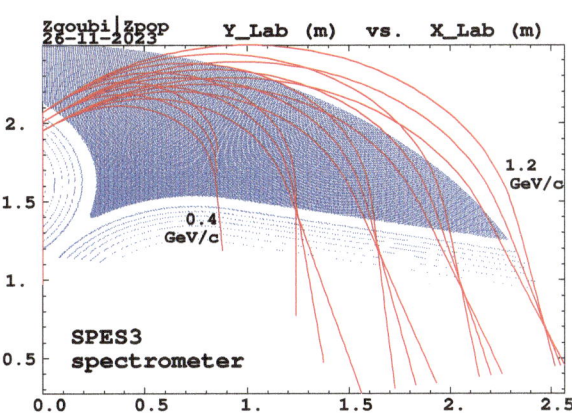

For additional details, including histograms of pion and decay muon beams at the focal plane, please refer to Zgoubi Users' Guide [13, PART C, Sect. 3].

13.3 A High-Resolution Mass Separator

(a) The input data file for the simulation of the 90 deg bend is given in Table 13.9, resulting from the parameters given in Table 13.3.

Transport matrix

Transport matrix for BEND, an excerpt from zgoubi.res:

Table 13.9 Simulation input data file: HRS 90 deg bend using BEND (top), or using DIPOLE (bottom).   The two problems are stacked, zgoubi allows that. This sequence defines the segments HRS-BEND_S:HRS-BEND_E and HRS-DIPOLE_S:HRS-DIPOLE_E for the purpose of INCLUDEs in subsequent exercises

Note: this file is available in zgoubi sourceforge repository at
https://sourceforge.net/p/zgoubi/code/HEAD/tree/branches/exemples/book/zgoubiMaterial/spectrometers/HRS-DESIR_massSeparator/

```
HRSDESIR  U180_V4. Bending dipole. HRS-BEND+DIPOLE.inc
'MARKER' HRS-BEND+DIPOLE_S                                    ! Just for edition purposes.
'OBJET'
405.179                                                                    ! Sn132+, 60 KeV.
5                                               ! Generate a 13 particle set, ready for use by MATRIX.
.01 .001 .01 .001 .0 .0001
0.  0.  0.  0.  0.  1.
'MARKER' HRS-BEND_S
'BEND'
0           ! BEND is straight magnet, theor. length=2*rho*sin(theta/2)=120.2081528 cm. Yet, length
120.41050  0.  4.76681176471          ! has to be part of YCE FIT so to preserve arc length=RM*pi/2.
20. 6. 0.62831853 ! Integration extent 20cm, fringe fall-off ~6 cm. Entrance EFB wedge angle 36 deg.
6  0.498959 1.911289 -1.185953 1.630554 -1.082657 0.318111
20. 6. 0.62831853     ! Integration extent20 cm, fringe fall-off ~6 cm. Exit EFB wedge angle 36 deg.
6  0.498959 1.911289 -1.185953 1.630554 -1.082657 0.318111
.4
3  0. 0.25258653 -0.7853981633
'MARKER'  HRS-BEND_E
'FIT2'   ! BEND requires matching YCE because of fringe fields as they cause deviation before EFBs.
2                                                                     ! 1 variables:
4 10 0 .1                                         ! magnet length, +/-10% allowed;
4 72 0 [-2.,2.]                                                          ! YCE.
3                                                                     ! 3 constraints:
3.1 1 2 #End 0. 1. 0                             ! Particle 1, final Y = Y0 at OBJET;
3.1 1 3 #End 0. 1. 0                                  ! final T = T0 at OBJET.
3   1 6 #End 133.517687778 1. 0                      ! Path length =RM*pi/2.
'MATRIX'                                    ! Compute transport matrix, from OBJET down to here.
1  0
'MARKER' HRS-BEND+DIPOLE_mid                                  ! Just for edition purposes.

! Using DIPOLE
'OBJET'
405.179                                                                    ! Sn132+, 60 KeV.
5
0.001 .001 0.001 .001 .001 .0001
0. 0. 0. 0. 0. 1.
'MARKER'  HRS-DIPOLE_S
'DRIFT'      ! This drift, first set to 20cm (hrd-edge hypothesis), compensates for extra DIPOLE angle
-20.282644        ! AT=90; it has to be part of the RE, RS FIT, to maintain s=RM*pi/2=133.517687778cm.
'DIPOLE'
0
116.48103048    85.                                     ! AT = 90 deg + 2*atan(20cm/RM); RM.
58.240519915 4.76681176471 0. 0. 0.                  ! ACNT=omega+ + atg(20cm/RM); B0; indices all null.
6. -1.                                          ! Fringe fall-off ~6 cm; no gap height dependence.
6     0.498959 1.911289 -1.185953 1.630554 -1.082657 0.318111 0.
45.   36.    1.E6  -1.E6  1.E6  1.E6                   ! Entrance EFB wedge angle 36 deg
6. -1.                                          ! Fringe fall-off ~6 cm; no gap height dependence.
6     0.498959 1.911289 -1.185953 1.630554 -1.082657 0.318111 0.
-45.  -36.   1.E6  -1.E6  1.E6  1.E6                         ! Exit EFB wedge angle 36 deg.
0. 0.
0 0.    0.   0.    0.    0. 0. 0.
0. 0.   0.   0.   0. 0. 0.
2  20.
.2
2 87.675288  -0.231090667196 87.675288  0.231090667196
'DRIFT'  ! This drift, first set to 20cm (hard-edge hypothesis), compensates for extra DIPOLE angle
-20.282644         ! AT-90; it has to be part of the RE, RS FIT, to maintain s=RM*pi/2=133.517687778cm.
'MARKER'  HRS-DIPOLE_E
'FIT2'   ! DIPOLE requires matching RE=RS due to fringe fields as they cause deviation beyond EFBs.
2                                                       ! 2 variables, coupled: RE & RS,
5 64 -5.66 .2                                               ! allowed range is 20%.
4 1  -6.1  .1                              ! Couples the two drifts, allowed variation range is 10%.
3  1e-15                                             ! 3 constraints, penalty 1e-15:
3.1 1 2 #End 0. 1. 0                             ! Particle 1, final Y = Y0 at OBJET;
3.1 1 3 #End 0. 1. 0                                   ! final T = T0 at OBJET.
3   1 6 #End 133.517687778 1. 0                       ! Path length =RM*pi/2.
'MATRIX'                                      ! Compute transport matrix, from OBJET down to here.
1  0
'MARKER' HRS-BEND+DIPOLE_E                                    ! Just for edition purposes.
'END'
```

```
Reference particle (#    1), path length :   133.51770    cm  relative momentum :   1.00000

          TRANSFER  MATRIX  ORDRE  1  (MKSA units)
     0.727662        0.847070        0.00000        0.00000        0.00000       0.846136
    -0.555454        0.727662        0.00000        0.00000        0.00000       1.72576
     0.00000         0.00000        -4.838890E-02    1.34063       0.00000       0.00000
     0.00000         0.00000        -0.744172       -4.838846E-02   0.00000       0.00000
     1.72576         0.846136        0.00000        0.00000        1.00000       0.489978
     0.00000         0.00000         0.00000        0.00000        0.00000       1.00000

     DetY-1 =      0.0000000448,    DetZ-1 =      0.0000000415
     R12=0 at   -1.164     m,      R34=0 at    27.71    m
  First order symplectic conditions (expected values = 0) :
     4.4756E-08    4.1481E-08     0.000         0.000        0.000        0.000
```

Transport matrix for DIPOLE, an excerpt from zgoubi.res:

```
Reference particle (#    1), path length :   133.51769    cm  relative momentum :   1.00000

          TRANSFER  MATRIX  ORDRE  1  (MKSA units)
     0.727646        0.847072        0.00000        0.00000        0.00000       0.846136
    -0.555462        0.727664        0.00000        0.00000        0.00000       1.72576
     0.00000         0.00000        -4.838656E-02    1.34063       0.00000       0.00000
     0.00000         0.00000        -0.744172       -4.839868E-02   0.00000       0.00000
     1.72574         0.846139        0.00000        0.00000        1.00000       0.489978
     0.00000         0.00000         0.00000        0.00000        0.00000       1.00000

     DetY-1 =     -0.0000023223,    DetZ-1 =      0.0000000165
     R12=0 at   -1.164     m,      R34=0 at    27.70    m
  First order symplectic conditions (expected values = 0) :
    -2.3223E-06    1.6482E-08     0.000         0.000        0.000        0.000
```

Note the agreement with theoretical values of horizontal transport coefficients in the hard edge model (Eq. 14.7 with $\phi = 90$ deg deviation and $\alpha = 36$ deg wedge angle):

$$r11 = \frac{\cos(\phi - \alpha)}{\cos \alpha} = 0.72654 \qquad r16 = RM(1 - \cos \phi) = 0.85$$
$$r21 = -\frac{1}{RM}\frac{\sin(\phi - 2\alpha)}{\cos^2 \alpha} = -0.55545 \qquad r26 = \sin \phi + (1 - \cos \phi)\tan \alpha = 1.7265$$

The agreement is not as good for the vertical coefficients, namely

$$r33 = 1 - \phi \tan \alpha = -0.14125 \qquad r34 = RM\phi = 1.3351$$
$$r43 = -\frac{\tan \alpha}{RM}(2 - \phi \tan \alpha) = -0.73402 \qquad r44 = 1 - \phi \tan \alpha = -0.14125$$

This should improve if the correction for the fringe field extent is accounted for. Equations 14.19 and 14.20 detail how to do that, it requires calculation of the I_1 integral, which can be performed using the Enge fringe field model for the fall-off (with coefficients $C_0 - C_5$ provided in the simulation data file, Table 13.9). From there, the first order correction ψ to the wedge angle can be included. This exercise is left to the reader. Note that zpop has an option to compute I_1 from a field fall-off profile, namely, menu 8 (Analysis/Graphic), sub-menu 18 (FRINGE-FIELD MATCHING).

Field along trajectories.

Set IL = 2 under BEND and DIPOLE (or, an alternate possibility, use OPTIONS[.plt, 2]). A graph of the field is given in Fig. 13.19.

(b) Adjusting the quadrupole doublet

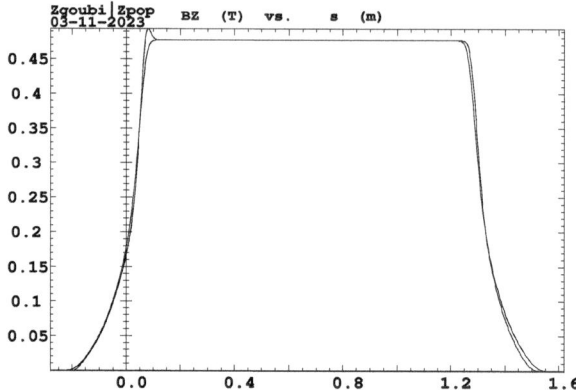

Fig. 13.19 Vertical magnetic field component $B_Z(s)$ along the reference axis and along a trajectory launched with $Z = 2$ cm, across BEND and DIPOLE. $B_Z(s)$ curves from both magnets superimpose well, they cannot be distinguished at this scale. The overshoot at the entrance EFB is for the $Z = 2$ cm case. A graph obtained using zpop, menu 7: 1/1 to open zgoubi.plt; 2/[6, 32] for B_Z versus s; 7 to plot

The input data file is given in Table 13.10. A FIT finds MQ1 ad MQ2 voltages for point-to-point imaging.

Transport matrix, field, trajectories

Transport matrix for point-to-point imaging, an excerpt from zgoubi.res:

```
Reference particle (#      1), path length :    116.50000     cm relative momentum :    1.00000

           TRANSFER  MATRIX  ORDRE  1   (MKSA units)
    -0.218728        1.179161E-08    -7.159944E-12     9.628594E-18     0.00000         0.00000
    -8.12640         -4.57208        -5.743011E-11     1.985314E-16     0.00000         0.00000
     6.864604E-17    6.231754E-17    -2.73607          2.690797E-08     0.00000         0.00000
     5.819981E-16    3.796530E-16    -6.80266         -0.365488         0.00000         0.00000
     0.00000         0.00000          0.00000          0.00000          1.00000         0.00000
     0.00000         0.00000          0.00000          0.00000          0.00000         1.00000

     DetY-1 =       0.0000447215,    DetZ-1 =       0.0000003068
     R12=0 at   0.2579E-08 m,        R34=0 at   0.7362E-07 m
   First order symplectic conditions (expected values = 0) :
     4.4722E-05    3.0682E-07    9.7321E-18    2.1246E-17    4.5622E-11    3.2735E-11
```

Fields along $Y = Z = 1$ cm lines are given in Fig. 13.20. Point-to-point imaging is shown in Fig. 13.21. These graphs are obtained from data logged in zgoubi.plt by setting ELMULT[IL = 2].

(c) Adjusting transport at middle of multipole M.

Radial point-to-parallel focusing requires $R_{22} = 0$. Axial point-to-point and parallel-to-parallel focusing requires $R_{34} = 0$ and $R_{43} = 0$.

Before matching, the transport matrix from object to middle of M multipole, 502.0177 cm downstream, is (an excerpt from zgoubi.res):

Table 13.10 Simulation input data file MQdoublet.inc: d1-MQ1-d2-MQ2-½d3 quadrupole doublet section. This list defines the segment HRS-MQ1-2_S:HRS-MQ1-2_E for the purpose of INCLUDEs in subsequent exercises

```
HRSDESIR U180_V4. Quadrupole doublet. MQdoublet.inc
'MARKER' HRS-MQDoublet_S                                ! Just for edition purposes.
'OBJET'
405.179                                                 ! Sn132+, 60 KeV.
5                                       ! Generate a 13 particle set, ready for use by MATRIX.
.01 .001 .01 .001 .0 .0001
0.  0.  0.  0.  0.  1.
'PARTICUL'          ! Nature of particle is necessary in order to solve motion, as electric
122957.21 1.602176487E-19 0.0 0.0 0.0                   ! fields are involved.
'MARKER' HRS-MQ1-2_S
'DRIFT'   d1
42.                                      ! Use SPLIT option to log particle data in zgoubi.plt.
'ELMULT'  MQ1
0
18.5 2. 0. -705.49787 0. 0. 0. 0. 0. 0. 0.
9. 4. 1. 0. 0. 0. 0. 0. 0. 0. 0.                        ! Quadrupole fringe fields are set.
6  .296471 4.533219 -2.270982 1.068627 -0.036391 0.022261
9. 4. 1. 0. 0. 0. 0. 0. 0. 0. 0.
6  .296471 4.533219 -2.270982 1.068627 -0.036391 0.022261
0. 0. 0. 0. 0. 0. 0. 0. 0. 0.
#60|30|60                                    ! 60 steps in fringe regions, 30 steps in body.
1  0. 0. 0.
'DRIFT'   d2
10.0                                     ! Use SPLIT option to log particle data in zgoubi.plt.
'ELMULT'  MQ2
0
18.5 2. 0. 816.41216  0. 0. 0. 0. 0. 0. 0.
9. 4. 1. 0. 0. 0. 0. 0. 0. 0. 0.                        ! Quadrupole fringe fields are set.
6  .296471 4.533219 -2.270982 1.068627 -0.036391 0.022261
9. 4. 1. 0. 0. 0. 0. 0. 0. 0. 0.
6  .296471 4.533219 -2.270982 1.068627 -0.036391 0.022261
0. 0. 0. 0. 0. 0. 0. 0. 0. 0.
#60|30|60                                    ! 60 steps in fringe regions, 30 steps in body.
1 0. 0. 0.
'DRIFT'   1/2d3
27.5                                     ! Use SPLIT option to log particle data in zgoubi.plt.
'MARKER' HRS-MQ1-2_E
'FIT2'
2                                        ! 2 variables: voltage in MQ1 and Mq2.
6  5 0 .2                                ! Allowed range 20%.
8  5 0 .2
2  1e-15                                 ! 2 constraints; penalty 1e-15:
1 1 2 #End 0. 1. 0                       ! R12=0,
1 3 4 #End 0. 1. 0                       ! R34=0.

'MATRIX'                                 ! Compute transport matrix, from OBJET down to here.
1   0
'MARKER' HRS-MQDoublet_E                 ! Just for edition purposes.
'END'
```

Fig. 13.20 Electric field component E_Z at constant $Y = Z = 1$ cm along MQ1 (bottom curve) and MQ2 (top curve). A graph obtained using zpop, menu 7: 1/1 to open zgoubi.plt; 2/[8, 35] for E_Z versus X; 7 to plot

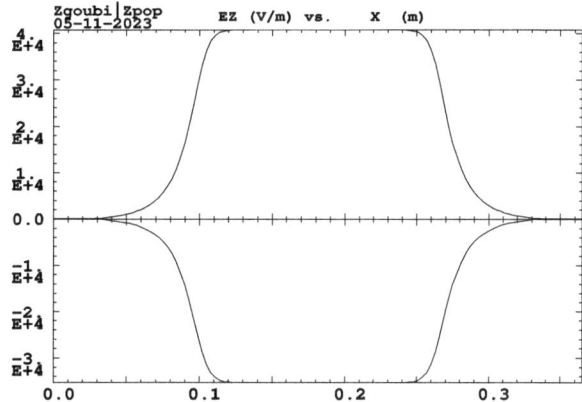

Fig. 13.21 Four trajectories leaving the object with $Y_0' = \pm 10$ mrad (red, largest excursion) and $Z_0' = \pm 10$ mrad (blue). A graph obtained using zpop, menu 7: 1/1 to open zgoubi.plt; 2/[6, 2] (and 2/[6, 4]) for Y (and Z) versus s; 7 to plot

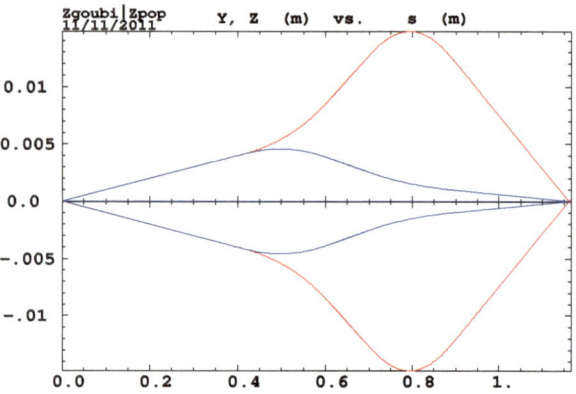

```
Reference particle (#    1), path length :    502.01770    cm  relative momentum :    1.00000

              TRANSFER  MATRIX  ORDRE  1  (MKSA units)
     -33.9960        -18.6071      -7.678697E-12     0.00000      0.00000       2.39932
     -8.998661E-03   -3.437068E-02  1.443290E-11     0.00000      0.00000       1.72576
     -2.289502E-16   -2.770760E-16  7.86076          0.164950     0.00000      -8.342532E-31
     -5.380441E-16   -1.719575E-16 -0.178453         0.123471     0.00000      -8.109384E-30
     -58.6475        -32.0288      -4.547474E-11     0.00000      1.00000       0.489978
      0.00000         0.00000       0.00000          0.00000      0.00000       1.00000
```

R_{22} is nearly zero, R_{34} and R_{43} are small, a good starting point for the matching procedure. This is done running the input data file, Table 13.11. The FIT status comes out to be (an excerpt from zgoubi.res):

```
               FIT  variables  and  constraints  in  good  order,  FIT  will  proceed.
STATUS OF VARIABLES  (Iteration #    0 /   999 max.)
LMNT VAR PARAM  MINIMUM    INITIAL     FINAL       MAXIMUM      STEP        NAME    LBL1
    5   1   5   -847.      -705.     -694.57599     -564.     1.672E-03   ELMULT   MQ1
    7   2   5    653.       816.      839.78984      980.     1.808E-02   ELMULT   MQ2
   15   3   5  -1.109E+03  -924.     -909.05344     -740.     1.448E-03   ELMULT   FQ
STATUS OF CONSTRAINTS (Target penalty =   1.0000E-10)
TYPE  I   J LMNT#    DESIRED      WEIGHT       REACHED        KI2     NAME   LBL1
  1   2   2   23   0.000000E+00  1.000E+00  -8.850032E-06  Infinity MATRIX   -
  1   3   4   23   0.000000E+00  1.000E+00   1.346251E-06  Infinity MATRIX   -
  1   4   3   23   0.000000E+00  1.000E+00  -2.077152E-06  Infinity MATRIX   -
Fit reached penalty value  8.4450E-11
```

It shows convergence toward the expected $R_{22} = 0$, $R_{34} = 0$ and $R_{43} = 0$. This is obtained with quadrupole voltages:

$$MQ1 : \ -694.57599\,\text{V}, \qquad MQ1 : \ 839.78984\,\text{V}, \qquad FQ1 : \ -909.05344\,\text{V}$$

The resulting transport matrix is (an excerpt from zgoubi.res):

```
Reference particle (#    1), path length :    502.01770    cm  relative momentum :    1.00000

              TRANSFER  MATRIX  ORDRE  1  (MKSA units)
     -35.3729        -19.4907      -1.766768E-11     0.00000      0.00000       2.39932
      5.135288E-02   -8.850032E-06  2.275957E-11     0.00000      0.00000       1.72576
     -1.259986E-16   -2.668953E-16  7.59199          1.346251E-06 0.00000      -8.342532E-31
     -6.770227E-16   -2.254787E-16 -2.077152E-06     0.131717     0.00000      -8.109384E-30
     -61.1686        -33.6362      -7.105427E-11     0.00000      1.00000       0.489978
      0.00000         0.00000       0.00000          0.00000      0.00000       1.00000
```

Table 13.11 Simulation input data file objetToM.dat: from the object plane to the middle of multipole M. The FIT procedure finds the proper MQ1, MQ2 and FQ1 voltages for $R_{22} = 0$, $R_{34} = 0$ and $R_{43} = 0$. The FS-FQ.inc file INCLUDEd here is comprised of (Table 13.3) ½d3-FS1-d4-FQ1-d5

```
objetToM.dat
'OBJET'
405.179                                                    ! Sn132+, 60 KeV.
5                                        ! Generate a 13 particle set, ready for use by MATRIX.
.01 .001 .01 .001 .0 .0001
0.  0.  0.  0.  0.  1.
'PARTICUL'                    ! Nature of particle is necessary in order to solve motion, as electric
122957.21 1.602176487E-19 0.0 0.0 0.0                      !   fields are involved.

'INCLUDE'
1
HRS-MQdoublet.inc[HRS-MQ1-2_S:HRS-MQ1-2_E]

'MATRIX'
1  0

'INCLUDE'
1
FS-FQ.inc[FS-FQ_S:FS-FQ_E]

'INCLUDE'
1
HRS-BEND+DIPOLE.inc[HRS-BEND_S:HRS-BEND_E]

'DRIFT'
75.0
'ELMULT'  M
0
15. 20. 0. 0. 0.0001 0. 0. 0. 0. 0. 0. 0.
0. 0. 0. 0. 0. 0. 0. 0. 0. 0. 0.
0 0. 0. 0. 0. 0. 0.
0. 0. 0. 0. 0. 0. 0. 0. 0. 0. 0.
0 0. 0. 0. 0. 0. 0.
0. 0. 0. 0. 0. 0. 0. 0. 0. 0.
1.
1 0. 0. 0.
'MATRIX'
1  0

'FIT2'
3                                             ! 3 variables: voltage in MQ1, MQ2, FQ1.
5  5 0 .2                                     !      Allowed range 20%.
7  5 0 .2
15 5 0 .2
3    ! 3 constarints:
1 2 2 #End 0. 1. 0                                          ! R12=0,
1 3 4 #End 0. 1. 0                                          ! R34=0.
1 4 3 #End 0. 1. 0                                          ! R43=0.

'MATRIX'                                      ! Matrix in the middle of M multipole.
1  0

'END'
```

(d) Adjusting transport at final focus.

A full HRS line is obtained by prolonging the upstream half, Table 13.11, by its reverse. Now constrain R_{11}, R_{22} to -1 and R_{33}, R_{44} to 1, using, as variables, MQ1, MQ2, FQ1 coupled with respectively MQ4, MQ3, FQ2. The FIT data needed for that are:

```
'FIT2'
3                                          ! 3 variables: voltage in MQ1, MQ2, FQ.
5  5 -36.5 .4                                      ! Allowed range 40%.
7  5 -34.5 .4                                      ! Allowed range 40%.
15 5 -28.5 .4                                      ! Allowed range 40%.
7                                                ! 7 constraints:
1 2 2 23 0. 10. 0          ! with a small weight, maintain earlier constraints at middle-M;
1 3 4 23 0. 10. 0
1 4 3 23 0. 10. 0
1 1 1 #End -1. 1. 0                                          ! R11=-1,
1 2 2 #End -1. 1. 0                                          ! R22=-1,
1 3 3 #End  1. 1. 0                                          ! R33=+1,
1 4 4 #End  1. 1. 0                                          ! R44=+1,
```

With these variables and constraints, the final status of the FIT comes out to be:

```
STATUS OF VARIABLES  (Iteration #      0 /   999 max.)
LMNT VAR PARAM  MINIMUM    INITIAL        FINAL      MAXIMUM     STEP        NAME   LBL1
  5   1    5  -1.667E+03   -695.       -694.57707     278.     3.396E-09  ELMULT    MQ1
 36   1    5  -1.667E+03   -695.       -694.57707     278.     3.396E-09
  7   2    5    504.        840.        839.79050    1.176E+03 7.467E-10  ELMULT    MQ2
 34   2    5    504.        840.        839.79050    1.176E+03 7.467E-10
 15   3    5  -1.273E+03   -909.       -909.05681    -545.     2.022E-09  ELMULT    FQ
 28   3    5  -1.273E+03   -909.       -909.05681    -545.     2.022E-09
STATUS OF CONSTRAINTS (Target penalty =   1.0000E-10)
TYPE  I   J LMNT#    DESIRED         WEIGHT        REACHED        KI2    NAME  LBL1
  1   2   2   23  0.000000E+00     1.000E+01    -1.196637E-05  1.04E-06 MATRIX   -
  1   3   4   23  0.000000E+00     1.000E+01     1.245864E-05  1.12E-06 MATRIX   -
  1   4   3   23  0.000000E+00     1.000E+01    -3.831489E-05  1.06E-05 MATRIX   -
  1   1   1   38 -1.000000E+00     1.000E+00    -1.000833E+00  5.03E-01 MATRIX   -
  1   2   2   38 -1.000000E+00     1.000E+00    -9.991717E-01  4.97E-01 MATRIX   -
  1   3   3   38  1.000000E+00     1.000E+00     9.999992E-01  4.80E-07 MATRIX   -
  1   4   4   38  1.000000E+00     1.000E+00     9.999994E-01  2.59E-07 MATRIX   -
Fit reached penalty value   1.3808E-06
```

as expected. It tells that this requires the following adjusted quadrupole voltages:

$$MQ1, MQ4/ - 694.58, \quad MQ2, MQ3/ + 839.79, \quad FQ1, FQ2/ - 909.05$$

Following from these voltage adjustments, the transport matrix at the first focus is still near point-to-point focusing, both planes:

```
Reference particle (#    1), path length :   116.50000    cm  relative momentum :    1.00000

          TRANSFER  MATRIX  ORDRE  1  (MKSA units)
   -0.358361     -8.350650E-02   -7.209470E-12    1.617389E-17    0.00000        0.00000
   -8.40677      -4.74958        -5.627678E-11    2.159291E-16    0.00000        0.00000
    7.779357E-17  8.032343E-17   -2.67238         5.825994E-02    0.00000        0.00000
    5.850307E-16  4.135978E-16   -6.67972        -0.228576        0.00000        0.00000
    0.00000       0.00000         0.00000         0.00000         1.00000        0.00000
    0.00000       0.00000         0.00000         0.00000         0.00000        1.00000
```

Transport matrix at the middle of the multipole still features near-zero R_{22}, R_{34}, R_{43}:

```
Reference particle (#    1), path length :   502.01770    cm  relative momentum :    1.00000

          TRANSFER  MATRIX  ORDRE  1  (MKSA units)
   -35.3731      -19.4908        -2.759947E-11    0.00000        0.00000        2.39932
    5.134701E-02 -1.196637E-05    1.332268E-11    0.00000        0.00000        1.72576
   -7.607480E-17 -2.503423E-16    7.59203         1.245864E-05   0.00000       -8.342532E-31
   -5.997751E-16 -1.885611E-16   -3.831489E-05    0.131716       0.00000       -8.109384E-30
   -61.1688      -33.6364        -7.389644E-11    0.00000        1.00000        0.489978
    0.00000       0.00000         0.00000         0.00000        0.00000        1.00000
```

The matrix at final focus features the expected $R_{11} = R_{22} = -1$ and $R_{33} = R_{44} = +1$:

```
Reference particle (#    1), path length :   1004.0354     cm  relative momentum :    1.00000

          TRANSFER  MATRIX  ORDRE  1  (MKSA units)
    -1.00083       4.562349E-04   -5.069283E-10   -1.355253E-16    0.00000     -67.2908
    -3.63122      -0.999172       -9.200092E-10   -2.117582E-16    0.00000     -122.115
     6.224330E-17  -6.475335E-17   0.999999        3.285171E-06    0.00000      3.477282E-15
    -7.653723E-15  -5.663523E-15  -5.421393E-04    0.999999        0.00000      4.949883E-15
    -122.092      -67.2728        -7.958079E-11    0.00000         1.00000      9.26076
     0.00000        0.00000        0.00000         0.00000         0.00000      1.00000
```

From this matrix, the mass dispersion (Eq. 13.5) comes out to be

$$D_m = \frac{1}{2} D_x = -33.6 \, \text{cm/\%}$$

fairly close to the expected $D_m \approx 31 \, \text{cm/\%}$ design value [8].

(e) A 6000 particle initial object. Minimizing second order aberrations.

To track a 6000 particle initial object, in the previous simulation for the transport coefficients, change OBJET[KOBJ = 5] to a Monte Carlo object definition. Its formatting and functioning are detailed in [13, cf. MCOBJET, Sect. 6.2]. Add as well FAISTORE[FNAME = zgoubi.fai FF], where FF is a LABEL1 (as in "'MARKER' FF") placed at the location of the image plane, end of the line. The resulting input data file is given in Table 13.12. It grabs from the HRSComplete.inc file (the INCLUDE statement) the segment [HRS-MQ1-2_S:FF] which is the complete HRS line sequence (the same as used in question (d)).

Outcomes are given in Fig. 13.22. These results are actually obtained with appropriate settings of the mid-plane multipole M, which reduce the sextupole aberration. Finding the necessary M component is left to the reader; FIT[IC = 3 or 4] can be used

Table 13.12 Input data file: tracking 6,000 Sn132$^+$ ions from object to final image plane

```
HRS-DESIR. Monte Carlo object.
'MCOBJET'
405.1790552244900274    ! (Sn132+, 60.58188 KeV)
2
6000
2        2        2        2        1    1
0.       0.       0.       0.       0.   1.
1        1        1        1        1    3
0.       0.       0.       0.       0.   0.0005
0.5e-3   2.e-3    0.5e-3   2.e-3    0.   .0
1        1        1        1        1    1
1 0. 0. 0. 0.
123456 234567 345678
! .5e-3   2.e-3    .5e-3    2.e-3    0.   .0    !!!005
'PARTICUL'
122957.208  1.602176487E-19  0.0 0.0 0.0    931.494028*118.710
'FAISTORE'
zgoubi.fai  FF
1

'INCLUDE'
1
HRSComplete.inc[HRS-MQ1-2_S:FF]

'FIN'
```

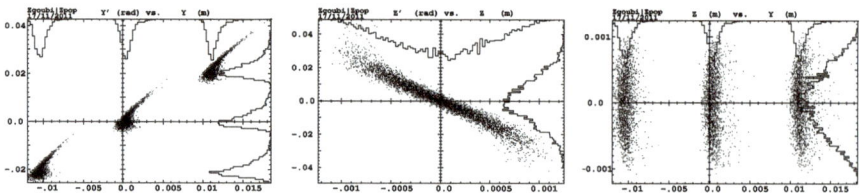

Fig. 13.22 At final focus of HRS-DESIR, for three different momenta, $p = 121.46962 \pm 5 \times 10^{-4}$ MeV/c: horizontal phase space (Y',Y), vertical phase space (Z'Z) and (Y, Z) transverse cross section. These results are obtained accounting for a second degree EFB curvature (a sextupole effect, Fig. 13.9) at both EFBs of both 90 deg bends

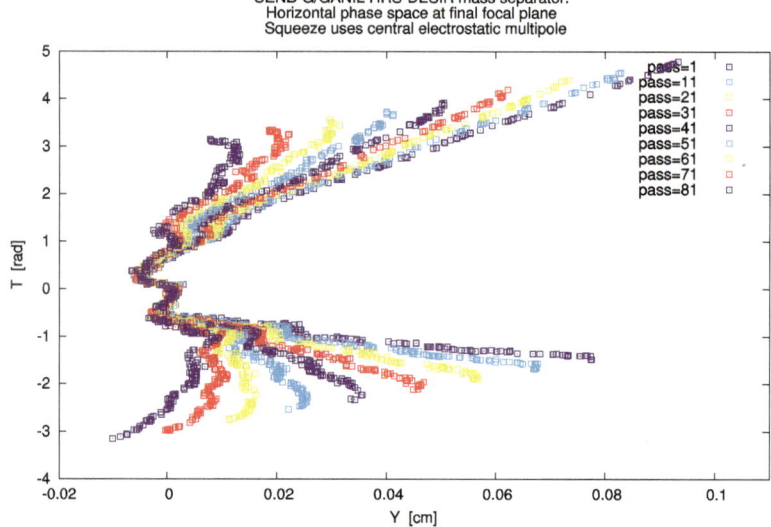

Fig. 13.23 Horizontal phase space for the central momentum, at final focus of HRS-DESIR, with high order aberrations progressively minimized. The process starts from initial condition where M is not used (curve with greater Y excursion). Minimization involves sextupole, octupole, decapole, dodecapole and 20-pole components of M

for that. Note that similar simulations may be found in `zgoubi` sourceforge repository, https://sourceforge.net/p/zgoubi/code/HEAD/tree/branches/exemples/ folder.

Minimization of high order aberrations is pushed further in Fig. 13.23: the sextupole, octupole, decapole, dodecapole and 20-pole components of M are set to minimize the image size. The initial object in this case is comprised of 300 particles, and only has horizontal divergence (initial Y, Z and P of particles are zero).

(f) Animation of the image squeeze.

An excerpt of an animation of the image squeeze is displayed Fig. 13.23. The squeeze relies essentially on the use of REBELOTE[IOPT = 1], which loops over varying values of a set of non-linear field components in the M multipole. Zgoubi input files can be found at [14]. The animation reads its data from a zgoubi.fai style file.

The files for that simulation are available in zgoubi sourceforge repository at https://sourceforge.net/p/zgoubi/code/HEAD/tree/branches/exemples/book/ zgoubiMaterial/spectrometers/HRS-DESIR_massSeparator/animation.

References

1. J. Saudinos, Spectrométrie magnétique. Ecole d'été des Houches "Méthodes expérimentales en physique nucléaire" (1974)
2. F. Bødker et al., Magnets and Wien filters for SECAR, in *Proceedings of IPAC2017* (Copenhagen, Denmark TUPVA051). Figure 13.1: copyrights under license CC-BY-3.0, https://creativecommons.org/licenses/by/3.0; no change to the material
3. Lezione 10: Magnetic Spectrometers. INFI LN-Legnaro, Italy. Figure 13.2: credit Dr. L. Corradi and Dr. E. Fioretto, INFN-LNL
4. L. Bianchi et al., SPEG: An energy loss spectrometer for GANIL. NIM A **276**(3), 509–520 (1989)
5. L. Perrot et al., Status of the CAVIAR detector at LISE-GANIL. https://accelconf.web.cern.ch/HIAT2009/papers/g-02.pdf Figure 13.3, LISE: copyrights under license CC-BY-3.0, https://creativecommons.org/licenses/by/3.0; no change to the material
6. R. Heine et al., Extension of the 3-spectrometer beam transport line for the KAOS spectrometer at MAMI and recent status of MAMI, in *MOPZ037 Proceedings of IPAC2011* (San Sebastián, Spain). Figure 13.4: copyrights under license CC-BY-3.0, https://creativecommons.org/licenses/by/3.0; no change to the material
7. A. Tkatchenko, F. Méot, Calculs optiques pour le spectromètre à kaons de GSI. Rapport Interne CEA/LNS/GT/88-07, Saclay (1988)
8. T. Kurtukian-Nieto et al., SPIRAL2/DESIR High Resolution Mass Separator. NIM B Volume 317, Part B, 2013, pp. 284–289. https://accelconf.web.cern.ch/ipac2021/papers/mopab264.pdf
9. J. Michaud et al., Commissioning of the DESIR high-resolution separator at CENB-G, in *12th International Particle Accelerator Conference IPAC2021* (Campinas, SP, Brazil). Figures 13.5, 13.13: copyrights under license CC-BY-3.0, https://creativecommons.org/licenses/by/3.0; no change to the material
10. J. Thirion, P. Birien, Le Spectromètre II. Rapport Interne DPh-N ME, CEA Saclay (1975)
11. H. Catz, *Le Spectromètre SPES II* (CEA Saclay, circa, Rapport Interne DPh-N ME, 1975)
12. SPES II input data files for Zgoubi simulations, using the early 1970s measured field maps of the three magnets, are available at https://sourceforge.net/p/zgoubi/code/HEAD/tree/branches/exemples/spectrometers/spes2_spectrometer/usingFieldMaps/
13. F. Méot, Zgoubi Users' Guide. https://www.osti.gov/biblio/1062013-zgoubi-users-guide. An up-to-date version of the guide can be found at: https://sourceforge.net/p/zgoubi/code/HEAD/tree/trunk/guide/Zgoubi.pdf
14. An animation of DESIR-HRS image squeeze using gnuplot, reading data from a zgoubi.fai style file, can be found at https://sourceforge.net/p/zgoubi/code/HEAD/tree/branches/exemples/CENBG/HRS-DESIR/animationAtFinalFocus/

Chapter 14
Optical Elements and Keywords, Complements

Abstract This chapter is not a review of the 60+ optical elements of zgoubi's library. They are described in the Users' Guide. One aim here is, regarding some of them, to briefly recall some aspects which may not be found in the Users' Guide and yet addressed, or referred to, in the theoretical reminder sections and in the exercises. This chapter is not a review of the 40+ monitoring and command keywords available in zgoubi, either. However it reviews some of the methods used, by keywords such as MATRIX (computation of transport coefficients from sets of rays), FAISCEAU (which produces beam emittance parameters), and others. This chapter in addition recalls the basics of transport and beam matrix methods, in particular it provides the first order transport matrix of several of the optical elements used in the exercises, in view essentially of comparisons with transport coefficients drawn from raytracing, in simulation exercises.

14.1 Introduction

Optical elements are the basic bricks of charged particle beam lines and accelerators. An optical element sequence is aimed at guiding the beam from one location to another while maintaining it confined in the vicinity of a reference optical axis.

Zgoubi library offers of collection of about 100 keywords, amongst which about 60 are optical elements, the others being commands (to trigger spin tracking, trigger synchrotron radiation, print out particle coordinates, compute beam parameters, etc.). This library has built over half a century, so it allows simulating most of the optical elements met in real life accelerator facilities. Quite often, elements available provide different ways to model a particular optical component. A bending magnet for instance can be simulated using AIMANT, or BEND, CYCLOTRON, DIPOLE[S][-M], FFAG, FFAG-SPI, MULTIPOL, QUADISEX, or a field map and TOSCA, CARTEMES or POLARMES to handle it. These various keywords have their respective subtleties, though, more on this can be found in the "Optical Elements Versus Keywords" Section of the guide [1, pp. 12, 227], which tells "Which optical component can be simulated. Which keyword(s) can be used for that purpose". For a

© The Author(s) 2024

F. Méot, *Understanding the Physics of Particle Accelerators*, Particle Acceleration and Detection, https://doi.org/10.1007/978-3-031-59979-8_14

complete inventory of optical elements, refer to the "Glossary of Keywords" found at the beginning of PART A [1, p. 9] or PART B of the Users' Guide [1, p. 229].

Optical elements in zgoubi are actually field models, or field modeling methods such as reading and handling field maps. Their role is to provide the numerical integrator with the necessary field vector(s) to push a particle, and possibly its spin, along a trajectory. The following sections introduce the analytical field models which the simulation exercises resort to.

Zgoubi's coordinate nomenclature, as well as the Cartesian or cylindrical reference frames used in the optical elements and field maps, have been introduced in Sect. 1.2 and Fig. 1.5.

14.2 Drift Space

This is the DRIFT, or ESL (for the French "ESpace Libre") optical element, through which a particle moves on a straight line. From the geometry and notations in Fig. 14.1, with L the length of the drift, coordinate transport satisfies

$$\left|\begin{array}{l} X_f - X_i = L \\ Y_f - Y_i = L \tan T \\ Z_f - Z_i = L \tan P / \cos T \\ \text{path length } d = L/(\cos T \cos P) \end{array}\right. \qquad (14.1)$$

Linear approach

Coordinate transport from initial to final position in the linear approximation is written (with z standing indifferently for x or y, subscripts i for initial and f for final coordinates) (Fig. 14.2)

Fig. 14.1 An L-long drift in zgoubi (O;X,Y,Z) frame, with origin at the upstream end of the drift. A particle flies from $A(Y_i, Z_i)$ to $B(Y_f, Z_f)$, at an angle P to the (X, Y) plane. Projection W of its straight path in (X, Y) plane is at an angle T to the X axis

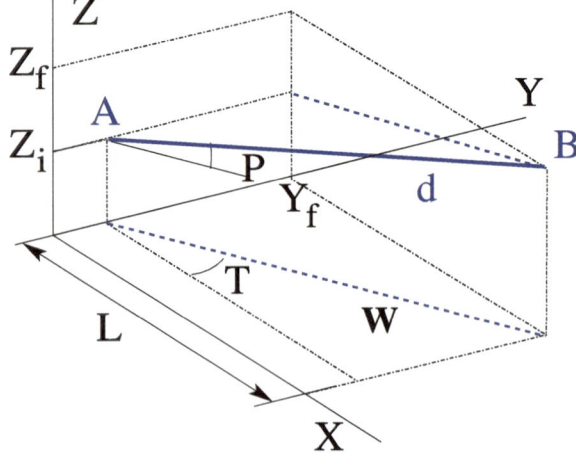

Fig. 14.2 A drift section
with length $L = s_f - s_i$, and
projection of a straight
trajectory in the (s, z) plane,
at an angle z' (standing for x'
or y') to the s axis

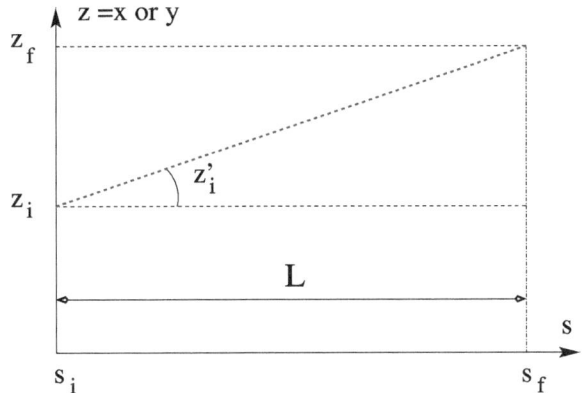

$$
\begin{vmatrix}
z_f = z_i + L z_i' \\
z_f' = z_i' \\
\delta l_f - \delta l_i = \beta c \delta t = \dfrac{L}{\gamma^2} \dfrac{\delta p}{p} \\
\delta p_f / p = \delta p_i / p
\end{vmatrix}
\quad \text{or,} \quad
T_{\text{drift}} =
\begin{pmatrix}
1 & L & 0 & 0 & 0 & 0 \\
0 & 1 & 0 & 0 & 0 & 0 \\
0 & 0 & 1 & L & 0 & 0 \\
0 & 0 & 0 & 1 & 0 & 0 \\
0 & 0 & 0 & 0 & 1 & \dfrac{L}{\gamma^2} \\
0 & 0 & 0 & 0 & 0 & 1
\end{pmatrix}
\quad (14.2)
$$

where βc is the particle velocity, $p = \gamma m \beta c$ its momentum, γ is the Lorentz relativistic factor.

14.3 Guiding

Beam guiding is in general assured using dipole magnets to provide a field vector normal to a bend plane. Gradient dipoles combine guiding and focusing in a single magnet, this is the case in cyclotrons where the field index is tailored to ensure isochronism, in scaling FFAGs where $B \propto r^k$ ensures the zero-chromaticity property. This may also be the case in strong focusing synchrotrons, for instance in the BNL AGS [2], in the CERN PS [3]. Dipole magnets sometimes include a sextupole component for the compensation of chromatic aberrations [4]. Non-linear optical effects may be introduced in addition by shaping entrance and/or exit EFBs, a parabola for instance for x^2 field integral dependence, a cubic curve for x^3 dependence (see Chap. 13).

Low energy beam guiding also uses electrostatic deflectors, shaped to provide a field normal to the trajectory arc, and possibly focusing properties. Plane condensers may be used as well for beam steering, including beam filtering in combination with a magnetic field, and at high energy in addition for such functions as pretzel orbit separation, extraction septa, etc.

Guiding optical elements are dispersive systems: trajectory deflection has a first order dependence on particle momentum.

14.3.1   Dipole Magnet, Curved

This is the DIPOLE element (an evolution of the 1972s AIMANT [1]) or variants: DIPOLES, DIPOLE-M. Lines of constant field in the magnet body are isocentric circle arcs. The magnet reference curve is a particular arc, at a reference radius R_0 for which the field value is B_0. The field in the median plane can be written

$$B_Z(r, \theta) = \mathcal{G}(r, \theta) \, B_0 \left(1 + N \, \frac{r - R_0}{R_0} + N' \left(\frac{r - R_0}{R_0} \right)^2 + N'' \left(\frac{r - R_0}{R_0} \right)^3 + \cdots \right)$$

(14.3)

$N^{(n)} = d^n N / dY^n$ are the field index and derivatives. $\mathcal{G}(r, \theta)$ describes the azimuthal shape of the field, from a plateau value in the body to zero away from the magnet. It can be written under the form [5]

$$\mathcal{G}(r, \theta) = G_0 \, F(d(r, \theta))$$

(14.4)

where G_0 a constant factor, and $F(d)$ a convenient model for the field fall-off, such as the Enge model discussed in Sect. 14.3.3. In that model take $d(r, \theta)$ the distance from particle location (X, Y, Z) to the magnet EFB, $\lambda(r)$ an r-dependent characteristic extent of the field fall-off (e.g., representing a radial dependence of dipole gap height gap(r), such that $\lambda(r) \approx$ gap(r)). The latter allows modeling the r-dependence of the flutter and its effect on vertical focusing.

Linear approach

The first order transport matrix of a sector dipole with curvature radius ρ, deflection α and index n, in the hard-edge model, writes

$$T_{\text{bend}} = \begin{pmatrix} C_x & S_x & 0 & 0 & 0 & \frac{r_x^2}{\rho}(1 - C_x) \\ C_x' & S_x' & 0 & 0 & 0 & \frac{1}{\rho} S_x \\ 0 & 0 & C_y & S_y & 0 & 0 \\ 0 & 0 & C_y' & S_y' & 0 & 0 \\ \frac{1}{\rho} S_x & \frac{r_x^2}{\rho}(1 - C_x) & 0 & 0 & 1 & \frac{r_x^3}{\rho^2}(\rho\alpha - S_x) \\ 0 & 0 & 0 & 0 & 0 & 1 \end{pmatrix} \quad \text{with} \quad \begin{bmatrix} C = \cos \frac{\rho\alpha}{r} \\ C' = \frac{dC}{ds} = \frac{1}{\rho} \frac{dC}{d\alpha} = \frac{-S}{r^2} \\ S = r \sin \frac{\rho\alpha}{r} \\ S' = \frac{dS}{ds} = \frac{1}{\rho} \frac{dS}{d\alpha} = C \\ (*)_x : r = \rho/\sqrt{1 - n} \\ (*)_y : r = \rho/\sqrt{n} \end{bmatrix}$$

(14.5)

or, explicitly,

$$T_{\text{bend}} = \begin{pmatrix} \cos\sqrt{1-n}\alpha & \frac{\rho}{\sqrt{1-n}}\sin\sqrt{1-n}\alpha & 0 & 0 & 0 & \frac{\rho}{1-n}(1-\cos\sqrt{1-n}\alpha) \\ -\frac{\sqrt{1-n}}{\rho}\sin\sqrt{1-n}\alpha & \cos\sqrt{1-n}\alpha & 0 & 0 & 0 & \frac{1}{\sqrt{1-n}}\sin\sqrt{1-n}\alpha \\ 0 & 0 & \cos\sqrt{n}\alpha & \frac{\rho}{\sqrt{n}}\sin\sqrt{n}\alpha & 0 & 0 \\ 0 & 0 & -\frac{\sqrt{n}}{\rho}\sin\sqrt{n}\alpha & \cos\sqrt{n}\alpha & 0 & 0 \\ \frac{1}{\sqrt{1-n}}\sin\sqrt{1-n}\alpha & \frac{\rho}{1-n}(1-\cos\sqrt{1-n}\alpha) & 0 & 0 & 1 & \rho\frac{\sqrt{1-n}\alpha-\sin\sqrt{1-n}\alpha}{(1-n)^{3/2}} \\ 0 & 0 & 0 & 0 & 0 & 1 \end{pmatrix} \quad (14.6)$$

Cancel the index in the previous sector dipole, introduce a wedge angle ε at entrance and exit EFBs, introduce the flutter term ψ to account for dependence of vertical focusing on fringe field extent (see Sect. 14.4.1, Eq. 14.20). The first order transport matrix, accounting for the entrance and exit EFB wedge focusing, then writes

$$T_{\text{bend}} = \begin{pmatrix} \frac{\cos(\alpha-\varepsilon)}{\cos\varepsilon} & \rho\sin\alpha & 0 & 0 & 0 & \rho(1-\cos\alpha) \\ -\frac{\sin(\alpha-2\varepsilon)}{\rho\cos^2\varepsilon} & \frac{\cos(\alpha-\varepsilon)}{\cos\varepsilon} & 0 & 0 & 0 & \frac{\sin(\alpha-\varepsilon)+\sin\varepsilon}{\cos\varepsilon} \\ 0 & 0 & 1-\alpha\tan(\varepsilon-\psi) & \rho\alpha & 0 & 0 \\ 0 & 0 & -\frac{\tan(\varepsilon-\psi)}{\rho}(2-\alpha\tan(\varepsilon-\psi)) & 1-\alpha\tan(\varepsilon-\psi) & 0 & 0 \\ \sin\alpha & 0 & 0 & 0 & 1 & \rho(\alpha-\sin\alpha) \\ 0 & 0 & 0 & 0 & 0 & 1 \end{pmatrix} \quad (14.7)$$

14.3.2 Dipole Magnet, Straight

This is the MULTIPOL element. Lines of constant field in the magnet body are straight lines. An early instance of a straight dipole magnet is the AGS main dipole (Fig. 9.2), which combines steering and focusing, and features in addition a small sextupole defect component [7]. The multipole components $B_n(X, Y, Z)$ [n = 1 (dipole), 2 (quadrupole), 3 (sextupole), ...] in the Cartesian frame of the straight dipole derive, by differentiation, from the scalar potential

$$V_n(X, Y, Z) = (n!)^2 \left(\sum_{q=0}^{\infty} (-1)^q \frac{\mathcal{G}^{(2q)}(X)(Y^2+Z^2)^q}{4^q q!(n+q)!} \right) \left(\sum_{m=0}^{n} \frac{\sin\left(m\frac{\pi}{2}\right) Y^{n-m} Z^m}{m!(n-m)!} \right) \quad (14.8)$$

where $\mathcal{G}^{(2q)}(X) = d^{2q}\mathcal{G}(X)/dX^{2q}$. In the case of pure dipole field for instance

$$V_1(X, Y, Z) = \mathcal{G}(X) Z - \frac{\mathcal{G}''(X)}{8}(Y^2+Z^2) + \frac{\mathcal{G}^{(4)}(X)}{512}(Y^2+Z^2) Z \ ... \quad (14.9)$$

and

$$B_X(X, Y, Z) = -\frac{\partial V_1}{\partial X} = \mathcal{G}'(X)\,Z - \frac{\mathcal{G}'''(X)}{8}(Y^2 + Z^2) \dots$$

$$B_Y(X, Y, Z) = -\frac{\partial V_1}{\partial Y} = -\frac{\mathcal{G}''(X)}{4}\,Y + \frac{\mathcal{G}^{(4)}(X)}{256}\,YZ \;..$$

$$B_Z(X, Y, Z) = -\frac{\partial V_1}{\partial Z} = \mathcal{G}(X) - \frac{\mathcal{G}''(X)}{4}\,Z + \frac{\mathcal{G}^{(4)}(X)}{512}(Y^2 + 3Z^2) \dots \qquad (14.10)$$

The longitudinal form factor $\mathcal{G}(X)$ accounts for the field fall-offs at the ends of the magnet, it is modeled using the Enge model discussed in Sect. 14.3.3.

14.3.3 Fringe Field, Modeling, Overlapping

A fringe field model is described here, which is resorted to in several optical elements of zgoubi's library.

Field shape at the EFBs of magnetic or electrostatic devices can be simulated using a hard-edge model (the field is assumed to change following a Heaviside step). When using stepwise ray-tracing techniques however, a smooth change of the field can accurately be accounted for. An efficient model is Enge's field form factor [6]

$$F(d) = \frac{1}{1 + \exp P(d)} \qquad (14.11)$$

$$P(d) = C_0 + C_1\left(\frac{d}{\lambda}\right) + C_2\left(\frac{d}{\lambda}\right)^2 + C_3\left(\frac{d}{\lambda}\right)^3 + C_4\left(\frac{d}{\lambda}\right)^4 + C_5\left(\frac{d}{\lambda}\right)^5$$

where d is the distance to the field boundary, and $\lambda \approx$ gap aperture is the extent of the fall-off. The latter is normally commensurate with gap aperture in a dipole, or $r_{\text{pole tip}}/(n-1)$ in a multipole ($n = 2, 3, \dots$ for quadrupole, sextupole...).

As an illustration, Fig. 14.3 shows $F(d)$ as matched to the measured end fields of BNL AGS main magnet [8, 9], using

$$\lambda = \text{gap aperture} \approx 10\,\text{cm} \quad \text{and} \qquad (14.12)$$

$$C_0 = 0.45473, \; C_1 = 2.4406, \; C_2 = -1.5088, \; C_3 = 0.7335, \; C_4 = C_5 = 0$$

These C_i coefficient values result from an interpolation to measured field data, which are also represented in the figure. The location of the EFB results from the following constraint, which is part of the matching: the field integral on the down side of the fall-off (the region from A to X = 0 in Fig. 14.3) is equal to the complement to 1 of the field integral on the rising side of the fall-off (X = 0 to B region in the figure), which writes

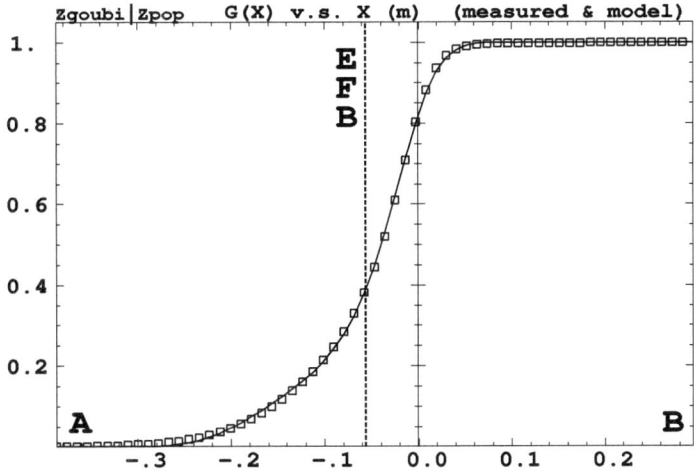

Fig. 14.3 Longitudinal field form factor $\mathcal{G}(X)$ (normalized to one) in BNL AGS main bend, taken along the magnet reference axis. Solid line: from Eqs. 14.11 and 14.12; square markers: measured field data. $X = 0$ is the origin in the field map frame, the vertical dashed line at $X_{\mathrm{EFB}} = -5.62\,\mathrm{cm}$ is the location of the EFB

$$\int_{X_A}^{X_{\mathrm{EFB}}} F(X)\,dX = \int_{X_{\mathrm{EFB}}}^{X_B} dX - \int_{X_{\mathrm{EFB}}}^{B} F(X)\,dX \quad \Rightarrow \quad X_{\mathrm{EFB}} = X_B - \int_A^B F(X)\,dX \tag{14.13}$$

A convenient property of this model is that changing the slope of the fall-off (i.e., changing λ) will not affect the location of the EFB.

Inward fringe field extents may overlap when simulating an optical element (Fig. 14.4). A way to ensure continuity of the resulting field form factor in such case is to use

$$F = F_E + F_S - 1 \quad \text{or} \quad F = F_E * F_S \tag{14.14}$$

where F_E (F_S) is the entrance (exit) form factor and follows Eq. 14.11. Both expressions can be extended to more than two EFBs (for instance 4, to account for the 4 faces of a dipole magnet: entrance and exit faces, inner and outer radial boundaries). Note that in that case of overlapping field extents, the field integral is affected, decreasing with more pronounced overlapping, it is therefore necessary to change the field value (B_0 in Eq. 14.4 for instance) to recover the proper integrated strength.

Overlapping Fringe Fields

Zgoubi allows a superposition technique to simulate the field in a series of neighboring magnets. The method consists in computing the mid-plane field at any location (r, θ) by adding individual contributions, namely [5]

Fig. 14.4 A sketch of overlapping entrance field form factor $F_E(d_E)$ (at the entrance "EFB-E") and exit $F_S(d_S)$ (at the exit "EFB-S"). The resulting form factor $F = F_E \times F_S$ is actually accounted for in modeling the field

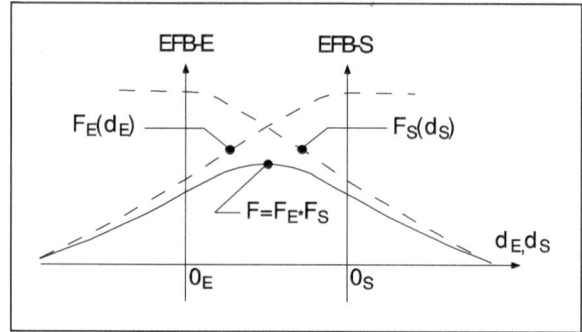

$$B_Z(r, \theta) = \sum_{i=1,N} B_{Z,i}(r, \theta) = \sum_{i=1,N} B_{Z,0,i}\, \mathcal{F}_i(r, \theta)\, \mathcal{R}_i(r)$$

$$\frac{\partial^{k+l} \mathbf{B}_Z(r, \theta)}{\partial \theta^k \partial r^l} = \sum_{i=1,N} \frac{\partial^{k+l} \mathbf{B}_{Z,i}(r, \theta)}{\partial \theta^k \partial r^l} \tag{14.15}$$

with $\mathcal{F}_i(r, \theta)$ and $\mathcal{R}_i(r)$ taken independently for each individual dipole in the series (for instance as per Eqs. 10.7 and 10.15). Note that, in doing so it is not meant that field superposition would apply in reality (if magnets are closely spaced, cross-talk may occurs), however it appears to allow closely reproducing magnet computation code outcomes.

Short Optical Elements

In some cases, an optical element in which fringe fields are taken into account (of any kind: dipole, multipole, electrostatic, etc.) may be given small enough a length, L, that it finds itself in the configuration schemed in Fig. 14.4: the entrance and/or the exit EFB field fall-off extends inward enough that it overlaps with the other EFB's fall-off. In zgoubi notations, this happens if $L < X_E + X_S$. As a reminder [1]: in the presence of fringe fields, X_E (resp. X_S) is the stepwise integration extent added upstream (resp. added downstream) of the actual extent L of the optical element.

In such case, zgoubi computes field and derivatives along the element using a field form factor $F = F_E \times F_S$. F_E (respectively F_S) is the value of the Enge model coefficient (Eq. 14.11) at distance d_E (resp. d_S) from the entrance (resp. exit) EFB.

This may have the immediate effect, apparent in Fig. 14.4, that the integrated field is not the expected value $B \times L$ from the input data L and B, and may require adjusting (increasing) B so to recover the required $\int B\, dl$.

14.3.4 Toroidal Condenser

This is the ELCYLDEF element in `zgoubi`. With proper parameters, it can be used as a spherical, a toroidal or a cylindrical deflector.

Motion along the optical axis, an arc of a circle of radius r normal to electric field **E**, satisfies

$$Er = v\frac{p}{q} = v(B\rho)$$

with $p = mv$ the particle momentum, q its charge and $(B\rho) = p/q$ the particle rigidity.

The first order transport matrix of an electrostatic bend writes

$$T_{condenser} = \begin{pmatrix} C_x & S_x & 0 & 0 & 0 & \frac{2-\beta^2}{p_x^2}r_0(1-C_x) \\ C_x' & S_x' & 0 & 0 & 0 & \frac{2-\beta^2}{r_0}S_x \\ 0 & 0 & C_y & S_y & 0 & 0 \\ 0 & 0 & C_y' & S_y' & 0 & 0 \\ -\frac{2-\beta^2}{r_0}S_x & -\frac{2-\beta^2}{p_x^2}r_0(1-C_x) & 0 & 0 & 1 & r_0\,\alpha\left[\frac{1}{\gamma^2} - \left(\frac{2-\beta^2}{p_x^2}\right)^2(1-\frac{S_x}{r_0\alpha})\right] \\ 0 & 0 & 0 & 0 & 0 & 1 \end{pmatrix} \quad (14.16)$$

$$\text{with} \quad \begin{bmatrix} \alpha = \text{deflection angle} \\ C = \cos p\alpha \\ C' = \frac{dC}{ds} = -\frac{p^2}{r^2}S \\ S = \frac{r}{p}\sin p\alpha \\ S' = \frac{dS}{ds} = C \\ (*)_x: \ p = p_x = \sqrt{2-\beta^2 - r_0/R_0} \\ (*)_y: \ p = p_y = \sqrt{r_0/R_0} \end{bmatrix}$$

14.4 Focusing

Particle beams are maintained confined along a reference propagation axis by means of focusing techniques and devices. Methods available in `zgoubi` to simulate those are addressed here.

14.4.1 Wedge Focusing

Wedge focusing is sketched in Fig. 14.5. A wedge angle ε causes a particle at local excursion x to experience a change $\int B_y\,ds = xB_y \tan\varepsilon$ in the field integral, com-

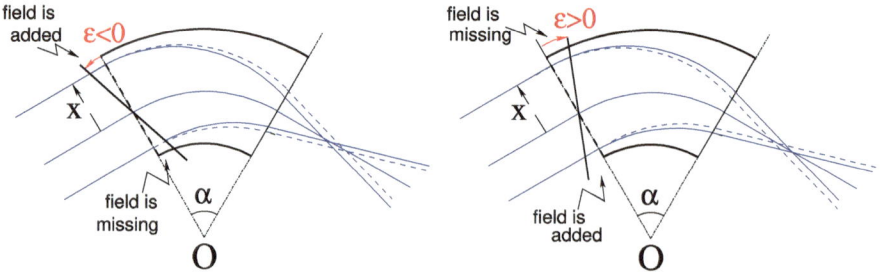

Fig. 14.5 Left: a focusing wedge ($\varepsilon < 0$ by convention); opening the sector increases the horizontal focusing. Right: a defocusing wedge ($\varepsilon > 0$); closing the sector decreases the horizontal focusing. The effect is the opposite in the vertical plane, opening/closing the sector decreases/increases the vertical focusing

pared to the field integral through the sector magnet. In the linear approximation this causes a change in trajectory angle

$$\Delta x' = \frac{1}{B\rho} \int B_y \, ds = x \frac{\tan \varepsilon}{\rho_0} \tag{14.17}$$

with $B\rho$ the particle rigidity and ρ_0 its trajectory curvature radius in the field B_0 of the dipole. Vertical focusing results from the non-zero off-mid plane radial field component B_x in the fringe field region (Fig. 14.7): from (Maxwell's equations) $\frac{\partial}{\partial y} \int B_x \, ds = \frac{\partial}{\partial x} \int B_y \, ds$ and Eq. 14.17 the change in trajectory angle comes out to be

$$\Delta y' = \frac{1}{B\rho} \int B_x \, ds = -y \frac{\tan \varepsilon}{\rho_0} \tag{14.18}$$

A first order correction ψ to the vertical kick accounts for the fringe field extent (it is a second order effect for the horizontal kick):

$$\Delta y' = -y \frac{\tan(\varepsilon - \psi)}{\rho_0} \tag{14.19}$$

with

$$\psi = I_1 \frac{\lambda}{\rho_0} \frac{1 + \sin^2 \varepsilon}{\cos \varepsilon} \quad \text{with} \quad I_1 = \int_{\text{edge}} \frac{B(s)\,(B_0 - B(s))}{\lambda \, B_0^2} \, ds \tag{14.20}$$

λ is the fringe field extent, I_1 quantifies the flutter (see Sect. 4.2.1); a longer/shorter field fall-off (smaller/greater flutter) decreases/increases the vertical focusing.

Fig. 14.6 Field components in the $B_y(s)$ field fall-off at a dipole EFB

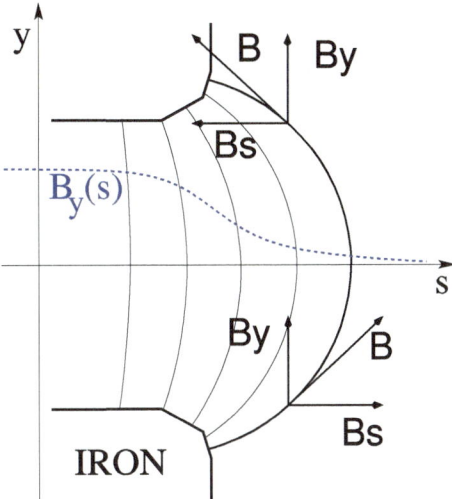

Fig. 14.7 Field components off mid-plane, in the fringe field region at the ends of a dipole ($y > 0$, here, referring to Fig. 14.6). $B_{//}$ parallel to the particle velocity has no effect. B_x pulls a positively charged particle away from the median plane, under the effect of a $\mathbf{v} \times \mathbf{B}_x$ force component. Inspection of the $y < 0$ region gives the same result: the charge is pulled away from the median plane

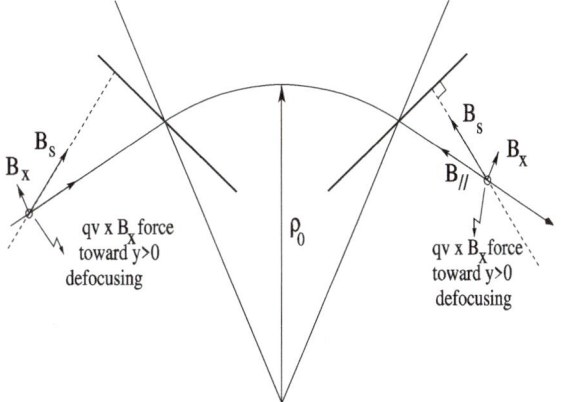

Linear approach

A wedge focusing first order transport matrix writes

$$T_{\text{wedge}} = \begin{pmatrix} 1 & 0 & 0 & 0 & 0 & 0 \\ \frac{\tan \varepsilon}{\rho} & 1 & 0 & 0 & 0 & 0 \\ 0 & 0 & 1 & 0 & 0 & 0 \\ 0 & 0 & -\frac{\tan \varepsilon}{\rho} & 1 & 0 & 0 \\ 0 & 0 & 0 & 0 & 1 & 0 \\ 0 & 0 & 0 & 0 & 0 & 1 \end{pmatrix} \tag{14.21}$$

Substitute $\varepsilon - \psi$ to ε in the R_{43} coefficient, when accounting for fringe field extent λ.

 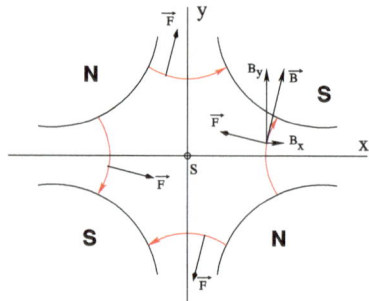

Fig. 14.8 Left: a quadrupole magnet [11]. Right: field lines and forces (assuming positive charges moving out of the page) over the cross section of an horizontally focusing/vertically defocusing quadrupole

14.4.2   Quadrupole

Quadrupoles are the optical lenses of charged particle beams, they ensure confinement of the beam in the vicinity of the optical axis. Most of the time in beam lines and cyclic accelerators, guiding and focusing are separate functions, focusing is assured by quadrupoles, magnetic most frequently, possibly electrostatic at low energy.

The field in quadrupole lenses results from hyperbolic equipotentials, $V = axy$. Pole profiles follow these equipotentials, in a $2\pi/4$-symmetrical arrangement for technological simplicity.

Magnetic Quadrupole

Magnetic quadrupoles are the optical lenses of high energy beams (Fig. 14.8).

The theoretical field in a quadrupole can be derived from Eq. 14.8 for the scalar potential, with $n = 2$ which yields

$$V_2(X, Y, Z) = \mathcal{G}(X)YZ - \frac{\mathcal{G}''(X)}{12}(Y^2 + Z^2)YZ + \frac{\mathcal{G}^{(4)}(X)}{384}(Y^2 + Z^2)^2 YZ - \cdots$$
$$(14.22)$$

and

$$B_X(X, Y, Z) = -\frac{\partial V_2}{\partial X} = \mathcal{G}'(X)YZ - \frac{\mathcal{G}'''(X)}{12}(Y^2 + Z^2)YZ + \cdots \quad (14.23)$$

$$B_Y(X, Y, Z) = -\frac{\partial V_2}{\partial Y} = \mathcal{G}(X)Z - \frac{\mathcal{G}''(X)}{12}(3Y^2 + Z^2)Z + \cdots \quad (14.24)$$

$$B_Z(X, Y, Z) = -\frac{\partial V_2}{\partial Z} = \mathcal{G}(X)Y - \frac{\mathcal{G}''(X)}{12}(Y^2 + 3Z^2)Y + \cdots \quad (14.25)$$

Fig. 14.9 Horizontal and
vertical projections of
particle trajectories across a
stigmatic quadrupole
doublet. The first quadrupole
(QF) is horizontally focusing
($K > 0$; thus vertically
defocusing), the second one
(QD) has reverse sign
($K < 0$) and reverse effect

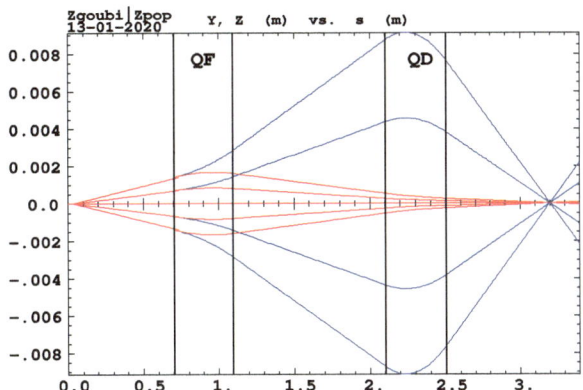

$G(X)$ is given by Eq. 14.4 whereas

$$G_0 = \frac{B_0}{a} \quad \text{and} \quad K = \frac{G_0}{B\rho} \tag{14.26}$$

define respectively the quadrupole gradient and strength, the latter relative to the
rigidity $B\rho$. The quadrupole is horizontally focusing and vertically defocusing if
$K > 0$, and the reverse if $K < 0$, this is illustrated in Fig. 14.9 which shows the
effect of a doublet of quadrupoles with focusing strengths of opposite signs.

Linear approach

The first order transport matrix of a quadrupole with length L, gradient G and strength
$K = G/B\rho$ writes

$$T_{\text{quad}} = \begin{pmatrix} C_x & S_x & 0 & 0 & 0 & 0 \\ C_x' & S_x' & 0 & 0 & 0 & 0 \\ 0 & 0 & C_y & S_y & 0 & 0 \\ 0 & 0 & C_y' & S_y' & 0 & 0 \\ 0 & 0 & 0 & 0 & 1 & \frac{L}{\gamma^2} \\ 0 & 0 & 0 & 0 & 0 & 1 \end{pmatrix} \quad \text{with} \quad \begin{bmatrix} C_x = \cos L\sqrt{K}; \ C_x' = \frac{dC_x}{dL} = -KS_x \\ S_x = \frac{1}{\sqrt{K}}\sin L\sqrt{K}; \ S_x' = \frac{dS_x}{dL} = C_x \\ C_y = \cosh L\sqrt{K}; \ C_y' = \frac{dC_y}{dL} = KS_y \\ S_y = \frac{1}{\sqrt{K}}\sinh L\sqrt{K}; \ S_y' = \frac{dS_y}{dL} = C_y \end{bmatrix} \tag{14.27}$$

$K > 0$ for a focusing quadrupole (by convention, in the (x, x') plane, thus defocusing
in the (y, y') plane). Permute the horizontal and vertical 2×2 sub-matrices in the
case of a *defocusing* quadrupole.

Electrostatic Quadrupole

The hypotheses are those of Sect. 2.2.2: paraxial motion, field normal to velocity, etc.
Take the notations of Eqs. 2.25 and 2.26 for the field and potential, case of electrodes

Fig. 14.10 A sketch of a solenoid, and quantities used to define it

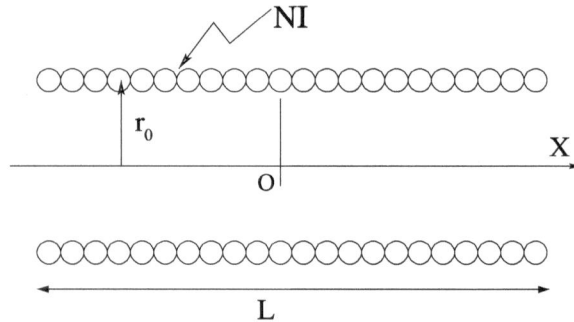

in the horizontal and vertical planes (Fig. 2.14). Electrode potential is $\pm V/2$, pole tip radius a, so that $K = -V/2a^2$ in Eq. 2.26. The equations of motion then write

$$\begin{bmatrix} \frac{d^2x}{ds^2} + K_x x = 0 \\ \frac{d^2y}{ds^2} + K_y y = 0 \end{bmatrix} \quad \text{with} \quad K_x = -K_y = \frac{-qV}{a^2 m v^2} = \pm\frac{V}{a^2} \underbrace{\frac{1}{|E\rho|}}_{\substack{\text{electrical} \\ \text{rigidity}}} \tag{14.28}$$

The transport matrix is the same as for the magnetic quadrupole, Eq. 14.27, taken for that K value.

14.4.3 Solenoid

Assume a solenoid magnet with longitudinal axis (OX). In a cylindrical frame $(O; X, r, \phi)$, Fig. 14.10 (r is the radial coordinate, the angle ϕ is taken in the X-normal plane), $B_\phi(X, r, \phi) \equiv 0$. Take solenoid length L, mean coil radius r_0 and an asymptotic field $B_0 = \mu_0 NI/L$, with NI = number of ampere-Turns, $\mu_0 = 4\pi \times 10^{-7}$ H/m. The asymptotic field value is defined by

$$\int_{-\infty}^{\infty} B_X(X, r < r_0)\, dX = \mu_0 NI = B_0 L \qquad \text{independent of r} \tag{14.29}$$

There is a variety of methods to compute the field vector $\mathbf{B}(X, r)$. Opting for one in particular may be a matter of compromise between computing speed and field modeling accuracy. A simple model is the on-axis field

$$B_X(X, r = 0) = \frac{B_0}{2} \left[\frac{L/2 - X}{\sqrt{(L/2 - X)^2 + r_0^2}} + \frac{L/2 + X}{\sqrt{(L/2 + X)^2 + r_0^2}} \right] \tag{14.30}$$

with $X = r = 0$ taken at the center of the solenoid. This model assumes that the coil thickness is small compared to its mean radius r_0. The magnetic length comes out to be

$$L_{\text{mag}} \equiv \frac{\int_{-\infty}^{\infty} B_X(X, r < r_0) dX}{B_X(X = r = 0)} = L\sqrt{1 + \frac{4r_0^2}{L^2}} > L \qquad (14.31)$$

so satisfying

$$\text{on-axis } B_X(X = r = 0) = \frac{\mu_0 NI}{L\sqrt{1 + \frac{4r_0^2}{L^2}}} \xrightarrow{r_0 \ll XL} \frac{\mu_0 NI}{L}$$

Maxwell's equations and Taylor expansions provide the off-axis field $\mathbf{B}(X, r) = (B_X(X, r), B_r(X, r))$. One has in particular in the $r_0 \ll XL$ limit,

$$B_X(X, r) = \frac{\mu_0 NI}{L} \quad \text{and} \quad B_r(X, r) = \frac{-r}{2} \frac{dB_X}{dX} \qquad (14.32)$$

An other way to compute the field vector $\mathbf{B}(X, r)$ is the elliptic integrals technique developed in [12], which constructs $B_X(X, r)$ and $B_r(X, r)$ from respectively

$$B_X(X, r) = \frac{\mu_0 NI}{4\pi} \frac{ck}{r} X \left[K + \frac{r_0 - r}{2r_0}(\Pi - K) \right] \qquad (14.33)$$

$$B_r(X, r) = \mu_0 NI \frac{1}{k}\sqrt{\frac{r_0}{r}} \left[2(K - E) - k^2 K \right]$$

wherein K, E and Π are the three complete elliptic integrals, X is an X- and L-dependent form factor, and

$$k = 2\sqrt{r_0 r}/\sqrt{(r_0 + r)^2 + X^2} \ ; \quad c = 2\sqrt{r_0 r}/(r_0 + r)$$

As an illustration, Fig. 14.11 displays a trajectory across a $L = 1$ m solenoid and its field fall-offs, and the field experienced along that trajectory, in the axial model of Eq. 14.30. In the paraxial approximation, a pitch requires a distance $l = 2\pi/K$, with $K = B_0/B\rho$ the solenoid strength, which is a condition satisfied here if the fringe field extent is short enough (solenoid radius r_0 is small enough).

Linear approach

The equations of motion write, to the first order in the coordinates, in respectively the central region (field B_s) and at the ends (at $s = s_{\text{EFB}}$),

$$\begin{vmatrix} x'' - K z' = 0 \\ z'' + K x' = 0 \end{vmatrix} \quad \text{and} \quad \begin{vmatrix} x'' - \dfrac{K}{2} z \delta(s - s_{\text{EFB}}) = 0 \\ z'' + \dfrac{K}{2} x \delta(s - s_{\text{EFB}}) = 0 \end{vmatrix} \qquad (14.34)$$

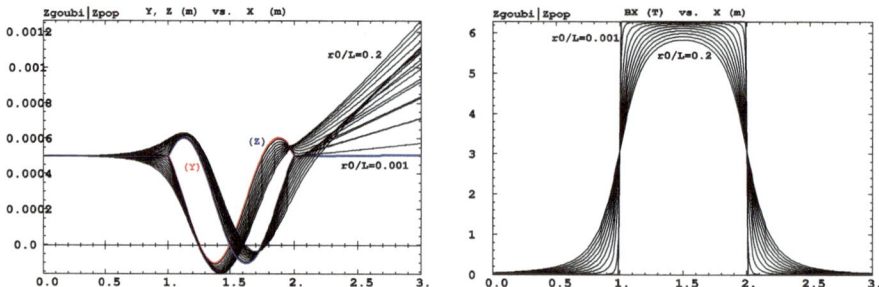

Fig. 14.11 Left: Horizontal (Y) and vertical (Z) projections of a particle trajectory across a $L = 1$ m solenoid, with additional 1 m extents upstream and downstream of the coil to account for the extended field fall-offs. The particle is launched with zero incidence, from transverse position $Y = Z = 0.5$ mm. Sample solenoid radius/length values in the range $0.001 \leq r_0/L \leq 0.2$ show that only for smallest $r_0/L = 0.001$ does the trajectory end with $Y = Z = 0.5$ mm and quasi-zero incidence (the thicker Y(X) and Z(X) curves), whereas greater r_0/L causes final Y(X) and Z(X) to be offset. Right: field $B_X(X, r)$ experienced along the trajectory for the various r_0/L values, the steep fall-off case is for $r_0/L = 0.001$

The first order transport matrix of a solenoid with length L writes

$$
T_{\text{sol}} =
\begin{pmatrix}
C^2 & \frac{2}{K}SC & SC & \frac{2}{K}S^2 & 0 & 0 \\
\frac{-K}{2}SC & C^2 & -\frac{K}{2}S^2 & SC & 0 & 0 \\
-SC & -\frac{2}{K}S^2 & C^2 & \frac{2}{K}SC & 0 & 0 \\
\frac{K}{2}S^2 & -SC & -\frac{K}{2}SC & C^2 & 0 & 0 \\
0 & 0 & 0 & 0 & 1 & \frac{L}{\gamma^2} \\
0 & 0 & 0 & 0 & 0 & 1
\end{pmatrix}
\quad \text{with} \quad
\begin{bmatrix}
K = \frac{B_s}{B\rho} \\
C = \cos\frac{KL}{2} \\
S = \sin\frac{KL}{2}
\end{bmatrix}
\qquad (14.35)
$$

A solenoid rotates the decoupled axis longitudinally by an angle $\alpha = KL/2 = B_s L/2B\rho$.

14.5 Data Treatment Keywords

14.5.1 Concentration Ellipse: FAISCEAU, FIT[2], MCOBJET, …

It is often useful to associate the projection of a particle bunch in the horizontal, vertical or longitudinal phase space with an *rms* phase space concentration ellipse (CE). Various keywords in zgoubi resort to concentration ellipses:

– FAISCEAU for instance prints out, in zgoubi.res, CE parameters drawn from particle coordinates,
– random particle distributions by MCOBJET are defined using CE parameters,

– ellipse parameters computed from CEs are possible constraints in FIT[2] procedures.

Transverse phase space graphs by zpop also compute CEs.

The CE method is resorted to in various exercises, for instance for comparison
of the ellipse parameters it gets from the *rms* matching of a bunch, with theoretical
beam parameters derived from first order transport formalism (such as computed
from rays by MATRIX, or TWISS).

The CE method used in these various keywords and data treatment procedures is
the following. Let $z_i(s)$, $z_i'(s)$ be the phase space coordinates of $i = 1, n$ particles
in a set observed at some azimuth s along an optical sequence. The second moments
of the particle distribution are

$$\overline{z^2}(s) = \frac{1}{n} \sum_{i=1}^{n} (z_i(s) - \overline{z}(s))^2$$

$$\overline{zz'}(s) = \frac{1}{n} \sum_{i=1}^{n} (z_i(s) - \overline{z}(s))(z_i'(s) - \overline{z'}(s)) \tag{14.36}$$

$$\overline{z'^2}(s) = \frac{1}{n} \sum_{i=1}^{n} (z_i'(s) - \overline{z'}(s))^2$$

From these, a concentration ellipse is defined, encompassing a surface $S_z(s)$, with
equation

$$\gamma_c(s)z^2 + 2\alpha_c(s)zz' + \beta_c(s)z'^2 = S_z(s)/\pi \tag{14.37}$$

Noting $\Delta = \overline{z^2}(s)\,\overline{z'^2}(s) - \overline{zz'}^2(s)$, the ellipse parameters write

$$\gamma_c(s) = \frac{\overline{z'^2}(s)}{\sqrt{\Delta}}, \quad \alpha_c(s) = -\frac{\overline{zz'}(s)}{\sqrt{\Delta}}, \quad \beta_c(s) = \frac{\overline{z^2}(s)}{\sqrt{\Delta}}, \quad S_z(s) = 4\pi\sqrt{\Delta} \tag{14.38}$$

With these conventions, the *rms* values of the z and z' projected densities satisfy

$$\sigma_z = \sqrt{\beta_z \frac{S_z}{\pi}} \quad \text{and} \quad \sigma_{z'} = \sqrt{\gamma_z \frac{S_z}{\pi}} \tag{14.39}$$

14.5.2 Transport Coefficients: MATRIX, OPTICS, TWISS, Etc.

Zgoubi does not know about matrix transport, it does not define optical elements by a
transport matrix, it defines them by electrostatic and/or magnetic fields in space (and
time possibly). Well, except for a couple of optical elements, for instance TRANS-
MAT, which pushes particle coordinates using a matrix, or SEPARA, an analytical

mapping through a Wien filter. Zgoubi does not transport particles using matrix products either, it does that by numerical integration of Lorentz force equation through these **E** and/or **B** fields.

However it is often useful to dispose of a matrix representation of an optical element or a beam line, or of paraxial parameters drawn from the first or second order one-turn mapping of a ring accelerator. Several commands in zgoubi perform the required treatment to derive these informations from particle coordinates. Examples are MATRIX: computation of matrix transport coefficients up to 3rd order, from initial and current coordinates of a particle sample. OPTICS transports a beam matrix, given its initial value using OBJET[KOBJ = 5.1]. TWISS derives a periodic beam matrix from a 1-turn mapping of a periodic sequence, and transports it from end to end so generating the optical functions along the sequence.

These capabilities are resorted to in the exercises. It may be required for instance to compare transport coefficients derived from raytracing, with the matrix model of the optical element(s) concerned. Or to compute a periodic beam matrix in a periodic optical sequence, this is how betatron functions are produced, often for the mere purpose of comparisons with matrix code outcomes, or with expectations from analytical models.

Coordinate Transport

In the Gauss approximation (i.e., trajectory angle $\theta \sim \sin\theta$), particles follow paths which can be described with simple functions: parabolic, sinusoidal or hyperbolic. A consequence is that a string of optical elements, and coordinate transport through the latter, can be handled with a simple mathematics toolbox. Taylor expansion (also known as transport) techniques are part of it, whereby a coordinate excursion v_{2i} (with index $i = 1 \to 6$ standing for x, x', y, y', δs or $\delta p/p$) from some reference trajectory at a location s_2 along the line is obtained from the excursions v_{1i} at an upstream location s_1, via

$$v_{2i} = \sum_{j=1}^{6} R_{ij} v_{1j} + \sum_{j,k=1}^{6} T_{ijk} v_{1j} v_{1k} + \sum_{j,k,l=1}^{6} v_{1ijkl} v_{1j} v_{1k} v_{1l} + \dots \quad (14.40)$$

This Taylor development can be written under matrix form, for instance to the first order in the coordinates, for non-coupled motion,

$$\begin{pmatrix} x \\ x' \\ y \\ y' \\ \delta s \\ \delta p/p \end{pmatrix}_2 = \begin{pmatrix} T_{11} & T_{12} & 0 & 0 & 0 & T_{16} \\ T_{21} & T_{22} & 0 & 0 & 0 & T_{26} \\ 0 & 0 & T_{33} & T_{34} & 0 & T_{36} \\ 0 & 0 & T_{43} & T_{44} & 0 & T_{46} \\ 0 & 0 & 0 & 0 & T_{55} & T_{56} \\ 0 & 0 & 0 & 0 & T_{65} & T_{66} \end{pmatrix} \begin{pmatrix} x \\ x' \\ y \\ y' \\ \delta s \\ \delta p/p \end{pmatrix}_1 = T(s_2 \leftarrow s_1) \begin{pmatrix} x \\ x' \\ y \\ y' \\ \delta s \\ \delta p/p \end{pmatrix}_1 \quad (14.41)$$

These are the quantities which such keywords as MATRIX [1, cf. Sect. 6.5] and
OPTICS [1, cf. Sect. 6.4] compute, from particle coordinates. Most of the time they
are resorted to for mere comparison with theoretical matrices such as recalled in
Sects. 14.2–14.4.

Beam Matrix

OPTICS and TWISS keywords cause the transport of a beam matrix. The former
requires initial beam ellipse parameters: these are provided as part of the initial
object definition, by OBJET. The latter first derives a periodic beam matrix from
initial and final particle coordinates resulting from raytracing throughout an optical
sequence. Basic principles are recalled here, regarding the way these keywords work
in zgoubi. They are resorted to quite often in the exercises.

In the linear approximation, the transverse phase space ellipse associated with a
particle distribution (for instance, the concentration ellipse, Sect. 14.5.1) is written
(with z standing for indifferently x or y)

$$\gamma_z(s)z^2 + 2\alpha_z(s)zz' + \beta_z(s)z'^2 = \frac{\varepsilon_z}{\pi} \tag{14.42}$$

in which the ellipse parameters

$$\beta_z(s), \quad \alpha_z(s) = -\frac{1}{2}\frac{d\beta_z}{ds}, \quad \gamma_z(s) = \frac{1+\alpha^2}{\beta_z} \tag{14.43}$$

are functions of the observation location s along the optical sequence. The surface
ε_z of the ellipse is an invariant if the beam travels in magnetic fields, however field
non-linearities, phase space dilution, etc. may distort the distribution and change the
surface of its *rms* matching concentration ellipse. In the presence of acceleration or
deceleration the invariant quantity is $\beta\gamma\varepsilon_z$ instead, with $\beta = v/c$ and γ the Lorentz
relativistic factor.

The ellipse Eq. 14.42 can be written under the matrix form

$$[z, \ z']\,\sigma_z^{-1}(s)\begin{bmatrix} z \\ z' \end{bmatrix} = 1 \tag{14.44}$$

with σ_z the beam matrix:

$$\sigma_z = \frac{\varepsilon_z}{\pi}\begin{pmatrix} \beta_z(s) & -\alpha_z(s) \\ -\alpha_z(s) & \gamma_z(s) \end{pmatrix} \tag{14.45}$$

The ellipse parameters can be transported from s_1 to s_2 using

$$\sigma_{z,2} = T\,\sigma_{z,1}\,\tilde{T} \tag{14.46}$$

with $T = T(s_2 \leftarrow s_1)$ the transport matrix (Eq. 14.41) and $\tilde{T}$ its transposed. This can also be written under the form

$$\begin{pmatrix} \beta_z \\ \alpha_z \\ \gamma_z \end{pmatrix}_2 = \begin{pmatrix} T_{11}^2 & -2T_{11}T_{12} & T_{12}^2 \\ -T_{11}T_{21} & T_{21}T_{12} + T_{11}T_{22} & -T_{12}T_{22} \\ T_{21}^2 & -2T_{21}T_{22} & T_{22}^2 \end{pmatrix}_{s_2 \leftarrow s_1} \begin{pmatrix} \beta_z \\ \alpha_z \\ \gamma_z \end{pmatrix}_1 \quad (14.47)$$

(subscripts 1, 2 normally hold for horizontal plane motion, $z = x$: change to 3, 4 for vertical motion, $z = y$). This beam matrix formalism can be extended to the longitudinal phase space and coordinates $(\delta s, \delta p/p)$. Thus a 6×6 beam matrix can be defined,

$$\sigma = \begin{pmatrix} \sigma_{11} & \sigma_{12} & 0 & 0 & 0 & \sigma_{16} \\ \sigma_{21} & \sigma_{22} & 0 & 0 & 0 & \sigma_{26} \\ 0 & 0 & \sigma_{33} & \sigma_{34} & 0 & \sigma_{36} \\ 0 & 0 & \sigma_{43} & \sigma_{44} & 0 & \sigma_{46} \\ 0 & 0 & 0 & 0 & \sigma_{55} & \sigma_{56} \\ 0 & 0 & 0 & 0 & \sigma_{65} & \sigma_{66} \end{pmatrix} \quad (14.48)$$

This can be generalized to non-zero anti-diagonal terms, if motions are coupled.

Periodic Structures

In the hypothesis of an S- periodic structure: a long beam line with repeating pattern, a cyclic accelerator, transverse motion stability requires the transport matrix over a period, from s to $s + S$ to satisfy

$$[T_{ij}](s + S \leftarrow s) = I \cos \mu + J \sin \mu \quad (14.49)$$

where $\mu = \int_{(S)} ds/\beta$ is the betatron phase advance over the period (independent of the origin),

$$I = \begin{pmatrix} 1 & 0 \\ 0 & 1 \end{pmatrix} \text{ is the identity matrix, } J = \begin{pmatrix} \alpha_z(s) & \beta_z(s) \\ -\gamma_z(s) & -\alpha_z(s) \end{pmatrix} \text{ (and } J^2 = -I)$$

$$\quad (14.50)$$

14.6 Exercises

14.1 Magnetic Sector Dipole
Solution 14.1.

(a) Simulate a $\rho = 0.5$ m radius, $\alpha = 60°$ sector dipole with $n = -0.6$ field index, in both cases of hard edge and of soft fall-off fringe field model. Find the reference

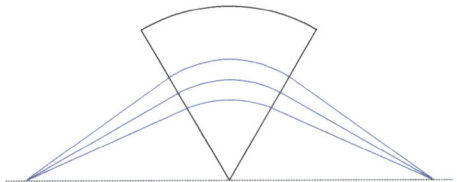 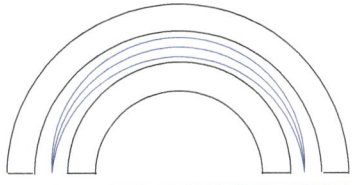

Fig. 14.12 Symmetric point to point focusing, case of a 60° or a 180° sector dipole

arc, such that $\int_{arc} B \, ds = BL$ with L the arc length in the hard-edge model and B the field along that arc.

Make sure that the reference arc has the expected length.

Produce the field along the reference arc, for a few different values of the fringe-field extent.

(b) A possible check of the first order: OBJET[KOBJ = 5], MATRIX[IORD = 1, IFOC = 0] can be used to compute the transport matrix from the rays. Compare what it gives with theory.

(c) Consider a sector dipole with parallel gap, uniform field. Show the well known geometrical property of point-to-point focusing represented in Fig. 14.12.

Produce the aberration curve $x'(x)$ in the horizontal phase-space at the image plane.

Test the convergence of the numerical solution versus integration step size.

(d) Transport a proton along the reference axis, injected with its spin tangent to the axis. Compare spin rotation with theory.

Test the convergence of the numerical solution versus integration step size.

14.2 Quadrupole Doublet
Solution 14.2.

Reproduce Fig. 14.9.

14.3 Solenoid
Solution 14.3.

An introduction to SOLENOID.

(a) Reproduce Fig. 14.11. Use both field models of Eqs. 14.30 and 14.33 and compare their outcomes, including the first order paraxial transport matrices, and some higher order coefficients as well (computed from in and out trajectory coordinates).

(b) Compare final coordinates in (a) with outcomes from the first order transport formalism (Sect. 14.4.3).

(c) Make a 1-dimensional (on-axis) field map of a $r_0 = 10$ cm, $L = 1$ m solenoid (namely, a map $B_{X,i}(X_i)$ of the field at the nodes of a X-mesh with mesh size $X_{i+1} - X_i$). Reproduce the trajectory in (a) (case $r_0 = 10$ cm) using that field map, with the keyword BREVOL. Check the convergence of the final particle coordinates, using the field map, depending on the mesh size.

14.7   Solutions of Exercises of This Chapter: Optical Elements and Keywords, Complements

14.1 Magnetic Sector Dipole
DIPOLE input data.

(a) A simulation of a $\rho = 0.5$ m radius, $60°$ sector dipole with n $= -0.6$ field index, in the hard-edge field model, is given in Table 14.1. A simulation which includes fringe fields is given in Table 14.2.

A major difference between the two is in the angular extent of the field domain, AT, in order to allow encompassing the fringe field extents, however there is more, as follows.

Hard edge model

The effective field boundaries (EFB) have to be placed on the angular opening limits, which means, in the representation of Fig. 14.13, and according to the users' guide [13, see DIPOLE],

$$\omega^+ = ACENT > 0, \quad \omega^- = -AT + ACENT < 0, \quad \omega^+ - \omega^- = AT > 0$$

Otherwise, in the case AT would be greater than the magnet deflection angle $\alpha = 60°$, particles would jump from zero field to plateau field value over the EFB, and so miss part of the field integral. Note that for mere code-specific, geometry computation reasons, it also requires that ACENT = AT/2, so that, *in fine*, $\omega^+ = -\omega^- = AT/2$.

Table 14.1 Input data file: definition of a dipole with index in the hard-edge field model. Definition of the [#S_60dSectDip_hardE:#E_60dSectDip_hardE] segment, mostly for the purpose of possible further INCLUDE. This file is used under the name sectorDIP_hardE.inc in subsequent exercises

```
! File sectorDIP.inc (hard-edege, here)
'MARKER'    #S_60dSectDip_hardE                              .! Label should not exceed 20 characters.
'DIPOLE'                                            ! Analytical definition of a dipole field.
2                        ! IL=2, only purpose is to log trajectories to zgoubi.plt, for further plotting.
60. 50.                                      ! Sector angle AT; reference radius RM.
30. 5. -0.6 0. 0.           ! Reference azimuthal angle ACN; BM field at RM; indices, N=-0.6 at RM=50cm.
0.  0.                                                    ! EFB 1 is  hard-edge,
4  .1455    2.2670  -.6395  1.1558  0. 0.  0.       ! hard-edge only possible with sector magnet.
30. 0.  1.E6  -1.E6  1.E6  1.E6
0.  0.                                                           ! EFB 2.
4  .1455    2.2670  -.6395  1.1558  0. 0.  0.
-30. 0.  1.E6  -1.E6  1.E6  1.E6
0. 0.                                                    ! EFB 3 (unused).
0  0.     0.     0.      0.      0. 0.  0.
0.  0.  1.E6  -1.E6  1.E6  1.E6 0.
4   10.
0.5                         ! Integration step size. The smaller, the more accurately the orbits close.
2  0. 0. 0. 0.                                   ! Magnet positioning RE, TE, RS, TS.
'MARKER'    #E_60dSectDip_hardE                      ! Label should not exceed 20 characters.
'END'
```

Table 14.2 Input data file: definition of a dipole with index in the soft-edge field model. The field extent in the Enge model (Eq. 14.11) is taken to be $g = 5$ cm ($\lambda_E = \lambda_S = g$ in Users' Guide's notations), so subtended by an angle $\text{atan}(g/RM) = 5.71059°$, thus well comprised in a $10°$ angular aperture. ACENT value is free, $30°$ as adopted here is arbitrary, it is just left to the value it was given in the hard edge settings (Table 14.1). This input includes the definition of the [#S_60dSectDip_softE:#E_60dSectDip_softE] segment. This file is used under the name sectorDIP_softE.inc in subsequent exercises

```
! File sectorDIP.inc (soft-edege, here)
'MARKER'    #S_60dSectDip_softE                      ! Label should not exceed 20 characters.
'DIPOLE'                                             ! Analytical definition of a dipole field.
2                        ! IL=2, only purpose is to log trajectories to zgoubi.plt, for further plotting.
80. 50.                           ! Sector angle AT=60 deg deflection+2*10deg for fringes; reference radius RM.
30. 5. -0.6 0. 0.   ! Reference angle ACENT (arbitrary value); field at RM; indices, N=-0.6 at RM=50cm.
5. 0.      ! Entry EFB: lambda~gap=5 cm, well comprised in RM*tan(10deg)=; same gap at all R -> nappa=0.
4 .1455 2.2670 -.6395 1.1558 0. 0. 0.                           ! Enge coefficients at entry.
20. 0. 1.E6 -1.E6 1.E6 1.E6 ! omega^+ = +20 deg from ACENT leaves 10deg room (8.8cm) for entry fringe.
5. 0.      ! Exit EFB: lambda~gap=5 cm, well comprised in RM*tan(10deg)=; same gap at all R -> nappa=0.
4 .1455 2.2670 -.6395 1.1558 0. 0.  0.                          ! Enge coefficients at exit.
-40. 0. 1.E6 -1.E6 1.E6 1.E6  ! omega^- =-40 deg from ACENT leaves 10deg room (8.8cm) for exit fringe.
0. 0.                                                           ! EFB 3 (unused).
0 0.       0.      0.      0.      0. 0. 0.
0. 0.  1.E6  -1.E6  1.E6  1.E6 0.
4   10.
0.5                      ! Integration step size. The smaller, the more accurately the orbits close.
2 0. 0. 0. 0.                                         ! Magnet positioning RE, TE, RS, TS.
'MARKER'    #E_60dSectDip_softE                      ! Label should not exceed 20 characters.
'REBELOTE'

'END'
```

Fig. 14.13 Parameters used to define the geometry of a dipole magnet with index, using DIPOLE [13, see DIPOLE]

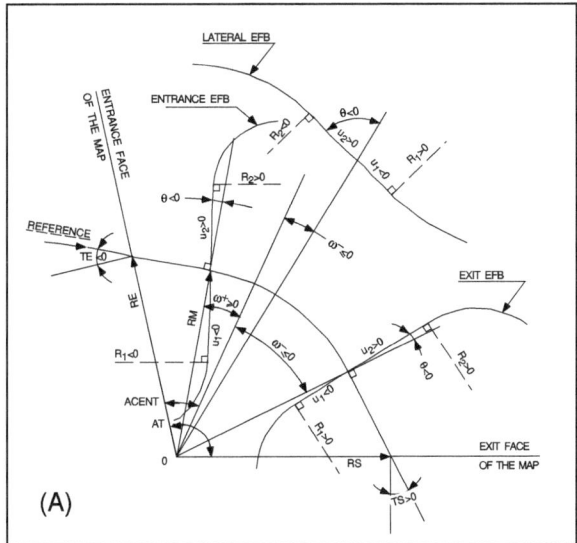

Soft edge model

AT has to be greater than the magnet deflection angle $\alpha = 60°$ in order to encompass the fringe field extent beyond the entrance and exit EFBs, so that, in the representation of Fig. 14.13, and according to the users' guide,

$$ACENT > \omega^+, \quad |\omega^-| < AT - ACENT$$

Integration-wise, particles will smoothly traverse the field fall-off regions, step by step, no field discontinuity there. Note that motion integration accuracy requires the step size to be small enough, compared to the fringe field extent. In the notations of Fig. 14.13, the resulting additional optical axis lengths l_E and l_S within the AT sector, on entrance and exit side respectively, to account for the field fall-offs, write

$$l_E = RM \times \tan(ACENT - \omega^+), \quad l_S = RM \times \tan[AT - (ACENT - \omega^-)]$$

Checking back one fortunately finds

$$\underbrace{\operatorname{atan}\left(\frac{l_E}{RM}\right)}_{\substack{\text{entrance} \\ \text{fringe field}}} + \underbrace{\omega^+ - \omega^-}_{\text{magnet body}} + \underbrace{\operatorname{atan}\left(\frac{l_S}{RM}\right)}_{\substack{\text{exit} \\ \text{fringe field}}} = AT$$

It also results from the fringe field modeling that the reference trajectory (which is ideally the trajectory that coincides with R = RM in the body of the magnet) enters the AT sector at radius RE, with an incidence TE. These two quantities have to be accounted for in setting the entrance and exit reference frames, however this is user's matter, regarding the choice of reference frames: most often (in synchrotron rings for instance) the reference curve is R = RM, so that Y and T coordinates of the reference particle are zero (the moving frame has its origin at the origin of the polar frame in which the field is defined, and rotates with the particle, clockwise in Fig. 14.13 representation). Thus, one has to set

$$TE = -(ACENT - \omega^+) < 0, \quad RE = RM/\cos TE$$

Note that, because of the small deflection due to fringe fields, RS and TS need be adjusted if the DIPOLE process has to end up with the reference particle featuring zero Y and T coordinates. Expectedly, that would be satisfied with RS and TS values near

$$TS = AT - (ACENT - \omega^-) > 0, \quad RS = RM/\cos TS$$

The radius R of the reference arc, such that $\int_{\text{arc}} B\,ds = BL$ with L the arc length in the hard-edge model, has to be found. Same thing for the arcs at $\pm 0.1\%$ momentum offset. FIT can be used for that (Table 14.3).

(b) First order transport.

This is left to the reader. Theoretical matrices are given in Eqs. 14.6 and 14.7. Refer to exercises in earlier chapters, such comparison is often performed.

Table 14.3 Input data file: find closed orbits, using FIT or FIT2, and log stepwise data in zgoubi.plt. Closed orbits are found for the reference particle (a particle with rigidity $B\rho = 5_{[kG]} \times 50_{[cm]}$ kG cm) and for particles with $\pm\delta p/p$ momentum offset. FIT starts with initial Y_0 radius values resulting from a hard edge model, i.e., $Y_0 = B\rho/B = 250_{[kG\,cm]}/5_{[kG]}$ and $\pm0.1\%$. This file produces the field along these trajectories, an effect of DIPOLE[IL = 2]. The [#S_60dSectDip_softE:#E_60dSectDip_softE] segment of Table 14.2 is INCLUDEd; simply substitute [#S_60dSectDip_hardE:#E_60dSectDip_hardE] (as defined in Table 14.1) to work with the hard edge model instead

```
Uniform field sector with index. Field on orbits at different momenta.
'MARKER'   DIPOLEField_S                                      ! Just for edition purposes.
!                                       First stage: find closed orbit at 1 MeV, for some k value.
'OBJET'
64.62444403717985                       ! Reference Brho ("BORO" in the users' guide) -> 200keV proton.
2                                                  ! Particles are defined one by one.
3 1                                        ! 3 particles, classified in a single momentum set.
50. 0. 0. 0. 0. 3.8685052339 'o'             ! Y_0=50cm is hard edge case -> 2.9886MeV proton.
50.125472 0. 0. 0. 0. 3.8723737392 'p'   ! +0.001 mom. offset. Circular orbit Y_0 is hard edge case.
49.875465 0. 0. 0. 0. 3.8646367287 'm'   ! -0.001 mom. offset. Circular orbit Y_0 is hard edge case.
1 1 1                           ! As many '1' as there are particles (that dates from programs on punched cards!
'INCLUDE'
1
./sectorDIP.inc[#S_60dSectDip_softE:#E_60dSectDip_softE]       ! DIPOLE with fringe, RM=50cm n=-0.6.
!./sectorDIP.inc[#S_60dSectDip_hardE:#E_60dSectDip_hardE]      ! DIPOLE with hard-edge, RM=50cm n=-0.6.
'FIT'                                  ! This matching procedure finds the closed orbit radius.
3    nofinal
2  30  0  .9                              ! Variable : Y_0. Variation allowed up to 90%.
2  40  0  .9                              ! Variable : Y_0. Variation allowed up to 90%.
2  50  0  .9                              ! Variable : Y_0. Variation allowed up to 90%.
3   1e-15  99                             ! Penalty; max numb of calls to the function.
3.1 1 2 #End 0. 1. 0                         ! Constraint : Y_final=Y_0, particle 1.
3.1 2 2 #End 0. 1. 0                         ! Constraint : Y_final=Y_0, particle 2.
3.1 3 2 #End 0. 1. 0                         ! Constraint : Y_final=Y_0, particle 3.
'MARKER'   DIPOLEField_E                            ! Just for edition purposes.
'END'
```

(c) Point-to-point focusing.

The hard-edge model DIPOLE of Table 14.1 can be used, with the following modifications and addenda in order to simulate the symmetric 60́deg sector and drifts configuration of Fig. 14.12:

– add OBJET[KOBJ = 1, IMAX = 41] so to generate 41 particles launched with $T_0 \in [-20, 20]$ mrad, like so:

```
'OBJET'
64.62444403717985
1
1   41  1  1  1  1
0.  1.  0. 0. 0. 0.
50. 0.  0. 0. 0. 3.8685052339
```

– following OBJET add a drift with length $RM/\tan(30°) = 86.6025403784$ cm,
– following DIPOLE add a drift with length $RM/\tan(30°)$,
– in DIPOLE: set the field index to zero,
– add AUTOREF[I = 3, I1 = 1, I2 = 2, I3 = 3] after DIPOLE: that will cause computation of the location of the waist formed by particles 1, 2 and 3,
– add FAISTORE[FNAME = zgoubi.fai, IP = 1] after AUTOREF, before END. This logs particle data at that location.

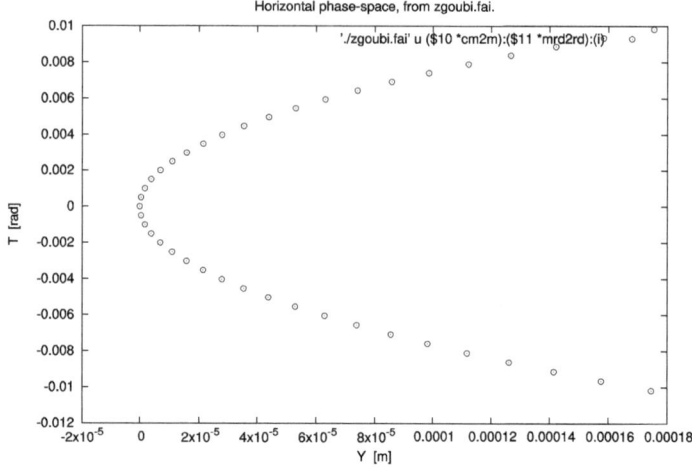

Fig. 14.14 Aberration curve at the focal point of a 180° uniform field dipole: a second order (sextupole) aberration, $Y \propto T^2$, typical of a bend non-linearities

In the execution listing zgoubi.res one finds:

```
      6  Keyword, label(s) :  AUTOREF
Change  of  reference,  horizontal, XC= -0.00011588 cm, YC = 49.999999 cm, A= -0. deg
TRAJ 1 IEX,D,Y,T,Z,P,S,time :  1    3.869  -1.1786E-16  0. 0. 0.   225.56   9.44931E-02
```

This indicates that AUTOREF confirms expectations: it found the waist

– at $XC = 0$, which means right at the end of the downstream drift,
– at a radial excursion $YC = 50$ cm as expected (the origin of the Y axis is at DIPOLE curvature center),
– with the reference frame X axis at an angle $A = 0$ to particle 1 direction of motion.

QED.

The following gnuplot script can be used to print the horizontal phase space $T(Y)$ at the image plane (Fig. 14.14)

```
cm2m = 1e-2; mrd2rd = 1e-3
plot './zgoubi.fai' u ($10 *cm2m):($11 *mrd2rd) w p ps .9 pt ; pause 2
```

In the case of an $\alpha = 180°$ dipole, the previous input data file can be used, changing DIPOLE angles to $AT = \omega^+ - \omega^- = 180°$ with for instance $\omega^+ = -\omega^- = 90°$. Remove the drifts in order to obtain the 180° sector configuration of Fig. 14.12.

Step size:
The method is the same as in Exercise 2.2 (b), case of a toroidal condenser, which can be referred to.

(d) Spin precession.

Add SPNTRK[KSO = 1] at the beginning of the input data file to track spin, starting aligned on the X axis. Tracking spin also requires PARTICUL, in order to define particle's mass, charge and anomalous magnetic moment.

The theoretical value of the spin precession angle in the moving frame is $G\gamma\alpha$ (Eq. 3.32), with $\alpha = \pi/3$ or $\alpha = \pi$ in the previous two deflection cases considered.

This is the value which the stepwise integration produces.

14.2 Quadrupole Doublet

The input data file for this problem is given in Table 14.4.

Table 14.4 Input data file: a double-focus quadrupole doublet

```
100 particles on an ellipse, through drift
'OBJET'
1000.
2
9  1
0.   0. 0.   0. 0. 1. 'o'
0.   1. 0.   0. 0. 1. 'a'
0.  -1. 0.   0. 0. 1. 'b'
0.   2. 0.   0. 0. 1. 'c'
0.  -2. 0.   0. 0. 1. 'd'
0.   0. 0.   1. 0. 1. 'e'
0.   0. 0.  -1. 0. 1. 'f'
0.   0. 0.   2. 0. 1. 'g'
0.   0. 0.  -2. 0. 1. 'h'
1 1 1 1 1 1 1 1 1
'FAISCEAU'

'MARKER'  dum .plt
'DRIFT'
70.  split 100 2
'QUADRUPO'        QF
2
40. 10.  4.7907188      I 11.1111
0. 0.
6  .1122 6.2671 -1.4982 3.5882 -2.1209 1.723
0. 0.
6  .1122 6.2671 -1.4982 3.5882 -2.1209 1.723
1.
1 0 0 0
'DRIFT'
100.  split 100 2
'QUADRUPO'        QD
2
40. 10. -4.7907188      I -11.1111
0. 0.
6  .1122 6.2671 -1.4982 3.5882 -2.1209 1.723
0. 0.
6  .1122 6.2671 -1.4982 3.5882 -2.1209 1.723
1.
1 0 0 0
'DRIFT'
70.   split 100 2
'MARKER'  dum .plt
'FAISCEAU'

! 'FIT'                                              ! This FIT procedure
! 2                                       ! varies QF and QD fields so to get
! 5 12 0 .4                 ! common focus point in both planes, 3.2 meters downstream of the object.
! 7 12 0 .4
! 4  1E-15
! 3   6 2 #End 0. 1. 0
! 3 11 2 #End 0. 1. 0
! 3   2 4 #End 0. 1. 0
! 3   3 4 #End 0. 1. 0

'IMAGE'
'IMAGEZ'

'DRIFT'
20.  split 100 2

'END'
```

Table 14.5 Input data file: a 1 m long solenoid, with 1 m upstream and downstream fringe field extents. The initial coil radius is $r_0 = 0.1$ cm, it is scanned (by REBELOTE) over the range $1 \leq r_0 \leq 20$ cm. For each r_0 a particle is launched with initial position $Y = Z = 1$ mm and initial angles $T = P = 0$

```
A 1 meter long solenoid.
'MARKER'  opticalLmntsProbSolenoA_S
'OBJET'
1000.
2                                               ! OBJET style KOBJ=2.
1 1
0.1  0. 0.1 0. 0. 1. 'o'         ! Initial coordinates Yo, To, Zo, Po, Xo, Do.
1
'SOLENOID'
200                          ! Log particle data to zgoubi.plt, every other 100 steps.
100.  .1 62.8318530718      ! length (cm); radius (cm); field (kG); [MODL=1] default.
100. 100.                    ! Extent of integration regions upstream and downstream of coil.
.01
1  0. 0. 0.
'FAISCEAU'
'REBELOTE'                              ! Used to repeat the sequence.
10 0.1 0  1                             ! Repeat 10 times.
1
SOLENOID 11 1.:20.              ! Vary parameter 11 (= R0) under SOLENOID.
'MARKER'  opticalLmntsProbSolenoA_E
'END'
```

14.3 Solenoid

(a) The paraxial trajectory pitch is $l = 2\pi B\rho/B_0$ (Sect. 14.4.3). Take $L = 1$ m (Fig. 14.11) and $B\rho = 1$ T m for simplicity, thus $B_0 = 2\pi$ T. Assume a particle launched from $Y = Z = 1$ mm with zero incidence. Scan the solenoid radius value in the range $1 \leq r_0 \leq 200$ mm to reproduce the figure. The data to be plotted (X, Y, Z, B_X) are read from zgoubi.plt.

The beam optics model is given in Table 14.5. Note the use of KOBJ = 2 in OBJET, which allows creating particles in an arbitrary number (just one, here), with arbitrary initial coordinates. REBELOTE[IOPT = 1] is used to repeat the sequence, varying the parameter R_0 under SOLENOID.

(b) To allow comparison, theoretical matrices (Eq. 14.35) must be computed for the theoretical length, L, of the matrix transport solenoid model. Tracking must extend upstream and downstream of the solenoid, over a distance much greater than the solenoid diameter (the latter determines the field fall extent, Eq. 14.30) (Table 14.7).

(c) A 1-dimensional (on-axis) field map of the solenoid field, $B_{X,i}(X_i)$, can simply be generated by tracking a particle along the solenoid axis. It has to extend upstream and downstream of the solenoid, over a distance much greater than the solenoid diameter. The integration step size will be the mesh size, take it in the centimeter range ($\lesssim r_0$), 5 cm here. An intermediate stage is necessary, which consists in reading $X, B_X(X)$ from zgoubi.plt and re-writing it in a dedicated ASCII file in a format proper for use by the keyword BREVOL.

The input file to generate the field and log to zgoubi.plt is given in Table 14.6.

Similar exercises, generating a 1D field map and using BREVOL, can be found be found in zgoubi sourceforge repository [14].

Table 14.6 Input data file: track a particle along the central axis of the solenoid, to generate a 3 m long, 1D field map, with mesh step 5 cm

```
! A 3 meter long solenoid field map.
'MARKER'  opticalLmntsProbSolenoC_S
'OBJET'
1000.
2                                              ! OBJET style KOBJ=2.
1 1
0.  0. 0.  0.  0.  1. 'o'              ! Initial coordinates Yo, To, Zo, Po, Xo, Do.
1
'SOLENOID'
200                          ! Log particle data to zgoubi.plt, every other 100 steps.
100.  .1  62.8318530718      ! length (cm); radius (cm); field (kG); [MODL=1] default.
100. 100.              ! Extent of integration regions upstream and downstream of coil.
5.
1 0. 0. 0.
'FAISCEAU'
'END'
```

Table 14.7 Input data file: track a particle in the solenoid, in a similar manner to the input data file of Table 14.6, using a field map model instead

```
A 1 meter long solenoid, 3 meter long field map.
'OBJET'
1000.
2
1 1
0.  0. 0.  0.  0.  1. 'o'
1
'BREVOL'
0 0
1. 1.
Test solenoid 1D field map
61                                      ! Number of nodes of the 1D mesh.
solenoid_1meter.map
0  0. 0. 0.
2
1.
1 0 0 0
'FAISCEAU'
'END'
```

References

1. Zgoubi Users' Guide, updated Sourceforge version (at revision 2037, here): https://sourceforge.net/p/zgoubi/code/HEAD/tree/trunk/guide/Zgoubi.pdf. Méot, F.: Zgoubi Users' Guide. Report BNL-98726-2012-IR, C-A/AP 470 (2012). https://www.osti.gov/servlets/purl/1062013
2. The AGS at the Brookhaven National Laboratory. https://www.bnl.gov/rhic/AGS.asp
3. The CERN PS. https://home.cern/science/accelerators/proton-synchrotron
4. J.T. Volk, Experiences with permanent magnets at the Fermilab recycler ring. James T Volk 2011 JINST6 T08003. https://iopscience.iop.org/article/10.1088/1748-0221/6/08/T08003/pdf
5. F. Méot, F. Lemuet, Developments in the ray-tracing code Zgoubi for 6-D multiturn tracking in FFAG rings. NIM A **547**, 638–651 (2005)
6. H.A. Enge, Deflecting magnets, in *Focusing of Charged Particles*, vol. II ed. by A. Septier (Academic Press Inc., 1967), pp. 203–264
7. Y. Dutheil, et al., A model of the AGS based on stepwise ray-tracing through the measured field maps of the main magnets, in *Proceedings of IPAC2012, New Orleans, Louisiana, USA, TUPPC101*, pp. 1395–1399. https://accelconf.web.cern.ch/IPAC2012/papers/tuppc101.pdf; F. Méot, et al., Modeling of the AGS using zgoubi - status, in *Proceedings of IPAC2012, New Orleans, Louisiana, USA, MOPPC024*, pp. 181–183. https://accelconf.web.cern.ch/IPAC2012/papers/moppc024.pdf

8. R.E. Thern, E. Bleser, The dipole fields of the AGS main magnets, BNL-104840-2014-TECH, 1/26/1996. https://technotes.bnl.gov/PDF?publicationId=31175

9. F. Méot, L. Ahrens, K. Brown, et al., A model of polarized-beam AGS in the ray-tracing code Zgoubi. BNL-112453-2016-TECH, C-A/AP/566 (2016). https://technotes.bnl.gov/PDF?publicationId=40470, https://www.osti.gov/biblio/1336073

10. G. Leleux, Accélérateurs Circulaires. Lectures at the Institut National des Sciences et Techniques du Nucléaire, CEA Saclay (1978). (unpublished)

11. Credit: Brookhaven National Laboratory. https://www.flickr.com/photos/brookhavenlab/8495311598/in/album-72157611796003039/

12. M.W. Garrett, Calculation of fields [...] by elliptic integrals. J. Appl. Phys. **34**(9) (1963)

13. F. Méot, Zgoubi Users' Guide. https://www.osti.gov/biblio/1062013-zgoubi-users-guide Sourceforge revision 1379 (2020-02-29). https://sourceforge.net/p/zgoubi/code/HEAD/tree/trunk/guide/Zgoubi.pdf

14. https://sourceforge.net/p/zgoubi/code/HEAD/tree/branches/exemples/KEYWORDS/BREVOL/

Index-Zgoubi Optical Elements

© The Editor(s) (if applicable) and The Author(s) 2024
F. Méot, *Understanding the Physics of Particle Accelerators*,
Particle Acceleration and Detection,
https://doi.org/10.1007/978-3-031-59979-8